CALORIMETRY IN HIGH ENERGY PHYSICS

CALORIMETRY IN HIGH ENERGY PHYSICS

XII INTERNATIONAL CONFERENCE

Chicago, Illinois *5 – 9 June 2006*

EDITORS
Stephen R. Magill
Rik Yoshida
Argonne National Laboratory

SPONSORING ORGANIZATIONS
Hamamatsu Photonics K.K.
Photonis Group
Northern Illinois Center for Accelerator and Detector Development (NICADD)
University of Chicago
Deutsches Elektronen-Synchrotron (DESY)
European Organization for Nuclear Research (CERN)
HEP Division, Argonne National Laboratory
Office of Science, U.S. Department of Energy

AIP
75 Years of Service
Melville, New York, 2006
AIP CONFERENCE PROCEEDINGS ■ VOLUME 867

Editors

Stephen R. Magill
Rik Yoshida

HEP Division, Bldg. 362
Argonne National Laboratory
9700 S. Cass Avenue
Argonne, IL 60439
E-mail: srm@anl.gov

L.C. Catalog Card No. 2006935205
ISBN 978-0-7354-0364-2
ISSN 0094-243X

Printed in the United States of America

Contents

SESSION III

ASTROPARTICLE AND NEUTRINO CALORIMETRY

Conveners: Karen Byrum and Bob Wagner

SESSION IV

MEDICAL APPLICATIONS OF CALORIMETRY

Convener: Karen Byrum

SESSION V

LHC CALORIMETRY—TECHNOLOGY

Conveners: Jim Freeman and Peter Schacht

SESSION VI

LHC CALORIMETRY—COMMISSIONING

Convener: Stanislav Nemecek

SESSION VII

LHC CALORIMETRY—TEST BEAM

Convener: Calin Alexa

SESSION VIII

LHC CALORIMETRY—CALIBRATION AND MONITORING

Convener: Ambreesh Gupta

SESSION IX

LHC CALORIMETRY—SIMULATION

Convener: Ren-Yuan Zhu

SESSION X

CALORIMETER SIMULATION

Conveners: Adam Para and Norman Graf

SESSION XI

CALORIMETRY AT THE ILC—PFAs

Convener: Jerry Blazey

SESSION XII

CALORIMETRY AT THE ILC

Conveners: Jerry Blazey and Harry Weerts

Preface

The 12th International Conference on Calorimetry in High Energy Physics was held from June 5th to 9th, 2006, in Chicago hosted by the HEP division of Argonne National Laboratory. The conference sessions were held in Wieboldt Hall of the Kellogg School of Management, Chicago Downtown Campus. The attendees were welcomed by Dr. Robert Rosner, the director of Argonne National Laboratory.

Over a hundred physicists from a dozen countries attended the conference. This series of conferences, which began at Fermilab in 1990, has, under the strong guidance of the International Advisory Committee, become the premier gathering of the experts on calorimetry in high energy physics and related fields where in-depth discussions of latest developments and innovations take place.

The conference program was designed both to address the chronological development of calorimetry as well as its differing applications. Thus the conference began with presentations of operating calorimeters in high energy physics, astroparticle and neutrino physics, medicine, and calorimeter technologies. Then the calorimeters for the Large Hadron Collider (LHC), presently under construction, were discussed, followed by presentations on simulations, first for the existing, and soon-to-exist LHC, calorimeters then moving on to simulation studies for the International Linear Collider (ILC). The final sessions of the conference were devoted to the R&D efforts on calorimeters for the ILC.

The different sessions in the conference started with introductory talks by invited speakers who were A. Ereditato (overall), S. Wakely (astroparticle), C.-T. Chen (medicine), H. Oberlack (LHC), and F. Sefkow (ILC). In addition to the invited talks, there were 81 contributed talks. The sessions were organized by J. Repond (calorimeter technologies and medicine), S. Kuhlmann (operating calorimeters), K. Byrum (astroparticle and neutrino calorimetry), J. Freeman, M. Oreglia, J. Proudfoot (LHC calorimetery), N. Graf (simulations), G. Blazey, and H. Weerts (ILC). Thus, it was a conference at which the vast majority of the attendees actively contributed to its success.

It was a pleasure to be in the city of Chicago during the sunny and warm spring week. We would like to thank Ms. Libby Mahoney, the Chief Curator of the Chicago History Museum, for giving us a most interesting talk about the city of Chicago. The complicated logistics of holding such a conference was ably handled almost single-handedly by S. Klepec, the conference manager, with help from P. Malhotra.

Last but not least, we would like to thank the sponsors, without whom this conference could not have been held. These were, Hamamatsu Photonics K.K., the Photonis Group, Northern Illinois Center for Accelerator and Detector Development (NICADD), the University of Chicago, Deutches Elektronen-Synchrotron (DESY), European Organization for Nuclear Research (CERN), the HEP division of Argonne National Laboratory, and the Office of Science of U.S. Department of Energy.

S. Magill and R. Yoshida (editors)

International Advisory Committee

G. Belletini	Pisa Univ.
J. Colas	LAPP/Annecy
A. Ereditato	INFN/Napoli
F.L. Fabbri	INFN Frascati
H.A. Gordon	BNL/Brookhaven
D. Green	FNAL/Batavia
P. Jenni	CERN/Geneva
T. Kobayashi	ICEPP/Tokyo
P. Lubrano	INFN/Perugia
A. Maio	FCUL & LIP/Lisbon
A. Menzione	INFN/Pisa
H. Oberlack	MPI/Munich
A. Para	FNAL/Batavia
K. Pretzl	Bern Univ.
J.P. Rutherfoord	Univ. of Arizona
R. Wigmans	Texas Tech/Lubbock
R. Yoshida	ANL/Argonne
R-Y. Zhu	Caltech/Pasadena

Local Organizing Committee

G. Blazey	NIU
K. Byrum	ANL
J. Freeman	FNAL
S. Kuhlmann	ANL
S. Magill	ANL
M. Oreglia	Univ. of Chicago
J. Proudfoot	ANL
J. Repond	ANL
H. Weerts	ANL
L. Xia	ANL
R. Yoshida	ANL
S. Klepec	ANL

Calorimeter issues today

Antonio Ereditato

Laboratory for High Energy Physics (LHEP), University of Bern, Switzerland

Abstract. After a few historical remarks on the Calorimeter Conference (CALOR series) the author attempts outlining a review of the main issues of modern calorimetry, with emphasis to the debate on the future high energy applications after the Large Hadron Collider at CERN (LHC): the International Linear Collider (ILC).

Keywords: Calorimeters, particle detectors, high energy physics, colliders
PACS: 29.40.Vj

INTRODUCTION AND HISTORICAL CONSIDERATIONS

The CALOR conference series started with a Workshop successfully organized by Adam Para at Fermilab in 1990. After that, the author of this paper, willing to hold a calorimetric meeting in Capri initiated discussions with Adam Para and Peter Jenni about the possibility of creating a more ambitious project, namely a series of conferences. This idea got momentum and the second edition of CALOR was actually organized in Capri in 1991. This meeting was then followed by conferences in Corpus Christi (1992), Elba (1993), Brookhaven (1994), Frascati (1996), Tucson (1997), Lisbon (1999), Annecy (2000), Pasadena (2002), Perugia (2004) and now in Chicago, under the leadership of Steve Magill and Rik Yoshida.

The Conference has successfully reached the XII edition, basically alternating US and EU venues with a 1-1.5 years periodicity. In more than 15 years the CALOR conference hosted the debate around SSC/LHC detectors, featured the design, prototyping, construction and commissioning of LHC calorimeters, included papers on the construction and operation of non-collider calorimeters, experienced the move from HEP to particle physics calorimetry, certified the growing relevance of astroparticle and neutrino physics applications, played the role of the main forum where new ideas and R&D studies are presented, is (will be) the place where design and tests of ILC calorimeters are (will be) discussed.

The conference format and program evolved together with the technology and under the pressure of the community. More than1000 (!) invited and contributed papers were submitted in 12 editions covering all aspects of basic detection principles, R&D studies, performance of running calorimeters and studies on future detectors [1]. Altogether a big success.

As far as this introductory paper is concerned, the idea is not to have another review on calorimetry, since excellent papers by distinguished colleagues are available, and not a politically-correct gallery of results and achievements from past and present calorimeters. Instead, I would like start recalling what I personally learned working for many years with these detectors, mostly in realizing neutrino physics experiments, and then

CP867, *XII International Conference on Calorimetry in High Energy Physics,*
edited by S. R. Magill and R. Yoshida

continue with a few selected topics among which the challenge of calorimetry for the future high energy applications (*i.e.* the ILC).

EXPERIENCE WITH CALORIMETERS FOR NEUTRINO EXPERIMENTS

During the 80's fine-grained gas calorimeters were fashionable and widely used for large mass (volume) and relatively cheap detectors. As an example I like to recall the CHARM II neutrino detector for which both granularity (200000 R/O channels), mass (700 ton) and low density were essential requirements fulfilled by employing limited streamer tubes and glass as absorber material. The experiment has been able to collect the largest world statistics of muon-neutrino off electron elastic-scattering (more than 2000 events) [2].

Around 1990 the subject of compensation was first developed and understood theoretically [3] and then successfully studied in several experimental applications. I like to mention our fruitful experience within the CERN SPACAL and RD1 Collaborations within which we conducted a systematic study of the performance of compensating lead-scintillating fiber calorimeters for integrated ECAL and HCAL detectors mainly meant for SSC and LHC experiments [4][5]. World record hadronic energy resolutions were obtained by the SPACAL prototype. However, the main implementations of the technique were not realized for proton colliders. In particular, the SPACAL technique was applied to another neutrino experiment, namely for the calorimeter of the CERN CHORUS experiment searching for muon- to tau-neutrino oscillations [6]. The excellent hadronic energy resolution had been a key issue for the kinematical preselection of the events.

Still talking about neutrino physics it is clear that large mass calorimeters are the best possible detector for this application, since they can provide adequate statistics despite the small neutrino cross section and good event reconstruction and energy measurement if sufficient granularity is provided. Obviously, this becomes a serious practical (and financial) problem if one is confronted to very large neutrino observatories, *e.g.* those designed for underground astroparticle physics experiments. The major breakthrough for the technique was represented by the idea of surface-instrumented detectors instead of volume-instrumented devices (sampling calorimeters). For the former, the actual signal readout devices are placed at the surface of a large volume usually occupied by a (cheap) large mass sensitive material. A typical example is represented by water Cerenkov detectors (Kamiokande, Super-Kamiokande), liquid scintillator detectors (Kamland) and, last but not least, Liquid Argon Time Projection Chambers (LAr TPC), developed in the last 10-15 years under the pioneering initiative of the ICARUS Collaboration.

The LAr TPC provides bubble-chamber like event reconstruction capabilities with excellent calorimetric features and constitute a relatively cost effective solution. In recent years the technique has got momentum in view of future applications in astroparticle physics and for the next generation of precision neutrino experiments [7], motivated and justified by the discovery of neutrino oscillations. For the latter application, LAr TPCs are the ideal detector low energy-threshold measurement of neutrino interactions as well as for beam normalization and background determination.

A completely different calorimetric technique for neutrino detection is the so-called Emulsion Cloud Chamber (ECC) featuring a 'standard' sampling calorimeter where the active layers are made of thin, high space resolution, nuclear emulsion films. This is actually a very 'old' technique, re-discovered in recent years thanks to new technological developments both concerning the industrial production and handling of photographic films, and automatic scanning devices (microscopes). The 'new' ECC technique is perfectly suited for extremely fine-grained and compact calorimeters for the detection and kinematical analysis of short lived particles. It provides detailed event topology, energy measurement *e.g.* by micro-track counting, particle ID by dE/dX and multiple scattering measurement, and the possibility of realizing detectors of large mass (kton).

The main, recent application is represented by the OPERA experiment at the Gran Sasso laboratory, aimed at the direct detection of the (short lived) tau-lepton from muon- to tau-neutrino oscillations in the CNGS long baseline beam [8]. The total 'ECC calorimeter' mass is 1800 ton built up with millions of lead plates and emulsion films arranged in individual units called 'bricks'. A parallel achievement of the OPERA project has certainly been the setting up of the required 'scanning power' within the Collaboration: more than 50 automatic microscopes are able to scan the emulsion films with a speed of tens of cm^2 per hour, a value completely unthinkable even 10 years ago. The OPERA experiment has very recently started running and interesting results are expected in the forthcoming years.

TOMORROW'S HEP CALORIMETRY: GENERAL CONSIDERATIONS

In my opinion the progress of calorimetry is mainly determined by two different approaches. On the one hand, there are the curiosity-driven R&D studies on new techniques and methods, often originated from an improved understanding of the underlying physics mechanisms. On the other hand, the motivation is also given by the 'next to come' frontier experiments, for which the increasing energy, luminosity, detector dimensions and mass, etc. represent a great challenge.

In this respect, I think that we have to pay here a tribute to all colleagues presently very busy in following the second approach, in their effort in building and operating calorimeters for the LHC experiments [9] and in the intense activity around the design and prototyping of future ILC detectors. We could start summarizing the not-exhaustive list of (met) milestones for LHC calorimetry

- achieve high energy, position and angular resolution;
- keep hermeticity and coverage down to high rapidity;
- realize radiation hard/tolerant devices;
- maintain uniformity and stability of response, build suitable calibration methods;
- sustain high-rate and handle signal pile-up;
- produce suitable electronics to actually exploit the detector performance;
- guarantee long term operation;
- undergo very long production phases;

- take into account engineering and integration issues;
- keep cost within budget !

Also for future calorimeters, an important aspect is the high-energy frontier to be explored by accelerator experiments. It is clear that in pushing further this frontier the importance of calorimetry will be increasingly high due to the improvement (or at least not worsening) of the energy resolution with the energy. The opposite occurs in the energy (momentum) measurement by magnetic spectrometers for which one has a linear worsening of the momentum resolution with increasing energy. It is also worth noting that whatever new physics will appear at the energy frontier, this will produce in the final state Standard Model particles, the same particles which are measured with today's calorimeters.

Calorimeters provide means to detect all known final state particles: electrons, photons, muons, tau's, hadrons, jets and also neutrinos through a missing energy measurement. In nowadays and future experiments as well, a suitable combination of electromagnetic (EM) and hadronic calorimetry allows to perform particle identification by exploiting differences in the X_0 and Λ values of the absorbers (e/π separation) or differences in the electromagnetic radiation cross-section (e/μ separation).

Concerning the calorimetric energy resolution one can easily show that jet, dimass and missing energy resolution are determined by the single particle resolution if the energy measurement is the only error. One can prove that it is possible to achieve high-energy resolution with EM sampling calorimeter with high sampling fraction, realistically yielding to stochastic terms as small as 5%. However, the best energy resolution is obtained by exploiting fully active devices such as crystals. In this case one has no sampling fluctuations, but noise, photon statistics, and light collection non-uniformity limit the achievable resolution.

Different processes limit the performance of hadronic (sampling) calorimeters. In particular, the large event-by-event fluctuations on the EM content of the shower induce consequently large fluctuations in the energy deposition in the active medium, severely limiting the energy resolution. Compensation is considered a possible solution to improve the resolution. It is well known, in fact, that the medium can respond differently to EM and hadronic shower components due to the differences in ionisation, the existence of nuclear binding energy, of neutrons and undetected fragments. A detector with e/h close to unit shows the best energy resolution since it induces a small constant term in the resolution formula. However, the effect of non-compensation reduces at increasingly high energy, where the energy resolution improves with ln(E).

Despite the many efforts in the technology of hadronic calorimeters, other intrinsic limitations remain limiting their performance. The detector transverse size set by the shower extent (X_0 and Λ) limits the smallest tower size. The minimal longitudinal depth is set by containment (about 20 X_0 and 10 Λ); the largest depth is set by jet leakage. The speed is limited by 25 ns bunch crossings at LHC; no reduction in pileup is expected if signals would be faster. The jet resolution is limited by final state radiation (FSR) and not really by the calorimeter energy resolution.

If one looks at the (next) future, namely at the high-energy frontier that could be explored by a hadron collider such as the LHC, one has to notice that both luminosity and CM energy increases are possible in principle. For particle masses of the order of the CM energy the required luminosity (L) rises rapidly, indicating that the energy is the most important issue; for masses much lower than the CM energy, L goes as the square of the mass. However, higher mass states or higher L in hadron colliders will require calorimetry that can withstand more than 10 (2) Mrad for the ECAL (HCAL) calorimeter for pseudorapidity values larger than 3 units. Scintillators would not survive for ECAL regardless the CM energy. Hermiticity will require a coverage to smaller angles as the CM energy increases and the 'plateau' extends. Forward calorimetry will then need to withstand more than 1 Grad. This dose is a real challenge and likely only gaseous or Cerenkov light detectors could be used.

THE CHALLENGE OF THE ILC

After the generation of forthcoming experiments at the LHC and the desirable discovery of Higgs and Super-Symmetry, precision studies will have to be performed at an electron machine, following what happened after the discovery of the W and Z bosons at the SPS and the precision Standard Model measurements at LEP. A future electron linear collider will have to face the challenge of very steeply falling cross-sections of the interesting processes: by large factors lower than the ones at LEP. In order to measure masses from 500 to 1000 GeV one will require very large luminosities in the range from 10^{33} to a few 10^{34} $cm^{-2}s^{-1}$. This will be the task of the International Linear Colliders (ILC), the machine that will be hopefully realized after the LHC and whose design has largely profited from the past R&D studies performed in Europe (CLIC, TESLA), in Japan (JLC) and in the USA (NLC) [10].

The main tool of the detectors in addressing the relevant physics issues at a future electron machine will be the detection of jets. This feature is complementary to what is going to be performed at the LHC: detectors at a linear collider have to do well what LHC finds difficult (detection of hadronic decays, multijets, etc.). The primary goal will be revealing the nature of the electroweak symmetry breaking through precision studies.

The main requirements for ILC calorimetry can be summarized as follows

- energy resolution for single particles (e/γ and hadrons);
- energy resolution for jets;
- linearity (1-100 GeV);
- hermeticity (no cracks);
- operation in strong magnetic field.

As compared to the specific needs of LHC detectors, we will require the inner vertex layer to be 3-6 times closer to the interaction point, a vertex pixel size 30 times smaller with a parallel reduction on the amount of material by a factor of 30, the global vertex detector material amount reduced by a factor 6, a track momentum resolution 10 times better and possibly an EM calorimeter granularity 200 times higher. The goal is then to obtain a jet energy resolution for physics events of about $0.3/\sqrt{E}$(GeV) with full hermeticity (required to veto background events and measure missing energy) down to 5-10 mrad. The typical benchmark measurement will be the ability to separate WW and ZZ in multi-jet events.

The above requirements will imply high energy resolution for hadrons. This could be naturally achieved *e.g.* by designing compensating calorimeters. This feature has been very well studied and established experimentally by the already mentioned SPACAL collaboration and by the ZEUS experiment at DESY, and already proposed for the JLC detector.

From the point of view of calorimetry the environment at an electron machine will be rather clean, with low occupancy apart from the presence of low transverse momentum electron pairs. Large B fields could be employed to 'open' the jets and to apply one of the main distinctive experimental methods at these colliders: the energy flow or particle flow analysis (PFA). This concept is based on the assumption that charged particles in a jet are more precisely measured with the tracker than with a calorimeter. In fact, the energy of a typical linear collider multi-jet event belongs for 64% to low momentum (suitable for tracking) charged particles, for 25% to photons and only for 11% to neutral hadrons. The energy flow method consists in using trackers for the energy measurement of charged particles and calorimeters for neutrals. One has to locate and remove charged particle's calorimetric energy by means of dense, highly-granular ECAL and HCAL. ECAL can be considered 'transparent' to hadrons: one detects hadrons in HCAL and photons in ECAL. The main calorimetric figures of merit are a 'large' BR^2/R_m, suitable for cluster detection in the ECAL, as well as a transverse segmentation of the order of R_m with a small X_0/Λ ratio.

The PFA largely determines the design of the calorimeters, and the latter, in turn, drive the design of the apparatus since the cost of ECAL, similar to that of HCAL, reaches for a typical ILC detector the cost of $ 50-100 M. Much more than at the LHC, they really constitute the heart of the experiment. Under these conditions, if the goal is measuring the energy of each individual particle, one necessarily needs extremely high longitudinal and transverse segmentation, *i.e.* high density absorber material.

However, one should mention that although the PFA concept is potentially attractive, in order to have a really working (and powerful) tool one will need to perform more extensive simulation studies to confirm the actual feasibility of the hardware, as well as proving that the method works experimentally by beam test results. Some trials made by re-analizing CDF data only yielded a small energy resolution improvement from $0.85/\sqrt{E}$ to $0.65/\sqrt{E}$. All in all, we will have to be convincing in showing that PFA is really what one needs and not just an expensive overkill...

Concerning the calorimeter technology at an electron linear collider, one option envisages for the ECAL large scale tungsten/silicon calorimetry with more than 1000 m^2 total Silicon surface and a 20 longitudinal-layer deep detector. Following the prescriptions of the energy flow concept, one can design a finely segmented calorimeter, such as the one already used, although at a much smaller scale, for luminosity monitors at SLC and LEP. The main drawback is the (presently) large or unaffordable (according to optimistic/pessimistic point of view) cost of the Silicon. However, we could hope of a sort of Moore's law for Silicon that by the year 2010 should bring the cost of this material to a reasonable level.

As far as the HCAL is concerned one should take into account that there is only a small contribution from neutral hadrons. This may allow only a modest HCAL energy resolution. People still consider scintillator tiles as a viable option, or the use of the so-called 'digital' hadronic calorimetry characterized by small segmentation (1 cm^2) with only 1-2 bit readout, possibly employing RPCs or similar cheap detectors.

Assuming the PFA method one is led to the following expressions for the energy measurement

$$E_{tot} = \Sigma p_{ch} + \Sigma E_{\gamma} + \Sigma E_{neut}$$

$$\sigma^2_{E\,jet} = \sigma^2_{ch} + \sigma^2_{\gamma} + \sigma^2_{neut} + \sigma^2_{confusion}$$

$\sigma^2_{confusion}$ is relatively important and is related to the discrimination between separate energy clusters and to track-cluster association.

As previously said, a strong magnetic field is required to bend charged-particle tracks as well as a fine segmentation (in volume) for both the ECAL and HCAL. This brings to designing detectors with large ECAL inner radius (for particle separation), small Moliere radius (in order to get small clusters) and large λ_{int}/X_0 ratio.

One more word on the possible use of digital HCAL readout. This option fits with the requirement of very fine granularity and segmentation (~1cm, ~50 layers) in order to build a cost effective device. The large number of channels is balanced by single-bit R/O whereas a simple hit counting can be successfully applied to estimate the total cluster energy. In addition, one may get rid of Landau fluctuations that could harm the analog readout.

Starting from these general considerations several groups have started designing and proposing actual calorimeters for ILC experiments. Among the common features we note the need of dense absorber material for both ECAL and HCAL. In the hypothesis of adopting PFA, Silicon is considered as a baseline option for the active ECAL medium. Under the same assumptions, neutron detection in HCAL is an issue. Therefore, the presence of Hydrogen would be desirable in the active medium, while gaseous active media might be suitable for large radius HCAL detectors.

Among the ECAL options (active-passive media) we can list: Silicon-Tungsten, scintillator/Silicon-Lead, scintillator/Silicon-Tungsten, scintillator-Lead and scintillator-Tungsten. For HCAL we mention: scintillator-stainless steel (analog), scintillator-Lead (analog), GEM-stainless steel (digital), RPC-stainless steel (digital), scintillator-stainless steel (digital) and scintillator-Lead-stainless steel (digital).

CALORIMETERS FOR THE ILC PROPOSED DETECTORS

As discussed above, the community interested in the ILC physics program is organizing itself by clustering around active proto-collaborations.

The Silicon Detector - SiD Collaboration [11] envisages the use of a calorimeter with the highest granularity, made of a Tungsten absorber and Silicon detectors. To make the detector affordable, a relatively small inner calorimeter radius of 1.3 m is chosen. Shower separation and good momentum resolution are achieved with a 5 T magnetic field and very precise Silicon detectors for charged-particle tracking. The fast timing of the Silicon tracker makes SiD a suitable detector to fight backgrounds.

The Large Detector - LDC Collaboration [12] has designed an apparatus derived from the detector described in the technical design report for TESLA and uses a somewhat larger radius of 1.7 m. It also features a Silicon-Tungsten calorimeter, possibly with a coarser granularity. For charged-particle tracking, a large TPC is planned to allow efficient and redundant particle reconstruction. The larger radius is needed to achieve the required momentum resolution.

The GLD detector [13] features a larger radius of 2.1 m to take advantage of a shower separation just by distance. It uses a calorimeter with even coarser segmentation and gaseous tracking similar to the LDC with a moderate 3 Tesla B field value.

Last but not least, a fourth Collaboration (actually called '4th' [14]) proposes an alternative detector design consisting of four basic subsystems: a pixel vertex detector, a Time Projection Chamber, a high precision multiple-readout fiber calorimeter, complemented with an EM dual-readout crystal calorimeter and an iron-free dual-solenoid muon system for the inverse direction bending of muons in a gas volume. The pixel vertex chamber, TPC and calorimeter are inside the solenoidal magnetic field.

The SiD detector design requirement, nearly analogous to those of the other proposals, well explains the case of the future detectors at the ILC

- two-jet mass resolution comparable to the widths of W/Z for a unambiguous identification of the final states;
- excellent flavor-tagging efficiency and purity (for b, c and s quarks);
- momentum resolution adequate to reconstruct recoil-mass to dimuons with resolution better than the beam-energy spread;
- high hermeticity to precisely determine the missing momentum;
- good timing resolution to tag bunch-crossings to suppress BG in calorimeter and tracker;
- very forward calorimetry that resolves each bunch in the train for veto capability.

Having in mind the goal of a jet energy resolution of $0.3/\sqrt{E}$(GeV) SiD chooses a dense, highly segmented, Si/W ECAL and HCAL. High magnetic field values limit the radius and the cost of calorimeters and solenoid (B = 5 Tesla). The detector is rather compact: a sort of cube with a side of about 12 m. The ECAL features 20 layers of 2.5 mm thick W followed by 10 layers of 5 mm thick W interleaved by 1 mm Silicon detector gaps. For the HCAL either scintillator strips, RPCs or GEMs are possible options.

The GLD Collaboration, instead, considers that the fine segmentation of the calorimeters is intrinsically limited by the Moliere radius of the passive material (W/scintillator 1.8 cm, Pb/scintillator 2.5 cm). High B field can spread hits but might be not efficient for dense jets. Therefore, they adopt a large ECAL radius (W-scintillator) with a few cm^2 granularity coupled to a large radius TPC for tracking and a relatively moderate B field.

This 'large' detector option has obviously vantages and disadvantages. Among the first we recall good jet PFA measurement (better cluster separation and charge particle separation at the entrance of ECAL due to large BL^2), easier pattern recognition (more hits?), good momentum resolution for charged particles, good dE/dx measurement for charged particles and smaller relative volume of the dead space.The main disadvantages are the larger solenoid (lower magnetic field) and hence the larger vertex detector inner-radius, as well as the calorimeter volume increase that then calls for cheap calorimeter options.

As said above, the approach followed by '4th' is substantially different from the other three detectors. As far as the calorimeter is concerned a new technology has been developed based on the 'dual' approach, somehow alternative to the PFA analysis. The calorimeter is a spatially fine-grained dual-readout fiber sampling calorimeter with the additional ability to measure the neutron content of a shower. The dual fibers are scintillation and Cerenkov for the separation of hadronic and electromagnetic components of hadronic showers.

CONCLUSIONS

The role of calorimetry is becoming increasingly important in high energy physics, since these detectors are well suited to face the challenge of the high energy frontier. After the planned generation of LHC experiments (and calorimeters) we do expect detector improvements and upgrades together with a better understanding of the physics scenario. However, it is also clear that intrinsic and conceptual limitations in the performance of calorimeters will remain. The issue of radiation hardness, in particular, will become fundamental in considering the possible luminosity upgrade of the accelerators as required to access higher-mass states.

Different considerations apply to the next generation detectors for the International Linear Collider, expected to be operational after the likely discovery of new physics at the LHC. From the technical point of view linear-collider (highly segmented) calorimeters could substantially contribute to the precision measurements by employing new methods such as the energy flow. However, both machines and experiments are in the R&D phase; further prototyping and simulations are needed. The main challenge will be the likely high cost of the detectors.

Since this is the introductory talk of this Conference I wish to conclude with a series of question that I would like be addressed in the following talks

- Is the calorimeter underlying theory completely understood?
- Are there new promising techniques from recent R&D studies?
- Astroparticle/neutrino and collider calorimeters: are they the two future main streams of the technique?
- What will be the lesson from LHC calorimetry?
- Electronics, software, calibration: are they adequate to meet the future experimental challenges?
- ILC calorimetry, the next big enterprise. How to assess the PFA issue in an objective way?
- What is the best (performing, affordable, cost effective) solution for ILC ECAL/HCAL?
- Specific ILC questions: what is the effect of material in front of the calorimeter? How much do we rely on simulations? Will beam tests give clear answers for the design of the final detectors? How relevant will be the 'association confusion' ?

I look forward triggering an interesting discussion!

ACKNOWLEDGMENTS

I am very glad to Steve Magill and Rik Yoshida for the kind invitation and for the excellent organization of the Conference.

REFERENCES

1. http://calor.pg.infn.it/
2. K. De Winter et al., A detector for the study of neutrino-electron scattering, Nucl. Instr. and Meth. A278 (1989) 670.
3. R. Wigmans, Nucl. Instr. and Meth. A259 (1987) 389.
4. D. Acosta et al., Electron, pion and multiparticle detection with a lead/scintillating fiber calorimeter, Nucl. Instr. and Meth. A308 (1991) 481.
5. J. Badier et al., Test results of an electromagnetic calorimeter with 0.5 mm scintillating fiber read-out, Nucl. Instr. and Meth. A337 (1994) 314.
6. E. Di Capua et al. ,Response to electrons and pions of the calorimeter for the CHORUS experiment, Nucl. Instr. and Meth. A378 (1996) 221.
7. A. Ereditato and A. Rubbia, Ideas for future liquid Argon detectors, Nucl. Phys. Proc. Suppl. 139 (2005) 301;
   A. Ereditato, A. Rubbia, The liquid Argon TPC: a powerful detector for future neutrino experiments and proton decay searches, Nucl. Phys. Proc. Suppl.154 (2006) 163;
   A. Ereditato, A. Rubbia, Conceptual design of a scalable multi-kton superconducting magnetized liquid Argon TPC, Nucl. Phys. Proc. Suppl.155 (2006) 233.
8. http://operaweb.web.cern.ch/operaweb/index.shtml
9. P. Checchia and A. Ereditato, Particle detectors for future collider experiments, IEEE Instrumentation and Measurement Magazine, Vol 8, No. 2, (2005) 27.
10. http://www.linearcollider.org/cms/
11. http://www-sid.slac.stanford.edu/
12. http://www.ilcldc.org/
13. http://ilcphys.kek.jp/gld/
14. http://physics.uoregon.edu/ lc/wwstudy/concepts/

SESSION I

OPERATING CALORIMETERS

Convener: Steve Kuhlmann

Performance of the DØ Calorimeter During Run IIA of the Tevatron

N. J. Buchanan

Florida State University, Tallahassee, FL

Abstract. The performance of the DØ liquid argon calorimeter during Run IIA of the Tevatron at Fermilab is described. Online calibration and data quality monitoring are included, as well as the effects of noise on the calorimeter and overall system maintenance during the run period.

Keywords: calorimetry
PACS: 07.20.Fw

THE DØ CALORIMETER

The DØ calorimeter is comprised of three liquid argon (LAr) sampling calorimeters that use uranium, or steel, for absorbers. A cryostat holds the central calorimeter which covers the pseudo-rapidity range $|\eta|<1$, and two forward cryostats hold the endcap calorimeters which further cover out to $|\eta|=4.2$. The calorimeters are sectioned into electromagnetic, fine hadronic, and coarse hadronic cells with the electromagnetic cells closest to the beam pipe and the fine hadronic and coarse hadronic cells following radially outward. The segmentation of the calorimeter cells is $\Delta\eta=0.1$ and $\Delta\phi=0.1$ except for the third (of four) electromagnetic layer where the segmentation is half as coarse for sampling the shower maximum. The calorimeters were installed for Run I of the Fermilab Tevatron and more details can be found in Ref [1]. Figure 1 shows the a cross-sectional view of one quarter of the Run II DØ calorimeter.

An additional scintillator-based inter-cryostat detector (ICD) is situated in the gap between the central and forward cryostats in order to recover some of the loss in energy resolution that results from the lack of active material in the region. The ICD was installed for Run II of the Tevatron. THe ICD can be seen in the space between the cryostats in figure 1.

The DØ detector was upgraded between Run I and Run II, and while the calorimeters were not modified, the calorimeter readout was upgraded to address the 396 ns bunch separation for Run II. A solenoid magnet and preshower detector, including a lead absorber were also installed between the beam pipe and the calorimeter. Figure 2 shows the location of material between the beam pipe and the calorimeter.

During the 2006 Tevatron shutdown that took place between February and May, a new level 1 calorimeter trigger was installed [2]. An upgrade was also made to the ICD that should greatly increase the sensitivity of this detector.

CP867, *XII International Conference on Calorimetry in High Energy Physics,*
edited by S. R. Magill and R. Yoshida

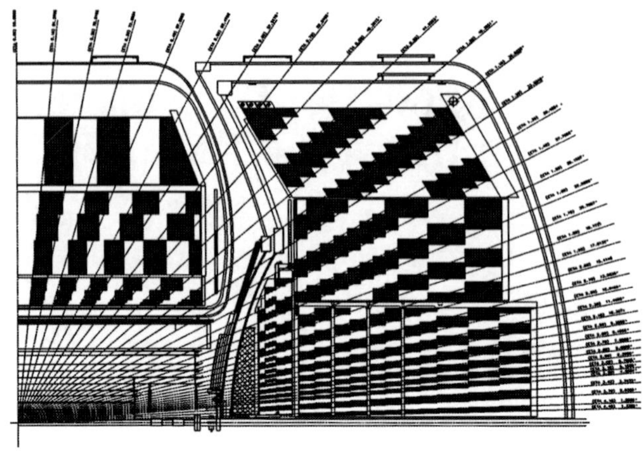

FIGURE 1. Cross-sectional view of one quarter of the DØ calorimeter.

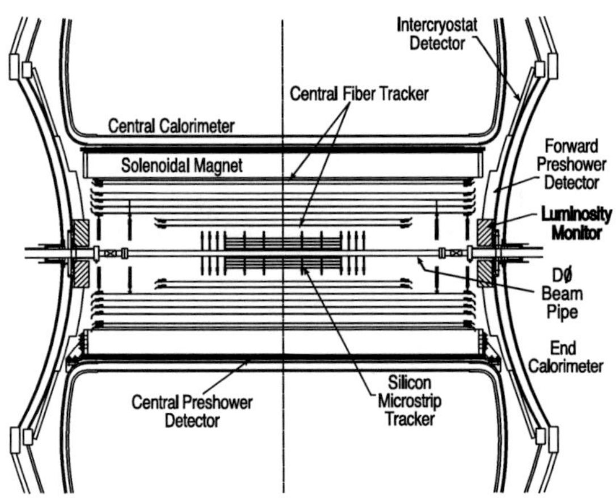

FIGURE 2. Zoomed in view of the region between the interaction point and the calorimeter.

Readout Electronics

The calorimeter readout electronics consists of 55,296 channels, of which 47,032 are connected to physical calorimeter modules. The signal into each channel is amplified and shaped with dual parallel-input JFETs and driven along a 115 Ω twisted-pair cable to baseline subtraction (BLS) cards. Several species of preamplifiers are used to account for the variation in calorimeter cell capacitance, and a special species is used to make the ICD signals similar to those of the LAr calorimeter channels. The remainder of the readout chain for the ICD is identical to that of the rest of the calorimeter.

The resulting signals are then shaped and stored, at 132 ns intervals, in switched-capacitor arrays (SCAs) while awaiting a level 1 trigger decision. Two gain paths, unity

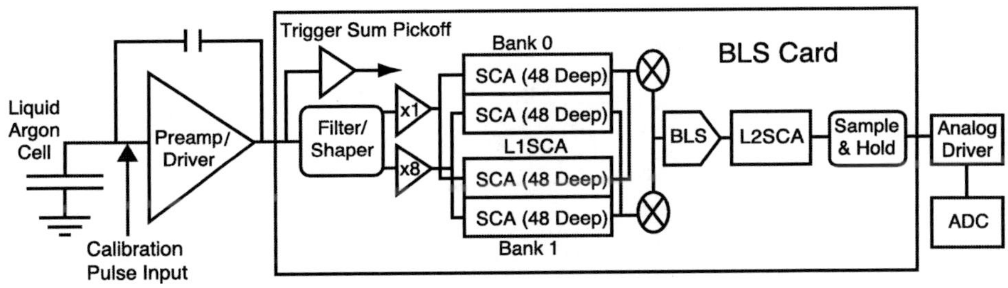

FIGURE 3. Schematic of the calorimeter readout chain.

and gain 8, are used to achieve the desired 15-bit dynamic range. Prior to the shaping stage copies of the signals are "picked off", summed into projective towers, and sent to the level 1 calorimeter trigger[1]. Signals selected by the level 1 trigger are passed to a second bank of SCAs for storage prior to a level 2 trigger decision. The BLS cards also provide baseline subtraction to remove low frequency noise and pile-up. Following a level 2 trigger accept, the signals are driven off to analog-to-digital (ADC) cards for digitization and optional pedestal subtraction and 1.5 σ zero suppression. Figure 3 shows a schematic of the calorimeter readout electronics. Further details can be found in ref [3].

Calibration

Pedestal calibrations are performed precedinging each store[2], or about once per 1-2 days. The pedestal procedure uses 10,000 events, for each gain path, and takes about 10 minutes. A level 3 filter is used to process the data to allow close to real time processing. Each set of pedestals is given a unique version number to ensure that the pedestals used offline during reconstruction match those that were loaded on the hardware when the data was taken. Fluctuations in the pedestal values are typically observed to be less than one ADC count over several months. A typical RMS value for an electromagnetic channel is between 3 and 4 ADC counts while coarse hadronic channels can have RMS values about twice this amount.

Gain calibrations are taken less often as the procedure requires about 30 minutes of time with no beam in the accelerator. For gain calibrations charge is injected onto the preamplifier inputs using a pulser system with a 17-bit digital-to-analog converter (DAC). The average current per DAC unit of 0.825 μA/DAC and the injected signals are similar in shape and duration to physics signals. Three separate runs, of varying DAC step granularity, are taken to fully sample the dynamic range of the readout. Figure 4 shows that the linearity of the system is generally better than 1% (plotted is

---

[1] Each level 1 calorimeter trigger tower is made up of 4 calorimeter towers giving a segmentation of $\Delta\eta$=0.2 and $\Delta\phi$=0.2

[2] A store is the period during which the Tevatron provides proton-antiproton collisions to the experiments. A store has a typical lifetime of 15-20 hours.

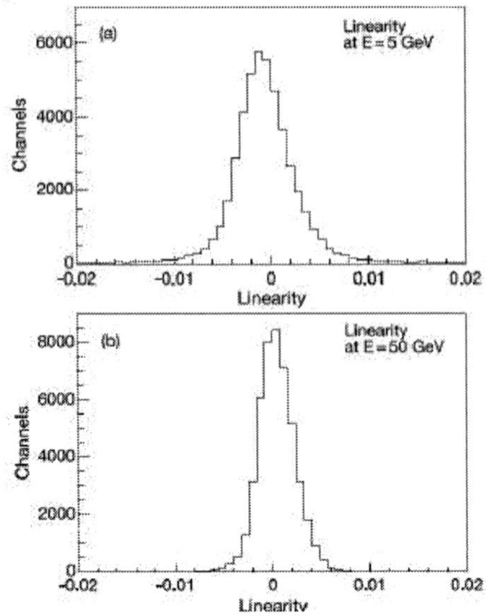

FIGURE 4. Plot of $ADC_{measured}/ADC_{fit}$)-1 that shows the good linearity of the calorimeter readout when pulsed. The top plot corresponds to an equivalent energy of 5 MeV and the bottom plot corresponds to an equivalent energy of 50 MeV.

($ADC_{measured}/ADC_{fit}$)-1).

Studies have shown that the SCAs saturate at the high and low ends of their dynamic range. To account for these non-linearities a method has been developed to empirically parametrize the non-linearities for off-line corrections. The same runs taken for the gain calibrations are used for the non-linearity corrections. A detailed description of the calorimeter calibration system and procedure is given in ref [3].

ONLINE MONITORING

Online monitoring of both the hardware performance and data quality is provided to the shifters and system experts through a variety of graphical interfaces (GUIs) and examination tools (Examines). All critical hardware parameters, such as voltages, currents, temperatures, status registers, etc..., are monitored in real time. A significant event system (SES) is set to trigger warnings and alarms if user-defined thresholds for the monitored parameters are exceeded. The SES provides a mechanism for displaying the warnings and alarms to the shifters in the control room and in addition, can be set to pause data taking in situations where hardware failures or performance could affect the data quality. All of the monitored parameters are also written to files that are archived weekly. This allows the reconstruction of conditions leading to failures for more detailed study.

TABLE 1. Typical maintenance of calorimeter hardware located in the collision hall. The platform is located below the detector and is readily accessible once in the collision hall, while access to the cathedral requires an additional 4 hours to open the muon toroid. The mean time between failures (MTBF) is the typical time between the failure of any one component, this is different from the MTBF for each individual component.

Component	Channels affected	MTBF (weeks)	Location	Repair (hours)
preamp power supply	4608	26	cathedral	8
BLS power supply	768-1536	8	platform	1
pulser power supply*	4608	8	platform	.5
BLS card	1-48	4	platform	.25
SCA card	1-12	1	platform	.25

* not critical to data taking

Data quality is monitored in real time using the outputs of trigger objects and reconstructed physics objects. Shifters in the control room and users connecting via a web interface can monitor the data quality as events are taken. References plots are periodically recorded for several different luminosity ranges to ensure that quantities such as jet multiplicity are as expected. We are currently moving toward a mode of operation where the data Examines will have the ability to trigger SES alarms. An example of this application would be to monitor jet trigger rates against reference sets to catch hot trigger towers when they first appear. The move toward more automated monitoring has led to increased efficiency.

OPERATION AND MAINTENANCE

The calorimeter has run with high efficiency and very little downtime during RunII. A modification to the BLS power supplies in 2004 has dramatically reduced the number of power supply failures, which accounted for the majority of downtime prior to that time.Since the 2004 Tevatron shutdown which took place between September and December of that year, there have been less than 10 stores, of about 200, where the calorimeter was not used in its entirety. A typical reason for removing the calorimeter from the experiment readout during data taking is the loss of a major hardware component, such as a BLS power supply, that resides in the collision hall. Access for such repairs is limited to the time between stores and thus as much as a day can pass before a repair is made. Most hardware failures affect between 1 and 48 channels and have minimal affect on the data quality. Such failures are typically repaired within one day. Table 1 gives some typical maintenance figures for hardware located inside the DØ collision hall. The preamplifiers and associated power supplies reside inside the cathedral, which is inside the outer layers of the muon system and the toroid magnet. It typically takes about 6 hours to open the detector to the point where the cathedral can be accessed. The remaining calorimeter electronics in the collision hall sit on the platform under the detector.

An important part of the maintenance procedure is providing failure and repair information to the offline data quality group so that data quality records can be adjusted as necessary. This ensures that the minimum amount of data is discarded. For example, if a power supply was to fail at the end of a 4 hour run we would not want to discard the data taken prior to the failure.

NOISE

Noise[3] is always a concern when making precise energy measurements and high gain systems, such as the DØ LAr calorimeter, are particularly susceptible. Prior to the 2004 shutdown the DØ calorimeter experienced periods of noise that were eventually correlated with an arc welder operating in the DØ assembly area. During the shutdown a grounding error was discovered and fixed. Following this repair the effects of external noise have been essentially eradicated. This experience led to the development of tools, such as special noise triggers, that are used periodically to monitor calorimeter noise. During shutdowns several days are allotted for calorimeter noise studies, and related maintenance if necessary.

While some sources of noise remain, and about 3% of unbiased data contain some form of coherent noise, significant effort has been spent in order to understand and permit the filtering of such noisy events. Events that are deemed noisy are categorized, flagged, and written into the data. This permits analyzers to remove the events from their data, while keeping the events available for further study. There is work currently underway to reduce the amount of coherent noise in the calorimeter readout chain.

During the past year studies have been performed to look for luminosity dependence of calorimeter noise. To date, no luminosity dependence, in the rate or severity of noise, has been observed.

PERFORMANCE

The performance of the calorimeter has been excellent and is continually improving. During 2005, 93% of the calorimeter data was of sufficiently high quality to be used in data analysis. This number continues to increase and we expect efficiencies of at least this level this year.

As examples of calorimeter performance, Figures 5 and 6 show plots of the reconstructed Z mass, in the di-electron channel, and the missing transverse energy (MET) spectrum for reconstructed Z's in di-electron channel respectively. Both plots agree with what is expected from Monte-Carlo simulations.

The missing transverse energy plot is consistent with the MET distributions observed for unbiased, with respect to trigger, data which have a typical width of 5-6 GeV. In the mass plot, the circles correspond to the case when the Z candidate was reconstructed with two electrons in the central cryostat and at a pseudorapidity less than 0.1, and the triangles correspond to the case where both electrons were in the central cryostat at a psudorpidity greater than 0.8. The difference between the two cases comes from

---

[3] In the context of this note, noise excludes pile-up.

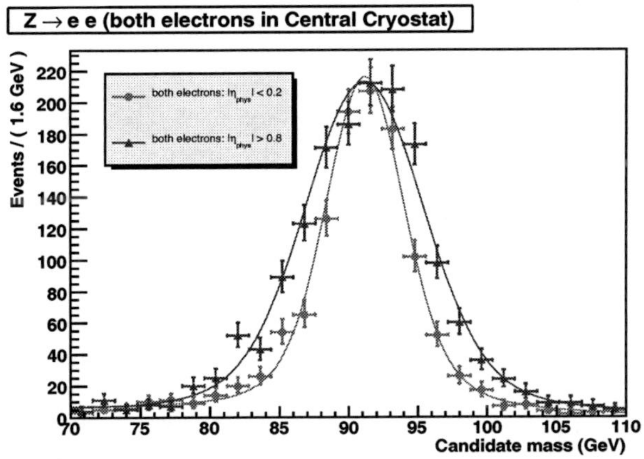

FIGURE 5. Reconstructed di-electron mass spectrum for the cases when both electrons are contained in the central calorimeter but for 2 different η ranges.

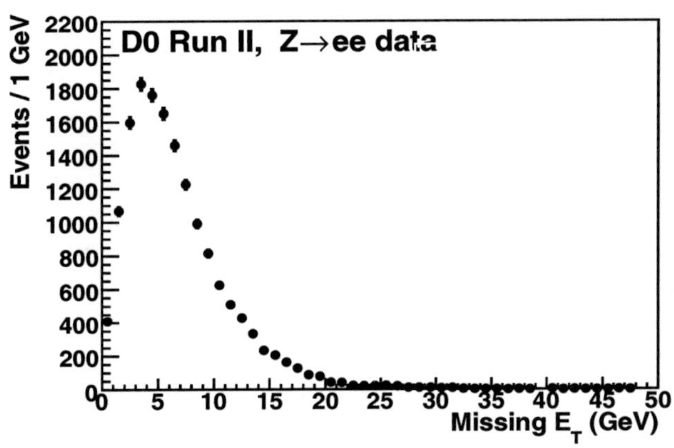

FIGURE 6. Missing transverse energy spectrum taken from di-electron Z decays.

an angular dependence in the shower maximum due to the increased dead material between the interaction point and the first active LAr layer of the calorimeter. Almost 4 radiation lengths of dead material sit between the interaction point and the first active layer of the LAr calorimeter. Work is currently underway to correct for this effect. In addition, an upgrade to the readout of the preshower detectors is expected to improve the dynamic range of these detectors which will compensate for some of the dead material. It is expected that with these improvements and increased statistics the DØ Run II performance goals should still be met.

Preliminary results from the new level 1 calorimeter trigger show improved turn-on curves, compared to those from Run IIA. Studies are still ongoing to to improve the performance of this trigger, these include tuning the jet, electron, and tau algorithms.

CONCLUSION

Following the upgrade of the DØ detector, including new calorimeter readout electroics, between Run I and Run II of the Tevatron the caloriemter has performed well. During the past year 93% of the data was of high enough quality to be used for physics analysis. A move over the next several months to further automate operation should further improve detector efficiency while offline algorithm work and improvements to the preshower detector readout will help to compensate for the amount of dead material in front of the calorimeter.

Careful attention to detector grounding, along with effective mitigation methods at the data quality level have greatly reduced noise concerns and there have been no indications of luminosity dependent noise.

While many studies are currently underway to understand the performance of the new level 1 trigger system, early indications are that the performace af the trigger is measureably better than the trigger used during Run IIA. All expectations are that the calorimeter and calorimeter trigger will meet the goals set out at the start of Run II.

ACKNOWLEDGMENTS

We thank the staffs at Fermilab and collaborating institutions, and acknowledge support from the DOE and NSF (USA); CEA and CNRS/IN2P3 (France); FASI, Rosatom and RFBR (Russia); CAPES, CNPq, FAPERJ, FAPESP and FUNDUNESP (Brazil); DAE and DST (India); Colciencias (Colombia); CONACyT (Mexico); KRF and KOSEF (Korea); CONICET and UBACyT (Argentina); FOM (The Netherlands); PPARC (United Kingdom); MSMT (Czech Republic); CRC Program, CFI, NSERC and WestGrid Project (Canada); BMBF and DFG (Germany); SFI (Ireland); The Swedish Research Council (Sweden); Research Corporation; Alexander von Humboldt Foundation; and the Marie Curie Program.

REFERENCES

1. S. Abachi *e*t al, *Nucl. Inst. Meth. Phys. Res.* **A 338**, 185 (1994).
2. M. Abolins *e*t al, *IEEE Trans. on Nucl. Sci.* **51** 340 (2004)
3. V. M. Abazov *e*t al, "The Upgraded DØ Detector", Accepted for publication in *Nucl. Inst. Meth. Phys. Res.*, (2006).

Status of the CDF II Calorimeters

Mark Mattson

Fermilab, P.O. Box 500, Batavia, IL 60510, USA

Abstract. The status of the CDF calorimeters was reported at the CALOR2002 conference, about a year after Run II started at the Tevatron Collider. I will review upgrades to the system since that conference, as well as the operation and performance of the calorimeters.

Keywords: CDF calorimeter
PACS: 01.30.Cc, 07.20.Fw

INTRODUCTION

The Collider Detector at Fermilab (CDF) is an ongoing experiment, collecting data from $\bar{p}p$ collisions at the Tevatron. Upgrades to the detector were made after the Run I data taking period ended in 1996, to account for increases in luminosity expected for Run II due to shorter time between bunch crossings and more particles per bunch. A description of the CDF II detector is available elsewhere [1]. The calorimeter is composed of systems built over the course of two decades. The last upgrade to the calorimeter hardware was completed in the fall of 2004. With the the determination of the jet energy scale for the Run II data [2], it is worthwhile to review the current status of the CDF II calorimetry.

This review does not cover the muon detectors, or the far forward calorimeter systems. The calorimeter timing system is discussed in a different talk by Max Goncharov.

CALORIMETER DESCRIPTION

The CDF calorimeter has five calorimeter compartments as shown in Fig. 1: central electromagnetic CEM [3], central hadronic CHA [4], plug (forward) electromagnetic PEM [5], plug hadronic PHA [5], and the wall hadronic WHA [4] in the region between the central and the plug calorimeter at $\theta \approx \pm 30°$ (see Fig. 1). The CDF detector is divided into two halves at the origin of the coordinate system, where the z axis points along the beam pipe, θ is the polar angle, ϕ is the azimuthal angle, and $\eta = -\ln\tan(\theta/2)$ is the pseudo-rapidity.

The central calorimeter consists of scintillator plates with wavelength-shifting (WLS) bars. It was used in Run I, and covers the region $|\eta| < 1.1$. It is segmented in towers of 15° in azimuth and 0.1 in η with lead-scintillator sampling for the electromagnetic measurements and steel-scintillator sampling for hadronic measurements. The material in the CEM has a depth of 18 radiation lengths. The energy resolution for high energy electrons and photons is $\frac{\sigma(E_T)}{E_T} = \frac{13.5\%}{\sqrt{E_T}} \oplus 1.5\%$. The CHA and WHA are of similar construction, with alternating layers of steel and scintillator and are 4.7 interaction lengths deep. Three of the WHA towers are situated behind the CEM/CHA and three

CP867, *XII International Conference on Calorimetry in High Energy Physics,*
edited by S. R. Magill and R. Yoshida

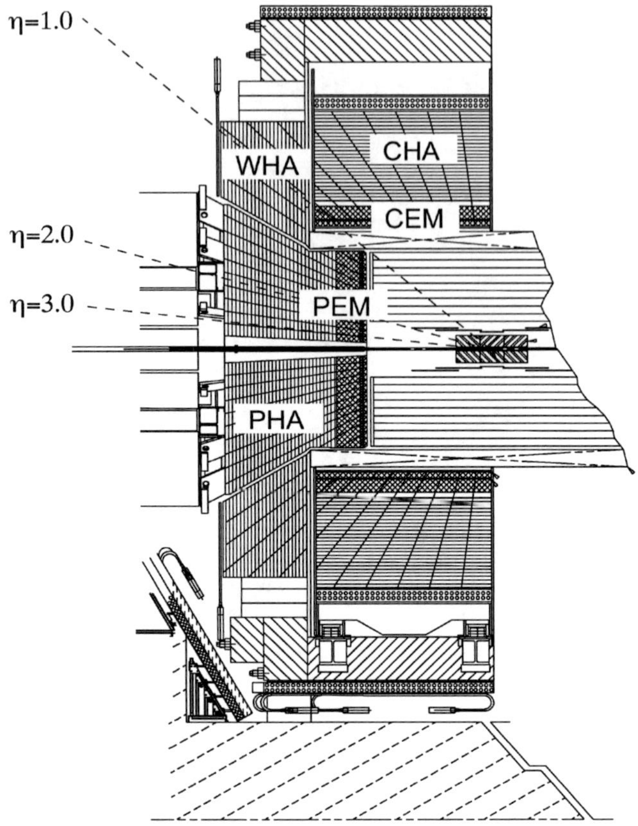

FIGURE 1. Elevation view of one half of the CDF detector displaying the components of the CDF calorimeter: CEM, CHA, WHA, PEM and PHA.

are behind the plug calorimeter. The energy resolution of the CHA is $\frac{\sigma(E_T)}{E_T} = \frac{50\%}{\sqrt{E_T}} \oplus 3\%$ and WHA is $\frac{\sigma(E_T)}{E_T} = \frac{75\%}{\sqrt{E_T}} \oplus 4\%$ for charged pions that do not interact in the CEM.

The forward "plug" calorimeters cover the angular range corresponding to $|\eta| > 1.1$. The Run I plug gas calorimetry would have too high occupancy under Run II conditions, and was replaced with scintillator tile/WLS fiber technology. The newer calorimeter is more compact and hermetic, with $|\eta|$ up to 3.6 (3° from the beam axis), as shown in Fig. 2. The PEM has a depth of 23.2 radiation lengths. The energy resolution for high energy electrons and photons is $\frac{\sigma(E)}{E} = \frac{16\%}{\sqrt{E}} \oplus 1\%$. The PHA has alternating layers of iron and scintillating tile, for a total of 6.8 interaction lengths. The energy resolution of the PHA is $\frac{\sigma(E)}{E} = \frac{80\%}{\sqrt{E}} \oplus 5\%$ for charged pions that do not interact in the PEM.

During test beam and the previous Run I data taking period (1992-1996) the time between two bunches was 3.5 μs and the integration time was 600 ns for all calorimeter

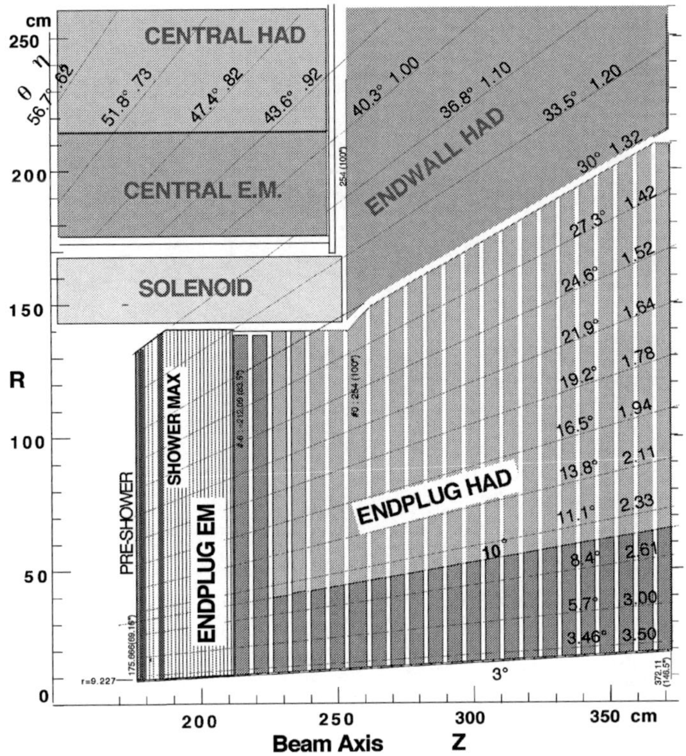

FIGURE 2. Diagram of the "plug" calorimeter. The different shaded region above 10° represents the location of the original Run I gas calorimeter.

compartments. The calorimeter readout electronics was upgraded [6] for the Run II data taking period to accommodate the 396 ns beam bunch spacing as well as a possible upgrade to 132 ns. The ADC integration gate for the charge collection is 120 ns wide and collects 94 to 98% of the signal depending on the calorimeter type. The fractional energy loss due to the shorter ADC integration gate in Run II is measured in muon, electron and jet data with an uncertainty of 1.5%.

The central and forward part of the calorimeter have their own shower profile detector positioned at the expected maximum of the lateral shower profile. The Central Electromagnetic Showermax (CES [7]) has proportional wire chambers for ϕ and scintillator strips for z position. The Plug Electromagnetic Showermax (PES [8]) detectors has scintillating strips with WLS fibers. The Central Pre-Radiator (CPR [9]) chambers are located at the inner face of the central calorimeter, and use the solenoid coil as a radiator to measure the shower development. The Plug Pre-Radiator (PPR) chambers are located at the inner face of the plug calorimeter, and have the same tile segmentation as the PEM. The showermax and pre-radiator detectors are mainly used for photon and electron identification.

The last major upgrade to the calorimeter was completed in the fall of 2004. The previous CPR gas system was replaced with a tile-fiber system similar to the PPR. Central "Crack" (CCR) detectors were installed at the same time, used to cover the structural edges of the calorimeter tower modules (about 7° of ϕ).

OPERATIONAL EXPERIENCE

Both central and plug calorimeters have been performing well, but sometimes it is more interesting to explore the problems that arise while operating the detector. At the beginning of Run II, the calorimeter was one of the few complete systems, so effort was made to have every tower and read-out channel (more than 4000) working all the time. In recent times, that has been more difficult. The accelerator is performing well enough that repair opportunities, which requires significant beam down time, is less frequent than in the past.

At the CALOR02 conference, the reported failure rate for the calorimeter front-end electronics was approximately 0.65% per year (30 units) for the CAFE daughter cards, and 3% per year (8 units) for the ADMEM mother boards. [6] Most of the failures were due to blown fuses for the CAFE cards, and worn out capacitors on the ADMEMs. Under stable operating conditions, the failure rate in recent years has been less than half of what was reported. The overall failure rate is still the same due to unusual conditions, like unplanned power outages or surges from work on nearby detector systems, that result in blown fuses.

After the installation of the PES, it was discovered that cross-talk between channels was 6% instead of the expected 0.5%. It was discovered that slight differences between the test stand set up and the full system resulted in the gap between the fiber cookie and the multi-anode PMT being 2.5mm, compared to 1mm for the test stand. After considering the options, the decision was made to install baffles to the existing MAPMT. This reduced cross-talk to a reasonable 1.4%, with only 8.5% signal reduction.

A more serious, and still partially unexplained, problem has been the decrease in the light energy response in the calorimeter. There is a light loss of about 1% per year, 2/3 of which is from scintillator attenuation (that has been going on since 1984). The remainder is an unexpected PMT gain loss, largest for towers near the beam line. This loss is seen in both laser and radioactive source calibrations, but not seen with the showermax or pre-shower detectors. There was a correlation with the integrated charge from the PMTs, so the highest eta phototubes had their gains lowered. In addition, the plug HV is set to standby (reduced but enough to keep the tube warm) during non-data taking periods. This did not eliminate the problem, but the gain loss is now acceptable for finishing the Run II schedule. For low eta towers away from the beam line, the gain loss was 1% in 2002 and 3% in 2003. The increase is due to higher colliding beam luminosity by the Tevatron. For the modified highest eta towers, the gain loss was 20% in 2002 and 8% in 2003.

SELECTED RESULTS

At the time the calorimeter upgrades were finished in September 2004, CDF had recorded events with 360 pb^{-1} of integrated luminosity. New results will use up to 1 fb^{-1}. I will list a few recent results that will demonstrate that the endplug calorimeter, built between Run I and Run II, is understood and functioning as well as the older central calorimeter.

Of primary importance to the analyses that use the calorimeter is the recent paper by the CDF jet energy corrections group. [2] The energy scale are the corrections from the measured jet energy back to the energy of the actual particle level jet. The CDF jet energy corrections are divided into different levels to accommodate different effects that can distort the measured jet energy, such as, response of the calorimeter to different particles, non-linearity response of the calorimeter to the particle energies, uninstrumented regions of the detector, spectator interactions, and energy radiated outside the jet clustering algorithm.

In the central calorimeter, the energy scale uncertainty is slightly better than the final Run I estimate, despite additional material in the Run II detector and a shorter ADC integration gate. As a result of having a better CDF simulation, the Run II jet energy scale uncertainties in the non-central regions have been decrease up to a factor of 5. From April 2002 through September 2004, the reconstructed $Z \rightarrow ee$ mass after corrections is stable versus time to within 0.3%, as seen in Fig. 3. The mean mass and width of the mass peak are the same whether both electrons are in the central calorimeter, or when an electron is in both the central and plug calorimeters.

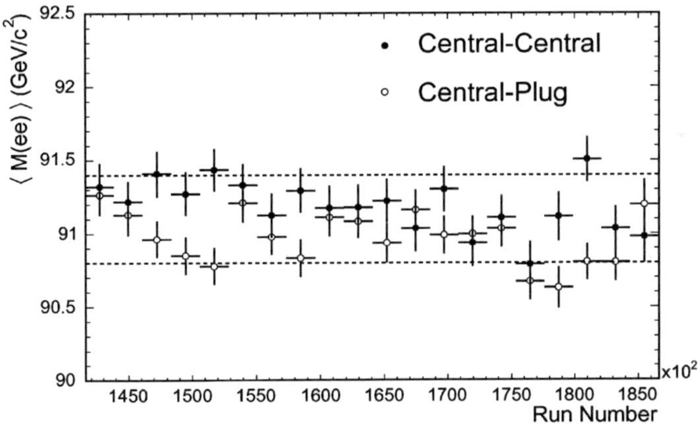

FIGURE 3. Corrected mass for $Z \rightarrow ee$, for data runs taken between April 2002 and September 2004.

One of the recent results is "Top Quark Mass Measurement Using the Template Method in the Lepton + Jets Channel at CDF II". [10] There are a number of Top results, but this one included W and b jets in the plug calorimeter. (Some analyses only use the central calorimeter, since the Central Outer Tracker wire chamber covers only $|eta| < 1.3$.) The jet energy scale uncertainty is the dominant systematic uncertainty on the CDF measurement of the top quark mass. A 1% uncertainty on the energy scale of jets results in an uncertainty of 10 % on the cross section for jet production at transverse momenta of 500 GeV/c [11] and in a 1 GeV/c^2 uncertainty on the top quark mass [12].

Another result is "Measurement of the forward-backward charge asymmetry of electron-positron pairs in pbar-p colisions at sqrt(s)=1.96 TeV". [13] The Z is recontructed with one lepton in the central calorimeter, while the other is allowed to be in either the central or plug calorimeter. The measured charge asymmetry is consistent with leading order calculations.

The last result to be mentioned is a currently unpublished preliminary result measuring the production cross section of $\sigma(p\bar{p} \to W) \times \mathscr{B}(W \to e\nu)$, where the electron is required to be in the plug calorimeter. [14] The cross-section is consistent with results that required the election to be in the central calorimeter [15] and theoretical estimates. Also, the missing E_T spectrum resembles the electron transverse momentum distribution, indicating good overall measurement of the recoil.

ACKNOWLEDGMENTS

The results presented here would not be possible without the continued operation of CDF and the Fermi National Accelerator Laboratory. In particular, I thank Willis Sakumoto, Larry Nodulman, Steve Kuhlmann, and the members of the CDF Calorimeter Group for their contributions and suggestions for the conference talk.

REFERENCES

1. D. Acosta *et al.*, Phys. Rev. D **71**, 032001 (2005).
2. A. Bhatti *et al.*, "Determination of the jet energy scale at the Collider Detector at Fermilab," arXiv:hep-ex/0510047. The article has been accepted by Nucl. Instr. Meth. A, but not yet published. `http://www-cdf.fnal.gov/physics/new/top/public/jets/cdfpublic.html`
3. L. Balka *et al.*, Nucl. Instrum. Meth. A **267**, 272 (1988).
4. S. Bertolucci *et al.*, Nucl. Instrum. Meth. A **301** 267 (1988).
5. G. Apollinari *et al.*, Proceedings of the Fourth International Conference on Calorimetry in High Energy Physics, World Scientific, Singapore, p. 200, (1994).
6. C. A. Nelson and T. M. Shaw, Prepared for 10th International Conference on Calorimetry in High Energy Physics (CALOR 2002), Pasadena, California, 25-30 Mar 2002.
7. S. Kuhlmann *et al.*, Nucl. Instrum. Meth. A **518**,39-41 (2004).
8. G. Apollinari *et al.*, Nucl. Instrum. Meth. A **412**, 515-526 (1998).
9. F. Abe *et al.*, Phys. Rev. Lett. **73**, 2662 (1994).
10. A. Abulencia *et al.*, Phys. Rev. D **72**, 032003 (2006).
11. T. Affolder *et al.*, Phys. Rev. D **64**, 032001 (2001).
12. T. Affolder *et al.*, Phys. Rev. D **64**, 032001 (2001).
13. D. Acosta *et al.*, Phys. Rev. D **71**, 052002 (2005).
14. Preliminary results from "Measurement of sigma(p-pbar->W) x BR(W->e nu) with electron identified by the Plug Calorimeter (1.2< |eta| < 2.8)". `http://www-cdf.fnal.gov/physics/ewk/2006/plugw`
15. D. Acosta *et al.*, Phys. Rev. Lett. 94, 091803 (2005).

Precision Calibration of the DØ HCAL in Run II

Krisztián Peters

School of Physics & Astronomy, University of Manchester, Manchester M13 9PL, UK

for the DØ Calorimeter Algorithm Group

Abstract. Most of the physics analyses at a hadron collider rely on a precise measurement of the energy of jets in the final state. This requires a precise *in situ* calibration of the calorimeter with the final detector setup. We present the basic procedure and results of the *in situ* gain calibration of the D0 HCAL in Run II. The gain calibration works on top of the pulser-based calibration of the readout electronics and is based entirely on collision data.

Keywords: calorimeter, calibration, hadronic, DØ detector
PACS: 01.30.Cc, 07.20.Fw, 29.40.Vj

INTRODUCTION

The detailed description of the DØ Calorimeter can be found in Run I and Run II instrumentation papers [1, 2]. Let us briefly summarize some of its basic aspects.

The DØ calorimeter is segmented into towers in η and ϕ. The precision towers divide the calorimeter in 64 segments in the ϕ and 72 segments in the η-direction. Each precision tower consists of four physical layers in the EM part and four or more physical layers in the hadronic part. The calorimeter is also divided into trigger towers, which are 2x2 arrays of precision towers, dividing the calorimeter into 32 segments in ϕ, and 37 in η. Trigger towers are the smallest calorimeter units seen by the Level 1 trigger.

We distinguish a central section covering pseudorapidities $|\eta|$ up to ≈ 1.1, and two end calorimeters (EC) that extend coverage to $|\eta| \approx 4.2$, with all three housed in separate cryostats, Fig.1. These are sampling LAr calorimeters with mainly Uranium absorber plates. A calorimeter basic unit consists of a 3 mm thick plate of an absorber material, 2.3 mm liquid Argon gap and a signal board consisting of a Copper pad surrounded by G10 insulator coated with high resistivity epoxy. The absorber is grounded and the pad is kept at positive voltage of 2000 V. Charge induced at the pads gives the physical signal. The electron drift time across the 2.3 mm gap is approximately 450 ns.

Although for Run II the calorimeter itself is unchanged from Run I, the charge integration time has been reduced from $\sim$2.2 μs in Run I to $\sim$260 ns in Run II, resulting in an enhanced sensitivity to the finite mechanical precision of the calorimeter. Additional mechanical boundaries like module edges, any non-uniformities in the LAr gap and Uranium plate widths or possible board bendings are directly related to the amount of charge collected and the response of the modules and cells. In adittion, the associated readout electronics have been largely redesigned to address the need for a shorter shaping and readout time and the need for analog buffering to store the data until a Level 1 Trigger decision becomes available.

CP867, *XII International Conference on Calorimetry in High Energy Physics*, edited by S. R. Magill and R. Yoshida

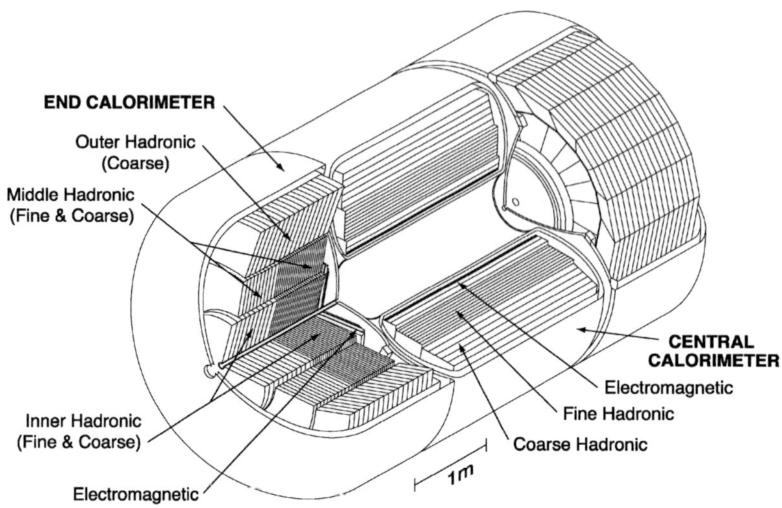

FIGURE 1. Isometric view of the DØ central and two end calorimeters in Run II.

The Run II upgrade strongly influenced the calorimeter in an indirect way as well. The amount of dead material in front of the calorimeter increased significantly with the upgrade and it is non-uniformly distributed. This material comes from pre-shower detectors, the solenoid, fiber tracker and silicon vertex detector. Together with the cryostat walls the particles traverse at least of 3.7 X_0 before they reach the calorimeter. The amount of dead material depends also significantly on the angle of incidence of the measured particles and increases with the pseudo-rapidity.

Due to this significant changes for the upgrade, the calorimeter had to be calibrated again. Moreover, it was essential to obtain a calorimeter Run II calibration *in situ* with the final detector setup.

CALIBRATION PROCEDURE

The calibration procedure for the DØ calorimeter contains two parts: calibration of the readout electronics using pulser data, and correction of non-uniformities due to mechanical variations in the detector using collision data.

The basic idea of the electronics calibration is to send a pulse of known charge into the readout, and to compare it to the measured charge. In this way we identify technical problems in the electronics, e.g. dead channels and correct for the channel-by-channel differences in the response. Pulses of different heights are used to probe the full dynamic range of every readout channel. In this way, the response of every single channel can be linearized, and the gains of the different channels can be equalized.

The gain calibration of the DØ calorimeter factorize into two parts: the calibration of the EM calorimeter and the calibration of the hadronic calorimeter. For this two parts of the calorimeter we determine the energy scale (i.e. a multiplicative correction factor), if possible per cell. Both parts of the calorimeter have been calibrated in two steps. First, the ϕ-intercalibration to reduce the number of degrees of freedom, where special triggered Run II data was used. Second, the η-intercalibration to get access to the remaining degrees of freedom, as well as the absolute scale of the EM calorimeter. For the η-intercalibration we used $Z \rightarrow e^+e^-$ events for the EM calorimeter and QCD dijet events for the hadronic calorimeter.

The best standard candle for the absolute calibration of the EM calorimeter is the Z-peak, which is well known from LEP measurements. However, we lack of statistics to use the Z-peak alone in calibrating on a tower or cell level. Therefore we used special triggered EM data to intercalibrate over rings of fixed η. Once the ϕ-degree of freedom is eliminated, the amount of Z events is sufficiently high to absolutely calibrate each intercalibrated η-ring.

For this purpose the reconstructed Z mass is written in terms of the electron energies and their opening angle. The electron energies are evaluated as the raw energy measurement from the calorimeter plus a parametrized energy-loss correction from a detailed detector simulation. Calibration constants are multiplicative to the raw cluster energy of each cell. A set of calibration constants is then determined that minimize the experimental resolution on the Z mass and gives the correct LEP measured value. After the fully calibrated EM calorimeter, we address the calibration of the hadronic part.

ϕ-INTERCALIBRATION OF THE DØ HCAL

Due to the fact that the $p\bar{p}$ beams in the Tevatron are unpolarized, the energy flow in the direction transverse to the beam should not have any azimuthal dependence. Based on this, we can use an energy flow method with the following basic principle:

Consider in each case a given η-bin of the calorimeter. Measure the density of calorimeter objects above a given E_T threshold as a function of ϕ. With a perfect detector this density would be flat in ϕ. Assuming that any ϕ-non-uniformities are due to energy scale variations, the uniformity of the detector can be improved by applying multiplicative calibration factors to the energies of the calorimeter objects in each ϕ-region in such a way that the candidate density becomes flat in ϕ.

A special trigger was designed to record efficiently data for the hadronic ϕ-intercalibration. It requires a transverse energy threshold of 5 GeV in the Trigger Tower at Level 1, then it requires at Level 2 that 5 GeV is in the hadronic part of the tower and finally it tightens the hadronic transverse energy cut for a Precision Tower at Level 3 to 7 GeV. Data for the ϕ-intercalibration was taken during normal physics running. The quality of the recorded data was studied in detail to separate failures in redout electronics from gain miscalibrations. Systematic uncertainties from trigger non-unifomities were avoided by placing trigger and offline cuts sufficiently above the trigger conditions.

The task of deriving ϕ-intercalibration constants for cells in towers at given η is divided into the two following steps:

- Finding a tower calibration constant, which is a multiplicative factor for all cells in the tower, such that tower occupancies above an E_T threshold are equalized in ϕ.

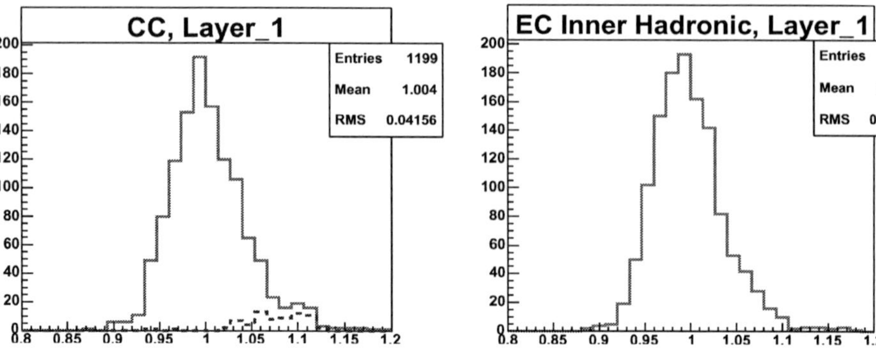

FIGURE 2. Spread of calibration constants for the first hadronic layer of the calorimeter. The constants are separately plotted for the central, EC inner-hadronic part.

- Fitting layer calibration constants, which then intercalibrate cells within the tower. This is performed using cell energy fraction distribution shapes which are compared to the ϕ-averaged reference shape.

Since the above two steps influence each other, the procedure of layer and tower calibration has to be iterated until stability is reached. The final ϕ-intercalibration constants are the products of these layer and tower constants.

With the calibration method described, calibration constants on cell level have been determined for the whole η region with available trigger information (up to $|\eta|$ of 3.2). Due to statistical limitations in our calibration data sample, for the inter cryostat region and for the region of $|\eta|$ above 2.4 a calibration on tower level is used only.

In Fig. 2, as an example, the spread of the calibration constants is plotted for the first hadronic layers of the detector separated into two regions: central calorimeter and EC inner hadronic calorimeter. The calibration constants are mainly in the range of 0.90-1.15 and the root mean squares are at the order of 0.05. Calibration constants are slightly smaller for the central region compared to the EC and for the first hadronic layer compared to the other hadronic layers. The plots of the spread of constants have tails resulting from outliers with higher constants. For the central calorimeter this is mainly due to a single module, whose contribution is plotted separately with dotted lines. In the EC inner hadronic part this outliers are from the region of higher η's with the lack of statistics. In these regions only a calibration on tower level is aimed anyway.

The error estimation was done with a MC method: we generate toy simulations of the data with known miscalibrations and compare to the fitted calibration constants of our calibration procedure. The central calorimeter is now calibrated with the precision of the order of 1%, for the high η-regions it is a few per cent.

In general the energy response of the modules is less uniform than it was in Run I. The dominant reason for this is the short integration time in Run II. This amplifies the effect of the finite precision of the calorimeter modules. The electron drift time across the 2.3 mm LAr gap is at the order of 450 ns. While in Run I the integration time was essentially "infinite" on the time scale of this drift time, with the shorter Run II integration time we

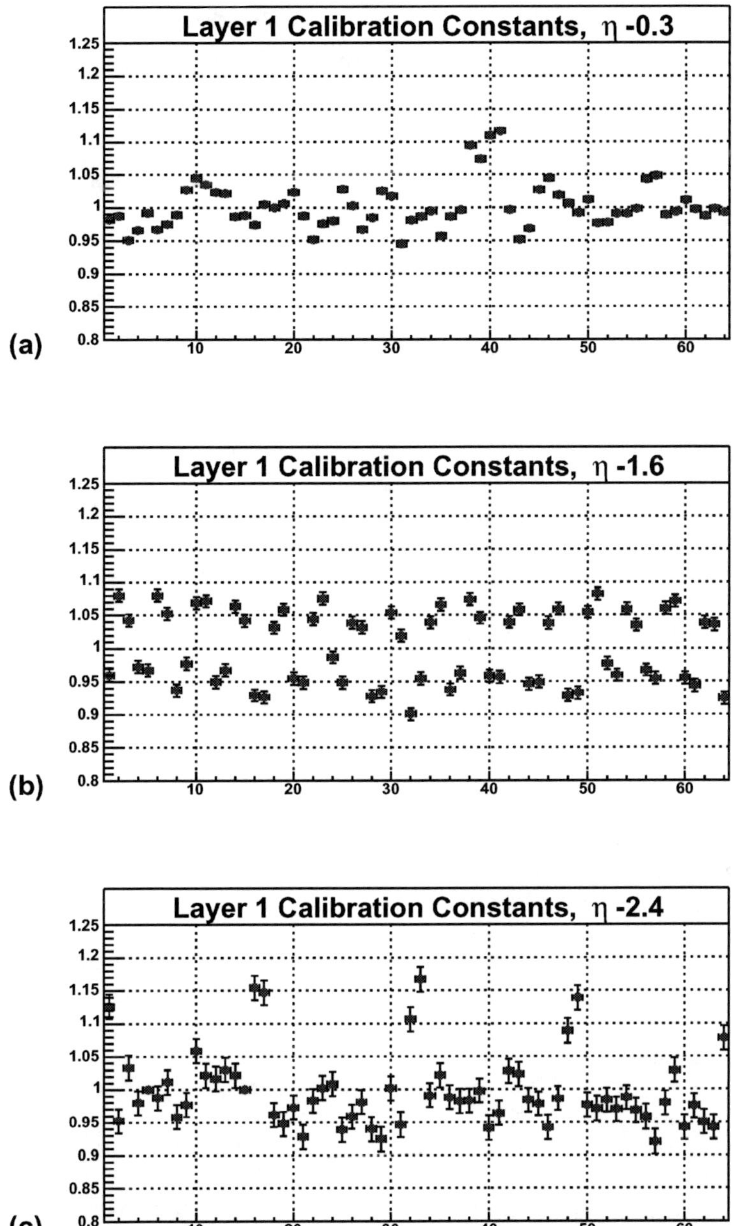

FIGURE 3. ϕ-intercalibration constants for the first hadronic layer at three different η-rings at η -0.3, -1.6 and -2.4 respectively. The 64 constants corespond to the 64 calorimeter segments in ϕ-direction.

cut into the signal. The ϕ-intercalibration accounts for these charge collection effects.

This is illustrated in three examples, where three extreme cases have been choosen. One is a whole module out of the 16 modules in the central calorimeter which has a low response, thus had to be boosted up. As an example we present Fig. 3(a), the calibration constants at η of -0.3, but the same pattern is visible in the whole fine hadronic central calorimeter. The spread of calibration constants due to this module is plotted in Fig.2, (dashed line).

In addition, the effect at the edges of this module is stronger and such cells need to be boosted more than at the center of the module. This kind of charge collection effects do not only concern this particular module, they are also visible throughout the hadronic calorimeter. At a closer look to Fig. 3(a), a similar pattern can be recognized for all the modules included in the discussed plot. The same inefficient charge collection is clearly visible in some cells of the end cap calorimeter which are at the boundaries of this calorimeter section. The effects are strongly enhanced by the additional calorimeter borders. An example is plotted in Fig. 3(b) for the η-ring at -1.6.

The inner hadronic calorimeter was built on one module, thus charge collection effects in the inner hadronic part are of different kind compared to the central part. The inner hadronic modules and absorption layers have been mounted together from half-circles. These modules and absorption layers are oriented within $\pm 90^{\circ}$ with respect to each other to obtain a structure without any gaps. Charge collection effects due to these rotated semi-circles are visible in the calibration constants. An example is plotted in Fig. 3(c) for the η-ring at -2.4. Four groups consisting of two constants need to be boosted up due to charge collection effects.

η-INTERCALIBRATION OF THE DØ HCAL

The next step in our calibration procedure is the η-intercalibration, where we determine overall calibration factors for each η-ring. These will be 64 constants for $-3.2 < \eta < 3.2$ on top of the ϕ-intercalibration constants. At this stage the EM layers of the calorimeter are already calibrated and the hadronic cells are equalized in ϕ. Our aim at the η-intercalibration is to determine a relative weight between the EM and the hadronic calorimeter which yields the best jet energy resolution. The necessary consequence of this procedure is well known [3], the jet response will be non-linear. This is due to the fact that the sampling fraction decreases considerably as the shower develops, because the calorimeter response is smaller for the soft γ component in the tail of the shower than for mips. The fraction of energy deposited in the hadronic part of the calorimeter will rise with the energy. The shower starts later and the sampling fraction will rise in this section for the above reasoning. Since a high sampling fraction gives less fluctuations, the hadronic part of the calorimeter demands a higher weight for the best energy resolution. Consequently there are no single optimal constants for all energies. The default constants have been chosen to be optimal for jet of 45 GeV which satisfies the vast majority of physics program at the DØ detector.

This discussed non-linearity does not imply that only a certain jet energy is measured correctly. With the presence of the amount of dead material in front of the calorimeter it is anyway non-linear regardless of the weights. The non-linear calorimeter response to

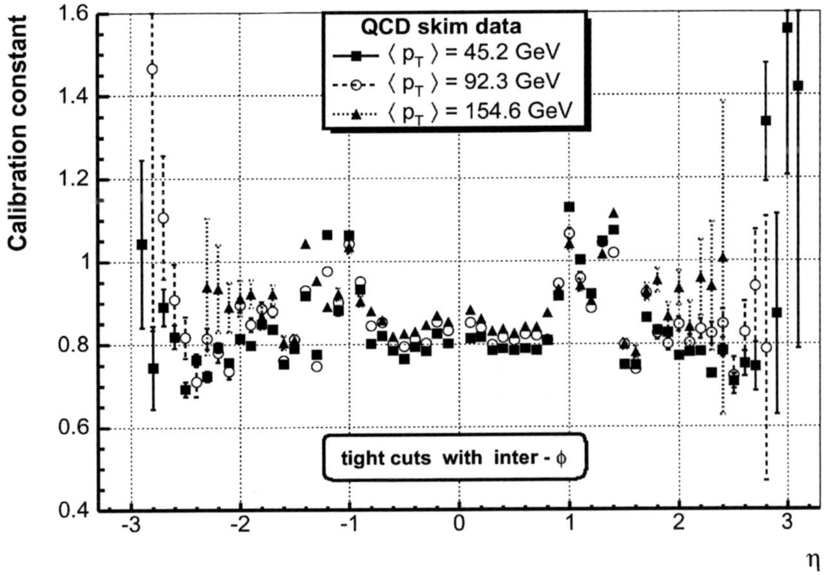

FIGURE 4. η-intercalibration constants for three different mean jet p_T values of 45.2 GeV, 92.3 GeV and 154.6 GeV respectively. Constants are on top of the older ones which roughly reproduced the right sampling fractions.

jets is corrected with an energy-dependent Jet Energy Scale [4].

For the η-intercalibration we used a sample of QCD dijet events where the total missing p_T-fraction of the events was minimized by weighting the hadronic calorimeter cells within the jets. Only well reconstructed back-to-back two jet events have been selected and an average jet p_T well above the trigger threshold was required.

Results of the η-intercalibration are plotted in Fig.4 for three different mean jet p_T's. As discussed, calibration constants rise with the jet p_T, however there was no appreciable dependence on the jet cone size. These are constants that are on top of the older ones which roughly reproduced the right sampling fractions. There is a discontinuity visible in the constants which is due to the fact that there are no EM cells for $1.2 < |\eta| < 1.4$. The large error bars at the high $|\eta|$ regions is due to a limited statistics. The correction factor for the regions of $2.0 < |\eta| < 2.7$ are however stable, thus we choose to extrapolate the mean values of this range to higher η-values rather than to use the constants with the large errors.

After the full hadronic calorimeter calibration the jet p_T-resolution were redetermined using dijets and the same 1 fb^{-1} sample to account for the improvements due to the hadronic calibration and Jet Energy Scale. The result is plotted in Fig.5 for the η-range of $0.0 < |\eta| < 0.4$. The dotted line is the earliest Run II result and the solid line is with the fully calibrated 1 fb^{-1} data set. The hadronic calibration lead to significant improvements in the central region (ca. 15% improvement at the energy range of Higgs and top decays).

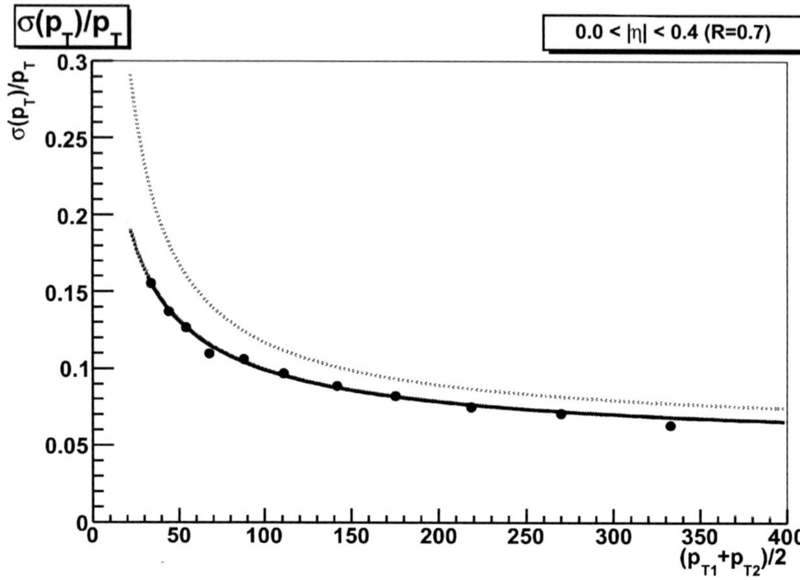

FIGURE 5. Jet p_T-resolution in dijet events from the earliest Run II result (dotted line) and with the fully calibrated 1 fb^{-1} data set (solid line) in the η-range of $0.0 < |\eta| < 0.4$.

REFERENCES

1. S. Abachi *et al.* [D0 Collaboration], "The D0 Detector," Nucl. Instrum. Meth. A **338**, 185 (1994).
2. V. M. Abazov *et al.* [D0 Collaboration], "The upgraded D0 detector," arXiv:physics/0507191, FERMILAB-PUB-05-341-E.
3. R. Wigmans, "Calorimetry: Energy measurement in particle physics," Oxford, UK: Clarendon (2000) 726 p.
4. J. Kvita, "Jet Energy Scale determination at DØ," These proceedings.

Calorimeter Timing System at CDF

Max Goncharov on behalf of CDF Collaboration

Texas A&M University

Abstract. We report on the design and performance of the electromagnetic calorimeter timing readout system (EMTiming) for the Collider Detector at Fermilab (CDF). The system is used in searches for rare events with high-energy photons to verify that the photon is in time with the event collision, and to reject cosmic-ray and beam-halo backgrounds. The installation and commissioning of all 862 channels was completed in Fall 2004 as part of an upgrade to the Run II version of the detector. Using in-situ data, we measure the energy threshold for a time to be recorded to be 3.8 ± 0.3 GeV and 1.9 ± 0.1 GeV in the central (CEM) and plug (PEM) portions of the detector. Similarly, for the central and plug portions we measure a timing resolution of 600 ± 10 ps and 610 ± 10 ps for electrons above 10 GeV (6 GeV). Pathologies such as noise and non-Gaussian tails are virtually non-existent.

Keywords: EMTiming, electromagnetic calorimeter, timing, CDF
PACS: 29.40.Vj

INTRODUCTION

Timing readout for the electromagnetic (EM) calorimeters [1, 2] was recently installed as part of an upgrade to the Run II version of the Collider Detector at Fermilab (CDF) [3]. The system, known as EMTiming [4] has a resolution of less than a nanosecond and covers the central (CEM, $|\eta| < 1.1$) and plug (PEM, $1.1 < |\eta| < 2.1$) portions of the calorimeter, where $\eta = -\ln\tan(\theta/2)$, and θ is the angle from the beam line.

The design of the EMTiming system is optimized for high-energy photons. The system records the time of arrival of the particles that deposit large amounts of energy in the EM calorimeter. With this information it is possible to verify that all photons (or other energy) are from the primary collision and to reject or estimate the rate of cosmic-ray and beam-related backgrounds.

In the following sections we describe the various components of the EMTiming system and its performance from the first several months of operation. In Sections 1 – 3 we give an overview of the system and describe the path of the calorimeter signals through the various components in the system, with particular emphasis on the differences between the CEM and PEM systems which were originally built with different readout. In Section 4 we describe the system performance as measured using collision data in-situ.

1. EMTIMING SYSTEM LAYOUT

The EMTiming system is shown schematically in Figure 1. Particles from the collision that deposit energy in the EM calorimeter create light within the scintillators [1, 2] that can be used for a timing measurement. Photo-multiplier tubes (PMTs) collect this light and convert it into an electrical analog signal. The EMTiming system routes a copy of the

CP867, *XII International Conference on Calorimetry in High Energy Physics*,
edited by S. R. Magill and R. Yoshida

PMT signal to a passive Transition Board (TB) and an Amplifier-Shaper-Discriminator board (ASD) that, in turn, converts analog signal into digital and sends it to a Time-to-Digital Converter (TDC) board for a timing measurement.

There are physical differences between the existing CEM and PEM detectors. While the PEM part of the EM calorimeter was designed with a timing upgrade in mind, the CEM was not. In Section 2 we describe the differences between the PMT outputs for the CEM and PEM. In Section 3 we describe the boards that collect the signals from the PMTs, amplify, shape, combine and use them as input to the discriminator, and output the results to a TDC. The exact system specifications can be found in [4].

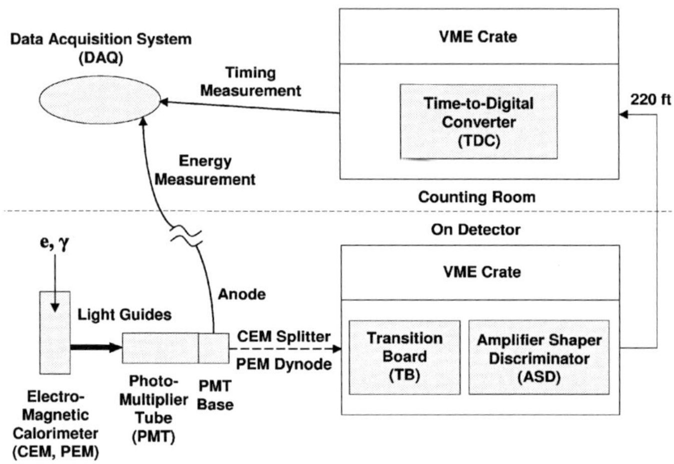

FIGURE 1. The EMTiming system layout.

2. CEM PMT SIGNAL SPLITTER

The PMT bases in the CEM and PEM have different designs. In the CEM, the bases are custom designed and only provide an anode output via a LEMO connector. In the PEM, the system was designed with the dynode signal already available. The rise times of the PMT output signals, as measured from 10% of the voltage-max to 90% of the voltage-max, are $\approx$10 ns and 6 ns for the CEM and PEM respectively.

In order to not disrupt the energy measurement in the CEM by modifying the bases, we designed an inductive signal "splitter" board that is placed between the PMT base and the original 25 ft RG174 cable that carries the anode signal for an energy measurement. The splitter routes a fraction of the PMT pulse energy for timing use, while not affecting the energy measurement. As shown in Figure 2 the anode cable is connected, via a LEMO connector, to a printed circuit board where the primary and shield wires are separated. The primary line is then passed through a small circular ferrite after which the wires are rejoined. A secondary wire is wound around the ferrite so that a signal from the PMT anode induces a voltage for timing use. This secondary signal is sent via an RG174 cable with a length that varies between 23 and 36 feet to the TBs depending on the PMT location on the detector. The splitter solution avoids potential ground-loop problems since there is no electrical contact between the timing and energy readout lines. All cables have LEMO connectors on each end to facilitate installation.

The secondary pulse used for the timing measurement has a voltage that is $\sim$15% of the primary signal as shown in Figure 3. Since the two lines are only inductively coupled, and the energy measurement is done with a charge integration device [5], in principle this solution should not affect the energy measurement since no charge can be lost. In order to verify this, test-stand comparisons of the integrated charge for a PMT pulse, with and without a splitter, were performed at various points over the full energy integration range. There was no observed (systematic or otherwise) effect on the linearity or resolution for any energy (see Figure 3), with a measurement uncertainty of approximately 10% of the 1σ variation in the charge-integration measurement itself (for a given energy).

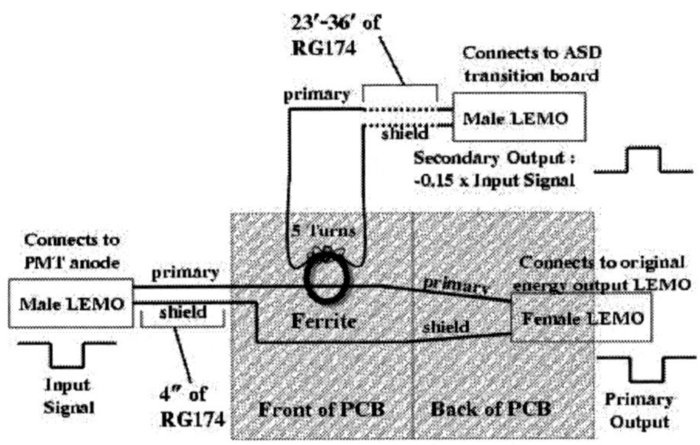

FIGURE 2. The splitter schematics.

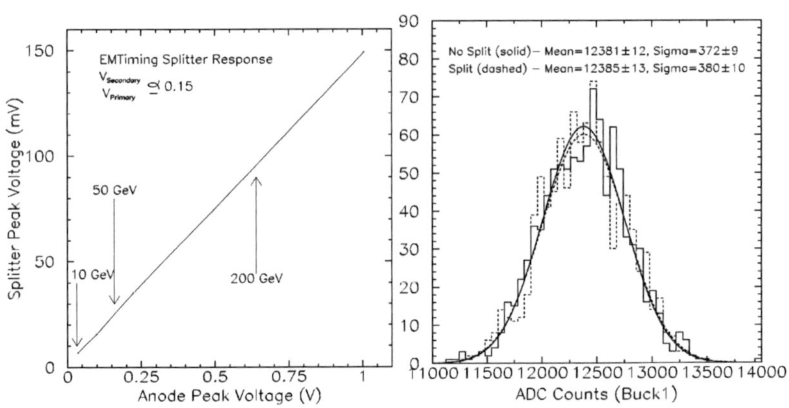

FIGURE 3. On the left: the splitter peak voltage as a function of anode peak voltage. The function is linear for almost all possible photon energies. On the right: the comparison between the energy readouts with and without the signal splitting. The energy readout is not affected by the splitter.

3. SIGNAL DISCRIMINATION AND TRANSFER TO THE TDC

Beyond the inputs to the transition boards the remaining system components (TBs, ASDs, the long cables and TDCs) are identical for the CEM and PEM. Each component is designed for low cost and large numbers of channels per board/cable.

Each TB is capable of receiving cables from 48 PMTs. As shown in Figure 4, processing begins with an *RC* circuit and a transformer. The *RC* circuit eliminates reflections back to the PMT that might otherwise cause a second, erroneous signal to be sent, and the transformer reduces ground-loop or potential DC-offset problems that might cause noise. Each line is passed, via the crate backplane, to the ASD.

On the ASD board the input analog signals from two PMT lines are summed and compared to a threshold with the digital result being sent to the TDC. As shown on the right-hand side of Figure 4, each PMT signal is amplified, then combined in an analog sum with a threshold offset, and sent to a comparator. This operates as a fixed-threshold discriminator with a 2 mV threshold. The resulting signal triggers a monostable with an output width of 70$\pm$10 ns that in turn feeds a National Semiconductor DS90C031 LVDS driver.

The output for each channel on an ASD board is placed on a single 220 ft multiwire twisted-pair cable (3M 3644B/50) that goes from the collision hall calorimeter crates, upstairs to the TDCs. At this length these cables have a rise time of $\sim$50 ns. However, test-bench studies show that as long as the input LVDS signal width is $>$50 ns, then we expect negligible data-transfer loss and timing jitter of $<$30 ps. The receiver chip set is located on one of the standard CDF 96 channel TDCs [6] that are a part of the data acquisition system (DAQ) and read out for each event.

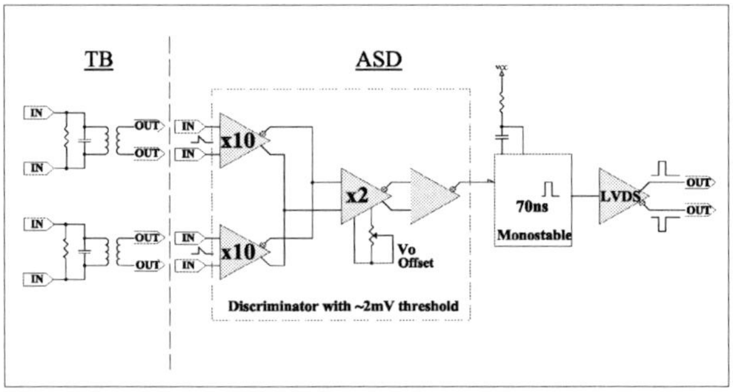

FIGURE 4. The ASD board schematics.

4. SYSTEM PERFORMANCE

The system operates at CDF when protons and anti-protons collide every 396 ns, on average, at the center of the detector. The variation in the time and position of the collision is $\sigma_{\text{time}} \approx$1.3 ns and $\sigma_{\text{position}} \approx$25 cm, respectively. If there is enough energy deposited, the TDC will record a time of arrival. Various corrections must be applied to remove channel-to channel differences and to eliminate the dependence on the particle energy (slewing effect).

The system operation can be described in terms of its efficiency to record a time, the noise rate, and the timing resolution after calibrations are applied. We study the system performance using data from the collisions that produce jets, photons, or electrons from $Z \to e^+e^-$ or $W \to e\nu$ events. The results presented here represent the data taken during the first four months of Tevatron running in 2005.

4.1. System Efficiency

The efficiency is defined as the ratio of the number of events with a time recorded to all events. The efficiency as a function of the deposited energy is shown in Figure 5 for one channel in CEM and one channel in PEM. As described in previous section, the ASD sums the analog signals from two input lines, so we actually use the sum of energies from individual PMTs and refer to it as the tower "energy". The data confirms that indeed one indeed should take a linear sum of energies from two PMTs. We plot the efficiency in a two dimensional view as a function of energy in each PMT. This is shown in Figure 6. One can see that the turn-on curve has a linear shape.

The efficiency curves are well described by a smeared step function. The top plot in Figure 7 shows the efficiency curves for all all towers (channels) together. We investigate each fully-instrumented tower separately and find that the efficiency plateaus at 100% and, as shown in the bottom parts of Figure 7, that the threshold and width values, as determined from a fit, are 3.8±0.3 GeV (1.9±0.1 GeV) and 1.4±0.2 GeV (0.6±0.1 GeV) in the CEM (PEM) calorimeters respectively.

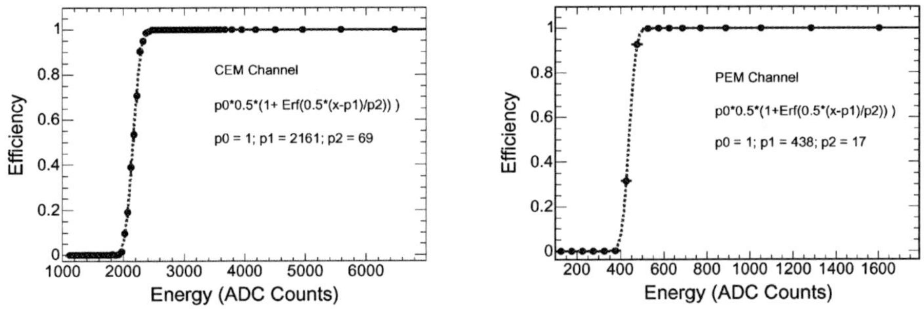

FIGURE 5. The efficiency curves for one channel in the CEM and one channel in the PEM. We fit the distribution with the modified error function. Typically the $p0$ parameter is referred to as an efficiency, $p1$ as a threshold, and $p2$ as a turn-on width.

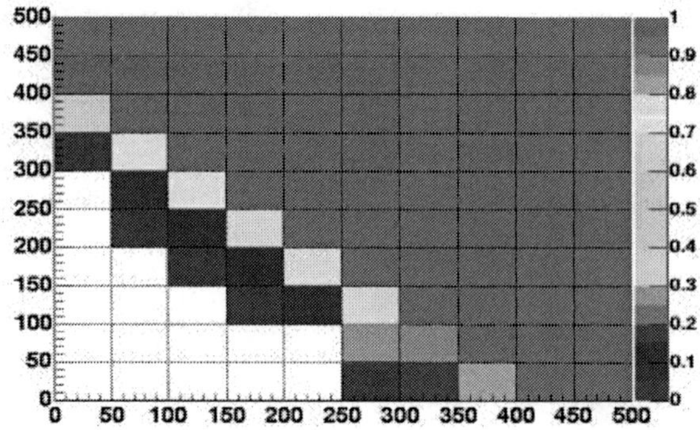

FIGURE 6. The efficiency as a function of energy in first and second PMTs. The turn-on has a linear shape confirming that the ASD responds to the analog sum of signals from both PMTs. The x-axis is the energy (in ADC counts) in the first PMT, and the y-axis is the energy in the second PMT.

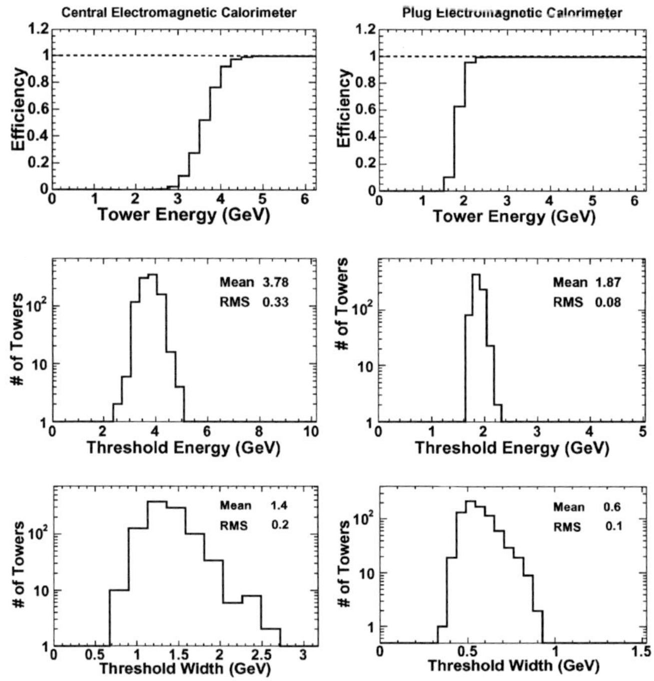

FIGURE 7. The plots on the top show the efficiency curves for CEM and PEM detectors for all towers. The plots on the bottom show the thresholds and turn-on widths for all channels. Here the ADC counts are translated into the GeV energy units. One can see that the system is 100% efficient above the threshold and is very uniform.

4.2. Noise Level

We define an event as noisy if there is timing information recored and there is no energy in the tower. We measure the noise rate to be typically less than one occurrence in 10^8 collisions, per tower. We note however that this rate can rise to one in 10^7 if we allow cases where there is some evidence that the energy is not properly read out by the DAQ. The rate at which multiple times are recorded for a single energy deposit, presumably from reflections, is roughly the same as the fake rate, but can be as high as one in 10^5 collisions for a few (<10) towers. However, this is easily corrected for in the final data analysis.

4.3. Calibrations

The time of arrival recorded by the TDC is a "raw" time and has a large RMS due to channel-to-channel differences and energy dependence typical for the systems that use fixed threshold discriminators (slewing). The uncorrected timing distribution is non-Gaussian with an RMS of ~6 ns and is not centered at zero. The plot on the left in Figure 9 shows the raw timing for one of the CEM towers. The distribution is non-Gaussian and is centered at an arbitrary value. The plot on the right shows the means of the timing distributions for all towers in the central detector. One can see that there are large channel-to-channel variations.

To calibrate the system we use a set of empirically derived functions for each channel given by:

$$C(x) = p_0 + \frac{p_1}{ln(x)} + \frac{p_2}{x^2}, \tag{1}$$

where p_0, p_1, and p_2 are constants and x is the sum of the energies from the two PMTs. The typical slewing curve is shown in Figure 8. The calibrations are performed by fitting each channel with Equation 1. In this way the coefficients p_1 and p_2 calibrate out the slewing effects, and the p_0 constant takes into account the channel-to-channel variations.

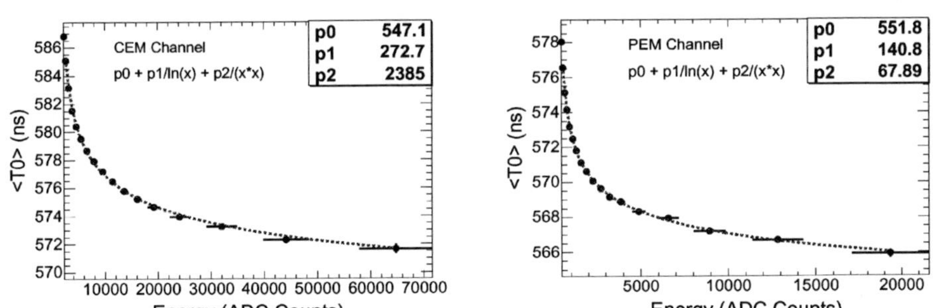

FIGURE 8. The raw time as a function of energy for one channel in CEM and PEM.

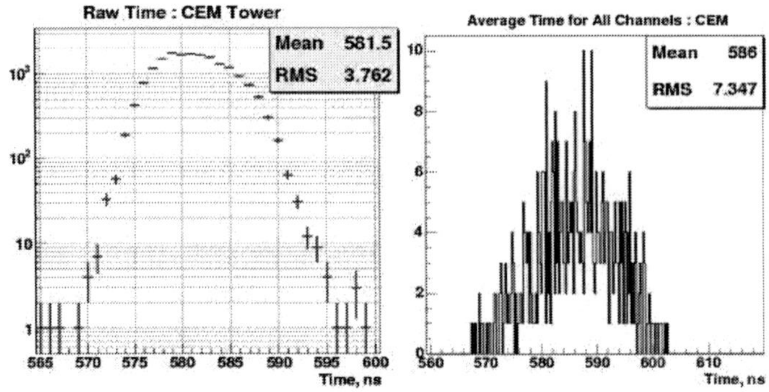

FIGURE 9. On the left: the raw timing distribution for a single CEM tower. The distribution is non-Gaussian with an RMS of ~4 ns and is centered at an arbitrary value. On the right: the means of all CEM towers.

4.4. Resolution

The system resolution can be measured in different ways. The most straightforward one is to use electrons from $Z \to e^+e^-$ events and compare the timing for both electrons. In this way several effects such as collision time and possible run dependence cancel automatically. The result is shown in Figure 10 for combinations of electrons in the CEM-CEM, and CEM-PEM. The RMS per electron is 600±10 ps and 610±10 ps for the CEM and PEM respectively, and is consistent between the measurements. This includes the irreducible TDC resolution of 288 ps (the TDC has a 1 ns binning output, which corresponds to an RMS of 1 ns/$\sqrt{12}$ = 288 ps). There is no offset between CEM and PEM detectors, therefore, even though there are physical differences between the two parts, the system looks uniform to the user.

In order to establish that there are no non-Gaussian tails we use single electrons from high statistics $W \to e\nu$ sample. Figure 11 shows the timing distribution for those electrons. The resolution is consistent with the $Z \to e^+e^-$ result; the slight increase in RMS is explained by the fact that for single electrons one has to subtract the collision time (measured by the COT tracking chamber [7]) that has an RMS of ~25 ps. The distribution in Figure 10 is well described by a Gaussian even out to roughly 5σ for the CEM.

The last point one has to show is that the resolution does not dependent on energy. Figure 12 shows the single tower energy resolution measured from a mixed sample of jets and electrons. While the resolution is slightly worse for low energies, the asymptotic resolution stabilizes at energies just above 10 GeV and 6 GeV in the CEM and PEM respectively to values that are very close to the electron resolution from W's (Figure 11).

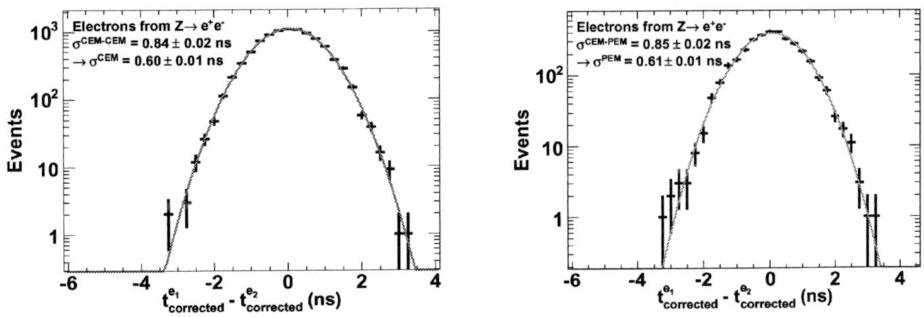

FIGURE 10. The corrected time distribution as estimated from a sample of $Z \rightarrow e^+e^-$ events in the CEM and PEM.

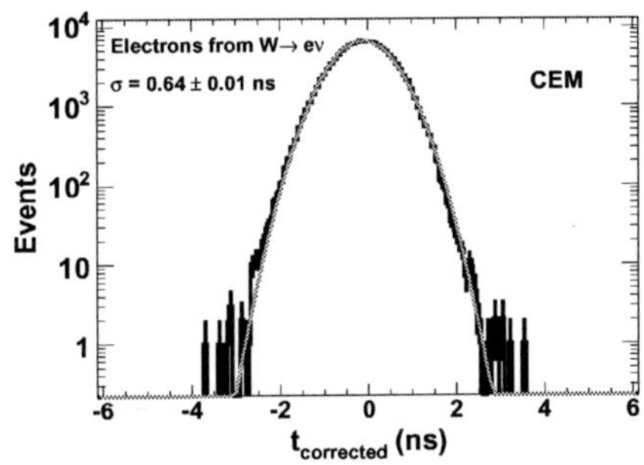

FIGURE 11. The corrected time distribution as estimated from a sample of CEM $W \rightarrow e\nu$ events. The distribution is well described by a Gaussian. There are no tails up to 5σ.

CONCLUSION

The EMTiming system provides timing readout for the CDF electromagnetic calorimeters with good uniformity and resolution. It has its 50% efficiency points at 3.8±0.3 GeV and 1.9±0.1 GeV in the CEM and PEM respectively, and is 100% efficient well above threshold. After a full set of corrections we find 600±10 ps and 610±10 ps timing resolutions, respectively, with only small deviations from a Gaussian distribution. There are very few pathologies observed in the data such as recording a time when no energy is deposited or recording multiple times for a single energy deposit. The system is well understood and is already heavily used in various analysis that involve photons.

ACKNOWLEDGMENTS

I would like to thank all people who contributed to the design and construction of the EMTiming system: D. Allen, W. Badgett, M. Campbell, M. Cordelli, H. Frisch, C. Gins-

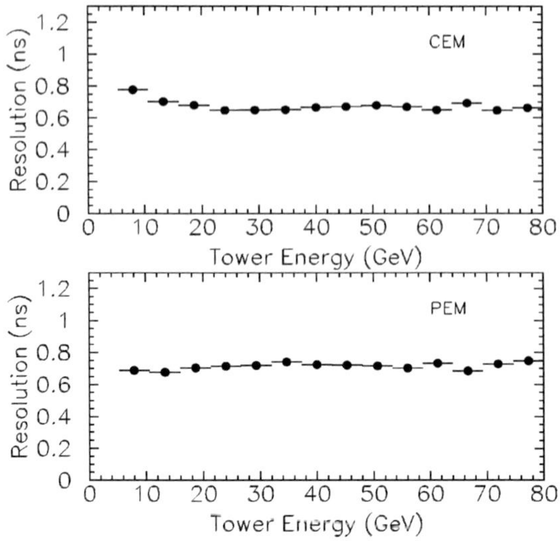

FIGURE 12. The resolution as a function of energy after all corrections are applied. The distribution is flat, as it should be.

burg, R. DeMaat, F. Happacher, E. James, M.J. Kim, v. Krutelyov, V. Khotilovich, S.W. Lee, M. Lindgren, N. Luzhetskiy, S. Miscetti, R. Roser, P. Ptohos, W. Sakumoto, R. Tesarek, D. Toback, P. Wagner, R. Wagner, and P. Wilson.

REFERENCES

1. L. Balka *et al.*, Nucl. Instr. and Meth. A267 (1988) 272.
2. M. Albrow *et al.*, Nucl. Instr. and Meth. A480 (2002) 524.
3. R. Blair et al., CDF Collaboration, CDF II Technical Design Report, FERMILAB-Pub-96/390-E (1996).
4. M. Goncharov *at al.*, arXiv:physics/0512171, accepted for publication in NIM.
5. R. Erbacher, Proc. of the 31st International Conference on High Energy Physics (ICHEP 2002), FERMILAB-CONF-02/251-E; T. Shaw, C. Nelson, and T. Wesson, IEEE Trans. Nucl. Sci. 47 (2000) 1834.
6. R. Moore, Proc. of the 2004 IEEE Nuclear Science Symposium and Medical Imaging Conference, FERMILAB-CONF-04-262-E.
7. The tracking chamber is described in T. Affolder *et al.*, Nucl. Instr. and Meth. A526 (2004) 249. A description of the algorithms designed to measure the collision time and *z*-position can be found in M. Goncharov, V. Krutelyov, D. Toback, and P. Wagner, CDF Note #8015, unpublished.

Electromagnetic and Hadron Calorimeters in the MIPP Experiment

T.S. Nigmanov (presenter), H.R. Gustafson, M.J. Longo, D. Rajaram (on behalf of MIPP collaboration).

University of Michigan, Ann Arbor, MI 48109, USA

Abstract. The purpose of the MIPP experiment is to study the inclusive production of photons, pions, kaons, and nucleons produced in π, K, and p interactions on various targets using beams from the Main Injector at Fermilab. The purpose of the calorimeters is to measure the production of forward-going photons and neutrons. The electromagnetic calorimeter consists of 10 lead plates interspersed with proportional chambers followed by the hadron calorimeter with 64 steel plates interspersed with scintillator. We collected data with a variety of targets with beam energies from 5 GeV/c up to 120 GeV/c. The energy calibration of both calorimeters with electrons, pions, kaons and protons is discussed. The performance of the calorimeters was tested on a neutron sample.

Keywords: MIPP; Calorimeters; Showers; Calibration; Photons; Electrons; Hadrons; Fermilab
PACS: 29.40.Vj

INTRODUCTION

The MIPP (Main Injector Particle Production) experiment (E907) [1] took place in the Meson area at Fermilab. The main purposes of the experiment were to verify the scaling law of hadron fragmentation, to measure particle production from the NUMI target to predict the NUMI neutrino flux, and to measure proton-nucleus cross sections for the purpose of proton radiography. The electromagnetic and hadron calorimeters allow us to measure the production of forward-going neutral particles – photons and neutrons – since they are invisible to the upstream detectors. The electromagnetic calorimeter was built for the MIPP experiment, while the hadron calorimeter was reused from the HyperCP (E871) experiment [2].

A schematic of the MIPP setup is shown in Fig.1. The detector consisted of two magnetic spectrometers. The target was placed in front of a time projection chamber (TPC) that sits in the magnetic field of the Jolly Green Giant magnet. The TPC can identify charged particles in the low energy region (<1 GeV/c). Immediately downstream, a series of drift chambers and Cerenkov counters provide tracking and particle identification for the medium energy region (1-10 GeV/c). The second spectrometer consists of the Rosie dipole magnet, a time of flight detector, a series of drift and multiwire chambers, and a RICH detector, for tracking and particle identification in the high energy region. Forward-going photons and neutrons were detected by the electromagnetic and hadron calorimeters placed at the downstream end of the second spectrometer. The data sample collected by MIPP is summarized in Table 1.

CP867, *XII International Conference on Calorimetry in High Energy Physics*,
edited by S. R. Magill and R. Yoshida

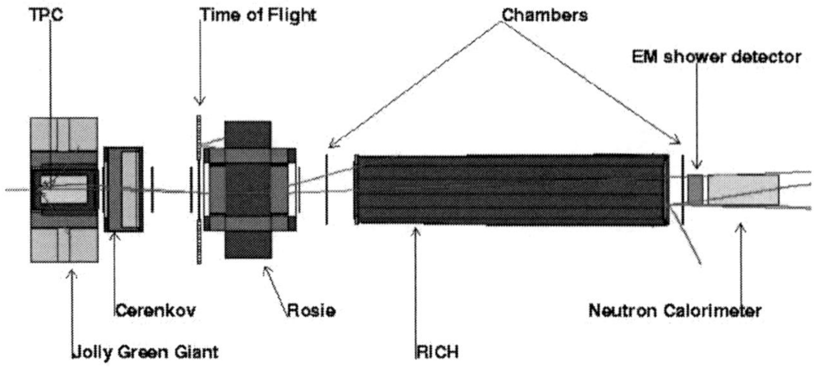

FIGURE 1. Experimental layout schematic.

TABLE 1. MIPP data sample.

Targets	Momentum (GeV/c)	Events x 10^6
LH2	5, 20, 60 and 85	7.08
Beryllium	35, 60 and 120	1.74
Carbon	20, 60 and 120	1.33
NuMI	120	1.78
Aluminum	35	0.1
Bismuth	35, 60 and 120	2.83
Uranium	60	1.18
K-mass (no target)	40 and 60	14.33
Total		31.38

CALORIMETER SPECIFICATIONS

A schematic of the calorimeters is shown in Fig. 2. The electromagnetic calorimeter (EMCAL) consists of 10 layers of 5.08 mm thick lead interspersed with planes of gas proportional chambers. The proportional chambers are made of aluminum extrusions. There are 64 anode wires with 25.4 mm spacing in each plane. The chambers used a

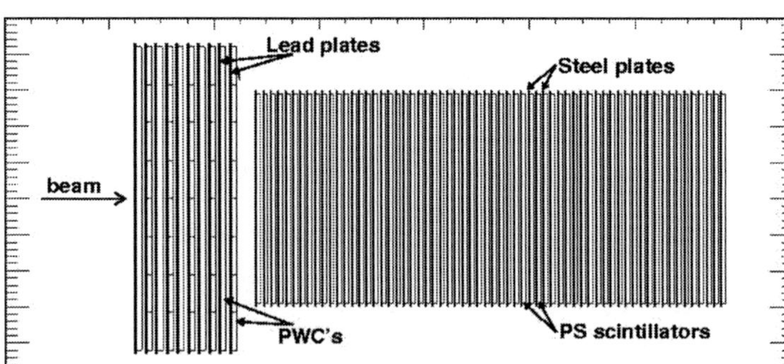

FIGURE 2. Calorimeters schematic.

gas mix of P10 (90% Argon and 10% Methane) and CF_4. The EMCAL active area was 1.6 m wide, 1.5 m high and 0.3 m long along the beam direction. The total thickness was ~10 radiation lengths. The EMCAL readout consisted of 640 channels.

The hadron calorimeter (HCAL) was composed of 64 layers of 24.1 mm thick Fe and 5 mm thick scintillator. It was originally built for the HyperCP (E871) experiment [2]. The total thickness of the HCAL was 88.5 radiation lengths (9.6 interaction lengths). Its active area was 0.99 m wide, 0.98 m high and 2.4 m long along the beam direction. For readout purposes it was subdivided into four longitudinal and two lateral sections, for a total of 8 cells.

ENERGY CALIBRATION

For the energy calibration of the calorimeters we assume that the response of calorimeters is a linear function of the incoming particle's energy. Then we can write

$$E = C_E \sum EMCAL + C_H \sum HCAL$$

where E is the particle's energy measured by the upstream magnetic spectrometers, ΣEMCAL and ΣHCAL are summed ADC counts representing the EMCAL and HCAL responses for the passage of given particle and C_E and C_H are proportionality constants for the calorimeters. We selected events with a single charged track in the upstream chambers and required a match of EMCAL shower position with the projection of the incoming track. C_E and C_H are expected to be almost independent of energy if the ADC counts in the calorimeters are proportional to energy. The ΣHCAL vs. ΣEMCAL dependence is illustrated in Fig. 3. In the figure on the left, the large fraction of events near EMCAL=0 is due to protons that deposit essentially all their energy in the HCAL. However, protons deposit some energy into EMCAL planes.

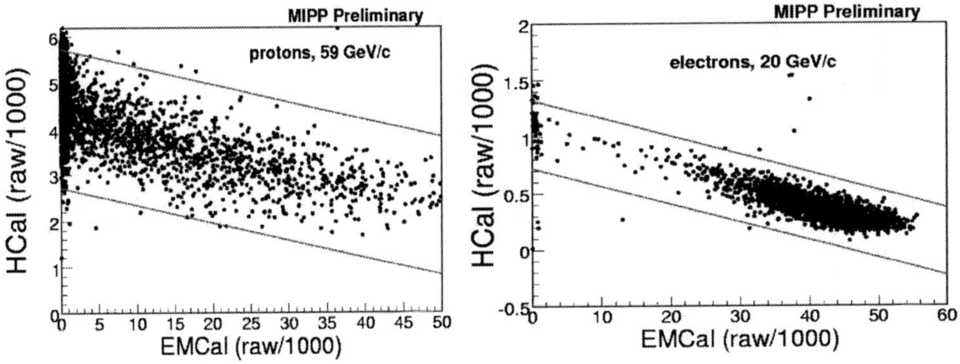

FIGURE 3. Scatter plot of HCAL vs. EMCAL responses for passage of 59 GeV/c protons (on left) and for 20 GeV/c electrons (on right). Lines above and below the data points indicate the boundaries for events used in the fitting procedure.

The deposited energies in these cases lie in wide energy ranges. In contrast with protons, as we can see in the plot on the right, electrons deposit most of their energy in the EMCAL and almost nothing in the HCAL. However, there is some fraction of events near EMCAL=0; these are presumably due to hadron contamination in the electron sample. The fit of HCAL vs. EMCAL responses is shown in Fig. 4.

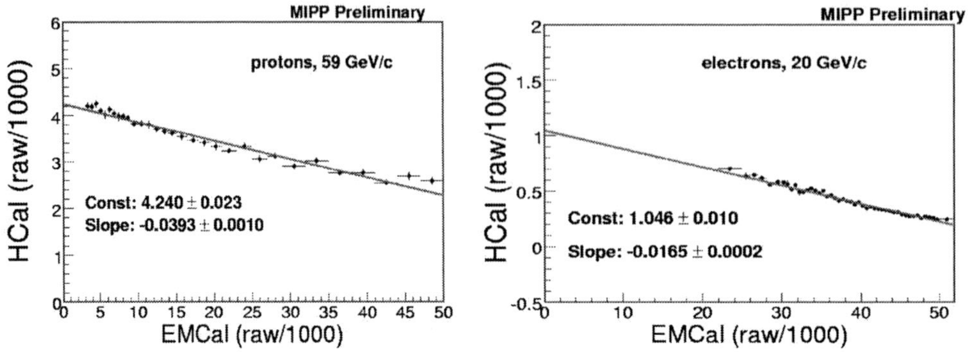

FIGURE 4. Fit of HCAL vs. EMCAL responses for 59 GeV/c protons (on left) and for 20 GeV/c electrons (on right). Lines represent the results of linear fit.

One can see on these figures that data in both plots demonstrate first-order linear dependence. This confirms that using the fit parameters we can derive the proportionality constants. Calculations were done in two ways: by least squares fit and by a maximum likelihood method. Both methods give consistent results. The final C_E and C_H constants are presented in Tables 2 and 3. The constants in Table 2 illustrate that the EMCAL and HCAL energy responses for π^+, p and K^+ particles are quite similar. Table 3 shows the corresponding constants for e^+ and e^- particles.

TABLE 2. The proportionality constants C_E and C_H for EMCAL and HCAL, respectively, calculated for 59 GeV/c π^+, p and K^+.

Particles	p (GeV/c)	C_E (MeV)	C_H (MeV)
π^+	59	0.52±0.01	13.2±0.1
p	59	0.52±0.01	13.9±0.1
K^+	59	0.57±0.02	13.2±0.1

TABLE 3. The proportionality constants C_E and C_H for EMCAL and HCAL, respectively, calculated for e^+ and e^- particles with different momenta.

Particles	p (GeV/c)	C_E (MeV)	C_H (MeV)
e^+	~5	0.43±0.02	12.8±0.1
e^+	~10	0.41±0.02	12.8±0.1
e^-	20	0.31±0.02	13.3±0.1

Applying the C_E and C_H constants one can calculate the ratio of the energy deposited in the calorimeters to the momentum of incoming particle. Fig. 5 illustrates the distribution of E_{e+h}/p ratios, where E_{e+h} is the calorimeters' response in energy units and p is the particle's momentum. The data in Fig. 5 shows that the mean values of E_{e+h}/p ratios are equal to 1. This means that the calibrated calorimeter's responses and the momentum measurements by the upstream magnet spectrometers are in good agreement. The widths of the distributions reflect the energy resolution of the system.

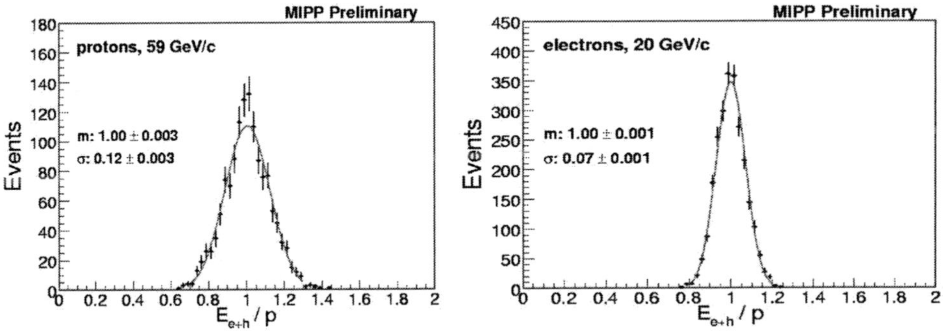

FIGURE 5. The distributions of E_{e+h}/p where E_{e+h} is the calorimeter response in energy units, p is the particle's momentum. The left plot represents 59 GeV/c protons data, right plot 20 GeV/c electron data. Curves represent the fit results using a Gaussian function.

ENERGY RESOLUTION and e/π RATIO

The energy resolution of the calorimeters derived from the E_{e+h}/p widths of the distributions are presented in Table 4. One can see that the combined EMCAL and HCAL energy resolution is in good agreement with what is expected for typical sampling calorimeters [3]. Table 5 presents the e/π ratios for our calorimeters.

TABLE 4. The combined energy resolution of both calorimeters.

Particle	p (GeV/c)	σ/p (%)
e^-	20	7±0.1
π^+	59	11±0.3
p	59	12±0.3
p (from E871)	70	9 (E871)

TABLE 5. The e/π ratios for EMCAL and HCAL calorimeters.

Particles	p (GeV/c)	e/π (EMCAL)	e/π (HCAL)
e^+	~5	0.83±0.02	0.97±0.02
e^+	~10	0.79±0.02	0.97±0.02
e^-	20	0.60±0.02	1.01±0.02

As is well known, the energy response of calorimeters is significantly different for electrons (or photons) compared to hadrons because hadrons produce more energy in the form of highly ionizing tracks that are not efficiently seen by the active medium. Also some of the energy from hadrons is carried out of the calorimeter by muons and neutrinos. This effect is embodied in the famous e/π ratio. This ratio generally determines the energy resolution of hadron calorimeters because of the large fluctuations in the fraction of the energy carried off by π^0's in hadron-induced showers. For optimal energy resolution the e/π ratio should be 1. In our case, with thin active layers in EMCAL, the e/π ratio appears to be less than 1. However, even with the low e/π ratio the energy resolution appears to be quite good. This suggests that one can use a lead calorimeter with the right ratio of active to passive layer thickness and get good energy resolution. Another argument for such a design is the relatively low cost.

MIPP is also performing calorimeter calibration studies on a plane-by-plane basis and studying the possibility of calibrating the EMCAL using hadrons [4].

STUDIES with NEUTRONS

Another way to study the calorimeter performance is by detecting neutrons. The expected signature of neutrons in the calorimeters is: no or very low energy contribution in the EMCAL (which is about 0.5 interaction lengths) and almost all the energy deposited into the HCAL. One possible source of neutrons is the proton to neutron elastic charge exchange reaction, p+n→n+p. For such neutrons we expect that the mean energy measured by the calorimeters to be consistent with the beam energy. We selected protons in the beam and used data from thin carbon and beryllium targets. For neutron selection we applied a veto on all events with any forward-going charged tracks downstream from the target. A neutron candidate event is presented in Fig. 6.

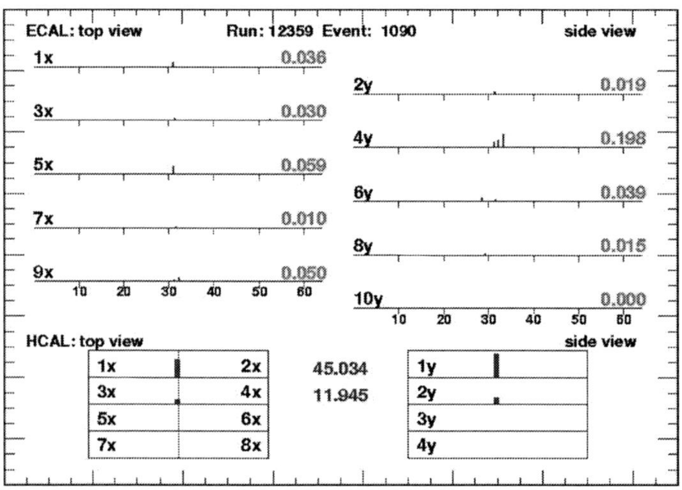

FIGURE 6. The calorimeter event display presenting a p→n elastic charge exchange candidate event. The beam momentum is 58 GeV/c, EMCAL energy 0.5 GeV, HCAL energy 57.0 GeV. The beam is incident from the top. Note that the HCAL is not to scale along the beam direction.

Fig. 7 represents the energy deposition of neutrons passing though the EMCAL and HCAL planes (on left plot) and the ΣHCAL distribution (on right plot). As expected there is minimal contribution in the EMCAL planes and almost all the energy is deposited into HCAL. The mean value of the ΣHCAL distribution (on right plot) is around HCAL=4500 counts. This value is consistent with proton data. (See HCAL values at EMCAL=0 on left plot in Fig. 3). Another way to check the consistency is to compare the neutron energy with the beam energy. Fig. 8 represents the E_{e+h}/p ratios for the neutron sample. The fit results suggest that the calorimeter responses are in good agreement with the incident beam momentum.

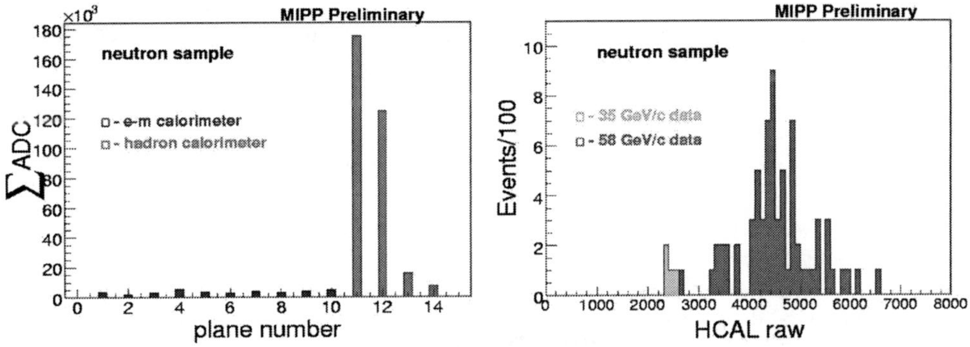

FIGURE 7. The left plot represents the energy deposition from neutrons while passing though EMCAL (planes 1 through 10) and HCAL (planes 11 through 14). The right plot represents the ΣHCAL distribution for both: 35 GeV/c and 58 GeV/c data.

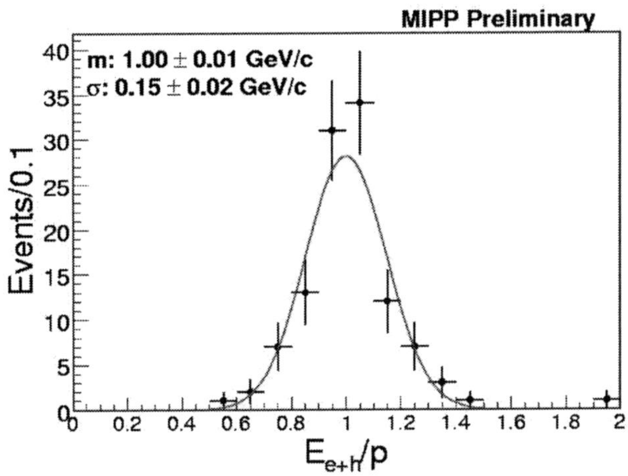

FIGURE 8. The E_{e+h}/p distribution, where E_{e+h} is calorimeter response to neutrons, p is the central value of beam momentum. The curve represents a fit with a Gaussian function.

CONCLUSIONS

The MIPP experiment collected about 31M events with hadron interactions on various targets with beam energies from 5 up to 120 GeV/c. Charged particles were identified in a wide range of momenta. Use of the electromagnetic and hadron calorimeters made it possible to detect photons and neutrons.

The calorimeters were energy calibrated using e^+, e^- and hadrons. The EMCAL and HCAL data demonstrate a linear response with respect to the incoming particle's energy. The calibration constants give an e/π ratio less than 1. However, the measured energy resolution was found to be quite good, and it is consistent with what was expected for the sampling calorimeters ($\sigma/E \sim 10\%$) [3].

The calorimeter performance was tested on a neutron sample. It demonstrates adequate response for neutrons: very low energy deposition in EMCAL and all the energy deposited into HCAL. The mean value of neutron energy was found to be consistent with the beam energy.

ACKNOWLEDGMENTS

The authors express their thanks to colleagues on the MIPP experiment for their efforts in running the experiment and data processing. We thank the committee for the pleasant atmosphere in the conference. This research was sponsored by the National Nuclear Security Administration under the Stewardship Science Academic Alliances program through DOE Research Grant DE-FG52-2006NA26182.

REFERENCES

1. P. D. Barnes Jr., et al. Proposal to measure particle production in the Meson area using Main Injector primary and secondary beams. see http://ppd.fnal.gov/experiments/e907/
2. R. A. Burnstein, et al., *Nucl. Instrum. and Methods* **541**, 516 (2005).
3. Review of Particle Physics. Physics Letters B **592** (2004) 265.
4.. Rajendran Raja. How to Calibrate an Electromagnetic Calorimeter Without Electrons. MIPP-NOTE-Gen-0061. 2005-03-11 at http://ppd.fnal.gov/experiments/e907/

Jet Energy Scale Determination at the DØ

Jiří Kvita

Institute of Particle and Nuclear Physics, Faculty of Mathematics and Physics, Charles University, V Holešovičkách 2, 180 00 Prague 8, Czech Republic

Abstract. Many physics measurements at a hadron collider critically depend on an accurate knowledge of the energy of jets resulting from the fragmentation of quarks and gluons generated in the hard scatter. The precise determination of the jet energy scale is a challenging project, involving corrections for physics, instrumental and jet algorithm-related effects. We present the most recent determination of the jet energy scale at the DZero experiment during Run II of the Tevatron proton-antiproton collider. We review the procedure followed, the estimated systematic uncertainties, as well as the validation studies performed. The experience gained by the Tevatron experiments in achieving percent-level determination of the jet energy scale should be of great value for the LHC.

Keywords: Jets, energy calibration, Dzero experiment.
PACS: 07.20.Fw, 13.87.-a

INTRODUCTION

With the 1fb^{-1} Run IIa dataset delivered by the Tevatron accelerator, we can explore a wealth of unique data with the DØ detector. As most physics analyses use jets as final hadron states, precision jet energy calibration is an essential input to already systematics limited measurements. The increased luminosity yields high statistics and potential to reach high accuracy. Here we present the preliminary JES determination on 150 pb^{-1}, the full $\approx 1\ \text{fb}^{-1}$ in progress. DØ has performed a full Calorimeter Calibration (ϕ-intercalibration as well as absolute calibration) and new and more advanced JES procedure and tools have been developed.

The DØ Detector

The DØ detector is in detail described in following Run I [1] and Run II [2] references. Here we shall mention only the essential parts. The main subdetectors relevant for the Jet Energy Scale determination are DØ Calorimeters (see Figure 1). These are Central and two End Cap Uranium-Liquid Argon Calorimeters providing uniform hermetic coverage up to $|\eta| \leq 4.2$ ($\eta \equiv -\ln\tan(\theta/2)$) and with fine segmentation up to $|\eta| < 3.2$: $\Delta\eta \times \Delta\phi = 0.1 \times 0.1$. With the DØ Calorimeter at the Tevatron Run II, we are challenged by the shorter bunch crossing (396 ns) and shorter charge integration time.

CP867, *XII International Conference on Calorimetry in High Energy Physics*, edited by S. R. Magill and R. Yoshida

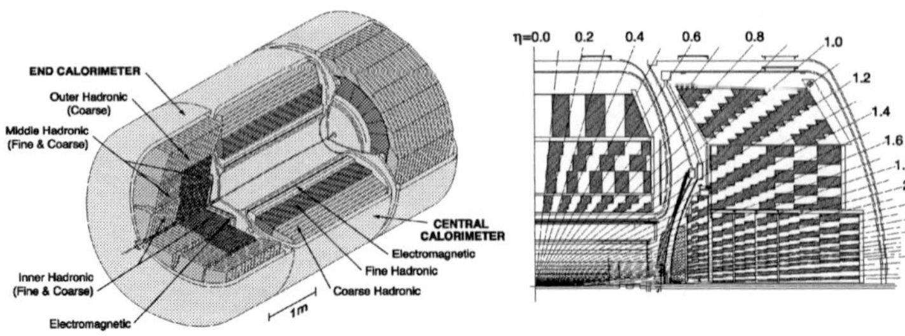

FIGURE 1. DØ Calorimeters in perspective and beam-parallel plane views.

Jet Algorithms

The same Run II cone jet algorithm is applied at three different levels on corresponding objects: at the parton level, Monte Carlo stable particles level and the detector level, where calorimeter clusters are used as seeds. The algorithm combines 4-vectors within a cone of radius R_{cone} (0.5 or 0.7) in $y \times \phi$ space. The jet axis is calculated and the procedure is iterated until a stable solution is found. The algorithm uses midpoints between jets to improve the infrared safety. At the end, identical solutions are removed and overlapping jets properly treated.

JET ENERGY SCALE DETERMINATION

The aim of the Jet Energy Scale (JES) procedure is to calibrate calorimeter jet energies to the particle level jets. There are various physics, instrumental and jet algorithm effects to be corrected for. As the agreement between simulation and Data is not precise enough to use only Monte Carlo samples, we need to employ in-situ calibration and derive JES separately for Data and Monte Carlo.

The basic idea is to assume calibrated EM calorimeter (using Z events) and use events with a well-measured EM object balanced by jet to determine the jet response.

The JES determination procedure involves the measurement of several subtle corrections, for each we also perform systematics cross checks and detailed error analysis. As a last point, we perform Closure tests where we compare the JES performance in Data and Monte Carlo to particle level jets. We calibrate separately Run II Algorithm Jets of Cone 0.5 and 0.7. The master JES formula is the following decomposition

$$E_{\text{new}} \equiv \frac{E_{\text{raw}} - \mathscr{O}}{\mathscr{F}_\eta \cdot \mathscr{R} \cdot \mathscr{S}},$$

where $\mathscr{O}$, $\mathscr{F}_\eta$, $\mathscr{R}$ and $\mathscr{S}$ stand for energy Offset, η-dependent correction, jet Response and Showering correction respectively.

In more detail, the corrections are derived sequentially, so the decomposition actually reads

$$E_{\text{new}} \equiv \frac{E_{\text{raw}} - \mathscr{O}(R_{\text{cone}}; \eta, \mathscr{L}, n_{\text{PV}})}{\mathscr{F}_\eta \cdot \mathscr{R}\left(R_{\text{cone}}; \frac{E_{\text{raw}} - \mathscr{O}}{\mathscr{F}_\eta}\right) \cdot \mathscr{S}\left(R_{\text{cone}}; E_{\text{raw}}\right)}.$$

First the energy Offset is measured as a function of η, inst. luminosity and number of Primary Vertices (multiple interactions dependence). Then the η-dependent correction is derived in several energy bins. Such η+Offset corrected energy serves as in input to Response measurement. As the last step, the Showering correction is derived.

The main data and Monte Carlo samples are γ+jets, dijet events and Z+jets.

Energy Offset

Offset ($\mathscr{O}$) is the energy not associated with the hard scatter process and can be decomposed as

$\mathscr{O}$ = Underlying Event + Noise + Multiple Interactions + Signal Pile-up.

In essence, we measure the energy density profile in detector projective towers, for which we use Minimum Bias events (luminosity monitor trigger) and bin Offset in number of primary vertices. Then, for each jet, the Offset energy is computed using the density for towers contained within the cone. The Offset energy in Central Calorimeter is typically 1 GeV $\times$ n_{PV}.

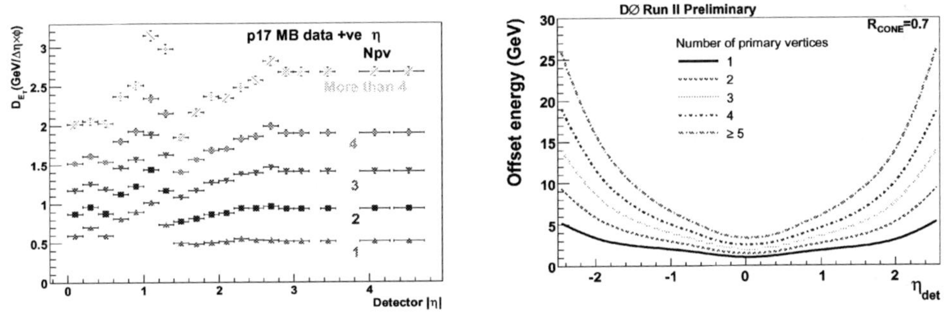

FIGURE 2. Energy density (left) and Offset energy for 0.7 Cone jets as functions of number of primary interactions vertices.

The MPF Method

The Missing E_{T} ($\not{E}_{\mathrm{T}}$) projection fraction method (MPF, see Figure 3) for the Response measurement uses back-to-back γ+jets events. Assuming a calibrated EM calorimeter ($Z \to e^+e^-$), we measure only the photon and missing transverse energy, so the method is jet algorithm independent. From the assumed p_{T} balance one can derive the MPF Response as

$$R_{\mathrm{MPF}} \equiv 1 + \frac{\not{E}_{\mathrm{T}} \cdot \vec{p}_{\mathrm{T}}^{\,\mathrm{EM}}}{(p_{\mathrm{T}}^{\mathrm{EM}})^2}.$$

Assuming back-to-back Tag object and the leading jet, the MPF response is expected to be the Jet Response. The same method is used for dijet events to derive the η correction.

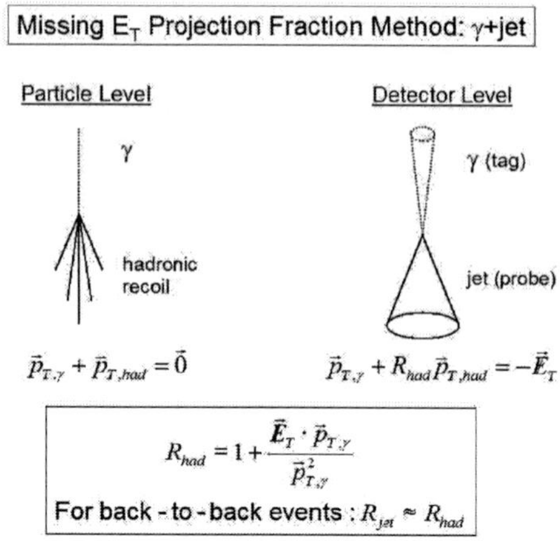

FIGURE 3. Illustration of the MPF method.

η-dependent Corrections

The purpose of the η correction is to make detector response uniform in η, which is especially important in the Inter Cryostat Region between Central and End Cap Calorimeters. We need to separate different instrumentation effects found over large η range. We measure jet Response relatively to Central Calorimeter Response using a Tag central object (jet or γ) and Probe jet anywhere.

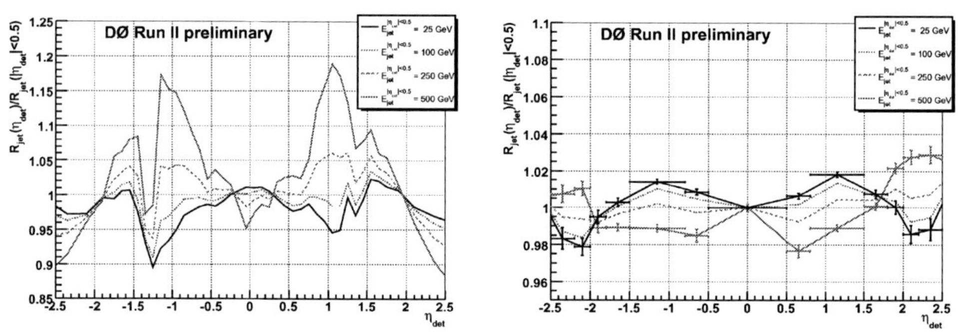

FIGURE 4. η-correction (left) and residuals in closure tests.

Response

As mentioned above, the Response measurement uses the MPF method. This particular correction is the largest correction (about 30%). An energy estimator combining well-measured photon energy and jet η is used to bin Response in terms of:

$$E' \equiv p_T^\gamma \cosh \eta_{\text{jet}} .$$

Measuring simultaneously Response and corrected jet energy as function of E', we finally remap $\mathscr{R}(E')$ to $\mathscr{R}(E_{\text{jet}})$. As can be seen in Figure 5, all η regions are on the same curve indicating the η correction work as desired.

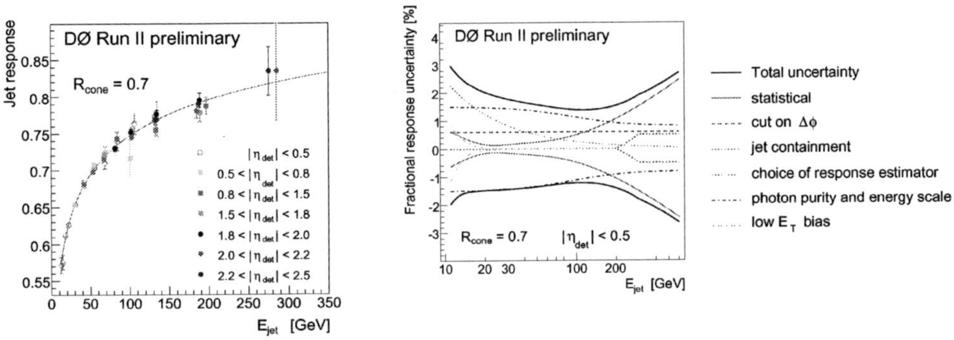

FIGURE 5. Response (left) as a function of (response corrected) energy in various calorimeter regions, and Response systematical error for Central Calorimeter jets (right).

Due to the preliminary dataset used in the report, large uncertainties appear at high energies. Another source of systematical error is the photon purity, since almost half of low p_T photons are actually background (jet faking an electromagnetic object), their energy undercorrected. It turns out that there is also an opposite effects, as we simultaneously overcorrect γ scale using the electron scale. Both effects partially cancel, while in the future we will specifically correct for both. Another error source are non-Gaussian tails in Response, especially in the ICR region and in Data.

Showering

The purpose of the Showering correction is to determine and correct for the fraction of jet energy inside the cone due to detector showering, being a result of multiple scattering, bending in the magnetic field etc.

Measuring the effect at detector level we observe effects of both detector+physics showering. Performing the same algorithm on MC particle jets we are sensitive only to physics showering effects. Thus, the ratio of these factors is the desired detector effect only.

In essence, we measure the energy density profile as a function of the distance from the jet center, subtract the fitted baseline due to Offset energy and determine the energy fraction inside and outside of the jet cone.

We have to treat properly not only particles leaking outside the jet, which would nominally be part of it, but also particles leaking inside the jet. Showering is a 1-2% correction, but a large uncentainty of the order of the correction itself is assigned due to the method.

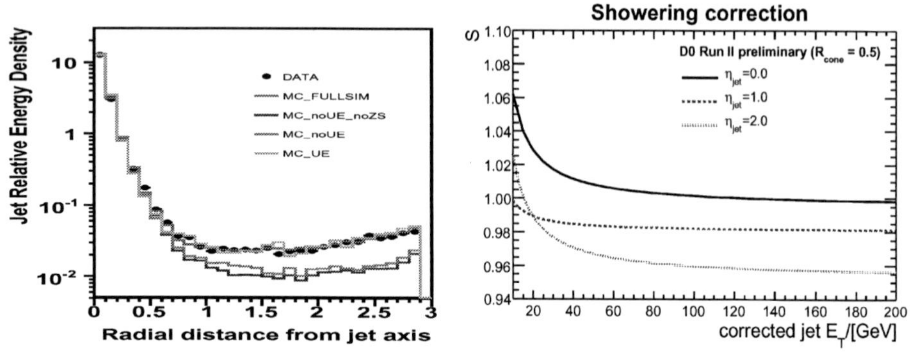

FIGURE 6. Jet profile, i.e. the energy density as a function of distance to jet center (left) and 0.5 Cone Showering correction itself (right).

JES Uncertainties

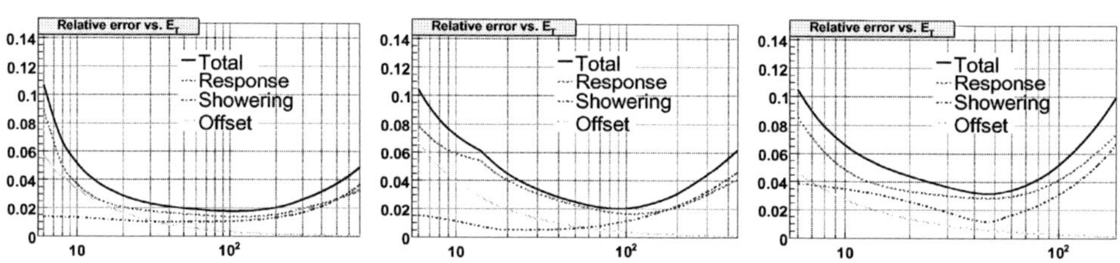

FIGURE 7. Fractional jet energy scale uncertainty for $R_{\text{cone}} = 0.7$ jets in data at $\eta = 0., 1., 2.$, as a function of uncorrected jet transverse energy.

Leading JES systematics is the Response. Offset uncertainties contribute mainly at low energies. η-dependent correction systematics comes from observed residuals after repeating the flattening procedure, being about 2% in the Inter Cryostat Region. Showering becomes important in forward region, i.e. at high energies, the uncertainty coming from statistics and instrumentation effects (detector edge), but mainly the method itself, which is now being significantly improved.

Closure Tests

It is important to test all corrections performance together, on broader range of samples and different event topologies. The goal is to validate JES within quoted uncertainties while we need to disentangle many effects from the possible JES failure itself. For many physics analyses, the relative Data-MC scale is most important, so it is useful to subtract Data and Monte Carlo, expecting points around 0.

In the so-called Hemisphere imbalance method (see Figure 8), we use a Tag object (EM-like: γ/Z, or a jet) and the highest p_T jet as our JES Probe, but we in fact combine information from all jets in the event. We project p_T's of all objects onto the Tag axis in the $x-y$ and compute separate scalar sums of projections lying in the EM Tag and Jet Probe hemispheres. We then define the Hemisphere imbalance as

$$\mathscr{H} \equiv \frac{\sum\limits_{\text{Probe hemi}} |\vec{p}_{\mathrm{T}} \cdot \vec{n}_{\mathrm{EM}}|}{\sum\limits_{\text{Tag hemi}} |\vec{p}_{\mathrm{T}} \cdot \vec{n}_{\mathrm{EM}}|} \approx 1.$$

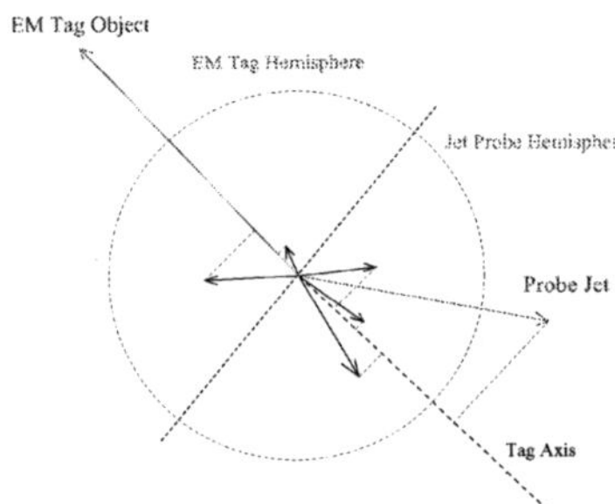

FIGURE 8. Hemisphere method illustration.

There are in fact many reasons why Hemi is not expected to be around 1. due to resolution, selection, kinematics and other biases. Therefore, we subtract Data and Simulation points, expecting to observe (in the case of the same biases present at both levels) points centered around zero.

Resulting Closure plots (see Figure 9) prove we correct jet energies in Data and Monte Carlo consistently to the same level and that all biases are present at both levels withing quoted uncertainties.

SUMMARY

Detailed understanding of JES at DØ has been achieved at the 2% level in Central Calorimeter at jet p_T in the 30-200 GeV range. This has only been possible due to the precision calorimeter calibration and new techniques to extract separate corrections and cross check their performance. There are ongoing efforts to further improvements in the areas of b-jets energy scale and semileptonic correction to correct semi-muonic b decays, as well as in improving standard corections themselves. Many algorithms of this in-situ JES technique can be of great importance for LHC experimets.

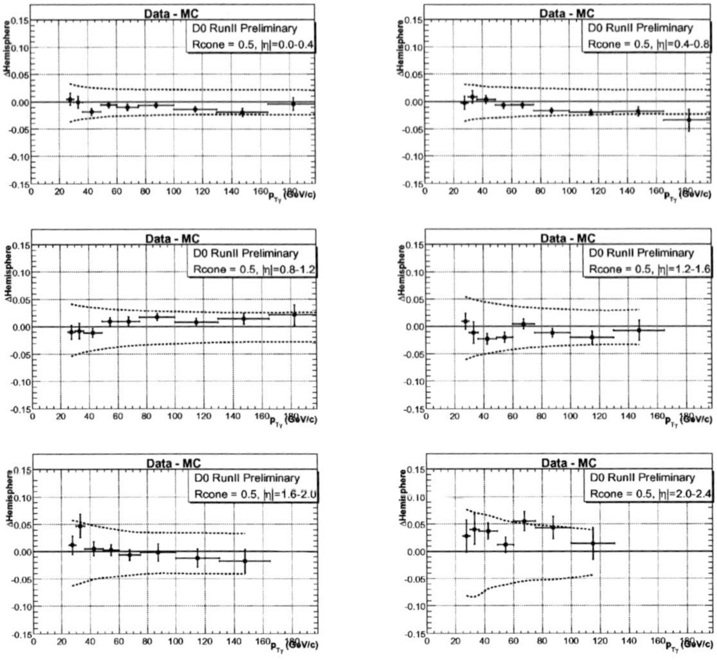

FIGURE 9. Hemisphere Method Closure, Data-MC, Cone 0.5

ACKNOWLEDGMENTS

This report is a result of long term efforts within the Jet energy Scale Group at the DØ experiment and would not be possible without the support of the whole DØ collaboration. The author would also like to thank the INGO grant 1P05LA257 of Ministry of Education of the Czech Republic, as well as the kind CALOR06 support of students.

REFERENCES

1. S. Abachi *et al.* [D0 Collaboration], "The D0 Detector," Nucl. Instrum. Meth. A **338**, 185 (1994).
2. V. M. Abazov *et al.* [D0 Collaboration], "The upgraded D0 detector," arXiv:physics/0507191, FERMILAB-PUB-05-341-E.
3. B. Abbott *et. al.* [D0 Collaboration], "Determination of the Absolute Jet Energy Scale in the DZERO Calorimeters", NUCL.INSTRUM.METH.A, **424**, 352 (1999).

Performance of the PrimEx Electromagnetic Calorimeter

M.Kubantsev*, I. Larin† and A. Gasparian**
for PrimEx Collaboration
*Institute of Theoretical and Experimental Physics, Moscow, Russia/Northwestern University, Evanston, Il 60208, USA
†Institute of Theoretical and Experimental Physics, Moscow, Russia
**Physics Department NC A&T State University Greensboro, NC 27411, USA

Abstract. We report the design and performance of the hybrid electromagnetic calorimeter consisting of 1152 $PbWO_4$ crystals and 576 lead glass blocks for the PrimEx experiment at the Jefferson Laboratory. The detector was built for high precision measurement of the neutral pion lifetime via the Primakoff effect. Calorimeter installation and commissioning was completed with the first physics run in fall of 2004. We present the energy and position resolution of the calorimeter. Obtained π^0 mass resolution of 1.3MeV/c^2 and its production angle resolution of 0.34mrad demonstrate the ability of the experiment to extract the π^0 lifetime on one percent level.

Keywords: calorimeter, photon, energy resolution

INTRODUCTION

Measurement of the partial width of the decay $\pi^0 \to \gamma\gamma$ provides an unique test of low energy QCD in confinement scale regime. The π^0 lifetime is arguably the most precise theoretical calculation in low energy QCD, which in the leading order depends on fundamental parameters only. The contribution of the next-to-leading order does not exceed 4% of the leading-order prediction [1]. The experiment PrimEx [2], [4] was designed to make a high precision (at one percent level) measurement of the π^0 decay width. The tagged photon beam at Hall B of the Jefferson Laboratory [3] was used for forward production of neutral pions in the Coulomb field of the target nucleus (Primakoff effect). The two photons from the pion decay are detected in the Hybrid Calorimeter (HYCAL). The HYCAL was designed to provide high efficiency detection of the photons with high resolution in energy and space. Here we describe the physics requirements for the calorimeter, the design and construction of the setup, and its calibration and performance during the physics data taking. We summarize calorimeter characteristics in the conclusion.

CP867, *XII International Conference on Calorimetry in High Energy Physics*,
edited by S. R. Magill and R. Yoshida

PHYSIC REQUIREMENTS

Precision measurement of an absolute value of the Primakoff π^0 production cross section requires a photon detector with high energy and space resolutions. Yet another condition is accurate monitoring of the energy and intensity of the tagged photon beam. The Primakoff production cross section is peaked at $\sim$ 0.3mrad for a nucleon target at energies of this experiment (E_γ = 4.9–5.5GeV). Therefore, high angular resolution is required for clear separation of this mechanism from the other processes at forward angles. Likewise, accurate reconstruction of the two-photon invariant mass is essential for rejection of background. Based on these requirements, we formulated the design concepts for the calorimeter on the Monte-Carlo level and then built prototype detectors that were tested in the beam [7]. Finally, we were able to construct the PrimEx detector whose design and performance are described below.

DESIGN OF THE CALORIMETER

To optimize the performance and cost of the detector, we have chosen a hybrid calorimeter design combining novel $PbWO_4$ scintillating crystals and conventional Cherenkov lead glass detectors. The central part of the calorimeter is built of 1152 $PbWO_4$ crystals surrounded with 576 lead glass blocks. The transverse dimensions of the calorimeter ($116x116cm^2$) are sufficient for a large geometrical acceptance for the π^0 mesons emitted at zero angle from the production target located 7.3 m upstream of the detector ($\sim$70%). Since the area around the photon beam is under a high rate of irradiation, the crystals used should be radiation hard. The $PbWO_4$ crystals fully meet the latter requirement.

In the past 10-15 years, the $PbWO_4$ crystal development resulted in improved light output that provides a good energy resolution. Furthermore, due to very strict requirements by LHC calorimetry groups [5], high quality and radiation hard $PbWO_4$ crystals became commercially available from the two basic manufacturers: BTCP, Russia [6] and Shanghai Institute of Ceramics (SIC), China. For the HYCAL calorimeter we have recieved 1250 crystals from SIC, China. The dimensions of the individual crystals are 20.5x20.5x180mm^3. To improve light collection, the crystals were wrapped in 100μm VM2000 reflective material [7] and then, for light isolation between the neighbours, in 36μm Tedlar. The crystals were viewed by Hamamatsu R4125HA photomultiplier tubes coupled to them with optical grease. As the $PbWO_4$ crystal light yield is highly temperature dependent ($\sim$-2%/^oC at room temperature), temperature stabilization of the calorimeter is mandatory. Throughout the experiment, the calorimeter was operated at 14 ± 0.1°C temperature, which was maintained by circulation of the cool liquid around the outer body of the calorimeter assembly.

The lead-glass modules of the calorimeter were provided by the IHEP (Russia) group [8]. The individual modules with the size of 38.2x38.2x450mm^3 (SF-2 type) were wrapped by 25μm aluminized mylar foil. The Cherenkov light produced in the lead glass radiator was detected by the Russian-made FEU-84-3 photomultiplier tubes.

FIGURE 1. Photograph of the HYCAL calorimeter during assembly: crystal and lead glass arrays are shown

The photograph of the detector during the assembly at the TestLab of the Jefferson laboratory is shown in Fig. 1.

The PrimEx experiment utilized the CODA DAQ system. The digital information was read out from over 2200 channels of FASTBUS based ADC and TDC. An advantage of this system is the ability to operate in fully buffered mode. Namely, events are buffered in the digitization modules themselves allowing the modules to be "live" during the readout process. This significantly reduced the dead time of the DAQ, which is essential for the cross-section measurement.

CALIBRATION

In order to maintain high performance of the calorimeter, a periodic energy calibration with tagged photon beam was required. During the experiment, HYCAL was calibrated by illuminating each module with a low-intensity tagged-photon beam. The entire assembly with weight of about five tons was movable in the transporter frame in both horizontal and vertical directions with a position accuracy of $\pm 0.7mm$. The calorimeter position was remotely controlled during the calibration.

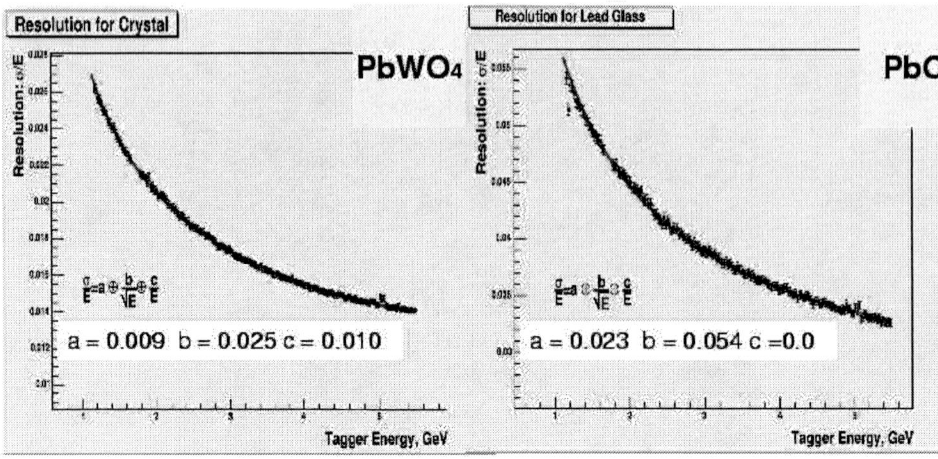

FIGURE 2. Energy resolution of the $PbWO_4$ crystals and the lead glass as function of the primary photon energy.

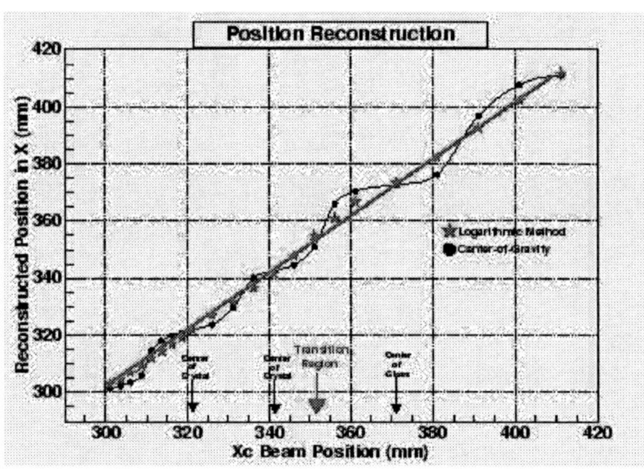

FIGURE 3. Shower position reconstruction in the calorimeter.

The tagged photon beam irradiated each module of the calorimeter using "a snake scan" during the calibration runs. Relative gains of the modules and energy dependence of the detector resolution, as determined through calibration, are shown in Fig. 2 for the $PbWO_4$ and lead-glass parts of the detector. The reconstruction of the γ cluster position in the calorimeter is illustrated in Fig. 3. The logarithmic method of coordinate reconstruction with fitted parameters was chosen. The final fine-tuning of the modules' gains was made using photons from decays of neutral pions observed in the experiment.

In between the beam calibrations the gain stability of each channel was monitored by a specially designed gain light monitoring system (LMS) [9]. The LMS consists of

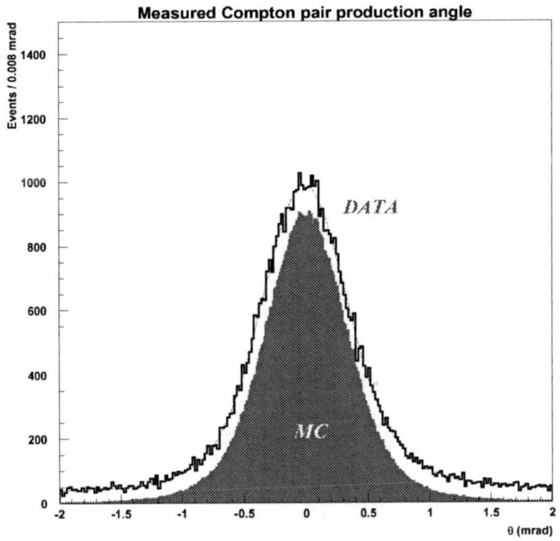

FIGURE 4. Measured angular distribution of Compton-pair clusters compared with the Monte Carlo simulations: $\sigma_{theta} = 0.34mrad$.

the following main parts: a light source, a light mixing box, a light distribution system, and reference detectors with a DAQ system. The light source comprises an assembly of NICHIA super bright blue LEDs (peak wavelength 470 nm). Light mixing is done using a six inch diameter integrating sphere (ORIEL) that provided 2000 distribution channels. The pulsed light was distributed to individual calorimeter modules via a bundle of plastic optical fibers. Each fiber is attached to the front face of a detector module. As a reference detector, a Hamamatsu PIN photodiode (S3399, 3mm diameter) with an AMPTEC low noise charge amplifier was used. Light from the light monitoring system was periodically injected into the detector modules between the data taking files ($\sim$1 hour period).

PHYSICS RUN PERFORMANCE

The experiment took physics data in the fall of 2004. Several exposures with beryllium, carbon, tin and lead targets were taken. The performance of the calorimeter was studied using several observed physics processes: quasielastic and inelastic photoproduction of neutral pions, Compton scattering and pair production.

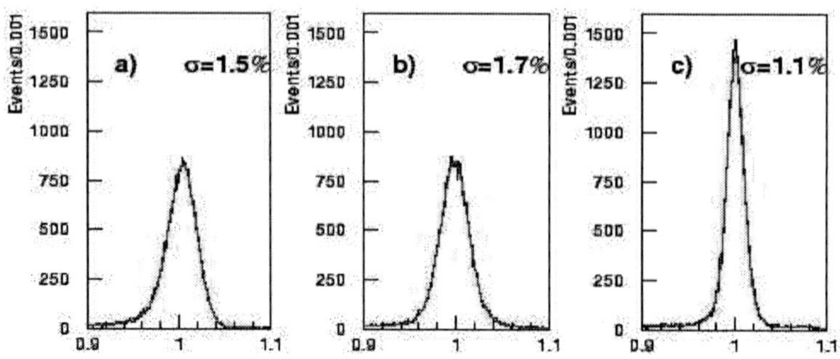

FIGURE 5. Ratio of sum of measured Compton pair energy and tagged γ energy at 5.2 ± 0.3 GeV. a - using reconstructed cluster energies; b - using cluster coordinates and Compton kinematical relations to calculate Compton pair energy; c - after applying the transverse momentum constraint.

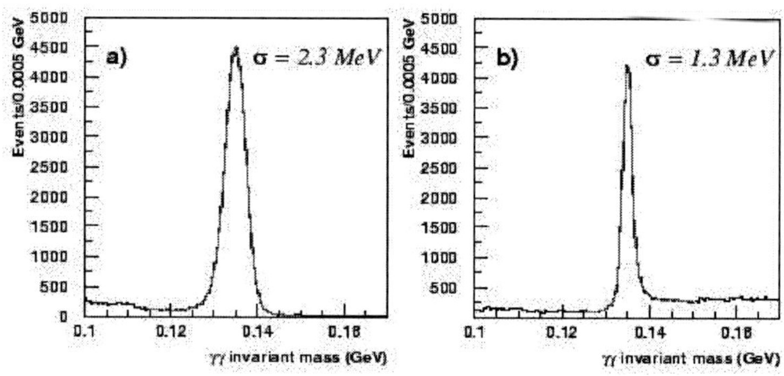

FIGURE 6. π^0 mass resolution for the $PbWO_4$ array. Left: after calibration on the nominal π^0 mass; Right: after correction on the tagged photon energy: the remaining width is mostly due to coordinate resolution.

First, we consider Compton scattering $\gamma + e \to \gamma + e$, which provides four constraints of the reaction kinematics and allows the study of energy and space resolutions of the detector, albeit in a limited range of small angles for the central $PbWO_4$ region of the calorimeter. Plotted in Fig. 5 is the ratio between the total energy as measured by the calorimeter and incident energy of the tagged photon ($\sim$5.2 GeV). The width of the distribution is $\simeq 1.5\%$ in (a) where complete information from the calibration is used, and $\simeq 1.7\%$ in (b) where the energy of the Compton pair is estimated only from the resonstructed shower coordinates using the kinematics of Compton scattering. The latter demonstrates the importance of good spatial resolution. As soon as total transverse momentum of the two showers is constrained to zero ($\vec{P_t} = 0$), the width of the distribution reduces to mere 1.1%, see Fig. 5c. As the Compton pairs are emitted at zero angle to the incident beam, the width of the distribution of the observed angle

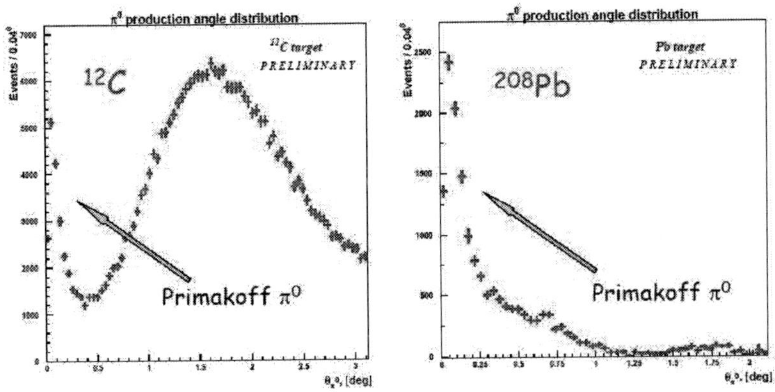

FIGURE 7. π^0 angular distribution for production on Carbon and Lead. The Primakoff production peak is clearly visible at small angles.

θ is a direct measure of the angular resolution of the detector. From the data of Fig. 4 we estimate the experimental resolution on θ as $\sigma_\theta = 0.34$ mrad, which agrees with the simulated value and is sufficient for measuring the Primakoff reaction.

Next, we estimate the detector resolution on the π^0 mass. Fig. 6 shows the $\gamma-\gamma$ invariant mass distribution for the central ($PbWO_4$) region of the calorimeter. The left-hand panel shows the distribution upon calibration on the nominal mass of the neutral pion (observed width of 2.3 MeV/c^2). In the right-hand panel, correction of the primary tagged photon energy has been applied. The observed width of 1.3 Mev/c^2 in the latter case reflects the coordinate resolution of the calorimeter. The constraint on the primary energy significantly improves the mass resolution. In addition, we see a high mass tail of the peak that is due to inelastic π^0 production. These pions have lower energy, and the beam-energy constraint shifts their corrected masses towards higher values.

The gaussian width of the π^0 peak increases from 1.3 MeV/c^2 for the $PbWO_4$ area of the calorimeter to 2.6 MeV/c^2 for the border area (on the $PbWO_4$ side of it), and to 4.5 MeV/c^2 for the middle of the lead-glass area.

The success of the experiment largely depends on accurate reconstruction of π^0 mesons emitted at small angles. Shown in Fig. 7 are the preliminary data on π^0 angular distributions for quasielastic production on carbon and lead nuclei. The Primakoff production peak is clearly seen for both data sets. The latter demonstrates good performance of the calorimeter, which should be adequate for a precision measurement of the π^0 lifetime.

CONCLUSION

The high precision hybrid electromagnetic calorimeter (HYCAL) has been designed, constructed, and run in the experiment aimed at measuring the lifetime of the neutral pion via the Primakoff effect. Measured characteristics of the detector have been shown to match the design parameters. The energy resolution of the calorimeter is about 1.5% for a photon energy of 5 GeV. Angular resolution of 0.34 mrad for Compton production is acheived. The π^0 mass resolution (upon constraining the beam energy) is $\sigma = 1.3MeV/c^2$ for the central PWO part of the calorimeter. High-quality data sets for extracting the π^0 lifetime have been collected. The first results are expected by the end of this summer. The authors thank all PrimEx collaborators for excellent design, construction, running, and data analysis of the experiment. Stable operation of the CEBAF accelerator is gratefully acknowledged. The work was supported by the US NSF grant (PHY-0079840) and the RFBR grant (04-02-17466).

REFERENCES

1. J.L.Goity, A.M.Bernstein, J.F.Donoghue, B.R.Holstein. Phys. Rev. D66:076014 (2002)

2. PrimEx Proposal (http://www.jlab.org/primex/primex_notes/PR99-014.ps)
PrimEx Collaboration, December 17, 1998

3. D.I.Sober et al. Nucl.Instrum.Meth.A440:263-284,2000

4. PrimEx Conceptual Design Report (http://www.jlab.org/primex/primex_notes/PrimEx_CDR.ps)
PrimEx Collaboration, March 3, 2000

5. The CMS Electromagnetic Calorimeter Project, CERN/LHCC 97-33,1997

6. A.Annenkov et al. Nucl.Instrum.Meth.A537:173-176,2005
D

7. A.Gasparian. Proceedings of 11th International Conference on Calorimetry in High-Energy Physics (Calor 2004), Perugia, Italy, 28 Mar - 2 Apr 2004. Published in *Perugia 2004, Calorimetry in particle physics* 109-115

8. M.Y.Balatz et al. Nucl.Instrum.Meth.A545:114-138,2005

9. S. Danagulian. Proceedings of 10th International Conference on Calorimetry in High EnergyPhysics (CALOR 2002), Pasadena, California, 25-30 Mar 2002.Published in *Pasadena 2002, Calorimetry in particle physics* 479-485

SESSION II

CALORIMETER TECHNOLOGIES

Convener: José Repond

Precision Crystal Calorimeters in High Energy Physics: Past, Present and Future

Ren-Yuan Zhu

California Institute of Technology, Pasadena, CA 91125, USA

Abstract. Precision crystal calorimeters traditionally play an important role in high energy physics experiments. In the last two decades, it faces a challenge to maintain its precision in a hostile radiation environment. This paper reviews the performance of crystal calorimeters constructed for high energy physics experiments and the progress achieved in understanding crystal's radiation damage as well as in developing high quality scintillating crystals for particle physics. Potential applications of new generation scintillating crystals of high density and high light yield, such as LSO and LYSO, in particle physics experiments is also discussed.

Keywords: Calorimeter, Crystal, Radiation Damage
PACS: 07.20.Fw; 29.40.Mc; 61.80.Fd

INTRODUCTION

Total absorption shower counters made of inorganic scintillating crystals have been known for decades for their superb energy resolution and detection efficiency [1]. In high energy and nuclear physics, large arrays of scintillating crystals have been assembled for precision measurements of photons and electrons. The discovery potential of

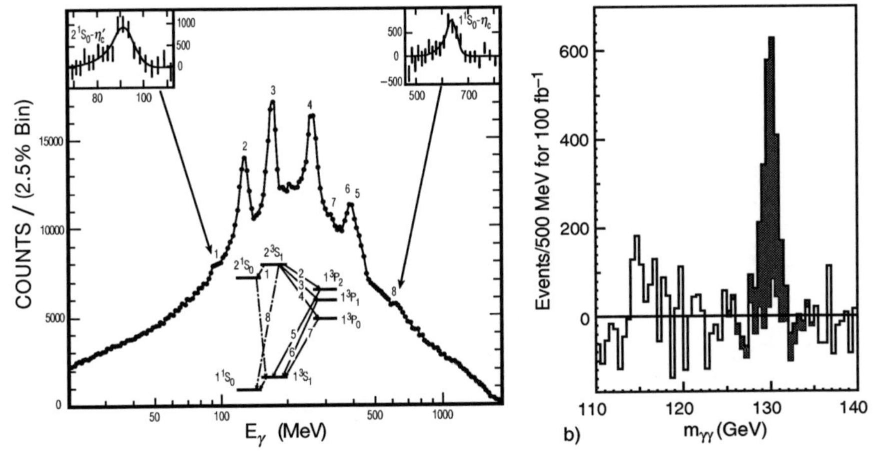

FIGURE 1. Left: An inclusive photon spectrum measured at the ψ' by the NaI(Tl) crystal calorimeter at SLAC [2]. Right: The expected background subtracted Higgs mass peak reconstructed from its two photon decays measured by the CMS $PbWO_4$ crystal calorimeter [3].

CP867, *XII International Conference on Calorimetry in High Energy Physics,*
edited by S. R. Magill and R. Yoshida

crystal calorimeters was early demonstrated by the Crystal Ball experiment through its study of radiative transitions and decays of the Charmonium family [2]. Figure 1 (Left) shows nearly all the principal radiative transition lines of the Charmonium system simultaneously measured by the NaI(Tl) crystal calorimeter. The designed goal of the CMS lead tungstate ($PbWO_4$) crystal calorimeter [3] is to maximize its discovery potential in searching for narrow resonances in photon and electron final states at LHC. Figure 1 (Right) shows the expected background subtracted Higgs peak reconstructed with its two decay photons by the CMS $PbWO_4$ calorimeter. The potential of the Higgs discovery via this decay channel is directly related to the energy resolution of the calorimeter.

Crystal calorimeters have been constructed, and their use has been a key factor in the successful physics programs of many experiments. With proper calibration and monitoring, crystal calorimeters usually achieve their designed resolution *in situ* [4]. Figure 2 (Left) shows energy resolution as a function of the electron energy obtained with the L3 BGO calorimeter in the CERN test beam, which is in a good agreement with the Bhabha electron resolutions measured *in situ* at LEP by using the RFQ calibration [5], as shown in Figure 2 (Right).

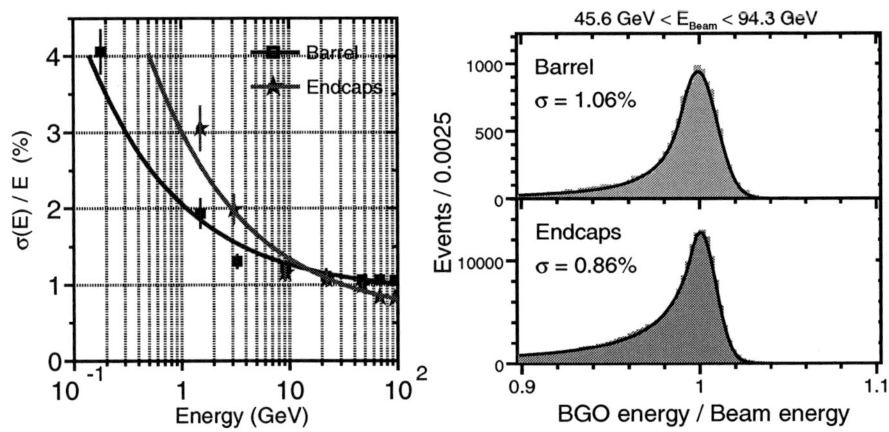

FIGURE 2. Left: The energy resolution of the L3 BGO calorimeter as a function of electron energy measured in the CERN test beam. Right: The energy resolution of Bhabha electrons observed by the L3 BGO calorimeter *in situ* at LEP by using the RFQ calibration.

Table 1 summarizes parameters of past and present crystal calorimeters in high energy physics. One notes that each of these calorimeters requires several cubic meters of high quality crystals. The most ambitious crystal calorimeter in Table 1 is presumably the CMS calorimeter which uses 11 m^3 $PbWO_4$ crystals. Its designed energy resolution [3] is

$$\sigma_E/E = \mathbf{2.5\%}/\sqrt{E} \oplus \mathbf{0.55\%} \oplus 0.2/E \tag{1}$$

for the barrel, and

$$\sigma_E/E = \mathbf{5.7\%}/\sqrt{E} \oplus \mathbf{0.55\%} \oplus 0.25/E \tag{2}$$

for the endcaps.

Figure 3 (Left) shows the CMS designed energy resolution as a function of energy. It can be decomposed to three contributions from photoelectron statistics (stochastic),

TABLE 1. Crystal Calorimeter in High Energy Physics: Past and Present

Experiment	C. Ball	L3	CLEO II	KTeV	*BaBar*	BELLE	CMS
Accelerator	SPEAR	LEP	CESR	Tevatron	PEP II	KEK	LHC
Date	75–85	80–00	80–00	90–10	94–10	94–10	95–20
Crystal Type	NaI(Tl)	BGO	CsI(Tl)	CsI	CsI(Tl)	CsI(Tl)	$PbWO_4$
B-Field (Tesla)	-	0.5	1.5	-	1.5	1.0	4.0
Inner Radius (m)	0.254	0.55	1.0	-	1.0	1.25	1.29
Number of Crystals	672	11,400	7,800	3,300	6,580	8,800	76,000
Crystal Depth (X_0)	16	22	16	27	16 to 17.5	16.2	25
Crystal Volume (m^3)	1	1.5	7	2	5.9	9.5	11
L. Yield (p.e./MeV)	350	1,400	5,000	40	5,000	5,000	2
Photosensor	PMT	Si PD	Si PD	PMT	Si PD	Si PD	APD†
Photosensor Gain	Large	1	1	4,000	1	1	50
Noise/Chan. (MeV)	0.05	0.8	0.5	Small	0.15	0.2	30
Dynamic Range	10^4	10^5	10^4	10^4	10^4	10^4	10^5

† Avalanche photodiode.

intrinsic shower leakage (stochastic and constant) and readout noise (noise). Figure 3 (Right) shows the energy resolution as a function of electron energy measured in the CERN test beam for two groups of 3×3 crystals, independent of their impact position on the crystal front face [6]. The measured resolution in the low and middle energy region agrees with the designed resolution. It also shows a smaller constant term because of the perfect calibration in the test beam.

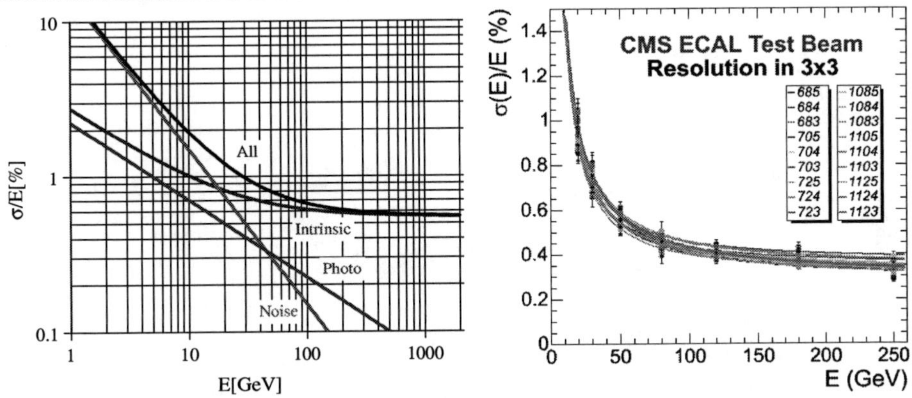

FIGURE 3. Left: The designed energy resolution of the CMS $PbWO_4$ calorimeter is shown as a function of energy and corresponding contributions [3]. Right: The energy resolution of two groups of 9 $PbWO_4$ crystals is shown as function of electron energy obtained in the CMS ECAL beam test [6].

Over the last two decades, however, crystal calorimeters faced a new challenge: the radiation damage caused by the increased center of mass energy and luminosity. To preserve crystal precision in a severe radiation environment crystal quality control is a crucial issue. The rest of this paper is devoted to a review of optical and scintillation properties of heavy crystal scintillators commonly used for the particle physics. The progresses achieved in understanding crystal's radiation damage and in developing high quality crystals is also be discussed.

COMMONLY USED CRYSTAL SCINTILLATORS

Table 2 lists the basic properties of commonly used heavy crystal scintillators: NaI(Tl), CsI(Tl), undoped CsI, BaF_2, CeF_3, bismuth gemanade ($Bi_4Ge_3O_{12}$ or BGO), lead tungstate ($PbWO_4$ or PWO), Ce-doped lutetium oxyorthosilicate ($Lu_2(SiO_4)O$ or LSO(Ce)) [7]. Mass production capabilities exist for all these crystals. All, except LSO(Ce) have been used in high energy physics experiments. LSO(Ce) is widely used in the medical industry.

TABLE 2. Properties of Heavy Crystal Scintillators with Mass Production Capability

Crystal	NaI(Tl)	CsI(Tl)	CsI	BaF_2	BGO	$PbWO_4$	LSO(Ce)
Density (g/cm^3)	3.67	4.51	4.51	4.89	7.13	8.3	7.40
Melting Point (°C)	651	621	621	1280	1050	1123	2050
Radiation Length (cm)	2.59	1.86	1.86	2.03	1.12	0.89	1.14
Molière Radius (cm)	4.13	3.57	3.57	3.10	2.23	2.00	2.07
Interaction Length (cm)	42.9	39.3	39.3	30.7	22.8	20.7	20.9
Refractive Index[a]	1.85	1.79	1.95	1.50	2.15	2.20	1.82
Hygroscopicity	Yes	Slight	Slight	No	No	No	No
Luminescence[b] (nm)	410	560	420	300	480	425	420
(at Peak)			310	220		420	
Decay Time[b] (ns)	230	1250	30	630	300	30	40
			6	0.9		10	
Light Yield[b,c]	100	165	3.6	36	21	0.29	83
			1.1	3.4		0.083	
d(LY)/dT[b,d] (%/°C)	-0.2	0.3	-1.3	-1.3	-0.9	-2.7	-0.2
				~0			

a At the wavelength of the emission maximum.
b Top line: slow component, bottom line: fast component.
c Relative and PMT quantum efficiency taken out.
d At room temperature.

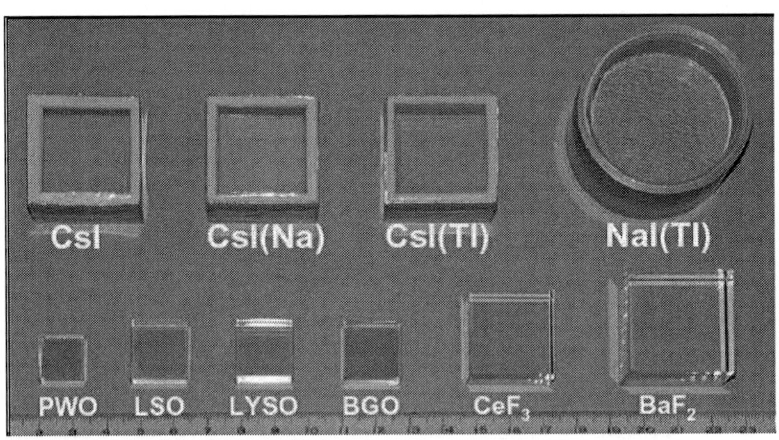

FIGURE 4. A photo shows ten crystal scintillators with dimension of 1.5 X_0.

The optical and scintillation properties were measured and compared for various crystal scintillators. Figure 4 is a photo showing ten crystal samples used in this comparative study. All samples are wrapped with white Tyvek paper as reflectors. NaI and CsI based samples are sealed in quartz window of 3 mm thick to avoid surface degradation caused by their hygroscopicity. To minimize sample size dependence all samples have a cube shape, $1.5 \times 1.5 \times 1.5\ X_0^3$, except the NaI(Tl) sample which is a cylinder with a length of 1.5 X_0 and an area of two ends equal to $1.5 \times 1.5\ X_0^2$ to match the 2 inch diameter of the PMT cathode.

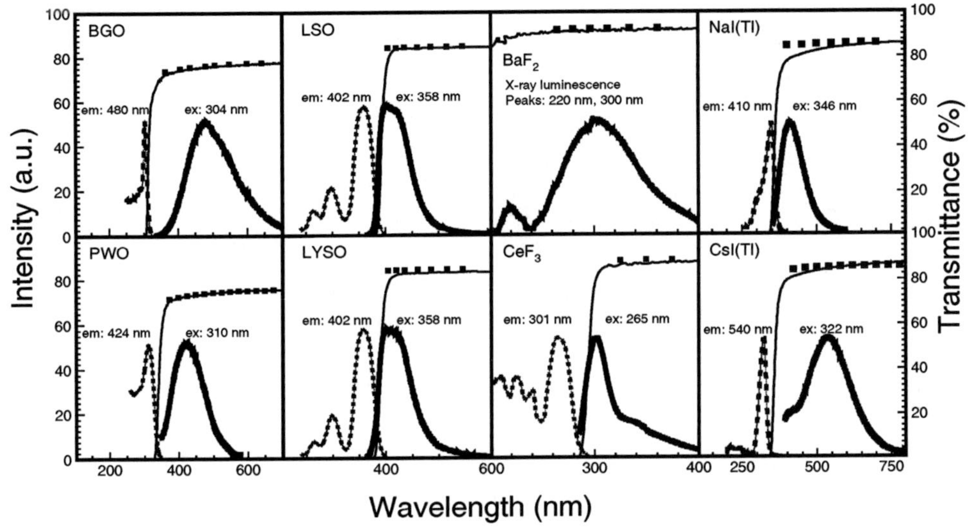

FIGURE 5. The excitation (dashed lines) and emission (thick solid lines) spectra (left scale) and the transmittance (thin solid lines) spectra (right scale) are shown as a function of wavelength for eight crystal scintillators.

Figure 5 shows a comparison of the transmittance (thin solid lines), emission (thick solid lines) and excitation (dashed lines) spectra as a function of wavelength for eight samples. The solid black square dots in these plots show the theoretical limit of the transmittance, which was calculated by using corresponding refractive index as a function of wavelength, taking into account multiple bounces between the two parallel end surfaces and assuming no internal absorption [8]. The measured transmittance approaches the theoretical limits, indicating negligible internal absorption. It is interesting to note that while the BGO, BaF_2, NaI(Tl), $PbWO_4$ and CsI(Tl) crystals have their emission spectra well within the transparent region, the UV absorption edge in the transmittance spectra of the LSO, LYSO and CeF_3 crystals cuts into the emission spectra and thus affects crystal's light output. This effect is more seriously observed for long LSO and LYSO samples [9].

Figure 6 shows light output in photoelectrons per MeV energy deposition as a function of integration time, measured by using a Photonis XP2254b PMT with multi-alkali photo cathode, for five fast crystal scintillators (Left): LSO, LYSO, CeF_3, undoped CsI and $PbWO_4$ and five slow (Right) crystal scintillators: NaI(Tl), CsI(Tl), CsI(Na), BaF_2 and BGO. The undoped CsI, $PbWO_4$ and BaF_2 crystals are observed to have two decay components as shown in Table 2. The LSO and LYSO samples have consistent fast decay time (~40 ns) and high photoelectron yield, which is 6 and 230 times of BGO

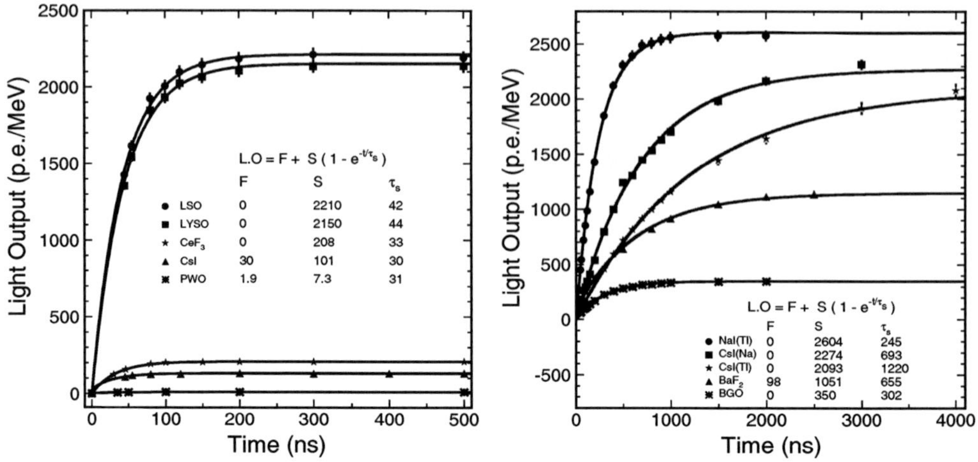

FIGURE 6. Light output measured by using the XP2254b PMT is shown as a function of integration time for five fast (Left) and five slow (Right) crystal scintillators.

and $PbWO_4$ respectively.

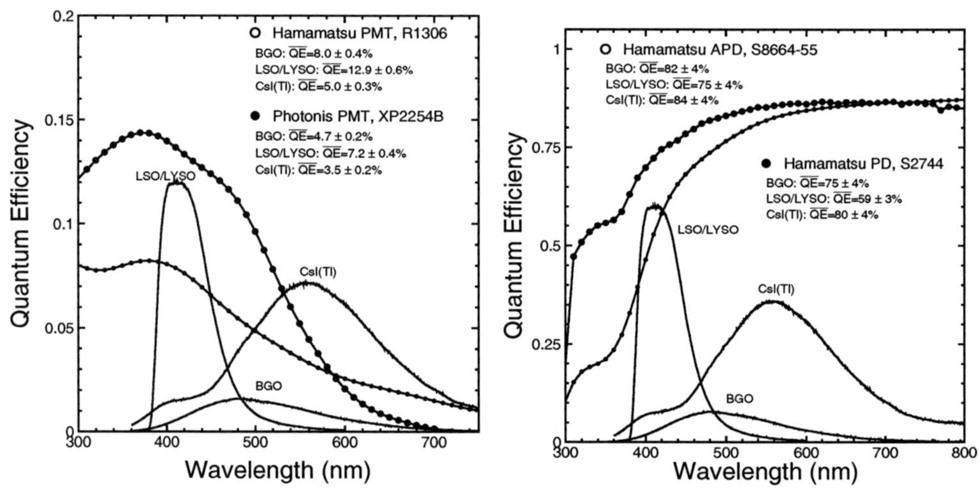

FIGURE 7. Left: The quantum efficiencies of a Hamamatsu 1306 PMT with bi-alkali cathode and a Photonis 2254B (solid dots)PMT with multi-alkali cathode (open circles) are shown as a function of wavelength together with the emission spectra of the LSO/LYSO, BGO and CsI(Tl) samples, where the area under the emission curves is proportional to their corresponding absolute light output. Right: The same for a Hamamatsu S8664 Si APD (open circles) and a Hamamatsu S2744 Si PIN diode (solid dots).

Since the quantum efficiency (QE) of the PMT used for the light output measurement is a function of wavelength, it must be taken out to compare crystal's photon yield. Figure 7 shows typical QE of two PMT with bi-alkali cathode (Hamamatsu R1306) and multi-alkali cathode (Photonis 2254B), a Si APD (Hamamatsu S8664) and a Si PIN PD

(Hamamatsu S2744), as well as the emission spectra of LSO/LYSO, BGO and CsI(Tl) crystals.

TABLE 3. Emission Weighted Quantum Efficiencies (%)

Emission	LSO/LYSO	BGO	CsI(Tl)
Hamamatsu R1306 PMT	12.9±0.6	8.0±0.4	5.0±0.3
Hamamatsu R2059 PMT	13.6±0.7	8.0±0.4	5.0±0.3
Photonis XP2254b	7.2±0.4	4.7±0.2	3.5±0.2
Hamamatsu S2744 PD	59±4	75±4	80±4
Hamamatsu S8664 APD	75±4	82±4	84±4

Table 3 summarized numerical result of the emission weighted average QE for several readout devices. Taking into account the PMT response, we conclude that the light output of the LSO and LYSO crystals is a factor of 4 and 200 of that of BGO and $PbWO_4$ respectively, as shown in Table 2.

Large size LSO and LYSO crystals with consistent optical and scintillation properties have been developed recently for the medical industry. Figure 8 shows four long crystal samples: SIC BGO, CTI LSO, CPI LYSO and Saint-Gobain LYSO of $2.5 \times 2.5 \times 20\,cm^3$ size. Their availability provides a new possibility for the precision crystal calorimeters.

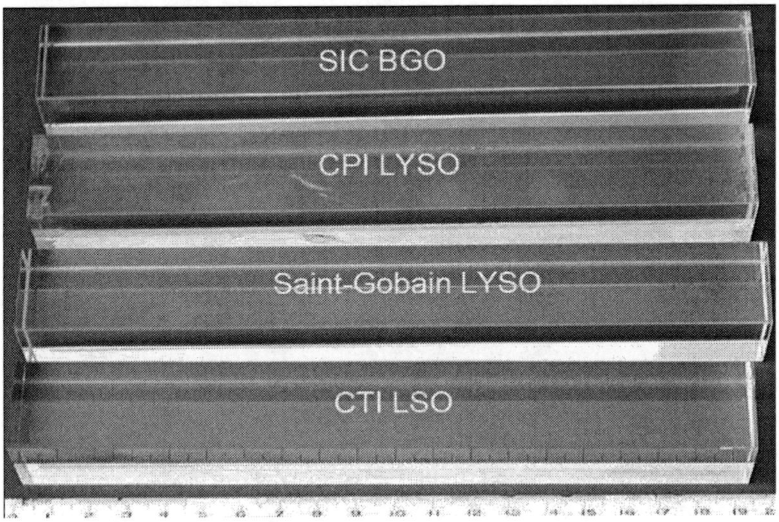

FIGURE 8. A photo shows four long crystal samples with dimension of $2.5 \times 2.5 \times 20\,cm^3$.

Figure 9 shows spectra of 0.51 MeV γ-rays from a ^{22}Na source observed by these samples with coincidence triggers. The readout devices used are a Hamamatsu R1306 PMT (Left) and 2 Hamamatsu S8664-55 APDs (Right). The FWHM resolution for the 0.51 MeV γ-ray with the PMT readout is about 12% to 13% for the long LSO and LYSO samples, which can be compared to 15% for the BGO sample. With APD readout, the γ-ray peaks are well visible for the long LSO and LYSO samples, but is much less distinguished for the BGO sample. The energy equivalent readout noise in our laboratory set up of APD readout is below 40 keV for the LSO and LYSO samples.

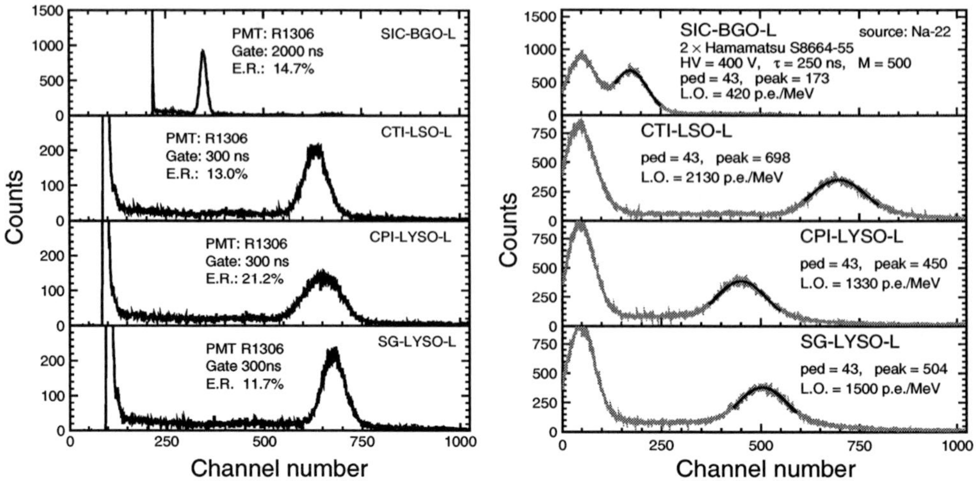

FIGURE 9. The spectra of 0.511 MeV γ-rays from a ^{22}Na source measured with a coincidence trigger using a Hamamatsu R1306 PMT (Left) and two Hamamatsu S8664-55 APDs (Right) for long BGO, LSO and LYSO samples of $2.5 \times 2.5 \times 20\,\text{cm}^3$ size.

CRYSTAL RADIATION DAMAGE

All known large size crystal scintillators suffer from radiation damage [10]. There are three possible radiation damage effects in crystal scintillators. First, radiation would induce internal absorption, caused by the color center formation, which would reduce the light attenuation length [8], and thus the light output, and may also cause a degradation of the light response uniformity.

Figure 10 shows the longitudinal transmittance spectra and their degradation under irradiation measured for full size CMS $PbWO_4$ (23 cm long, Left) and *BaBar* CsI(Tl) (30 cm long, Right) crystal samples respectively. The radiation induced absorption and corresponding color center formation are clearly observed in these samples. Second, radiation would induce phosphorescence (afterglow), which would cause an increase of the readout noise. Last, radiation may also reduce scintillation light yield. If so, both the light output and the light response uniformity would be degraded since the radiation dose profile *in situ* is usually not uniform.

Radiation induced absorption may also recover under application temperature through a process called color center annihilation. If so, the damage would be dose rate dependent [10]. Figures 11 (Left) shows light output normalized to that before irradiation (solid dots with error bars) as a function of time under irradiation for a full size CMS $PbWO_4$ sample. Measurements were made step by step for different dose rates: 15, 100, 500 and 1,000 rad/h, as shown in these figures. The degradation of the light output shows a clear dose rate dependence.

If no recovery or the recovery speed is very slow, however, the color center annihilation process would be less important, the color center density would not reach an equilibrium under certain dose rate rather continuous increasing until all defect traps are

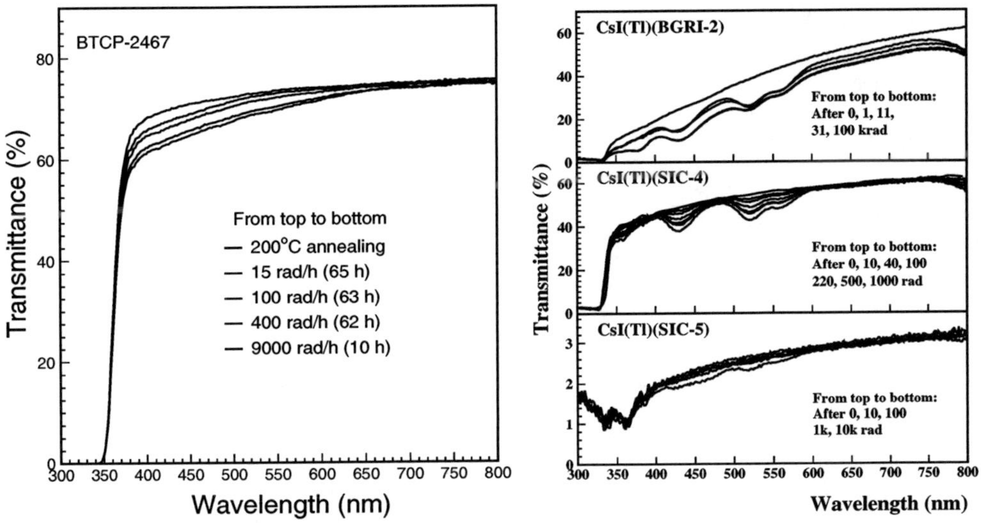

FIGURE 10. Longitudinal transmittance of $PbWO_4$ (Left) and CsI(Tl) (Right) samples, showing radiation induced absorption bands.

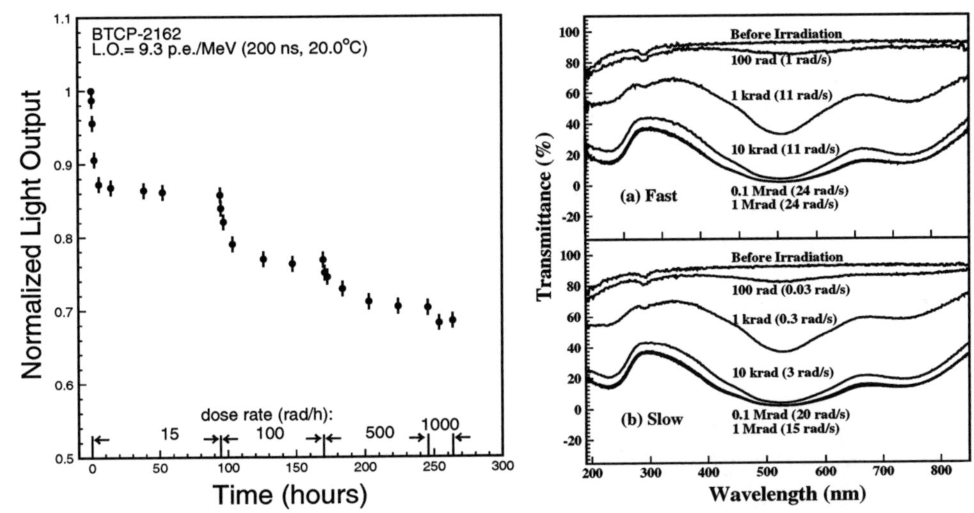

FIGURE 11. Left: Normalized light output is plotted as a function of time under irradiations for a CMS $PbWO_4$ sample, showing dose rate dependent radiation damage. Right: No dose rate dependence was observed for the longitudinal transmittance spectra measured for a GEM full size BaF_2 sample.

filled. This means no dose rate dependence. Figure 11 (Right) shows the transmittance as a function of wavelength for a GEM full size (25 cm) BaF_2 sample before and after 100, 1k, 10k, 100k and 1M rad irradiations (from top to bottom) under a fast (a) and a slow

(b) dose rates. While the fast dose rate is up to a factor of three higher than the slow rate, the damage levels for the same integrated dose are identical. This was expected, since no recovery at room temperature was observed for BaF_2.

TABLE 4. Radiation Damage in Crystal Scintillators

Item	CsI(Tl)	CsI	BaF_2	BGO	PWO
Color Centers	Yes	Yes	Yes	Yes	Yes
phosphorescence	Yes	Yes	Yes	Yes	Yes
Scintillation Damage	No	No	No	No	No
Recover @RT	Slow	Slow	No	Yes	Yes
Dose Rate Dependence	No	No	No	Yes	Yes
Thermally Annealing	No	No	Yes	Yes	Yes
Optical Bleaching	No	No	Yes	Yes	Yes

Crystal radiation damage may also be cured by either thermal annealing or optical bleaching [11]. Table 4 summarizes radiation damage observed in various crystal scintillators which were used in the high energy and nuclear physics experiments.

For LYSO(Ce) crystals our initial investigation on $2.5 \times 2.5 \times 20$ cm^3 bar samples indicates that they suffer less radiation damage than other crystals, such as BGO, CsI(Tl) and $PbWO_4$ [9]. Figure 12 (Left) shows an expanded view of the longitudinal transmittance spectra measured for a long LYSO samples before and after each step of γ-ray irradiations at 2, 100 and 9,000 rad/h. An immediate increase of the longitudinal transmittance after the first irradiation under 2 rad/h was observed, which was followed by small degradation under higher dose rate. Because of LYSO's high light yield its natural phosphorescence and radiation induced phosphorescence have small effect in readout noise. Fig. 12 (Right) shows the γ-ray induced PMT anode current for two long LYSO

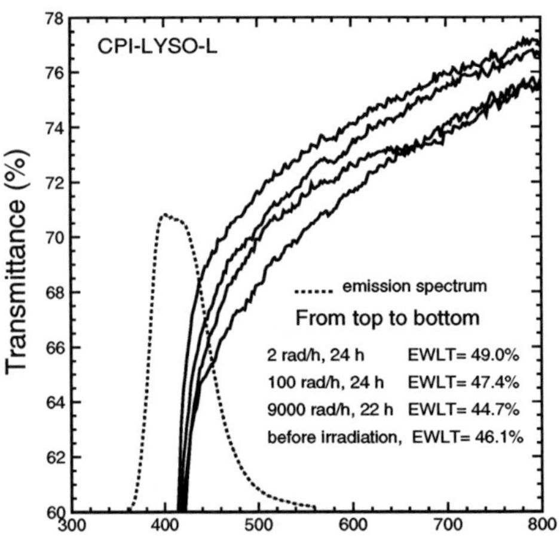

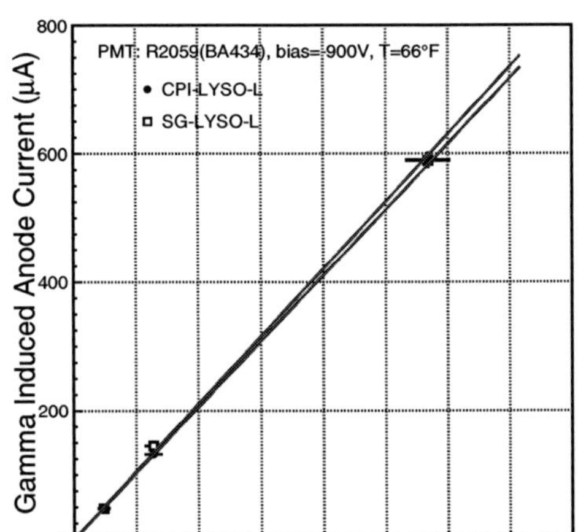

FIGURE 12. Left: The longitudinal transmittance spectra before and after 2, 100 and 9,000 rad/h irradiations are shown as a function of wavelength for the Saint-Gobain long LYSO sample. Right: The γ-ray induced anode current is shown as a function of the dose rate for the CPI and Saint-Gobain long LYSO samples.

samples and a linear fit. The radiation induced phosphorescence related readout noise with 100 ns integration time is estimated to be about 0.2 MeV and 1 MeV equivalent respectively in a radiation environment of 15rad/h and 500 rad/h for LYSO samples of $2.5 \times 2.5 \times 20$ cm^3 size.

CRYSTAL DEVELOPMENT AND QUALITY IMPROVEMENT

Commercially available mass produced crystals usually do not meet the quality required for precision crystal calorimeters. A research and development program is usually needed to systematically study the correlations between crystal's radiation hardness and its impurities and point defects. By removing harmful impurities from the raw materials and developing an approach to effectively reduce the density of defects in the crystal during the growth and processing, the quality of mass produced crystals may be improved. This approach has been successfully carried out for BGO [12], BaF_2 [13], CsI(Tl) [10] and $PbWO_4$ [10, 14]. Two examples are given below in this section.

CsI(Tl) Development

Figure 13 (Left) shows the light output as a function of accumulated dose for full size ($\sim$ 30 cm) CsI(Tl) samples produced at the Shanghai Institute of Ceramics (SIC), and compared to the *BaBar* radiation hardness specification (solid line) [10]. While the late

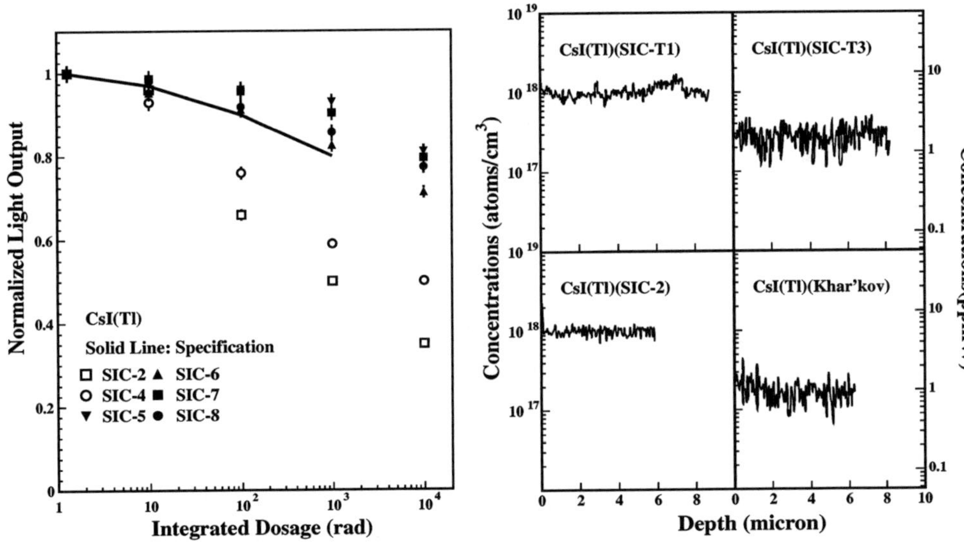

FIGURE 13. Left: The progress of CsI(Tl) radiation hardness is shown for full size ($\sim$30 cm) CsI(Tl) samples from SIC together with the rad-hard specification of the *BaBar* experiment. Right: The depth profile of oxygen contamination is shown for two rad-soft CsI(Tl) samples (SIC-T1 and SIC-2) and two rad-hard samples (SIC-T3 and Khar'kov).

samples SIC-5, 6, 7 and 8 satisfy the *BaBar* specification, early samples SIC-2 and 4 did not. This improved radiation hardness of CsI(Tl) crystals was also observed by *BaBar* and BELLE experiments [15].

The improvement of CsI(Tl) quality was achieved following an understanding that the radiation damage in halide crystals is caused by the oxygen or hydroxyl contamination. The identification of oxygen contamination was achieved by the Secondary Ionization Mass Spectroscopy (SIMS) analysis carried out at Charles Evans & Associates. A Cs ion beam of 6 keV and 50 nA was used to bombard the CsI(Tl) sample. All samples were freshly cleaved prior before being loaded into the UHV chamber. An area of 0.15×0.15 mm^2 on the cleaved surface was analyzed. To further avoid surface contamination, the starting point of the analysis is at about 10 μm deep inside the fresh cleaved surface. Figure 13 (Right) shows the depth profile of oxygen contamination for two radiation soft samples (SIC-T1 and SIC-2) and two radiation hard samples (SIC-T3 and Khar'kov). Crystals with poor radiation resistance have oxygen contamination of 10^{18} atoms/cm^3 or 5.7 ppmW, which is 5 times higher than the background count (2×10^{17} atoms/cm^3, or 1.4 ppmW). The practical solution at SIC is to use a scavenger to remove oxygen. This leads to the development shown in Figure 13 (Left).

$PbWO_4$ Development

Figure 14 shows the light output as a function of time under various dose rates for CMS full size (23 cm) $PbWO_4$ samples produced at SIC [10]. Samples produced late

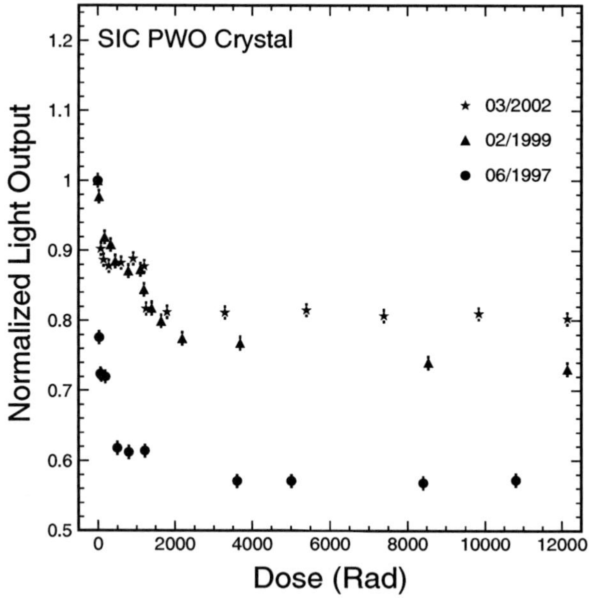

FIGURE 14. The progress of $PbWO_4$ radiation hardness is shown for full size (23 cm) CMS $PbWO_4$ samples from SIC.

2002 is much more radiation hard than the early samples. This improved radiation hardness of $PbWO_4$ crystals was also confirmed by an evaluation of mass produced $PbWO_4$ crystals [16].

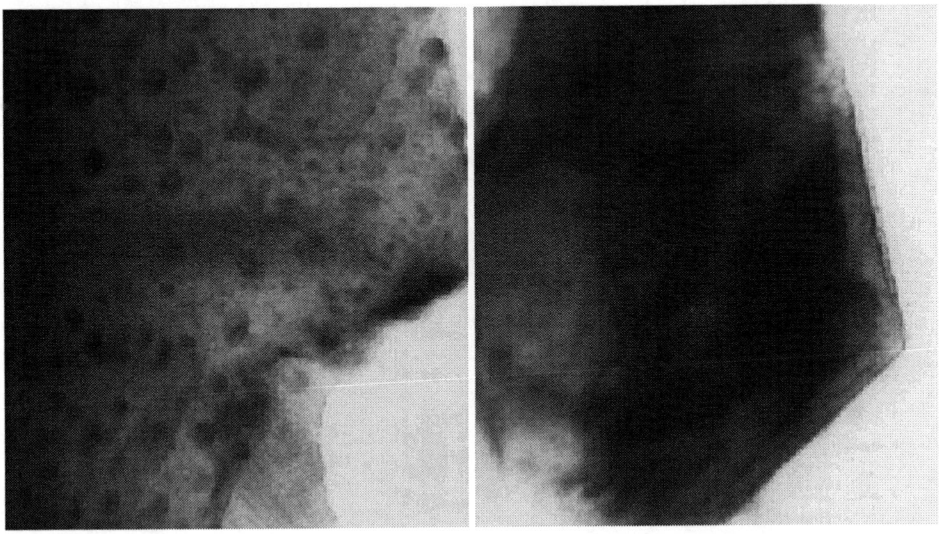

FIGURE 15. TEM pictures of a $PbWO_4$ crystal of poor (Left) radiation hardness, showing clearly the black spots of ϕ5–10 nm related to oxygen vacancies, as compared to that of a good one (Right).

The improvement of $PbWO_4$ quality was achieved following an understanding that the radiation damage in oxide crystals is caused by the oxygen or lattice structure vacancies. By using Transmission Electron Microscopy (TEM) coupled to Energy Dispersion Spectrometry (EDS), a localized stoichiometry analysis was used to identify oxygen vacancies. A TOPCON-002B Scope was first used at 200 kV and 10 μA. Samples were made to powders of an average grain size of a few μm, and then placed on a sustaining membrane. With a spatial resolution of 2 Å, the lattice structure of $PbWO_4$ crystals was clearly visible. Figure 15 (Left) shows a TEM picture taken for a sample with poor radiation hardness. Black spots of a diameter of 5 – 10 nm were clearly seen in the picture. On the other hand, samples with good radiation hardness show stable TEM picture with no black spots, as shown in Figure 15 (Right).

By employing a TEM with EDS system, a localized stoichiometry analysis was carried out at SIC [17]. The system is a JEOL JEM-2010 scope and a Link ISIS EDS. The spatial resolution of this system allows a localized stoichiometry analysis in a region of a diameter of 0.5 nm. An as grown sample was first analyzed, and black spots were observed. Points inside and surrounding the black spots were analyzed as well as points far away from the black spots. The uncertainty of the analysis is typically 15%. The resultant atomic fractions (%) at these areas are listed in Table 5. A clear deviation from the atomic stoichiometry of O:W:Pb = 66:17:17 was observed in the center of these black spots, pointing to a severe deficit of the oxygen component. In the peripheral area, the oxygen deficit was less, but still significant. There was no oxygen deficit observed in the area far away from the black spots. As a comparison, the same sample after oxygen compensation was re-analyzed. No black spot was found. The result of the analysis is also listed in Table 5. In all randomly selected points no stoichiometry deviation was observed. This analysis thus clearly identified oxygen vacancies in $PbWO_4$ samples of poor radiation hardness.

TABLE 5. Atomic Fraction (%) of O, W and Pb in $PbWO_4$ Samples Measured by TEM/EDS [17]

As Grown Sample

Element	Black Spot	Peripheral	$Matrix_1$	$Matrix_2$
O	1.5	15.8	60.8	63.2
W	50.8	44.3	19.6	18.4
Pb	47.7	39.9	19.6	18.4

The Same Sample after Oxygen Compensation

Element	$Point_1$	$Point_2$	$Point_3$	$Point_4$
O	59.0	66.4	57.4	66.7
W	21.0	16.5	21.3	16.8
Pb	20.0	17.1	21.3	16.5

Various approaches were tried to compensate oxygen vacancies by annealing $PbWO_4$ crystals in an oxygen-rich atmosphere [10] and by doping [14]. Significant improvement of radiation hardness was observed in both cases. The practical solution at SIC is to dope $PbWO_4$ crystals with yttrium. This leads to the development shown in Figure 14.

AN LSO/LYSO CRYSTAL CALORIMETER

As discussed in previous sections LSO(Ce) and LYSO(Ce) crystals are a new type of crystal scintillators with light yield 4 and 200 times of BGO and $PbWO_4$ respectively and a decay time about 40 ns. The LYSO crystals are also known to suffer less radiation damage as compared to other commonly used crystal scintillators. Mass production capability of LSO/LYSO crystals has been established for the medical industry. Crystals of size sufficient for building a crystal calorimeter are routinely grown. Assuming the same readout scheme as the CMS $PbWO_4$ calorimeter, the expected energy resolution of an LSO/LYSO based crystal calorimeter would be

$$\sigma_E/E = \mathbf{2\%}/\sqrt{E} \oplus \mathbf{0.5\%} \oplus 0.001/E, \tag{3}$$

which represents a fast calorimeter over large dynamic range with very low noise. Such calorimeter would provide excellent physics potential for high energy physics experiment in the International Linear Collider [18] or in a super B factory [19].

SUMMARY

Precision crystal calorimeters have been an important part of high energy physics detector. Its energy resolution, position resolution and photon identification capability has been a key factor in many physics discoveries. In the last two decades, however, it faces a challenge: the radiation damage in scintillation crystals. Progresses have been made in understanding crystal's radiation damage and developing high quality crystals for high energy physics experiments.

Recent availability of mass production capability of large size LSO and LYSO crystals provides an opportunity to build a crystal calorimeter with unprecedent energy resolution over a large dynamic range down to MeV level. This crystal calorimeter, if built, would greatly enhance the discovery potential for future high energy and nuclear physics experiments.

ACKNOWLEDGMENTS

Work supported by U.S. Department of Energy Grant No. DE-FG03-92-ER40701.

REFERENCES

1. G. Gratta *et al.*, *Annu. Rev. Nucl. Part. Sci.* **44** 453 (1994).
2. E. Bloom and C, Peck, *Ann. Rev. Nucl. Part. Sci.* 33 143-197 (1983).
3. *The CMS Electromagnetic Calorimeter Project*, CERN/LHCC 97-33 (1997).
4. R.Y. Zhu, *Nucl. Instr. and Meth.* A537 344 (2005).
5. U. Chaturvedi *et al.*, *Nucl. Instr. and Meth.* A461 376 (2001).
6. A. Zabi, in *Proceedings of the 12th International Conference on Calorimetry in Particle Physics*, Chicago, (2006).
7. C. Melcher, V.S. Patent 4958080 (1990) and 5025151 (1991).
8. D.A. Ma and R.Y. Zhu, *Nucl. Instr. and Meth.* A333 (1993) 422.
9. J.M. Chen *et al.*, *IEEE Trans. Nucl. Sci.* **52** (2005) 3133.
10. R.-Y. Zhu, *Nucl. Instr. and Meth.* **A413** (1998) 297–311.
11. D.A. Ma and R.Y. Zhu, *Nucl. Instr. and Meth.* A332 113 (1993) and D.A. Ma *et al.*, *Nucl. Instr. and Meth.* A356 309 (1995).
12. Z.Y. Wei *et al.*, *Nucl. Instr. and Meth.* A297 163 (1990).
13. R.Y. Zhu, *Nucl. Instr. and Meth.* A340 442 (1994).
14. X.D. Qu *et al.*, *Nucl. Instr. and Meth.* **A480**, 470 (2002),
15. T. Hryn'ova, in *Proceedings of the 10th International Conference on Calorimetry in Particle Physics*, World Scientific, Ed. R.-Y. Zhu, 175 (2002).
16. R.H. Mao *et al.*, *IEEE Trans. Nucl. Sci.* **NS-51** (2004) 1777.
17. Z.W. Yin *et al.*, in *Proceedings of SCINT97 Int'l Conf.*, Ed. Zhiwen Yin *et al.*, CAS Shanghai Branch Press, (1997) 191.
18. R.-Y. Zhu, *An LSO/LYSO Crystal Calorimeter for the ILC*, talk presented in 2005 ILC Workshop, Snowmass. See http://www.hep.caltech.edu/ zhu/talks/ryz_050818_lc.pdf.
19. W. Wisniewski, *Consideration for Calorimetry at a Super B Factory*, in *Proceedings of Tenth International Conference on Calorimetry in Particle Physics*, Ed. R.-Y. Zhu, World Scientific (2002).

The Philosophy and Feasibility of Dual Readout Calorimetry

John Hauptman

Physics, Iowa State University, Ames IA 50011

Abstract. I will discuss the general physical ideas behind dual-readout calorimetry, their implementation in DREAM (Dual REAdout Module) with exact separation of scintillation and Cerenkov light, implementation with mixed light in DREAM fibers, anticipated implementation in $PbWO_4$ crystals with applications to the 4th Concept detector and to CMS, use in high energy gamma-ray and cosmic ray astrophysics with Cerenkov and N_2 fluorescent light, and implementation in the 4th Concept detector for muon identification.

Keywords: dual readout, calorimetry
PACS: 29.40.Vj, 29.40.Ka, 29.40.Mc

INTRODUCTION

The idea of dual readout calorimetry has been around for many years [1], but only recently has this idea been successfully implemented in the DREAM (Dual REAdout Module) beam test, and with spectacular results [2]. The essential idea is simple: as a hadronic shower develops there are large fluctuations in the numbers of charged and neutral pions produced in hadronic interactions with nuclei, in addition to smaller numbers of neutral and charged kaons and nulceons. The neutral pion decays immediately to photons, $\pi^0 \to \gamma\gamma$, which in turn initiate localized electromagnetic showers, while the charged pions lose energy by ionization and reinteract hadronically continuing the hadronic shower cascade. In general, the response of a calorimeter to the electrons and positrons of an electromagnetic shower is different from the response to charged pions, characterized by the single number $(e/h) \neq 1$ and, therefore, fluctuations in the π^0 content, or EM fraction f_{em}, lead directly to fluctuations in response, *i.e.,* poor energy resolution.

Indeed, these large fluctuations in π^0 content often result in a non-Gaussian response function and a non-linear response to incident energy. The latter problem can be far more serious than poor energy resolution in a high energy experiment that must probe physics beyond the energies of available test beams.

The DREAM module has tested the principle of dual-readout calorimetry using Cerenkov and scintillation light in an optical calorimeter, and for this type of calorimeter there are five ways to separate the light,

1. *physically*, by producing light in separate media, such as in distinct scintillating and Cerenkov (clear) fibers. This is the approach taken by DREAM and discussed below.

CP867, *XII International Conference on Calorimetry in High Energy Physics,*
edited by S. R. Magill and R. Yoshida

2. *by direction*: Cerenkov light is produced on a forward cone at an angle θ_C with respect to the charged particle direction, whereas scintillation light is isotropic.
3. *by wavelength*: Cerenkov light is produced constant in $\nu d\nu$, or in wavelength as $d\lambda/\lambda^2$, as thus mostly in the blue and violet. The scintillation wavelength is a property of the scintillating mechanisms and material, but typically in the green and therefore at higher wavelengths.
4. *by time*: Cerenkov light is essentially instantaneous, whereas scintillation light depends on the material, but can range from a fraction of a nanosecond to hundreds of microseconds.
5. *by polarization*: Cerenkov light is polarized radially outward from the charged particle axis $\hat{v}$, and the electric field vector $\vec{E}$ is defined by $\vec{E} \times \vec{B} = \hat{k}$ where $\vec{B}$ is the curl field produced by the moving charged particle, $\hat{k}$ is the Cerenkov photon vector, and Cerenkov angle is given by $\hat{k} \cdot \hat{v} = cos\theta_C$ [3].

Dual readout need not be just with light, but is a general principle applying to any signal that contains components with large fluctuations. There are other conceptually equivalent schemes, such as the weighting of EM clusters in a fine-grained calorimeter by the WA1 experiment that improved both hadronic linearity and energy resolution, Abramowicz, H. *et al.*, *Nucl. Instrum. and Methods* **180** (1981) 429; the dual measurement of ionization energy loss, dE/dx, and the production of Cerenkov light in liquid Argon, although the scintillation light is 100-fold brighter, by Winn, D.R. and Worstel, W.A., *IEEE Trans. Nucl. Sci.*, **NS-36** (1989) 334; Particle Flow Analysis (PFA) of very finely three-dimensionally segmented calorimeters advocated by the SID, GLD and LDC concept groups on the ILC is an extreme form that could be called "everything"-readout [4]; the truncated mean dE/dx analysis in a TPC is a form of dual readout through EM cluster weighting in which a weight of 1 is assigned to samples below a fixed percentile, and zero above; and, a TPC truncated mean analysis might be used in the MediPix analysis with micropattern grid devices for TPC endcaps to discriminate against δ-rays that with their longer meandering ranges degrade the spatial resolution of a cluster.

We discuss successful applications of these ideas in the DREAM fiber geometry, and we speculate on dual readout in high energy astrophysics and in muon identification in DREAM .

DREAM(1): physical separation of scintillation and Cerenkov light

In the DREAM beam test, in which scintillation and Cerenkov light were generated in separate fibers, the electromagnetic fraction, f_{em}, was measured each event and used to estimate the hadronic shower energy. The resulting energy resolution on single pions and "interaction jets" was about $20\%/\sqrt{E}$ plus a constant term of 1-2% when leakage fluctuations are suppressed. Four paper have been published on this test [5, 6, 7, 8]

We go to some length here to explicitly define the employment of dual readout in this simplest of cases. All other cases are straightforward extends or modifications of this. In the simplest formulation of the response of a calorimeter to hadronic energy, the relative EM-to-hadronic responses for the scintillating fibers and the Cerenkov fibers in

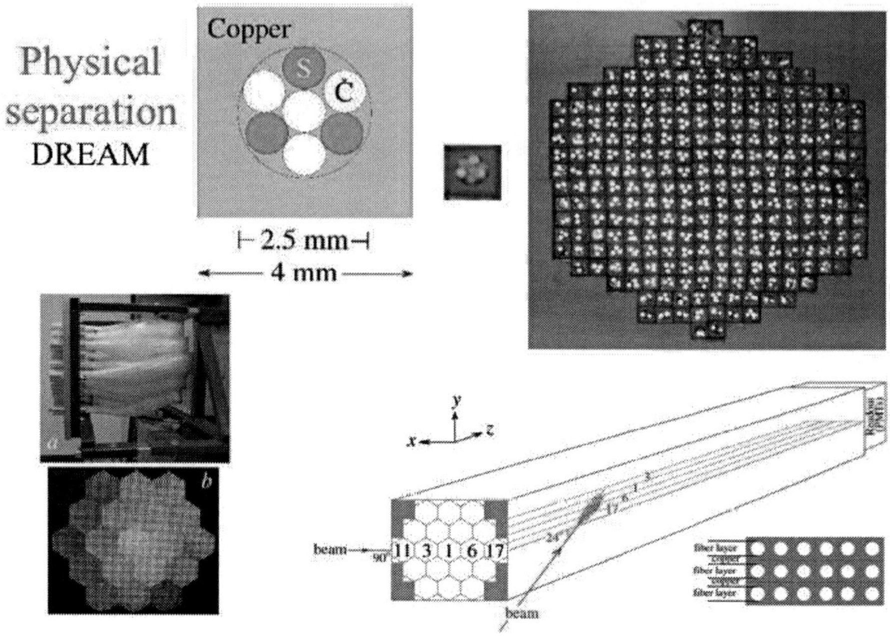

FIGURE 1. Fibers and geometry of the DREAM module.

this module are approximately

$$(\frac{e}{h})_C \equiv \eta_C \approx 5 \quad \text{and} \quad (\frac{e}{h})_S \equiv \eta_S \approx 1.4 \tag{1}$$

and the energy response of each calorimeter to incoming hadronic energy, E, can be written directly for the Čerenkov fibers and the scintillation fibers, respectively, as

$$C = [f_{em} + \frac{1-f_{em}}{\eta_C}]E \quad \text{and} \quad S = [f_{em} + \frac{1-f_{em}}{\eta_S}]E. \tag{2}$$

The electromagnetic fraction, f_{em}, is well correlated with the Cernkov signal as can easily be seen in the data by writing C of Equ. (2) as

$$\frac{C}{E} = \frac{1}{\eta_C} + f_{em}(1 - \frac{1}{\eta_C}),$$

and since $1/\eta_C \approx 0.20$, the Čerenkov signal for events with zero EM activity should be 20% of the shower energy. This is seen directly in Fig. 2 where the f_{em}=0 intercept is at 40 GeV out of 200 GeV, *i.e.*, $C/E \approx 0.20$.

Separately, the scintillation fibers and the Cerenkov fibers constitute very poor calorimeters with their large values of (e/h), and in particular the scintillation response shown in Fig. 2(a) has a poor resolution of 14%, it is highly non-Gaussian, and it has the wrong energy. Now we combine these independent measurements. The electromagnetic fraction can be calculated from the Cerenkov signal in Equ. 2 as

$$f_{em} \approx (\frac{C}{E_{\text{shower}}} - \frac{1}{\eta_C})(\frac{\eta_C}{\eta_C - 1}),$$

and using this estimate of f_{em} in Equ. 2 for S, the energy resolution improves from 14% to

$$\sigma_E/E \approx 5.1\% \quad \text{(Fig.2(b))}.$$

This estimate still includes lateral leakage fluctuations in this one tonne Cu module of about 4% from f_{em}. However, in this beam test it is possible to make a better estimate of f_{em} each event by calculating f_{em} as

$$f_{em} \approx (\frac{C}{E_{\text{beam}}} - \frac{1}{\eta_C})(\frac{\eta_C}{\eta_C - 1})$$

in which the denominator does not fluctuate and, in fact, is exact. *This effectively eliminates the effects of leakage fluctuations.* Of course, f_{em} still contains the *pe* fluctuations in C and the estimated energy contains the fluctuations in both S and C. The resulting resolution is

$$\sigma_E/E \approx 2.15\% \quad \text{(Fig.2(c))}.$$

Dual readout in the DREAM module has achieved good energy resolution, a Gaussian response, and an absolute hadronic energy linearity in this beam test from 20 to 300 GeV, as shown in Fig. 3. It is the point of this paper to suggest that these extraordinary features can be used elsewhere is high energy calorimetry, for purposes other than calorimetery, and possibly in other fields.

DREAM(2): Scintillation and Cerenkov light in the same fibers

The DREAM module was augmented to include a few fibers with deliberately mixed scintillation and Cerenkov light into the same fibers so that the efficacy of at least two of the five separation techniques could be attempted. We found that both direction and timing allowed an acceptable separation of the two kinds of light. A fifth paper has been publish on these measurements [9] and some data are shown in the conference talk.

DREAM(3): Scintillation and Cerenkov in a single crystal

There is no reason that the calorimeter medium cannot be a continuous medium such as a crystal. In the near future we will test the dual readout of scintillation and Cerenkov light in $PbWO_4$ crystals. A bench test with cosmic rays indicates that about 15% of the light is Cerenkov light, and that both direction and timing are possible means of separating the light. This is a proposed technique for the 4th Concept detector [10] for the International Linear Collider, and should also be a possible augmentation or upgrade for the $PbWO_4$ crystals of the CMS experiment at the LHC.

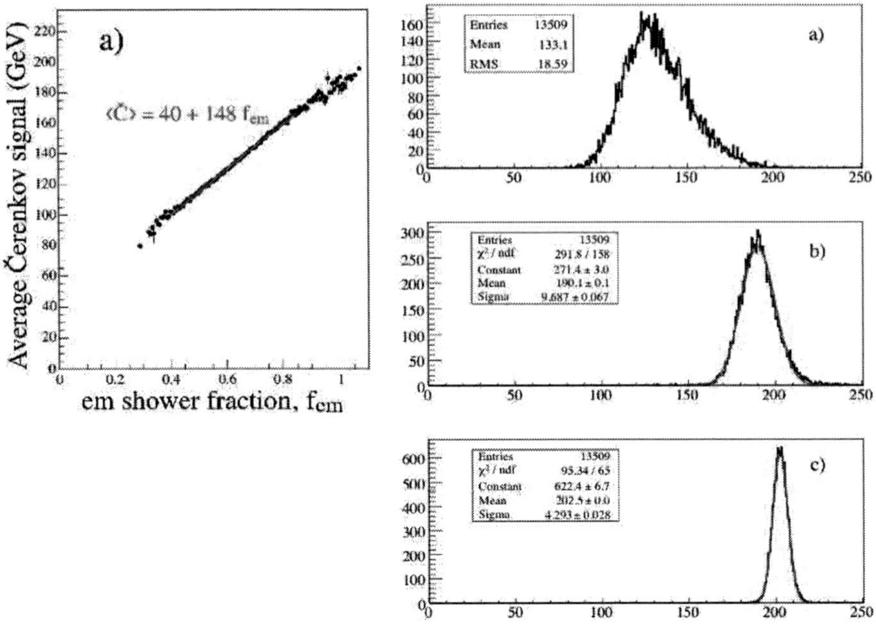

FIGURE 2. Direct linear correlation of the Cerenkov signal with the electromagnetic fraction , f_{em}, in hadronic showers at 200 GeV. Frames (a-c) display the energy resolution of DREAM for (a) scintillation fibers only, (b) scintillation and Cerenkov fibers, and (c) scintillation and Cerenkov with f_{em}calculated using E_{beam}.

UHE Events above 10^{20} eV: Cerenkov plus N_2 fluorescent light

The field of high energy γ-ray and cosmic ray astrophysics (including present detectors HESS, Whipple, and future detectors VERITAS, Auger, *etc.*) has spotted several sources of highly energetic photons by detecting the 1-meter thick pancake of Cerenkov light that is aligned with the electromagnetic shower in the atmosphere and that reaches the surface of earth. Two current experiments (Fly's Eye, AGASA) have detected ultra-high energy showers by the blue fluorescence of the N_2 molecule in the 300-400 nm range. There is a mystery as to how some showers above 5×10^{19} eV reach earth, since the Greisen (GZK) limit that is based on well-understood physics does not allow a proton of those energies to reach earth due to the cross secton for a proton to scatter from the primordial cosmic microwave background (CMB) radiation, $p + \gamma_{cmb} \to \Delta^+ \to p\pi^0, n\pi^+$ thereby degrading the energy of the protons. The simultaneous "dual readout" of the pancake of Cerenkov light and the isotropic scintillation light might be a means to better measure the energies of these showers. AGASA and Fly's Eye have seen events up to 32×10^{19} eV, a clear factor of 6 beyond GZK.

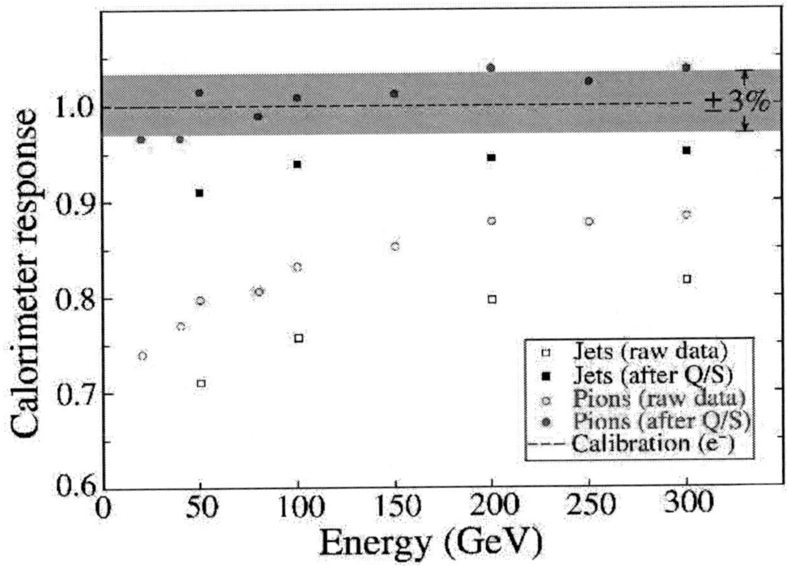

FIGURE 3. Demonstration of absolute hadronic energy linearity in DREAM beam test.

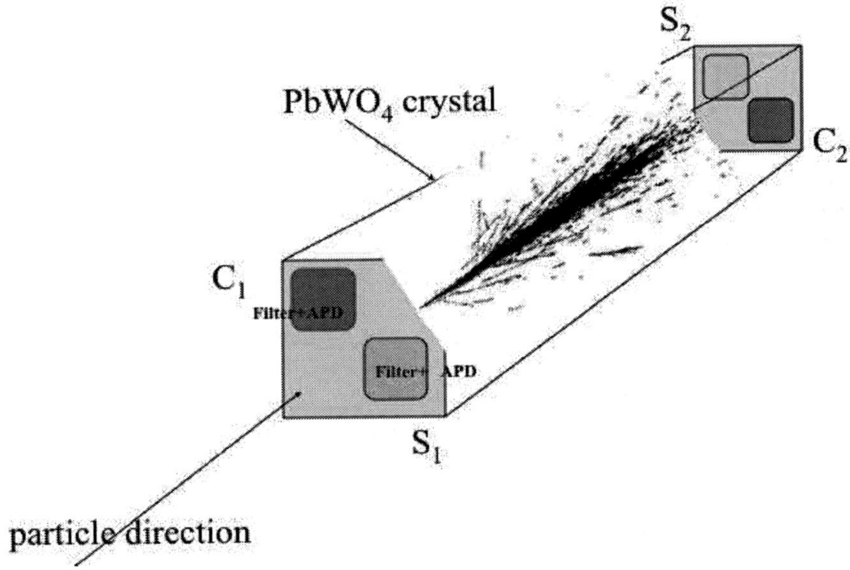

FIGURE 4. A possible way to instrument a crystal for dual-readout that allows separation by wavelength and direction, and possibly time.

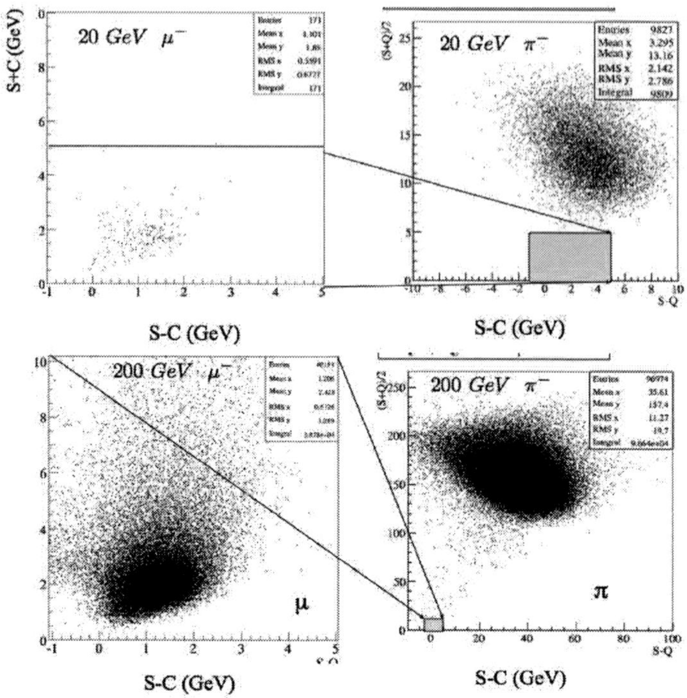

FIGURE 5. Muon separation from pions in the DREAM module at 20 and 200 GeV.

Muon Identification: Separation of Ionization and Radiative Energy Losses of Muons

A dual readout calorimeter provides a unique and powerful identity tag of a muon passing through the calorimeter volume [7, 10]. The main idea is that an approximately aligned muon will generate Cerenkov light at θ_C all of which is outside the numerical aperture of a clear plastic fiber, and therefore the ionization signal of a muon is zero. The electromagnetic showers of radiative energy losses are statistically equal in the scintillation and Cerenkov fibers, so that $(S-C)$ is just the ionization energy loss of about 1.2 GeV in DREAM , and $(S+C)/2$ is approximately the shower energy. These quantities for both muons and pions in DREAM data at both 20 and 200 GeV are shown in Fig. 5 in which dramatic separation of pions from muon is evident. This powerful feature in incorporated into the design of the 4th Concept detector for the ILC [10].

SUMMARY

Any signal that is the sum of widely varying statistical components (hadronic showers, dE/dx samples, *etc.*) can be better measured by isolating, by some means, the separate components of the signal. This has been easy to implement in high energy hadronic calorimetry, and it opens the door to obvious extensions, such as triple-readout in which the MeV neutrons of a hadronic shower are measured to suppress the binding energy loss fluctuations.

ACKNOWLEDGMENTS

The work was supported by DoE grant DE-FG02-01ER41155, and the DREAM module was designed and built at TTU by R. Wigmans, H. Kim and N. Akchurin funded by the DoE Advanced Detector Research (ADR) program administered by Saul Gonzalez. Lastly, I thank Richard Wigmans for his deep understanding of calorimetry.

REFERENCES

1. The first mention of this dual-readout as a means to improve hadronic energy resolution was made by Paul Mockett, "A Review of the Physics and Technology of High-Energy Calorimeter Devices", proceedings of the *11th SLAC Summer Institute on Particle Physics*, July 1983, SLAC Report No. 267. Offline compensation was achieved in the WA1 experiment by weighting EM clusters, Abramowicz, H., *et al.*, *Nucl. Instrum. and Methods* **180** (1981) 429. David Winn proposed readout of Cerenkov light and ionization in LAr, IEEE *Trans. Nucl. Sci.* **NS-36** (1989) 334. A detailed discussion and calculation of a dual readout calorimeter was made by Erik Ramberg, "Recent Developments in High-Speed, Non-Sampling Electromagnetic Calorimetry", March 1990, Fermilab-Conf-90/46. The main ideas were first discussed in detail by Wigmans, "Quartz Fibers and the Prospects for Hadron Calorimetry at the 1% Resolution Level", Proceeding for the 7th International Conference on Calorimetry in High Energy Physics, Tucson, 1997. A coherent discussion of all these issues is in *Calorimetry: Energy Measurement in Particle Physics*, Oxford University Press, 2000, by R. Wigmans.
2. "Dual-Readout Calorimetry for High-Quality Energy Measurements", October 2001, proposal to Advanced Detector Research program of DoE, R. Wigmans, *et al.*. The website containing the proposal, all papers and figures is http://www.phys.ttu.edu/dream, which can be linked from http://highenergy.phys.ttu.edu.
3. The polarization was calculated to be a few percent averaged over an electromagnetic shower in DREAM (Hauptman) and searched for and possibly seen in the data (Akchurin).
4. These concepts are described and available at http://physics.uoregon.edu/ lc/wwstudy/concepts/.
5. "Hadron and Jet Detection with a Dual-Readout Calorimeter", N. Akchurin, *et al.*, *Nucl. Instrum. and Methods* **A537** (2005) 537-561. All DREAM papers and all figures are accessible at http://www.-phys.ttu.edu/dream.
6. "Electron Detection with a Dual-Readout Calorimeter", N. Akchurin, *et al.*, *Nucl. Instrum. and Methods* **A536** (2005) 29-51.
7. "Muon Detection with a Dual-Readout Calorimeter", N. Akchurin, *et al.*, *Nucl. Instrum. and Methods* **A533** (2004) 305-321.
8. "Comparison of Scintillation and Čerenkov Light in High-energy Electromagnetic Profiles", N. Akchurin, *et al.*, *Nucl. Instrum. and Methods* **A548** (2005) 336-354.
9. "Separation of Scintillation and Čerenkov Light in an Optical Calorimeter", N. Akchurin, *et al.*, *Nucl. Instrum. and Methods* **A 550** (2005) 185-200.
10. The 4th Concept detector for the International Linear Collider is described in detail and is available on the WWS website: http://physics.uoregon.edu/ lc/wwstudy/concepts/.

Rad Hard Active Media For Calorimeters

E. Norbeck, J. E. Olson, A. Moeller, and Y. Onel

Department of Physics and Astronomy, University of Iowa, Iowa City, IA, USA

Abstract. Zero-degree calorimeters have limited space and extreme levels of radiation. A simple, low cost, radiation hard design uses tungsten metal as the absorber and a suitable liquid as the Čerenkov radiator. In other applications a PPAC (Parallel Plate Avalanche Counter) operating with a suitable atmospheric-pressure gas is an attractive active material for a calorimeter. It can be made radiation hard and has sufficient gain in the gas that no electronic components are needed near the detector. It works well even with the highest concentration of shower particles. For this pressure range, R134A (used in auto air conditioners) has many desirable features.

Keywords: calorimeter, Cherenkov detectors, avalanche counters.
PACS: 29.40.Ka, 29.40.Cs, 29.40.Vj

INTRODUCTION

Some of the CMS detectors will suffer unacceptable radiation damage with the proposed 10 times luminosity upgrade for the LHC (the SLHC). The zero-degree calorimeters (ZDC) have limited space and the most extreme levels of radiation [1]. We consider here a design for a ZDC that is simple and inexpensive but highly resistant to radiation damage. We also present some recent measurements with a simple PPAC (parallel-plate avalanche counter). A PPAC is the "hydrogen atom" for gas detectors with gain. We show that it can be fast with excellent energy resolution, as well as being resistant to radiation damage.

ČERENKOV LIQUID WITH TUNGSTEN METAL

A ZDC for CMS (and also ATLAS) must fit in a space 9.6 cm wide, 1.0 m long, and 61 cm deep with the beam located 6.7 cm above the bottom of the space. The current design of the ZDCs at the LHC and at RHIC makes use of tungsten metal (ALICE uses tantalum) as the absorber and quartz fibers as the active material. Quartz fibers have served well at RHIC and may be adequate at LHC energies and normal luminosities. A more radiation hard alternative uses a suitable liquid, which can be easily replaced, instead of quartz.

CP867, *XII International Conference on Calorimetry in High Energy Physics,*
edited by S. R. Magill and R. Yoshida

The proposed ZDC design uses blocks of tungsten metal at the bottom of a polished aluminum tank that is open at the top. The tungsten is polished and flashed with aluminum to increase its reflectivity. The blocks have hooks on the top so that they can be easily lifted out of the tank. The photocathodes of the photomultiplier tubes (PMT) are immersed in the liquid to provide the maximum optical coupling. The liquid also serves as a radiation shield for the PMTs.

Because the ZDC is so narrow (it must fit between the two beam pipes) that part of the energy of showers from 2.75 TeV neutrons will leak out the sides of the active region causing the size of the signal to depend on the horizontal position of the incoming neutron. Our latest design uses liquid both between the tungsten blocks and at the sides. Most of the light is from the core of the shower in the space between the blocks, but shower leakage out the side will produce light in the liquid at the side to compensate for lost energy. Figure 1 is a schematic view of this arrangement looking at the tungsten blocks from the top. Careful simulations using GEANT-4 will be made to determine the optimum dimensions of the spaces.

The tungsten-liquid ZDC is easily subdivided horizontally in both the longitudinal and transverse directions. This is done with thin, polished aluminum sheets that go from the bottom of the tank up to the level of the PMTs.

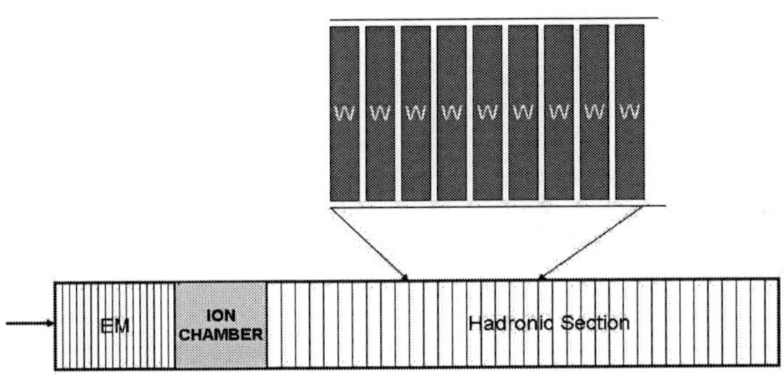

FIGURE 1. Interior of ZDC seen from above.

In principle, almost any liquid could be used, but there are properties that affect the size of the signal. The liquid needs to be transparent over wavelengths accepted by PMTs. A larger refractive index will increase the light output, but it is also useful to have the refractive index match that of the PMT entrance window. Our simulations have shown a small increase with reduced density, because shower particles have greater range in the liquid.

For convenience, the liquid should not be corrosive, too volatile, a health or fire hazard, or acquire any such noxious characteristics under radiation. It must also be available in adequate purity at a reasonable price. If the liquid is damaged by excessive radiation it can be easily replaced, provided it does not form polymers that coat the tungsten blocks. Water is a possibility, but it has a low refractive index, n = 1.33, and things grow in it. An interesting possibility is ethylene glycol, n = 1.43, which is antifreeze without additives. Another candidate is mineral oil with n = 1.46 and a density of only 0.78 g/cm^3. MiniBooNE at FNAL has 10^6 l of it.

A large increase in light could result from adding a wavelength shifting material to the liquid because most of the Čerenkov photons are produced in the UV at wavelengths too short to be seen by PMTs. The number of photons produced is proportional to $1/\lambda^2$. Finding a suitable wavelength shifting material and liquid will require considerable study. The liquid with the wavelength shifting material should have the desirable qualities given above. Because most wavelength shifting materials are large molecules, there will be some difficulty finding a material that is not too sensitive to radiation and does not polymerize.

PPAC STUDIES

In some applications a PPAC (parallel plate avalanche counter), operating with a suitable atmospheric-pressure gas is an attractive active material for a calorimeter. It can be made radiation hard and has sufficient gain in the gas that no electronic components are needed near the detector. It works well even with the highest concentration of shower particles. For this pressure range, the gas R134A (used in auto air conditioners) has many desirable features.

A PPAC is the simplest type of gas detector with gain. It consists of two conducting plates with a gas between them. Much of what we learn about PPACs is useful in understanding the operation of more complicated gas detectors.

Our study of PPACs is basic research with possible applications related to CMS. The primary motivation for starting this project was to find a replacement for some of the scintillators in HE (the endcap calorimeter in CMS) for use with the SLHC. A PPAC can be made of materials that are not affected by radiation and can have excellent time and energy resolution. Our findings will have applications with other, similar detectors, such as resistive plate chambers (RPCs), which are used in the muon system of CMS. We are working with a group at Argonne National Laboratory on the application of RPCs for use in a digital calorimeter for the ILC.

We have studied a dozen different gasses with pressures from 5 torr to one atmosphere. We looked at the fast electron and slow ion signals as a function of voltage and the area and separation of the plates and found that a PPAC can have excellent time and energy resolution [2]. The best gas we have found for use at one atmosphere is R134a.

For the HE upgrade, PPACs would operate at atmospheric pressure. Small gas leaks are then unimportant if the gas is non-toxic and nonflammable. A minimum ionizing particle (MIP) will make about a dozen primary electrons with a spacing of 0.5 mm using the heavy gas, R134a. This may be compared with two electrons from the photocathode for a MIP in the present HE scintillators [3]. The gain, corresponding to the photodetector gain, is the increase in the number of electrons caused by the avalanche. As with photodetectors the gain depends on the applied voltage. The number of primary electrons is easily estimated. The ideal gas law gives 0.2 mg/cm^2 for R134A, mw 102, with a plate separation of 0.5 mm. Assuming a MIP loses 2 MeV in 1 g/cm^2 and that one electron-ion pair requires 30 eV, the number of primary electrons is 13. Because of the factor of 6 improvement, compared to scintillating tiles, HE would still have adequate energy and position resolution with some of the PPACs turned off.

The small, 0.5 mm, plate spacing is used to keep the voltage under 4 kV. For small area detectors the width of the large electron signal is between one and two ns. The slow ion signal is so small that it can be ignored in most situations. To maintain a constant gas gain the high voltage may need small adjustments to compensate for small temperature and barometric pressure changes, which change the number of gas molecules between the plates.

Compared to the scintillating tiles, PPACs provide better energy resolution, a better time structure, and can be made entirely of radiation hard materials. Any small leaks of the slightly above atmospheric pressure gas are harmless because the gas is not toxic or flammable. The signals are put directly into 50 Ω coax cable so that no sensitive electronics are needed near the detector. The 3 kV required for the detector is not a hazard because the currents are only at the microamp level. The cost of a PPAC system is less than for the original scintillating tile system. It is basically a simple device, two flat plates with a gas between them–a proportional counter having plane parallel symmetry.

The gas in the PPACs should be constantly replenished to keep it pure and free of decomposition products caused by radiation. This can be done by a flow rate that replaces the gas once each hour. Because the plate spacing is only 0.5 mm, the total amount of gas in the PPACs is small.

For large area detectors the width, W, of the electron signal is determined by the detector capacitance and is approximately W = RC, R = 50 Ω and C is the capacitance of the PPAC. The circuit for a PPAC is simple.

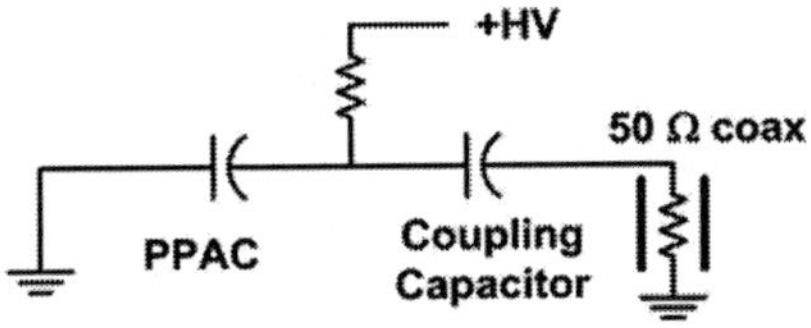

The signal divides between the coupling and detector capacitances, but the energy available to a spark is the sum of the two capacitances. For many applications the two capacitances should be set equal. This provides half of the maximum signal size to the coax with a minimum of stored energy.

With the present scintillating tiles, the signal from HE extends over three beam-crossing intervals of 25 ns each. PPACs with area of up to 100 cm^2 provide a signal short enough so that the entire signal fits into a single 25 ns time interval. A 100 cm^2 area provides better position resolution than is needed for HE. To get the same time structure for a larger area, signals from individual 100 cm^2 plates can be brought out of HE on small coax cables and combined by a simple summing amplifier.

PPACs are not affected by the large magnetic fields in HE because the electron motion in the PPAC is parallel to the magnetic field. In any case the effect of a magnetic field is small because the electrons travel less than 0.5 mm in an electric field of 60,000 V/cm.

We have studied the performance of PPACs with many different gasses at pressures from 5 torr to one atmosphere with a variety of plate spacings and configurations. Most recently we have concentrated on PPACs operating at atmospheric pressure. For large systems, other pressures complicate the design, increase the cost, and reduce the reliability. At atmospheric pressure a MIP gives a useful signal with even the smallest plate spacings.

A 1.0 GHz oscilloscope (Tektronix TDS 5104) is used to observe the signals directly from the detector. Because of the noise in the scope we sometimes use a fast 10 X amplifier to improve the signal to noise ratio for small signals near the operating voltage threshold of the PPAC. Much information can be obtained from the study of the signals.

The signal that is seen on the scope is caused by charges moving between the plates, not, as might be expected, by the charge that is being collected. With normal operation a PPAC signal has a fast electron part with a width of a few nanoseconds and a small ion signal that is constant for about the first 2/3 of the time and linearly approaches zero for the last third. During the constant part of the signal the ions are moving between the plates. The signal goes toward zero as the ions are collected at the cathode. Because of the exponential growth of the avalanche, most of the charges are generated close to the anode. The electron signals are short because the electrons have a large velocity, because of their small mass, and travel only a short distance. The last electrons formed in the avalanche fall freely in the electric field without any additional collisions with gas molecules. The ions travel slowly over a much longer distance with many collisions on the way. The presence of ions does not interfere with formation of a new avalanche. The fractional change in the potential between the plates is insignificant and the cross section of a single avalanche is minute.

The ion transit times with R134a gas are long, even longer than one would expect for such a heavy ion (MW: 102). This is caused by the highly polar nature of the R134a molecule. R134a is 1,1,1,2-tetrafluoroethane, which means that three of the fluorine atoms are at one end of the molecule. Since the fluorine strongly attracts electrons, this arrangement results in a large electric dipole moment for the molecule. The dipole moment is about the same as for water, which owes many of its peculiar properties to its large dipole moment. The symmetric R134 (1,1,2,2-tetrafluoroethane), with two fluorine atoms on each end and no dipole moment, gives electrical signals that are identical except for the ion part, which is faster than for R134a, but with the same area. The attraction of the polar R134a molecules interferes with the motion of the ions, which results in a longer ion-collection time. The counter used for these measurements consists of two circular plates, each with an area of 1.0 cm^2, separated by 0.5 mm and operating at 700 torr and 2120 V. The ion signal is constant for a time t_0 and then goes linearly to zero at time t_1. The values of t_0 and t_1 are 1.3 μs and 1.8 μs for R134a, but only 0.8 μs and 1.3 μs for R134. These are not precise times because the signals are very small and the values depend on the location of the primary ion formation (from a ^{137}Cs γ source). During the constant part of the signal the ions are moving between the plates. The signal goes toward zero as the ions are collected at the cathode. For both gasses the large signal from the electrons is fast with a full width at half maximum of only 1.0 ns.

REFERENCES

1. O. A. Grachov et al., Proceedings of this conference.
2. E. Norbeck, J. E. Olson, and Y. Onel, "A New look at the PPAC" in *Proc. 21st Winter Workshop on Nuclear Dynamics*, edited by W. Bauer, R. Bellweid and S. Panitkin, Feb. 5-12, 2005, Breckenridge, CO, USA, pp285-290.
3. CMS TDR 2, CERN/LHCC 97-31, p.224.

On the Calibration of Segmented Calorimeters

Richard Wigmans

Department of Physics, Texas Tech University, Lubbock TX79409-1051, USA

Abstract. Most calorimeter systems used in particle physics experiments are longitudinally subdivided into several compartments. The intercalibration of these compartments is highly non-trivial, as a result of two effects: 1) The dependence of the calorimeter response on the type (and energy) of the showering particle, and 2) The depth dependent sampling fraction of showers.

The problems arising from this are illustrated with examples from the scientific literature. I also review some of the methods applied in practice to calibrate segmented calorimeters, and discuss the (lack of) merit of these methods.

Keywords: Calorimetry, calibration, shower sampling
PACS: 29.40Ka, 29.40Mc,29.40Vj

INTRODUCTION

One of the most important, and most underestimated, aspects of working with calorimeters is the calibration, *i.e.* the determination of the relationship between the signals produced by the detector (in terms of photoelectrons or picoCoulombs) and the energy deposited by the showering particles. Most calorimeters used in particle physics experiments are of the *sampling type*. The calibration of such detectors is greatly complicated by the fact that the sampling fraction changes as the shower develops in the absorbing structure. This leads to considerable problems when the calorimeter is segmented. In this talk, I illustrate these problems with practical examples, and I discuss methods that are being used to deal with them.

CALIBRATING SEGMENTED CALORIMETERS

Origin of the problems

The signal from a sampling calorimeter is the sum of all the signals generated by individual shower particles in the active layers. When these devices were first built, one naively expected that these secondaries constituted a collection of minimum ionizing particles ("mips") and that, therefore, the total calorimeter signal would be the same as that for N mips, N being the ratio of the shower energy and the energy deposited by a single muon traversing the detector. However, this turned out to be a fundamentally incorrect oversimplification.

Let us first consider a high-energy electromagnetic shower. It turns out that in typical sampling calorimeters, which use high-Z absorber material such as lead, $\sim 40\%$ of the energy is deposited by electrons that carry less than 1 MeV kinetic energy[1]. Such electrons are produced in Compton scattering and photoelectric effect. The cross

CP867, *XII International Conference on Calorimetry in High Energy Physics,*
edited by S. R. Magill and R. Yoshida

sections for these processes strongly favor the absorber material. This combined with the fact that the typical range of these electrons is much smaller than the thickness of a sampling layer (see Figure 1) leads to a very inefficient sampling of this shower component, much less efficient than for a mip. Most of these low-energy electrons are

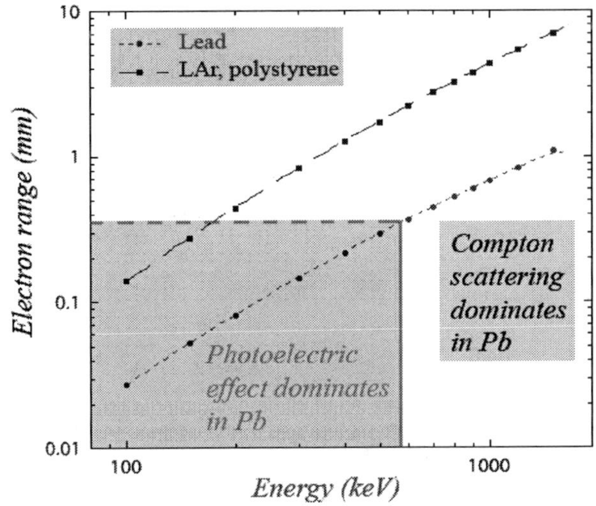

FIGURE 1. Range of electrons in lead and in liquid argon, in the 0.1 - 1 MeV range. The energy region where photoelectric effect is the dominant process for photons in lead is shaded.

produced late in the shower, beyond the shower maximum. Early in the shower, the energy deposit is dominated by fast e^+e^- pairs produced by energetic Bremsstrahlung photons generated by the primary particle. These secondaries are sampled in a much more mip-like manner, since they may traverse several sampling layers.

As a result, the effective sampling fraction of an em shower in a sampling calorimeter with high-Z absorber material is smaller than that of a mip ($e/mip < 1$), and decreases rather considerably as the shower develops. In lead/plastic-scintillator calorimeters, used in a variety of experiments, this decrease in sampling fraction amounts to about 30% (Figure 2).

A consequence of this phenomenon is that if the calorimeter is subdivided into different longitudinal segments, the relationship between deposited energy and resulting signal is different for each of these segments. Moreover, these differences are energy dependent, since the domination of the mip-like component will extent to greater depth if the energy of the incoming particle is larger. Also, these differences will depend on whether the incoming particle was an electron or a photon. At equal energy, photon-induced showers develop, on average, deeper inside the calorimeter[2].

A given signal corresponds thus to a larger energy deposit if it is generated in a deeper segment of the calorimeter. This is an inevitable consequence of the fact that only a small (and gradually decreasing) fraction of the charged secondaries produced in the shower development actually contribute to the calorimeter signals. The overwhelming majority of these secondaries are stopped in the same absorber layer in which they are generated and thus do *not* contribute to the signals. However, this situation greatly complicates the

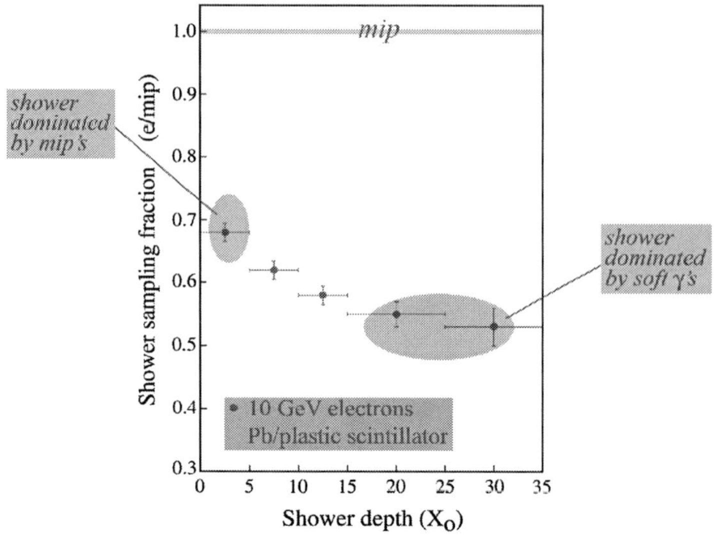

FIGURE 2. Development of the sampling fraction of em showers in a lead/plastic-scintillator calorimeter, as a function of age, or depth inside the calorimeter. The sampling fraction for mips in the same detector structure is used for normalization purposes.

calibration issues, since it is not at all clear how to determine the relationship between measured signals and deposited energy for individual calorimeter segments. As is shown in the following, any choice one makes in this respect has undesirable consequences.

An example of the practical importance of this effect is shown in Figure 3, which concerns the em calorimeter of the AMS experiment[3]. This lead/scintillating-fiber calorimeter is subdivided into 18 longitudinal segments of $\sim 1X_0$ each. Each of these segments was calibrated with mips, and the energy equivalent of a mip traversing one segment was established to be 11.7 MeV.

However, the total depth of this calorimeter ($17X_0$) is not sufficient to fully contain high-energy electron showers, as can be seen in the top diagram of Figure 3. As a result, the total signal is not proportional to the beam energy. The larger the beam energy, the larger the fraction that leaks out (bottom diagram). The authors tried to compensate for this effect by fitting the measured signals in the 18 segments to a standard shower profile. By integrating this profile to infinity, they expected to determine the average shower leakage. However, as illustrated by the circles in the bottom diagram, this method only led to a partial recovery of the missing energy. This can be understood from the fact that the measured signals beyond the shower maximum corresponded to a significantly larger energy than the signals from the early part of the shower development. By using the same signal-to-energy conversion throughout the module, the energy leaking out the back is thus systematically underestimated. Therefore, the reconstructed shower energy is systematically too small, more so if the leakage fraction is larger. The practical consequences of the depth-dependent sampling fraction may thus be very substantial (up to 20% in this example).

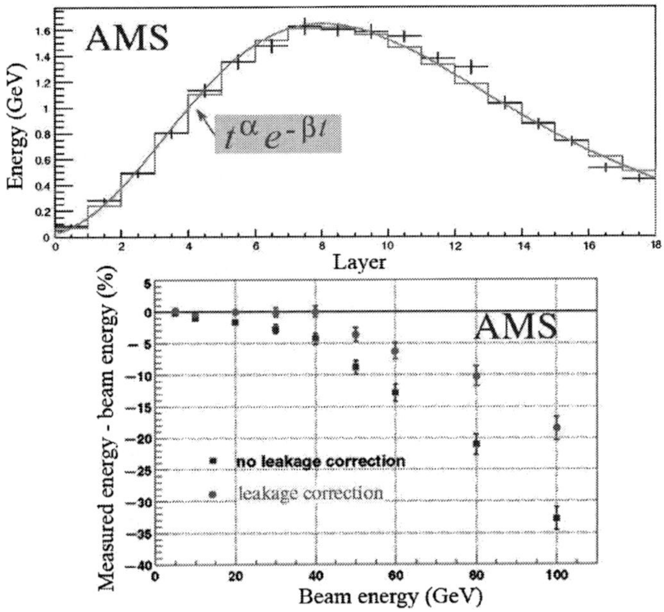

FIGURE 3. Average signals for 100 GeV electrons in the 18 longitudinal sections of the AMS lead/scintillating fiber calorimeter (top). Average difference between the measured energy and the beam energy, before and after leakage corrections based on extrapolation of the fitted shower profile (bottom).

Two longitudinal segments

In practice, experiments with calorimeters segmented into two longitudinal sections often try to (inter-)calibrate these segments (*e.g.*, with calibration constants A and B) by minimizing the width of the total signal distribution, *i.e.* by minimizing the quantity

$$Q = \sum_{j=1}^{N} \left[E - A \sum_{i=1}^{n} S_{ij}^{\mathrm{em}} - B \sum_{i=1}^{n} S_{ij}^{\mathrm{had}} \right]^2 \qquad (1)$$

where E is the beam energy and $\sum S^{\mathrm{em}}$ and $\sum S^{\mathrm{had}}$ are the sums of all the signals in the towers i of the em and hadronic calorimeter sections that contributed to the measured signal for event j. With this method, values for A and B and, more importantly, for the intercalibration constant B/A can be determined for each calorimeter tower.

However, several fundamental difficulties are encountered when such calibration methods are applied[2]:

1. The values of A, B and B/A are found to be energy dependent.
2. The values of B/A differ considerably (typically 20%) from those found for muons.
3. If a value B/A different from that found for muons is used as the basis for calibration, then the calorimeter becomes non-linear for em shower signals.

All these effects are a direct consequence of the depth-dependent sampling fraction. The relationship between signal and energy is different in the two segments. By giving the signals from the deeper segment (where the sampling is less efficient) a relatively smaller weight, the total resolution is seemingly improved. However, as the energy increases, a larger fraction of the energy is deposited in this deeper segment. By giving the signals from this segment systematically a smaller weight, the *total* energy is thus increasingly *underestimated*, which leads to signal non-linearity and systematic energy mismeasurement. Also, as the shower energy is increased, more energetic particles reach the deeper segment, and as a result the sampling fraction of this segment increases. The optimal intercalibration ratio B/A thus becomes gradually larger as the energy increases.

The only way to avoid such effects is to set the energy scale of the different segments in exactly the same way, *e.g.,* with muons traversing both segments[4].

Lateral segmentation, hadron showers

The phenomena described in the previous sections are not unique for electromagnetic showers, nor are they an exclusive feature of longitudinal shower development. This is illustrated in Figure 4, which shows experimental results from the SPACAL R&D project [5].

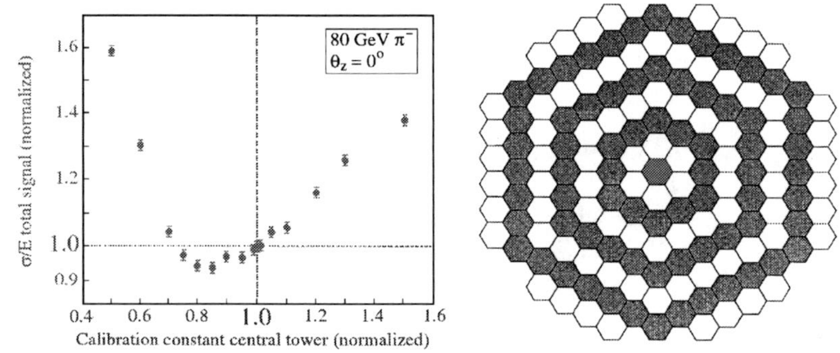

FIGURE 4. Total width of the signal distribution for 80 GeV π^- showers in SPACAL, as a function of the calibration constant of the central detector tower (left). The SPACAL tower geometry (right).

SPACAL was a compensating lead/scintillating-fiber calorimeter consisting of 155 longitudinally unsegmented hexagonal towers. Each tower was calibrated with electrons. The electron showers were almost completely (95%) contained in one tower. When high-energy pions were sent into this detector, typically half of the shower energy was deposited in the central (hit) tower, the other half was distributed among the other towers. Figure 4 shows that the width of the total signal distribution could be improved (by as much as 8%) by reducing the calibration constant of the central tower with respect to the other towers. However, also in this case minimizing the width of the total signal distribution led to wrong calibration constants, with the associated undesirable consequences (*e.g.,* signal non-linearity). These were avoided in the original calibration scheme (which, therefore, did *not* produced the narrowest total signal distribution).

Hadron showers in an ECAL/HCAL system

The calibration problems resulting from longitudinal segmentation are even worse for hadron showers than for em ones. Hadron showers consist of em and non-em components, which are typically sampled very differently (when $e/h \neq 1$). *On average*, the em shower components (initiated by π^0s) occur early on in the shower development, while the shower tails are dominated by the products of the numerous nuclear interactions, where the "invisible" energy needed for nuclear breakup causes important detection inefficiencies. Some experiments base their calibration strategy on this characteristic: The energy scale for the most upstream calorimeter segment (the em calorimeter) is set with em showers, which are contained in this segment, while the scale for the downstream segment(s) is set with hadrons that penetrate the em calorimeter section without starting a shower. It has been demonstrated that schemes of this type lead to serious biases and systematic energy mismeasurement for hadrons that start showering in the em calorimeter, typically the majority of the events[6, 7].

It has also been demonstrated that such effects can be avoided by treating the individual segments of the calorimeter system in exactly the same way for what concerns the calibration, *e.g.*, with (contained) electron showers or with muons traversing all segments. The resulting overall discrepancy between this "electromagnetic energy scale" and the hadron energies can be eliminated with a correction factor derived from the e/h value of the calorimeter system[7].

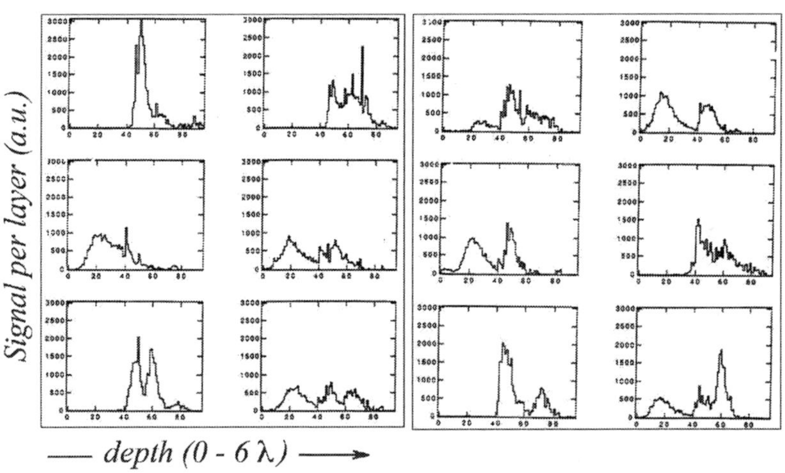

FIGURE 5. Longitudinal energy deposit profiles for 270 GeV π^- showers in lead.

Whatever the method chosen to determine the calibration constants for such segmented calorimeters, they may be very wrong for individual hadronic events. This is a result of the fact that the starting point of em showers initiated by π^0s produced in the shower development may fluctuate wildly. This is illustrated in Figure 5, which shows the longitudinal energy deposit profiles for randomly selected showers induced by 270 GeV π^- in lead. Em showers are indicated by the peaks in these profiles[8]. There is thus often very significant em activity beyond the ECAL.

More than two longitudinal segments

The discussion in the previous sections illustrates that calibration of the individual sections of a longitudinally segmented calorimeter is a delicate issue. The mentioned strategies (minimizing the width of the total signal distribution, each section its own particles, *etc.*) each produce only good results for a specific *subset* of events (electrons of a certain energy, pions penetrating the ECAL), while yielding negative consequences for the rest of the events (systematic mismeasurement of energy, biases depending on the starting point of the shower, signal non-linearity).

A more correct approach seems to be to treat all sections in exactly the same way, *e.g.,* with muons that traverse the entire calorimeter, or with em showers contained in the individual segments. This method gives good results for systems consisting of only two longitudinal segments[4, 7]. However, the example shown in Figure 3 illustrates that also this method has its limitations. So what to do in such situations?

One approach that has produced reasonable results in practice comes from the ATLAS Collaboration. Their Pb/LAr ECAL consists of 3 longitudinal segments ($4.3X_0$, $16X_0$ and $2.0X_0$ deep, respectively). Figure 6 shows how the sampling fraction for electron showers of different energies decreases as a function of depth in this detector (*cf.* Figure 2).

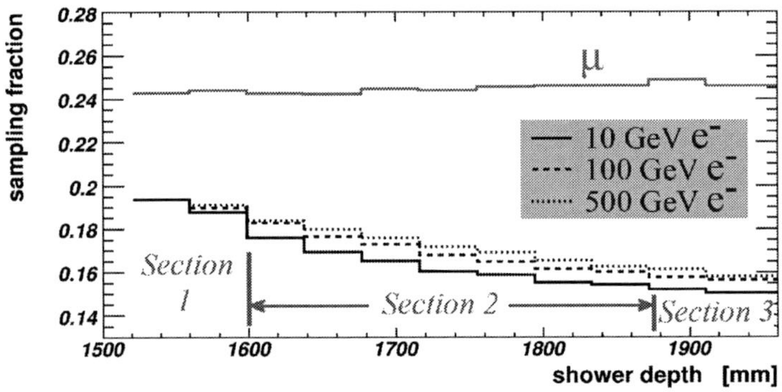

FIGURE 6. Energy sharing between the 3 longitudinal segments of the ATLAS em calorimeter, for electron showers of different energies.

The calibration constants were determined on the basis of detailed Monte Carlo simulations, which were optimized such as to achieve simultaneously good energy resolution and signal linearity[9]. The reconstructed energy was derived from the measured signals with a formula that contains at least 4 parameters that depend in a non-linear way on the energy of the incoming electrons. With this formula, the authors achieved excellent linearity for the energy range 15 - 180 GeV. Impressive as these results may be, one has to realize that the parameter values are only valid for one particular pseudorapidity, and that the parameter values will also have to be different when the signals are produced by photons instead of electrons. It is also unclear how these results extend beyond the energy range for which they were obtained: At 10 GeV, a very significant deviation from linearity was observed. Life will not be easy at the LHC.

SOME FINAL THOUGHTS

The calibration of a longitudinally segmented calorimeter system is a nightmare. The more segments, the more complicated the problems become. I personally don't know how to do it, and I even don't know if there exists a fundamentally correct way of doing it. What I do know is that all methods I have seen lead to unintended, undesirable consequences: Systematic mismeasurement of the energy, signal non-linearities, dependence of the reconstructed energy on the starting point of the showers, *etc.*

There are several ways to avoid or reduce the problems introduced by longitudinal segmentation:

- *No* longitudinal segmentation. This is the trivial solution. Especially if the calorimeter is compensating, this is extremely attractive, since all one has to do in this case is to send an electron beam in each and every tower, and the calibration is finished: No ifs, ands or buts. SPACAL (Figure 4) is an example of such a calorimeter. The experience with this calorimeter proved that electron identification (one of the main reasons to have a separate em calorimeter section) could very effectively be achieved without longitudinal segmentation[10, 11].
- Use *homogeneous* calorimeters. This would eliminate the problems deriving from the depth-dependent sampling fraction (Figures 2,6). By simultaneously measuring the Čerenkov and scintillation light produced in the shower development, one could possibly also obtain excellent results for hadrons (based on a calibration with electrons) in such a device[12].
- If sampling calorimeters are to be used, one should choose passive and active materials with similar Z values. Since the effects depicted in Figures 2,6 derive from the *difference* between these Z values, they could be eliminated in this way. However, the problems deriving from $e/h \neq 1$ would of course remain.

REFERENCES

1. R. Wigmans, *Calorimetry, Energy Measurement in Particle Physics*, International Series of Monographs on Physics, Vol. 107, Oxford University Press (2000).
2. R. Wigmans and M.T. Zeyrek, Nucl. Instr. and Meth. **A485** (2002) 385.
3. AMS Collaboration, Nucl. Instr. and Meth. **A490** (2002) 132.
4. T. Åkeson *et al.* , Nucl. Instr. and Meth. **A262** (1987) 243.
5. D. Acosta *et al.* , Nucl. Instr. and Meth. **308** (1991) 481.
6. O. Ganel and R. Wigmans, Nucl. Instr. and Meth. **A409** (1998) 621.
7. M. Albrow *et al.* , Nucl. Instr. and Meth. A487 (2002) 381.
8. D. Green, *Proc. 4th Int. Conf. on Calorimetry in High Energy Physics*, La Biodola, Italy, 1994, eds. A. Menzione and A. Scribano, (Singapore: World Scientific), p. 1.
9. M. Aharrouche *et al.* (ATLAS Collaboration), *Energy Linearity and Resolution of the ATLAS Electromagnetic Barrel Calorimeter in an Electron Test-Beam*, Nucl. Instr. and Meth. (in press, 2006). See also contribution of W. Lampl to these Proceedings.
10. D. Acosta *et al.* , Nucl. Instr. and Meth. **A302** (1991) 36.
11. D. Acosta *et al.* , Nucl. Instr. and Meth. **A305** (1991) 55.
12. J. Hauptman, contribution to these Proceedings.

Silicon Photomultipliers, A New Device For Low Light Level Photon Detection

Hans-Günther Moser

Max-Planck-Institut für Physik (Werner Heisenberg Institut), Föhringer Ring 6, 80805 München, Germany
MPI Halbleiterlabor, Otto-Hahn-Ring 6, 81739 München, Germany

Abstract. Silicon Photomultipliers (SiPM) are novel detectors for low level light detection based on arrays of avalanche photodiodes operating in Geiger mode. Offering good characteristics (fast response, high gain, photon counting capability, insensitivity to magnetic fields, low voltage operation) they have the potential to replace classical photomultipliers (PMT) in many applications. Drawbacks are dark rate and optical cross talk. Though their quantum efficiency is already comparable or better than that of bialkali PMT it is still limited by the structures on the light sensitive front surface. A new concept, presently developed at the Max-Planck semiconductor laboratory, allows boosting the efficiency to almost 100%. Using a fully depleted substrate the light enters through the unstructured backside. A drift diode structure collects the electrons on a small "point like" avalanche structure for multiplication. Engineering the thin entrance window at the backside using antireflective layers a high efficiency can be achieved in a wide wavelength range (300–1000nm). The paper will summarize the status of front illuminated SiPMs and report on the development of the backside illuminated devices.

Keywords: Photon Detection, Avalanche Photon Counters, Backside Illumination
PACS: 29.40.Wk

INTRODUCTION

The Silicon Photomultiplier (SiPM) is a novel type of photon detector which allows for compact, robust, low cost devices with photon counting capability, high quantum efficiency, large gain and fast response (<100 ps). In contrast to photo tubes SiPMs are insensitive to magnetic fields. The development started about a decade ago, mainly in Russia [1,2,3]. When the first prototypes became available they triggered large interest for many applications, like calorimeters, air-shower Cerenkov telescopes, medical applications etc.

A SiPM is an array of avalanche photodiodes operating in limited Geiger mode. Figure 1 shows a cross section of an individual pixel and the corresponding electrical field. Each pixel has an individual quenching resistor and the outputs of all pixels are connected together. The SiPM is operated at a voltage above the breakdown voltage. If a photon converts in the sensitive (=depleted) volume of such a pixel, the electron or hole drifts in the high field region and triggers the avalanche multiplication, which discharges the diode capacitance C_{pixel}. The multiplication process stops if the voltage drops below the breakdown voltage. Hence per pixel a charge of $Q=C_{pixel}$ $(U_{op}$ –

CP867, *XII International Conference on Calorimetry in High Energy Physics,*
edited by S. R. Magill and R. Yoshida

$U_{breakdown}$) is released. The quenching resistor R limits the current recharging the diode which is needed not to reinitialize the avalanche process. The pixels is active after a recovery time $\tau \sim C_{pixel} R$.

Hence a single cell is a binary device giving the same signal if one or more photons convert. In a matrix the signals of all cells are added. For low photon flux, when the probability for a single cell to be hit very low, the output is simply proportional to the number of photons. However, for higher photon fluxes some cells might be hit by more than one photon leading to a nonlinear response and finally to saturation. In a crude approximation the dynamic range is given by the number of cells.

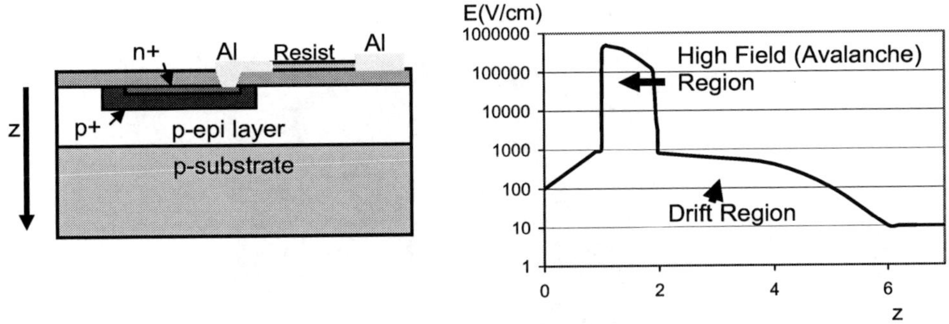

FIGURE 1. Cross section through a SiPM single cell and configuration of the electric field. The high field region (E ~ $5\text{x}10^5$ V/cm) is built up in the highly doped n^+p^+ junction.

FEATURES OF SILICON PHOTOMULTIPLIERS

Dark Rate

Due to thermally generated currents SiPMs have a substantial dark rate, a typical value is about 1 MHz for a 1 mm^2 device operated at room temperature. This causes a problem especially for large area devices. Dark rates increase with operation voltage and therefore with gain. The leakage current has a strong temperature dependence of:

$$I(T) = I(T_0)\frac{T^2}{T_0^2}\exp\left(-\frac{E_g}{2kT}\right) \Big/ \exp\left(-\frac{E_g}{2kT_0}\right) \tag{1}$$

With E_g = 1.1 eV (band gap in Silicon), k is the Bolzmann constant and (T_0) T the (reference) temperature in ^{0}K. Lowering the temperature by ~7°K reduces the dark rate be a factor of two.

Since leakage currents depend largely on the quality of the fabrication process, careful processing will be one tool to produce devices with low dark rates.

Crosstalk

The hot carriers (electrons and holes) in an avalanche emit light, about 1 photon per 10^5 electrons [4], see Fig. 2. These photons may travel to another pixel of the matrix and initiate an avalanche breakdown there. As a result the average number of pixel fired per incident photon is larger than one and the dark count distribution does not follow Poisson statistics. Practically this leads to non-linearities, additional noise, and calibration problems. The amount of cross talk depends on many parameters, like pixel size, dead space between the active area and gain. One way to reduce cross talk is to optically insulate the pixels from each other by trenches etched around each pixel. A reduction by a factor of ~100 has been achieved [5].

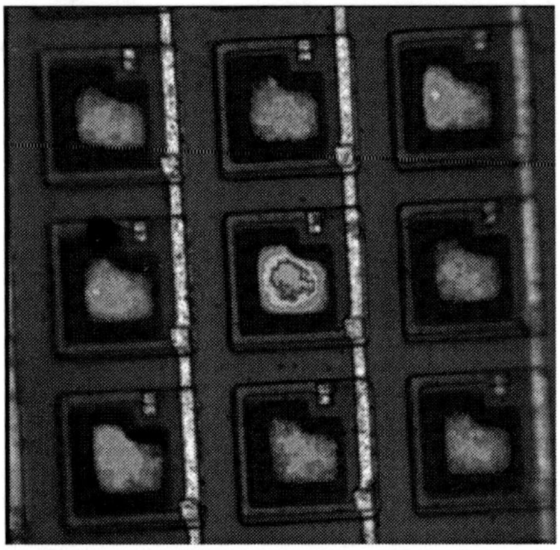

FIGURE 2. Light emitted by avalanche breakdowns in a SiPM. Picture taken using a Hamamatsu Phemos 1000 emission microscope. SiPM from MEPhI with 42 μm pitch. The cells fire due to dark currents, the light is integrated over several seconds.

Quantum Efficiency

At sufficiently high electric fields the efficiency for an electron or hole to trigger an avalanche breakdown is almost 100%. Hence it should be possible to produce SiPMs with a Quantum Efficiency (QE) close to 100%. However, available devices reach at most 35% QE. The limiting factors are:

- Surface transmission: A fraction of the light is reflected at the surface of the entrance window or absorbed in the undepleted surface layer. To achieve high efficiency this layer should be as thin as possible. The reflection losses can be reduced applying special antireflective coatings.
- Fill Factor: The major limitation of the present front illuminated devices is the fact that a substantial area of the pixel is inactive: space is needed for guard rings separating pixels electrically, bias resistors and Al-traces for electrical contacts. Furthermore the high field regions of each pixel have to be separated

from the neighbors using guard rings and gaps. Therefore the resulting fill factor is always considerably lower than 100% and can be as low as 16% for small pixels (Fig. 3). It generally becomes better for larger pixel sizes, however, such devices may develop other problems (e.g. small dynamic range because of limited number of pixels). Backside illuminated devices, as discussed in the last section do not have this problem.

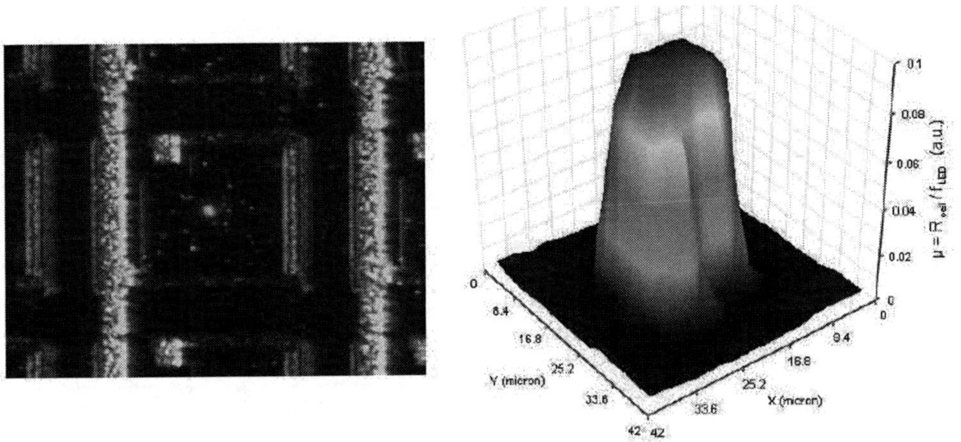

FIGURE 3. Measurement of the Fill Factor. Left figure: a light spot of ~3 μm diameter (visible in the centre of the image) is scanned across the SiPM surface, while the SiPM response is measured. Right Figure: Response map of a single cell. The active area corresponds to 16% of the cell area [6]. Note that the sensitive area corresponds exactly to the light emitting area of Fig. 2 (same SiPM measured).

Sensitivity for Blue and UV Light

The peak sensitivity of SiPMs is in the range of 450-600 nm. Especially at short wavelengths their sensitivity drops considerably which limits the performance for the detection of Cerenkov Light or blue/UV scintillator light. The wavelength dependence of the SiPM sensitivity can be understood from the light absorption curve of Silicon (Fig. 4). Three regions can be identified:

- $\lambda > 450$ nm: the light converts deep in the silicon below the high field region. Due to the field configuration in p-type devices (to date standard) electrons drift back into the high field region. Electrons trigger avalanches efficiently; hence in this wavelength region the efficiency is high.
- 350 nm $< \lambda <$ 400 nm: the light is absorbed in the first microns of the device. Here holes drift into the high filed region and trigger the avalanche. Since holes have (at the same field) a lower probability to trigger an avalanche (Fig. 5), the QE is reduced in this wavelengths region.

- $\lambda < 350$ nm. The light is absorbed in a thin surface layer. The electron/holes tend to recombine before they diffuse into the drift and finally high field region. In this region the QE drops sharply.

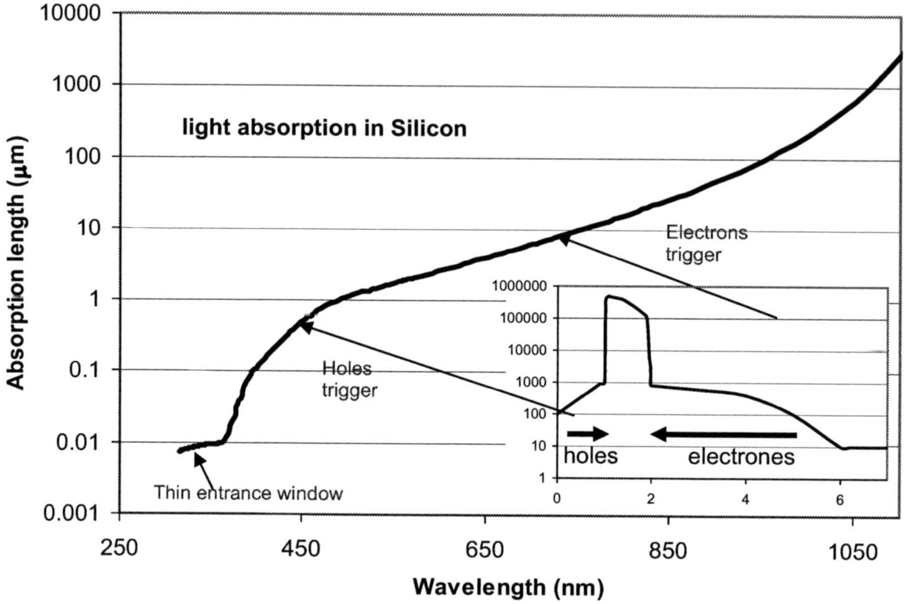

FIGURE 4. Light absorption in Silicon. The insert indicates the field configuration in a SiPM (see Fig. 1) and the drift direction of electrons and holes.

Several methods exist to increase the sensitivity for short wavelength: Operation at very high voltage increases the avalanche probability for holes. However, in practice the maximal operation voltage is limited due to the increase of cross talk and dark rate. Clearly, entrance windows have to be made as thin as possible. An interesting approach is the inverted structure: the n-substrate is used instead of p-type and the doping of the high field junction is reversed. Hence the role of electrons and holes are reversed resulting in an avalanche trigger by electrons for short wavelengths.

Overall Optimization

As described above the performance of a SiPM depends on many parameters and it is difficult to optimize all simultaneously. For instance, large QE needs large pixels with small insensitive gaps, a feature which necessarily enhances cross talk. Or, enhancing the operation voltage leads to higher (enhanced avalanche probability) but also increases cross talk and dark rate. It is therefore important to design a SiPM carefully for a given application and most likely compromises must be made.

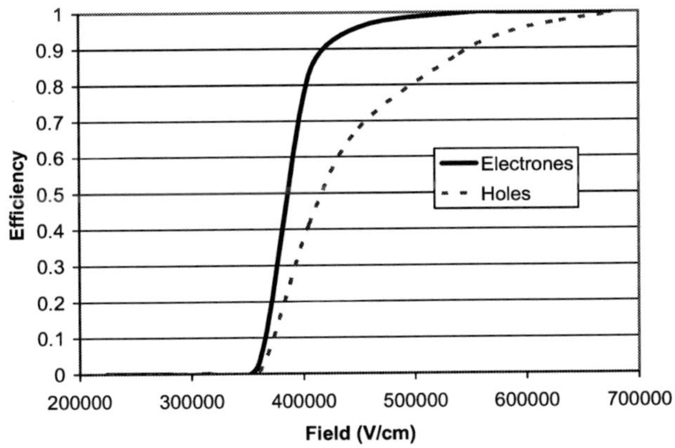

FIGURE 5. Efficiency for a single electron or hole to initiate an avalanche breakdown. Qualitative Monte Carlo calculation based on ionization coefficients from [7].

EXISTING DEVICES

Various prototypes have been produced by several companies and research institutes (CPTA, Moscow; MEPhI/Pulsar, Moscow; Dubna/Micron, Moscow; Hamamatsu, Japan; and Sensl, Irland). Sizes range from 0.5x0.5 mm^2 to 5x5 mm^2, with 100 – 10000 pixels. Typical pixel sizes are 25x25 μm^2 to 100x100 μm^2. To date no device seems to be commercially available, only prototypes have been distributed for evaluation, but this should change in the near future.

DEVELOPMENT OF BACKSIDE ILLUMINATED DEVICES

Driven by the need for photon detectors with very high QE for use in the MAGIC air Cerenkov telescope [8] MPI Munich started to develop SiPM with backside illumination [9]. The basic structure of a single cell is shown in Fig. 6: Light enters through the unstructured backside and photons convert in the fully depleted silicon bulk. Assisted by field shaping drift rings the electrons drift in an electric field into a small high field region where the avalanche multiplication occurs. While the cell may have a size of 100 μm, the avalanche region can be as small as a few micron. The advantages of this approach are:

- The backside is unstructured and can be made very thin. This leads to 100% fill factor and high conversion efficiency even at short wavelengths. In addition it is also possible to coat the backside with antireflective layers enhancing the QE further [10].
- The avalanche is always triggered by electrons giving high and uniform QE.

- Since the avalanche region can be kept small resulting in small capacitance and low gain (which must still be kept high enough to have detectable signals). These can help to reduce cross talk and recovery time.

However, these advantages are counterbalanced by some disadvantages:

- The whole silicon volume must be depleted, which leads to larger dark currents and rates. Possible solutions are: thinning of the device form nominal 450 μm to about 50 μm, using processes with inherent low dark currents, and cooling.
- The large sensitive volume may increase the cross talk. It is not yet known if this overcompensates the reduction expected due to the small capacitance. Again, thinning should help to reduce the crosstalk.
- The additional drift of the electrons through the bulk towards the high field regions increases time jitter. From simulations [11] it is expected that the time resolution can still be kept below 2ns.

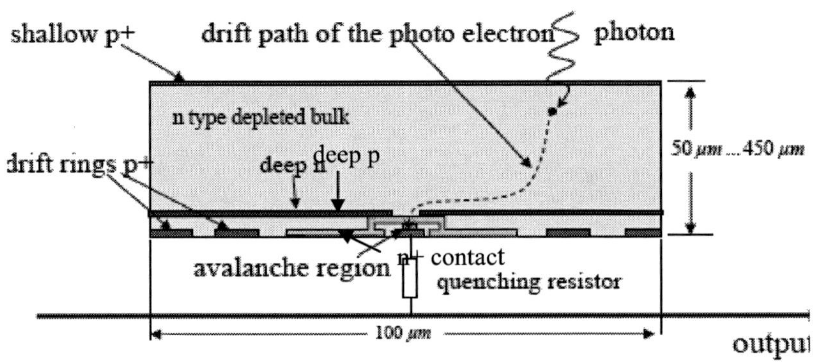

FIGURE 6. Schematic cross section of a back illuminated SiPM cell [9]. The high field region is between the deep p implantation and the n+ contact (which is connected to the quenching resistor).

Such devices have been designed in the MPI Semiconductor laboratory and production of first structures is ongoing. Since such devices need a more complicated production process (double sided processing) their price will be higher. Nevertheless they will be very interesting for photon hungry applications where the need for high QE outweighs other requirements.

CONCLUSIONS

Since several years SiPMs have been produced and proved to be suitable photon detectors with a performance comparable or even better than photomultiplier tubes. Noteworthy are single photon resolution, good efficiency, insensitivity to magnetic fields and (probably) low costs. This makes them especially interesting for large scale applications like in calorimeters. Drawbacks are the high dark rate, cross talk and poor UV/blue sensitivity. Ongoing R&D may improve the devices further. A new concept using back illumination has the potential to boost the quantum efficiency close to 100%, which will be interesting for applications where extremely high QE is needed.

ACKNOWLEDGMENTS

I would like to thank B. Dolgoshein and E. Popova for discussions and providing material. I am also indebted to my colleagues from the MPI Semiconductor laboratory for preparing this paper.

REFERENCES

1. Z.Y. Sadygov, et al., IEEE Trans. Nucl. Sci, NS-43 (3), p. 1009 (2005)
2. Z.Y. Sadygov, et al. Proc SPIE 1621, p 158 (1991)
3. B. Dolgoshein, Nucl.Instrum.Meth.A504, p. 48-52 (2003)
4. A. Lacaita et al., IEEE Trans. Electron Devices, ED-40, No 3, p. 577 (1993)
5. B. Dolgoshein, private communication.
6. M. Laatiaoui, Master Thesis, TU Munich, Germany (2005)
7. R. van Overstraeten and H. De Man, Solid State Electron (GB) vol 13, p. 583 (1970)
8. J. A Barrio et al., Design Report on MAGIC, MPI-PhE/98-5 (1998)
9. G. Lutz et al., IEEE Trans. Nucl. Sci. 52, p. 1156 (2005)
10. R. Hartmann et al., Proc. SPIE 5903, N1-N9 (2005)
11. C. Merck et al, Proceedings Beaune05, to be published in NIM A.

The Electromagnetic Calorimeter of the future PANDA Detector

Rainer Novotny* for the PANDA Collaboration

*2nd *Physics Institute, Justus-Liebig-University Giessen, Heinrich-Buff-Ring 16, D-35392 Giessen, Germany*

Abstract. Experiments with a cooled antiproton beam at the future accelerator facility FAIR at GSI, Darmstadt, will be performed with the 4π detector PANDA comprising a high resolution, compact and fast homogeneous electromagnetic calorimeter to detect photons between 10MeV and 10GeV energy inside a superconducting solenoid (2T). The target calorimeter comprises more than 20,000 $PbWO_4$ crystals of significantly enhanced quality read-out with large area avalanche photodiodes at an operating temperature of -25°C. The paper describes the quality of PWO-II and illustrates the future performance based on response measurements with high-energy photons.

Keywords: Electromagnetic Calorimetry, $PbWO_4$, PWO-II, Inorganic Scintillator.
PACS: 07.20.FW, 29.40.MC, 85.60.Dw

INTRODUCTION

The LHC program initiated a variety of R&D projects to make possible the design and the construction of unprecedented detectors. The lead tungstate - $PbWO_4$ (PWO) - scintillator is a result of one of these successfully performed projects [1]. Currently, PWO is used to construct the **E**lectromagnetic **CAL**orimeter (ECAL) of CMS [2], **PHOS**, the **PHO**ton **S**pectrometer of ALICE [3], **HYCAL** for the PrimEx experiment [4] and **DVCS** [5] as part of the CLAS detector, both installed at Jefferson Laboratory.

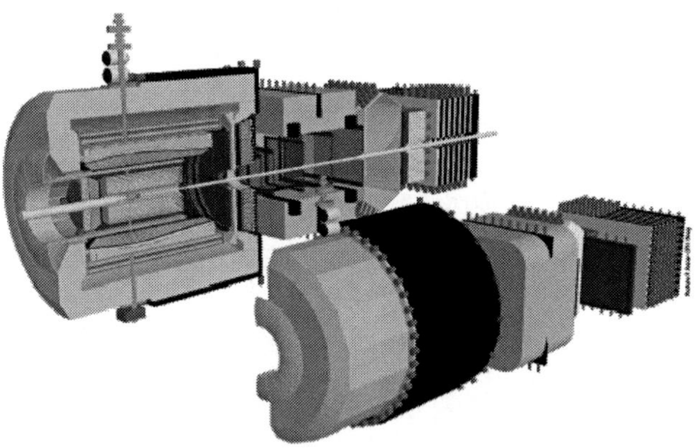

FIGURE 1. Schematic view of the proposed PANDA detector at the future FAIR facility at GSI.

CP867, *XII International Conference on Calorimetry in High Energy Physics,*
edited by S. R. Magill and R. Yoshida

In this paper we report on a future application of lead tungstate for the electromagnetic section of the 4π detector PANDA to be installed at the antiproton storage ring HESR at the proposed *International Accelerator Facility for Beams of Ions and Antiprotons FAIR* at the GSI in Darmstadt (Germany) [6]. The extreme requirements on compactness, count rate capability, fast response, position and energy resolution can be met by the new generation of $PbWO_4$ crystals with enhanced luminescence yield, identified as PWO-II, combined with a highly efficient read-out within a strong magnetic field using large area avalanche photo diodes (LAAPD).

THE PANDA DETECTOR

The proposed new accelerator at GSI will deliver hadron and heavy ion beams up to energies of 30 GeV/u provided by the double ring synchrotron SIS 100/200. Among several additional facilities antiprotons will be produced, collected and finally stored in the HESR (**H**igh **E**nergy **S**torage **R**ing). Depending on the requested quality of the beam momentum, achieved by stochastic or electron cooling, luminosities up to $2 \cdot 10^{32} cm^{-2} s^{-1}$ will become available for a momentum of antiprotons up to 15GeV/c. The envisaged hadron physics program will focus on non-perturbative processes to study the structure of hadrons and their interactions. High resolution experiments will perform charmonium spectroscopy, search for gluonic excitations such as charmed hybrids or glueballs and will address open problems of in-medium modifications of hadrons.

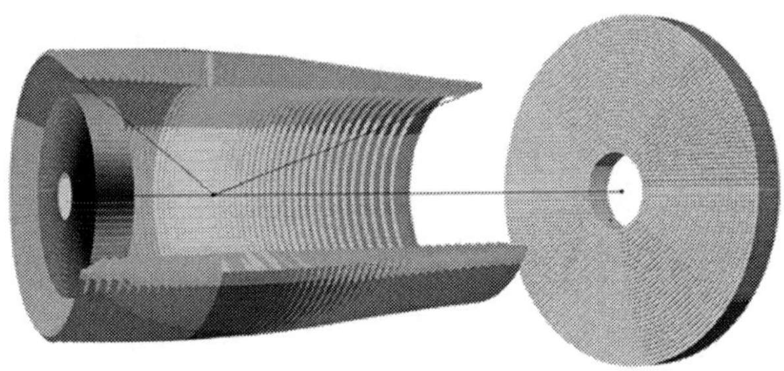

FIGURE 2. Schematic view of the EM calorimeter in the target region of the PANDA detector.

In one of the straight sections of the storage ring, the multipurpose PANDA detector will be installed for fixed target experiments with antiprotons, anticipating an annihilation rate of 10^7/s. The general schematic layout of the detector is shown in Figure 1. The central part houses inside a superconducting solenoid of 2T magnetic field the target system, vertex reconstruction (D, K_s, Λ,), charged particle tracking ($e^{\pm}$, $\mu^{\pm}$, $\pi^{\pm}$, p,), particle identification and a homogeneous electromagnetic calorimeter for photon and neutral meson detection. The most forward part below a scattering angle of 5^0 relative to the beam axis is covered by a dipole magnet (3.5T) and contains similar detector components to identify in particular leading particles.

The electromagnetic calorimeter has to provide a nearly 4π coverage for multi-photon and multiple meson detection, a fast response to cope with the high rate of annihilations and sufficient granularity and energy resolution for position and energy reconstruction over a large dynamic range of photon energies from ~ 8GeV down to as low as 10MeV to enable the planned physics program. Figure 2 illustrates the geometrical arrangement of the calorimeter, comprising nearly 20,000 tapered crystals of 20cm length (22.5 X_0) and a front face diameter close to the Molière-radius of 2.2cm.The asymmetric structure is adapted to the fixed target kinematics anticipating end-caps of different size. The barrel part is subdivided into 16 slices containing 720 modules in a carbon fiber construction. The overall concept of the PANDA detector requires a dense, fast and sufficiently bright scintillator coupled to a fast photo sensor to be operated within the strong magnetic field.

THE SCINTILLATOR MATERIAL PWO-II

In close collaboration with the group of M. Korzhik (Research Institute for Nuclear Problems, Minsk, Belarus) and the Bogoroditsk Technical Chemical Plant (BTCP), Bogoroditsk (Russia) a research program has been initiated to improve the crystal quality and to establish the technology for future mass production [7].

Based on R&D work at BTCP the highest light yield and optimum radiation hardness can be achieved for crystals simultaneously doped with La- and Y-ions. Using pre-synthesized and stoichiometric raw material of very high purity point structure defects can be efficiently compensated along the axis of crystal growth by introducing the ions at different stages of the production chain. Performed EPR studies confirmed that only single RE-$(WO_4)^{3-}$ (RE = La,Y) centers are formed, if the concentration of the dopants stays below 20ppm for each ion, respectively. As a result, these crystals contain at least two times less defects of Frenkel type leading to an increase of light of ~80% compared to standard type material, the mass produced crystals for CMS/ECAL. The kinetics of this enhanced material, named PWO-II, remains fast and allows integrating 97% of the light within a 100ns wide time gate at room temperature. The technology has been established and a large quantity of up to 200mm long straight or tapered crystals has been produced with nearly constant quality as illustrated in Figure 3.

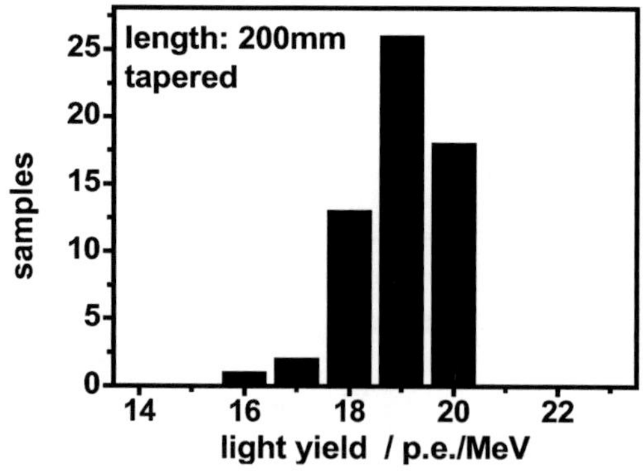

FIGURE 3. Distribution of the light output of 60 tapered PWO-II crystals in PANDA geometry. The light yield has been measured with a 2 inch photomultiplier tube with bialkali photocathode.

Due to the thermal quenching of the luminescence process at room temperature, cooling of the crystals can effectively increase the light yield. However, the obviously low concentration of deep traps in PWO-II does not distort the scintillation kinetics even down to a temperature of -25°C. Figure 4 shows for different temperatures the scintillation light obtained for a ^{60}Co-source of 1.2 MeV mean γ-energy by integrating over a gate width between 75ns and 1μs, respectively. Even at the lowest temperature, 95% of the light can be collected within less than 300ns.

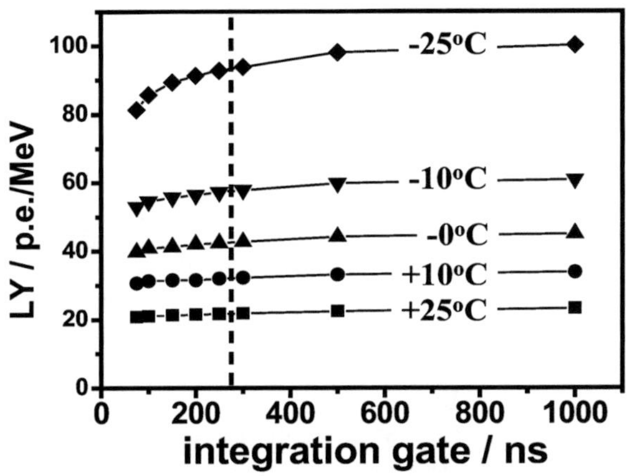

FIGURE 4. Light yield of a 200mm long PWO-II crystal in PANDA geometry measured at different temperatures and integration gates using a 2 inch photomultiplier tube with bialkali photocathode.

Figure 5 documents the strong linear correlation between the light yield measured at room temperature and the corresponding value obtained at –25°C, the envisaged operating temperature of the calorimeter. That allows performing all measurements for quality assurance at room temperature.

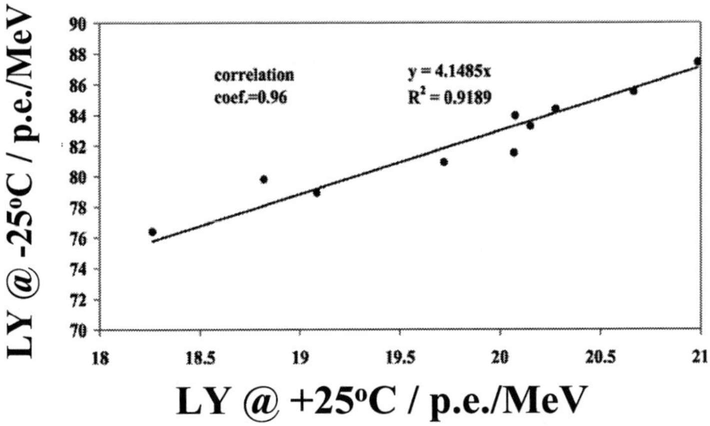

FIGURE 5. Correlation of the light yield of 200mm long PWO-II crystals in PANDA geometry measured at two extreme temperatures using an integration gate of 100ns width.

First investigations of the radiation hardness using a high dose of γ-rays as well as up to 10^{13} protons/cm^2 of 90MeV energy, delivered by the cyclotron at KVI (Groningen, The Netherlands), have shown a performance similar to CMS crystals indicating a slight reduction of the optical transparency but no permanent defects.

RESPONSE MEASUREMENTS TO HIGH-ENERGY PHOTONS

First prototype tests were performed with quasi-monochromatic photons delivered by the tagged photon facility at the electron accelerator MAMI at Mainz (Germany) exploiting the tagging of bremsstrahlung produced by a monoenergetic electron beam (E_e = 855 MeV). After bremsstrahlung emission the momenta of the slowed down electrons were analyzed by the magnetic spectrometer of the Glasgow-Mainz tagger [8], requiring a time coincidence of the detected bremsstrahlung photon with the corresponding electron identified in the focal plane. The typical energy width per tagging channel varies between ΔE = 2.3 MeV at 50 MeV to ΔE = 1.5 MeV at 500 MeV photon energy, respectively. Eight photon energies in the range between 64MeV and 520MeV, respectively, were chosen. The $PbWO_4$-arrays were located at a position 12 m downstream from the radiator. A set of collimators limited the beam spot to a diameter of d = 8 mm at the front face of the crystals.

In both presented cases, the blocks of crystals, which were individually wrapped in TEFLON foil, were mounted in an insulated container, which could be cooled down to a well stabilized ($\Delta T < 0.1^0$C) temperature by circulating cold and dry air. A thin plastic scintillator covering the total front of the calorimeter served as a charged particle veto to reject already converted photons. The crystal matrix could be moved remote controlled in two dimensions perpendicular to the axis of the collimated photon beam by stepping-motors, to perform a relative calibration of each detector element under beam conditions.

3x3 matrix read-out with photomultiplier tubes

The 3x3 test matrix of rectangular crystals (200mm long, 20x20mm^2 cross section), manufactured by BTCP, was read-out individually by photomultiplier tubes (Philips XP1911). The experiment provides data taken with a minimum of electronic noise and allows a direct comparison to previous measurements with crystals of former quality [9, 10]. The individual detector signals were digitized by means of a charge-sensitive ADC and a timing signal was deduced to determine the response time relative to the central module. Coincidences of the central calorimeter module with one of the timing signals of the chosen tagger channels were accepted under the condition that no charged particle was identified in the plastic veto. The response function of the crystal matrix has been measured at T=+10°C and T=-25°C, respectively, to study the crystal quality as well as the impact of cooling.

The lineshape of the response function can be reconstructed by summing up event-wise the energy deposited into the matrix. Figure 6 shows as an example at the photon energy of 63.8MeV the measured energy depositions integrated individually in the central unit, the sum of all neighboring modules and finally in the whole array. The energy resolution at T=-25°C of σ/E=4.6% represents the best ever achieved result with $PbWO_4$.

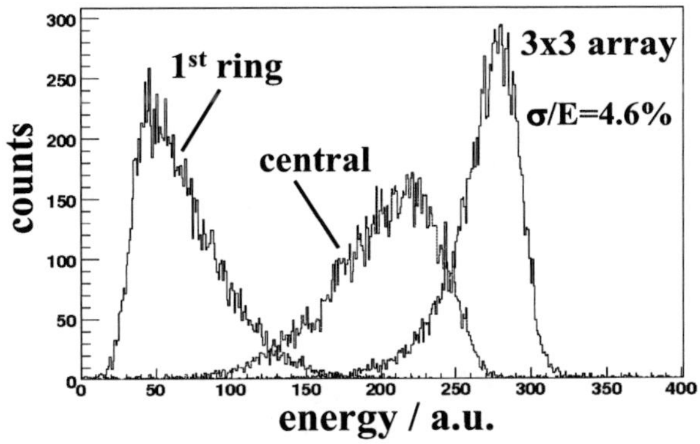

FIGURE 6. The energy response of a 3x3 matrix to photons of 63.8MeV energy. The energy deposition into the central and the surrounding modules are shown in comparison to the total matrix. The response has been obtained at T=-25°C with photomultiplier readout.

The experimental resolution obtained in the investigated range of photon energies is summarized in Figure 7 at both operating temperatures. In particular, the results obtained at the lower temperature of T=-25°C document the significant overall improvement of the photon statistics compared to previous measurements and leads to an extremely low stochastic term of 0.95%. In spite of the fact that there is a significant lateral leakage of the electromagnetic shower indicated by the relative large constant term, an energy resolution well below 2% can be extrapolated for 1GeV incident photon energy.

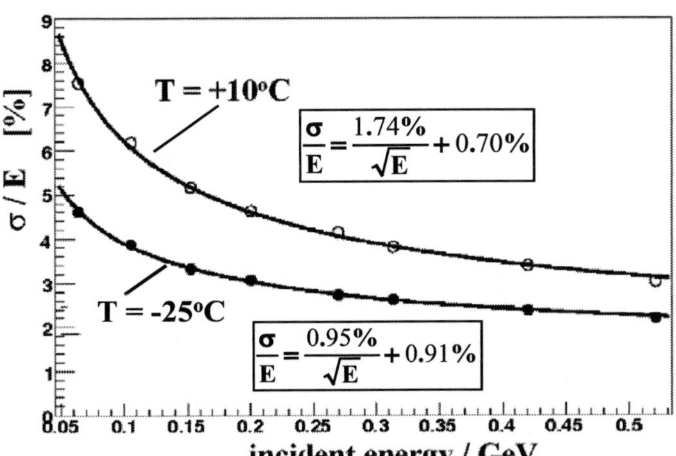

FIGURE 7. Energy resolution of a 3x3 matrix of 200mm long PWO-II crystals measured at two operating temperatures with energy marked photons between 64MeV and 520MeV, respectively. The scintillation light has been read-out via photomultiplier tubes.

3x3 matrix read-out with photomultiplier tubes

Since avalanche photo diodes (APD) are compact photo sensors, which are insensitive to high magnetic fields, the CMS/ECAL group has developed reverse biased diodes in collaboration with Hamamatsu Photonics with small overall thickness (~200μm), a quantum efficiency of 70-80% at 420nm, sufficient radiation hardness and a narrow amplification section (<20μm) to minimize the response to directly ionizing particles. However, the achieved active area was limited to $5x5mm^2$.

In continuation, a new large area avalanche photo diode (LAAPD) of $10x10mm^2$ active size has been developed for PANDA (Hamamatsu S8664-1010). Detailed tests have established excellent performance including radiation hardness with respect to low energy γ-rays as well as protons of ~90MeV energy up to an integral dose of 10^{13} protons, tested at room temperature and at -25°C [11]. In order to meet the electrical and geometrical constraints a new low noise and low power charge sensitive pre-amplifier (LNPP) has been designed with a charge sensitivity of 0.5V/pC and a short feedback time constant. Cooling in addition decreases the noise contribution of the LNPP by 15-20%. For the input capacitance of the LAAPD the equal noise charge (ENC) drops from $2000e_{RMS}$ at room temperature down to ~ $1600e_{RMS}$ measured at –25°C.

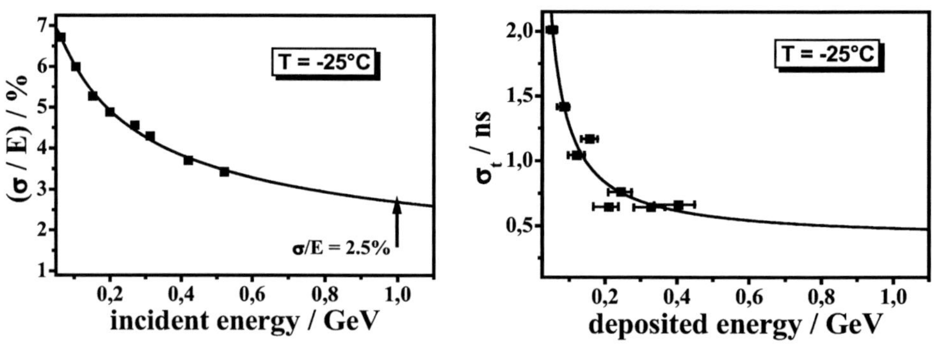

FIGURE 8. (Left) Energy resolution of a 3x3 matrix of 150mm long PWO-II crystals measured at an operating temperature of –25°C with energy marked photons between 64MeV and 520MeV. The scintillation light has been read-out via avalanche photo diodes. (Right) Measured time resolution of the central PWO-module as a function of the mean deposited photon energy. The crystal has been read-out via an avalanche photo diode.

In a first test, a 3x3 matrix of 150mm long crystals was equipped with LAAPD-read-out. The bias voltage was chosen to achieve a moderate intrinsic amplification of M~100-150. The output signals of the pre-amplifiers were passively split to serve a timing and an analog circuit. A spectroscopy amplifier treated the analog pulse with 0.5μs Gaussian shaping. The timing signal was derived by a constant-fraction-discriminator after fast shaping. In particular, the coincidence timing of the central crystal with one of the timing signals of the fast tagger was used to estimate the achievable time resolution. The response function to high-energy photons has been measured at T=-25°C only. Figure 8 (left) summarizes the energy resolutions obtained in the investigated range of photon energies. In spite of the fact that there is a significant lateral leakage of the electromagnetic shower an energy resolution of $\sigma/E=2.5\%$ can be extrapolated for 1GeV incident photon energy.

The obtained timing information of the central detector allows getting a first estimate of the time resolution (see Figure 8 (right)). The shown error bars indicate the most probable range of the energy deposited in the central module for the different incident energies. The given time resolutions have not been corrected for the contribution of the reference detector (σ_t<400ps). In spite of the not optimized read-out chain for the timing information, an excellent and very promising result has been obtained. Even the present data indicate time resolutions <1ns above 150MeV photon energy and a value near 500ps at 1GeV, respectively.

CONCLUSION AND OUTLOOK

The very complex and ambitious detector project is presently focusing on the design and construction of a large prototype matrix comprising 60 detector modules including read-out with LAAPDs and a new low power and low noise quad-preamplifier in order to study primarily the mechanical stability, cooling, thermal insulation and temperature stabilization. In a second step, a complete slice of the barrel calorimeter will be manufactured and assembled housing 720 PWO-crystals in carbon fiber containers. The PANDA detector should be completed and ready for commissioning in 2011/12.

ACKNOWLEDGMENTS

The project is supported by the European Community-Research Infrastructure Activity under the FP6 "Structuring the European Research Area" programme (HadronPhysics contract number RII3-CT-2004-506078), the Bundesministerium für Forschung und Technologie (BMBF), Germany and Gesellschaft für Schwerionenforschung (GSI) at Darmstadt, (Germany).

REFERENCES

1. A. A. Annenkov et al., *Lead Tungstate Scintillation Material*, Nucl. Instr. Methods, vol. A490, pp. 30 (2002).
2. CMS, Technical Design Report, CERN-LHCC-97-033.
3. ALICE, Technical Proposal, CERN/LHC 9-17.
4. A. Gasparian, *A High Performance Hybrid Electromagnetic Calorimeter at Jefferson Lab,* Proc. of the 11th Int. Conf. on Calorimetry in Particle Physics, Perugia, Italy, March 29, - April 2, 2004, p. 109.
5. J. Pouthas, Orsay, private communication.
6. H. Gutbrod et al., *Conceptual Design Report : An International Accelerator Facility for Beams of Ions and Antiprotons,* http.//www.gsi.de/GSI-Future/cdr.
7. A. Borisevich et al., Nucl. Instr. Methods, vol. A537, p. 101-104 (2005).
8. I. Anthony et al., Nucl. Instr. Methods, vol. A301, pp. 230 (1991).
9. K. Mengel et al., *Detection of Monochromatic Photons Between 50 and 790MeV With a $PbWO_4$-Scintillator Array,* IEEE Trans. Nucl. Sci., vol. 45, no. 3, p. 681-685 (1998).
10. R. Novotny et al., *Electromagnetic Calorimetry With $PbWO_4$ in the Energy Regime Below 1GeV,* IEEE Trans. Nucl. Sci., vol. 47, no. 4, p. 1499-1502 (2000).
11. B. Lewandowski, GSI, private communication.

The Electromagnetic Barrel Calorimeter for the GlueX Experiment

M. Barbi

Department of Physics, University of Regina
3737 Wascana Parkway, Regina, SK, S4S 0A2, Canada

Abstract. The electromagnetic barrel calorimeter is one of the main components of the planned GlueX experiment. It will consist of 48 modules made of consecutive layers of 4 m long lead sheet and fast green scintillator fibers for an overall number of approximately 3000 readout channels with silicon-photomultiplier-based photo-sensors for light collection. The calorimeter is expected to achieve energy and time resolution better than $5\%/\sqrt{E} \oplus 2\%$ and 200 ps, respectively.

In this contribution we present an overview of the calorimeter design and some preliminary studies of its performance using Monte Carlo simulations and beam test measurements.

Keywords: GlueX, hybrid mesons, calorimeter, silicon-photomultipliers, scintillator fiber
PACS: 07.20.F, 29.40.–n, 12.38.-t, 12.38.Aw, 12.39.Mk,

INTRODUCTION

The Standard Model (SM) is the most successful theory to date to explain the fundamental building blocks which make all the existing matter in the Universe and the forces through which they interact. Within the SM, the Quantum Chromodynamics theory (QCD) describes the strong interaction that bounds quarks and gluons together inside the nucleons. QCD has succeeded in explaining the quark-quark interactions at short distance scale, where the strong coupling constant α_s is small and perturbative QCD calculations apply. However, at large distance scales, α_s is large, perturbative calculations no longer apply and the complicated QCD Lagrangians are difficult to solve. However, it is at these scales that the QCD confinement mechanism which prevents quarks from escaping the strong potential generated by the gluonic fields plays its role in the nucleon structure. One way of working around the non-perturbative regime is by using Lattice QCD calculations [1]. Another way is to rely on physics models in an attempt to approximately reproduce the quark-gluon dynamics at large α_s values. In either cases, the difficult in solving the QCD Lagrangian in the non-perturbative regime leads to one of the more challenging and outstanding problems still to be solved in QCD, that of unfolding the real nature of the QCD confinement mechanism.

The Flux-Tube Model [2] is one of the most accepted models to describe the quark-gluon dynamics at large distance scales, in which quarks are bounded to each other through gluonic flux tubes. Regular quark bound states happen when the flux tubes are in their ground state. However, when in excited mode, the gluonic flux tubes can add

CP867, *XII International Conference on Calorimetry in High Energy Physics,*
edited by S. R. Magill and R. Yoshida

degrees of freedom to the quark system, consequently leading to final states where gluons also contribute to the total quantum numbers. These states, where not only quarks but also gluons effectively participate in the full dynamics, are called hybrid states. Quark-gluon bound states are allowed by QCD, though they are not permitted in the regular Quark Model.

At the Jefferson Thomas Laboratory (JLAB), a dedicated detector is being projected and developed to map the hybrid meson states: the GlueX experiment [3] (Figure 1). The data will be collected from fixed-target photoproduction in γp collisions using 9 GeV coherent Bremssthalung photons generated from a primary 12 GeV electron beam source. The final states will be fully established using Partial Wave Analysis (PWA) techniques. To be effective and unbiased, PWA requires GlueX to have a very large acceptance to fully contain the particles produced in γp reactions, and to have high energy, momentum and position resolution to allow for reconstruction of the four-momentum of individual particles to high accuracy.

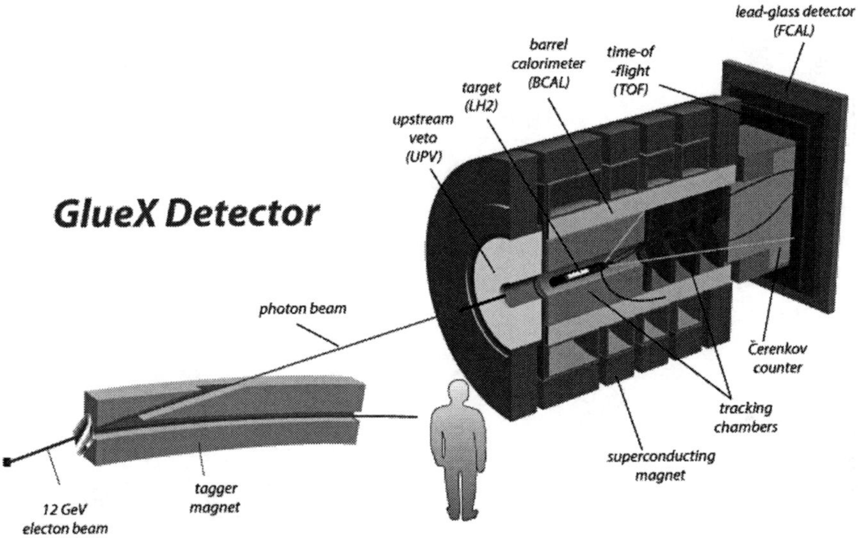

FIGURE 1. The GlueX detector.

The electromagnetic barrel calorimeter (BCAL) is one of main components of the GlueX detector. Its primary goal is to detect photons from decays of π^0 and η particles. The calorimeter is made of consecutive layers of 4 m long lead (Pb) sheet and scintillator fibers (SciFi) acting as passive and active material respectively, and will consist of 48 modules as depicted in Figure 2. The SciFi's are laid into grooves machined longitudinally in the lead sheets relative to the incoming photon beam. The tracking chambers and superconducting solenoid constrain the inner and outer radius of the BCAL to 65 cm and 90 cm respectively, corresponding to an overall acceptance of $12° < \theta < 135°$.

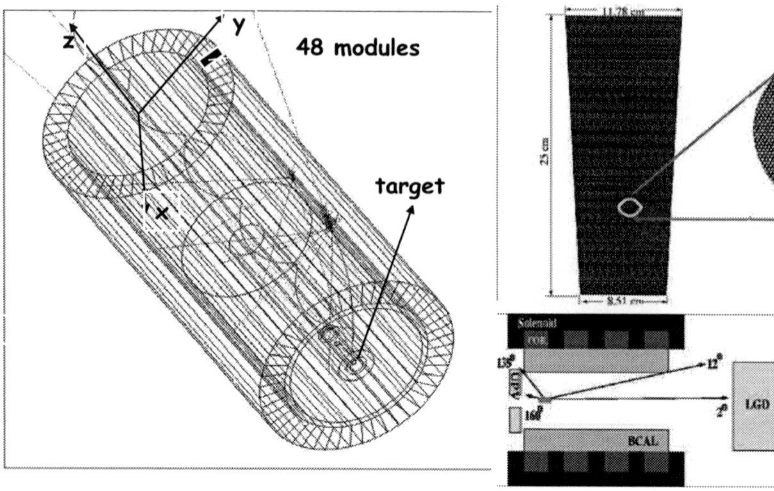

FIGURE 2. An overview of the full calorimeter is shown on the left with a simulation of a γp reaction using GEANT-3, while the top right figure is a GEANT-3 simulation of the cross-section of one of the calorimeter modules. The bottom right sketches the calorimeter acceptance region.

The GlueX-BCAL is a non-compensating sampling calorimeter with a Pb:SciFi:glue ratio of 37:49:14. A section of a module is represented in Figure 3 showing the geometrical dimensions of the lead-scintillator layers including the gluing material used to fix the SciFi's into the grooves of the lead sheets.

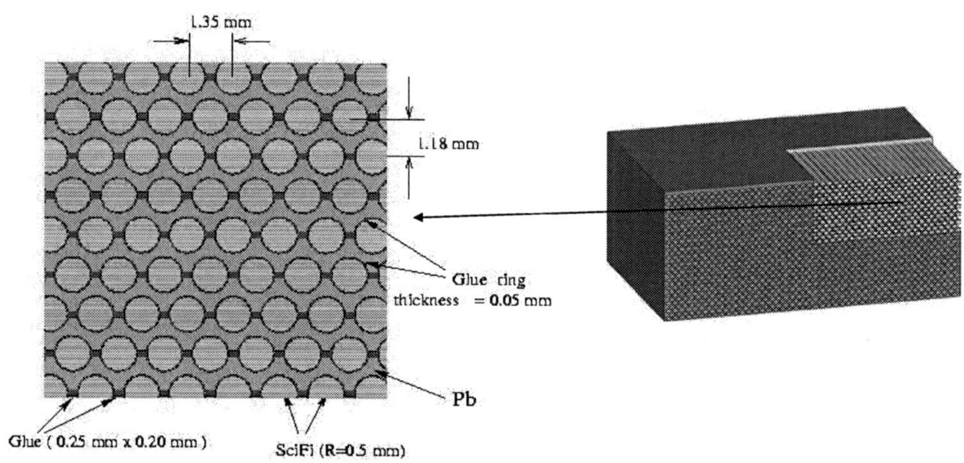

FIGURE 3. The inner geometrical structure of the calorimeter.

The BCF-20 double-clad fast green scintillator fibers from BICRON with emission peak at 492 nm, 1 mm diameter, attenuation length greater than 300 cm, decay time of ~ 3 ns and 8000 photons/MeV were chosen as the best candidate to date for the active material of the BCAL. A total of ~3500 Km of fiber equivalent to a density of 63 fibers/cm^2 will be needed for the full construction of the calorimeter.

THE GLUEX-BCAL REQUIREMENTS

The designed BCAL energy, position and time resolution are given below:

$$\frac{\sigma(E)}{E} = \frac{5\%}{\sqrt{E}} \oplus 1\% \,; \qquad \sigma(x) = 1\ cm\,; \qquad \sigma(t) = 200\ ps. \tag{1}$$

The geometrical and material composition of the calorimeter ensure a thickness of ~15.6 X_0 radiation lengths needed to fully contain electromagnetic showers produced by incident particles with energy up to 3 GeV. The resulting Molière radius $R_M \approx 4\ cm$ requires fine readout segmentation area of at most 4 cm^2 to accomplish good shower separation and therefore fulfill the requirements of excellent particle identification and high energy resolution. A possible readout scheme will have the most inner layers of each BCAL module, corresponding to the first section of 10 cm in depth where most of the energy of the income particles is deposited, segmented into 20 readout channels of ~2x2 cm^2 area each, while the outer layers corresponding to the second section of 14 cm in depth will have 12 readout channels with area of ~3.5x4 cm^2 each, for a total of 64 readout channels per module, 32 for each BCAL side, leading to the overall number of 3072 readout channels for the whole calorimeter.

The GlueX-BCAL will be inserted and operated in a ~2 T magnetic field required for the GlueX experiment. This high magnetic field is a constraint to the choice of the photo-sensors for the calorimeter. Under this condition, regular vacuum PMT's are very sensitive to the orientation of the magnetic field. On the other hand, preliminary studies [4] have shown that silicon photomultipliers (SiPM's) are suitable for the requirements of the calorimeter readout operation, i.e., up to ~1 MHz counting rate, 20 MeV – 3 GeV dynamic range, and high energy and time resolution. These devices are also insensitive to magnetic fields up to at least ~4 T [5], and can deliver high resolution single photo-electron detection at low noise level.

A detailed description of SiPM's can be found elsewhere [6]. Some of their main properties are: high gain (10^6-10^7 times); low bias voltage normally in the range of 20-60 V; high dynamic range; fast recovery time at the order of few tenths of nanosecond; short rise time at the order of 1-2 ns. Also, currently available SiPM's have higher sensitivity in the green region of the light spectrum. This is most desirable since preliminary studies carried out by the GlueX collaboration (unpublished) have shown that even using long blue scintillator fibers, the light that survives to the ends of the BCAL is in the green region of the spectrum. It has been therefore decided that a better approach would be to use green scintillator fibers coupled to green sensitive photo-sensor.

PRELIMINARY STUDIES

Several Monte Carlo simulations have been carried out in order to optimize the calorimeter response and test its potential to reconstruct electromagnetic showers in GlueX. Figure 4 is a GEANT-3 simulation of the process $\gamma\, p \rightarrow \eta\, \pi^0\, p \rightarrow \gamma\gamma\gamma\gamma\, p$, where 9 GeV photons collide with fixed-target protons. The first three distributions show the ratio between the total energy $E_{\gamma,p}$ of the final state and the energy deposited in the lead (E_{dep}^{Lead}), SciFi (E_{dep}^{SciFi}) and glue (E_{dep}^{Glue}) respectively, while the fourth distribution shows the fraction of energy E^{Leak} leaking from the calorimeter. 97% of the total energy produced in this reaction is fully contained in the calorimeter.

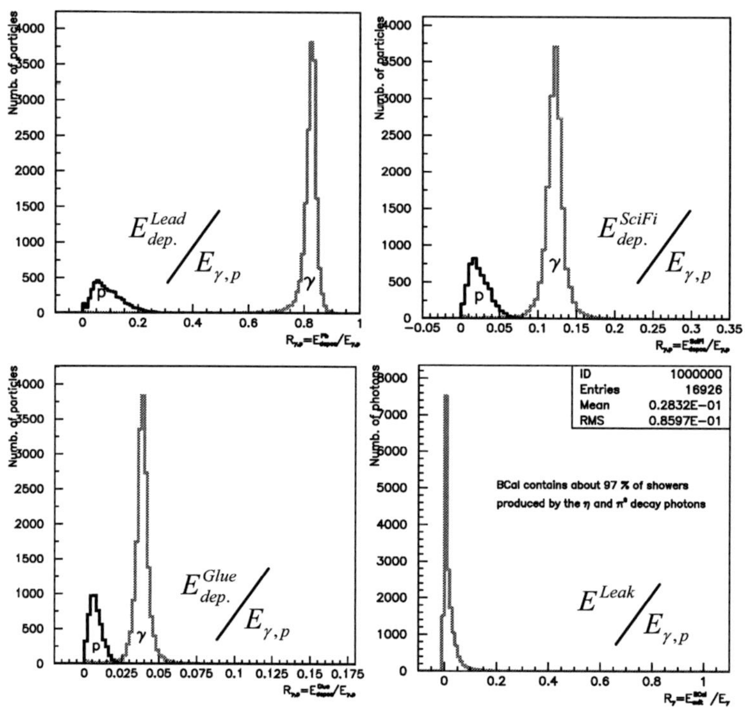

FIGURE 4. Distribution of the energy fraction deposited in the lead, scintillator and glue material. The fraction of energy not captured by the calorimeter is also shown in the bottom right distribution.

The reconstructed invariant mass of π^0 particles simulated with energy of 2 GeV using the proposed BCAL readout scheme as discussed in the previous section implemented into GEANT-3 is shown in Figure 5. The π^0's enter the BCAL at random angles. There is no electronics simulation included and the reconstruction software is a very preliminary version which is still being optimized for events with more than one electromagnetic shower. The large background at lower invariant masses is due to photons being associated to wrong showers. This background is expected to significantly improve with the use of optimized reconstruction software.

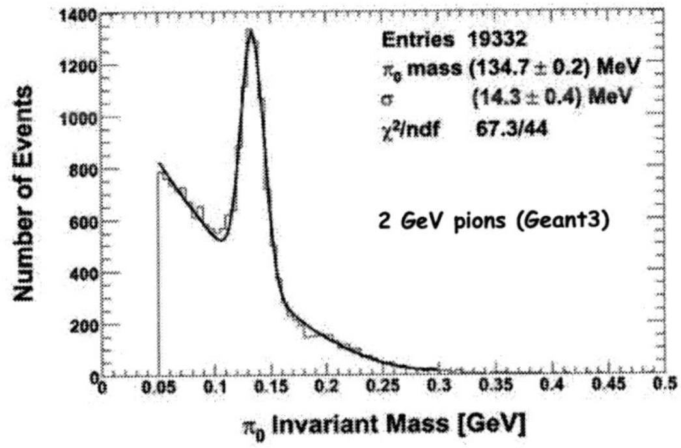

FIGURE 5. Invariant mass of simulated π^0's using GEANT-3. The pions are generated at random angle with respect to the calorimeter.

Studies using real beam conditions were also carried out at the M11 pion facility at TRIUMF. Beams of pions, muons and electrons with energy ranging from 120 MeV to 300 MeV were aimed at a full-scale 4 m long prototype of a BCAL module[1]. Trigger counters placed between the source of the beam and the detector were used for timing and trigger. Regular Burle PMT's 8575 photo-sensors were used to collect light from the prototype. Figure 6 shows some preliminary results of an analysis to estimate the attenuation length of the fibers and the time resolution achieved in the tests[2]. While the attenuation length of 309±13 cm is consistent with the manufacturer specifications, it is expected that the time resolution of 320±2 ps will significantly improve with the use of the final version of the photo-sensors (possibly SiPM's), fast green scintillator fibers and improved readout electronics.

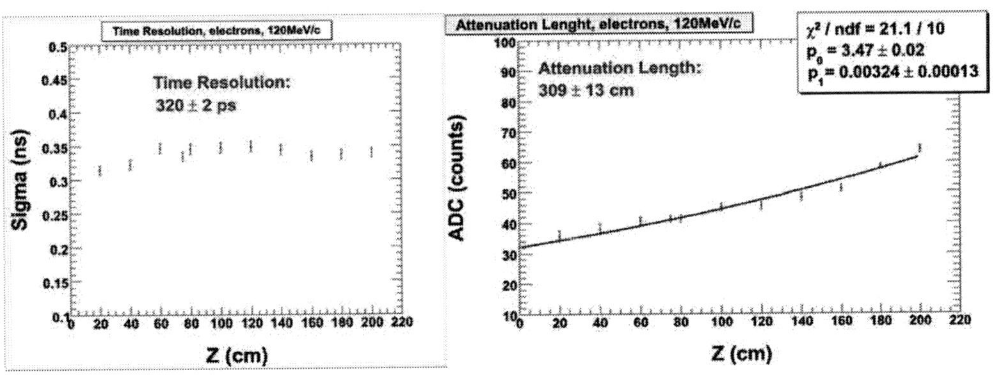

FIGURE 6. Preliminary measurements of the attenuation length and time resolution with data from tests using 120 MeV electrons from the M11 beam facility at TRIUMF.

---

[1] Pol.Hi.Tech. double-clad blue scintillator fibers were used for these tests.
[2] These measurements are part of the Master thesis being prepared by Mrs. Gergana Koleva at the University of Regina.

As discussed before, SiPM's have being considered as the best candidates to fulfill the requirement of operation in a ~2 T magnetic field, and deliver high energy and time resolution. Currently, only small devices with active area of 1x1 mm^2 are available in the market. Given the proposed BCAL readout scheme, as discussed in the previous sections, these small devices cannot be used to instrument the calorimeter. However, the GlueX collaboration has entered into an agreement with the SensL Company[3] for the development of customized large area photo-sensors consisting in arrays of SiPM's for the BCAL-GlueX detector. Each photo-sensor is a matrix of 3x3 mm^2 SiPM's. A study on the dependence of the minimum number of detectable photoelectrons as a function of the photo-sensor area and dark rate is shown in Figure 8 based on simulations of the SiPM parameters. This study was carried out by the SensL group using two values for the SiPM dark rate at one photoelectron level: 3 MHz and 5 MHz. The total photo-sensor dark rate is the sum of the dark rate of each SiPM in the array. The SiPM's were not optimized to low optical cross-talk effect and no cooling system was considered into the simulation. A minimum number of 170 photoelectrons is expected to reach the lower 20 MeV energy threshold level with the calorimeter. As can be observed, 4x4 arrays of 3x3 mm^2 SiPM's, equivalent to photo-sensors with 12x12 mm^2 active area, can deliver a minimum of 160 photoelectrons at 80 MHz dark rate level (5 MHz/SiPM). These parameters will be significantly improved with the optimization of the SiPM's for low optical cross-talk effect and the use of a cooling system which will further decrease the dark rate to even lower levels with consequent increase in the minimal number of detectable photoelectrons.

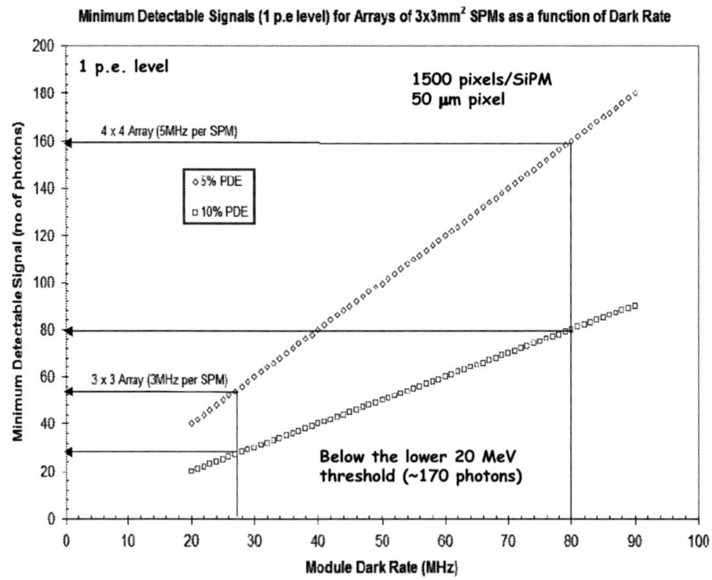

FIGURE 7. Minimum number of detectable photoelectrons as a function of the photo-sensor array area and total dark rate.

---

[3] The Low Light Sensing Company (SensL), http://www.sensl.com/.

In order to collect light from individual calorimeter readout areas of 20x20 mm^2 (inner layers), each photo-sensor will be coupled to Winston cones with maximal geometrical reduction factor of 4 times.

CONCLUSIONS

This contribution summarizes some of several studies for the development of the electromagnetic calorimeter for the GlueX detector. These studies have shown that the project of a large acceptance non-compensating SciFi-Pb calorimeter using silicon-photomultiplier based photo-sensors for readout is feasible. High energy and time resolution can be achieved with the proposed geometrical configuration, leading to electromagnetic shower reconstruction to high accuracy as required for the success of the GlueX physics programs.

ACKNOWLEDGMENTS

The author would like to thank the GlueX Collaboration for the material presented in this contribution. Particular thanks goes to Mrs. Gergana Koleva for providing some of her master thesis results to be included in this paper, to Dr. Rafael Hakobyan, Dr. Nikolay Kolev and Mr. Blake Leverington for the Monte Carlo simulations, to Dr. Zisis Papandreou and Dr. George Lolos for several calculations and discussions and finally to SensL Company for the information concerning the development of the photo-sensors. This research is supported by the National Sciences and Engineering Research Council of Canada and U.S. Department of Energy.

REFERENCES

1. C.J. Morningstar and M. Peardon, Phys. Rev. D 60, 034509 (1999).
2. G.S. Bali, K. Schilling and C. Schlichter, Phys. Rev. D 52, 5165 (1995) .
   T. Burns, F.E. Close, Phys. Rev. D 74, 034003 (2006).
3. A.R. Dzierba, "QCD Confinement and the Hall D Project at Jefferson Lab" in e^+e^- Physics at Intermediate Energy Worshop-2001, eConf C010430 T04, 2001. Also in e-Print Archive hep-ex/0106010.
   GlueX Collaboration Homepage: http://gluex.org.
4. V.D. Kovalchuck, G.J. Lolos, Z. Papandreou, K. Wolbaum, Anucl. Instrum. Meth. A538, 408-415 (2005).
5. E. Garutti, M. Groll, A. Karakash and S. Reiche, ALC-DET-2004-025, (2004).
6. P. Buzhan et al, ICFA Instrum. Bull. 23, 28-41 (2001).

Silicon-Tungsten Tracking Calorimeters for the Forward Direction in the PHENIX Experiment at RHIC

E. O'Brien*, V. Dzhordzhadze†, D. Karmanov▼, E. Kistenev*, Z. Li*, A. Litvinenko▲, M. Merkin▼, I. Nakagawa•, V. Peresedov▲, R. Seto†, A. Takesani, A. Sukhanov*, V. Volkov▼, A. Voronin▼, V. Vrba♦, L. Zolin▲

*Brookhaven National Laboratory, Upton, NY 11973, USA,
†University of California at Riverside, CA 92521, USA,
▼D.V. Skobeltyn Institute of Nuclear Physics, M.V. Lomonosov Moscow State University, 1/2 Leninskie Gory, Moscow 119992 Russia,
▲Joint Institute for Nuclear Research, Dubna, Russia,
•RIKEN-BNL Research Center at Brookhaven National Laboratory, Upton, NY 11973, USA,
♦Institute of Physics, Academy of Sciences of the Czech Republic, Prague

Abstract. The PHENIX detector at RHIC has been designed to study hadronic and leptonic signatures of the Quark Gluon Plasma in heavy ion collisions and spin dependent structure functions in polarized proton collisions. The baseline detector measures muons in two muon spectrometers located forward and backward of mid-rapidity, and measures hadrons, electrons, and photons in two central spectrometer arms. Further progress requires extending rapidity coverage for hadronic and electromagnetic signatures by upgrading the functionality of the PHENIX muon spectrometers with calorimeters to include photon and jet measurements. In this paper we present the basics of the calorimeter design and argue that combining highly segmented calorimeter with high resolution preshower and shower maximum detectors allows to resolve overlapping showers separated by ~X0/4 and properly reconstruct effective masses even under constrains of extremely compact experiment.

Keywords: Calorimeter, forward spectrometer, mechanical design, analog and digital readout electronics, triggering.
PACS: 24.85.+p, 29.40.Vj, 29.40.Wk

INTRODUCTION

The PHENIX [1,2] detector (Fig.1) is designed to perform a broad study of nucleus-nucleus (A-A), proton or deuteron nucleus (p-A or d-A), and proton-proton (p-p) collisions to investigate nuclear matter under extreme conditions and spin dependent structure functions. The PHENIX detector utilizes global detectors to characterize the collisions, a pair of central spectrometers at mid rapidity to measure electrons, hadrons, and photons, and a pair of forward spectrometers to measure muons. Each

CP867, *XII International Conference on Calorimetry in High Energy Physics*, edited by S. R. Magill and R. Yoshida

spectrometer has a geometric acceptance of about one steradian and excellent particle identification.

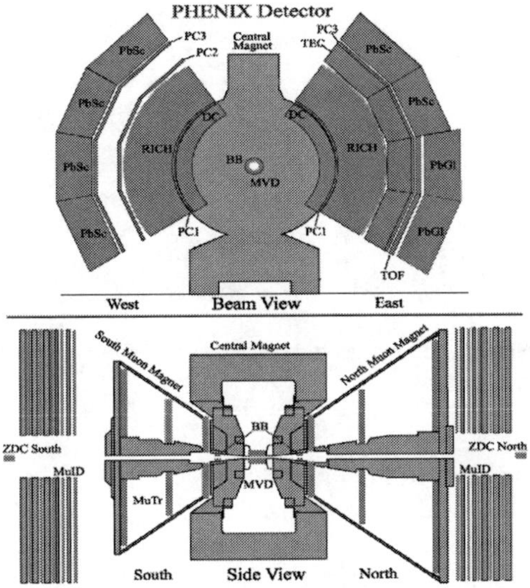

FIGURE 1. A beam view (top) and side view (bottom) of the PHENIX detector in its most recent configuration. MVD - Multiplicity Vertex Detector; BB-Beam-Beam trigger counters; DC-Drift Chambers; RICH - ring imaging Cherenkov counters; PC-Pad Chambers; TEC - Time Expansion Chambers; PbSc/PbGl – Electromagnetic Calorimeters; MuTr/MuId -Muon Tracking and Muon Identification; ZDC - Zero Degree Calorimeters. There is an axial magnetic field in the central regions; the muon magnets produce a radial magnetic field.

The forward spectrometer upgrade will introduce new detector elements in the forward direction which will result in a ten-fold increase in rapidity coverage for photons and charged hadrons. This upgrade will give greater coverage for the direct photon-jet measurements, critical to the measurement of the spin dependent structure functions, for photon decays of the charmonium states such as the $\chi C \rightarrow J/\psi+\gamma$ for the study of charmonium suppression in heavy ion collisions, for the forward production of inclusive jets, direct photons or Drell-Yan pairs[3, 4]. Below we discuss the constrains imposed by the geometry of existing experiment, present design we have chosen and PHENIX specific performance results based upon simulation data. The first proof-of-principle prototype of the future detector was built in 2005 in Russia and tested in particle beams at IHEP (Protvino). Details of the prototype construction and beam test results are subject of a separate presentation to this Conference [5].

PHENIX FORWARD CALORIMETERS (NCC)

The upgraded PHENIX Forward Spectrometer is shown diagrammatically in Fig. 2.

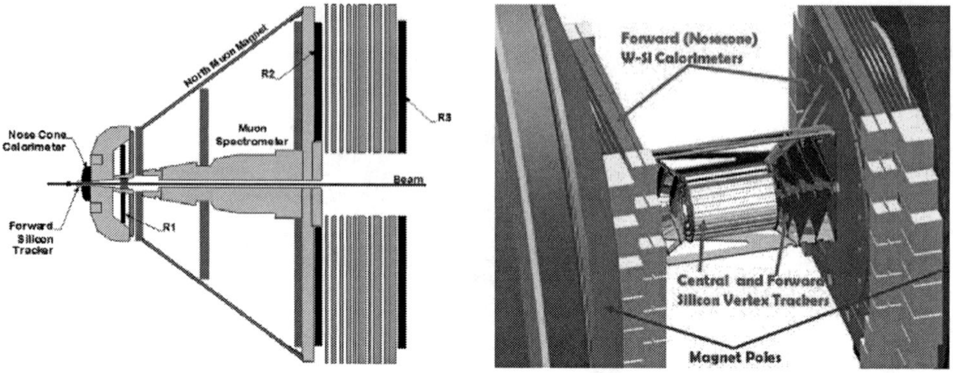

FIGURE 2 Left panel: Schematic rendering of the new PHENIX Forward Spectrometer. R1-R3 – resistive plate chambers built for triggering on muons. Right panel: 3-d rendering of the new detectors employing silicon sensors in the central PHENIX region.

The principal components to the upgrade are: 1) an enhanced muon trigger (trigger RPC chambers; 2) a forward Si vertex detector (FSVT); and 3) a compact W-Si sampling calorimeter (NCC). The NCC occupies space along the collision axis which is currently used by supplementary muon filters (brass disks 19 cm deep and ~ 1m in diameter). Two identical devices are proposed for the North and South Arms of PHENIX (see Fig. 2 right panel).

The NCC must provide or contribute to the following:

- precision measurements of individual electromagnetic showers;
- $\gamma/\pi 0$ discrimination similar to the central PHENIX electromagnetic calorimeters;
- photon/hadron discrimination;
- jet finding , jet energy and impact position measurements;
- direct lepton identification with isolation measurements;
- data for fast triggering.

Located 40 cm from the nominal collision point the NCC's are a combination of a highly segmented electromagnetic and hadronic compartments supplemented by high resolution position detectors: downstream of active converter (at 2 X0, preshower detector) to count the photons, and at a shower max position to measure decay asymmetries. The NCC discriminates between electromagnetic and hadronic showers measuring shower development over the total calorimeter depth of ~40 X0. To that effect it is longitudinal segmented into a relatively shallow EM compartment (16 X0, two depth segments) and a nearly 1 Labs deep leakage segment. Main calorimeter parameters are briefly summarized in Table 1.

TABLE 1. Major design parameters of the Nose Cone Calorimeter.		
Parameter	Value	Comment
Distance from collision vertex	40 cm	
Radial coverage	50 cm	
Geometrical depth	~19 cm	
Absorber	W + Pb	42 X0 or 1.6 Labs
Readout	Si pads (15x15 mm2) and pixilated strips (0.5x0.5 mm pixels grouped into 60 mm long strips)	
Calorimeter	EMC: 14 sampling cells 4mm W + 2.5 mm readout each. Longitudinally structured into two identical nonprojective segments. Hadronic: 6 sampling cells 4 mm W + 10 mm Pb + 2.5 mm readout each.	
Preshower detector (PS)	Stripixel layer (0.5 mm strips) with 2-d readout	
Shower max detector (SMD)	Stripixel layer (0.5 mm strips) with 2-d readout	
Expected EM energy resolution	$15\%/\sqrt{E}$ stochastic term , ~4% constant term	
Expected jet energy resolution	$100\%/\sqrt{E}$	
Two showers resolved at — in calorimeter	3 cm	
Two showers resolved at — in preshower	2 mm	
Two showers resolved at — in shower max.	4 mm	

A single calorimeter tower is sketched in Fig.3

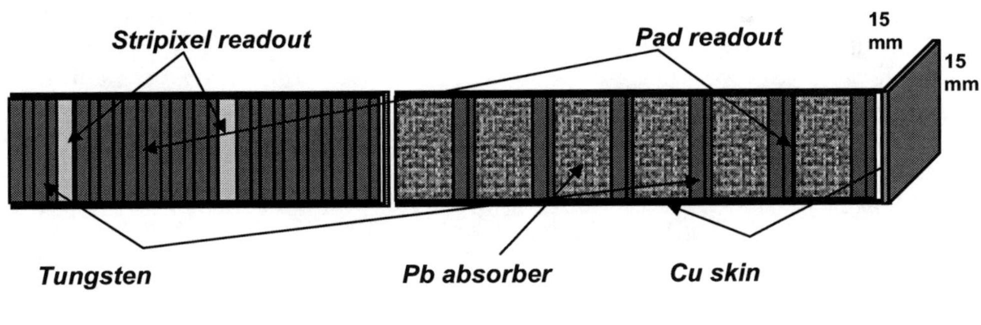

FIGURE 3. Longitudinal structure of a single calorimeter tower.

SILICON DETECTORS

The calorimeter is built of 20 pad-structured Si layers organized into towers in readout and two strip-structured Si layers (pixilated strips) used to track individual charged particles and resolve close showers. The pad size chosen for NCC is

1.5x1.5cm2, the strip size is 0.5x60 mm2, all silicon detectors in NCC are 62x62 mm2 diced of 4" wafers 525 mm thick. The estimates based upon total energy flow simulation predict the total fluence of ~100 kRad (in Si) or about $3*10^{12}$ equivalent neutrons per cm2 over the period of 10 years (assuming pp interactions at 200 GeV with luminosity $\sim 2*10^{32}$ s-1cm-2). For a pad size of the order of a few cm^2 and 500 μm sensor thickness, published data predicts the leakage current of the order of 5 – 10 mkA per pad after such an exposure. In our design of sensors and selection of raw material we were guided by the experience of D0 (FNAL) and ALICE (LHC) projects as well as published data on the research and development in the radiation hard silicon technology. We have chosen the single–sided single-metal p+ on n-bulk silicon devices. Two kinds of sensors are used to construct NCC's : pad and pixilated strip structured. The pad-structured sensors have many prototypes, this technology is well tested and we already have three established vendors ready to manufacture the 6x6 cm2 pad structured sensors using common design files (corrected for specifics of technology). The pixilated strips (stripixels) are novel detectors developed at BNL and currently selected as a base design for silicon sensor for the PHENIX central silicon tracker. The design chosen by PHENIX are interleaved stripixel detector where each pixel is divided into two parts: X-cell (or pixel) and Y-cell. X-strips and Y-strips connect X-cells and Y-cells, respectively, in a strip detector readout scheme. In this detector, two-dimensional (2D) position sensitivity is achieved with single-sided processing. Both kinds of sensors have been prototyped at ELMA (Russia) and are currently investigated.

THE PERFORMANCE OF THE NCC

The design of the NCC relies heavily on past experience in Si-W calorimetry [6,7,8] it is technologically challenging and nontraditional for PHENIX does not measure momenta of charged particles emitted forward. The NCC ability to discriminate between single and overlapping electromagnetic showers (mostly from π^0 decays) is fully defined by choice of absorber material and its granularity. In our design we rely on assumption that given the 50% containment of energy in electromagnetic showers within the radius of 0.25 $L_{Moliere}$ we should be capable of reaching a similar limit for the two-shower separation.

Shower Identification and Hadron Rejection

A shower identification procedure based upon longitudinal shower shape measurements was used to tune NCC performance. Lateral shape analysis will be added at a later stage and is expected to improve results considerably.

The momentum dependences of the average energy deposited by electrons in calorimeter segments and its variance were parameterized using simple polynomial functions. The χ^2 value computed ignoring the correlations between segments was used to compare visible energy distribution in calorimeter to the parameterized one. In Fig. 5 we show the ratio between the energy spectra of fake electromagnetic showers and initial hadrons. The points corresponding to different W thickness in EM segments are color-coded and connected by lines to guide the eyes.

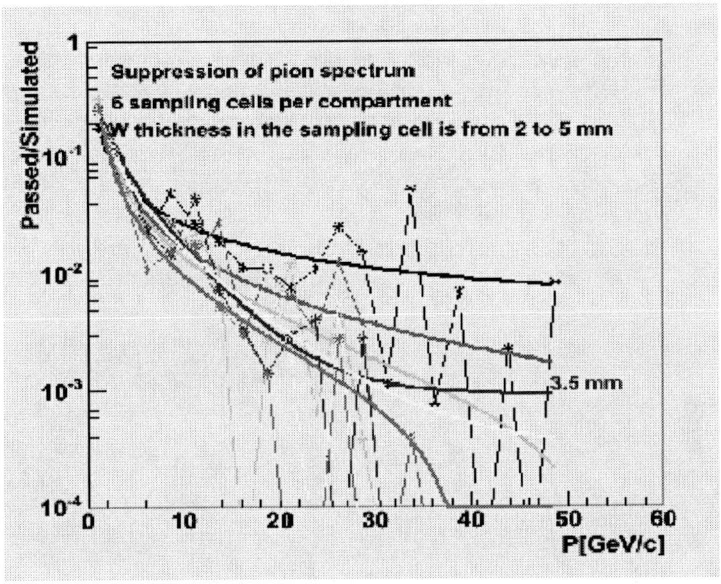

FIGURE 4. Results of the NCC hadron rejection procedure applied to the shower spectrum due to PYTHIA simulated hadronic background. Above 30 GeV the fake electromagnetic showers are ~10-3 of the inclusive hadronic cross section at a similar momenta.

The saturation effect at large momenta (results independent on EM plate thickness) is the natural consequence of the constraints imposed by cost (number of Si readout layers) and limited total depth of the calorimeter. Residual hadronic contamination to electromagnetic shower spectra will amount to ~0.1% of inclusive hadronic cross section. The latter value is to be compared to the expected direct photon yield estimated from PHENIX published data on π^0 and direct photon production in p+p interactions at 200 GeV [9]. In the pT range above 6 GeV/c it exceeds 10% of π^0 yield (~5% of the charge hadron yield). The fake electromagnetic showers produced at a level of 1% of the total hadronic yield will contribute 20% to residual direct photons and could be accounted for using simulation.

π^0 reconstruction

Given the characteristic slope of the shower core ~0.2 $R_{Moliere}$ the separation of the maxima due to two photons with similar energies should be possible down to a similar value of $\Delta \sim 0.2\ R_{Moliere}$. Reaching this limit requires high granularity lateral shower shape measurements and proper algorithms to filter out the fluctuations. The solution adopted for NCC employs two layers of identical 2-d position sensitive detectors (0.5 mm pitch) inserted into the calorimeter at two depths: ~2 X0 and ~8X0. The first (PreShower) is a hit counting layer where position resolution is limited by the strip width to ~150 μm. The second (ShowerMaximum) layer will serve to measure the decay asymmetry through shower shape analysis at a depth where the shower expands to radius ~$R_{Moliere}$.

GEANT simulated π^0's with momenta up to 30 GeV/c were used to test this approach. A simple clustering algorithm was applied to the hits seen in the longitudinal segments of the calorimeter. Clusters in calorimeter segments were combined into ``tracks" originating from collision vertex (track energy is equal to the sum of contributing cluster energies). Whenever the algorithm produced more then one track in NCC, an attempt was made to compute an effective mass. The comparison of the effective mass distributions for π^0 events and for single-electron events in multi-track configurations is shown in the left panel in Fig. 5.

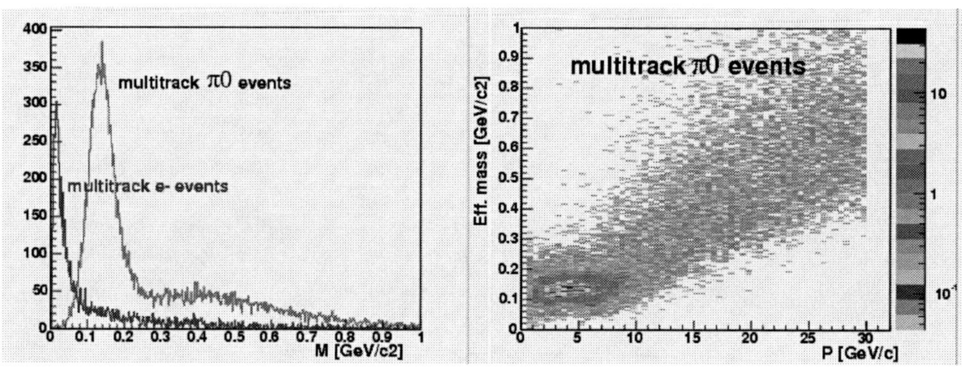

FIGURE 5. Two-shower effective mass distribution in the events with two distinct maxima found in the pattern of deposited energy in NCC.

A plot of the effective mass vs π^0 momentum in ``multitrack" events is shown in the right panel. A clear π^0 signal with a width of ~25 MeV/c2 is seen in π^0 events while the distribution for electron events is totally consistent with background. This approach works reasonably well for π^0 momenta below 7 GeV/c and fails at higher momenta where the showers begin to merge. The single ``track" events were further subjected to a reconstruction procedure relying on hits in PS and SM detectors (see example in Fig. 6).

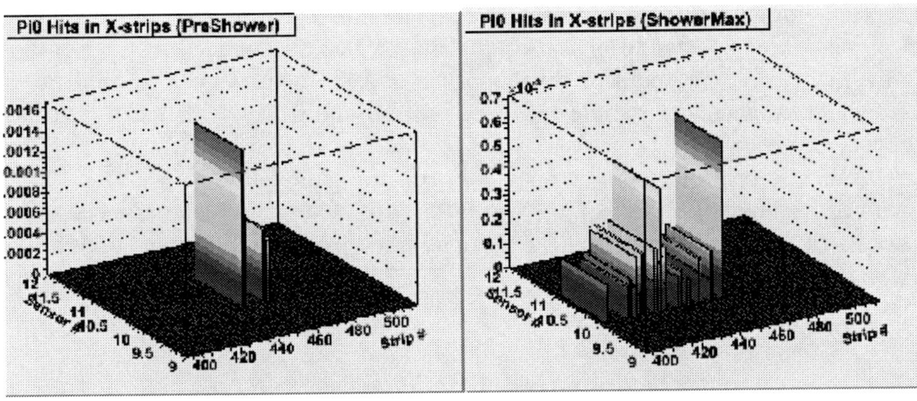

FIGURE 6. Energy patterns in the PreShower and ShowerMax detectors due to two overlapping photons from 12 GeV/c π^0 decay. (vertical axis is an energy per strip in GeV).

The reconstruction algorithm is based upon decay opening angle measurements using hits in the PS detector. The decay asymmetry is measured fitting shower profile in the SM detector with the sum of two Gaussians constrained by PS data. Once total energy, decay angle and decay asymmetry are known, the effective mass can be computed using formulas based upon decay kinematics. The comparison of the distributions of effective masses reconstructed using this procedure in electron and π^0 events is shown in Fig. 7.

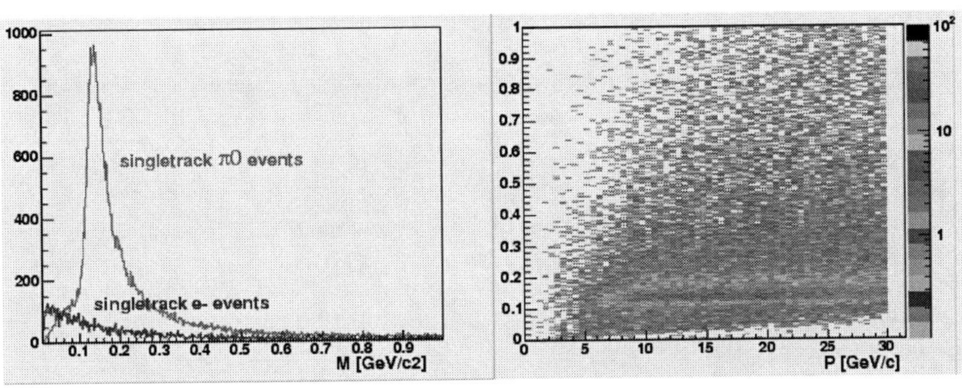

FIGURE 7. Left: reconstructed effective mass distribution in the events with a single track in calorimeter and multiple hits in PS. Right: scatterplot of reconstructed masses vs π^0 momentum in a single-track π^0 events.

The π^0 peak is present in π^0 events only, whereas the distribution for e-events is totally consistent with background. The shape of as it is reconstructed by this procedure is a subject to further improvements. The π^0 signal is characterized by σ_{sim}~20 MeV/c^2 consistent with detector energy resolution. The background under the signal reflects correlations between the decay asymmetry and the effective mass resolution intrinsic to this approach. Considerable improvements to the algorithm performance are expected when test beam data will become available.

If a mass cut of 2 standard deviations around the central peak is applied to π^0 masses reconstructed in ``single track" events only about 50% of real π^0 events are retained.

All other algorithm failures combined result in additional 10% loss of π^0's. The combined estimated efficiency of algorithm to reconstruct π^0 produced 40 cm upstream of NCC using both multi- and single-track π^0 events is shown in Fig.8.

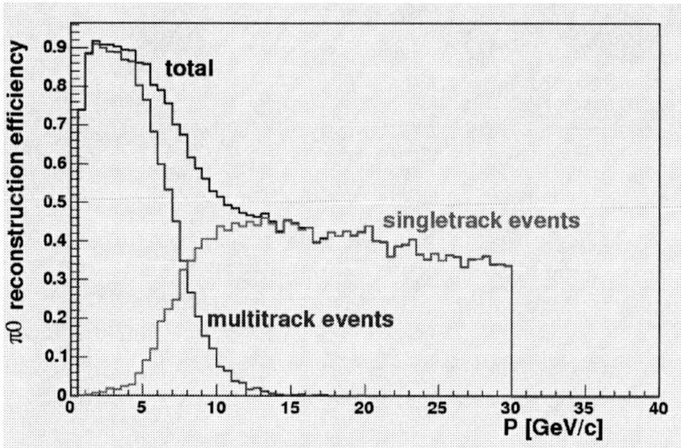

FIGURE 8. Probability for π^0 in NCC to be reconstructed within 2 standard deviations mass window.

Even at this early stage in the development the efficiency stays above 40% to 30 GeV. For comparison the π^0 reconstruction efficiency in the PHENIX central calorimeters drops to zero around 15 GeV/c.

ACKNOWLEDGMENTS

We thank our PHENIX collaborators for the support and encouragement of the proposed upgrade and fruitful discussions.

REFERENCES

[1] PHENIX Conceptual Design Report, PX20, BNL48922 internal report, 1993
[2] The Relativistic Heavy Ion Collider Project: RHIC and its Detectors, Edited by M. Harrison, T. Ludlam and S. Ozaki, NIMA499 , Issues 2-3, 235-880, 2003
[3] Seto, R., Future measurements at RHIC, Proceedings of Workshop on Nuclear Dynamics, Bahamas, January 2002, World Scientific 2002, nucl-ex/ 0204003.
[4] L. McLerran, R. Venugopalan, Phys. Rev. D49, 1994, 2233; L. McLerran, R. Venugopalan, Phys. Rev. D49, 1994, 3352; L. McLerran, R. Venugopalan, Phys. Rev. D50, 1994, 2225; see also E. Iancu, A. Leonidov, and L. McLerran hep-ph/0202270 and hep-ph/0202025 for a nice review of the saturation and RHIC physics.
[5] M.Merkin, Performance of the PHENIX NCC Prototype, talk presented to this Conference.

[6] I. Golutvin et al.. A Silicon Hadron Calorimeter module operated in a strong magnetic field with VLSI read out for LHC, CERN-DRDC-91-54, CERN-DRDC-P-34, Jan 1992
[7] J.H. Adams et al., The Silicon Matrix as a Charge Detector for the ATIC Experiment, Instrum.Exp.Tech. 44, 455-461, 2001, Prib.Tekh.Eksp. 2001 N4, 38-44, 2001
[8] V. Bonvicini et al., New concepts in silicon calorimetry for space experiments, NIM.A518:186-187,2004
[9] S.S.Adler et al, Mid-rapidity direct-photon production in p+p collisions at s**(1/2) = 200-GeV, Phys.Rev.D71: 071102,2005

Performance of the PHENIX NCC Prototype

E.O'Brien[*], V.Dzhordzhadze[†], D.Karmanov[▼], E. Kistenev[*], Z. Li[*], A.Litvinenko[▲], M. Merkin[▼], I.Nakagawa[•], V.Peresedov[▲], R. Seto[†], A.Sukhanov[*], V.Volkov[▼], A.Voronin[▼], V.Vrba[♦], L.Zolin[▲]

[*]*Brookhaven National Laboratory, Upton, NY 11973, USA,*
[†]*University of California at Riverside, CA 92521, USA,*
[▼]*D.V. Skobeltyn Institute of Nuclear Physics, M.V. Lomonosov Moscow State University, 1/2 Leninskie Gory, Moscow 119992 Russia,*
[▲]*Joint Institute for Nuclear Research, Dubna, Russia,*
[•]*RIKEN-BNL Research Center at Brookhaven National Laboratory, Upton, NY 11973, USA,*
[♦]*Institute of Physics, Academy of Sciences of the Czech Republic, Prague*

Abstract. The first prototype of NCC Si-W electromagnetic calorimeter have been built and tested at U-70 accelerator (IHEP, Protvino). Tests have been performed for 10 GeV electrons and 70 GeV protons.
This paper describes design and construction of the prototype and tests results.
Final prototype energy resolution is about 11% at 90% CL.

Keywords: Electromagnetic sample calorimeter, silicon sensors, tungsten (W)
PACS: 29.40.Vj, 29.40.Wk

I. INTRODUCTION

The Nose Cone Calorimeter (NCC) increases the calorimetric coverage of the PHENIX detector to forward and backward rapidities so that the total coverage is increased by more than an order of magnitude.

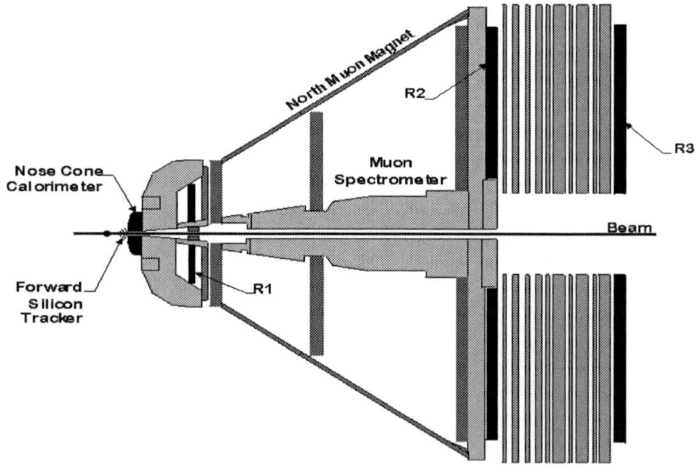

FIGURE 1. Schematic rendering of the new PHENIX Forward Spectrometer.

CP867, *XII International Conference on Calorimetry in High Energy Physics,* edited by S. R. Magill and R. Yoshida

The NCC would add substantial rapidity coverage to the PHENIX detector for the detection of photons and π_0 mesons, covering $1 < |\eta| < 3$, and increasing the overall coverage PHENIX acceptance about a factor of 10. Jet physics and energy loss studies using both photon-tagged jets as well as leading π_0 mesons will be possible with the NCC. This additional pseudorapidity coverage will allow for the study of physics observables away from central rapidity. Because the effective energy density of the created matter decreases as a function of pseudorapidity, when combined with measurements in the central arms, the NCC observables will provide a robust picture of the created matter for a variety of energy densities within a single colliding system.

This impacts the entire PHENIX physics program in nucleus-nucleus (A-A), proton or deuteron nucleus (p-A or d-A), and proton-proton (p-p).

The design requirements imposed on the detector from physics considerations are:

- the ability to measure electromagnetic showers with a precision comparable to the sum of all systematic uncertainties characteristic of similar measurements with PHENIX central electromagnetic calorimeters (π^0 effective mass resolution better then 20 MeV in the p_T range above 5 GeV/c),
- the ability to discriminate between electromagnetic and hadronic showers in the calorimeter allowing extraction of the direct photon signal down to ~5% of the π^0 yield,
- the ability to reconstruct π^0's via effective mass measurements and shower shape analysis to the p_T extent allowed by the calorimeter acceptance and RHIC luminosity,
- the ability to measure shower impact vectors with resolution sufficient for efficient matching to charged tracks reconstructed by Forward Silicon Tracking when it becomes available,
- the ability to measure the total jet energy and jet angle with precision sufficient to reconstruct kinematics of the g-q hard scattering resulting in γ-jet events with jets in the forward direction,
- ability to measure energies inside the jet cone around high p_T lepton candidates for isolation testing.

II. TEST BEAM STUDY OF THE PROTOTYPE NCC CALORIMETER

The prototype NCC was designed as a proof-of-principle device for the Collaboration to acquire the experience in silicon detector production, and to test at least one possible assembly scheme and the robustness of technology. The prototype calorimeter was designed of 22 sampling layers, with the first 16 cells in groups of 6 and 10. These were designed as the as electromagnetic segments (2.5 mm W + 2.5 mm readout). The last 6 cells (15 mm W and 2.5 mm readout) serving as a hadronic segment. The calorimeter was built to allow installation up to 4 sensors per layer (6x12 cm^2). We used one position in every layer to install prototype DC coupled sensors. Fig. 2 is the photograph of a single assembled ROU (readout unit) connected to the preamplifier board.

FIGURE 2. Single assembled readout layer for the prototype NCC connected to readout electronics.

FIGURE 3. Prototype of NCC with readout electronics.

For the test beam experiment we used preamplifier boards developed at MSU and equipped with CR1P4 16 channels preamplifier chips.

To make an optimal use of available electronics the number of geometrically matching pads from individual layers ganged together was limited to 4. Ganged and amplified signals (7 in total) were digitized using off-the-shelf multi-channel ADC unit. The assembled prototype is shown in Fig. 3. Fully assembled prototype was structured into 5 electromagnetic segments and 2 hadronic segments. The prototype calorimeter was exposed to the electron and proton beams from IHEP U70 proton synchrotron in Protvino (Russia) in November 2005. In what follows we present data accumulated exposing the calorimeter to 70 GeV/c protons and 10 GeV/c positrons.

III. DATA ANALYSIS

The amplitude distributions recorded in proton runs were used to establish values of the pedestals, the shape of the signal, minimum ionizing peak values and the signal/background ratio. The result is illustrated in Fig. 4 which shows the pedestal and MIP signal. The signal/noise ratio is 4-5 with a good MIP resolution.

In the case of the electron data the individual pedestals computed using data from proton run were subtracted to get signal amplitudes in every segment. MIP peaks varied segment-to-segment but was nearly independent of hit position in the plane. The recorded data are strongly affected by a common mode noise. Pads outside the base matrix of 3x3 towers centered at the tower with maximum amplitude were used to measure the common mode noise value which was subtracted from all 16 measured amplitudes prior to normalization. In addition the following cuts were applied to clean the data of hadronic contamination present in the electron beam:

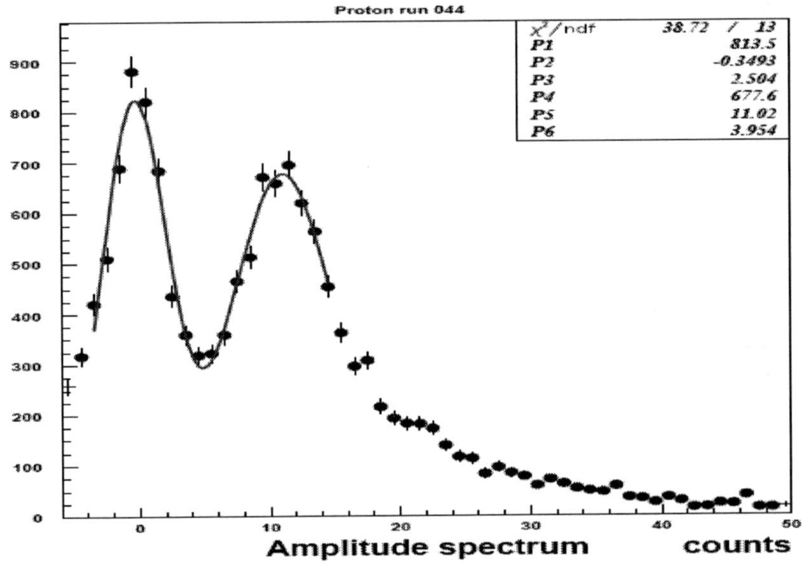

FIGURE 4. Amplitude distribution of the signal from the pad near beam impact.

• Amplitude sum in segment 1 was required to be smaller then amplitude sum in segment 2 (to exclude tails of showers started upstream of the prototype detector);

• The amplitude sum in segment 1 was required to be smaller then 600 cnts (to exclude events with multiple hits).

A typical event in the prototype calorimeter exposed to positrons of 10 GeV/c momenta cleaned of common mode noise is shown in Fig. 5.

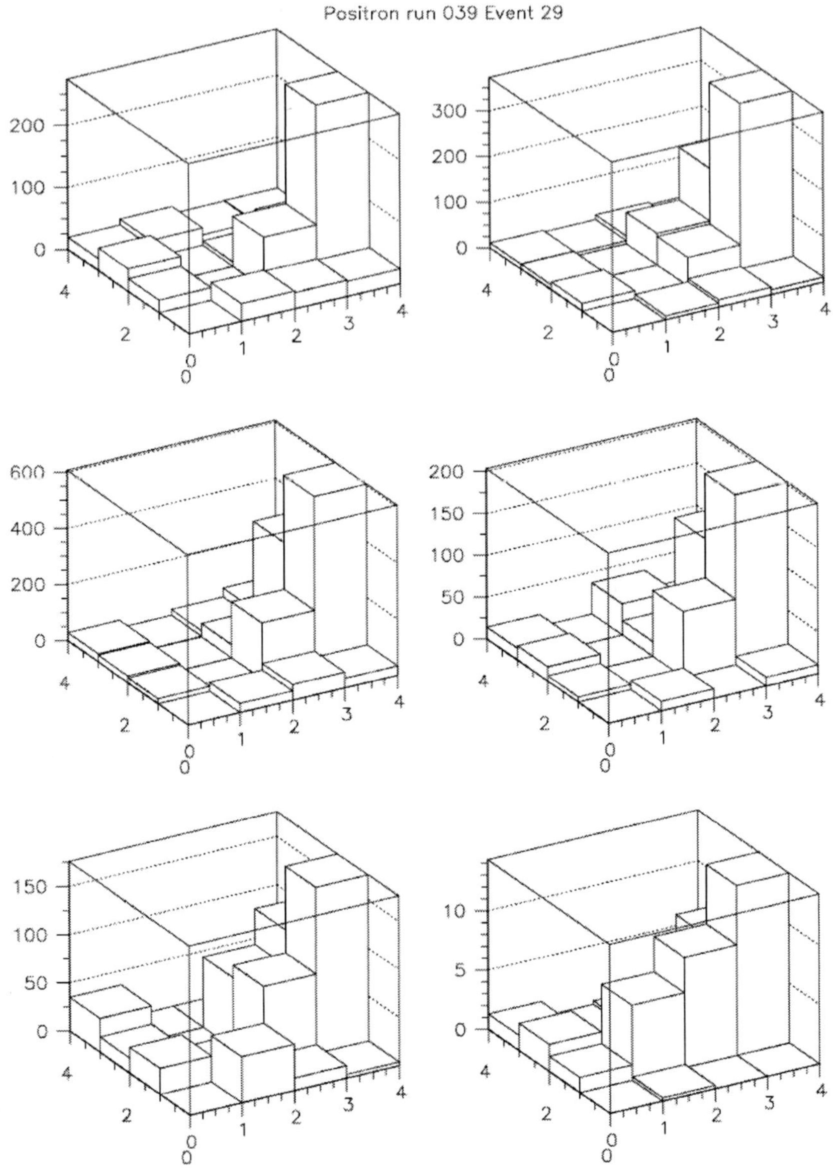

FIGURE 5. Example of the electromagnetic shower in prototype W-Si calorimeter.

Average energy in the segment when exposed to positrons is plotted as function of the calorimeter depth at the geometrical center of the segment in Fig. 6.

The observed behavior is consistent with the expected electromagnetic shower shape. The maximum in the distribution of deposited energy is reached at a depth of ~ 7 X_0 in the calorimeter.

The mean longitudinal profile was fitted with the functional form $k \times X_0^{(a-1)} \times exp^{-bt}$ which describes the data with good accuracy.

In order to improve the experimental resolution an additional weighting procedure based upon measured longitudinal shower shape was applied to positron data. The final distribution of the weighted amplitude sum with is shown in Fig. 7. The data were fitted by sum of the Gaussian peak and polynomial background resulting in measured resolution of 11% (90% CL) for the 10 GeV/c positron beam consistent with predictions in earlier chapters of this CDR i.e. ~ 18% /√(E) + 4%

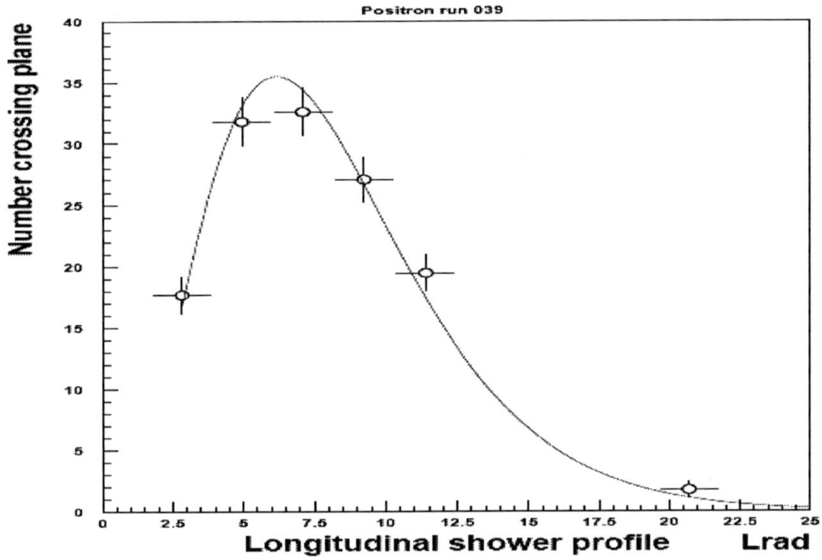

FIGURE 6. Longitudinal e+ shower profile.

The positron data recorded in every segment of the prototype were further used to study the dependence of position resolution measured in individual segments on the depth position of the segment in calorimeter. In the absence of beam tracking resolution was computed subtracting measured coordinates from those computed using straight line fit to all measured points (see Fig. 8).

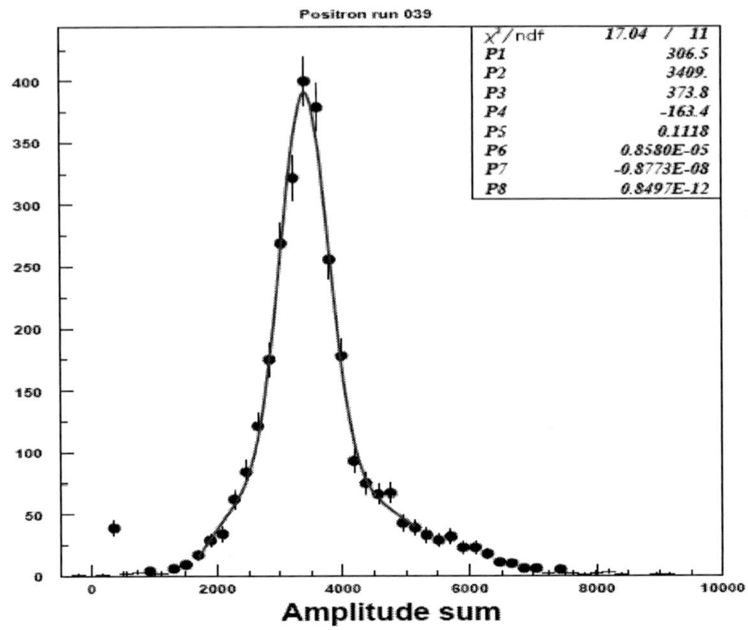

FIGURE 7. Energy resolution for 10 GeV e^+.

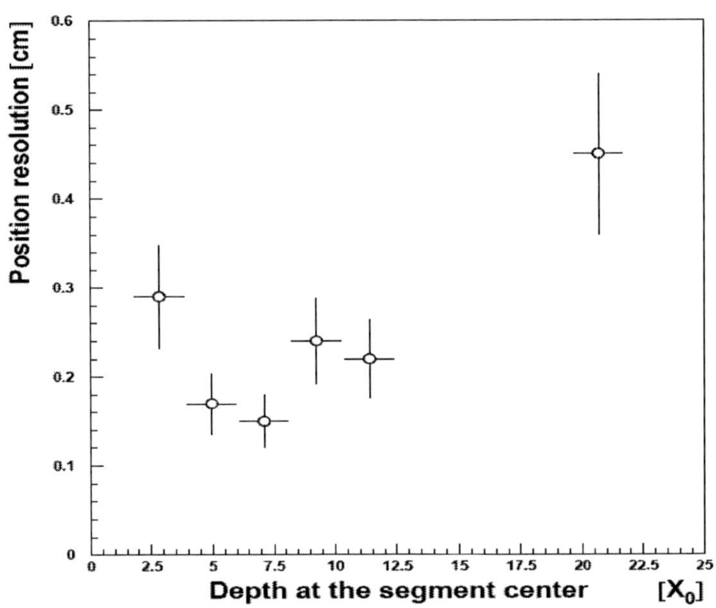

FIGURE 8. Precision of the impact point measurements in the segments of prototype detector. Location of the segment geometrical center in the prototype is plotted along X.

The Universe as Calorimeter

Scott P. Wakely

Enrico Fermi Institute, Dept. of Physics, and Kavli Institute for Cosmological Physics
University of Chicago, Chicago IL 60637, USA
E-mail: wakely@ulysses.uchicago.edu

Abstract. The maturation of very high energy astrophysics has enabled many novel physics applications. Among these is the investigation of extragalactic photon fields, which in some cases encode specific calorimetric information about such things as the formation of large-scale structure and the total high-energy luminosity of the universe. The origins of these backgrounds, and techniques for probing them will be discussed.

Keywords: Calorimetry, Gamma-rays, Infra-red, Background Radiations
PACS: 07.20.Fw, 98.70.Vc, 95.85.Pw, 95.85.Hp

I: INTRODUCTION

Applications of calorimetry are found virtually everywhere in modern high-energy astrophysics. In many cases, these applications are rather straight-forward translations of traditional high-energy physics instrumentation, adapted for deployment on balloon or space-based platforms. This includes traditional homogeneous and sampling calorimeters for both electromagnetic (EGRET[1], GLAST[2]) and hadronic (SEZ[3], SOKOL[4]) primary particles, as well as advanced "thin" calorimeters designed to maximize geometric factors for the rapidly decreasing particle flux (ATIC[5], CREAM[6]). Other calorimetric applications utilize the atmosphere itself as the absorber volume, with various ground-level observables (e.g., air-shower electron[7,8], Cerenkov[9] or fluorescence light densities[10, 8]) being used to infer the properties of the primary particle.

This talk, however, will step back from these specific pursuits – which will be well-covered in other talks - and take a larger viewpoint on calorimetry in astrophysics and cosmology. Specifically, it will address ways in which we might consider the universe itself as a calorimeter. "Universe" in this case can be considered a (somewhat fanciful) shorthand term referring to both matter (specifically dusts and metals) in the universe, as well as to the distribution of universal background photon fields, such as the cosmic microwave background (CMB), and the cosmic infrared background (CIB).

CP867, *XII International Conference on Calorimetry in High Energy Physics,*
edited by S. R. Magill and R. Yoshida

II: VHE GAMMA RAYS AND AGN AS COSMIC BEAM-LINES

Over the past 20 years or so, VHE astrophysics has undergone a period of tremendous growth. At TeV energies, the imaging atmospheric Cerenkov technique has truly come of age, now boasting a catalog of high-confidence sources surpassing some 30 objects, and with a source variety and physics potential that continues to grow as well[11]. In the GeV regime, the EGRET instrument has populated the sky with over 270 sources, and motivated the construction of a follow-up instrument, GLAST, which will further increase this count, perhaps to over 5000 objects. Some of the most intriguing sources seen at both of these wavelengths belong to the class of extragalactic objects known as active galactic nuclei (AGN). The standard radio-loud AGN model includes a central supermassive ($\sim 10^8$ – 10^{10} solar-mass) black hole surrounded by an accretion disk, with relativistic jets emanating along the spin axis of the black hole[12]. Within this jet, particles are accelerated to high energies, presumably through shock acceleration. When this jet is aligned to the observer, broad-band photon emission extending into the TeV range can be observed, along with rapid variability in the overall flux. AGN aligned in this way are often called "blazars".

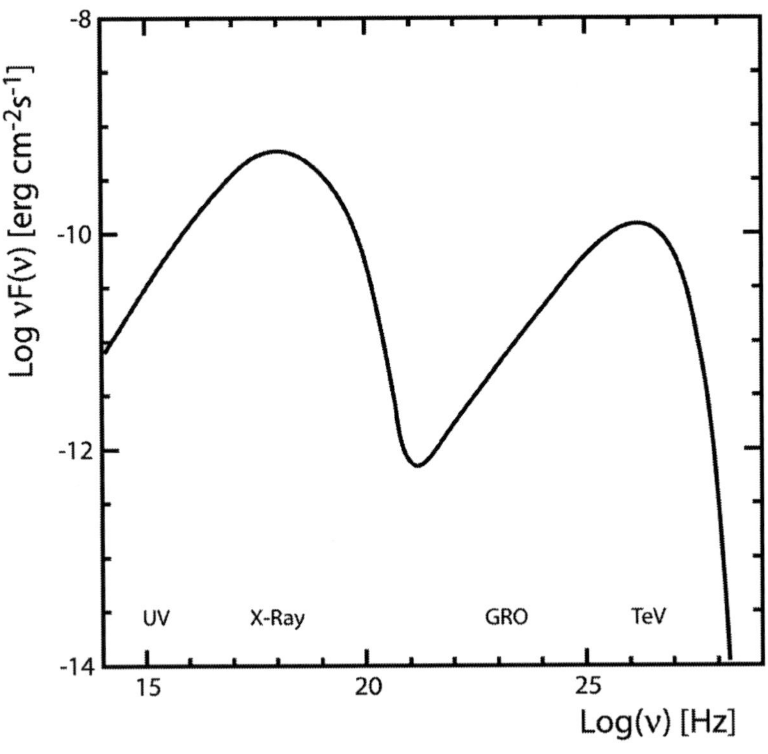

FIGURE 1. Photon spectrum from a typical SSC blazar model. See text for details[13].

The photon spectrum of blazars extends from the radio to the TeV, with a characteristic two-humped appearance, as shown in Figure 1. In the generally favored Synchrotron Self-Compton model, the lower-energy hump is produced by synchrotron radiation from energetic electrons produced in the jet. The higher-energy hump is due to the inverse Compton up-scattering of those synchrotron photons by the same population of energetic electrons. As shown in Figure 1, at TeV energies, the form of the flux spectrum approaches a power-law, typically with a differential spectral index of around -2.5.

At TeV energies, blazars have been observed to redshifts at least as large as z~0.186. Coupled with relatively simple high-energy source spectra (the high-energy side of the inverse Compton peak approaches a simple power law), these large distances make blazars and their observed spectra useful probes of low-probability or long-baseline interaction mechanisms, such as Lorentz Invariance Violation[14]. One such mechanism is the interaction of high energy TeV primaries with low-energy background photons, a process which can attenuate the blazar flux at high energies.

III: PAIR ATTENUATION

Soon after the cosmic microwave background radiation was discovered, it was realized[15] that due to interactions with this background field, high energy photons would have a limited range in the universe. The relevant interaction process is called Breit-Wheeler pair production[16] and it differs from the more widely known Bethe-Heitler process - which occurs in the vicinity of atomic nuclei. With Breit-Wheeler pair production, two real photons interact to form an electron/positron pair.

As a result of this process, the spectrum of photons observed from distant AGN differs from the source spectrum of the photons by the addition of an exponential attenuation term, which varies with the optical depth ,$\tau_{\gamma\gamma}$, as shown in Eq (1.1).

(1.1) $$J_{obs}(E) = J_{src}(E)e^{-\tau_{\gamma\gamma}(E,z)}$$

Here J_{obs} is the observed gamma-ray energy spectrum, J_{src} is the source spectrum, E is the primary photon energy and z is the redshift of the source. The optical depth, $\tau_{\gamma\gamma}$, can be shown to vary as seen in Eq. (1.2):

(1.2) $$\tau_{\gamma\gamma}(E) \propto \int_{\varepsilon_{thr}}^{\infty} n(\varepsilon)\,\sigma_{\gamma\gamma}(E,\varepsilon)d\varepsilon$$

Here E is the energy of the primary photon, ε is the energy of the background photon, ε_{thr} is the background photon energy at the interaction threshold, $\sigma_{\gamma\gamma}$ is the Breit-Wheeler cross-section, and $n(\varepsilon)$ is the spectral density of the background photons.

The cross-section $\sigma_{\gamma\gamma}$ is very narrowly peaked around $s \sim 2E\varepsilon \sim 8m_e^2$, close to the CM threshold energy of the interaction. As a result, the process effectively "couples" primary photons at energies of ~1 TeV to background photons of around ~1 eV. The result of this coupling is that the optical depth, and hence the severity and location of the cutoff in the AGN spectrum, will encode information about the density and distribution of the low-energy background photons. This turns out to be especially useful because the 1eV background photons probed by VHE gamma-rays lie in the ~1μm range, which places them in one of the most cosmologically interesting extragalactic background radiation fields, the cosmic infrared background.

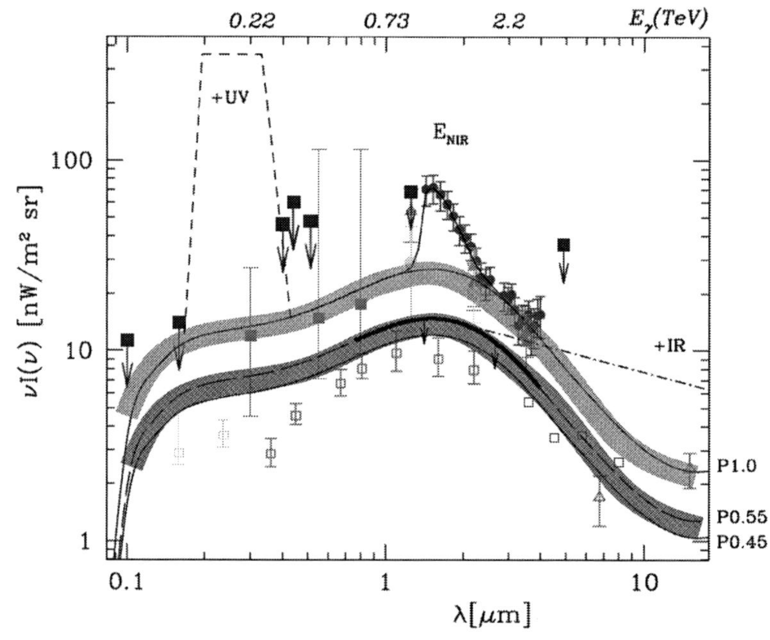

FIGURE 2. Spectral energy distribution of the cosmic infrared background field; taken from [17] - see that work for complete description and full references. The numbers on the upper axis correspond to the energy of the primary photons "coupled" to the background photons. Open symbols correspond to the integrated light from galaxy counts, and thus must be considered lower limits for the CIB: in the UV–O range, from Hubble data (green, red); in the NIR, from Spitzer (blue) and ISO data. The upper limits (purple) are 2σ estimates. Direct measurements are shown as filled symbols: IRTS data from the NIR spectrometer (blue), and data from COBE/DIRBE (green, magenta and red triangles). Red squares correspond to tentative detections in the optical. The curves show the EBL shapes used to reconstruct the intrinsic spectra. P1.0 (thick red) gives 26, 23 and 14 $nWm^{-2}\ sr^{-1}$ at 1.25, 2.2 and 3.5μm, respectively. The heavy black line shows the range most effectively constrained by the HESS data. In the long-dashed regions, higher fluxes than P0.55 (thick blue) would not be in conflict, as long as the flux in the 1–3μm range is within or around the limit.

IV: CALORIMETRY OF THE COSMIC INFRARED BACKGROUND

The cosmic infrared background field represents the second highest energy density (the CMB is the highest) in the so-called "grand unified photon spectrum" of cosmic background fields. It is produced by red-shifted stellar emission and thermally-reprocessed radiation from the early universe, and hence it encodes much information about the formation and evolution of pre-galactic stars, galaxies, quasars, black-holes, decaying relic particles, etc. See [18] for a comprehensive review.

In this sense, the CIB is a calorimetric record of many cosmologically important processes which have occurred since the time of matter/radiation decoupling. In a nutshell, any luminosities which could contribute to the heating of ambient matter in the early universe must have an imprint in the CIB. Hence, measurements of this field can provide important information about a period in the evolution of the universe not easily addressed with other observables. Unfortunately, the CIB is notoriously difficult to measure directly because of the very large foregrounds associated with galactic emission and especially zodiacal photon fields. As a result, flux estimates around $\sim 1\,\mu m$ tend to be uncertain and limited by systematics.

It was realized nearly 15 years ago[19] that TeV gamma-rays could provide a valuable service in this regime. The process is a simple one: find distant AGN and measure the cutoffs in their spectra. Make reasonable assumptions about the shape of the source spectra, and infer - using the interaction cross-section - the spectral energy density of background photons needed to generate the observed flux. In this way, limits can be obtained which are free from the systematic issues resulting from large foreground subtraction[*]. Furthermore, with an ensemble of AGN at different distances, the redshift-dependence of the CIB can possibly be mapped out.

The most spectacular application of this method has been performed by the HESS Collaboration, who recently reported results[20] on two newly-discovered blazar sources, 1ES1101-232, and H2356-309. The measured cutoffs in the spectra of these sources, located at redshifts of z=0.186 and z=0.165, respectively, allowed HESS to place limits on the flux of CIB photons, especially in the range of 0.8-4μm. Figure 2 shows the results of this work. The central result of this work is that, due to the unexpectedly hard (i.e., under-attenuated) spectra, the inferred CIB density is much lower than previously assumed. This invalidates some previous measurements and places the CIB very close to the lower limits provided by the integrated light from resolved galaxy counts. This greatly constrains the contributions to the CIB from additional sources, and may impact, for example, some theories of massive metal-free Population III stars.

---

[*] To be fair, however, this method has its own systematic issues which it must face - most notably, the lack of definitive knowledge of the source production mechanism and the subsequent energy spectrum. This situation may be ameliorated in the future with the additional information which could be provided by GLAST observations.

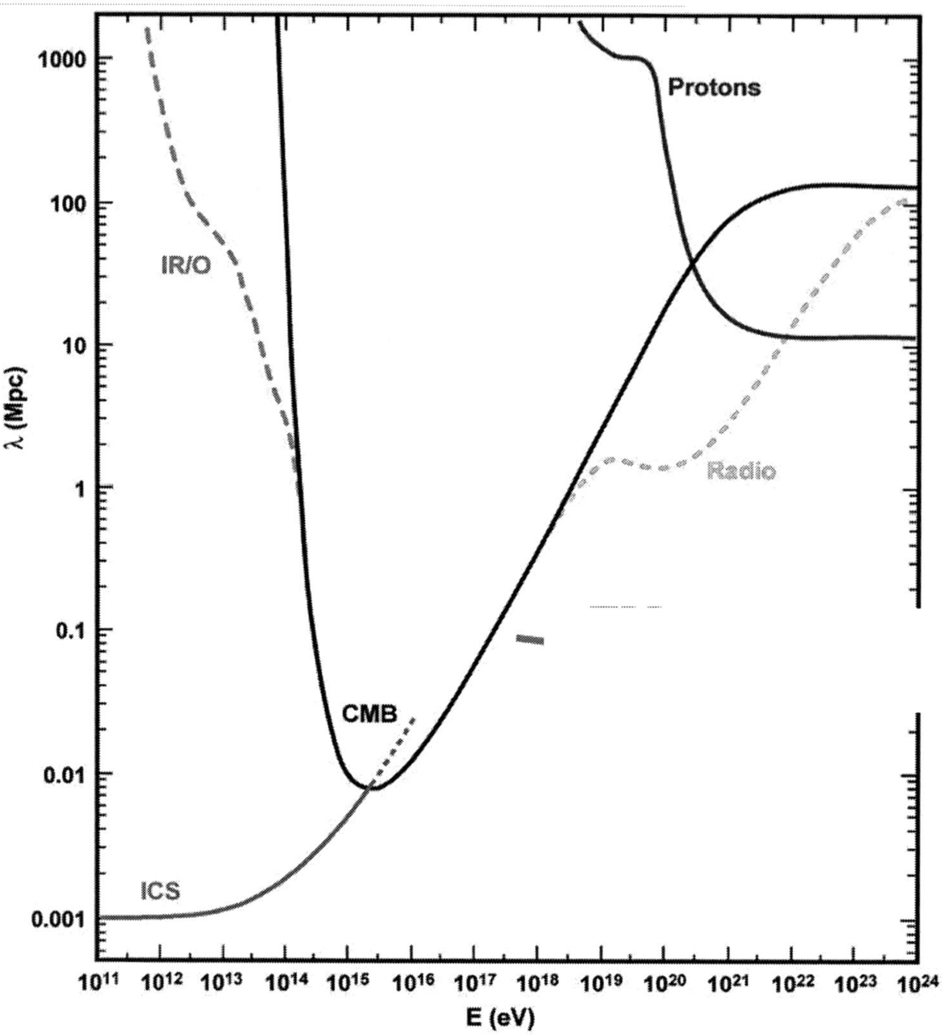

FIGURE 3. Particle interaction lengths versus energy. – after [21]. The lines labeled "IR/O", "CMB", and "Radio" show the interaction length for primary photons interacting with those background fields through simple pair production. The line labeled "Protons" shows the interaction length for primary protons on the CMB via GZK mechanisms. The line labeled "ICS" shows the interaction length of primary electrons due to inverse Compton scattering. Synchrotron losses for electrons will also be important in most astrophysical environments and even in the complete absence of magnetic fields, triplet pair production (i.e., $e + \gamma \rightarrow 3e$) will limit electron ranges at energies above $\sim 10^{17}$ eV.

V: PARTICLE HORIZONS AND GAMMA-RAY BACKGROUNDS

One of the obvious consequences of the pair-production process is the existence of a gamma-ray horizon. That is, there should be a distance beyond which gamma-rays cannot propagate, because of their destructive interactions with universal background fields.

The distance to this horizon depends on the energy of the primary, but in most cases it is small, by cosmological standards - as shown in Figure 3. In this figure, the CIB is labeled "IR/O", and shows that TeV photons tend to interact within a few hundred megaparsecs or so. At PeV energies, the huge density of CMB photons produces a very local cutoff at tens of kiloparsecs, while radio backgrounds limit large-scale propagation beyond 10^{18} eV. The existence of horizons is not limited to gamma-rays, however. Also shown in the figure are interaction lengths for other particles. Protons, heavier nuclei, and electrons are all attenuated through their own processes - resulting in a kind of generic opacity for high-energy particles in the universe. One obvious implication of this is that the very-high or ultra-high-energy universe must be more or less "local".

Another implication, however, concerns the actual by-products of the attenuation processes. For example, the photon energy absorbed through pair production doesn't simply disappear. It must be re-appear somewhere. In fact, what happens is that this energy is channeled into longer wavelength photons, through the process of pair cascading. A pair cascade is be generated, for example, when the daughter electrons produced in a Breit-Wheeler interaction inverse Compton up-scatter CMB photons. As shown in Figure 3, this up-scattering occurs on very short distance scales, and acts to cool the high-energy electrons relatively quickly. Some of the up-scattered photons undergo additional pair-production interactions, producing more electrons. The result is the reprocessing of the initial particle into a large flux of photons at lower energies. Figure 4 shows a typical daughter-particle energy spectrum from a single PeV photon propagating through 10 Mpc of open space, containing only CMB and CIB photons. The large blue histogram shows daughter photons produced by the smaller population of daughter electrons, shown in red.

In many ways, this pair-cascade reprocessing is identical to the thermal reprocessing which helps to generate the cosmic infrared background. Instead of thermal photons, however, sub-TeV photons are produced - to which the universe is largely transparent (see Figure 2). Hence, there should be a universal gamma-ray background, which encodes calorimetric information about the very high energy luminosity of the universe[22,23]. This includes such things as ultra-high-energy cosmic rays, high-energy AGN emissions, decays of exotic particles, etc. Also folded into this is the evolution of the CMB and CIB backgrounds, which provide the target photons for producing the gamma-ray backgrounds. Accurately measuring this isotropic extragalactic sub-TeV flux will be difficult with ground-based arrays, but may be possible with the next-generation space-based instrument, GLAST[24].

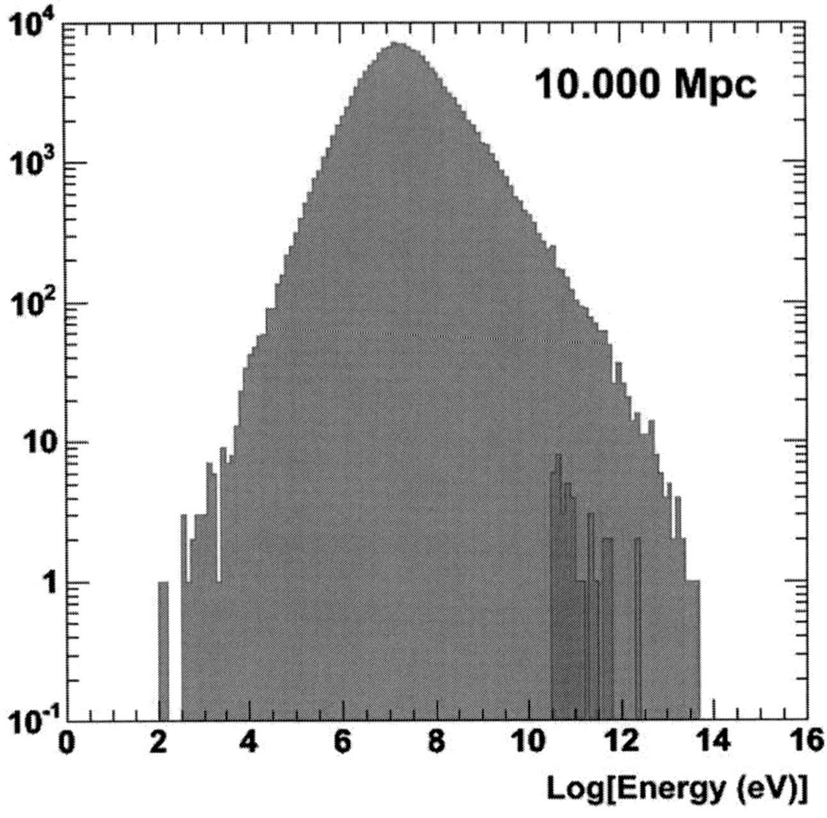

FIGURE 4. Energy histogram of daughter particles from a single 1 PeV gamma-ray propagating through 10 Mpc of space, populated only with CMB and CIB photons. The larger (blue) histogram shows daughter photons and the smaller (red) histogram shows daughter electrons. The large flux of photons is produced by the inverse Compton up-scattering of CMB photons by the electrons.

VI: SUMMARY

The advent of the imaging atmospheric Cerenkov technique has revolutionized the field of TeV gamma-ray astrophysics and enabled a large variety of new physics topics to be addressed. One such topic involves using TeV gamma-rays from active galactic nuclei as probes of the cosmic infra-red background. This is possible due to pair-production interactions which attenuate the gamma-ray flux, generating signatures which can be used to infer the spectral energy density of the CIB. This is important because the CIB, which is difficult to measure directly through traditional methods, contains photons thermally reprocessed from early-universe emission mechanisms, and hence encodes calorimetric information about the formation and evolution of galaxies, black holes, stars, etc.

A consequence of the CIB and CMB photon fields is that the universe is not transparent to very high energy photons (and many other particles). However, this opacity provides another opportunity to extract calorimetric information from the universe, as these background fields act as absorbers for the VHE photons, reprocessing it via pair cascades to lower energies, while increasing their overall flux. Similar to the CIB case, this flux of photons contains information about the high-energy luminosity of the universe, and could be measured with a future space-based gamma-ray mission such as GLAST.

REFERENCES

1 http://cossc.gsfc.nasa.gov/docs/cgro/egret/
2 http://www-glast.stanford.edu/
3 N. Grigorov, Rapoport I., Nesterov V. et al. Izv. AN SSSR, ser. Phys. 33, 1469 (1969).
4 I. Ivanenko, Shestoperov V., Rapoport I. Proc. 23th ICRC, Calgary (1993) 2, 17.
5 http://atic.phys.lsu.edu/aticweb/
6 http://cosmicray.umd.edu/cream/cream.html
7 http://www-akeno.icrr.u-tokyo.ac.jp/AGASA/
8 http://www.auger.org/
9 http://veritas.sao.arizona.edu/
10 http://www.cosmic-ray.org/
11 Ong, R. A.\ ,ArXiv Astrophysics e-prints, arXiv:astro-ph/0605191 2006
12 Urry and Padovani, PASP 107, 715 (1995)
13 Konopelko, A., et al., Astrophysical Journal 597, p851 (2003)
14 Biller, S.D., et al., Physical Review Letters 83, p2108 (1999).
15 Gould & Schreder Phys Rev Lett 16 (1966), J. Jelley, Phys Rev Lett 16 (1966)
16 Breit, G and Wheeler, J., Phys Rev 46 (1934)
17 Aharonian, F. et al., Nature 440, 1018-1021 (2006)
18 Hauser, M.G. and Dwek, E., Annual Review of Astronomy and Astrophysics 39, p249 (2001)
19 Stecker, F. W., De Jager, O. C. & Salamon, M. H, Astrophys. J. Lett. 390, 49 (1992).
20 Aharonian, F. et al., Nature 440, 1018-1021 (2006)
21 Coppi, P. and Aharonian, F., ApJ Lett 487 (1997)
22 Coppi, P. and Aharonian, F., ApJ Lett 487 (1997)
23 Halzen, F. et al., Phys. Rev. D41 (1990)
24 http://www-glast.stanford.edu/

SESSION III

ASTROPARTICLE AND NEUTRINO CALORIMETRY

Conveners: Karen Byrum and Bob Wagner

The Status Of GLAST CsI Calorimeter

A.Chekhtman
on behalf of GLAST collaboration

George Mason University/ Naval Research Laboratory
4555 Overlook Ave, SW, Washington, DC 20375

Abstract. The GLAST Large Area Telescope (LAT) CsI calorimeter, which consists of 16 flight modules and 2 spare modules, was assembled in 2004-2005 by an international team from the USA, France and Sweden. Each module contains 96 CsI crystals supported by a carbon fiber composite structure and read out from both ends with silicon PIN photodiodes. Signals from the array of photodiodes are processed by custom analog ASICs and commercial ADCs. After assembly, each module underwent a full environmental test program including electromagnetic interference and compatibility, vibration, and thermal-vacuum test to flight-acceptance levels. The functional performance of each module was verified before and after each test, and calibration with cosmic muons and charge injection was performed throughout the test sequence. All 18 modules showed stable functioning over the few months of the assembly and test program. None of the 1728 crystals experienced mechanical or optical degradation. Integration of the calorimeter modules with the other detector and electronics subsystems into the complete Large Area Telescope began at SLAC in April 2005 and was completed in December 2005. The environmental testing of full LAT instrument is planned for summer 2006. Simultaneously the beam test of calibration unit made of spare modules is planned to be done at PS and SPS beam lines at CERN.

Keywords: crystal calorimetry, gamma-ray telecope, instrumentation for gamma astronomy.
PACS: 95.55.Ka, 95.85.Pw, 29.40.Vj

INTRODUCTION

GLAST [1] is the next generation space based gamma ray telescope in the photon energy range 30 MeV - 300GeV. It consists of a spacecraft and two instruments: The Gamma Ray Burst Monitor (GBM) and the Large Area Telescope (LAT).

The LAT (Fig 1) contains three main detector subsystems: a tracker (TKR) with active silicon strip detector layers and thin passive tungsten radiators, a hodoscopic CsI crystal calorimeter (CAL), and a plastic anticoincidence detector system (ACD). The TKR is responsible for determining incident photon direction, and also plays a role in energy determination for incident photons with energies less than 1 GeV. The CAL supplies incident photon energy. In addition, its hodoscopic configuration allows it to determine photon direction, albeit with poorer resolution than the TKR.

CP867, *XII International Conference on Calorimetry in High Energy Physics*,
edited by S. R. Magill and R. Yoshida

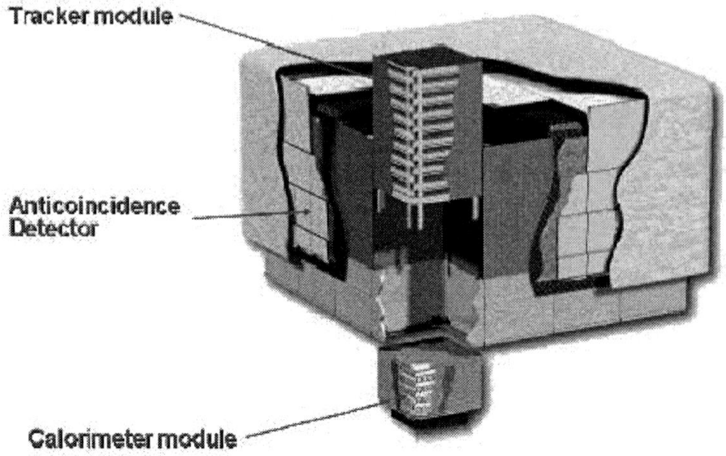

FIGURE 1. View of the LAT science instrument with one Tracker and one Calorimeter module pulled away from the grid.

CALORIMETER DESIGN

The Calorimeter [2] is built as a 4x4 array of identical modules. Each module contains 96 CsI crystals with dimensions 19.9x26.7x326 mm3 arranged in 8 horizontal layers of 12 crystals, as shown on Fig. 2.

FIGURE 2. Calorimeter module after insertion of crystals without electronics boards installed.

Crystals in even layers (0,2,4,6; also known as X layers) are orthogonal to crystals in odd layers (also known as Y layers), forming a hodoscopic arrangement supported by a carbon fiber composite structure.

Scintillation light is detected at each end of a crystal by two silicon PIN photodiodes (Figs. 3 & 4).

FIGURE 3. Dual PIN photodiode.

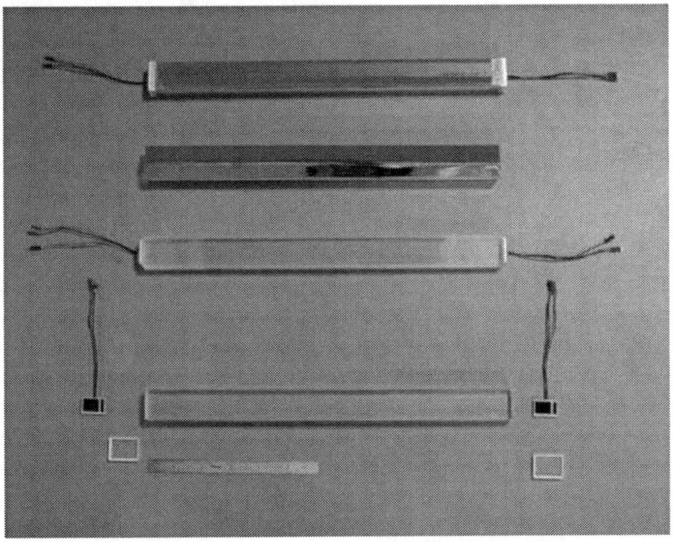

FIGURE 4. Crystal detector element and its components.

The large diode (1.5 cm^2) is used to measure energy deposits from ~2 MeV to ~1.1 GeV, while the small diode (0.25 cm^2) covers energy deposits up to 70 GeV. The PIN diodes are mounted in a ceramic carrier (Fig. 3) and bonded with an optical bond to either end of a CsI crystal, which is then given a light-tight, reflective wrap (process shown in Fig. 4). The resulting unit is referred to as a Crystal Detector Element (CDE).

Two of four side surfaces of each crystal are roughened to provide known attenuation of light collection along the crystal. This allows measurement of the longitudinal coordinate using the relative signal ("light asymmetry") from the two ends of a crystal.

CDEs are inserted into the carbon fiber structure using a precision insertion tool (Fig. 5).

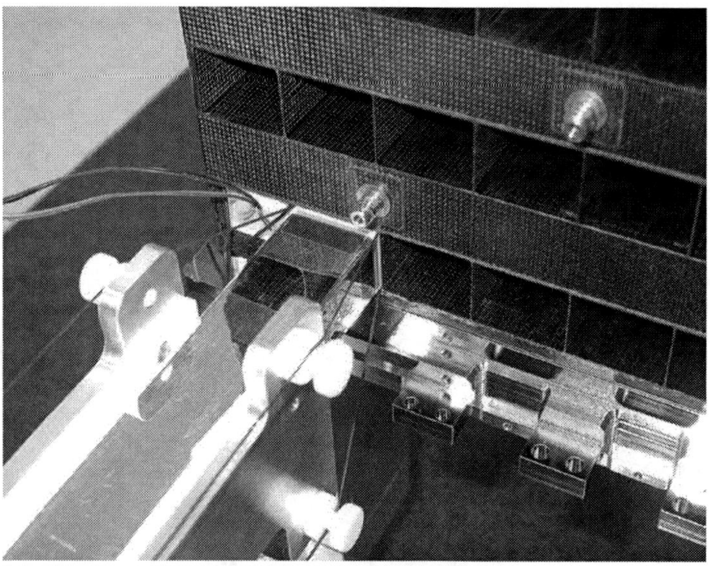

FIGURE 5. Crystal insertion into carbon fiber structure.

Longitudinal elastic bands keep the CDEs in place the their cells. Fig. 2 shows a module with all CDEs inserted.

The front-end electronics boards are mounted over each of the four vertical faces of the CAL module. The PIN diode leads are passed through feedthroughs in the boards and soldered in place, after which closeout plates are attached to complete the module.

The signal from each photodiode is processed by an individual electronics chain (Fig. 6) including preamplifier, shaper with ~3.5 μs shaping time, and two track-and-hold stages to produce nominal "x1" and "x8" output signals to cover the large dynamic range (the measured signal ratio between these two outputs, ~1:9, is somewhat different from the nominal value). In total, each crystal end produces 4 output signals ("energy ranges"; 2 for each diode) with ratio 1:9:60:540, digitized by a single 12-bit ADC. Programmable "range selection logic" selects the lowest unsaturated energy range for readout. Four range readout is available for calibration purposes.

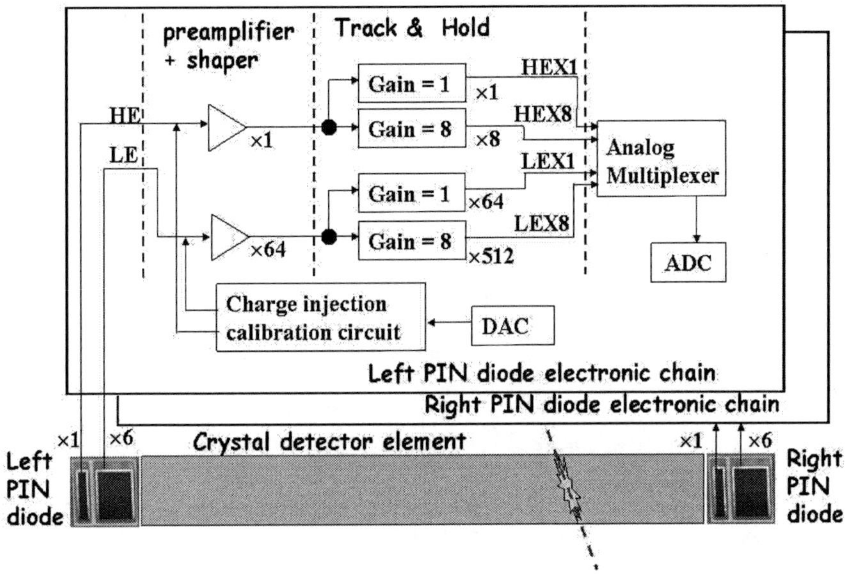

FIGURE 6. Front-end electronics chain of one crystal.

To decrease the data volume it is possible to suppress the transfer of the digitized data for crystal ends with signals below the Log Accept (LAC) threshold which can be set individually for each crystal end. In flight the LAC thresholds will be ~1.5 MeV.

The electronic chain of each photodiode also includes a trigger circuit, consisting of a fast shaper with ~0.3 μs shaping time and a discriminator (FLE for low energy diode and FHE for high energy diode). The FLE and FHE discriminator thresholds can be set individually for each crystal end. The logical OR of all FLE and of all FHE outputs form the CAL_LO and CAL_HI triggers respectively. In flight the FLE discriminator threshold will be set to ~100 MeV and FHE discriminator threshold to ~1 GeV per crystal. To trigger from cosmic muons during ground calibration, the FLE discriminator thresholds were set to ~10-15 MeV.

The front-end electronics boards are controlled and read out by the Tower Electronic Module (TEM), installed below the calorimeter module. The TEM provides collection of trigger signals and formation of the trigger decision; commands to the front-end boards to digitize the signals; buffering of the digitized data; optional collection of the diagnostic information on the status of LAC, FLE, and FHE discriminators on the front-end board; and data transfer to the online Ground Support Equipment (GSE) processor or LAT central processor.

CALORIMETER PRODUCTION RESPONSIBILITIES

Eighteen calorimeter modules were assembled in 2004-2005. This number includes 16 modules which have been integrated into the LAT instrument, and two additional modules that function as flight-qualified spares and/or post-launch reference units plus a third identical spare module not flight-tested.

AMCRYS (Kharkov, Ukraine) manufactured 1850 CsI crystals for the GLAST calorimeter (1728 required for 18 modules plus some spares) and sent them to Kalmar University, Sweden, for acceptance testing. The custom PIN diode assemblies (Fig. 3), containing two PIN diodes on one ceramic carrier, were produced by Hamamatsu (Japan). The crystals were assembled into detector elements (CDEs, see Fig. 4) at Swales Aerospace (USA).

The carbon fiber supporting structures (Fig. 5) were produced by LLR (Paris, France).

The production of the front-end electronics boards and the final assembly of the calorimeter modules (Fig. 7) were done at Naval Research Laboratory.

FIGURE 7. Fully assembled calorimeter module.

ENVIRONMENTAL TESTING

After production each module underwent a full environmental test program: vibration, electromagnetic interference (EMI) and compatibility (EMC), and thermal-vacuum. The functional performance of each Module was verified before and after each test, and calibration with cosmic muons and charge injection was performed throughout the test sequence. Then modules were delivered to SLAC for integration together with other detector and electronics subsystems into the complete LAT device. The assembled LAT was tested at SLAC using cosmic muons, charge injection and low energy photons and in April 2006 was delivered to NRL for thermal-vacuum test. After LAT-level environmental tests, the LAT will be delivered to General Dynamics for the integration with the GLAST spacecraft.

CALORIMETER PERFORMANCE MONITORING

A variety of calorimeter performance figures have been monitored over the course of environmental testing. For the most part, performance has remained stable, although small changes have been observed. For example, the energy calibration has changed by a small amount in some cases.

Fig. 8 shows the comparison of the energy scale (in ADC units per MeV) for all 96 crystals of one module, measured right after the module was assembled (horizontal axis) and 2 months later after the end of environmental testing (vertical axis). Each point corresponds to one crystal. In the ideal case of no change in energy scale, all points should be along the blue line. The spread along the line corresponds to ~10% crystal to crystal non-uniformity within a module. However, the distribution demonstrates ~1.5% signal decrease after environmental testing.

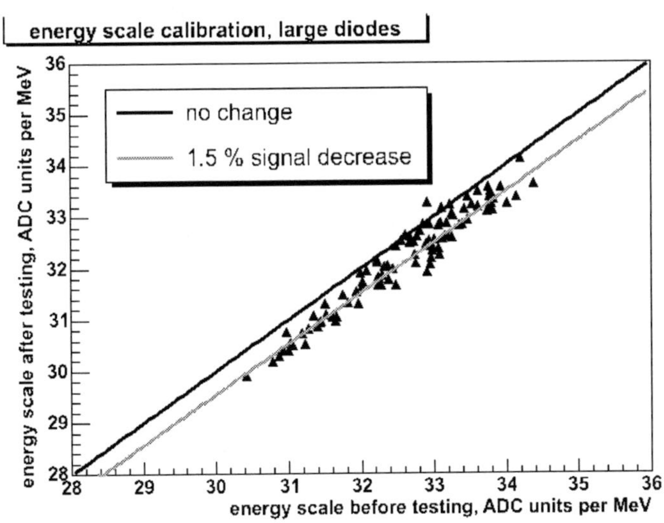

FIGURE 8. Energy calibration stability for all crystals of one calorimeter module.

The change is small relative to calorimeter energy resolution (~6%) and will be taken care of by future calibrations. This change in energy scale is more than likely caused by small drifts in the electronics chain due to differences in measurement setup and/or environment.

CONCLUSION

The assembly and test phase for the GLAST/LAT calorimeter modules resulted in delivery of 16 flight modules, 3 spare modules (of which 2 underwent full testing) and one engineering module to SLAC for integration with the LAT and further testing. These modules were extensively monitored during and after the assembly and test period. None of them showed evidence of significant degradation. Integration at SLAC was completed in December 2005. After further testing with flight software, the LAT has been shipped to NRL for instrument environmental tests, during which further monitoring and calibration of the calorimeter modules will be undertaken. Simultaneously the beam test of calibration unit made of three spare calorimeter modules and two spare tracker modules is planned to be done at PS and SPS beam lines at CERN.

REFERENCES

1. P. E. Michelson, " GLAST: A detector for high-energy gamma rays", *Proc.SPIE Conf. Gamma-Rays and Cosmic-Ray Detectors, Techniques and Missions*, **2806**, 31-40 (1996).
2. J.Ampe, A.Chekhtman, P.Dizon, J. E. Grove, W. N. Johnson, B. E. Leas, D. P. Sandora, M.S. Strickman, "The calibration and environmental testing of the Engineering Module of GLAST CsI calorimeter", *IEEE Trans. Nucl. Sci.,* **51**,No.5,2008-2011 (2004).

The PAMELA electromagnetic calorimeter: performances

E. Mocchiutti*, M. Albi*,†, M. Boezio*, V. Bonvicini*, J. Lundquist*, A. Vacchi*, G. Zampa*, N. Zampa*, J. Lund** and M. Pearce**

**INFN Sezione di Trieste, Padriciano 99, 34012 Trieste, Italy.*
†Dipartimento di Astronomia dell'Università di Trieste, via G.B. Tiepolo 11, 34131 Trieste, Italy.
***The Royal Institute of Technology (KTH), Department of Physics, Albanova University Centre, 10691 Stockholm, Sweden.*

Abstract. A sampling silicon–tungsten imaging calorimeter has been designed and built for the PAMELA satellite–borne experiment. The calorimeter has been designed to identify antiprotons from an electron background and positrons in a background of protons with a high efficiency and rejection power. In this work we present the identification capabilities of the calorimeter obtained using both Monte Carlo and test beam data. We show that the calorimeter provides a proton rejection factor of at least 10^5 while keeping a high efficiency in selecting electrons and positrons. Hence, the calorimeter will fulfill the identification power needed to reach the primary scientific objectives of PAMELA, that are the measurement of the flux of antiprotons, positrons and light isotopes in the cosmic radiation.

Keywords: PAMELA, cosmic rays, silicon–tungsten calorimeter, electron, hadron
PACS: 07.20.Fw, 96.50.S–, 95.30.Cq, 07.87.+v

INTRODUCTION

The PAMELA apparatus [2] (a Payload for Antimatter Matter Exploration and Light–nuclei Astrophysics) is a satellite–borne experiment which is primarily designed to study antiparticles (antiprotons and positrons) in the cosmic radiation.

PAMELA is housed on–board the Resurs–DK1 earth–observation satellite which has been launched into space by a Soyuz rocket on June 15th 2006 from the Russian cosmodrome of Baikonur. The satellite orbit is elliptical and semi–polar, with an altitude varying between 350 km and 600 km, at an inclination of 69.9°. Now PAMELA has passed the preliminary testing phase and is collecting data continuously. Detectors are performing in a nominal way and the analysis of data has started. Till now PAMELA has collected more data than the ones taken by a balloon borne experiment like CAPRICE98 [1]. The mission is foreseen to last for at least three years.

The main scientific goal of the experiment is the precise measurement of the cosmic–ray antiproton and positron energy spectra. An accurate determination of these spectra allows cosmic ray propagation [4] and solar modulation [5] to be studied. Antiparticles could also be produced from exotic sources such as primordial black holes [6] or the annihilation of supersymmetric [7] or Kaluza–Klein [8] dark matter particles. Almost all contemporary cosmic ray antiparticle data have been obtained by balloon–borne experiments for which the data–taking time (usually 24 hours) and the presence of a residual overburden of atmosphere (~5 g/cm^2) limit the accuracy of the measurements. Recent

CP867, *XII International Conference on Calorimetry in High Energy Physics*, edited by S. R. Magill and R. Yoshida

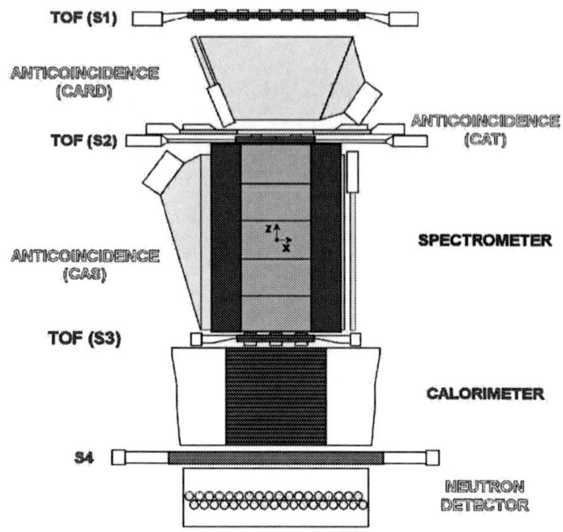

FIGURE 1. A schematic overview of the PAMELA satellite experiment. The experiment stands ~1.2 m high and, from top to bottom, consists of a time–of–flight (ToF) system, an anticoincidence shield system, a permanent magnet spectrometer, a silicon–tungsten electromagnetic calorimeter, a shower tail scintillator (S4) and a neutron detector. The experiment has an overall mass of 450 kg.

measurements for antiprotons include BESS00 [9], AMS [10] and CAPRICE98 [3]; and recent positron measurements include AMS [11], CAPRICE98 [12] and HEAT00 [13].

PAMELA will determine the flux of antiprotons and positrons in the cosmic radiation across a wide energy range and with unprecedented statistics. The measurements are complicated by the substantial background from protons (for the positron flux) and electrons (in the case of antiprotons). The ability of the PAMELA electromagnetic calorimeter to separate electrons from hadrons has been studied using particle beam tests and Monte Carlo simulations.

In the next section a general overview of the PAMELA apparatus is reported. Next an accurate description of the PAMELA electromagnetic calorimeter is given. Particle separation with the PAMELA experiment is then presented, describing also the analysis of simulations, tests at particle beams and ground data. Results are reported and discussed in the last two sections.

THE PAMELA EXPERIMENT

A schematic view of the PAMELA apparatus is shown in figure 1. PAMELA is built around a 0.43 T permanent magnet spectrometer ('tracker') equipped with 6 planes of 300 μm thick double–sided silicon detectors [14]. The tracker is surrounded by a plastic scintillator veto shield [15] (AC). An electromagnetic calorimeter mounted below the tracker measures the energy of incident electrons and allows topological discrimination between electromagnetic and hadronic showers (or non–interacting particles). This discriminatory ability is the focus of this paper. The calorimeter is described more fully in the next section. Planes of plastic scintillator mounted above and below the tracker form a time–of–flight (ToF) system which also provides the primary experimental trigger [16]. The volume between the upper two time–of–flight planes is bounded by an additional plastic scintillator anticoincidence system. A plastic scintillator system (S4) mounted beneath the calorimeter provides an additional stand–alone trigger for high energy electrons and is followed by a neutron detection system (ND) comprising ^{3}He–filled tubes within a polyethylene moderator for the selection of very high energy electrons and positrons (up to 3 TeV) which shower in the calorimeter but do not necessarily pass through the spectrometer.

THE ELECTROMAGNETIC CALORIMETER

The sampling electromagnetic calorimeter comprises 44 single–sided silicon sensor planes (380 μm thick) interleaved with 22 plates of tungsten absorber [17]. Each tungsten layer has a thickness of 0.26 cm, which corresponds to 0.74 X_0 (radiation lengths), giving a total depth of 16.3 X_0 ($\sim$0.6 nuclear interaction lengths). Each tungsten plate is sandwiched between two printed circuit boards upon which the silicon detectors and associated read–out electronics are mounted. The 8×8 cm^2 silicon detectors are segmented into 32 read–out strips with a pitch of 2.4 mm. The silicon detectors are arranged in a 3×3 matrix and each of the 32 strips is bonded to the corresponding strip on the other two detectors in the same row (or column), thereby forming 24 cm long read–out strips. The orientation of the strips of two consecutive layers is orthogonal to provide two–dimensional spatial information ('views'). A self–trigger system was implemented in the calorimeter read–out electronics to measure high–energy electrons in the cosmic radiation. Requiring only that the particles enter from one of the first four planes and cross at least 10 radiation lengths in the calorimeter, the overall acceptance ($\sim$600 cm^2 sr) becomes about a 30 factor larger than the normal acceptance of PAMELA. Figure 2 shows the display of a calorimeter self–trigger event taken in flight.

PARTICLE SELECTION AND ELECTRON–HADRON SEPARATION

The main stages for the identification of electrons and antiprotons are as follows: the Time–Of–Flight system is used to select down–going singly–charged particles to reject nuclei and the albedo component of cosmic rays. The magnet spectrometer measures the deflection, thus the rigidity of the particles and the sign of the charge separating negative particles from positives ones. The calorimeter, supported by the Neutron Detector, separates electromagnetic showers from non–interacting particles and hadronic showers.

Protons and electrons dominate the positively and negatively charged components of the cosmic radiation, respectively. Hence, positrons must be identified from a background of protons and antiprotons from a background of electrons. To achieve its scientific goals, the PAMELA system must separate electrons from hadrons at a level of 10^5

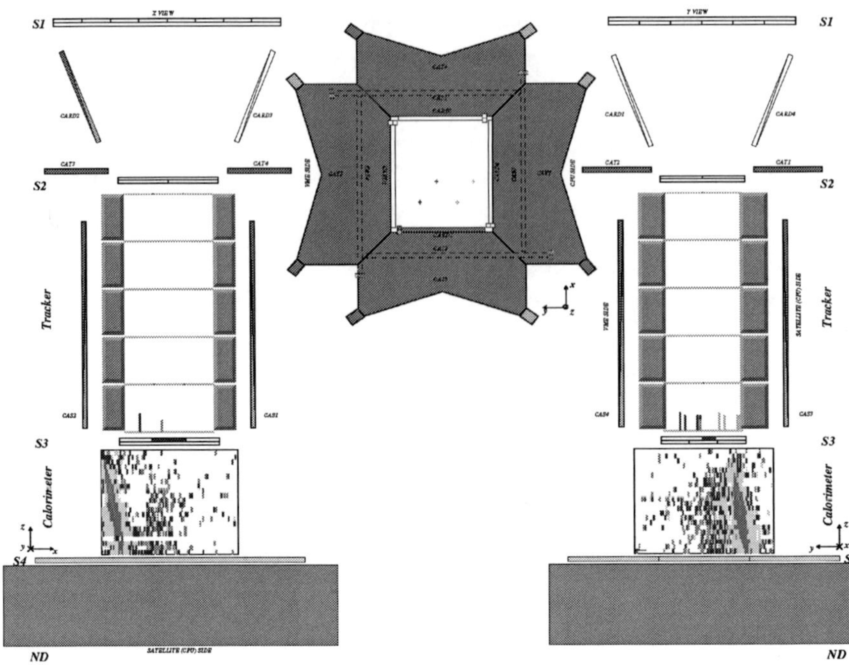

FIGURE 2. Display of a calorimeter self–trigger event from flight data. PAMELA is shown in the x–view on the left, in the y–view on the right and from above in the center. In this event anti–coincidence were hit and one neutron was detected. Notice that the trajectory of the event is out of the PAMELA acceptance.

– 10^6. Much of this separation must be provided by the calorimeter, i.e.: electrons must be selected with an acceptable efficiency and with as small a hadron contamination as possible.

The use of a simple variable, such as the total energy deposited in the calorimeter, allows an effective separation between electrons and hadrons. For incident hadrons of a given energy, the distribution of total energy deposited in the calorimeter is essentially flat with a sharp peak at low energies for non–interacting hadrons. For electrons, the total energy deposited in the calorimeter for a given incident energy is normally distributed, as long as most of the shower is contained. For higher electron energies, a tail to low energies can, however, decrease the efficiency for electron identification. The use of this variable alone allows a protons reduction of 99.98% while keeping the electrons efficiency at about 96%.

Additional variables have been used to further separate electromagnetic and hadronic showers in the calorimeter. These variables characterize the shower inside the calorimeter taking into account its starting point, its longitudinal and transverse profiles, and its topological development.

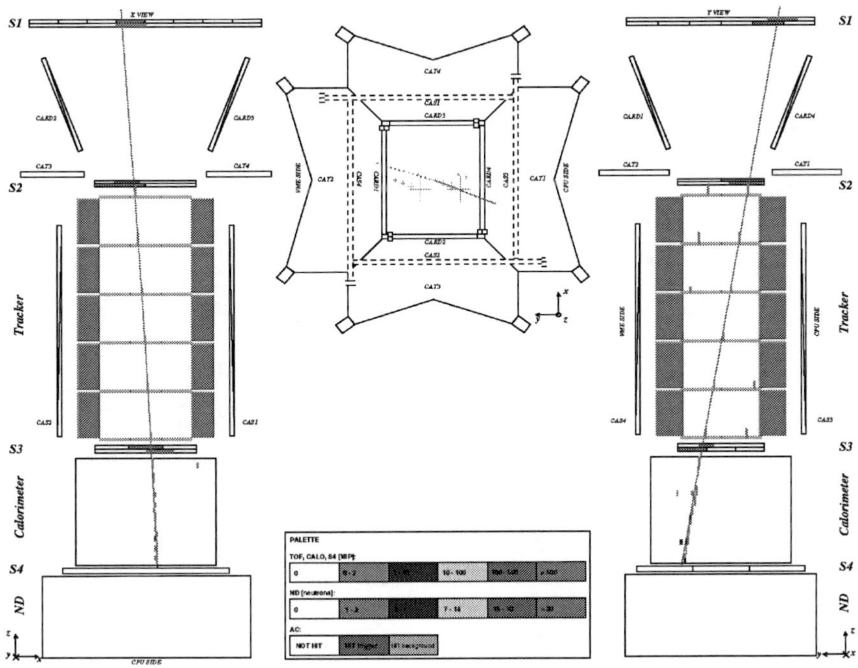

FIGURE 3. Display of a muon detected at ground. The solid line represent the fit of the track as performed by the tracker, notice the typical signature of a non–interacting particle in the calorimeter and the low level noise. In this figure S4 and the ND are not shown in same x and y scale.

DATA, PARTICLE BEAM TESTS AND SIMULATIONS

Three large sets of data exist to study the calorimeter performances: data from ground acquisition performed prior to the launch in the integration phase, data from tests with particle beams and Monte Carlo simulations. Data taken at ground consist mainly of low energy muons and some protons. Two events are displayed in figures 3 and 4.

Since ground data consist only of low energy particles and very few electrons they can not be used to study the calorimeter performances in the electron–hadron separation at high energy.

Hence, the calorimeter capabilities has been evaluated using data from tests with particle beams at CERN and Monte Carlo simulations.

The test beam data used in the analyses were collected at three different occasions at the CERN PS and SPS test beam facilities during 2002 and 2003. In each test, the calorimeter was equipped with approximately half of the total number of silicon planes. The performance results presented in the next section are therefore conservative. Monte Carlo simulations of the calorimeter performance used the GEANT framework, version 3.21 [18]. The default GHEISHA hadron shower package was used to simulate the interactions of hadrons with the nuclei of the matter traversed [19].

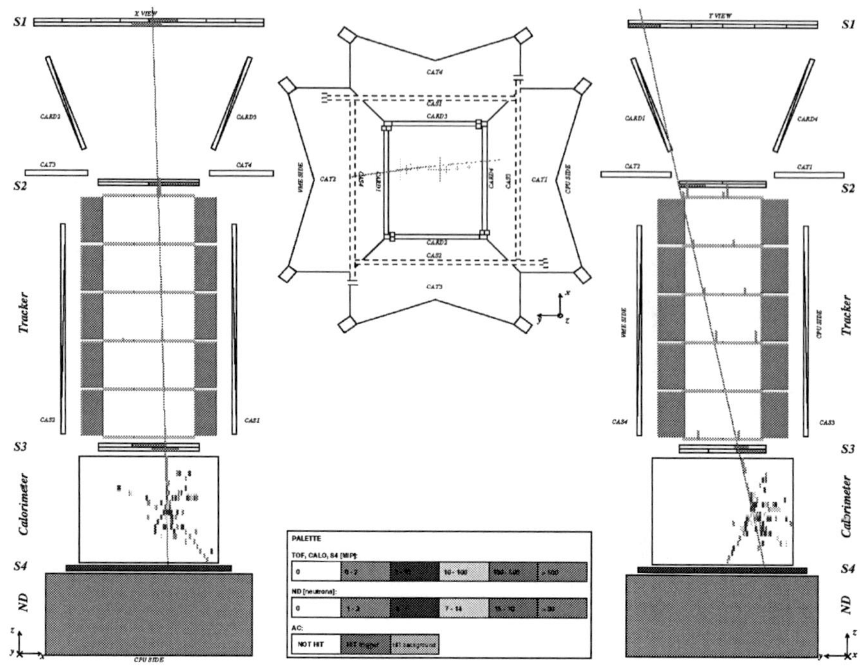

FIGURE 4. Display of an interacting proton detected at ground. The solid line represent the fit of the track as performed by the tracker, notice the typical signature of an interacting particle in the calorimeter and the detection of one neutron. In this figure S4 and the ND are not shown in same x and y scale.

RESULTS AND DISCUSSION

Combinations of the variables described previously were used to provide electron–hadron separation at a given incident particle energy. The selections were optimized using only the data collected during particle beam tests. Monte Carlo simulations were not used for this purpose. The complete set of selections used for each incident particle energy is listed elsewhere [19].

Figure 5 shows the efficiency (upper panel) and contamination (lower panel) of the final electron selection as a function of momentum. The full circles represent test beam data and the open squares represent simulated data. When tuning the data selections, the electron efficiency was kept above 90% whenever possible. The agreement between simulation and experimental data is acceptable. In the lower panel of figure 5 the proton contamination is shown as a function of momentum for both beam test data (full circles) and simulated data (open squares). The arrows show the 68% confidence levels if no events survived the selection. The data points at 3 GeV/c and 10 GeV/c are for electron/pion separation and therefore do not directly apply to the electron/proton separation study because pions with a momentum of 3 GeV/c have $\sim$1 GeV more kinetic energy that can be deposited in the calorimeter, than protons of the same momentum. This will result in a higher contamination for pions compared to protons. However, the pion data can usefully be used to compare simulation and beam test results. Notice that results for proton contamination at high energy are limited by the statistics of the collected data at the beam tests.

Good agreement between simulations and beam test data can be seen. Fluctuations in the measured efficiencies as a function of energy are due to different calorimeter configurations (i.e. number of planes equipped and read out [19]), energy dependent selections and different beam conditions (direction and impact point on the calorimeter). These results refer to a partially equipped calorimeter and should be considered as lower limits of the calorimeter performance.

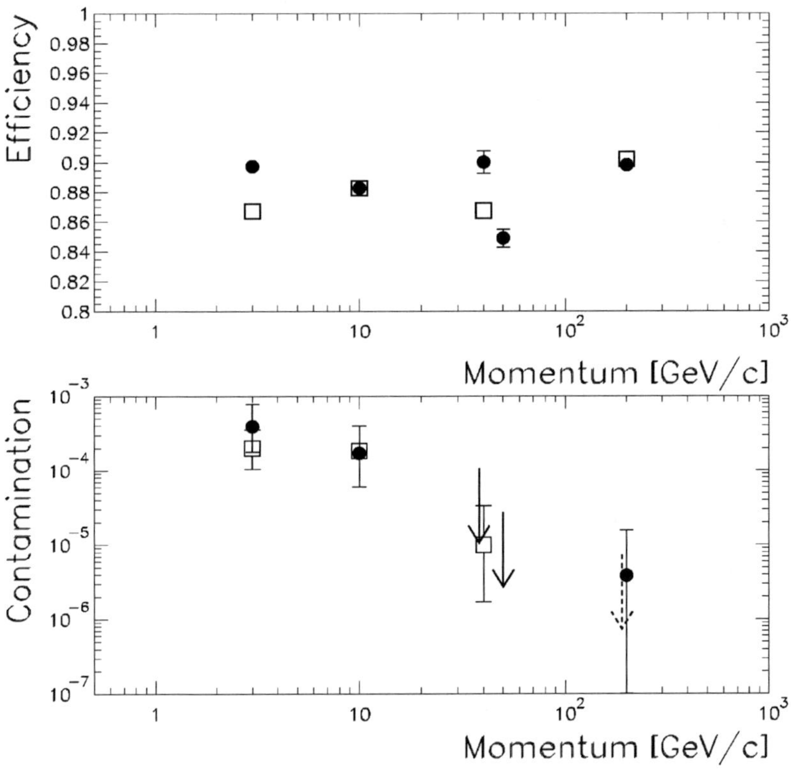

FIGURE 5. Upper panel: The electron identification efficiency, shown as a function of incident particle momentum. The full circles and open squares represent beam test and simulated data, respectively. Lower panel: The proton contamination in the electron selection, shown as a function of incident particle momentum. The full circles and open squares represent beam test and simulated data, respectively. The arrows indicate the 68% confidence level for test beam data if no events survive the selection. The dashed arrow is for simulated protons at 200 GeV/c. The data at 10 GeV/c, 40 GeV/c and 200 GeV/c are displaced for clarity.

Results obtained from test beam data indicate that even with a partially equipped calorimeter it is possible to achieve a proton rejection factor of at least 10^5 above 10 GeV/c while maintaining an electron selection efficiency of $\sim$90%.

CONCLUSIONS

The PAMELA calorimeter has been extensively tested at particle beam facilities and studied with simulations. Combining these data, it has been shown that the calorimeter is able to provide the rejection power needed by PAMELA to fulfill the scientific goals for which it was designed.

ACKNOWLEDGMENTS

The authors thank the PAMELA Collaboration for their support during the writing of this paper. In particular, we would like to acknowledge Francesco Cafagna and Marialuigia Ambriola for their help with the PAMELA simulation package, GPAMELA; Francesca Volpe for her help in setting up simulation codes; and Elena Vannuccini for her help with the tracker data. The KTH group thanks The Swedish National Space Board (Rymdstyrelsen) and The Swedish Research Council (Vetenskapsrådet) for financial support.

REFERENCES

1. M.L. Ambriola et al., *Nucl. Phys. B (Proc. Suppl.)*, **78**, 32 (1999).
2. M. Boezio, "The PAMELA Space Experiment", *Proc. 29th International Cosmic Ray Conference*, Pune, India, 2005.
3. M. Boezio et al., *Astrophys. J.* **561**, 787 (2001).
4. I.V. Moskalenko et al., *ApJ.* **586**, 1050 (2003).
5. Y. Asaoka et al., *Phys. Rev. Lett.* **88**, 051101 (2002).
6. K. Maki et al., *Phys. Rev. Lett.* **76**, 3474 (1996).
7. L. Bergström et al., *Phys. Rev. D.* **59**, 43506 (1999).
8. D. Hooper et al., *Phys. Rev. D.* **71**, 083503 (2005).
9. Y. Asaoka et al., *Phys. Rev. Lett.* **88**, 051101 (2002).
10. M. Aguilar et al., *Phys. Rept.* **366**, 331 (2002).
11. J. Alcaraz et al., *Phys. Lett. B* **484**, 10 (2000).
12. M. Boezio et al., *Advances in Space Research* **27**, 669 (2001).
13. J.J. Beatty et al., *Phys. Rev. Lett.* **93**, 241102 (2004).
14. O. Adriani et al., *Nucl. Instr. and Meth. A* **511**, 72 (2003).
15. S. Orsi et al., *Adv. Space Research.* In press (2005).
16. G. Osteria et al., *Nucl. Instr. and Meth. A* **535**, 152 (2004).
17. M. Boezio et al., *Nucl. Instr. and Meth. A* **487**, 407 (2002).
18. R. Brun et al., Detector description and simulation tool, CERN Program Library (1994).
19. J. Lund, PhD thesis, KTH, Stockholm (2004). Available online at <http://www.particle.kth.se>.

The CREAM Calorimeter: Performance In Tests And Flights

M.H. Lee[a], H.S. Ahn[a], P. Allison[b], M.G. Bagliesi[c], L. Barbier[d], J.J. Beatty[b], G. Bigongiari[c], P. Boyle[e], T.J. Brandt[b], J.T. Childers[f], N.B. Conklin[g], S. Coutu[g], M.A. DuVernois[f], O. Ganel[a], J.H. Han[a], J.A. Jeon[h], K.C. Kim[a], L. Lutz[a], P. Maestro[c], A. Malinine[a], P.S. Marrocchesi[c], S. Minnick[i], S.I. Mognet[g], S.W. Nam[h], S. Nutter[j], I.H. Park[h], N.H. Park[h], E.S. Seo[a,k], R. Sina[a], S. Swordy[e], S. Wakely[e], P. Walpole[a], J. Wu[a], J. Yang[h], Y.S. Yoon[k], R. Zei[c], S.Y. Zinn[a]

(a) Inst. for Phys. Sci. and Tech., University of Maryland, College Park, MD 20742, USA
(b) Dept. of Physics, Ohio State University, Columbus, Ohio 43210, USA
(c) Dept. of Physics, University of Siena and INFN, Via Roma 56, 53100 Siena, Italy
(d) Goddard Space Flight Center, Greenbelt, MD
(e) Enrico Fermi Institute and Dept. of Physics, University of Chicago, Chicago, IL 60637, USA
(f) School of Physics and Astronomy, University of Minnesota, Minneapolis, MN 55455, USA
(g) Dept. of Physics, Penn State University, University Park, PA 16802, USA
(h) Dept. of Physics, Ewha Womans University, Seoul, 120-750, Republic of Korea
(i) Dept. of Physics, Kent State University Tuscarawas, New Philadelphia, OH 44663, USA
(j) Dept. of Physics and Geology, Northern Kentucky University, Highland Heights, KY 41099, USA
(k) Dept. of Physics, University of Maryland, College Park, MD 20742, USA

Abstract. The Cosmic Ray Energetics And Mass (CREAM) balloon-borne experiment, designed to directly measure cosmic-ray particle energies from $\sim 10^{11}$ to $\sim 10^{15}$ eV, had two successful flights since December 2004, with a total duration of 70 days. The CREAM calorimeter is comprised of 20 layers of 1 radiation length (X_0) tungsten interleaved with 20 active layers each made up of fifty 1 cm wide scintillating fiber ribbons. The scintillation signals are read out with multi pixel Hybrid Photo Diodes (HPDs), VA32-HDR2/TA32C ASICs and LTC1400 ADCs. During detector construction, various tests were carried out using radioactive sources, UV-LEDs, and particle beams. We will present results from these tests and show preliminary results from the two flights.

Keywords: Cosmic-ray calorimeter; High energy cosmic rays.
PACS: 07.20.Fw, 07.87.+v, 95.55.Vj

1. INTRODUCTION

The CREAM experiment [1], a balloon borne instrument, is designed to study high energy cosmic rays by directly measuring nuclei from H to Fe, in the energy range from $\sim 10^{11}$ to $\sim 10^{15}$ eV. One of the scientific goals is to test a cosmic ray acceleration theory which predicts an acceleration limit of $\sim Z \times 10^{14}$ eV, where Z is the incident particle charge number. By measuring ratios of primary to secondary nuclei, the propagation of cosmic rays can also be studied. At its high end, CREAM's energy

CP867, *XII International Conference on Calorimetry in High Energy Physics*,
edited by S. R. Magill and R. Yoshida

range overlaps the range of ground based experiments, providing directly measured data to anchor their model-based calibration.

The CREAM instrument, described in detail elsewhere [1, 2], is comprised of several complementary detector systems to measure the charge and energy of cosmic rays, including a sampling tungsten/scintillating-fiber calorimeter. The CREAM calorimeter is comprised of 20 layers of 1 radiation length (X_0) tungsten absorber interleaved with 20 active layers of fifty 1 cm wide scintillating fibers ribbons made of 0.5 mm diameter fibers as shown in Fig. 1. Alternate layers are oriented orthogonally to each other, providing 10 measurements in X-Z and 10 alternate measurements in Y-Z. Scintillation light is generated in the scintillating fibers when an incident particle generates a shower of charged particles inside the calorimeter. These signals are transmitted from the fiber ribbon through a light mixer and a bundle of clear fibers before reaching the window of a HPD (Hybrid Photo Diode). In the HPD, photo-

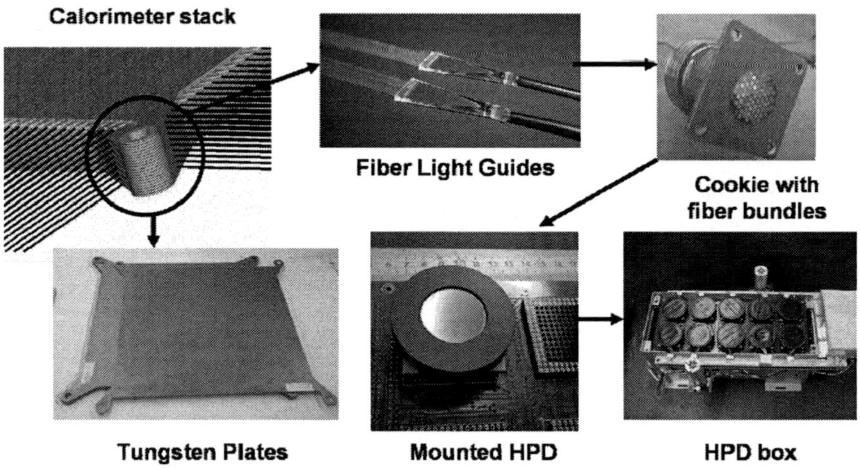

FIGURE 1. CREAM calorimeter components.

electrons are accelerated by a high voltage potential between photo-cathode and pixelated diode with a gain of about 2500 at 12 kV. The accelerated photo-electrons generate electron-hole pairs in the fully depleted region of the silicon diode of about 380 μm thickness by 100 V reverse bias. The charge signals are read out by a charge sensitive amplifier (VA32-HDR2) and digitized in 12-bit ADC chips (LTC1400) with a trigger circuit (TA32C) forming a trigger logic signal above a threshold value. A detailed explanation on the design of the calorimeter is provided elsewhere [3]. In this paper we present the performance of the calorimeter in various tests carried out during design, construction, and two successful flights.

2. OPTICS

The scintillation light generated by charged shower particles travels in the fibers to a ultra-violet transparent (UVT) acrylic light mixer and is transferred to a jacketed

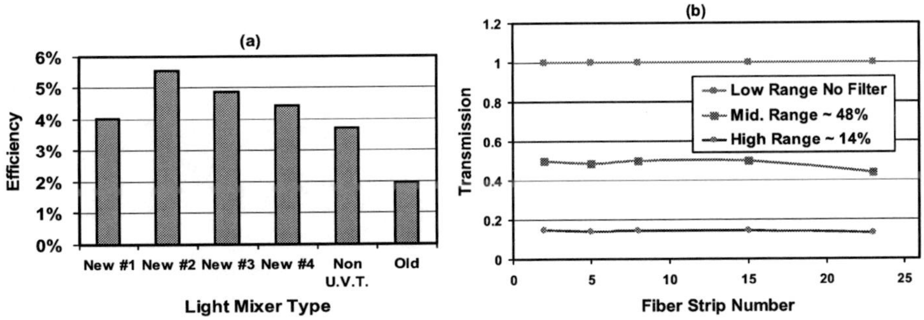

FIGURE 2. Transmission efficiencies with (a) various types of light mixers, "New" are those used in flight and are ~4.7 % on average, (b) neutral density filters in different fiber strips of a cookie.

bundle of forty eight 256 μm diameter clear plastic fibers. The transmission efficiency of light mixer and clear fibers, including all glue-joints were measured using a blue LED (Light Emitting Diode) source and a PMT (Photo Multiplier Tube) read out with an oscilloscope as shown in Fig. 2 (a). The average light transfer efficiency for flight assembly is ~4.7%. Also shown are overall transmission efficiencies for a previous design, as well as the flight design with non-UVT acrylic. The clear fiber bundle carries the light to the face of the HPD, where the bundle is split into three groups each with a different number of thin fibers, and through a different neutral density (ND) filter. The transmission efficiency of those ND filters are measured and shown in Fig. 2 (b). This arrangement forms three readout ranges to match the dynamic range of

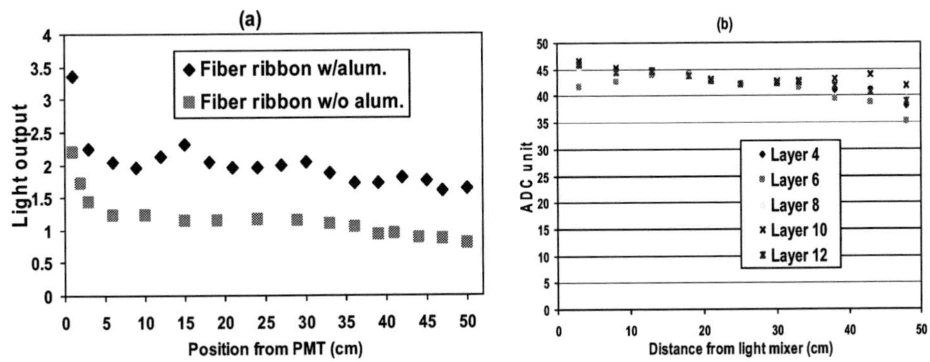

FIGURE 3. (a) Light output of fiber ribbons with and without aluminized non-readout ends, from UV LED light source, (b) Light output of aluminized fiber ribbons from electron showers.

the front end electronics, while covering the 1:200,000 range needed to read out the lateral tails of the shower for ~1 TeV showers, as well as the shower core maximum for ~1000 TeV showers. Aluminization on the non-readout side of each ribbon improves light collection in each ribbon, and increases the effective attenuation length in the ribbon as shown in Fig. 3 (a) by measuring UV-light-excited scintillation light from an UV LED. The non-uniformity caused by attenuation is measured using showers in different layers and are shown in Fig. (b). The measured non-uniformities were less than 15 % and are compensated by reading alternate (even vs. odd numbered) ribbons of a layer on opposite ends by two different HPDs.

3. READOUT ELECTRONICS

The calorimeter front end electronics are housed in four HPD boxes as shown in Fig. 1. Each box is instrumented with 10 HPDs socket-mounted in a Front board at the top of the box. Each HPD reads out 55 channels, of which 25 are low-range, 25 mid-range, and 5 high-range, based on the calorimeter optical division scheme [4]. At the bottom of the box a pair of Motherboards holds the connectors to the data sparsification system, the trigger system, the command system, the calibration system, and the power system. The Motherboards also provide the HPDs with bias voltage. A set of 10 vertically mounted ASIC boards connects the Front board and the two Motherboards. On the outside of the box, a pair of High Voltage Power Supplies (HVPSs) provides filtered high voltage, up to 12 kV (adjustable) at up to 1 μA each. Details of the high voltage system are explained elsewhere [4]. Each calorimeter HPD is read out by a pair of VA32-HDR2/TA32C ASICs, for a total of 80 such ASICs. Each pair of VA/TAs is socket-mounted in an ASIC board, allowing convenient

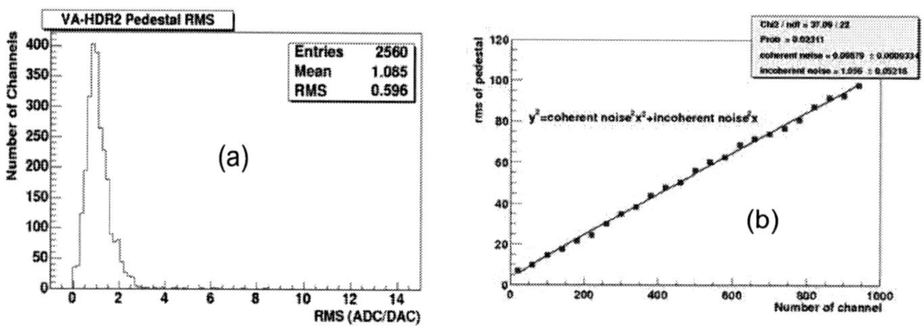

FIGURE 4. (a) Pedestal RMS of all 2560 calorimeter VA/TA channels, (b) Incoherent and coherent noise measurement from pedestal noise as a function of the number of channels summed.

replacement in case an ASIC is damaged before flight. Each VA/TA ASIC has 32 signal input channels, a 32-channel serial output and one trigger output channel. The VA receives the input signals from the HPD, and passes them through a Gaussian shaping circuit and, after an adjustable-delay peak hold circuit, into an ADC. The average VA peak delay time was set to ~1.5 μs. The digitized signals are sent through the Motherboard to a sparsification system that discards those signals that are below sparsification thresholds (adjustable on a channel-by-channel basis). This reduces the bandwidth required for data telemetry to the ground to manageable levels. In parallel to signal processing in the VA, the TA passes the input signals through a fast semi-Gaussian shaper, and compares the levels to an adjustable preset threshold. If any of the 32 shaped signals exceeds the threshold, the TA sends out a trigger signal. The low-range signals are used to form half-layer triggers, which are then combined with the other half-layer to form an overall OR of all ribbons in the layer. The overall calorimeter trigger is formed by requiring trigger signals from N consecutive layers, where N is adjustable in-flight by command from the ground to 4, 6, 8, or 10 layers.

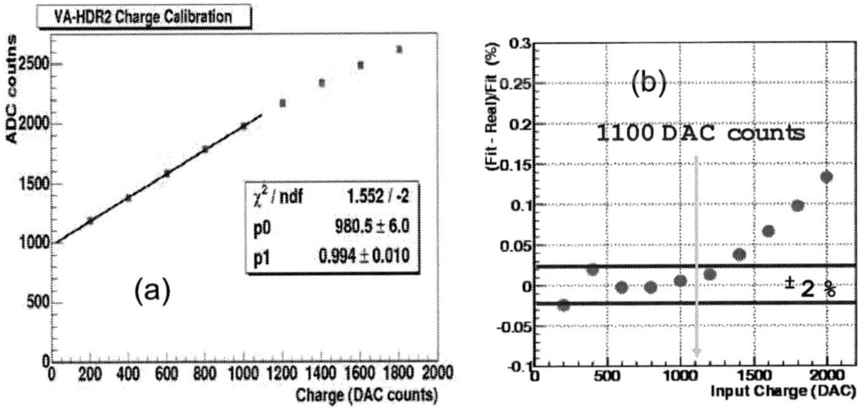

FIGURE 5. (a) Multi Charge calibration run averaged over 32 channels of a VA chip with a linear function fit, (b) Differences between measured values and fitted values, linear up to ~1100 DAC counts (~1 pC) within 2 %.

The characteristics of the VA/TA were tested in the lab. Figure 4 (a) shows a distribution of pedestal RMS noise for all 2560 VA/TA channels. The mean value of 1.085 ADC counts is ~1.23 fC (~7700 e) or ~54% higher than the nominal VA/TA noise specification of ~5000 e. Figure 4 (b) shows the measured coherent and incoherent noise from pedestal events by summing different numbers of channels. The estimated incoherent and coherent noise levels are about 1 and 0.1 ADC counts, respectively. The mean gain value of 0.98 (ADC/DAC) obtained from charge calibration corresponds to about 0.88 mV/fC as measured in the lab. Figure 5 shows the dynamic range of the VA, averaged over its 32 channels, is linear over 1:1000 (1100 ADC counts above pedestal, or ~1 pC with ~1.1 ADC counts noise). The trigger part of the VA/TA was tested and found to have a gain of ~2.85 DAC-counts-at-threshold/ADC-counts, as shown in Fig. 6. The fitted noise level of the TA circuit is ~13 DAC counts (~ 4.5 ADC counts).

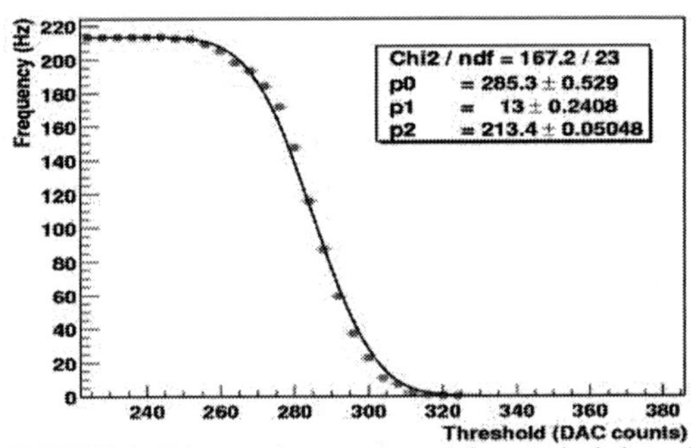

FIGURE 6. Trigger gain measurement for a typical channel of a VA/TA chip with a charge input of 100 DAC counts (~0.1 pC), showing a gain of about 2.85 DAC/ADC.

4. BEAM TESTS

With a total of 1000 fiber ribbons in 20 layers, each with three readout ranges, calibration is of critical importance in correctly reconstructing shower energy [5]. There are two steps to this calibration. First, one must equalize the low ranges of the different ribbons. Then, one must inter-calibrate the mid- to low-range and the high- to mid-range signals in each ribbon separately. The equalization process required scanning the surface of the calorimeter with beams of 150 GeV electrons, and for each ribbon, comparing the measured signal in ADC counts for those events where that ribbon was in the middle of the beam profile, with the energy deposit expected from simulated 150 GeV electron events with the same incident positions. The CREAM calorimeter is designed for multi-TeV showers, and thus has an electron sampling fraction of ~0.5%. This results in an energy deposit too low to be of use in calibrating the ribbons of the first and last few layers when using 150 GeV electrons. To address the calibration of the first layers, 2.5 cm thick (~5 X_0) lead bricks were placed in front of the graphite targets for a second scan to move the shower maximum closer to the top of the calorimeter. To calibrate the last few layers, a third scan was carried out with the detector rotated such that the beam was incident through its bottom. The shower maximum was moved to the bottom portion of the calorimeter in this configuration by placing the same lead bricks before the aluminum honeycomb pallet holding the stack. To compensate for the missing $2X_0$ of the targets and the tungsten layer above the first active layer, two additional tungsten plates (1 X_0 thick each) were placed between the lead bricks and the pallet in this configuration. A mean calibration constant of about 0.9 MeV/ADC counts was obtained in the beam test for the CREAM-I flight calorimeter. After applying the calibration constants for the 1000 low range readout channels, the beam data showed very good agreement with the simulation results as shown in Fig. 7 (a). The simulation included such effects as photon statistics, incoherent noise, and coherent noise. Figure 7 (b) shows energy

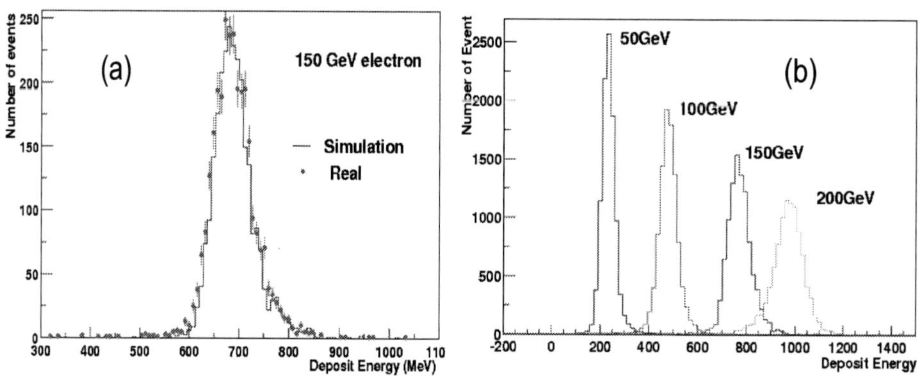

FIGURE 7. (a) Calorimeter energy distributions for 150 GeV electrons from simulation and beam data, (b) Energy distributions for 50, 100, 150, and 200 GeV electron beam data.

distributions for 50, 100, 150, and 200 GeV electron beam data, exhibiting a good linear response.

4. FLIGHTS

During the 42 days of CREAM-I's 2004 flight [6], the calorimeter operated quite satisfactorily [7]. Figure 8 (a) shows the mean noise levels in the calorimeter as measured by pedestal runs throughout the flight. The mean calorimeter noise value of

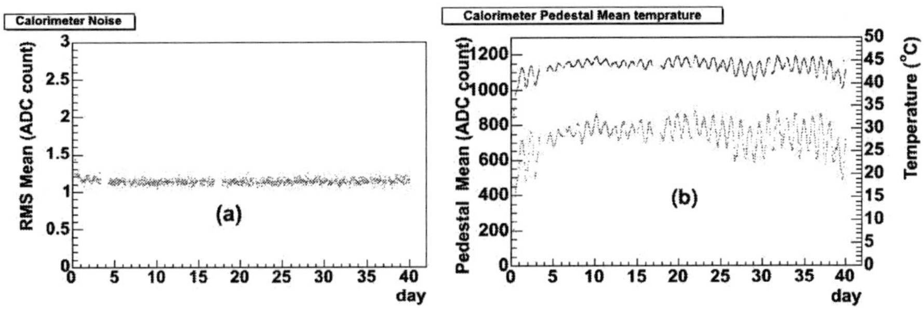

FIGURE 8. (a) Pedestal noise in the calorimeter, (b) Pedestal mean values and temperature for one calorimeter MB during CREAM-1 flight.

~1.15 ADC counts is comparable to the best achieved in the lab at full atmospheric pressure. During the flight, the calorimeter remained in a temperature range of 15°C – 35°C, well within its operational limits. Figure 8 (b) shows the time variation of the pedestal mean values and temperature for one calorimeter motherboard. To assure proper sparsification thresholds and pedestal subtraction despite this thermal drift, pedestal runs were performed once every five minutes throughout the flight, leading to a pedestal value drift of less than 1 ADC count between two

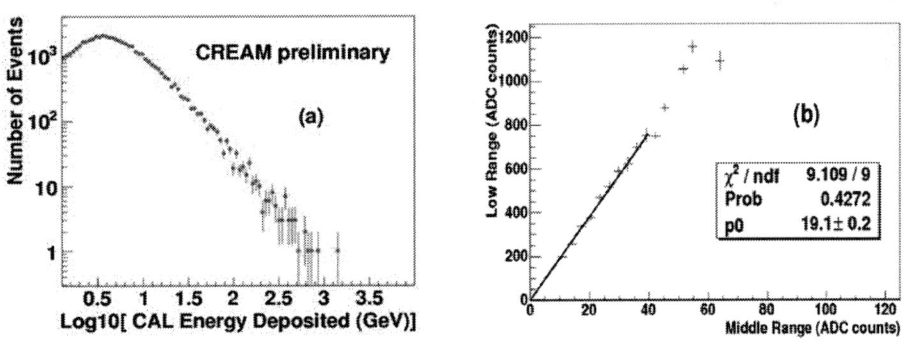

FIGURE 9. (a) Preliminary distribution of reconstructed calorimeter energy deposit for calorimeter triggers during flight, (b) An example of the mid-range vs. low-range signals measured in flight from the same ribbon, with a fit of the linear region.

consecutive pedestal runs. During the flight, the trigger was set to require at least 6 consecutive layers, with a threshold of ~60 MeV. This provided nearly 100% efficiency for showers of protons above 3 TeV (estimate based on a MC study). Figure

9 (a) shows a preliminary result of the calorimeter energy deposit distribution for events recorded with a calorimeter trigger. The energy deposit was reconstructed using a preliminary set of calibration constants from beam calibration, LED-based HV gain corrections, and flight measurements of the ratios between different readout ranges. Figure 9 (b) shows an example of such a ratio of mid- and low-range signals for one ribbon. Such plots were used to inter-calibrate the different ranges for each ribbon separately, providing measurements from the mid-range wherever the low-range was saturated.

The CREAM-II instrument, incorporating a calorimeter with the same design as that of CREAM-I was flown for ~29 days from December 2005 to January 2006, completing 2 circumnavigations of the Antarctic continent, exhibiting very stable operation. About 57 GB of data were collected during this flight. About 117 GB of data has been collected to date from the combined 70 days flights of CREAM-I and CREAM-II. Analysis of flight data is now in progress. At the same time, the CREAM-I calorimeter is being refurbished for flight in December 2007, and its calibration at a CERN test beam is planned for October of this year.

ACKNOWLEDGMENTS

This work was supported by NASA. The authors thank NASA/WFF, the Columbia Scientific Balloon Facility, the NSF Office of Polar Programs, and Raytheon Polar Services Company for the successful balloon launches, flight operations, and payload recovery. The authors also thank CERN for the use of their excellent beam test facilities and operations.

REFERENCES

1. E. S. Seo et al., *Advances in Space Research*, **33**/10, 1777-1785 (2004).
2. S. Coutu et al., “Performance of the transition radiation detector and the timing charge detector in the first flight of the CREAM instrument”, Proc. 29th Int. Cosmic Ray Conf., Pune, 2005, 3, pp. 393-396.
3. M. H. Lee et al., “Design and tests of the CREAM calorimeter”, Proc. 29th Int. Cosmic Ray Conf., Pune, 2005, 3, pp. 413-416.
4. M. H. Lee et al., “Electronics for the CREAM calorimeter and hodoscopes”, Proc. 29th Int. Cosmic Ray Conf., Pune, 2005, 3, pp. 409-412.
5. Y. S. Yoon et al., “Performance of the CREAM calorimeter in accelerator beam test”, Proc. 29th Int. Cosmic Ray Conf., Pune, 2005, 8, pp. 371-374.
6. E.S. Seo et al., “New Observations with CREAM”, Proc. 29th Int. Cosmic Ray Conf., Pune, 2005, 10, pp. 185-198.
7. M. H. Lee et al., “Performance of the CREAM calorimeter module during its first flight of 42 days”, Proc. 29th Int. Cosmic Ray Conf., Pune, 2005, 3, pp. 417-420.

The Fluorescence Detector of the Pierre Auger Observatory - a Calorimeter for UHECR

B. Keilhauer for the Pierre Auger Collaboration

Universität Karlsruhe, Institut für Experimentelle Kernphysik, Postfach 3640, 76021 Karlsruhe, Germany and
Observatorio Pierre Auger, Av. San Martin Norte 304, 5613 Malargüe, Argentina;
http://www.auger.org/auger-authors.pdf

Abstract. The Pierre Auger Observatory is a hybrid detector for ultrahigh energy cosmic rays (UHECR) with energies above $10^{18.5}$ eV. Currently the first part of the Observatory nears completion in the southern hemisphere in Argentina. One detection technique uses over 1600 water Cherenkov tanks at ground where samples of secondary particles of extensive air showers (EAS) are detected. The second technique is a calorimetric measurement of the energy deposited by EAS in the atmosphere. Charged secondary particles of EAS lose part of their energy in the atmosphere via ionization. The deposited energy is converted into excitation of molecules of the air and afterwards partly emitted as fluorescence light mainly from nitrogen in the wavelength region between 300 and 400 nm. This light is observed with 24 fluorescence telescopes in 4 stations placed at the boundary of the surface array. This setup provides a combined measurement of the longitudinal shower development and the lateral particle distribution at ground of the same event. Details on the fluorescence technique and the necessary atmospheric monitoring will be presented, as well as first physics results on UHECR.

Keywords: Cosmic Rays, Extensive Air Showers, Fluorescence Emission
PACS: 96.50.S, 96.50.sb, 96.50.sd, 33.50.Dq

INTRODUCTION

Ultrahigh energy cosmic rays (UHECR) are a continuous particle flux arriving at the Earth's atmosphere with strongly decreasing rate at highest energies. At energies above 10^{18} eV, their origin is expected to be extragalactic. However, even after several years of measurements, the major questions remain unanswered: where and how do these particles obtain such high energies and what is their elemental composition. For resolving the enigma about UHECR, experiments are designed for measuring the primary energy and type of cosmic rays as well as their arrival directions.

Entering the Earth's atmosphere, cosmic rays initiate extensive air showers (EAS). The primary particle of an EAS, which is the actual cosmic ray particle, interacts with a nucleus of air molecules. This hadronic interaction yields new secondary particles which will interact themselves with nuclei of air molecules. A growing cascade of secondary particles develops in the Earth's atmosphere, which is called an extensive air shower. After an initial phase of particle multiplication, the number of charged particles in the shower falls due to energy losses. For inclined showers with moderate primary energy, only some of the overwhelming amount of particles reaches the Earth's surface. These particles can be observed by particle detectors at ground. For UHECR with nearly vertical incidence, the maximum of shower development is reached close to ground or

CP867, *XII International Conference on Calorimetry in High Energy Physics,*
edited by S. R. Magill and R. Yoshida

even "below" ground.

Another detection technique arises from the energy losses of EAS in the atmosphere. The energy deposited in the air excites molecules of the atmosphere where particularly the excitation of nitrogen molecules is of interest. The de-excitation of nitrogen molecules is partly proceeding through the channel of fluorescence emission. So a light track of the air shower in the UV-wavelength region between 300 and 400 nm can be detected by adequate telescopes. This enables a calorimetric measurement of the deposited energy of EAS which can be used to obtain the primary energy of the cosmic ray.

THE PIERRE AUGER OBSERVATORY

The Pierre Auger Observatory is currently the only data-taking experiment for the detection of UHECR but still under construction. The first part nears completion in the Pampa Amarilla, near Malargüe, Argentina [1]. The second part will be installed in Colorado, USA. Each site of the observatory is designed as a hybrid detector, which consist of particle detectors at ground and fluorescence telescopes. This configuration ensures at least in dark nights the possibility of simultaneous detection of air showers with two different detection techniques.

In May 2006, about 1100 of the total 1600 water Cherenkov tanks, as particle detectors, have been deployed in the Pampa. The spacing of the tanks is 1.5 km and finally an area of 3000 km^2 will be covered. Within these detectors, energetic electrons and muons induce flashes of Cherenkov light which are recorded by three photomultipliers inside the water-filled tank. The duty cycle of this system is nearly 100 % [2].

The second detector component is fluorescence telescopes. Four stations are located at the boundary of the surface array each consisting of six telescopes. A single telescope has a field of view of roughly $30° \times 30°$ with a minimum elevation of 1.5° above horizon. Each station in total has 180° in azimuth observing the atmosphere above the surface array. The telescopes are built with a *Schmidt* optics with an UV-filter for reducing the background light. This system is operated during the one week before and after new moon at night. Therefore, the duty cycle reaches only 15 % of the surface detectors. The camera in the focal point consists of 440 photomultipliers each covering 1.5° [3].

Data taking started in January 2004 with a continuously increasing detector. Until June 2005, an exposure of 1750 km^2 sr yr could be achieved [4].

EXTENSIVE AIR SHOWERS

The development of extensive air showers in the atmosphere is strongly fluctuating from shower to shower. Nevertheless, some systematic behavior is correlated with the energy and type of the primary particle. High-energetic showers penetrate deeper in the atmosphere than lower-energetic ones. Vertical showers develop deeper in the atmosphere than more inclined ones. And also proton-induced air showers develop deeper than air showers induced by more heavy nuclei like iron. The shower development is described according to the amount of traversed air which is the *atmospheric depth* X given by

$$X(h_0) = \int_{h_0}^{\infty} \rho_{air}(h)\, dh, \qquad (1)$$

with h as geometrical height and ρ_{air} as the altitude-dependent air density[1].

As the density of air is quite low at higher altitudes, the height of first interaction is strongly fluctuating. The incoming particles represent the projectiles and the resting air nuclei are the target similar to the setup in fixed-target accelerator experiments. In the case of nuclei, not all nucleons of the projectile interact with the target, most of them are only spectators [5].

In general, the description of EAS can be divided into three particle components: the electromagnetic, muonic, and hadronic. If the incoming particle is a nucleus, high energy hadronic interactions produce mostly nucleons, charged and neutral pions, and kaons. The $\pi^{\pm}$ and K mesons have relatively long lifetimes of 10^{-8} s, thus they can interact with air nuclei or decay depending on their energy [6]. If the $\pi^{\pm}$ decay before interacting, they give rise to the muonic component accompanied by neutrinos: $\pi^{\pm} \rightarrow \mu^{\pm} + \nu$. Many of the muons reach the Earth's surface due to the relativistic time dilatation of their lifetime of 2.2×10^{-6} s and small energy loss. Those which decay, produce parts of the electromagnetic component: $\mu^{\pm} \rightarrow e^{\pm} + \nu + \bar{\nu}$.

If the incoming particle is an electron or positron, energy loss due to Bremsstrahlung is one of the most dominant processes. The emitted high-energetic photons create new pairs of electrons and positrons. In one radiation length, their energy is attenuated by the factor $1/e$, which is approximately the same as the $1/e$ attenuation of a gamma-ray beam due to pair production. In air, the radiation length X_0^{air} is about 37 g/cm^2. The particle multiplication is large and the electromagnetic component is the most numerous part in EAS. This electromagnetic part of an air shower can also be induced by the decay of a π^0 into two photons [7]. Losing more and more energy, these particles reach the critical energy which describes the equality of the energy loss due to radiation and the loss due to ionization. Since ionization energy loss is about 2.2 MeV/g/cm^2, the critical energy in air is 81 MeV. The interplay of particle production and energy loss yields in a shower maximum and closer to ground the shower starts to die out. Only some of the electrons and positrons reach the Earth's surface.

The overall longitudinal shower development, as simulated with CORSIKA[2] [8], is shown in Fig 1, where a proton-induced shower with 10^{19} eV is plotted. On the left side, the development in terms of energy deposit of a vertical shower is displayed and on the right side, the particle number for a 45° inclined shower.

The position of the shower maximum is a good indicator for the properties of the primary particle. The particle number at shower maximum is proportional to the primary energy E_0 and the atmospheric depth of the position of shower maximum is proportional to $ln(E_0/A)$, where A is the mass of the primary.

Using the fluorescence technique, the energy deposit of an air shower is measured by the amount of emitted fluorescence light. However, integrating the whole energy deposit

---

[1] Here written for a vertical particle trajectory.

[2] **CO**smic **R**ay **SI**mulations for **K**ascade and **A**uger

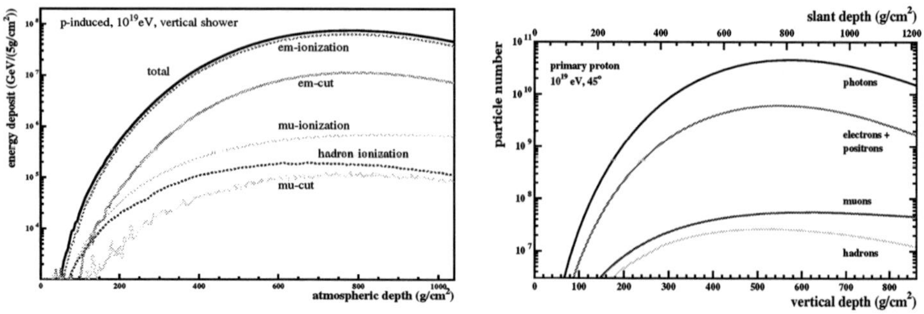

FIGURE 1. Simulated proton-induced extensive air shower with $E_0 = 10^{19}$ eV. Left: The overall deposited energy versus atmospheric depth is shown for the vertical case. The contributions from the electromagnetic, muonic, and hadronic part of air showers are separated. The contributions labeled by 'cut' are due to simulation limitations but can be added to the local energy deposit [9]. Right: The particle numbers of major contributions vs. atmospheric depth is shown. The shower inclination angle is 45 °, therefore the slant depth is indicated at the top of the frame [10].

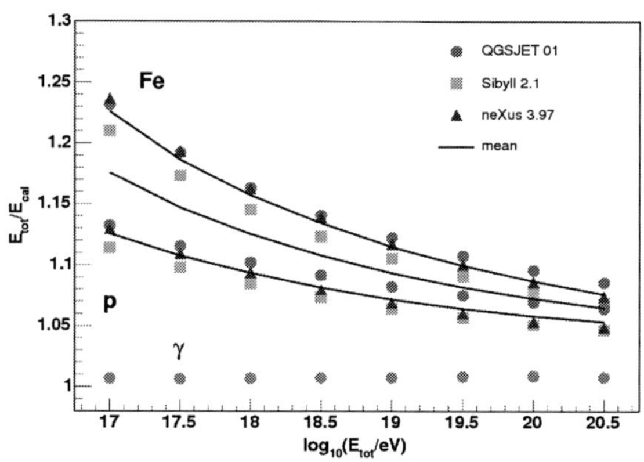

FIGURE 2. Correction factor for the energy which is not detected by fluorescence telescopes in dependence on primary energy and type of air shower. The missing energy is related to mainly neutrinos and muons as secondary particles in the air shower cascade [11].

will not result in the total primary energy. The energy deposit is only that of the charged particles but the energy carried away by neutrinos is not detected. Therefore, one has to perform a correction to obtain the total primary energy, which depends on the energy and type of cosmic rays, see Fig. 2.

CALIBRATION SYSTEMS

The atmosphere is an integral part of any detector system using extensive air showers. The longitudinal and lateral development is affected by local atmospheric conditions

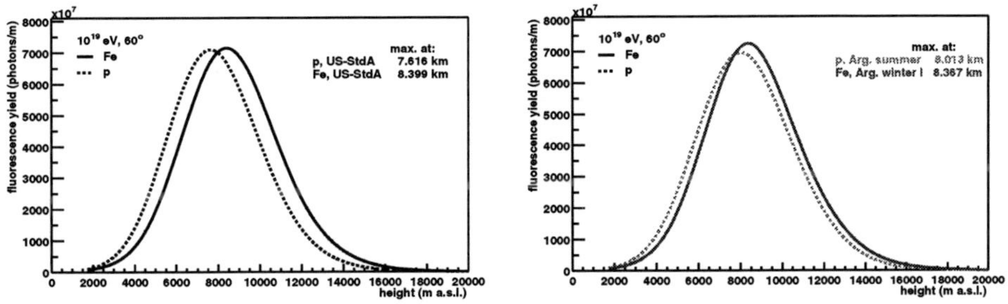

FIGURE 3. Left: Fluorescence yield profiles for p- and Fe-induced EAS in the US Standard Atmosphere with $E_0 = 10^{19}$ eV and 60° inclination. The fluorescence yield is the sum of all emitted photons between 300 and 400 nm [9]. Right: Fluorescence yield profiles for p-induced EAS in Argentine summer and Fe-induced EAS in Argentine winter both with $E_0 = 10^{19}$ eV and 60° inclination. The fluorescence yield is the sum of all emitted photons between 300 and 400 nm [9].

such as layering of air density and temperature. The optical signal of extensive air showers is even more influenced by the atmosphere, in the emission and transmission towards the detector.

Several calibration systems have been installed at the site of the Pierre Auger Observatory in Argentina. The *Central Laser Facility* [12] and the *Lidar stations* at each telescope station [13] are detecting the optical transmission conditions of the atmosphere. The absolute *drum* calibration determines the detector response [14]. Atmospheric variables like air temperature, pressure, and humidity are obtained by meteorological radio soundings [15]. From these data, the altitude and time dependent air density can be derived which is important for the reconstruction of the longitudinal air shower profiles of the calorimetric measurement using air fluorescence emission. Varying atmospheric conditions affect the conversion of atmospheric depth, as the characteristic quantity for the air shower development, to geometrical altitude, as reconstructed from fluorescence telescope data. Systematic variations of the position of shower maximum as seen by the telescopes due to changing air density profiles have been found [16]. In Fig. 3, the position of shower maximum for simulated proton- and iron-induced showers are shown on the left side for conditions like in the US Standard atmosphere [17]. On the right side, the same simulated showers are plotted for the case that the proton shower would develop in the Argentine summer atmosphere and the iron shower in winter conditions. The separation between the positions of maximum in the US Standard atmosphere as seen by the fluorescence telescopes is significantly reduced.

The uncertainties on the reconstructed energy of air showers from the fluorescence measurement are still quite large. Currently, known uncertainties for the signals in the PMT camera are 5 % for the light collection and 12 % from the absolute detector calibration [3]. However, the latter will improve in the near future. For the reconstructed photons at the telescope, the Auger experiment has to cope with 2 % uncertainty in the geometry reconstruction for hybrid events and with 10 % from aerosols in the atmosphere. The correction for the missing energy yields 3 % uncertainty and the

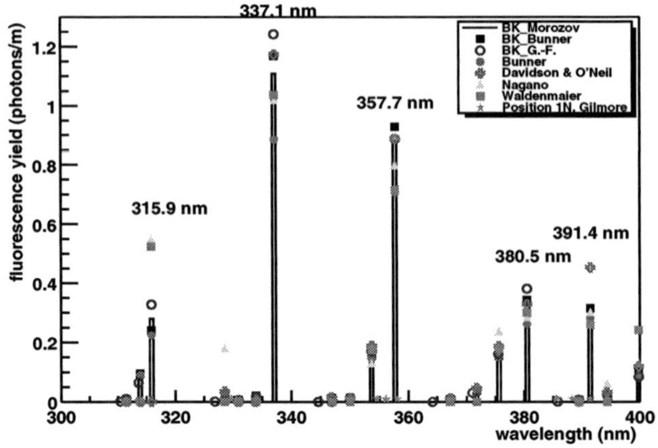

FIGURE 4. Fluorescence yield spectra of several calculations and measurements for 0.85 MeV electrons as exciting particles. The bars indicate the preferred calculation by the authors of the referenced article [18]. All calculations by the authors are labeled with "BK_*name*", where *name* indicates the authors of the input parameters.

varying air density profiles 2 % for monthly models for the site of the Auger experiment.

A very important quantity for the energy reconstruction is the fluorescence yield. The emitted light is mainly coming from 18 strong emission bands in the 2P system of nitrogen molecules for the wavelength range between 300 and 400 nm and one emission band in the 1N system. It is assumed that the fluorescence yield is proportional to the local energy deposit of an air shower

$$Fl.Yield_\lambda = \varepsilon_\lambda(p,T) \cdot \frac{\lambda}{hc} \cdot \frac{dE}{dX} \cdot \rho_{air} \left[\frac{\text{photons}}{\text{m}}\right], \tag{2}$$

where $\varepsilon_\lambda(p,T)$ is the fluorescence efficiency in dependence on air pressure and temperature. Strong efforts in improving laboratory measurements of the main variables of the fluorescence emission are ongoing as well as of theoretical descriptions of the molecular processes. A comparison of current experimental data and calculations for the fluorescence yield is given in Fig. 4 [18]. The yield is given for a 0.85 MeV electron as the exciting particle in US Standard atmosphere conditions at sea level. The uncertainties for individual wavelengths are large. For the overall wavelength region, some of the discrepancies cancel out and the uncertainty for the total fluorescence yield between 300 and 400 nm is about 15 % for air shower applications. However, the wavelength-dependent light transmission coefficients and detector response necessitate a good knowledge of the fluorescence yield of the individual emission bands. A world-wide cooperation of several groups will improve the accuracy of the most important variables linked to nitrogen fluorescence emission in air in the near future [19].

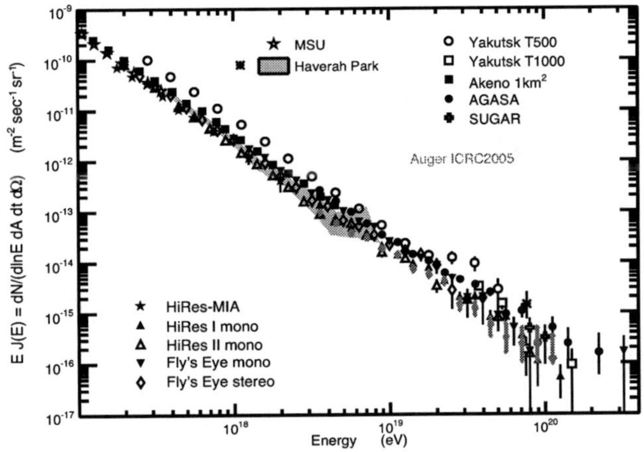

FIGURE 5. Primary cosmic ray flux [20]. Shown is a selection of previous measurements [21] together with the data from the Pierre Auger Observatory as presented at the ICRC 2005 [4].

COSMIC RAY ENERGY SPECTRUM

Based on the data taken with the Pierre Auger Observatory between January 2004 and July 2005, a cosmic ray energy spectrum could be derived [4]. All events with zenith angle between 0° and 60° have been taken into account. The *time average area* of the continuously growing array has been 660 km^2 which is about 22 % of the final size of the southern part of the experiment. Applying several quality cuts, a full efficiency could be reached above 3 EeV and 3525 events above $10^{18.5}$ eV could be collected by the surface array. The analysis has been performed using the following procedure. Since the surface array could collect much more event data, this higher statistics should be used. The zenith angle correction for these data is done with the *constant intensity cut* method [4]. Without fluorescence data, the energy reconstruction of the surface array events would depend on air shower simulation with its considerable uncertainties arising from different hadronic interaction models. Therefore, the energy is determined from the sub-sample of hybrid events. An energy conversion factor is derived by comparing the reconstructed energy of hybrid events, based on the calorimetric energy measurement with the fluorescence telescopes, with the surface detector signal at 1000 m from the shower core. Finally, the exposure of the growing array is calculated. The estimated energy spectrum can be seen in Fig. 5 together with previous measurements.

Statistics is still limited for the first data taking period and systematic uncertainties are quite large. So, based on that data, no decision can be made about the characteristics of the spectrum in the GZK cutoff region. Whether there is a systematic difference in the energy reconstruction for the fluorescence method and the surface detector method will be subject of detailed investigations. The Pierre Auger Observatory with its hybrid technique intends to resolve this uncertainty together with the investigations of the fluorescence processes.

ACKNOWLEDGMENTS

I would like to thank the organizers of the CALOR 2006 conference in Chicago for the invitation and the financial support. I gratefully acknowledge the fruitful cooperation and discussions with my colleagues from the Pierre Auger Collaboration. My research is supported by the German Research Foundation (DFG) under contract No. KE 1151/1-1.

REFERENCES

1. J. Abraham et al., Pierre Auger Collab., Nucl. Instr. and Meth. Phys. Res. **A523**, 50, (2004)
2. X. Bertou for the Pierre Auger Collab., Proc. 29th Int. Cos. Ray Conf., Pune, India, Vol. 7, 1, (2005)
3. J. Bellido for the Pierre Auger Collab., Proc. 29th Int. Cos. Ray Conf., Pune, India, Vol. 7, 13, (2005)
4. P. Sommers for the Pierre Auger Collab., Proc. 29th Int. Cos. Ray Conf., Pune, India, Vol. 7, 387, (2005)
5. J. Knapp, Vergleich hadronischer Wechselwirkungsmodelle zur Simulation ausgedehnter Luftschauer der kosmischen Strahlung, Report FZKA 5970, Forschungszentrum Karlsruhe, (1997)
6. G. B. Khristiansen et al., Cosmic Rays of Superhigh Energies, Verlag Karl Thiemig, München, (1980)
7. P. Sommers, C. R. Physique 5, 463, (2004)
8. J. Knapp, D. Heck, Extensive Air Shower Simulation with CORSIKA: A User's Guide, Report KfK 5196B, Kernforschungszentrum Karlsruhe, (1993); Auger technical note GAP-1998-012, (1998); for an up-to-date version see http://www-ik.fzk.de/~heck/corsika/usersguide/corsika_tech.html
9. B. Keilhauer, Investigation of Atmospheric Effects on the Development of Extensive Air Showers and their Detection with the Pierre Auger Observatory, Report FZKA 6958, Forschungszentrum Karlsruhe, (2004)
10. M. Risse, private communication, (2006)
11. T. Pierog et al., Proc. 29th Int. Cos. Ray Conf., Pune, India, Vol. 7, 103, (2005)
12. F. Arqueros et al. for the Pierre Auger Collab., Proc. 29th Int. Cos. Ray Conf., Pune, India, Vol. 8, 335, (2005)
13. R. Cester et al. for the Pierre Auger Collab., Proc. 29th Int. Cos. Ray Conf., Pune, India, Vol. 8, 347, (2005)
14. P. Bauleo et al. for the Pierre Auger Collab., Proc. 29th Int. Cos. Ray Conf., Pune, India, Vol. 8, 5, (2005)
15. J. Blümer et al. for the Pierre Auger Collab., Proc. 29th Int. Cos. Ray Conf., Pune, India, Vol. 7, 123, (2005)
16. B. Keilhauer et al., Astropart. Phys. **22**, 249, (2004), astro-ph/040548, (2005)
17. National Aeronautics and Space Administration (NASA), U.S. Standard Atmosphere 1976, NASA-TM-X-74335, (1976)
18. B. Keilhauer et al., Astropart. Phys. **25**, 259, (2006), astro-ph/0511153, (2005); for an updated version see http://www.particle.cz/conferences/floret2006/Keilhauer.ppt, (2006)
19. Series of *International Workshop on Air Fluorescence*,
    http://www.physics.utah.edu/ fiwaf/done/index.html, (2002)
    http://www.auger.de/events/air-light-03, (2003)
    http://lappweb.in2p3.fr/IWFM05/index.html, (2005)
    http://www.particle.cz/conferences/floret2006/Keilhauer.ppt, (2006)
20. R. Engel, private communications, (2005)
21. R. Engel, H. Klages, C. R. Physique 5, (2004)

Air Fluorescence Calorimetry with the High Resolution Fly's Eye and Telescope Array Experiments

Charles C.H. Jui, for the HiRes and TA Collaborations

Department of Physics, University of Utah Salt Lake City, Utah, USA

Abstract. The air fluorescence technique was first successfully deployed on the Fly's Eye Experiment (1981-1993) by the University of Utah. Its successor, the High Resolution Fly's Eye (HiRes) experiment has further exploited this technique, first in hybrid mode with the MIA muon array (1993-1996), and then in monocular and stereoscopic modes (1997-2006). Results from HiRes will be presented, including evidence for the Greisen-Zatsepin-K'uzmin (GZK) Effect predicted 40 years ago. Most recently, members of the HiRes are collaborating with groups from Japan, led by University of Tokyo, to construct and operate the Telescope Array (TA) experiment, which will deploy a large scintillation-based ground array in combination with fluorescence detectors. Funding for TA in the US has already been approved by NSF. TA will begin operation in 2007.

Keywords: ultrahigh energy cosmic rays astro-particle physics
PACS: 98.70.Sa, 95.85.Ry, 95.55.Vj

INTRODUCTION

The High Resolution Fly's Eye (HiRes) Experiment consists of two air fluorescence detectors located on the U.S. Army Dugway Proving Ground in the West Desert of Utah. The project is a collaboration between University of Utah, Columbia University, Rutgers University, University of New Mexico, University of Montana,the Los Alamos National Laboratory (LANL), University of Tokyo, and the Institute for High Energy Physics (IHEP) in Beijing, China. The two detector sites are placed 12.6 km apart. The HiRes-1 site, in operation since 1997, sits atop Little Granite Mountain and comprises 22 mirrors covering $3° - 17°$ in elevation. The HiRes-2 site began routine observations at the end of 1999. It consists of 42 mirrors which view elevation angles in the range $3° - 31°$.

Each of the two detectors serve as sampling calorimeters, using the Earth's atmosphere as the scintillation medium. As in the case of the original Fly's Eye experiment [1], the mirrors collect and focus the UV (*fluorescence*) light emitted by nitrogen molecules in the wake of the charged particles in the extensive air shower. Photomultiplier tubes (PMTs) are placed as pixel elements at the focal plane of the ~ 2 m diameter, spherical mirrors. In this way, HiRes, like Fly's Eye, images the longitudinal development of air showers. Unlike the original Fly's Eye experiment, full-sky coverage is not needed because HiRes detectors are able to detect showers up to 40 km away, whereas the scale height of the atmosphere itself is only about 8 km.

During 1993-1996, a group of 14 prototype mirror units were operated in a "tower" configuration, with the five highest mirrors pointing up to 70° in elevation. The tower

CP867, *XII International Conference on Calorimetry in High Energy Physics,*
edited by S. R. Magill and R. Yoshida

had a limited azimuth ($\sim 30^\circ$) in its field of view, which is centered on the MIA muon array located ~ 3.5 km away. This combination of detectors represented the World's first hybrid (fluorescence + ground array) detector system, preceding the AUGER experiment by more than a decade. The composition results from the HiRes-MIA combination are included later along with those of HiRes.

Each HiRes mirror units uses the same basic optical design. Accounting for obscuration by the PMT cluster, each mirror has an effective collection area of 3.72 m^2. Each mirror is instrumented with a 16×16 array of PMTs in a hexagonal close-packed cluster. Each PMT pixel covers a 1° cone in the sky. Each mirror unit thus covers about 16° in azimuth, and about 14° in elevation angle. The 1° pixel size gives HiRes a factor of five improvement in resolution. Since air showers are approximately linear, this five-fold improvement results in a corresponding five-fold increase in the signal-to-noise ratio for the individual pixels. The increase in sensitivity gives HiRes an order-of-magnitude improvement in the aperture over that of the Fly's Eye. Additional details of the detectors components and readout electronics can be found elsewhere [2, 3]). The HiRes experiment ceased routine observation in April of 2006.

HIRES RESULTS

Energy Spectrum

The HiRes experiment was designed to measure the energy spectrum, composition, and anisotropy of ultrahigh energy cosmic rays above 10^{18} eV. The combined monocular spectrum (from HiRes-1 and HiRes-2 monocular measurements) was first published in 2004 [4]. We show here a recent spectrum (As of June 2006) in Figure 1, along with the stereo spectrum reported by the Fly's Eye [5] and the HiRes-MIA hybrid detector [6]. This spectrum clearly shows two spectral features expected from attenuation from the cosmic microwave background: (a) the GZK suppression (from photo-pion production) [7] near $10^{19.8}$ eV and the associated *pile-up* just below, and (b) the ankle structure at $10^{18.5}$ eV. The latter is consistent with previous measurement shown.

Composition and Cross-Section Measurements

The depth of shower maximum (X_{MAX}) from the HiRes stereo data was used in a composition study of UHE cosmic rays above 10^{18} eV. The stereoscopic observation of HiRes offers the advantage of redundancy. The difference between the measurements from the two sites can be used to check the accuracy of the Monte Carlo simulations, especially in its ability to model the detector resolutions. Figure 2 shows the relative difference between the X_{MAX} values measured from HiRes-1 and HiRes-2 detectors, shown for both data (left) and Monte Carlo (right). The vertical axes in these graphs are in log scale. The two histograms have the same width. This excellent agreement serves to validate the X_{MAX} resolution of ~ 30 g/cm^2 obtained from simulations [8].

The average X_{MAX} vs. energy, as measured by HiRes, has been published [8]. These are shown, with the measurements at lower energies from the HiRes/MIA hybrid, in

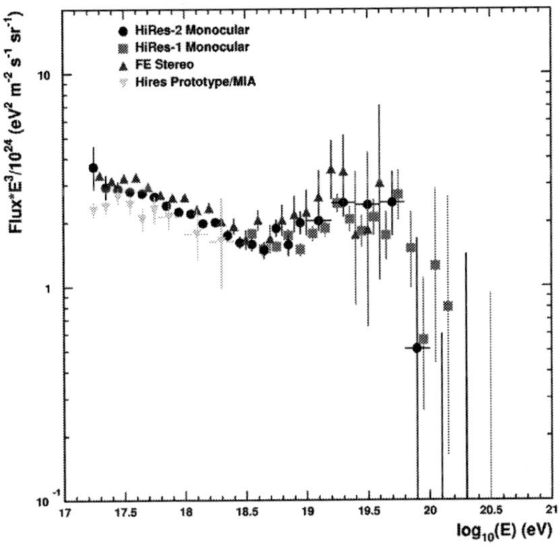

FIGURE 1. The HiRes monocular spectrum (6/2006) plotted with the Fly's Eye stereo [5] and HiRes/MIA [6] spectra. The HiRes measurement shows clearly an ankle structure at $10^{18.5}$ eV consistent with previous measurements, and a suppression at $10^{19.8}$ eV consistent with the GZK Effect [7].

Figure 3. Also included are model predictions for proton and iron. The HiRes results, which spans the energy range of $10^{18.0} - 10^{19.3}$ eV, are consistent with an apparently constant, and predominantly light composition. The elongation rate, $dX_{MAX}/d\log E$, roughly parallel to the model lines, is also indicative of constant composition. In contrast, the HiRes/MIA results show a transition from predominantly heavy composition at $\sim 10^{17}$ eV to a predominantly light one by 10^{18} eV.

Next, the stereo data has also been analyzed to extract the proton-air cross-section by fitting the slope of the exponential tail of the X_{MAX} distribution for values greater than 700 g/cm^2. This analysis uses a reverse Monte-Carlo deconvolution to fit for the exponential slope, λ_{p-air}, of the distribution of the interaction depth of the primary cosmic rays in the atmosphere. The result of the de-convolution fit applied to the HiRes stereo data (up to Jan. 2004) is shown in Figure 4. From a fitted value of $\lambda_{p-air} =$ 52.7 cm^2/g, a cross-section of 456 ± 17(stat.)+39(sys.)-11(sys.) mb was obtained for $E = 10^{18.5}$ eV. Details of this analysis can be found elsewhere [9].

The proton-air cross-section was translated to a proton-proton cross-section [10] and this result is shown in Figure 5. The figure also shows accelerator-based pp data selected by M. Block and fitted to a parametric model [10]. The extrapolation of this model gives a prediction that lies within the error bars of the HiRes result. The figure also shows the results from the Fly's Eye experiment and the old Akeno Array, renormalized with a k-factor of 1.32 obtained from CORSIKA simulations (using both QGSJet and Sybill) [10].

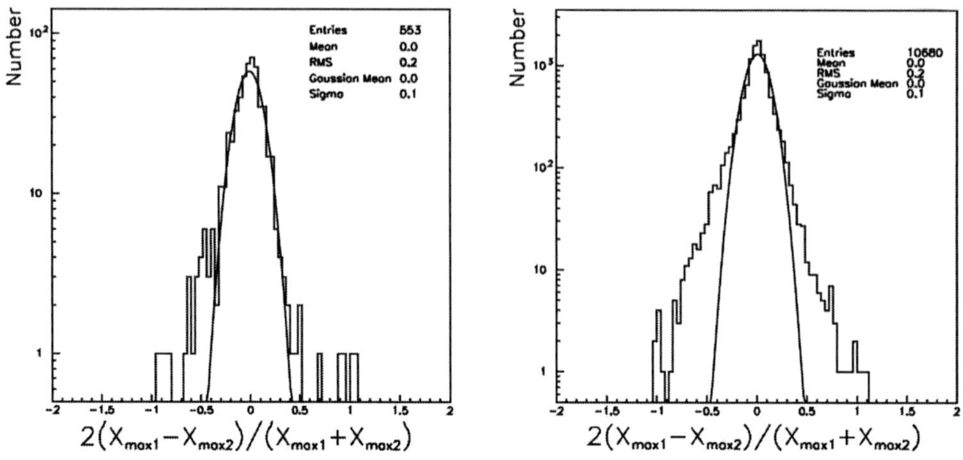

FIGURE 2. Histograms of the relative difference between the X_{MAX} values measured from HiRes-1 and HiRes-2 detectors, shown for data (left) and Monte Carlo (right).

Anisotropy

Several studies have been performed to search for both small and large-scale anisotropy using the HiRes-1 monocular data. These have yielded null results. They include searches for point-sources [11] and cross-correlation with the AGASA doublets and triplet [12, 13]. Null results were also obtained in a search for dipole enhancements in the direction the Galactic Center, Centaurus A, and M87 [14]. In particular, HiRes did not observe any small-scale clustering such as that reported by AGASA [15, 16].

Anisotropy studies were also performed on the HiRes stereo data, which has better than 0.6° angular resolution. One study showed the stereo events above 10^{19} eV to be consistent with isotropy at all small angular scales [17]. A search for coincidence with AGASA doublets and triplets did find one event in coincidence with the AGASA triplet. However, a Monte Carlo study of 10,000 isotropic datasets with yielded 47 sets which gave the same coincidence with equal or greater statistical significance. This is consistent with a $\sim 2.5\sigma$ fluctuation. Moreover, the chance probability of $\sim 5\times 10^{-3}$ does not take into consideration the penalty associated with the *a posteriori* AGASA cuts, which would further weaken the significance of the observed coincidence.

An apparent correlation between HiRes stereo events and BL-Lacertae (BL-Lac) objects was first noted by Gorbunov *et al.* [18]. Using a binned analysis, the authors compared the positions of HiRes stereo events above 10^{19} eV extracted from previous HiRes publications to those of the 156 “BL" objects in the Veron Catalog [19], and found 11 events to lie within 0.8° of a “BL" object, with a chance probability estimated at $\sim 10^{-3}$. An independent analysis by the HiRes group using an unbinned maximum-likelihood method reproduced the apparent correlation with a chance probability of $\sim 10^{-4}$.

The HiRes BL-Lac correlation has been published [20] with data up to January of

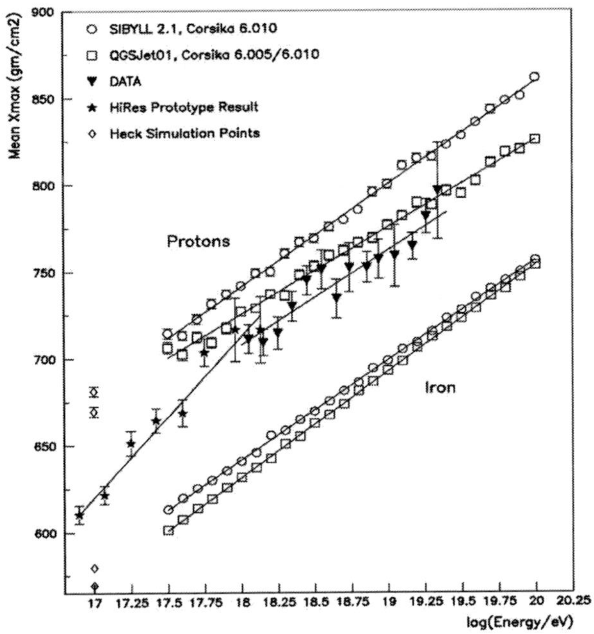

FIGURE 3. Average measured X_{MAX} for UHE cosmic rays plotted as a function of energy for both HiRes stereo data [8] and for the HiRes/MIA hybrid [6].

2004. An illustration of this correlation is shown in Figure 6, which plots the the number of pairs of HiRes events and known BL-Lac objects as a function of θ, their angle of separation. They are histogrammed in one degree angular bins and the pairs weighted by $1/\sin(\theta)$ to remove the sharp rise from zero of ordinary phase-space. We note that this plot was not used for the actual statistical analysis in the publication. A peculiar feature of this observation is that the correlation is consistent with the HiRes stereo angular resolution.

TELESCOPE ARRAY

Currently, a new hybrid experiment is under construction in Millard County, Utah. This new project is a Japan-U.S. collaboration that will deploy a ground array of 576 plastic scintillation counters in a rectangular grid of 1.2 km spacing. Three new fluorescence detectors will initially be deployed at the periphery of the ground array. Each of these stations will provide a field of view of $3° - 17°$ in elevation and $108°$ in azimuth, which is sufficient to provide full coverage of the ground array at energies above 10^{19} eV, which is about 8 times the size of AGASA.

The existing 64 HiRes mirror units will be redeployed at two additional sites inside the ground array at ~ 6 km from two of the new TA fluorescence stations. These new stereo pairs will extend stereo fluorescence coverage down to $\sim 10^{17.5}$ eV in order to

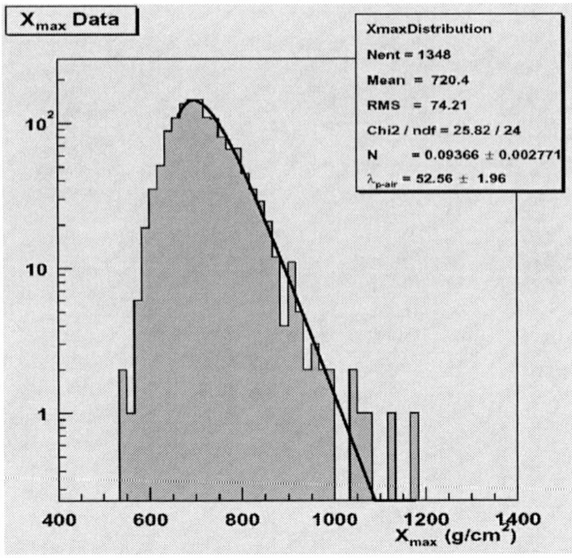

FIGURE 4. Histogram showing the measured X_{MAX} distribution and the result of the de-convolution fit to the exponential tail. A value of $\lambda_{p-air} = 52.7$ cm^2/g was obtained, which corresponds to a proton-air cross-section of 456 ± 17(stat.)+39(sys.)$-$11(sys.) mb.

provide a reliable stereo measurement of the energy spectrum over the ankle region. In addition, this configuration of detectors will also extend hybrid stereo coverage (where each event is measured by at least two fluorescence detectors and the ground array) over the entire ground array for energies above 10^{19} eV. In stereo-hybrid mode, the angular resolution will approach $\sim 0.1^\circ$. This attribute will make TA an extremely powerful cosmic ray anisotropy detector.

TA will also construct 15 additional detector units with 3 m diameter mirrors. These will be placed at one of the 6 km sites overlooking an infill ground array. With elevation coverage comparable to the original HiRes prototype tower, larger mirrors, and broader azimuthal coverage, the new tower detector will overlap with the stereo TA measurements and extend it down to $10^{16.5}$ eV.

The construction and deployment of TA detector elements are well under way. The ground array and the peripheral fluorescence sites will be completed in the spring of 2007. The TA project has also received funding support in the U.S. from the National Science Foundation (NSF). The interior fluorescence detectors and infill array should begin construction in early 2008.

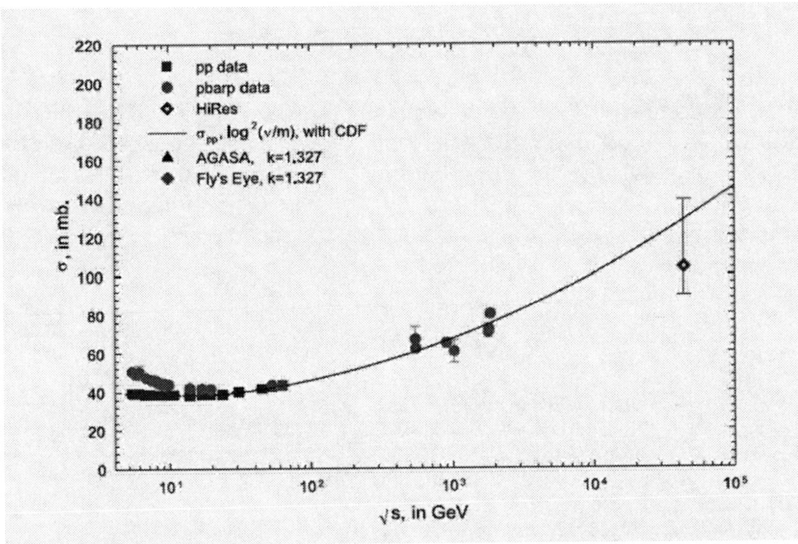

FIGURE 5. The HiRes proton-air inelastic cross section converted to a proton-proton cross-section [10]. The HiRes data point is shown with an extrapolation of lower-energy pp data [10], which gives a prediction that lies within the error bars of the HiRes result. Also shown are the Fly's Eye and Akeno results re-normalized using a k factor of 1.32 obtained from CORSIKA simulations (with QGSJet and Sybill) [10].

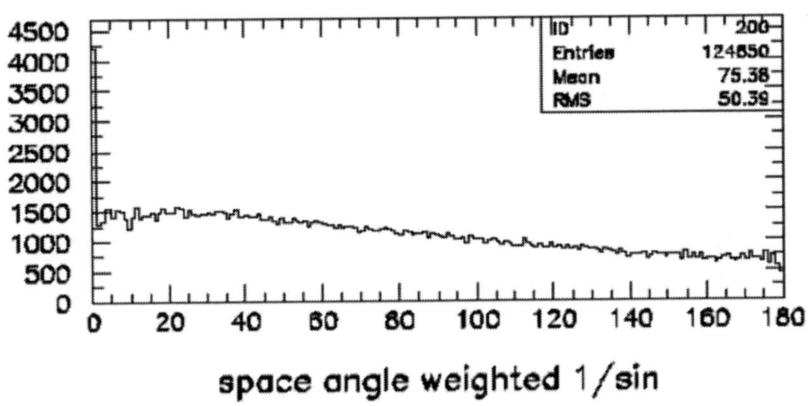

FIGURE 6. The number of pairs of HiRes events and known BL-Lac objects as a function of θ, their angle of separation. They are histogrammed in one degree angular bins and the pairs weighted by $1/\sin(\theta)$ to remove the sharp rise from zero of ordinary phase-space. We note that this plot was not used for the actual statistical analysis in the HiRes publication

ACKNOWLEDGMENTS

This work is supported by US NSF grants PHY-9321949, PHY-9322298, PHY-0098826, PHY-0245428, PHY-0305516, PHY-0307098, by the DOE grant FG03-92ER40732. We gratefully acknowledge the contributions from the technical staffs of our home institutions and the Utah Center for High Performance Computing. The cooperation of Colonels E. Fischer and G. Harter, the US Army, and the Dugway Proving Ground staff is greatly appreciated.

REFERENCES

1. R. Cady *et al.*, Proc. 18th ICRC (Bangalore), **9**, 351, (1983).
2. T. Abu-Zayyad *et al.*, Proc. 26th ICRC (Salt Lake City), **5**, 349, (1999).
3. J. Boyer *et al.*, NIM **A482**, 457 (2002).
4. R. U. Abbasi *et. al.*, Phys. Rev. Lett. **92** 151101, (2004).
5. D.J. Bird *et al.*, Phys. Rev. Lett. **71**, 3401, (1993).
6. T. Abu-Zayyad *et al.*, Phys. Rev. Lett. **84**, 4276, (2000).
7. K. Greisen, Phys. Rev. Lett. **16**, 748 (1966); G.T. Zatsepin and V.A. K'uzmin, Pis'ma Zh. Eksp. Teor. Fiz. **4**, 114 (166) [JETP Lett. **4**, 78 (1966)].
8. R. U. Abbasi *et. al.*, Astrophys. Journal **622**, 910-926, (2005).
9. K. Belov for the HiRes collaboration, Nucl. Phys. B (Proc. Suppl.) **151** (2006) 197-204.
10. To appear in Proc. Physics At The End Of The Galactic Cosmic Ray Spectrum Conference.
11. R. U. Abbasi *et. al.*, submitted to Astroparticle Phys.
12. M. Teshima *et al.*, Proc. of the 28th ICRC (Tsukuba, 2003) 437.
13. R. U. Abbasi *et. al.*, submitted to Astroparticle Phys.
14. R. U. Abbasi *et. al.*, Astroparticle Phys. **21** (2004) 111-123.
15. M. Takeda *et al.*, Astrophys.J. **522** (1999) 225-237
16. R. U. Abbasi *et. al.*, Astroparticle Phys. **21** (2004) 111-123.
17. R. U. Abbasi *et. al.*, Astrophys. J. **610** (2004) L73-76.
18. D. S. Gorbunov, P. G. Tinyakov, I. I. Tkachev, and S V. Troitsky, JETP Lett. **80** (2004) 145-148; Pisma Zh.Eksp.Teor.Fiz. **80** (2004) 167-170
19. Veron-Cetty,M.-P., and Veron, P. 2000, A Catalogue of Quasars and Active Nuclei (9th ed.; Garching: ESO) A&A, **374**, (2001) 92.
20. R. U. Abbasi *et. al.*, Astrophys. Journal **636** (2006) 680-684.

VERITAS: An Atmospheric Imaging Calorimeter

Elizabeth Hays

Enrico Fermi Institute, 5640 S. Ellis Ave, University of Chicago, Chicago, Illinois 60637
Argonne National Laboratory, Argonne, Illinois 60439

Abstract. VERITAS is one of a new generation of imaging air Čerenkov telescope arrays built to study astrophysical sources emitting gamma rays in the GeV to TeV energy range. The catalog of objects that can be studied by these ground-based instruments is expanding rapidly. The observations now possible make a critical contribution to understanding some of the most energetic phenomena in the universe. The imaging Čerenkov technique makes use of atmospheric calorimetry. The types of objects studied and the optical nature of ground-based gamma-ray telescopes places them solidly within the field of astronomy. However, the strong connections to particle processes, the electronic instrumentation, and reconstruction methods are those familiar to high energy particle physics. The advances in instrumentation that have brought about the current success of the field are reviewed and prospects for future improvements discussed.

Keywords: ground-based gamma-ray instrumentation, imaging air Čerenkov technique
PACS: 95.55.Ka

INTRODUCTION

Ground-based gamma-ray observatories fill a critical role in study of the nonthermal universe. A variety of astrophysical sites, including the locations of shock fronts from supernovae, pulsar wind interactions, and relativistic particle jets caused by massive black holes, emit very high energy (VHE, $E_\gamma > 100$ GeV) gamma rays. The observation of these extreme environments allows the study of accretion and explosion processes that drive galactic evolution and have brought about the present-day universe. Gamma-ray telescopes probe the sites of high energy particle acceleration both within and externally to our galaxy. Since the discovery of TeV emission from the Crab Nebula, a supernova remnant, using the Whipple 10-m imaging Čerenkov telescope[1], the instruments and techniques have advanced and the catalog of objects visible at energies from hundreds of GeV to tens of TeV has grown in both size and variety. Detected gamma-ray sources include BL Lacertae-type active galactic nuclei, shell-type and plerionic supernova remnants, microquasars, a pulsar binary system, diffuse emission from the galactic plane, the center of our galaxy, and a number of sources that do not have known counterparts at other wavelengths [2, 3, 4] .

The detectors used to study these objects have recently entered a new regime in sensitivity. To understand the expansion of the TeV source catalog and the potential for further advancements in the field, we explore the basis of the imaging technique and the underlying air shower physics that allows a calorimetric approach to be taken. The rest of the paper is organized as follows: section I presents the status of VERITAS; section II reviews aspects of the imaging Čerenkov technique; and section III discusses some

CP867, *XII International Conference on Calorimetry in High Energy Physics,*
edited by S. R. Magill and R. Yoshida

limitations of the current instruments and possible avenues for improvement.

VERITAS

VERITAS [5, 6] consists of four imaging Čerenkov telescopes with 12-m dimeter reflectors currently under construction at the Fred Lawrence Whipple Observatory in southern Arizona (111° W, 32°N). The optics is f/1.0 Davies Cotton design. Each camera covers a 3.5° diameter field of view and is made up of 499 PMTs. A specially designed light-collecting cone is mounted on each PMT to improve efficiency. Telescope motion is controlled by an alt-az mount capable of slewing at $1^\circ/sec$. The accuracy for tracking sources is better than 20 arcsec.

The trigger is applied in three successive levels. At the first level each PMT channel is monitored by a constant fraction discriminator. The second level is applied at the telescope level and is a pattern trigger requiring a coincidence of neighboring channel triggers. The third level trigger operates at the array level and requires a coincidence of telescope triggers. The thresholds, multiplicites, and time windows at each stage of the trigger are programmable and will be optimized to maximize the performance of the array.

When the array trigger condition is met, the telescopes participating in the array receive a signal indicating data should be recorded. The PMT signals are digitized using 500 MHz FADCs with a 64 μs memory buffer. This leaves ample time to make a trigger decision at the array level and then communicate the trigger signal to the telescopes. It also allows for the development of specialized trigger logic for specific applications. Two gain inputs are used to avoid saturation by large pulses. Additionally, the FADC data can be suppressed on a channel-by-channel basis at the hardware level to reduce the raw data size and data acquisition deadtime. At an array trigger rate of 150 Hz, the deadtime is less than 10%.

Commissioning observations made with the first two telescopes detected signals from the Crab Nebula, Mrk 421, and Mrk 501. The full array will begin operation in late 2006 and the planned science program will begin in 2007. In addition to further study of known gamma-ray emitters and observations of new objects in known classes, new source classes will be actively pursued in searches including gamma-ray bursts and candidate sites of dark matter annihilation.

ATMOSPHERIC IMAGING CALORIMETRY

The imaging air Čerenkov technique, which is employed in VERITAS, relies on the use of the atmosphere as a calorimeter for incident VHE photons. The goal of optical observations of extensive air showers is to collect the Čerenkov light generated by the passage of relativistic charged particles through the atmosphere, image it, and thus, record the track of the shower through the atmosphere. The properties of this track are used to indicate the primary particle type, the location of origin in the sky, the axis of shower development, and the energy of the incident particle. Figure 1 shows a gamma-like image recorded using two VERITAS telescopes.

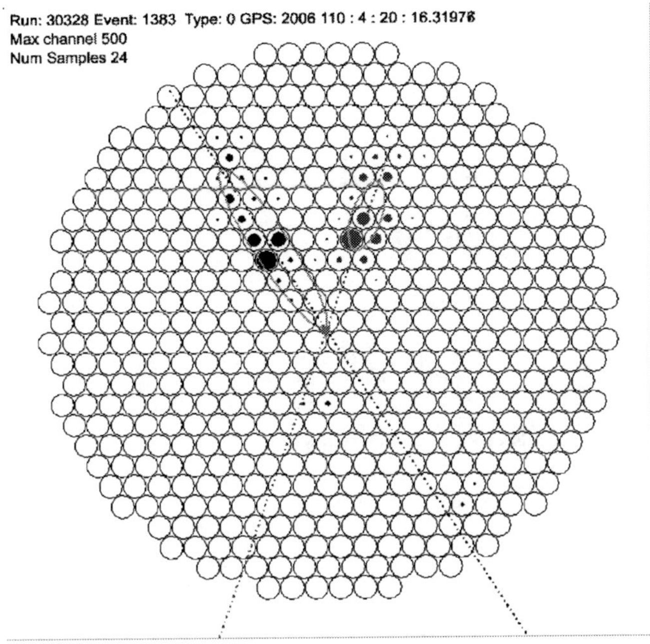

FIGURE 1. A gamma-like shower image in the VERITAS telescopes. The camera display shown for this event superimposes plots of the summed charge per channel from two telescopes. Each channel maps to an angular position in the field of view. The ellipses correspond to the Hillas parameter fits for each image with a dashed line indicating the major axes. The star marks the best fit location of the shower origin using both images.

Little of the energy dissipated in the electromagnetic cascade at these energies goes into muons, and so the Čerenkov light yield can be used as a reliable indicator of the primary energy if the primary type and the relative location of the shower development axis to the telescope are known. The energy resolution is limited by the reconstruction in accurate particle identification and good knowledge of the shower core location, by simulations in correctly portraying the atmosphere and the detector response, and ultimately by the fluctuations in the shower development.

To image an air shower, a telescope must be located somewhere within the region illuminated by the shower-generated Čerenkov light. The majority of the light is produced by the core of the cascade and at shower maximum, which occurs at an altitude of around 10 km for typical gamma-ray induced showers. The resulting light pool is roughly 250 m in diameter on the ground. Figures 2 and 3 show the lateral distributions of the average photon density for gamma-ray and proton induced air showers. Of note is the difference in shape of the profiles and the lower relative light yield of proton-induced showers. This is partly because more of the shower energy can escape into muon and neutrino channels and partly due to the difference in interaction heights for gamma rays and protons of the same energy. The differences between the hadronic and electromagnetic cascades are exploited to reject the cosmic ray nuclei background. This is important both for de-

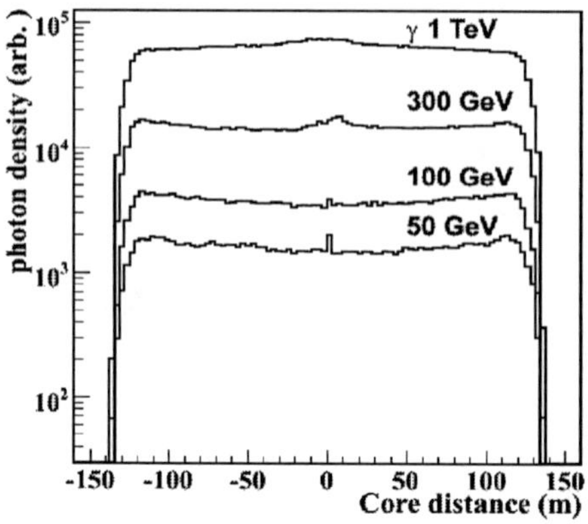

FIGURE 2. Average lateral density of Čerenkov photons at ground level for different incident gamma-ray energies.

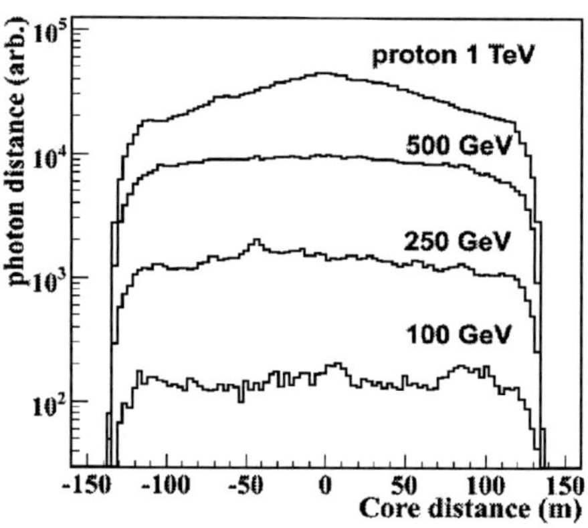

FIGURE 3. Average lateral density of Čerenkov photons at ground level for different incident proton energies.

tection sensitivity and for energy reconstruction since a contamination of the signal will introduce a low energy bias.

Hadron Rejection

The biggest complication for the technique is that the dominant background from impacting cosmic ray nuclei (primarily protons) has a flux that is orders of magnitude larger than that of gamma-ray sources at these energies. The signal to background ratio for events triggering a telescope is of order 1:1000. This background must be

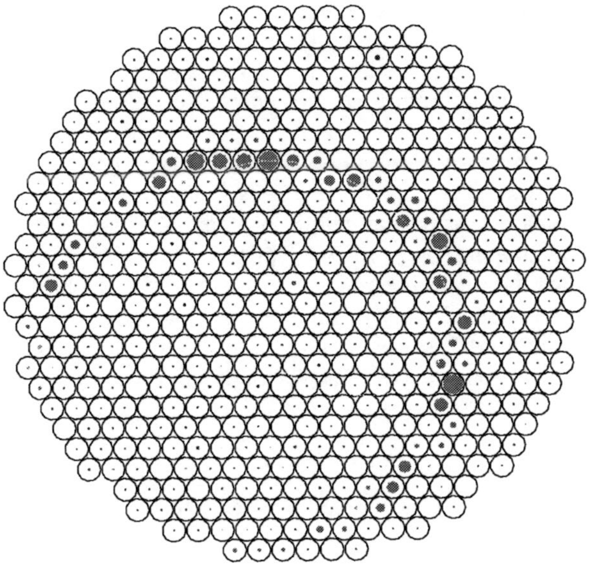

FIGURE 4. A muon image in one of the VERITAS telescopes. A nearby muon producing a nearly full ring is easily rejected by analysis cuts, but partial muon rings have a similar appearance to gamma images. Data from multiple telescopes allow partial rings to be distinguished from gamma events. This background is greatly reduced in an array at the trigger level by requiring multiple coincident telescopes. Further analysis cuts remove muon events passing the trigger requirements.

rejected with a very high efficiency to allow the detection of a gamma-ray signal. The differences in development between hadronic and electromagnetic showers are manifest in the shower image and can be exploited to achieve rejection power in excess of 99% efficiency.

The first detection of a TeV gamma-ray source was made using a single 10-meter telescope[1]. The reconstruction developed for analyzing this source is based on characterizing the shape and orientation of the shower image using Hillas parameters [7]. After image cleaning is applied to remove spurious hits, the moments of the shower image are used to construct the width, length, and orientation of the shower in the field of view. The major axis defined by this fit indicates the direction of the shower and provides one of the most powerful rejection tools. While cosmic ray showers come from all parts of the field of view, a gamma-ray point source generates showers originating from a single point in the sky. The angular difference of the measured shower track and the line from the centroid of the image to the center of the camera is known as the alpha parameter.

[1] The Whipple 10-m has since undergone several upgrades and is still in use at the Fred Lawrence Whipple Observatory.

Requiring alpha to be small eliminates the large fraction of the cosmic-ray background. Additional use of the ratio of the length of the shower image to the integrated charge, or size, helps to reduce a secondary background caused by atmospheric muons passing near the telescope. The image of a muon event recorded in one of the VERITAS telescopes is shown in Figure 4.

The shape of the shower image provides some simple, yet powerful tools for rejecting the cosmic-ray background. Hadronic cascades are broader than the electromagnetic cascades. This is apparent in the width of the shower image. With knowledge of the shower core provided by stereoscopic observations, the image width can be adjusted to take into account the relative position of the shower axis and the participating telescopes. Referred to as a mean-scaled width, this parameter is tested against expectations for the gamma ray shape determined from simulations. A similar treatment can be made of the image length. These are examples of rejection parameters that require multiple telescope images for their effectiveness.

The Power of Arrays

VERITAS is designed to trigger most efficiently on showers from photons with energies above $\sim$ 100 GeV. Achieving a low energy threshold requires a sufficient mirror area of the individual telescopes, optimal baselines between telescopes, and a coincidence trigger of multiple telescopes. The first increases the light collected per shower and allows detection of lower energy primary photons. The last allows a decrease in trigger threshold by rejecting single telescope backgrounds from the night sky and muons. The lower threshold translates into improved detector efficiency for lower energy gamma-ray primaries.

The use of multiple telescopes greatly enhances reconstruction of shower parameters [8, 9] because views of the shower from different locations on the ground are only partially correlated. The photon density for individual gamma showers remains relatively uniform while that of hadronic showers has nonuniformities and asymmetries. Detectors placed at different positions within the Čerenkov light pool provide strong constraints on the location of the shower core and the primary particle direction.

Viewing showers stereoscopically allows a more accurate reconstruction of the shower geometry than is possible with a single telescope. This improves the energy resolution dramatically in comparison to single telescope observations. Table 1 compares typical resolutions for several telescope configurations. The use of multiple telescopes is the biggest factor in improving angular and energy resolution. The use of larger reflectors enhances the sensitivity to lower energy primaries, which have lower photon densities, and reduces the energy threshold.

FUTURE DIRECTIONS

The current generation of IACTs are producing substantial results, but further improvements in the technique are possible and beneficial to the science that can be done at VHE energies. Future instruments could involve some combination of larger arrays,

TABLE 1. Comparison of performance characteristics of different IACT telescope configurations. These are typical values and can vary depending on properties of an individual source or analysis.

Number Telescopes	Mirror Area Per Telescope	Angular Resolution	Energy Resolution	Energy Threshold
1	80 m^2	0.4°	30%	> 300 GeV
5*	8.5 m^2	0.1°	20%	> 500 GeV
4	100 m^2	0.1°	15%	> 100 GeV

* Based on the HEGRA experiment[10].

larger telescopes, higher altitudes, and improved technologies. The single factor with the largest potential impact on detector performance is the quantum efficiency of the photo-detectors. The current cameras use photomultiplier tubes that are only around 20% efficient.

Analysis based on image shape and orientation has proved effective and robust. There are additional methods that seek to exploit some other properties of air showers to improve reconstruction and background rejection. The abundance of extended galactic sources makes it particularly desirable to find methods of background rejection that are effective over large areas of the field of view. Extragalactic sources require further reductions in the energy threshold to achieve detections of more distant galaxies and sensitivity to critical regions of the energy spectrum.

There are several features of air showers that may be useful in reducing backgrounds and could play a role in future reconstruction methods.

- Inhomogeneity in light density
- Shower asymmetry
- Timing properties of the Čerenkov light

To take practical advantage of any of these properties, however, data and simulation studies at lower energy thresholds are needed.

CONCLUSION

The combination of using 10-m class reflectors in stereoscopic arrays has produced ground-based gamma-ray observatories that provide valuable and reliable astronomical observations. The enhanced sensitivity and energy resolution of these detectors is largely due to applying calorimetry methods to air showers. The VERITAS experiment is now nearing completion. It has been designed to operate at low energy thresholds and makes use of electronics and triggering capabilities that are unique among IACT arrays. Future advancements beyond the current telescope systems are desirable to reach sensitivity to 50-100 GeV photons and to improve observations of large or diffuse sources. This will require new techniques and technologies and will benefit from continued overlap in these areas with high energy physics.

ACKNOWLEDGMENTS

This research is supported by grants from the U.S. Department of Energy, the U.S. National Science Foundation, the Smithsonian Institution, by NSERC in Canada, by Science Foundation Ireland and by PPARC in the UK.

REFERENCES

1. T. C. Weekes, et al., *ApJ* **342**, 379–395 (1989).
2. R. A. Ong, "Gamma ray astronomy, c 2005 (OG-3)," in *proc. 29th ICRC*, 2005, vol. 10, p. 329.
3. D. Horan, and T. C. Weekes, *New Astronomy Review* **48**, 527–535 (2004).
4. F. Aharonian, et al., *ApJ* **636**, 777–797 (2006).
5. T. C. Weekes, et al., *Astroparticle Physics* **17**, 221 (2002).
6. J. Holder, et al., *Astroparticle Physics* **25**, 391–401 (2006).
7. A. M. Hillas, "Cerenkov light images of EAS produced by primary gamma," in *proc. 19th ICRC*, 1985, vol. 3, p. 445.
8. F. A. Aharonian, W. Hofmann, A. K. Konopelko, and H. J. Völk, *Astroparticle Physics* **6**, 343–368 (1997).
9. H. Krawczynski, D. A. Carter-Lewis, C. Duke, J. Holder, G. Maier, S. Le Bohec, and G. Sembroski, *Astroparticle Physics* **25**, 380–390 (2006), astro-ph/0604508.
10. G. Pühlhofer, et al., *Astroparticle Physics* **20**, 267–291 (2003).

Detection of TeV Gamma-Ray Emission from the Cygnus Region of the Galaxy with Milagro Using a New Background Rejection Technique

A. A. Abdo for the Milagro Collaboration

Michigan State University, Dept. of Physics and Astronomy, East Lansing, MI, 48824

Abstract. Milagro is a TeV gamma-ray observatory that utilizes a large water Cherenkov detector to observe extensive air showers (EAS) produced by high energy particles impacting the Earth's atmosphere. Milagro's distinct advantage compared to other TeV gamma-ray detectors is that it views a wide field (2 steradian overhead sky) and it continuously operates (> 90% live time). These factors give Milagro the potential for discovery of new sources with unknown positions and times, such as gamma-ray bursts, flaring AGNs, and observation of diffuse and extended sources like the Galactic plane or large supernova remnants. Here we present a new technique for improving background rejection in Milagro as well as a gamma-ray image of the Cygnus region at energies near 12.5 TeV. We report the detection of both an extended source and a large area of diffuse gamma-ray emission from this region of the galaxy. This new extended source has an extent of 0.32 ± 0.12 degrees and an integral flux above 12.5 TeV of $(1.71 \pm 0.24_{stat} \pm 0.34_{sys}) \times 10^{-13} cm^{-2} s^{-1}$ assuming a differential source spectrum of $E^{-2.6}$. The best-fit location for the source is RA= $304.66 \pm 0.13_{stat} \pm 0.25_{sys}$ degrees and declination $36.96 \pm 0.08_{stat} \pm 0.2_{sys}$ degrees.

Keywords: gamma-ray astronomy; extensive air-shower detectors; cosmic-ray detectors
PACS: 95.55.Ka; 95.55.Ka; 95.85.Pw; 98.70.Rz

1. INTRODUCTION

Milagro is a water-Cherenkov detector at an altitude of 2650m capable of continuously monitoring the overhead sky. The Milagro detector is composed of a central 80m x 60m x 8m (depth) pond with a sparse 200m x 200m array of 175 "outrigger" tanks surrounding it. The central pond is instrumented by 723 photo multiplier tubes (PMTs) that are submerged in 24 million liters of purified water which acts as the detection medium for the secondary particles in an air shower. The 723 PMTs are split into two layers: the top "air shower" layer consists of 450 PMTs and is under 1.4m of water, and the bottom "muon" layer has 175 PMTs and is located 6m under the surface of the water. The shallow depth of the air shower layer allows the accurate measurement of shower particles' arrival times used for direction reconstruction and triggering. The greater depth of the muon layer is used to detect penetrating muons and hadrons which are mostly presented in hadronic air showers. The outrigger array extends the physical area of Milagro from 5000m^2 to 40,000m^2, thus improving the core location and angular resolution of the detector by providing a longer lever arm with which to reconstruct events. The angular resolution improves from 1.05° to 0.72° when outriggers are used in the reconstruction.

CP867, *XII International Conference on Calorimetry in High Energy Physics*,
edited by S. R. Magill and R. Yoshida

1.1. Identification and Rejection of Hadronic Events

It is well known that extensive air showers induced by hadronic cosmic rays contain many more muons (from pion decay) and hadrons than EAS induced by gamma rays of comparable energies. The atmospheric overburden at the Milagro site is 20.5 radiation lengths and 8.3 interaction lengths. The muon layer in Milagro is located under 6m of water (corresponding to 17 more radiation lengths and 7.2 interaction lengths). This means that most EM charged particles that enter the pond get absorbed before reaching this layer. On the other hand, muons with energies as low as 1.2 GeV can penetrate and shower near the PMTs of the muon layer. These penetrating muons and hadrons will result in bright compact clusters of light in this layer. Using Monte Carlo simulations we estimate that 79% of all proton showers that trigger Milagro contain a muon and/or a hadron that enters the pond, while only 6% of gamma ray induced air showers contain a muon and/or a hadron that enters the pond.

A new parameter has been developed to parametrize the "clumpiness" in the muon layer. This parameter, A_4, is given by:

$$A_4 = \frac{(fTop + fOut) \times nFit}{mxPE} \tag{1}$$

where

- $fTop$ is the fraction of the air shower layer PMTs hit in an event.[1]
- $fOut$ is the fraction of the outriggers hit in an event.[2]
- $nFit$ is the number of PMTs that entered in the angle fit.
- $mxPE$ is the number of photoelectrons (PEs) in the muon layer PMT with the highest hit.

The first part in the numerator of A_4 carries information about the size of the shower, while nFit carries information about how well the shower was reconstructed. mxPE carries information about the clumpiness in the muon layer.

Figure 1 shows A_4 distribution for gamma Monte Carlo, proton Monte Carlo, and data. We see a clear difference between the Monte Carlo gamma ray showers and the proton showers, while there is a good agreement between the data and the proton Monte Carlo. Figure 2 shows the efficiencies for retaining data, proton Monte Carlo, and gamma Monte Carlo as a function of A_4.

The relative increase in the significance of a signal for a given selection criterion is given by the Quality factor (Q) of the cut. For a large number of events, Q is given by:

$$Q = \frac{\varepsilon_s}{\sqrt{\varepsilon_b}} \tag{2}$$

[1] $fTop = \frac{nTop}{450}$, where nTop is the number of air shower PMTs hit in an event
[2] $fOut = \frac{nOut}{175}$, where nOut is the number of outriggers hit in an event

Where ε_s and ε_b are the efficiencies for retaining the signal and background events, respectively.

Figure 3 shows the Q-factor as a function of the minimum value of A_4 required to retain an event. Requiring events to have $A_4 \geq 3.0$ and $nFit \geq 40$ rejects 97.3% of the simulated proton induced air showers that trigger Milagro and 98.4% of the data (for this data sample), and retains 35% of the gamma-ray induced air showers. This results in a predicted Q-factor of 2.2 comparing Monte Carlo proton events to Monte Carlo gamma-ray events, and 2.7 comparing the data to Monte Carlo gamma-ray events.

1.2. Tests of A_4 on the Crab Nebula

The Crab Nebula acts as a standard candle for gamma-ray astronomy due to its long-term, unchanging energy emission through many wavelengths. Figure 4 shows a map of the statistical significance around the Crab Nebula with the $A_4 \geq 3.0$ and $nFit \geq 40$ cuts applied for 1.7 years of data. In this map the Crab Nebula is seen at 8.02 σ. Figure 5 shows a map of the statistical significance around the same region for a harder A_4 cut of 12 and $nFit \geq 40$. The Crab Nebula is seen at 5.58 σ. Although there was a $\approx 30\%$ loss in statistical significance of the Crab Nebula when the harder A_4 cut was applied, the main advantage of applying this cut is the higher S/B ratio (60.00%) achieved with this cut compared to that with the soft A_4 cut (3.42 %).

1.3. γ-Hadron Relative Weighting

The fact that one can achieve a higher S/B ratios with no major loss in statistical significance when applying harder A_4 cuts led to the development of an alternative analysis method that weights events based on the relative probability that the event was due to a gamma-ray. Events with large A_4 values are assigned high weights, while events with low A_4 are assigned lower weights. Application of this method increases sensitivity by a factor of $\approx 30\%$. Figure 6 shows a map of the statistical significance around the Crab Nebula for the same data set as in figures 3 and 4 with the new analysis method applied. The statistical significance of the Crab Nebula in this map is 10.55 σ. The median energy for gamma ray events with this new anaslysis method is around 12.5 TeV. The angular resolution improves from 0.72° to 0.3° when this method is used.

2. SURVEYING THE TEV GAMMA-RAY SKY

The new weighting analysis method was applied on 5.4 years of data, starting on July 20th 2000 and ending on March 2nd 2006. In total, 1914.56 days of on-time data were included. Sources are only observable as they transit through Milagro's field of view (FOV), so a day of data represents approximately 4-6 hours of exposure for any given northern hemisphere source. Figure 7 shows the northern hemisphere as viewed in TeV gamma-rays. The most prominent feature of this map is the Galactic ridge spanning declinations 0 to 50 degrees and right ascensions 285 to 320 degrees. The Crab Nebula is visible in this map at 14 standard deviations[3].

2.1. Cygnus Region of the Galactic Plane

The Cygnus region is the most luminous source of High Energy gamma rays in the northern sky at energies between 30MeV and 100 GeV as reported by EGRET[1]. High Energy gamma-ray emission from this region is complex and is likely due to the combination of both diffuse and point sources. A large contribution to the VHE emission must come from the interaction of cosmic rays with the relatively dense matter distribution in the region. The HEGRA experiment has detected a localized unidentified TeV source in the Cygnus region with flux of 30 mCrab[2]. A previous Milagro measurement[3] of the diffuse flux from the Galactic longitude region 40 - 100 degrees, normalized to the EGRET measurements in the same region, showed an excess in the region near Galactic longitude 80 degrees - the Cygnus region of the Galaxy[4].

Figure 8 shows the Cygnus region in Galactic coordinates. The contours on the figure show the EGRET model for the diffuse gamma ray emission in the region. At 12 standard deviations this is the second brightest region of TeV emission (after the Crab Nebula) in the northern hemisphere. The crosses in the figure show the location and location error of the EGRET sources in this region (none of which have been definitively identified with counterparts at other wavelengths). While the average angular resolution of Milagro for a single gamma ray is 0.5 degrees, the highest energy gamma rays detected have substantially better angular resolution (0.3 degrees) and background rejection. An examination of the arrival directions of the higher-energy photons shows that this source is most likely an extended source of TeV gamma rays. This new extended source has an extent of 0.32 ± 0.12 degrees and an integral flux above 12.5 TeV of $(1.71 \pm 0.24_{stat} \pm 0.34_{sys}) \times 10^{-13} cm^{-2} s^{-1}$ assuming a differential source spectrum of $E^{-2.6}$. The best-fit location for the source is RA= $304.66 \pm 0.13_{stat} \pm 0.25_{sys}$ degrees and declination $36.96 \pm 0.08_{stat} \pm 0.2_{sys}$ degrees.

ACKNOWLEDGMENTS

We acknowledge Scott Delay and Michael Schneider for their dedicated efforts in the construction and maintenance of the Milagro experiment. This work has been supported by the National Science Foundation, the US Department of Energy, Los Alamos National Laboratory, the University of California, and the Institute of Geophysics and Planetary Physics.

---

[3] The significance of the Crab Nebula on the 5.4 years is less than what one may expect by scaling the significance from the 1.7 years data set, namely $10.55 \times \sqrt{5.4/1.7} = 18.8$. This is due to the fact the larger data set contains data with lower quality (no outriggers) than the 1.7 years data set.

REFERENCES

1. Hartman R.C., Bertsch D.L., Bloom S.D. et al., *Ap. J. Suppl.* **123**, 79 (1999)
2. Aharonian F. et al., *Astron. and Astrophys.* **431**, 197 (2005)
3. Atkins. R.A., et al., *phys. rev. Lett.* **95**, id. 251103 (2005)
4. Hunter, S.D. et al., *Astrophys. J.* **481**, 205-240 (1997)
5. Atkins. R.A., et al., *Astrophys. J.* **595**, 803-811 (2003)

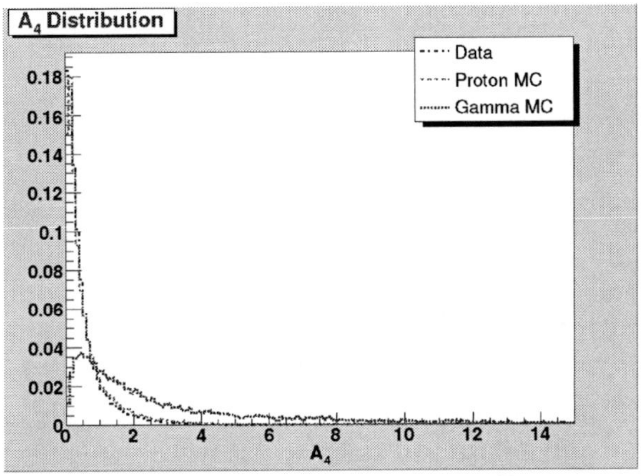

FIGURE 1. Distribution of A_4 for gamma Monte Carlo, proton Monte Carlo, and data.

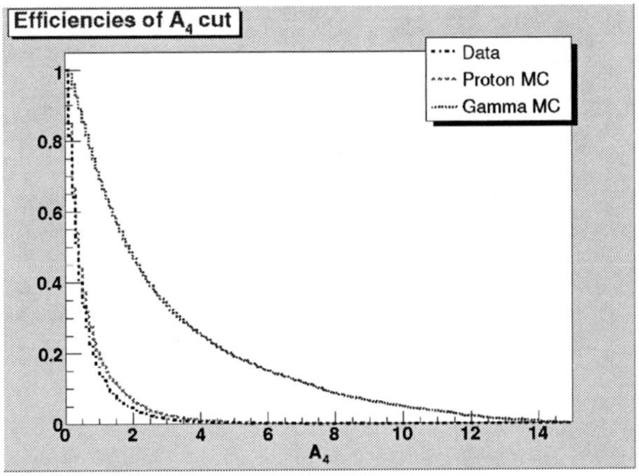

FIGURE 2. Efficiencies for retaining data, proton Monte Carlo, and gamma Monte Carlo as a function of A_4.

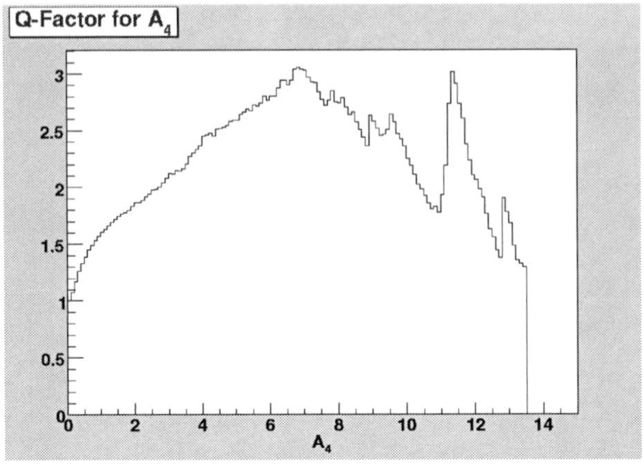

FIGURE 3. Q-factor as a function of the minimum value of A_4 required to retain an event.

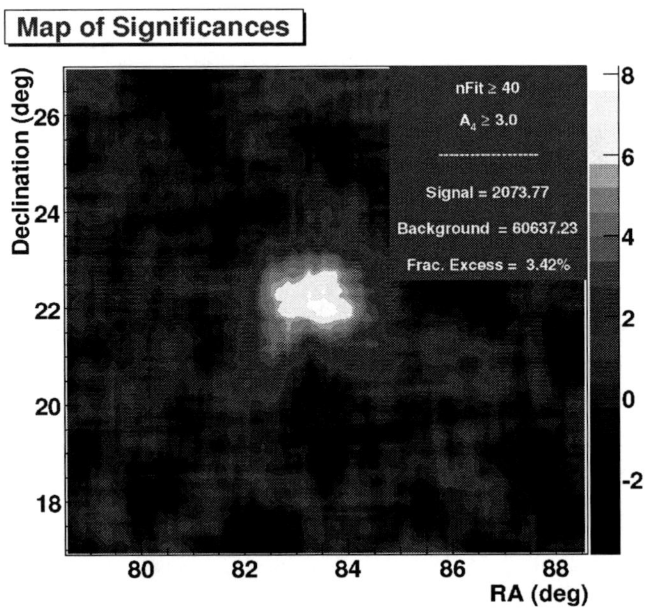

FIGURE 4. Map of the statistical significance around the Crab Nebula with the $A_4 \geq 3.0$ and $nFit \geq 40$ cuts applied.

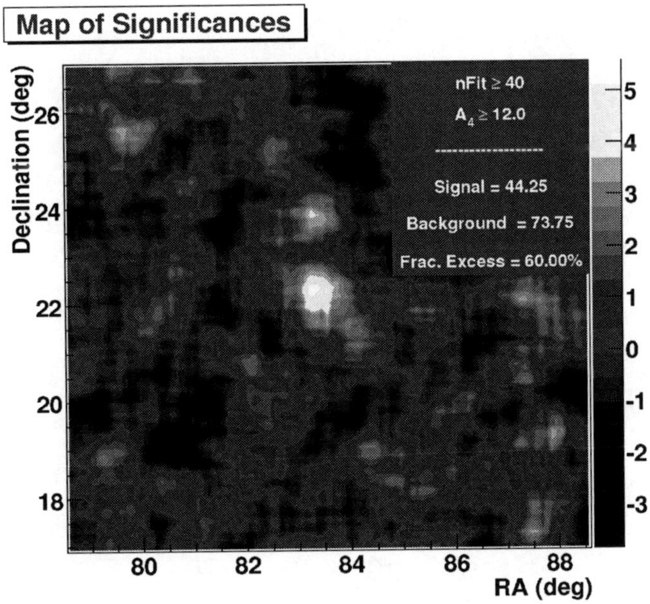

FIGURE 5. Map of the statistical significance around the Crab Nebula with the $A_4 \geq 12.0$ and $nFit \geq 40$ cuts applied.

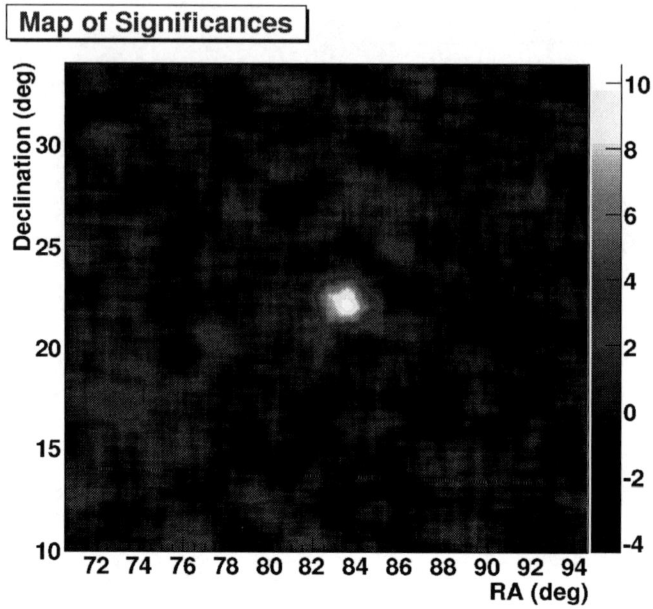

FIGURE 6. Map of the statistical significance around the Crab Nebula with the new γ-Hadron weighting analysis method applied.

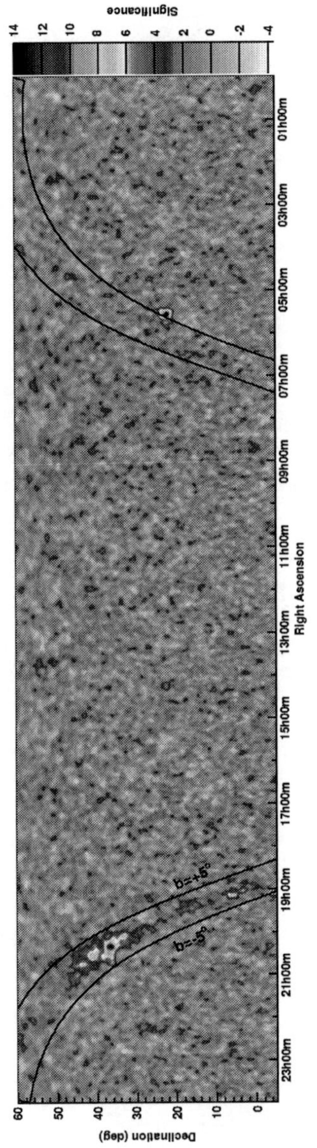

FIGURE 7. The northern hemisphere in TeV gamma rays. This figure shows the statistical significance of the observed excess (or deficit) at each point in the sky (points below 2 standard deviation have been suppressed for clarity). The Galactic ridge is clearly visible in this map. The Cygnus Region is within this ridge near a declination of 37 degrees. The Crab Nebula is the most significant TeV gamma ray source in the northern hemisphere (R.A. 5h 44m and declination 22 degrees), seen here at 14 σ. The dark lines mark a ± 5 degree region around the Galactic plane.

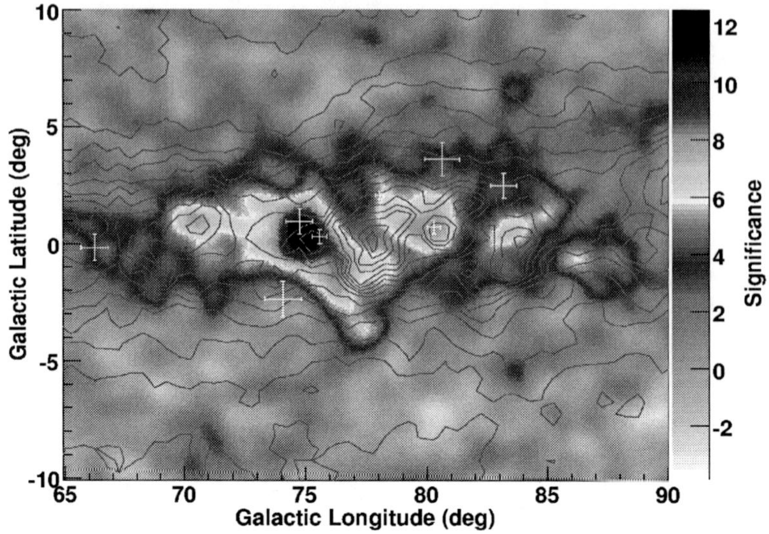

FIGURE 8. The Cygnus Region of the Galaxy as seen in TeV gamma rays. This figure shows the statistical significance of the observed excess (or deficit) at each point in the sky (points below 2 standard deviation have been suppressed for clarity). Superimposed on the image are contours showing the EGRET diffuse gamma-ray emission model for this region of the Galaxy. There is a good correlation between the TeV emission and the EGRET model. The white crosses show the location of the EGRET sources and their corresponding location errors.

The IceCube Neutrino Telescope

Mark Krasberg
for the IceCube Collaboration [1]

IceCube 5th floor, 222 West Washington, Madison, Wisconsin, 53713

Abstract. IceCube is the world's largest neutrino telescope, and it is still in the early stages of construction. In the last 19 months 604 Digital Optical Modules (DOMs) have been deployed, roughly 10% of the planned total. They comprise 540 modules on 9 strings located between 1.5km and 2.5km in the deep ice beneath the South Pole surface, and 64 modules located in a 16-station surface air shower array called IceTop. IceCube's instrumented volume will eventually reach 1 km^3 in the ice and an area of 1 km^2 on the surface.

Each DOM is an independently running detector looking for Cherenkov radiation from charged particles which are produced when neutrinos interact near or in the detector. The deep ice under the South Pole is an excellent calorimetric medium and makes it possible for the array of modules to not only determine the direction of the neutrinos, but to measure their energies as well.

The operation of the DOMs will be discussed, and results will be shown.

Keywords: neutrinos;astrophysics;cosmic rays
PACS: 95.85.Ry,95.55.Vj,98.54.-h

INTRODUCTION - NEUTRINO ASTROPHYSICS IN ICE

Neutrinos are the only known particles that we can observe that carry information about high energy particle production at the far reaches of the Universe. Charged cosmic rays are bent by magnetic fields and are known to be absorbed by resonant scattering with Cosmic Background Radiation photons. High energy gamma rays are absorbed by the interstellar medium. However, since neutrinos are not charged and interact only weakly with matter, they typically reach the earth with the same energy and direction they originated with.

Observation of very high energy neutrinos (in the TeV to PeV enery range) requires enormous target volumes, and large ice or water Cherenkov detectors are a practical means by which to instrument large volumes ([2]-[5]). The density of these media provide a large number of possible target nuclei for neutrino interactions to take place, and the transparency of these media make it possible to detect the Cherenkov light which is emitted by the charged particles (muons, electrons and taus) which can result from these neutrino interactions. Theoretical models predict low fluxes of cosmological neutrinos, which result in neutrino interaction rates of the order of a few events per cubic kilometer per year [2], and hence the need for cubic kilometer scale detectors.

---

[1] For the Collaboration list see http://www.icecube.wisc.edu/pub_and_doc/conferences/calor06/

CP867, *XII International Conference on Calorimetry in High Energy Physics,*
edited by S. R. Magill and R. Yoshida

THE ICECUBE DETECTOR

The Antarctic Muon and Neutrino Detector Array (AMANDA) has provided much of the technical foundation for operating large, deep-ice neutrino telescopes [6]. AMANDA's 677 modules (a combination of both analog and digital optical modules) were deployed between 1996 and 2000 and will continue operation during the construction of IceCube[1]. When fully completed, IceCube will comprise approximately five thousand DOMs (Digital Optical Modules) and will have a 1 km^3 instrumented volume, fifty times that of AMANDA. Figure 1 is a schematic of the planned IceCube detector, which is scheduled to be completed in 2011 [7].

Each DOM is an independently running detector looking for Cherenkov radiation from charged particles which travel near or in the IceCube volume. Of primary interest is the Cherenkov radiation from muons which are produced from the interaction of comological neutrinos in the rock or ice surrounding the detector volume. In a typical cosmological neutrino event approximately one hundred DOMs trigger within a time window of a few microseconds. By examining the amount and timing of light received at each DOM, IceCube will be able to reconstruct the event energy and direction. IceCube will be able to distinguish between three types of neutrino events: single track muons, electron cascades, and tau "double-bang" events when the tau has a large enough energy to make the cascades separable. Figure 2 shows a simulated "double-bang" event which is signified by two isolated cascades and a track between them (the first cascade is the result of recoiling hadronic material in the initial neutrino interaction, and the second cascade is the result of the decay of the tau after it has traveled through the detector).

Meanwhile, the IceTop DOMs instrument 1.8 m diameter ice tanks at the surface and comprise an air shower array which will be used for angular resolution calibration, atmospheric muon veto and for studying the cosmic ray composition. The IceTop array will enable the exploration of the cosmic-ray spectrum using the combined measurement of the electromagnetic and muon components of showers at energies up to 10^{18}eV with high statistics and good energy resolution.

Additionally, the technology used in IceCube will increase the angular and energy resolutions of the array beyond those of AMANDA [2] and therefore will result in improvements in excess of simple geometric gains (see figures 3 and 4). Reference [7] is a detailed report on the physics goals and expected performance of the array. When the IceCube array detects Cherenkov radiation from muons with energy over 1 TeV, the amount of light the detector sees grows with the logarithm of the muon energy (because of the additional Cherenkov light from bremsstrahlung and other fluctuations). For the 10-string AMANDA detector, the energy resolution for $\log_{10}(E)$ was approximately 0.4 [8]. With IceCube, waveform digitization allows for more precise counting of the number of photoelectrons which a DOM detects, leading to better energy resolution. Also, the large dynamic range of a DOM means that saturation will not occur (this is true even for electron cascades which occur near a DOM).

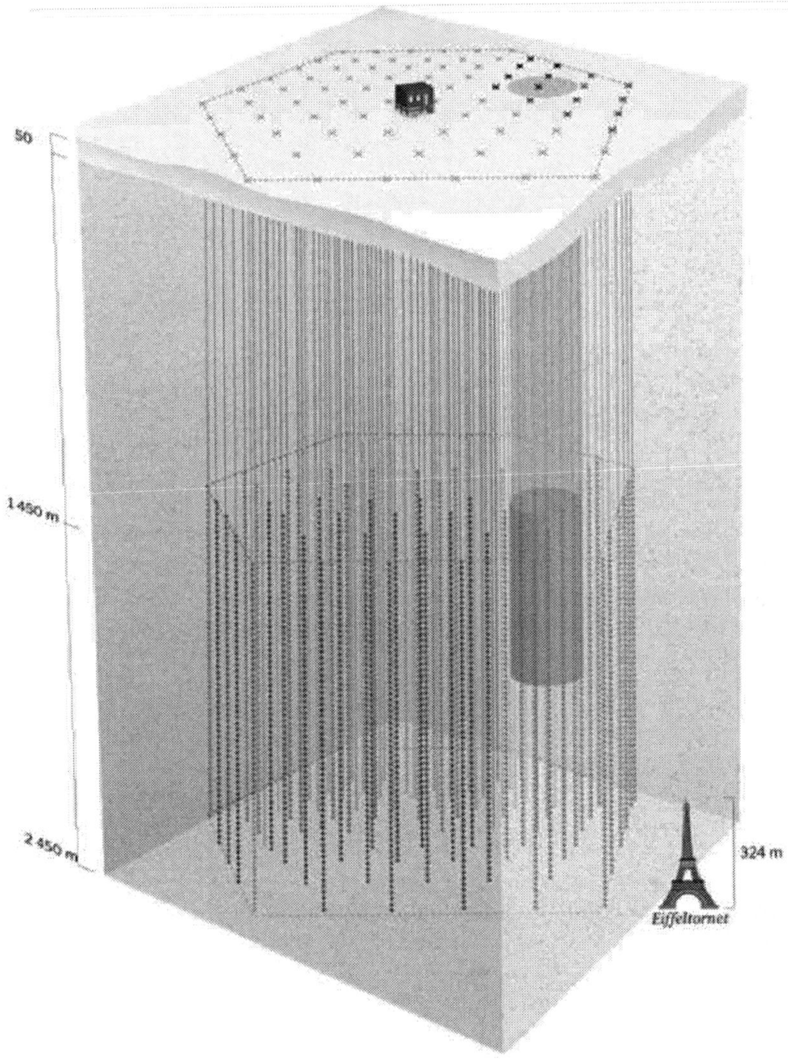

FIGURE 1. The IceCube detector. The vertical lines represent the strings which are separated by approximately 125 meters, and the 60 dots per string represent the in-ice DOMs depicted at depths ranging from 1450 to 2450 meters. At the surface the four-DOM IceTop stations are depicted as crosses above the strings, while the IceCube Lab is positioned at the center of the detector. AMANDA is shown as a cylinder in the upper right-hand corner of the larger IceCube array.

THE DIGITAL OPTICAL MODULE

The DOM (figure 5) is the main component of the detector. It contains a 10-inch Hamamatsu photomultiplier tube (PMT), enclosed in a pressure resistent sphere, which

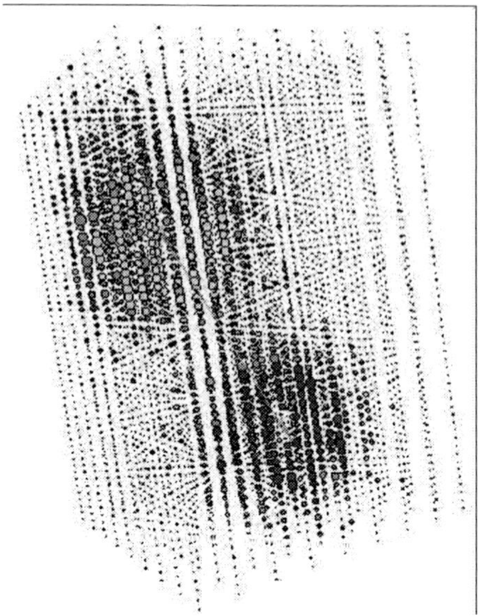

FIGURE 2. Simulated "double-bang" event in the 80-string detector. Electromagnetic cascades are visible from the neutrino interaction that created the tau, and its subsequent decay.

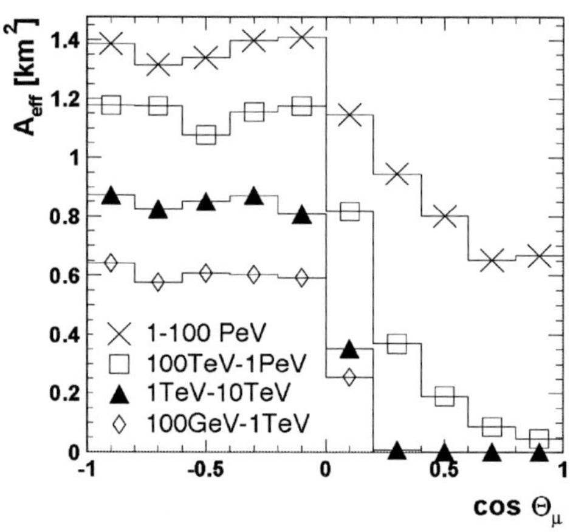

FIGURE 3. The simulated effective area of IceCube as a function of the zenith angle, for four separate energy ranges.

enables it to withstand the enormous pressures of the deep glacial ice. The DOM mainboard contains the analog and digital data acquisition circuitry, and is in effect an autonomous data collection unit. An Analog Transient Waveform Digitizer, or ATWD, records the PMT pulse with three different gains (x1/4, x2 and x16). The ATWD records 128 samples over approximately 425 ns with a rate of 3.3 ns/sample. The same pulse

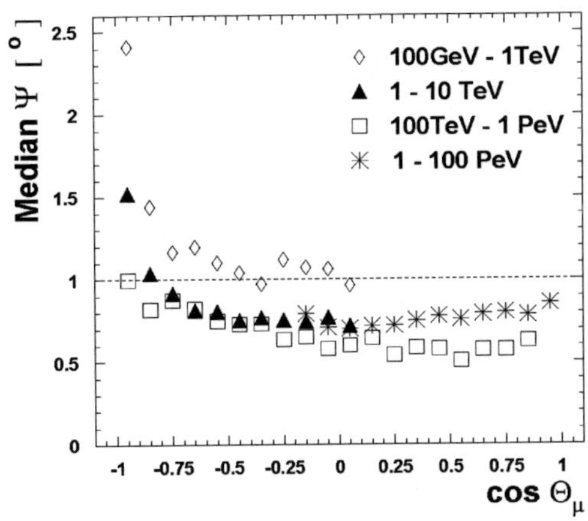

FIGURE 4. The simulated angular resolution of IceCube for neutrino-induced muon events. A additional improvement can be expected from using the digital waveform information in the filtering and reconstruction algorithms.

is also digitized by a 40MHz Flash Analog-to-Digital Converter (FADC) which records 255 samples over 6.4μs with a rate of approximately 25 ns/sample. The linear dynamic range of a DOM is 400 photo-electrons (p.e) per 15 ns during the recording period of the ATWD. Neighboring DOMs are connected to each other electronically and the ATWD is only digitized after a local triggering condition is met. Each waveform is time stamped locally, and the time stamps are synchronized to the GPS time providing better than 5 ns time resolution. The DOM mainboard contains an FPGA (Field Programmable Gate Array) that can be re-programmed from a computer at the surface. A flasherboard containing 12 LEDs sits just above the DOM mainboard and can be used to produce light pulses into the ice for calibration purposes.

DETECTOR CONSTRUCTION

During the 2004-2005 austral summer, the first deployed IceCube string and 4 IceTop stations were installed and commissioned. 60 cm diameter holes are drilled into the ice with a hot water drill, which takes approximately 40 hours to reach the target depth of 2450 meters. During the following 2005-2006 South Pole season, an additional 8 strings and 12 IceTop stations were deployed, bringing the total number of deployed DOMs to 604. More than 98% of these DOMs survived the freeze back and they are successfully taking data.

DETECTOR PERFORMANCE

The results for one year of operation of the first string and four IceTop stations are summarized in [9]. Many down-going muon tracks have been reconstructed, with an

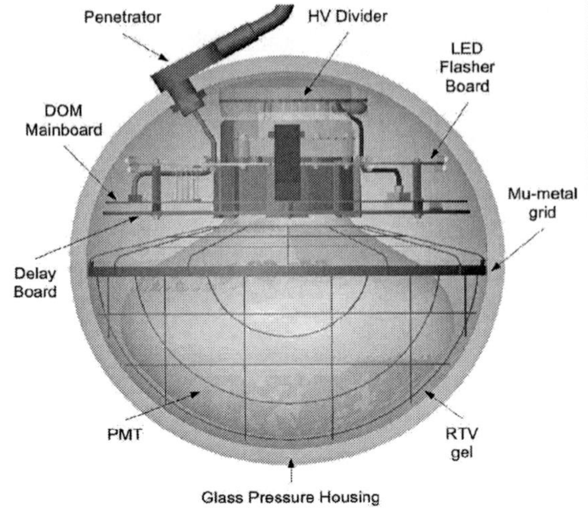

FIGURE 5. Schematic of a DOM

angular distribution consistent with expectations for atmospheric cosmic ray showers above the detector. A small number of up-going muon tracks have also been seen, as expected from interaction of neutrinos produced in cosmic ray showers in the northern hemisphere. Preliminary analyses indicates that the current 9-string in-ice trigger rate of approximately 140 Hz on a request for 8 DOMs to be hit in a time window of $5\mu s$ is also in agreement with expectations.

In order to perform the optimal direction and energy reconstruction, the properties of the ice must be carefully understood. Figure 6 shows the measured scattering coefficient (1/length) as a function of depth ([10]). The peaks correspond to dust layers in the ice. Figure 7 shows the local coincidence trigger rates in the deep ice when a DOM is required to trigger in coincidence with a neighboring DOM. Here the peaks correspond to the clearer ice. As can be seen, the overall DOM response is very uniform across all strings, and the trigger rate is very low. Coincidence triggers are mostly due to down-going muons from atmospheric cosmic ray showers. The rate is higher in clear ice because light from passing muons reaches these DOMs more easily.

A low noise rate helps simplify the reconstruction of high energy neutrino events, and is also beneficial for searches for low energy neutrinos from supernovae ([11]). Using a $51\mu s$ dead-time suppression of after-pulses, the dark-noise rate for each DOM in the deep ice is approximately 350 Hz. Supernova searches rely on increased counting rates in the whole IceCube array over a few seconds, because the Cherenkov light from each low energy neutrino interaction is typically seen as a single hit in one DOM.

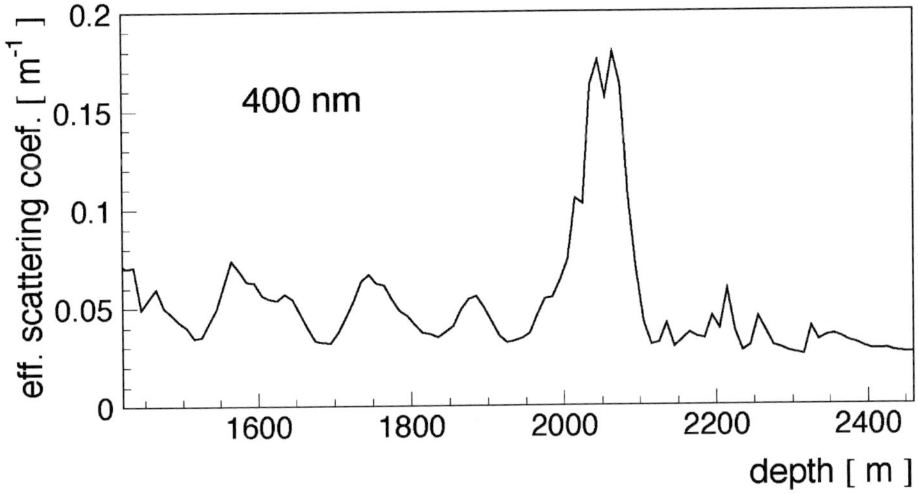

FIGURE 6. The effective scattering coeffi cient for light at 400nm, the sensitivity maximum for IceCube DOMs, as a function of depth in the ice. The typical scattering length of the extremely clear ice is 20 meters, while the typical absorption length is over 100 meters.

CONCLUSIONS

The IceCube detector construction is proceeding well. We expect to install 12-14 strings per year leading to a complete 1 km^3 detector during the 2010-2011 austral summer.

Physics analyses have commenced. As the detector grows in size we will be able to test the theoretical models which predict a handful of events per year [2], and in a few years the discovery of the first astrophysical neutrinos is a realistic possibility.

ACKNOWLEDGMENTS

We acknowledge the support from the following agencies: National Science Foundation-Office of Polar Program, National Science Foundation-Physics Division, University of Wisconsin Alumni Research Foundation, Department of Energy, and National Energy Research Scientific Computing Center (supported by the Office of Energy Research of the Department of Energy), the NSF-supported TeraGrid system at the San Diego Supercomputer Center (SDSC), and the National Center for Supercomputing Applications (NCSA); Swedish Research Council, Swedish Polar Research Secretariat, and Knut and Alice Wallenberg Foundation, Sweden; German Ministry for Education and Research, Deutsche Forschungsgemeinschaft (DFG), Germany; Fund for Scientific Research (FNRS-FWO), Flanders Institute to encourage scientific and technological research in industry (IWT), Belgian Federal Office for Scientific, Technical and Cultural affairs (OSTC); the Netherlands Organisation for Scientific Research (NWO); M. Ri-

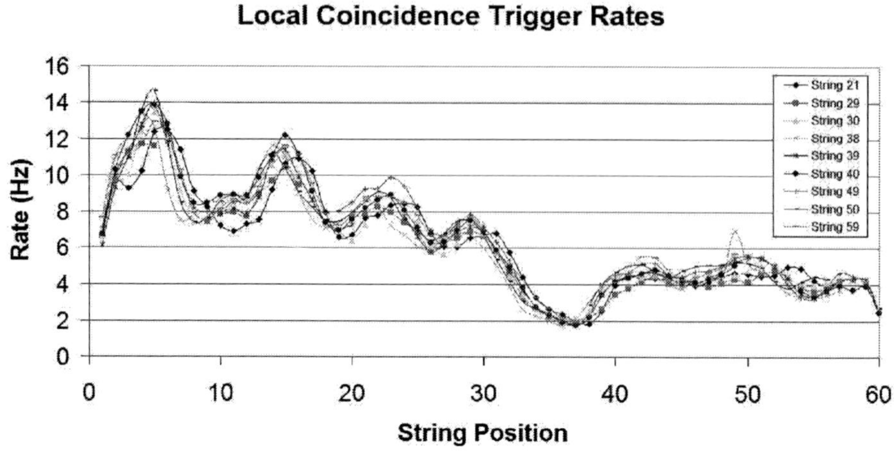

FIGURE 7. Readout rates when the deep ice DOMs are required to trigger in coincidence with a neighboring DOM. Valleys correspond to volcanic dust layers from years past. String-to-string uniformity is excellent.

bordy acknowledges the support of the SNF (Switzerland); J. D. Zornoza acknowledges the Marie Curie OIF Program (contract 007921).

REFERENCES

1. http://icecube.wisc.edu/
2. J. Ahrens et al. (Icecube Collaboration) Astropart. Phys. 20 (2004) 507. *Sensitivity of the IceCube Detector to Astrophysical Sources of High Energy Muon Neutrinos*
3. J.A. Aguilar et al. (ANTARES Collaboration), Astropart. Phys. 23 (2005) 131.
4. A. Capone et al. (NEMO Collaboration), Nucl. Instr. Meth. A 487 (2002) 423.
5. G. Aggouras et al. (NESTOR Collaboration), Nucl. Instr. Meth. A 552 (2005) 420.
6. A. Achterberg et al. (IceCube Collaboration), Contributions to the 29th International Cosmic Ray Conference (ICRC 2005), Pune, India, Aug. 2005, astro-ph/0509330.
7. IceCube Project Preliminary Design Document, Ahrens et al. (IceCube Collaboration), http://icecube.wisc.edu/
8. J. Ahrens et al. (IceCube Collaboration), Nucl. Instr. Meth. A 524 (2004) 169.
9. M. Ackermann et al.: 2006b *First year performance of the IceCube neutrino telescope, Astropart. Phys.* , in press.
10. M. Ackermann et al. (AMANDA Collaboration) (2006), J. Geophys. Res., In Press, doi:10,1029/2005/JD0068.
11. J. Ahrens et al. (AMANDA Collaboration), Astropart. Phys. 16 (2002) 345.

First data from the ANTARES neutrino telescope

V. Niess

(on behalf of the ANTARES Collaboration)

CPPM –IN2P3/CNRS – U. Méditerranée
163, av. de Luminy – case 902 –13288 Marseille cedex 09 - France

Abstract. This contribution reviews the recent progress achieved towards building the ANTARES neutrino telescope. The first results obtained by the operation of a Mini Instrumentation Line with Optical Modules, "MILOM", and the first complete detector line are highlighted.

Keywords: High energy neutrino; Underwater detector.
PACS: 95.55.Vj ; 91.50.Yf.

THE ANTARES DETECTOR

The European Collaboration ANTARES aims at building and operating a large undersea neutrino telescope located at a depth of 2500 m in the Mediterranean Sea, offshore from Toulon in France[1]. Neutrinos will be detected through their interaction in the matter surrounding the detector, producing muons radiating Cherenkov light while propagating in the sea water. Photons are recorded by a lattice of Optical Modules (OMs)[2], consisting of 10" hemispherical photomultiplier tubes (PMTs)[3] housed in pressure resistant glass spheres, installed on a set of mooring lines. The reconstruction of the muon track direction, pointing to a fraction of a degree towards the direction of the parent neutrino source for high energy neutrinos, is achieved from the measurement of the arrival times of the Cherenkov photons on the OMs, as well as their position in space.

The complete ANTARES detector will consist of 12 lines of 25 storeys, each storey being equipped with a triplet of Optical Modules looking at 45° downward and an electronics container mounted on a titanium frame, giving a grand-total of 900 OMs. On each storey, the local electronics container includes the front-end electronics of the PMTs, an Ethernet board for the data acquisition and the detector Slow Control, electronics boards for clock distribution and for Dense Wavelength Division Multiplexing of the Ethernet transmission, and a tiltmeter-compass board measuring the local tilt and orientation of the storey. Some storeys also support a hydrophone for acoustic positioning or an LED Optical Beacon used for inter-string time calibration.

CP867, *XII International Conference on Calorimetry in High Energy Physics,*
edited by S. R. Magill and R. Yoshida

The vertical distance between storeys is 14.5 m, the first one being placed at 100 m above the sea bed, leading to a total height of the detector strings of 480 m. Each string is anchored on the sea floor at a distance of 70 m from its neighbours. Every line is individually connected to a Junction Box by an interconnection cable a few hundred metres long. The Junction Box is itself linked to the shore by a 40 km long electro-optical cable equipped with 48 optical fibres.

The construction of the ANTARES detector started in October 2001 with the deployment of the main electro-optical cable from the ANTARES site, located in the Mediterranean Sea (42°48'N-6°10'E) offshore Toulon (France) at a depth of 2475 m, to the beach of La Seyne-sur-Mer where the shore station of the experiment is situated. In December 2002, the Junction Box was connected at the end of this cable and immersed on the site. In Spring 2003, two small test lines, the Prototype Sector Line (PSL) and the Mini Instrumentation Line (MIL), were installed, connected and operated for a few months. The genuine operation of the ANTARES neutrino telescope really started in April 2005 after the connection of the Mini Instrumentation Line with Optical Modules (MILOM) and more recently the installation of the first full complete detector line, Line 1, in March 2006. The second line, Line 2, was in integration phase in June 2006, at the time of the Calor'06 Conference, it has been successfully deployed on the ANTARES site at the end of July 2006 as scheduled. Thanks to two assembly sites running in parallel, the ANTARES detector is foreseen to be fully deployed and operational by the end of 2007 for several years of physics data taking.

The main performance parameters expected for the complete ANTARES neutrino telescope are summarized in figure 1. The left plot shows the effective area for neutrinos as a function of the neutrino energy for various incident angles. The effective area reaches 1 m^2 for $E_\nu > 100$ TeV, the energy above which the Earth shadowing starts to be of some importance. The right plot shows the angular resolution for the reconstructed muon compared to the true muon direction and to the parent neutrino direction, as a function of the neutrino energy. While the angular resolution is dominated by the physical angle between the muon and the parent neutrino at low energy, it is dominated by the reconstruction for $E_\nu \geq 10$ TeV and is expected to be as good as 0.3°. In this regime, the angular resolution is mainly dominated by two effects: the scattering and the chromatic dispersion of the Cherenkov light during its travel into the sea water, contributing for a time arrival spread of $\sigma \sim 1$ ns; the transit time spread (TTS) of the PMT signals being $\sigma \sim 1.3$ ns.

To achieve this good angular resolution, the ANTARES detector is designed such as additional electronics contributions to the time calibration contribute for less than ~0.5 ns to the time-stamping of the detected photons. In addition, the relative position reconstruction of the OM has to be controlled with a precision of ~10 cm.

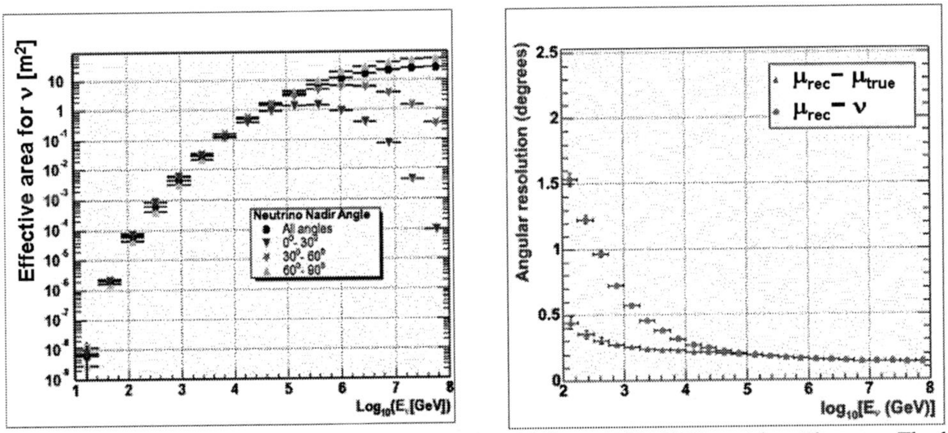

FIGURE 1. Expected performance for the complete 12 lines ANTARES neutrino telescope. The left figure shows the effective area for neutrinos as a function of the neutrino energy for several incident angles. The right figure shows the angular resolution of the reconstructed muon compared to the true muon direction and to the parent neutrino as a function of the neutrino energy.

RESULTS FROM THE FIRST ANTARES LINES

The MILOM line

The current data taking of the ANTARES detector started in March 2005 with the operation of the MILOM[4]. This instrumentation line, partly devoted to multi-disciplinary and environmental studies, consists of an instrumented releasable anchor, the Bottom String Socket (BSS), and three storeys respectively located at 100 m, 117 m and 169 m above the sea bed. The middle storey is a standard ANTARES storey housing a triplet of Optical Modules. The main other devices are a water current profiler located on the top storey, an LED Optical Beacon held on the bottom storey, an acoustic positioning transducer attached to the BSS and a seismometer buried into the sea floor 50 m away from the MILOM.

Optical background measurements

The operation of the MILOM allowed a continuous monitoring of the background rates of the Optical Modules. A typical OM counting rate display exhibits a baseline of ~60-80 kHz largely dominated by optical background due to ^{40}K decays and bioluminescence activities coming from bacteria, as well as bursts of a few seconds duration produced by bioluminescent emission of macro-organisms[5]. Figure 2 (left) shows the baseline rates recorded by the three OMs of the MILOM during a period of three months in Autumn 2005. A seasonal variation of the bioluminescence component of the baseline is clearly observed. The 15% higher counting rate of OM1 during the full period is due to a lower threshold set on the readout of this Optical

Module. Figure 2 (right) shows the burstfraction, defined as the fraction of the time when the counting rate is higher than the baseline by 20% during a 15 min interval, as a function of the water current intensity. A strong correlation of these two quantities is clearly observed.

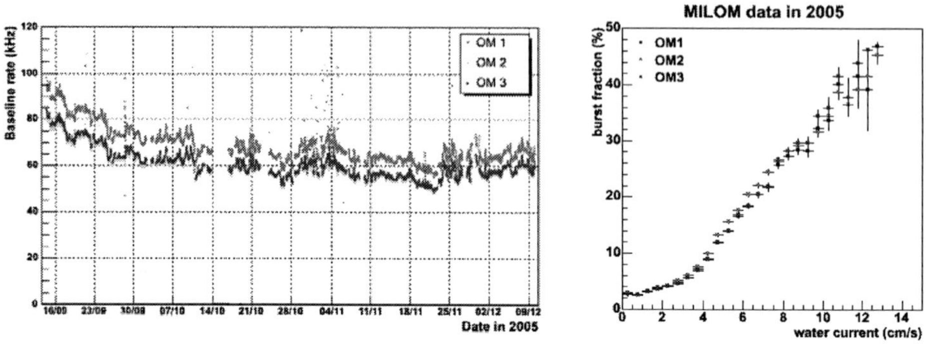

FIGURE 2. Baseline rates recorded by the three OMs of the MILOM during Autumn 2005 (left) and burstfraction as function of the water current intensity (right).

The time coincidences between pairs of neighbouring Optical Modules have also been studied. These distributions exhibit a flat background of random coincidences and a Gaussian peak of few ns width due to genuine coincidences of ^{40}K radioactive decays producing two detected photons. The ^{40}K coincidence rate is measured to be 13.0±0.5 Hz and is in good agreement with a simulation the signals induced by ^{40}K decays which leads to a coincidence rate of 12 Hz with a 4 Hz systematic error due to uncertainties in the effective area and angular response of the OMs.

Time calibration with the LED Optical Beacon

The time calibration of the MILOM Optical Modules has been checked by flashing the LED Optical Beacon located on the bottom storey. This device consists of a glass cylinder container containing 36 blue LEDs synchronised in time in order to produced intense light flashes with a time dispersion < 0.5 ns. The time calibration of the OMs can be checked either by looking at the arrival time of the signal on the PMT relative to the time of the flash, or by the time difference of the flash arrival time measured by two adjacent OMs. Due to the large intensity of the light flashes and the short 15 m distance of propagation of the light into the water, the time stamping of the OM signal is dominated by its electronics contribution and not by the TTS of the PMT in this case. The measured distributions confirm that the electronics contribution to the time calibration is ≤ 0.5 ns as expected.

The Line 1

The first complete line of the ANTARES neutrino telescope, Line 1, has been deployed on the site on February 14th 2006 and connected two weeks later on March

2nd by using the Remote Operated Vehicle Victor of IFREMER. This line is made of a BSS and of 25 storeys, holding a total of 75 OMs. It also includes 4 LED Optical Beacons and 5 acoustic positioning hydrophones spread along the line, as well as an acoustic transducer on its BSS.

Time calibration with the MILOM LED Optical Beacon

The time calibration of the Line 1 OMs can also be checked by flashing the LED Optical Beacon located on the MILOM bottom storey. Figure 3 (left) shows the detection time spread of the LED Optical Beacon flashes by an OM of Line 1 located on a storey at the same altitude than the MILOM LED Optical Beacon. In this case, the measured distribution width is $\sigma = 0.7$ ns for a "horizontal" travel path of ~80 m of the light in the sea water. Figure 3 (right) shows in comparison the detection time spread for an OM of Line 1 located at a higher altitude corresponding to a "diagonal" travel path of the flash light of ~150 m. A wider distribution is clearly observed due to the smaller intensity of the detected signal, as well as a tail of delayed photons coming from scattering light. All measured distributions have been found in good agreement with expectations.

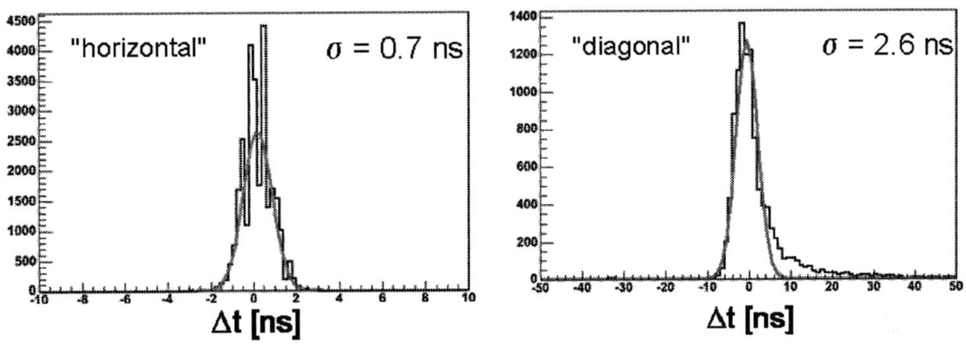

FIGURE 3. Detection time spread of the MILOM LED Optical Beacon flashes by an OM of Line 1 located at the same altitude as the LED Optical Beacon at an "horizontal" distance of ~80 m (left) and by an OM at a higher altitude at a "diagonal" distance of ~150 m (right).

Acoustic positioning measurements

The reconstruction of the detector geometry in real time is primarily based on acoustic triangulation of a small number of hydrophones scattered along every line. The triangulation is performed from distance measurements of each hydrophone to several fixed acoustic emitters located either on every line anchor base (transducers) or on autonomous pyramidal structures anchored around the detector field (transponders). A relative positioning of the hydrophones in space with a precision of few cm is necessary in order to obtain a precision of ~10 cm on the OM positions as a result of the line shape reconstruction performed by the addition of the tilts and heading measurements of every storey. The concomitant operation of MILOM and

Line 1 has allowed a check of the acoustic system performance by performing the Line 1 hydrophone triangulation based on the acoustic emission of the MILOM transducer and two autonomous transponders. The good resolution of the acoustic system, found to be well within the specification, can be appreciated on figure 4 which shows the radial displacement of the lowest hydrophone of Line 1, located on its bottom storey at 100 m above the sea bed, with respect to the line axis during a period of two weeks.

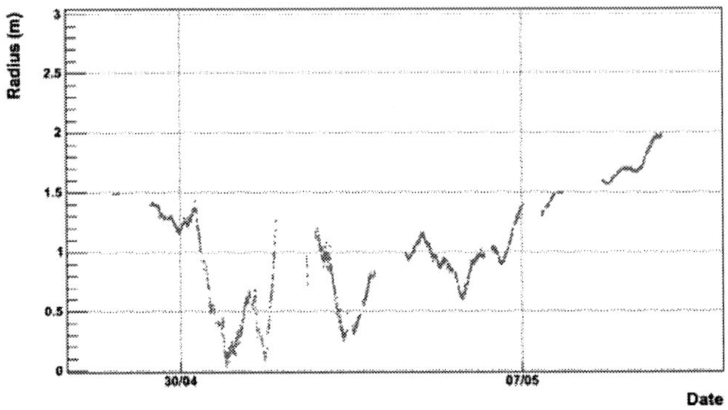

FIGURE 4. Radial displacement of the Line 1 lowest hydrophone with respect to the line axis measured by the acoustic relative positioning system during a period of two weeks.

Reconstruction of atmospheric muons

Although the OMs point at 45° downwards, the ANTARES detector has a none negligible efficiency for the detection and the reconstruction of downward-going atmospheric muons. The event selection is first performed by an online filter algorithm, running at the shore station, which looks for a set of ≥ 4 local coincidence hits, the triggered hits, causally compatible with a muon track passing through the detector during a 4 μs time window. The muon reconstruction is then performed with a χ^2 fit of the hit times as function of their altitudes, to a hyperbola corresponding to the intersection of the muon Cherenkov light front with the line plane, in order to determine the zenith angle of the muon track. An example of such a muon track fit is shown on figure 5. Several thousand of atmospheric muons have already been reconstructed after a few weeks of operation of the Line 1, the first being detected only two days after its connection. The study of the muon angular distribution is in progress. The preliminary results already show that the muon reconstruction is working well and that the hunt for the first undersea neutrino can be started.

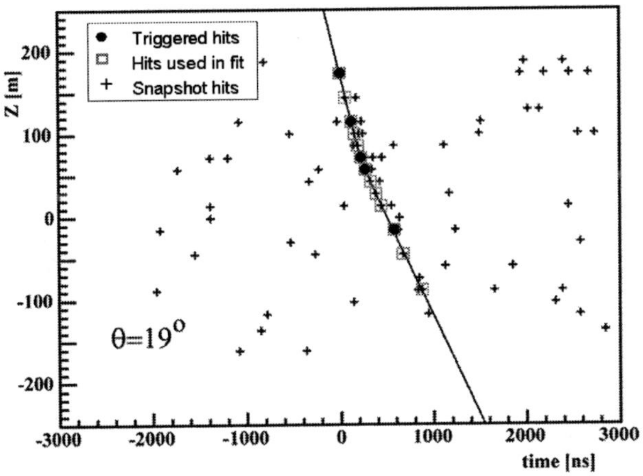

FIGURE 5. Example of a downward-going atmospheric muon track reconstruction with the ANTARES Line 1.

CONCLUSION

The ANTARES Collaboration has made a major step forward during the last year by the operation of a Mini Instrumentation Line with Optical Modules, the MILOM, for more than a year, and the installation of the first complete line of the detector in Spring 2006. All studies performed with these two lines show that the detector behaves well within the design specification and that all technical problems are solved. The detector should be fully installed by the end of 2007 and in operation for science during at least five years. It is also considered as a milestone towards the building of a km^3 underwater detector for which a design study is under preparation.

ACKNOWLEDGMENTS

We thank the ANTARES Collaboration for the opportunity to report this work and particularly the CPPM team. We also thank the Organizing Committee for financial support.

REFERENCES

1. E. Aslanides *et al.*, ANTARES Collaboration, "the ANTARES Proposal" (1999), *Preprint* astro-ph/9907432
2. P. Amram *et al.*, ANTARES Collaboration, *Nucl. Inst. Meth.* A **484** (2002) 369 (*Preprint* astro-ph/0112172)
3. J.A. Aguilar *et al.*, ANTARES Collaboration, *Nucl. Inst. Meth.* A **555** (2005) 132 (*Preprint* physics /0510031)
4. J.A. Aguilar *et al.*, ANTARES Collaboration, "First results of the Instrumentation Line for the deep-sea ANTARES neutrino telescope" (2006), *Preprint* astro-ph/0606229
5. P. Amram *et al.*, ANTARES Collaboration, *Astropart. Phys.* **13** (2000) 127 (*Preprint* astro-ph/9910170)

Calorimetry with Nonequilibrium Phonon Sensors for Dark Matter Direct Detection

Michael Dragowsky

Physics Department, Case Western Reserve University, 10900 Euclid Ave., Cleveland, OH, U.S.A.

Abstract. Low-temperature calorimetric particle detectors are applied to the problem of dark matter detection. These semiconductor crystal-based devices operate at sub-Kelvin temperatures. Nuclear-recoil interactions can be distinguished from electron-recoils by comparing the thermal energy yield with that in another channel, such as ionization or scintillation, essential in the search for dark matter and potentially quite useful in measuring neutrino–nucleus scattering cross-sections. The Cryogenic Dark Matter Search collaboration has demonstrated in phase two (CDMS II) that athermal phonon sensors instrumented with direct-current SQUID arrays achieve additional background rejection encoded in the pulse shape, particularly associated with leading edge features. This talk will present the general ideas of applying calorimetry to dark matter detection. It will also present CDMS II detector design and performance, and discuss future plans for dark matter direct detection with this approach.

Keywords: nonequilibrium phonon sensor, transition edge sensor, superconductivity, solid state calorimeter
PACS: 14.80.Ly, 95.35.+d, 95.30.Cq, 95.30.-k, 85.25.Oj, 29.40.Wk

WHAT IS DARK MATTER?

Astrophysical observations over the galactic and extra-galactic scales tell us there is more matter present than meets the eye. Historically, these studies charted the motion of stars around the centers of galaxies, and galaxies within galaxy clusters. In recent decades, experimental cosmology, principally big bang nucleosynthesis and the cosmic microwave background radiation, as well as advanced techniques for exploiting gravitational lensing have provided independent evidence for this cosmic dark matter.

The nature of the dark matter remains a mystery. Astrophysics, cosmology and particle physics have generated a compelling plausibility argument that fundamental particles having interactions at the weak scale and GeV–TeV scale mass would naturally account for the dark matter. It is important to keep in mind that light fundamental particles are also viable candidates, as is the possibility of new physics acting on extra-galactic scales of distance, time or acceleration, however, testing the weakly-interacting, massive particle hypothesis is very well motivated. Cryogenic semiconductor crystal-based detectors are playing a leading role in the search for dark matter.

An interesting aspect of the WIMP hypothesis is that it is independently motivated by work at the frontier of particle physics theory and by foundational ideas from cosmology. Big bang cosmology provides a framework that allows for new, heavy, neutral fundamental particles that would freeze out early in the expansion history of the universe. Particle mass and cross sections at the electroweak scale are well matched to explain dark matter. From particle theory, supersymmetry has emerged from efforts to

CP867, *XII International Conference on Calorimetry in High Energy Physics,*
edited by S. R. Magill and R. Yoshida

reconcile the heirarchy problem, providing a specific framework for calculation and delivering 'neutralino' into the dark matter lexicon. The details of the WIMP hypothesis for dark matter is not the subject of this paper, and so the interested reader is encouraged to choose from amongst these recent reviews [1].

CALORIMETRIC DETECTION STRATEGY

Dark matter particle interactions will occur by elastic scattering on atomic nuclei. A generic statement of the strategy for dark matter detection is then to distinguish nuclear and electron recoil events. Cryogenic detectors provide a differential measurement, in which the full energy of the interaction is manifested in lattice excitations, and in another channel, such as ionization charge or scintillation light, in which the share of energy depends on the particle type. Early implementations of these techniques relied on detecting the temperature rise of the crystal. This technique has a distinguished track record, and achieves good sensitivity to nuclear recoils, and very good energy resolution at low energy. Improved sensitivity can be had from sensing particle-induced nonequilibrium phonons. The challenge is to efficiently couple these carriers into a sensor that preserves the intrinsic microsecond timing of this energy channel.

The issue of whether to emphasize phonon readout at the microsecond time scale is influenced by the choice of alternate energy channel. Ionization is a fast process, having sub-microsecond collection times, while scintillation may be fast or slow. The CRESST [3] experiment (Cryogenic Rare Event Search with Superconducting Thermometers), for example, using $CaWO_4$ scintillating crystals with millisecond optical decay constants, deemphasizes microsecond phonon readout, and uses superconducting phase transition sensors that maximize phonon energy collection. In addition to the dark matter application, the CRESST devices have been used to measure for the first time the α-decay half life of ^{180}W [2].

The Cryogenic Dark Matter Search approach in phase two (CDMS II)[1] [4] [5] emphasizes fast phonon readout. Aluminum thin film structures (having a high superconductor band gap) are used for phonon collection, so that predominantly the fastest phonons are collected from the kg-scale crystals used in dark matter search experiments. As the particle-induced phonon population equilibrates via scattering, the mean frequency rapidly decreases until few phonons are capable of dissociating Cooper pairs in the Al collection structures. These structures are linked to tungsten (W) transition edge sensors (TES) by bilayer W/Al quasiparticle traps. Both the equilibrium and nonequilibrium phonon collection approaches are vetted as having high sensitivity to low-energy nuclear recoils and have the potential for ton-scale cryogenic experiments.

---

[1] CDMS I employed the equilibrium phonon approach

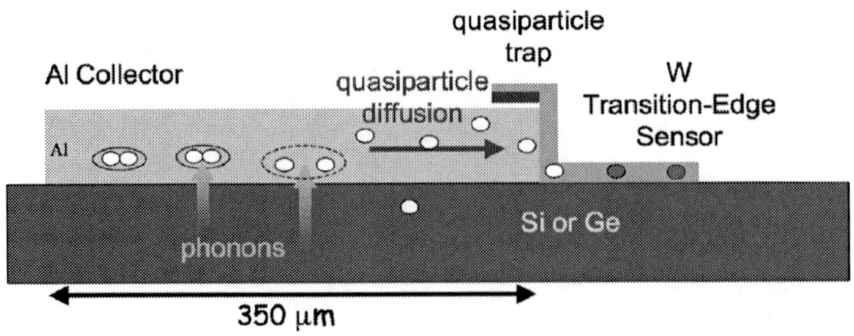

FIGURE 1. Cartoon depicting the collection of nonequilibrium phonons via extended Al quasiparticle trapping structures that feed this energy into W transition edge sensors.

CDMS II: NONEQUILIBRIUM PHONON-BASED DETECTORS

The CDMS II collaboration has created z-sensitive ionization and nonequilibrium phonon (ZIP) sensors. Two distinct mechanisms underlie the phonon sensors: collection of nonequilibrium phonon energy at the crystal surface and conversion of this energy into an output signal. Two superconducting thin-film systems are associated with these mechanisms, and these are depicted in cartoon form in Fig. 1. The output signal arises from tungsten (W) thin-films patterned into a parallel circuit with 1036 elements. The heat capacity of W is low which yields a rapid response. The parallel circuit allows voltage biasing promoting electrothermal feedback that decreases the pulse falltime and enhances stability. Readout is accomplished with 100-element direct current SQUID array preamplifiers operated at 1K. Impedence matching the SQUID array calls for low-impedence in the W circuit, which in turn limits the W surface area coverage. Phonon energy is collected then by features patterned in Al. Nonequilibrium phonons having sufficiently high energy promote Cooper pairs above the superconducting bandgap as quasiparticles that diffuse through the large-scale Al features. A W/Al bilayer links the Al collection structure to the W sensor, and because the bilayer has a superconducting bandgap intermediate to Al and W, quasiparticles that reach the bilayer preferentially move into the W.

Ionizing particle interactions ultimately carry the energy from electron motion into vibrations of the crystal lattice and ultimately into an increased temperature of the bulk crystal after this energy achieves equilibrium. The time scale for this process is dependent on the crystal size, so that making larger crystals increases the duration of a phonon pulse.

The cryogenic instrumentation needed to operate and readout the detectors must achieve continuous cooling to 50 mK or below, and provide noise sensitivity at 10 pA$\sqrt{Hz}$. The cryogenic environment is provided by means of a ^{3}He–^{4}He dilution refrigerator. The principle of operation was first realized by Fritz London in 1951, whereby a mixture of ^{3}He:^{4}He (optimally about 15% ^{3}He) cooled below 0.87 K will separate into phases dilute and enriched in ^{3}He, respectively. The dilute phase concentration will remain non-zero even as T$\rightarrow$0, thereby providing a system in which ^{3}He can be 'evaporated' from the concentrated phase across a static phase boundary to provide continuous

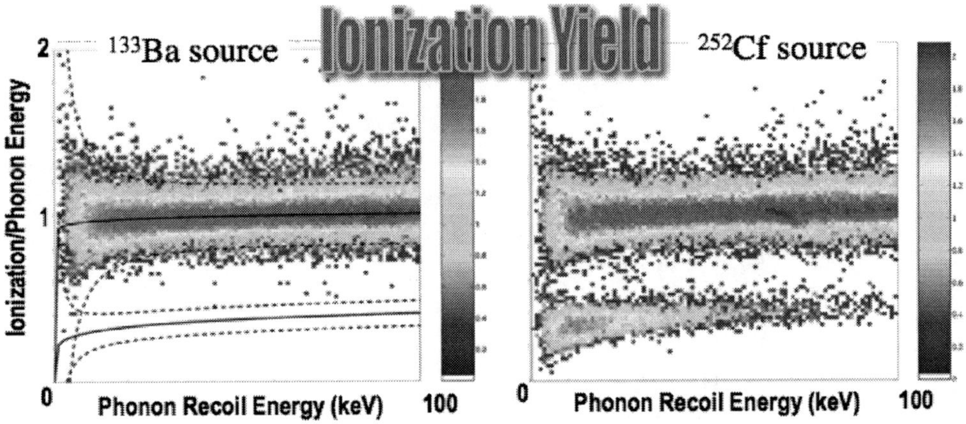

FIGURE 2. The response in ionization yield (ratio ionization to phonon signal) for the gamma-ray source ^{133}Ba and the fission neutron source ^{252}Cf. The latter source also generates a gamma-ray flux. The discrimination power in this parameter is approximately 10,000:1.

cooling.

Phonon sensor readout is accomplished via 100-element SQUID series arrays. The SQUID arrays are biased in flux-locked loop mode to act as low-impedence preamplifiers with high bandwidth. The intrinsic SQUID bandwidth is much higher than our readout requirements (GHz), and the circuit is organized to achieve a maximum of 2 MHz bandwidth through the SQUID amplifier. When the QET sensor is also biased, the input and feedback circuit bandwidth is 100-200 kHz. This bandwidth is quite adequate to reproduce the microsecond leading edge of the phonon pulse.

Energy Resolution

In principle, the collection of sub-meV phonons should provide excellent energy resolution at a few keV, as fluctuations in the number of phonons would be expected to be very small. In practice, very high resolution is obtained when collecting the equilibrium phonon energy. In the CDMS II implementation of athermal phonon collection, energy resolution is about 5% at 10 keV, primarily due to inefficiencies coupling the athermal phonons into the extended Al quasi-particle structures. Measures taken to improve this performance in detectors for the next phase experiment will be discussed later.

Setting the energy scale for the dark matter experiment is done using an external gamma-ray source with photons having sufficient energy to penetrate the copper cans of the dilution refrigerator. The radionuclide ^{133}Ba is used, as it provides peaks at 276-, 303- , 356- and 384-keV. A plot showing the data compared with Monte Carlo generated detector response is shown in Fig. 3. The phonon sensor response is nonlinear above about 200 keV, therefore, the phonon channels are calibrated against the charge channels. The accuracy of this method is checked following neutron calibration studies. In Ge-

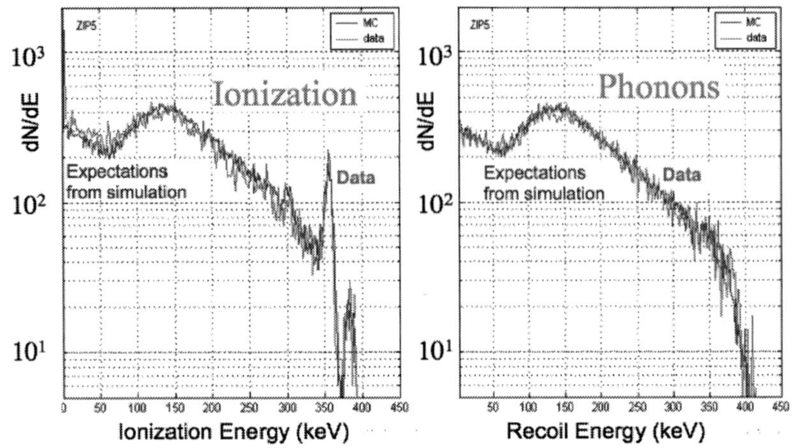

Source: L. Baudis, RWTH Aachen, Germany

FIGURE 3. The energy scale derived from ^{133}Ba gamma-ray calibration data for a Ge-based detector.

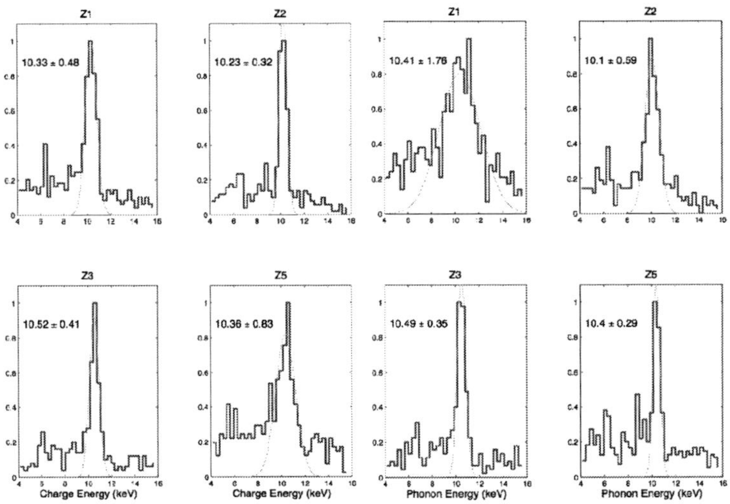

Source: V. Mandic, California Institute of Technology

FIGURE 4. The response in ionization (left) and phonons (right) to the decay of ^{71}Ge that arises from neutron capture on ^{70}Ge during neutron calibration.

based detectors, radioactive ^{71}Ge is produced via neutron activation, and decays with an 11.8 d half life, producing a flux of 10.36 keV gamma-rays distributed throughout the crystal. Analysis from data taken in 2003 is shown in Fig. 4. This analysis checks the accuracy of the energy scale at the level of one percent.

Phonon Pulse Timing Parameters

Each ZIP has four phonon sensors occupying symmetric quadrants one detector surface. The relative arrival times of phonon energy in each sensor provides the means to determine where an event has occurred in the plane of the sensors. Three dimensional position reconstruction can in principle be achieved through phonon transport modeling. Preliminary work has been reported in this reference [6] and work along these lines is ongoing.

Further discrimination between nuclear- and electron-recoil events is obtained by analysis of the temporal detector response. The fundamental property exploited in the timing analysis is that events at the detector surface generate a different phonon frequency distribution than that generated by events in the bulk. This leads to a faster signal for events at the surface. Two timing parameters are used in the principle analysis of the most recent CDMS II data. These are the delay between the ionization signal and start time of the local quadrant phonon pulse, and the risetime of the same local quadrant phonon pulse. The starttime is referenced to 10% of the peak pulse height on the rising edge of the pulse. The risetime is calculated as the time difference between the 40% and 10% pulse heights.

The timing parameter used in the principal analysis was the sum of these two parameters. The joint use of ionization yield and this timing parameter provide a high-acceptance, zero background signal region for the WIMP search analysis. Such a background-free region of parameter space is highly desirable for conducting a rare event search. For more details, please see reference [4], Fig. 1 and associated text.

COMPLEMENTARITY

In the outline of the talk, I promised to explain the buzzword of the day: complementarity. In the context of dark matter, there are really two questions. Firstly, are there dark matter particles in the galaxy/solar system that we can study in the laboratory? Secondly, if so, what do we need to understand to test whether these particles can account for the cosmological dark matter? One appealling aspect of the WIMP hypothesis is that the predicted heavy, neutral particle can account for the behaviour seen on galactic and cosmologic scales. A view has emerged that dark matter direct detection addresses the former question, and that accelerators such as the LHC or an ILC will be able to address the latter question. Understanding dark matter is then an investigation jointly pursued by particle physics and astrophysics where the two approaches bring complimentary information to bear on the matter.

The WIMP-search parameter space is shown in Fig. 5. Depicted in the figure are the regions for which WIMPs have already been ruled out by direct detection (arcs spanning the space horizontally) and by accelerator data (vertical lines). Discovery requires moving these lines respectively downward, and to the right. The image further illustrates a central truth about experiment: there are limits to our reach. For direct detection, the challenge is to achieve sensitivity for a few events per ton per year. For accelerators, the challenge is mass approaching a TeV/c^2. Fortunately, or perhaps naturally, there is theory suggesting that the phase space accessible now and within

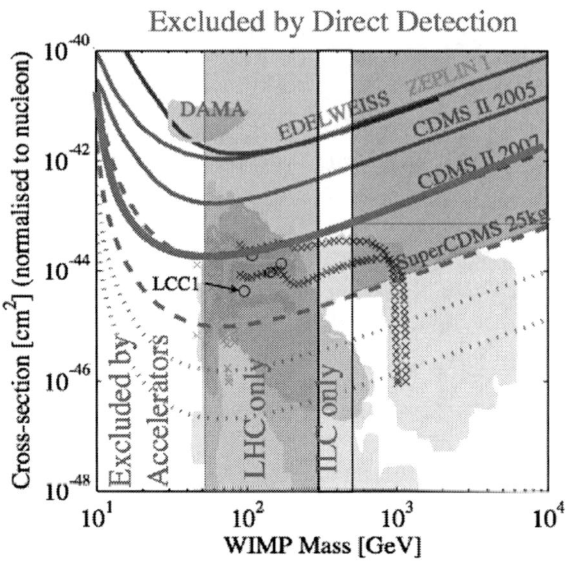

FIGURE 5. WIMP search parameter space. Regions of high cross-section are excluded above the horizontal arcs. Existing direct detection results appear as solid lines. Not all experiments are listed for clarity in the figure. The near-term expectation for CDMS II is the arc labeled CDMS II 2007. WIMP mass exclusion limits appear as vertical lines, with the shaded regions indicating the reach of LHC and ILC.

reach of upcoming experiments will be of great interest.

The attractive possibility of a fruitful interplay between accelerator and nonaccelerator techniques for understanding dark matter can be further extended. Accelerator measurements of WIMP annihilation and scattering cross sections can be used to understand the results from present and future high energy gamma-ray and neutrino astronomy efforts. Bringing these resources to bear allows not only testing that the WIMP hypothesis accounts for the galactic and cosmologic dark matter, but also holds the promise of mapping the galactic dark matter distribution. The detailed study of dark mater in our galaxy and in the local group will perhaps lead to a microscopic understanding that can be extended to larger scales.

ROOM FOR IMPROVEMENT

The utility of athermal phonon sensors for dark matter direct search is clear. The SuperCDMS collaboration intends to extend the CDMS II program by building a new array of detectors having 25 kg of target mass. The effectiveness of the SuperCDMS detectors to reject the surface events will be enhanced with a few simple alterations. These changes are expected to achieve a factor of 40 improvement in suppressing surface events relative to bulk events, comfortably meeting the requirements to conduct a background-free WIMP search.

Firstly, the devices will be made thicker by a factor of 2.5, enhancing the ratio of bulk to surface regions. Secondly, the phonon collection structures – the quasiparticle traps – will be expanded in area, improving phonon energy collection and timing statistics. Thirdly, fabrication will incorporate hydrogenated amorphous silicon into the charge carrier blocking layer (h-α-Si), reducing charge lost to the opposite contact by a factor of two according to previous results. Fourthly, exposure of the devices to Rn will be controlled at least as well as the latest towers produced for CDMS II, achieving a factor of two reduction largely from suppressed ^{210}Pb plate-out. Fifth, and finally, we expect improved analysis techniques to yield another factor of two enhancement.

Summarizing, device enhancements should provide a factor of 10 reduction, with factors of 2 coming from improved isolation from Rn and enhanced analysis, respectively. While a combined factor of 40 reduction is expected, only a factor of 20 reduction is needed to operate in the zero-background mode for a 50 kg-y exposure.

SUMMARY

Cryogenic non-equilibrium phonon sensors have demonstrated high sensitivity rejection of electron-recoil backgrounds with high efficiency to detect nuclear-recoils, precisely the characteristics need for dark matter direct detection. Understanding if the nature of dark matter lies in the particle sector requires direct detection and accelerators. CDMS II detectors are coming online for a 5 kg-years exposure that will substantially improve sensitivity for WIMP discovery compared to the latest results based on 74 kg-days. The collaboration is moving forward to build 25 kg target mass of new detectors to push the sensitivity frontier deep into parameter space favored by supersymmetric extensions to the standard model. New collaborators are welcome, and enthusiasm for solid state cryogenic detectors applied to dark matter direction detection and other rare event applications is very high.

REFERENCES

1. R. J. Gaitskell, Annu. Rev. Nucl. Part. Sci. 2004. **54**:315D59;
   Wendy L. Freedman and Michael S. Turner, Rev. Mod. Phys. **75**, 1433 (2003).
2. C. Cozzini *et al*, Phys. Rev. C **70**, 064606 (2004).
3. G. Angloher *et al* Astroparticle Physics **23** (2005) 325 Ð 339.
4. D.S. Akerib *et al* Phys. Rev. Lett. **96**, 011302 (2006).
5. D.S. Akerib *et al* Phys. Rev. D**72**, 052009 (2005).
6. G. Wang, Ph.D. dissertation, Case Western Reserve University, January 2005.

SESSION IV

MEDICAL APPLICATIONS OF CALORIMETRY

Convener: Karen Byrum

Calorimetry in Medical Applications: Single-Photon Emission Computed Tomography and Positron Emission Tomography

Chin-Tu Chen

Department of Radiology and Committee on Medical Physics, The University of Chicago, Chicago, Illinois 60637, USA

Abstract. Positron emission tomography (PET) and single-photon emission computed tomography (SPECT), two nuclear medicine imaging modalities broadly used in clinics and research, share many common instrumentation, detector, and electronics technology platforms with calorimetry in high-energy physics, astronomy, and other physics sciences. Historically, advances made in calorimetry had played major roles in the development of novel approaches and critical technologies essential to the evolution of PET and SPECT. There have also been examples in which PET/SPECT developments had led to new techniques in calorimetry for other application areas. In recent years, several innovations have propelled advances in both calorimetry in general and PET/SPECT in particular. Examples include time-of-flight (TOF) measurements, silicon photomultipliers (SiPMs), etc.

Keywords: PET, SPECT, time-of-flight, photosensors, semiconductor detectors
PACS: 87.58.-b; 87.58.Ce; 87.58.Fg

INTRODUCTION

Nuclear medicine imaging, in which radiopharmaceuticals (radioisotope-labeled tracers), often designed in such a way that they can reveal specific functions and physiologic processes inside a life system (for example, a human patient or an animal model representing a particular disease model), are introduced into the life system so that images or dynamic image sequences can be obtained. These nuclear medicine images represent signatures of the particular functions and physiology that can be employed to derive quantitative information relevant to the life processes (for example, metabolism or blood flow) that are indications in certain medical diagnostic procedures or some needed measurements useful in basic biomedical research.

There are three major imaging modalities in nuclear medicine imaging: planar scintigraphy, single-photon emission computed tomography (SPECT), and positron emission tomography (PET). In planar scintigraphy, 2-D projection images of a 3-D object with radiotracers distributed inside are acquired, with overlapping organs and tissues projected into a planar view. SPECT and PET are tomograpghic imaging modalities that acquire many projection views from different angles so that volumetric 3-D images of the radioactivity distributions within the object under study can be reconstructed from these projection data, eliminating the overlap issues encountered in planar scintigraphy.

CP867, *XII International Conference on Calorimetry in High Energy Physics,*
edited by S. R. Magill and R. Yoshida

As major applications of calorimetry in medicine, SPECT and PET share many common elements of physical processes, system design, and event detection and determination with those in calorimetry in general. There is a source distribution of radiation or particles that constitutes the origin of the physical processes. This source generates radiation or particles as a result of physical events such as nuclear decay or particle collision. Detection of the resulting radiation or particles is achieved by utilizing detectors such as scintillation crystals to capture them. Secondary photons or radiation generated in the detectors can then be analyzed with enhanced signals using photosensors such as photomultipliers (PMTs) or photodiodes. These photon signals are then converted to electrical/electronic signals for subsequent processing by the special circuits and electronics designed specifically for handling the signals under study.

This common chain of physics elements and their associated instrumentation requirements provide SPECT/PET, and calorimetry in other physical science areas such as high-energy physics and astronomy, a rich ground for cross fertilization of detector and electronics technologies, as well as event processing algorithms and computations. Historically, advances made in calorimetry had played major roles in the development of novel approaches and critical technologies essential to the evolution of PET and SPECT. There have also been examples in which PET/SPECT developments had led to new techniques in calorimetry for other application areas. In recent years, several innovations have propelled advances in both calorimetry in general and PET/SPECT in particular. Examples include time-of-flight (TOF) measurements, semiconductor detectors, etc.

SPECT AND PET FUNDAMENTALS

In SPECT, single-photon radionuclides that emit their respective characteristic gamma rays are employed in labeling the radiopharmaceuticals. These radioisotopes are, in general, more readily available and can be easily acquired through commercial channels. In order to determine the direction of incoming gamma rays, physical collimators are usually utilized in front of the detector scintillation crystals, thus reducing the photon flux, hence the system sensitivity, substantially. In addition, since there are two sets of unknown parameters in the reconstruction problem, i.e., the radioactivity distribution itself, and also the photon attenuation and scattering effects each photon experiences before detection, the resulting reconstructed images require a more challenging task to achieve high quality and accurate quantitative measurements.

One the other hand, PET utilizes positron-emitting radionuclides in producing the relevant radiotracers. These positron isotopes are, in general, more short-lived, therefore, require on-site cyclotron to produce them, making the overall cost of performing PET studies higher than that of SPECT. However, its underlying physics uses coincidence detection of two annihilation photons in recording an event. This mechanism of electronic collimation determines the direction of the line along which two annihilation photons travel, thus no physical collimators are required, making its system sensitivity relatively high. The coincidence detection also reduces the complexity of the effect of photon attenuation in PET, resulting in a relatively easier task in order to ensure high quantitative accuracy and image quality in reconstruction.

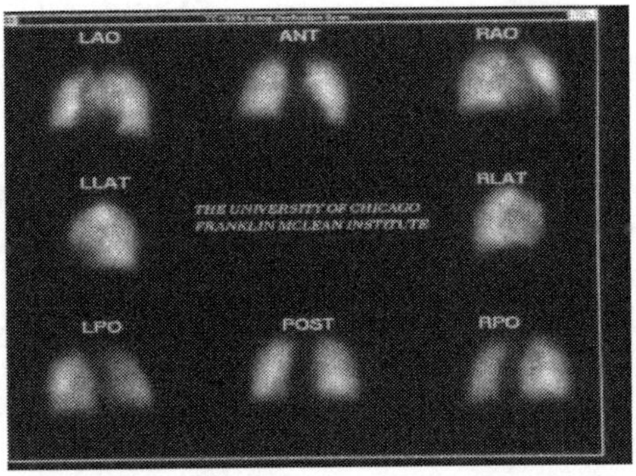

FIGURE 1. Example of Single-Photon Imaging. These are lung perfusion images acquired by planar scintigraphy using Tc-99m labeled radiotracer. Projection views from various perspectives are shown here to illustrate the collapsing of 3-D overlapping tissues onto various 2-D planes.

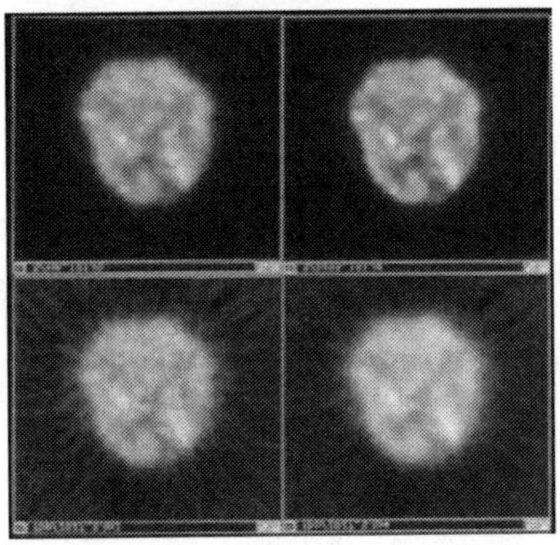

FIGURE 2. Example of PET Imaging. These are tomographically reconstructed brain images using different image reconstruction strategies. These images represent one single slice of the 3-D volumetric brain. F-18 labeled radiotracer was used to reveal brain glucose metabolism. A lesion at the top-right corner of the brain can be clearly seen.

Figures 1 and 2 illustrate examples of single-photon planar imaging and PET imaging respectively. Details on SPECT and PET can be found, for example, in [1].

TIME-OF-FLIGHT PET (TOFPET)

One of the recent advances in calorimetry that finds its use in both high-energy physics and PET is in the area of time-of-flight (TOF) measurement. Picosecond level of TOF measurements is now feasible in high-energy physics for which improved TOF measurements will allow improved identification of particles and their related physical characteristics [2].

The concept of time-of-flight PET (TOFPET) is not new. In the 1980s, there were some major efforts in developing TOFPET systems for improving imaging quality [3]. These initiatives also led to the discoveries of several novel uses of scintillation detectors such as BaF_2. TOF measurements on the order of few hundred psec were achieved at the time. However, the fast-responding scintillation crystals such as BaF_2 that is useful for TOF measurements at the time do not have an adequate stopping power to yield similar high detector resolution that some dense scintillation crystals such as BGO can provide. This issue, in essence, delayed the development of TOFPET systems for nearly two decades.

Since the late 1990s and early 2000s, a new class of scintillation crystals such as LSO, LYSO, LaBr3 has generated a renewed interest in developing TOFPET systems (see, foe example, [4], [5]) because, not only they are rapid-responding, therefore, suitable for TOF measurements, these scintillation crystals are also relatively dense, making high detector resolution also feasible. These new initiatives have created new avenues for developing the next-generation PET imaging devices that incorporate TOF and other novel features in system design for improving image quality and quantitative accuracy, and for reducing the complexity of the imaging procedures and perhaps even the radiation dose. Commercial TOFPET systems offering 600-700 psec TOF resolution are now available on the market.

Improvements in PET Images Using TOF

In TOFPET, two annihilation photons arrive at the opposing detectors with a timing difference of t, which can be measured by fast-responding scintillation crystals with reasonable timing resolution of, say, a few hundred psec. This can be translated into a spatial resolution in the direction of the line of response (LOR), along which the two annihilation photons travel, on the order of a few centimeters. Although an order of magnitude poorer that that of the detector resolution, this new piece of positional information can result in substantial improvement in PET image quality.

Figure 3 illustrates this TOF concept in PET. If Δt can be measured with a known amount of uncertainty, then the positional information, x_1 (or x_2), can be estimated from the total distance between the two opposing detectors d, the TOF measurement Δt, and the known speed of the light C, according to a formula: $x_1 = (C{*}\Delta t + d)/2$. The timing resolution (uncertainty associated with Δt measurements) can, therefore, be translated into spatial resolution along the LOR. A 200 psec TOF resolution is equivalent to 3 cm of spatial resolution in the direction of two traveling annihilation photons (i.e., LOR). Figure 4 demonstrates this broader spatial resolution in the LOR direction, when compared to the detector resolution.

A typical commercial human PET scanner currently available on the market offers a detector spatial resolution of approximately 4 - 5 mm. A TOF timing resolution of 200-300 psec can be translated into 3 - 4.5 cm spatial resolution in the direction of LOR, which is approximately 6 - 10 times of that of the detector spatial resolution. Figure 5 shows that this range of TOF resolution, although limited, can indeed produce improvements in the overall spatial resolution given by a TOFPET system, especially in the tail regions of the point spread function (PSF).

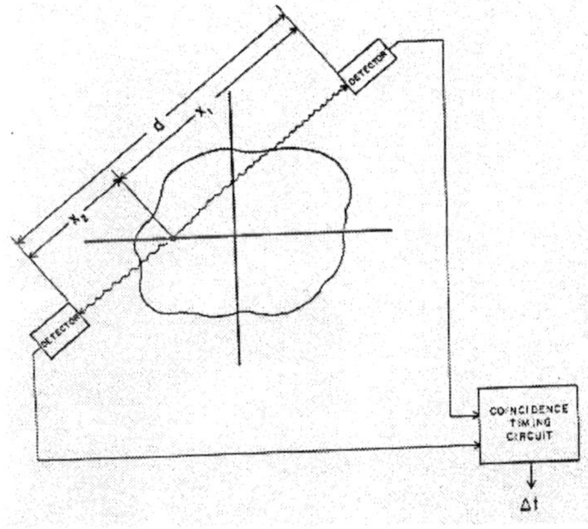

FIGURE 3. Schematic Illustration of the Concept of TOFPET. TOF is the difference in the arrival times of the two annihilation photons reaching the two opposing detectors.

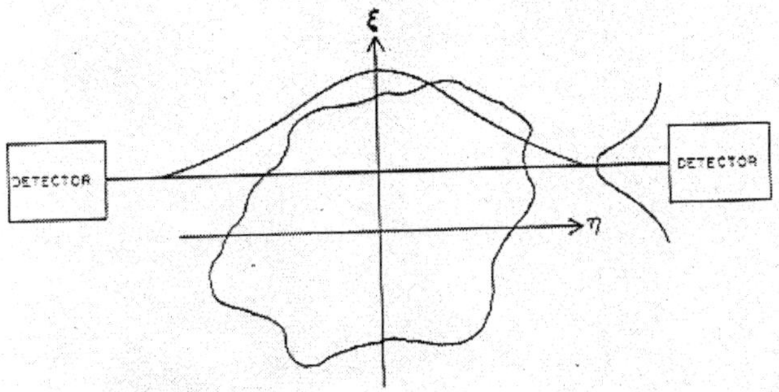

FIGURE 4. Detector Resolution vs. TOF Resolution. TOF measurements can be translated into limited spatial resolutions along the LOR.

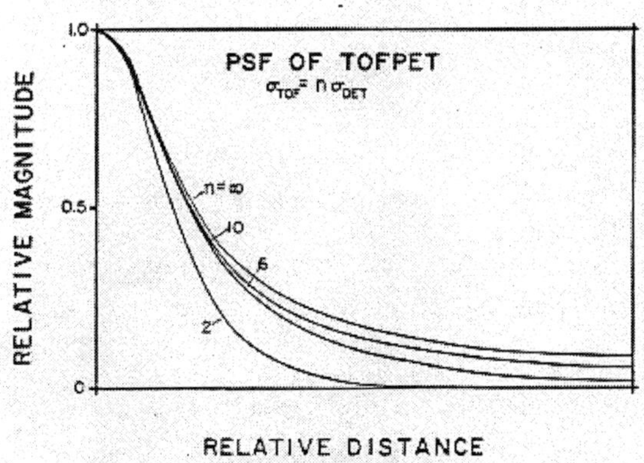

FIGURE 5. Improvements of Overall Spatial Resolution Using TOF Measurements. TOF information provides limited spatial resolution in the LOR direction that results in improvements in the overall spatial resolution of a TOFPET system..

Potential Advantages Offered by TOFPET

The goal of recent developments in TOF for high-energy physics applications is on the order of 1 psec timing resolution is equivalent to a PET LOR spatial resolution of 0.15 mm, which is more than 10 times better than the detector resolution offered by the commercial human PET systems currently available, as well as the dedicated PET systems for imaging of small animals.

If 10 psec TOF can be realized in PET with reasonable cost addition, which provides a LOR spatial resolution of 1.5 mm that is comparable to the detector resolution of current animal PET systems, there would no longer be a need for image reconstruction. PET images can be generated by simply histogramming the incoming events and registering at its location determined by TOF. With 30 psec TOF measurements available in a human PET system, which offers a LOR spatial resolution of 4.5 mm and is comparable to the system's detector resolution, again, image reconstruction is no longer required in such a circumstance.

Therefore, TOF information not only can provide substantial improvements in PET image quality as shown in Figure 5, when the targeted goal of picosecond-level TOF is achieved in high-energy physics and such technology is translated to applications in PET, image reconstruction will no longer be required in producing the PET images, thus eliminating the complexity and hardware/software requirements posed by the need of image reconstruction processes.

SEMICONDUCTOR DETECTORS

Another example of recent developments in calorimetry, originated in high-energy physics but also useful in nuclear medicine imaging including both PET and SPECT, is in the area of developing novel semiconductor detectors or photosensors such as silicon photomultiplier (SiPM). Development of SiPM technology has received considerable attention in recent years primarily because of its potential to replace the costly and bulky PMTs to provide a relatively inexpensive and compact solution for photon detection.

SiPM utilizes semiconductor technology to fabricate miniaturized photosensors that can provide high gain on the order of 10^6, similar to that achieved by using PMTs. Because of its small size, the voltage required to achieve such a high gain does not need to be very high, usually on the order of 40 – 50 volts. Its size is very compact, only of millimeters, with hundreds of photosensor cells on a single chip. SiPM also has an excellent single-photoelectron resolution and is relatively stable. In addition, SiPM is relatively fast in responding, on the order of 100 psec, making TOF measurements possible. Since it is fabricated with regular semiconductor technology, when this technology becomes more mature, its cost can be relatively inexpensive. All these potentials and relative advantages have generated tremendous interests in using this new technology in high-energy physics, astronomy, and other physical sciences applications (see, for example, [6] and [7]).

However, there are a number of limitations of SiPM's performance characteristics with the current technology. The nature of SiPM design allows only very limited dynamic range. Because of its relatively small size, cross-talks can be a significant problem. In addition, there exist substantial amount of dark counts, making some of the potential applications relatively limited; for example, this dark-count issue may prohibit its use in applications such as SPECT. All these issues need to be investigated further and new designs and fabrication strategies that may lead to relaxing these limitations remain to be active research areas for the leading SiPM groups.

Using SiPM technology in designing and building the next-generation PET and TOFPET system has also become a focus of active research in the last couple of years (see, for example, [8] and [9]). SiPM's characteristics offer great potentials in designing a very compact PET system with ultra-high resolution and TOF capability at a relatively low cost. In addition, the physical dimensions of SiPM also allow opportunities to utilize this new technology to measure new information such as depth-of-interaction (DOI), to explore novel designs of new detector configurations, and to experiment innovative system geometries and non-conventional scanning trajectories. It is expected that SiPM technology will be incorporated into the next-generation PET and TOFPET systems and become more broadly available and utilized.

Another potential advantage and novel application of SiPM technology is its use in the presence of magnetic fields. As a semiconductor device, SiPM is relatively insensitive to magnetic fields and radio-frequency interference. Therefore, it is possible to build PET or SPECT modules to be inserted into MRI scanners to become hybrid MRI/PET or MRI/SPECT imaging systems. With MRI providing anatomic information and structural details, the exact locations of the functions and physiology revealed by PET and SPECT can be more readily identified. As multi-modality imaging is becoming the choice of technology of the future, SiPM technology will expand the dimension of novel multi-modality approaches and make PET/MRI and SPECT/MRI hybrid systems more feasible and less expensive.

SUMMARY AND CONCLUSION

SPECT and PET share many common instrumentation, detector, and electronics technology platforms with calorimetry in high-energy physics, astronomy, and other physics sciences. Calorimetry technologies, originally developed for applications in high-energy physics and other physical sciences, have advanced developments of medical imaging technologies. In recent years, development of TOF technology has renewed the interest in designing and building TOFPET system that can offer improved image quality and other advantages, including the potential to eliminate the need of image reconstruction if certain level of TOF resolution can be reached. Development of SiPM technology offers great potentials to design and build the next-generation PET and SPECT devices that utilize these semiconductor photodetectors with more compact configuration at a reduced cost. SiPM also provides opportunity to design and build hybrid PET/MRI or SPECT/MRI systems that would have no or very minimal interference from the magnetic fields.

ACKNOWLEDGMENTS

The author wishes to thank the following colleagues and collaborators for their devoted work and continuing support: Drs. Chien-Min Kao, Qingguo Xie, Jeffrey Souris, Patrick La Riviere, Xiaochuan Pan, and Mr. Yun Dong of the Department of Radiology at The University of Chicago; Dr. Henry Frisch, Mr.Harold Sanders, and Mr. Fukun Tang of the Department of Physics at The University of Chicago; Drs. Alexandre Vaniachine, Robert Wagner, Karen Byrum, and Mr. Gary Drake of the High-Energy Physics Division of Argonne National Laboratory; and Dr. Christian Wietholt of the Division of Medical Engineering Research of the National Health Research Institutes in Taiwan. This work is supported in part by two seed grants for collaborative research projects conducted jointly at The University of Chicago and Argonne National Laboratory.

REFERENCES

1. *Emission Tomography: The Fundamentals of PET and SPECT*, Editors: M. Wernick and J.Aarsvold, Academic Press, San Diego, 2004.
2. http://hep.uchicago.edu/psec/
2. *Proceedings of Workshop on Time-of-Flight Tomography*, edited by M.M. Ter-Pogossian, et al., St. Louis, May 1982, IEEE Computer Society Press, Silver Spring, Maryland, 1982.
4. M. Moszyski, et al. "New Perspectives for Time-of-Flight PET with LSO Scintillators," *2005 IEEE Nuclear Science Symposium and Medical Imaging Conference Record,* J03-18.
5. A. Kuhn, et al. *IEEE Trans. Nucl. Sci.*, **51**, 2550-2557, (2004).
6. http://www.istc.ru/istc/sc.nsf/html-print/projects.htm?open&id=1275
7. V. Golovin and V. Saveliev, *NIM A*, **518**, 560–564 (2004).
8. S. Moehrs et al. *Phys. Med. Biol.* **51**, 1113-1127 (2006).
9. Q. Xie, C.-M. Kao, K. Byrum, G. Drake, A. Vaniachine, R. Wagner, G. Blazey, V. Rykalin, C. T. Chen, "Characterization of Silicon Photomutipliers for PET Imaging," *2006 IEEE Nuclear Science Symposium and Medical Imaging Conference Record,* October 29 – November 4, 2006, San Diego (in press).

SESSION V

LHC CALORIMETRY—TECHNOLOGY

Conveners: Jim Freeman and Peter Schacht

Crystal Properties in the Electromagnetic Calorimeter of CMS

Riccardo Paramatti on behalf of CMS ECAL group

INFN - Rome1

Abstract. The Compact Muon Solenoid (CMS) is a multi-purpose detector for LHC. The electromagnetic calorimeter (ECAL) contains 75848 lead tungstate crystals allowing a very accurate energy measurement of electrons and photons in the GeV - TeV energy range. More than two thirds of the ECAL Barrel has been already assembled.

In this paper an updated analysis on the optical and scintillation properties of about 50000 crystals and an overview on the construction status of the calorimeter are presented. Furthermore, the use of crystal production measurements for the calorimeter precalibration is discussed.

Keywords: electromagnetic calorimeter, crystal, light yield, optical transmission
PACS: 29.40.Vj - 42.25.Bs

INTRODUCTION

The Large Hadron Collider will allow the study of pp interactions at a center of mass energy of 14 TeV. The main physics goal of the CMS experiment [1], one of the four detectors at LHC, is the quest for Higgs and SUSY particles. The maximum design luminosity foreseen at this machine is 10^{34} cm^{-2} s^{-1} and the crossing rate will be 40 MHz. The production of around 25 events per bunch crossing will result in about 1000 tracks. The radiation environment is expected to be particularly severe. In ten years of running (corresponding to a total integrated luminosity of around 10^6 pb^{-1}) detector parts will be exposed to a maximum neutron fluence of 10^{17} n/cm^2 and γ doses of 10^6 Gy. These extreme conditions have imposed a long R&D phase in order to obtain high granularity, radiation resistant, fast and selective detectors.

The electromagnetic calorimeter of the CMS experiment [2] is composed of 75848 Lead Tungstate crystals organized in a barrel covering the central rapidity region $|\eta|<1.48$ and two endcaps which extend the coverage up to $|\eta|=3$. A preshower detector placed in front of the endcaps improves the γ - π^o separation in this region. To increase hermeticity, the barrel has a nearly pointing geometry both in ϕ and η.

In this paper an analysis on the optical properties of more than 50000 CMS crystals produced in Russia by the Bogoroditsk Techno-Chemical Plant is presented.

LEAD TUNGSTATE CRYSTALS

Lead tungstate ($PbWO_4$) is a very dense (ρ = 8.28 g/cm^3) scintillating material. The most appealing properties of $PbWO_4$ crystals are: the scintillation light decay time (80% of light is emitted within 25 ns), the basic radiation resistance, the small radiation

CP867, *XII International Conference on Calorimetry in High Energy Physics,*
edited by S. R. Magill and R. Yoshida

length (X_0 = 0.89 cm) and Molière radius (R_M = 2.2 cm). These properties allow the construction of a compact calorimeter with a granularity suitable for the multiplicity expected at LHC.

Nevertheless this material presents major drawbacks: the low level of light yield (100 photons/MeV, $\sim$ 0.2% with respect to NaI:Tl) imposes a photodetector read-out with internal gain, the strong dependence on temperature of the light yield (-2%/^{o}C around 18 ^{o}C) and the high refractive index (2.29 at peak emission wavelength λ = 420 nm) which makes the light extraction from the crystal very difficult.

During R&D phase (1995-1998), a huge effort was done to improve the light yield of these crystals without spoiling their fast response and to guarantee their radiation hardness at sufficient level to preserve the requested energy resolution.

The $PbWO_4$ crystals show a damage induced by electromagnetic radiation through the creation of color centers that reduce the crystal transparency. The damage amount, depending on the dose rate, reaches a stable level after a small administered dose. A partial recovery of the crystal transparency happens in few hours. The scintillation mechanism remains untouched and the observed light loss is tolerable and can be followed with a monitoring system by injecting light through the crystals.

Photodetectors

Avalanche photo-diodes (APD) were developed for the barrel part of the calorimeter. These silicon devices are insensitive to the 4 T magnetic field of the experiment and have an internal gain (M = 50 foreseen for CMS) essential for $PbWO_4$. The quantum efficiency around the wavelength of the $PbWO_4$ emission peak is about 75%-80%. The amplification is obtained in a small region ($d_{eff} \sim$ 6 mm) of high electric field. Two APDs of 25 mm^2 area are coupled to each crystal. In this configuration 4000 photo-electrons per GeV of deposited energy are produced.

In the endcap regions, radiation levels are too high to use APDs; the longitudinal magnetic field allows there the use of vacuum photo-triodes (VPT). These are single stage photo-multiplier tubes with a fine metal grid anode. Their active area is 280 mm^2, quantum efficiency is about 20% at 420 nm and a gain of 10 in the 4 T magnetic field is expected.

CONSTRUCTION

The calorimeter construction is a distributed process. The barrel construction started in 2002. Parts, as crystals, capsules, mechanical elements for the support structure etc. are produced under the responsibility of different Institutions taking part to the project and then sent in two Regional Centers located in CERN and Rome.

The calorimeter has a modular structure. The basic sub-unit of the barrel, composed by a crystal (whose dimensions vary with η, being approximately of 2 cm x 2 cm x 23 cm) and a capsule (hosting two APDs read in parallel) glued together, is inserted in a glass fiber alveolar structure. Ten sub-units fit in this structure that is closed by an aluminum tablet and constitute a "sub-module". Sub-modules are of 17 different types (as well as crystals) depending on η position. A "module" is made by 40 or 50 sub-modules mounted on a 3 cm thick aluminum grid. Modules are of 4 η types. A "super-module" is a set of 4 modules. The barrel consists of two identical halves made by 18 super-modules.

Endcaps will have a simpler structure; 138 "super-crystals", each made of 25 crystals, constitute the so-called "dee" and two dee make an endcap. Endcap crystals are larger with respect to the barrel; the transversal dimensions are $\sim$ 3 cm x 3 cm.

In the Regional Centers, all the elements are assembled and step by step checked, following a well defined quality control protocol. In particular, each crystal is subject to the systematic test of its geometrical shape, longitudinal transmission (related to radiation hardness), transversal transmission, light yield and uniformity of light collection. Both Centers are equipped with similar automatic machines to insure crystal quality. A detailed description of the machines installed in INFN/ENEA and CERN Regional Centers can be found in [3] and [4] respectively. The strict control of quality at each step of the construction will insure a fully operational calorimeter at best of its potentiality and reduce the risk of dead or malfunctioning channels presence. The "bare" supermodule assembly and its subsequent dressing are done at CERN. Around 75% of the barrel is currently (June 2006) assembled.

OPTICAL PROPERTIES

Longitudinal Transmission

The spectrum of transmission along the crystal axis (longitudinal transmission, LT) is measured at different wavelenghts for the acceptance tests of the crystal. In Rome, transmissions are measured through a system of mirrors with a single-beam CCD-diode array spectrophotometer (S2000 from Ocean Optics2) directly coupled to an integrating sphere (Labsphere of 150 mm diameter). Data are taken every 4 nm from 300 to 700 nm. At CERN, transmissions are measured by two separate spectrophotometers using a large area photodiode as detector. The light wavelengths are selected using a set of narrow band pass filters nominally centered at: 330, 340, 350, 360, 380, 405, 420, 450, 492, 620 and 700 nm. The crystal transmittance along its longitudinal axis (LT), in particular around 360 nm, has been proven to be connected with the presence of defects which are correlated with the radiation tolerance of the crystals. Based on this correlation, very strict requirements on the shape and level of the LT were set.

In Figures 1 and 2, the distribution of the LT360, LT420 and LT620 measurements are shown together with the acceptance cuts. The higher dispersion of LT360 measurements compared with the others, is due to the extreme sensitivity of the band-edge region to the production conditions. However, this dispersion corresponds to a spread in the edge position of less than 10 nm.

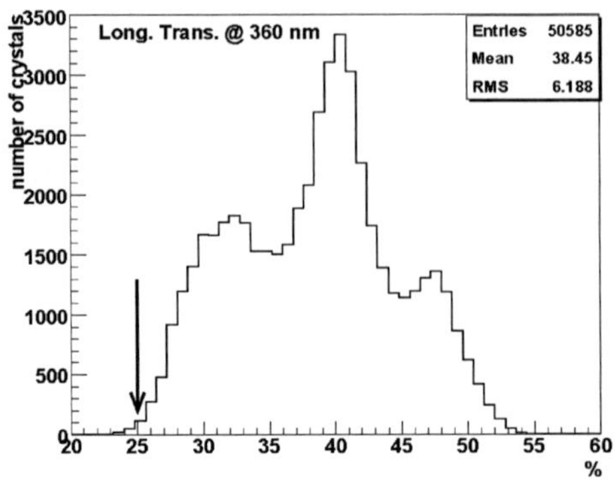

FIGURE 1. Longitudinal Transmission at 360 nm; the arrow shows the acceptance cut.

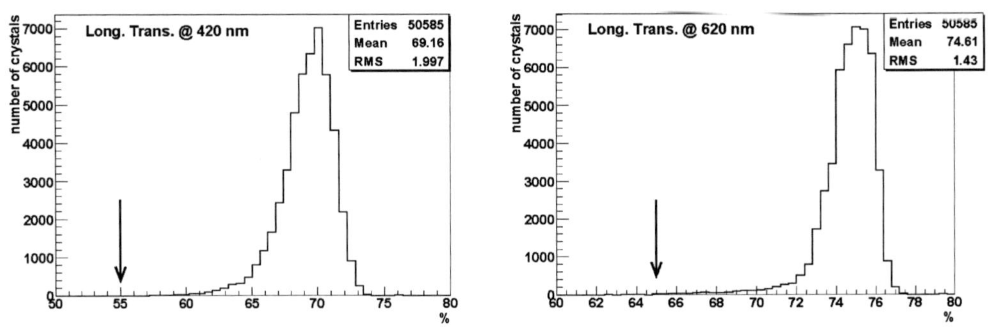

FIGURE 2. Longitudinal Transmission at 420 nm (left) and at 620 nm (right).

Light yield and front non-Uniformity

The Light Yield (LY) is important for its contribution to the stochastic term of the energy resolution of the calorimeter but also to the noise term. Crystals are accepted if their LY is higher than 7.2 pe/MeV when measured at 18°C, with a photomultiplier Philips XP2262B in optical contact with the large base of the crystal and the other surfaces wrapped with tyvek. This threshold value corresponds to around 4 pe/MeV when read with two APDs in the test beam. The required precision on the LY measurement is of the order of a few percent. For what concerns the LY measurement, both centers use a radioactive source to excite the scintillating light, driven by a motor to move the source along the crystal. The measurements, one per cm along the crystal axis, take place in a dark room and care has been taken to eliminate any possible spurious light source. The light signal itself is quite small, since the available radioactive sources with a long lifetime produce γ rays of energies around 1 MeV and $PbWO_4$ crystals have a rather low LY.

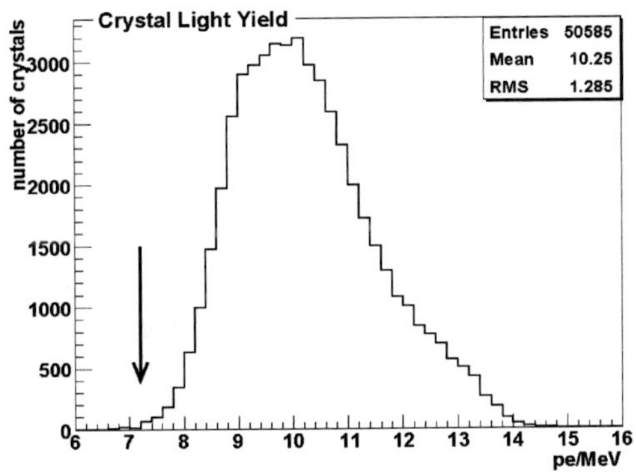

FIGURE 3. Light Yield measurements distribution; the arrow shows the acceptance cut.

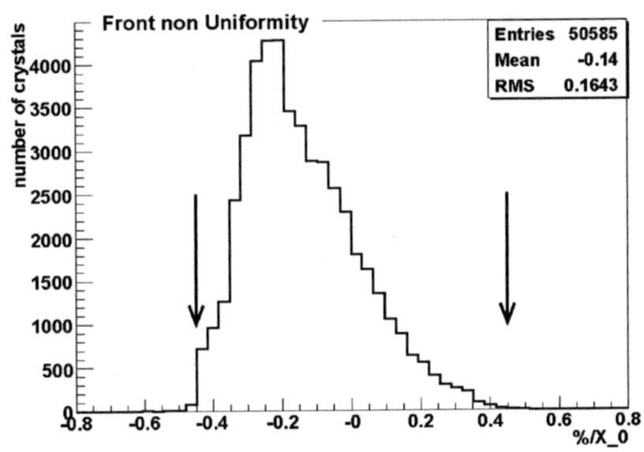

FIGURE 4. Front non-uniformity measurements distribution; the arrows show the acceptance cuts.

The tapered shape of crystals, required to obtain an hermetic pointing geometry, induces a focusing effect in the light collection. The same amount of energy deposited far or near the photo-detector will produce a higher or lower signal respectively. The uniformity of the light collection can be controlled by de-polishing with a given roughness one lateral face of the crystal. The ideal situation is to have a uniform light collection in the region of the maximum of the electromagnetic shower energy deposition. Due to the fact that the shower maximum position fluctuates, deviations from uniformity in this region will produce additional contribution to the constant term of the energy resolution. To maintain this contribution below 0.3% the maximum allowed deviation from uniformity in the crystal region 3 X_0 to 13 X_0 (front non-uniformity) was set to:

$$\frac{1}{LY}\frac{dLY}{dX} = \pm\, 0.45\%/X_0 \tag{1}$$

The LY and front non-uniformity distributions of Russian crystals are shown in Figures 3 and 4.

PRECALIBRATION

In the recent past, a strong correlation between the crystal light yield and their longitudinal transmission in the range of 350-370 nm has been noticed on a sample of around 6000 crystals in Rome Regional Center [5]. This correlation, verified on CERN data as well, is shown in Figure 5 on the full sample of 50000 Russian crystals measured up to now. The correlation factor between LY and LT360 is 77.3%. On the contrary there is no correlation between the LY and the transmission at 420 nm and 620 nm.

It is worthwhile to note that, assuming a simple model in which the emission spectrum is the same for all crystals and considering only the variation in the transmission spectrum, the maximum LY variation is expected to be below 10%, being the observed variation larger than 50% (Figure 5). This fact indicates that, in these $PbWO_4$ crystals, beside the natural correlation existing between light collection and absorption, there should be also a correlation between the amount of light produced and the transmission curve edge. This correlation might be due to a variation of absorption and emission center concentration from one crystal to another with consequent different positions of the crystal transmission band-edge and correspondingly different amount of light produced.

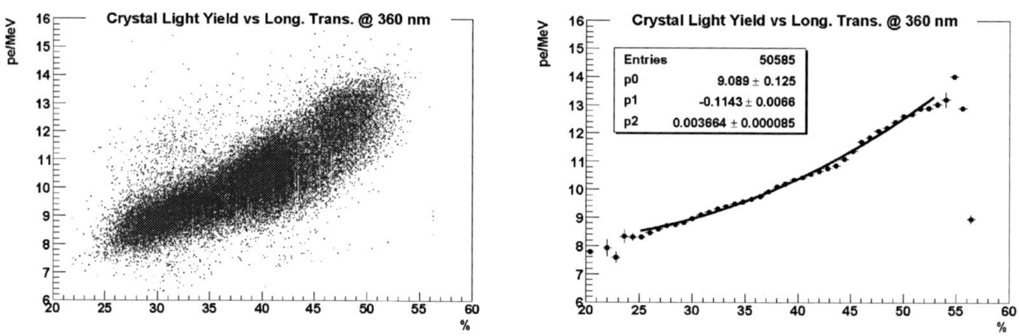

FIGURE 5. Light Yield vs Longitudinal Transmission at 360 nm; the profile on the right is fitted by a second order polynomial function.

Exploiting the observed correlation, the transmission measurement can be combined with the direct light yield determination to improve the intercalibration of the crystals. LY best estimation comes from the average between direct LY and LY obtained from LT360 using the fit shown in Figure 5. This has been verified with calibration from CERN-SPS test beam. The agreement between this laboratory "best" intercalibration and the test beam intercalibration with electron of 120 GeV is at level of 4% as shown in Figure 6; this is an excellent result considering the five orders of magnitude in energy between the MeV scale of the laboratory source and the electron energy in the test beam.

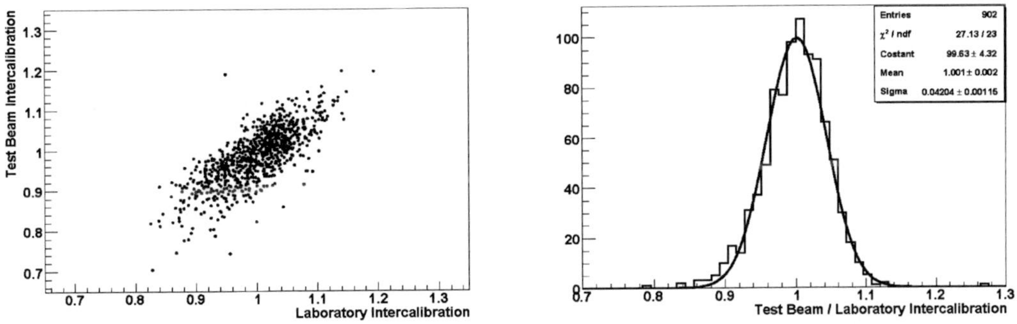

FIGURE 6. Comparison between the intercalibration coefficient obtained from CERN-SPS test beam with 120 GeV electrons and from laboratory measurements. On the right, the ratio distribution is fitted with a gaussian function.

CONCLUSIONS

The CMS electromagnetic calorimeter is a challenging project. After an extended R&D effort, the ECAL barrel construction is in the final phase in CERN and Rome Regional Centers while the endcap construction is starting now. Crystal properties are monitored in the ECAL Regional Centers continuously. Crystals are very uniform thanks to the precise depolishing of one lateral face; this contribute to reach the foreseen energy resolution. A very interesting correlation between Light Yield and Longitudinal Transmission is observed. This correlation provides an additional and independent measurement of the crystal LY, leading to a LY resolution improvement and to a 4% intercalibration precision of the calorimeter .

ACKNOWLEDGMENTS

My acknowledgments go to E. Auffray, F. Cavallari and I. Dafinei for the precious help in collecting data from Cern and Rome Regional Centers.

REFERENCES

1. CMS Collaboration, "The CMS Technical Proposal", CERN/LHCC 94-38, 1994.
2. CMS Collaboration, "The CMS Electromagnetic Calorimeter Technical Design Report", CERN/LHCC 97-33, 1997.
3. S.Baccaro et al. , "An automatic device for the quality control of large-scale crystal's production", Nucl. Instr. & Methods A 459 (2001) 278.
4. E. Auffray et al. , "Performance of ACCOS, an automatic crystal quality control system for the PWO crystals of the CMS calorimeter", Nucl. Instr. & Methods A 456 (2001) 325.
5. L.M. Barone et al. , "Precise measurements of light yield and transmission of $PbWO_4$ crystals at the INFN-ENEA regional centre.", Nucl. Instr. & Methods A 562 (2006) 76

A Radiation Damage and Recovery Study for Lead Tungstate Crystals from BTCP and SIC

Jianming Chen, Rihua Mao, Liyuan Zhang and Ren-Yuan Zhu

California Institute of Technology, Pasadena, CA 91125, USA

Abstract. This paper presents result of a study on radiation damage and recovery for lead tungstate crystals produced at BTCP and SIC. Correlations were observed between initial light output and initial longitudinal transmittance at 360 nm, between the loss of longitudinal transmittance at 440 nm and the loss of light output, and between radiation damages levels at different dose rates. No correlations, however, were found between crystal's initial optical properties and radiation hardness. Excellent linearity was observed between the variations of crystal's light output and its longitudinal transmittance at 440 nm in several cycles of irradiation followed by recovery, indicating these PWO crystals can be monitored *in situ* at LHC.

Keywords: Calorimeter; Lead Tungstate; Crystal; Radiation Damage; Correlation
PACS: 07.20.Fw; 29.40.Mc; 61.80.Fd

INTRODUCTION

Because of its high density and fast decay time, lead tungstate ($PbWO_4$ or PWO) crystal was chosen by the compact muon solenoid (CMS) experiment to construct a precision electromagnetic calorimeter (ECAL) of 76,000 crystals at the large hadron collider (LHC). After extensive R&D in the past decade, PWO crystals are now in mass production [1]. 40,000 crystals have been produced at Bogoroditsk Techno Chemical Plant (BTCP) in Tula, Russia, for the CMS experiment at CERN. Shanghai Institute of Ceramic (SIC) produced a few thousands crystals for the PrimEx experiment at the Jefferson Laboratory, and started to produce PWO crystals for CMS since 2003.

A total of 54 full size PWO samples were studied: 32 from SIC and 22 from BTCP. All samples went through a 200°C thermal annealing, which removes any residual absorption, followed by γ-ray irradiations at 15, 400 and 9000 rad/h until equilibrium [2]. Properties measured are longitudinal transmittance, emission and excitation spectra, light output, decay kinetics, light response uniformity and their degradation. Correlations between initial optical properties and with radiation damage were investigated for all 54 samples. In addition, two samples each from SIC and BTCP in a batch went through several irradiation and recovery cycles at 100 and 400 rad/h to study the linearity between the variations of the longitudinal transmittance and the variations of the light output.

CORRELATIONS

Correlations were observed between the initial light output and the initial longitudinal transmission at 360 nm, and samples from SIC and BTCP show different slopes, as shown in Fig. 1 (Left). This correlation may be attributed to the self-absorption of

CP867, *XII International Conference on Calorimetry in High Energy Physics*,
edited by S. R. Magill and R. Yoshida

the very blue part of the emission spectrum, as evidenced by the cross-over of the transmittance and emission spectra at 360 nm. This correlation was also observed by others [3].

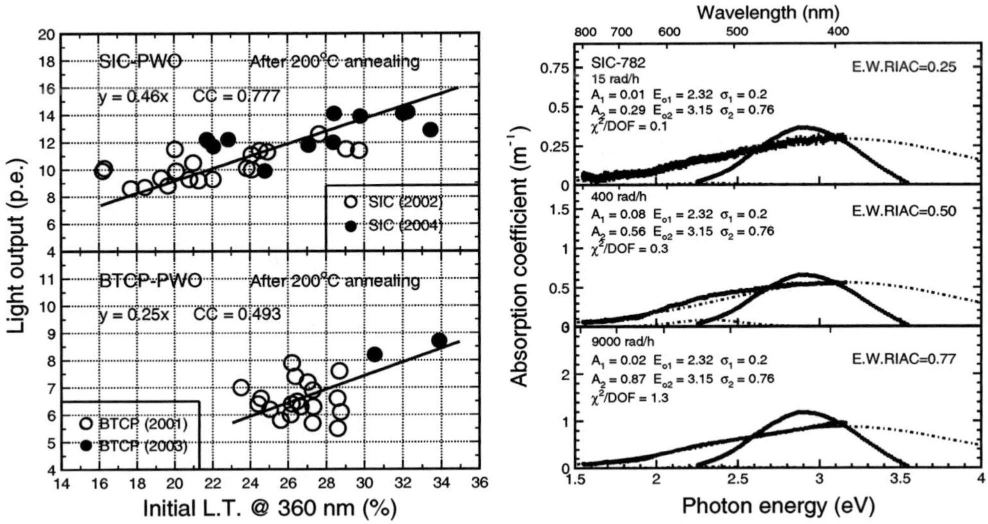

FIGURE 1. Left: Correlations are observed between initial light output and initial longitudinal transmission at 360 nm for BTCP and SIC PWO samples. Right: Radiation induced absorption coefficients are shown as a function of photon energy for SIC-782. Also shown in the figure are emission spectrum, decomposed color centers and EWRIAC in units of m^{-1}.

PWO crystals are known to suffer from radiation induced absorption which is dose rate dependent [2]. Fig. 1 (Right) shows radiation induced absorption coefficient (RIAC) as a function of photon energy for a sample SIC-782 under γ-ray irradiations with dose rate of 15, 400 and 9,000 rad/h in equilibrium [2]. Also shown in the figure is a decomposition of the RIAC to two color centers and crystal's emission spectrum with its peak between two color centers. The numerical values of the emission weighted radiation induced absorption coefficients (EWRIAC) [1] are also shown in the figure. Since the PWO scintillation mechanism is not affected by γ-ray irradiations, the loss of the light output is related to the EWRIAC. The linear correlation coefficient was measured to be 0.48 between the light output loss and the EWRIAC at 15 rad/h although samples have different preexisting absorption.

The EWRIAC values of all samples are less than 1 m^{-1} up to 400 rad/h, indicating no damage to the light response uniformity [2] for PWO crystals used in the CMS ECAL barrel if the LHC luminosity is increased by a factor of ten, i.e. the SLHC. The EWRIAC values measured at 9,000 rad/h, however, are diverse. Some samples show 3 m^{-1}, indicating damages in the light response uniformity and thus the energy resolution [2]. Rigorous quality control on radiation hardness thus is required for PWO crystals used in the endcaps. It is also found that SIC samples have relatively deeper color centers as compared to BTCP samples and thus longer recovery time constants.

Fig. 2 (Left) shows a correlation between the EWRIAC and the variations of the longitudinal transmission at 440 nm for all samples at all three dose rates. This correlation

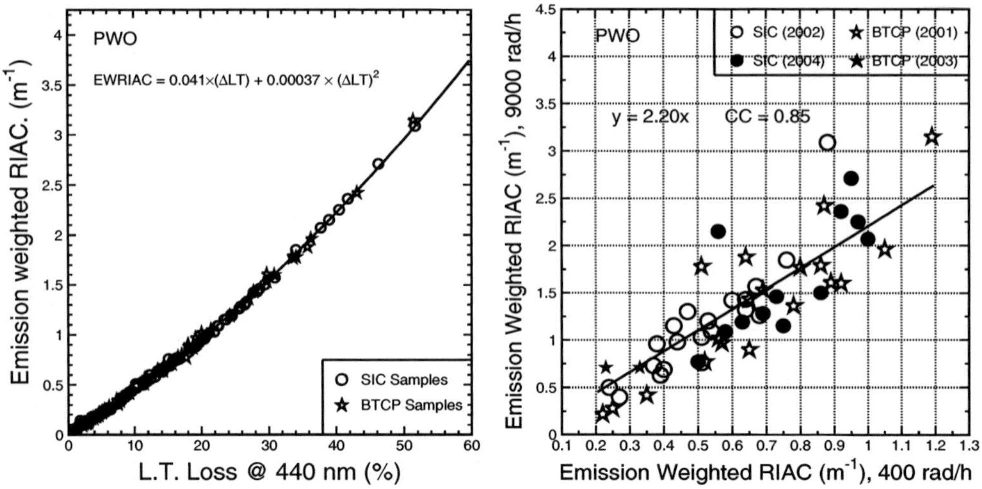

FIGURE 2. Left: Correlations are observed between the EWRIAC and the variations of the longitudinal transmission at 440 nm for all samples in equilibrium under all three dose rates. Right: Correlation between the EWRIAC values measured at 400 and 9,000 rad/h.

can be fit to a 2nd order polynomial, as shown in the figure. This result is consistent with previous investigation where 440 nm was chosen for the monitoring wavelength because of its best linearity [4].

Fig. 2 (Right) shows correlations between the EWRIAC values measured at 9,000 rad/h and 400 rad/h. The linear correlation coefficient is 0.85, which is much larger than 0.69 observed between 400 and 15 rad/h. A weaker correlation observed between low dose rates can be explained by the influence of preexisting absorptions. The strong correlation observed between high dose rates provides a foundation to transfer or compare radiation damage data measured at different laboratories, so is a useful tool for crystal quality control. This correlation was also observed by others [3].

No correlations were observed between initial optical properties and the EWRIAC. The numerical values of the linear correlation coefficient measured between the initial longitudinal transmittance at 360 nm versus the light output loss at 15 rad/h, the EWRIAC at 15 rad/h and the EWRIAC at 9,000 rad/h were found to be 0.19, -0.19 and 0.04 respectively, indicating that preexisting absorption in PWO crystals has no effect on the radiation induced absorption.

DAMAGE AND RECOVERY CYCLES

Two SIC samples (570 and 572) and two BTCP samples (2482 and 2531) went through long term irradiation and recovery cycles in a batch at 100 and 400 rad/h. Figs. 3 shows histories of the longitudinal transmission at 440 nm and the light output as a function of time for sample SIC-572.

The observed variations are consistent with our model on color center kinetics based

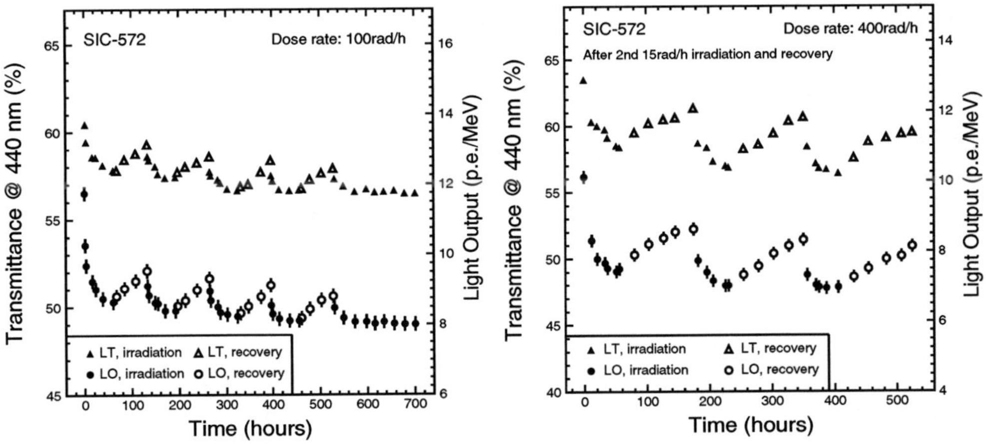

FIGURE 3. Variations of longitudinal transmission and light output are shown as a function of time for sample SIC-572 in multi-cycles of irradiation at 100 rad/h (Left) and 400 rad/h (Right) followed by recoveries.

on color center formation (damage) and annihilation (recovery) [2]. Variations of the light output and the longitudinal transmittance are also observed to be correlated. In addition, a longer time is found to be needed to reach equilibrium under 100 rad/h as compared to 400 rad/h, which can be explained by a longer damage time constant at low dose rate [2].

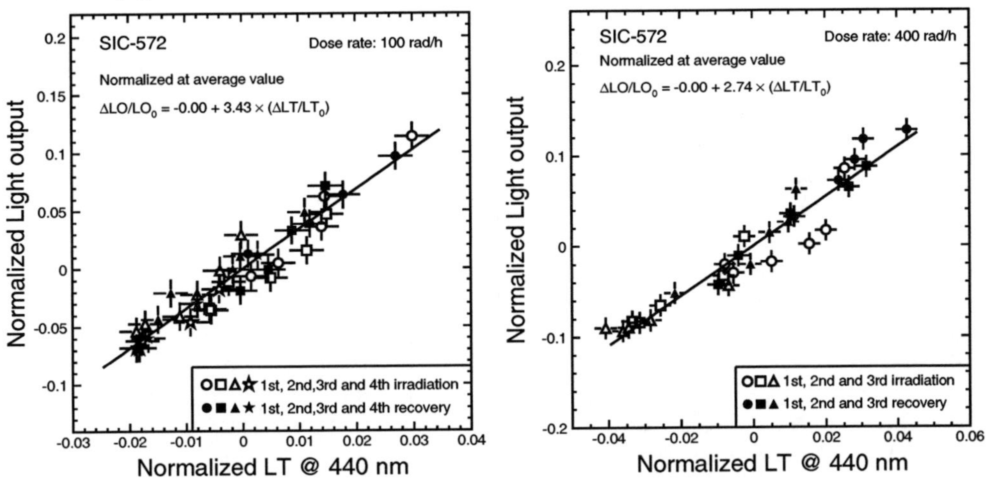

FIGURE 4. Normalized variations of the light output are shown as a function of normalized variation of the longitudinal transmission at 440 nm for sample SIC-572 under 100 rad/h (Left) and 400 rad/h (Right).

Fig. 4 presents linear fits to the data shown in Fig. 3 (Left) and (Right), respectively. The excellent linearity indicates that the variation of the light output may be calculated by using the variations of the longitudinal transmittance at 440 nm. The CMS ECAL monitoring system will use laser pulses at 440 nm to provide crucial monitoring information *in situ* at LHC [5]. It is also interesting to note that the slopes are 3.43 and 2.74 for dose rates of 100 rad/h and 400 rad/h respectively, indicating that this slope is damage level dependent, in addition to the wavelength dependence observed in our previous investigation [4]. The other SIC sample shows similar result.

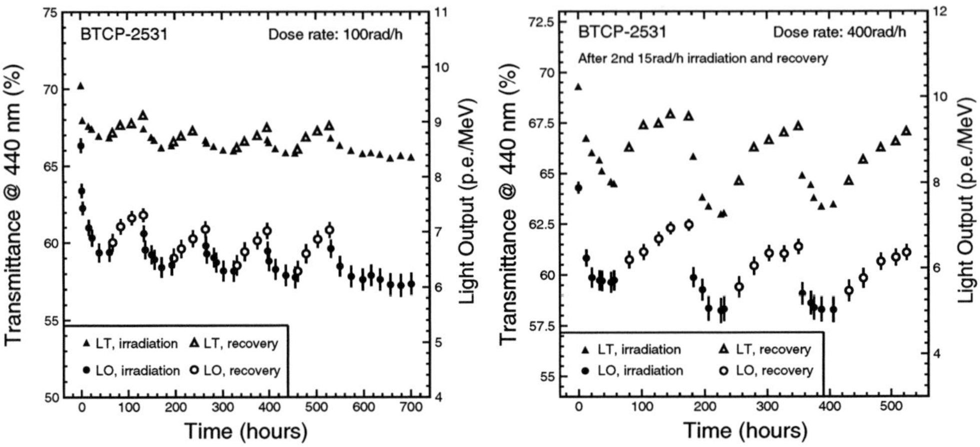

FIGURE 5. Variations of longitudinal transmission and light output are shown as a function of time for sample BTCP-2531 in multi-cycles of irradiation at 100 rad/h (Left) and 400 rad/h followed by recovery (Right) followed by recoveries.

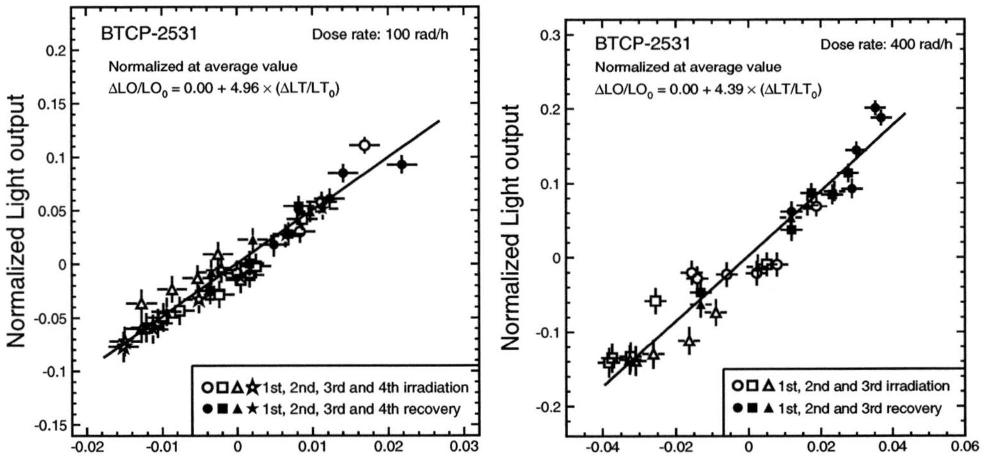

FIGURE 6. Normalized variations of the light output are shown as a function of normalized variations of the longitudinal transmission at 440 nm for sample BTCP-2531 under 100 rad/h (Left) and 400 rad/h (Right).

Similarly, Fig. 5 (Left) and (Right) show histories for sample BTCP-2531. Since SIC samples have relatively deeper color centers as compared to BTCP samples they take a longer time to reach equilibrium at low dose rates. Fig. 6 presents the linearity fits for sample BTCP-2531, showing slopes of 4.96 and 4.29 at 100 rad/h and 400 rad/h respectively. Again, the slope depends on the damage level. The result for the other BTCP sample is also consistent.

The BTCP samples have a slope of a factor of 1.5 larger as compared to the SIC samples. This observation is in a good agreement with the CMS beam test data. Based upon this observation, the variation amplitude of the light output for BTCP crystals *in situ* at LHC thus would be a factor of 1.5 larger as compared to the SIC crystals with the same variation amplitude in the longitudinal transmittance.

CONCLUSION

By using 54 PWO samples, correlations were observed between initial longitudinal transmittance at 360 nm and initial light output, between the loss of transmittance at 440 nm and EWRIAC, and between EWRIAC measured at different dose rates. No correlations, however, were found between crystal's initial optical properties and radiation hardness.

Excellent linearity was observed between the variations of crystal's light output and its longitudinal transmittance at 440 nm in several cycles of radiation at 100 and 400 rad/h followed by recoveries, indicating PWO crystals can be monitored *in situ* at LHC.

It is also noticed that the ratio between variations of the light output and the longitudinal transmittance is damage level dependent, indicating a necessity to extract this ratio *in situ* at LHC. The BTCP samples have also found to have these ratios of a factor 1.5 of SIC samples, indicating that the former is relatively more difficult to be monitored *in situ* at LHC.

ACKNOWLEDGMENTS

Discussions with CMS ECAL colleagues are acknowledged. This work was supported in part by the U.S. Department of Energy Grant No. DE-FG03-92-ER40701.

REFERENCES

1. Rihua Mao et al., IEEE Trans. Nucl. Sci. 51 (2004) 1777.
2. Ren-Yuan Zhu, Nucl. Instr. and Meth. A 413 (1998) 297.
3. I. Dafinei, in Proceedings of SCINT 2005, ed. A. Gektin and B. Grinyov (2005) 327.
4. Xiangdong Qu et al., IEEE Trans. Nucl. Sci. 47 (2000) 1741.
5. Liyuan Zhang et al., IEEE Trans. Nucl. Sci. 52 (2005) 1123.

Status of Zero Degree Calorimeter for CMS Experiment

O. A. Grachov, M. J. Murray

Department of Physics and Astronomy, University of Kansas
Lawrence, KS, USA

A. S. Ayan, P. Debbins, E. Norbeck, Y. Onel

Department of Physics and Astronomy, University of Iowa,
Iowa City, IA, USA

D. d'Enterria

CERN PH/EP, CH-1211 Geneva

CMS Collaboration

Abstract. The Zero Degree Calorimeter (ZDC) is integral part of the CMS experiment, especially, for heavy ion studies. The design of the ZDC includes two independent calorimeter sections: an electromagnetic section and a hadronic section. Sampling calorimeters using tungsten and quartz fibers have been chosen for the energy measurements. An overview of the ZDC is presented along with a current status of calorimeter's preparation for Day 1 of LHC.

Keywords: zero degree calorimeter, electromagnetic, hadronic.
PACS: Calorimeters, 29.40.Vj

INTRODUCTION

The two Zero Degree Calorimeters (ZDCs) are a major hardware contribution of the US Heavy Ion group to the CMS experiment. The ZDCs will measure neutrons and very forward photons for the heavy-ion and low-luminosity (up to 10^{33} $cm^{-2}s^{-1}$) pp collisions. The design is robust, as simple as possible and based on the Relativistic Heavy Ion Collider (RHIC) ZDC design [1], which showed highly successful performance for the four RHIC BNL experiments: BRAHMS, PHOBOS, PHENIX and STAR.

The ZDCs will provide reliable measurements of real-time luminosity, beam tuning and accelerator monitoring during A+A, p+A, and p+p running. In the A+A mode, the measured coincidence rate of forward-backward neutron signals emitted in the decay

CP867, *XII International Conference on Calorimetry in High Energy Physics,*
edited by S. R. Magill and R. Yoshida

of the Coulomb excited colliding ions can be used to determine the absolute A+A luminosity. The ZDCs, in addition, will help to discriminate effectively against beam-residual gas backgrounds and complement other available methods for commissioning the proton and ion beams at the LHC.

The event-by-event determination of the heavy-ion collision centrality is a basic prerequisite for any physics measurement of A+A reactions. The centrality of A+A reactions determines the degree of overlapping of the colliding nuclei and, thus, the maximum energy density reached in the interaction. Because the ZDCs are sensitive to "spectator" neutrons, they are important detectors for measurements of reaction centrality. The combination of ZDCs measuring the energy of forward neutrons, together with other multiplicity detectors at more central rapidity, has become the standard method for determining reaction centrality in A+A collisions at RHIC.

The ZDCs have been proven at RHIC to be of paramount importance for triggering and tagging UPC (ultra peripheral collision) events by measuring the forward neutrons issuing from the Coulomb-excited-nucleus dissociation.

For the photon measurement, it is necessary to have an electromagnetic zero-degree calorimeter in front of the neutron zero-degree calorimeter. The "standard" Pomeron-Pomeron diffractive A+A collision leading to central particle production and forward nucleus dissociation will not only be a background for $\gamma+\gamma$ and γ+A processes but also a signal in their own right, accessible to study with the ZDC.

MECHANICAL AND OPTICAL DESIGN

In order to measure neutrons and very forward photons the calorimeter will be placed downstream of the first beam dipole magnet in the straight section and between the two beam pipes. In CMS this region is located ~140 m on each side of the interaction vertex inside the neutral particle absorber (TAN), designed to protect LHC magnets by limiting their absorbed dose. This neutral particle absorber was built with a detector slot of 1 m length, 96 mm width and 607 mm height, into which are inserted 90 cm length copper absorber bars and a transversally segmented ionizing chamber (proton-proton luminosity monitor). The design of each individual ZDC includes two independent calorimeter sections, the electromagnetic section (EM section) and the hadronic section (HAD section). These sections replace the copper absorbers (see Fig. 1). The ZDC calorimeter sections are sampling calorimeters with the core of each structure consisting of a tungsten-plate/quartz-fiber-ribbon stack. A significant advantage of this technology is that the calorimeter is compact, extremely fast, and radiation hard. Quartz fibers were chosen as the active media of the ZDC calorimeters because of their unique radiation hardness features and the intrinsic speed of the Cherenkov effect. The quartz-quartz fibers can withstand up to 30 GRad with only a few percent loss in transparency in the wavelength range of 300 nm to 425 nm. The Cherenkov light generated by charged particles passing through the fibers is brought to photomultiplier tubes (PMT).

The HAD section consists of 24 layers of 15.5 mm thick tungsten plates and 24 layers of 0.7 mm diameter quartz fibers. The EM section is made of 33 layers of 2 mm thick tungsten plates and 33 layers of 0.7 mm diameter quartz fibers. The hadronic section is longitudinally segmented into 4 readout segments of ~1.4 nuclear interaction length each. The tungsten plates are tilted by 45°.The fibers of each individual tungsten/quartz fiber cell are grouped together to form readout bundles. The four fiber bundles are coupled to four photomultiplier tubes via long air core lightguides which provide optical mixing and matching of the fiber bundle to the photocathode.

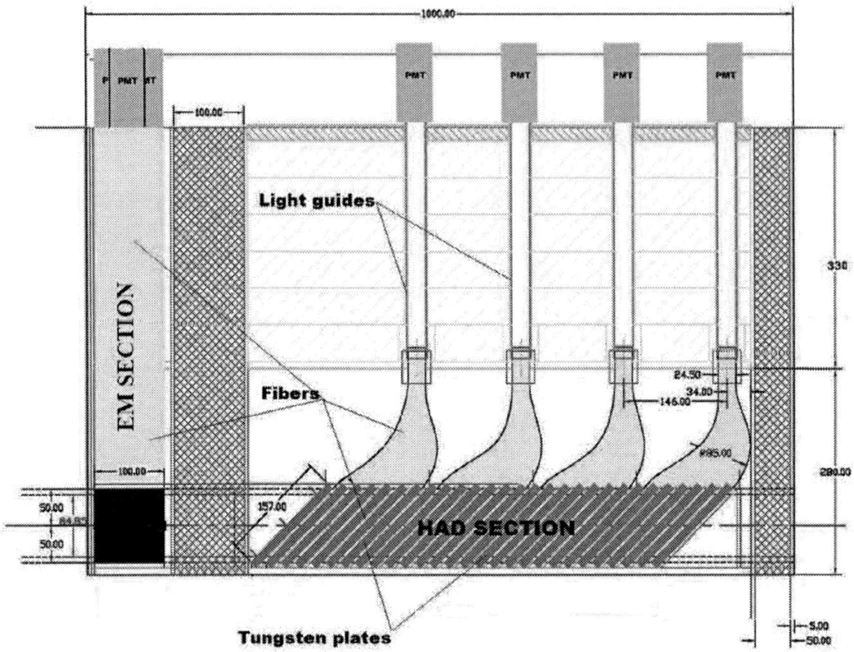

FIGURE 1. Schematic illustration of the TAN detector slot showing the location of the p-p luminosity monitor and the ZDC (EM section and HAD section).

The electromagnetic section is segmented into 5 horizontal (transverse) individual readout towers. The tungsten plates are oriented vertically. The fibers from each individual tower are grouped together to form a readout tower bundle. These are each coupled directly to the PMT. All calorimeter sections will be instrumented with the same type of PMT, R7525, used in the HF (forward hadronic) calorimeter. The physical characteristics of the ZDC are summarized in Table 1.

TABLE 1. The Physical Characteristics of ZDC.

	Hadronic Section	Electromagnetic Section
Sampling Ratio	15.5 mm W/0.7 mm QF	2 mm W/0.7 mm QF
Number of Cells	24	33
Interaction (Rad.) Length	$5.6\lambda_o$	$(19X_o)$
Number of Channels	4 longitudinal segments	5 horizontal divisions
Module Size (W x L x H), mm	92 x 711 x 705	92 x 116 x 705
Weight of Module, kg	~400	~65

Tungsten/quartz sampling calorimeters of this type have been successfully used in a number of colliding beam and fixed target experiments, and as a consequence, the technology is well known [2]. Recent examples of detectors incorporating this technology include the zero degree calorimeters for the four heavy-ion experiments at RHIC. Much of the technology employed in the optical and mechanical design of CMS ZDC has its origins in the original research and development conducted for the ZDCs of RHIC and the HF calorimeter of the CMS experiment at the LHC, including the preparation of the optical components (quartz optical fibers, optical lightguides) and the characterization of the phototubes.

RADIATION ENVIRONMENT

The LHC will have 7 TeV proton beams with nominal luminosity L= 10^{34} $cm^{-2}s^{-1}$ and with an interaction rate of $8\times10^{8}s^{-1}$. The power density, power dissipation, particle fluxes and spectra, accumulated dose and residual dose rates for the TAN neutral absorber have been evaluated by means of the DPMJET event generator for p-p collisions based on the MARS code [3]. During the pp runs with luminosity L= 10^{33} $cm^{-2}s^{-1}$ the expected average absorbed radiation dose will be of about 18 GRad/year. Whereas, for a one-month PbPb run the expected averaged absorbed dose will be ~30 MRad. To avoid radiation damage of the optical fibers, the ZDC will only be installed during the heavy-ion runs and the first low-luminosity p-p runs.

The phototubes will be located about ~500 mm above the beam line and ~140 m from the interaction point. According to estimates, the radiation environment will be on the same level as for the PMTs of the HF calorimeter (~10 krad/yr for the design p-p luminosity of 10^{34} $cm^{-2}s^{-1}$). This dose is permitting the use of the same Hamamatsu R7525 PMTs used by the HF calorimeter.

PMT SYSTEM

The PMT systems of the HF calorimeter have been adopted for the ZDC. The air-core light guide of the hadronic section design is based on studies performed by the HF group at the University of Iowa. These studies show that the diffusion of light due to the mixer is optimum. Thus the 20% variation in quantum efficiency that might be expected as a result of the non-uniform response of various fibers is reduced to a few percent as a result of utilization of the mixer. The PMTs for the ZDC are eight-stage Hamamatsu R7525 phototubes with bi-alkali photocathodes, resulting in an average quantum efficiency for Cherenkov light of about 10%. The resistive chain high voltage base (ratio "B") is optimized to achieve a high gain and the maximum dynamic range of linearity (linearity is on the level of 2% for dynamic range of up to 10^{3}). The PMT and base are housed in a shielding enclosure (housing). All housing components, except for the μ-metal shields, are fabricated in the University of Kansas (KU) physics machine shop.

PROTOTYPING

An engineering mechanical prototype was designed and constructed at KU. The test module design was virtually identical to one quarter of the HAD section design (one longitudinal tower). The module was used to establish assembling protocol and to verify the assembly tolerances and clearances. Fiber routing and fiber ribbons preparation was also studied with the first mechanical prototype. Experience gained through these studies was important in establishing the final fiber routing scheme and fiber ribbon design.

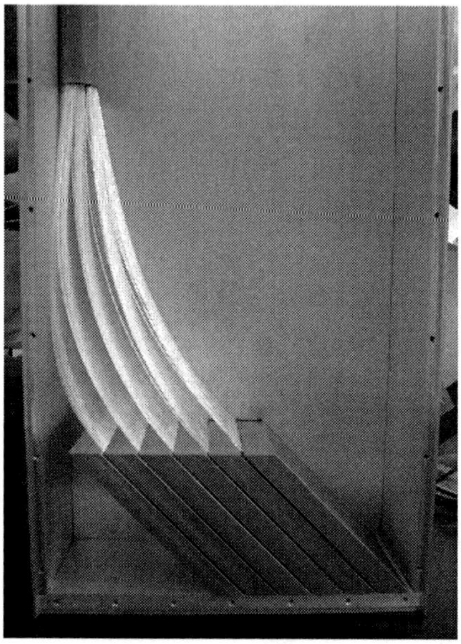

FIGURE 2. Picture showing the mechanical prototype of one longitudinal tower of the HAD section.

In consultation with the LHC integration group, many of the details of how to install the ZDC were worked out this year. A dummy ZDC was made from plastic and successfully inserted into the two TANs that will be used for CMS. Finally a wooden full scale copy of a TAN has been constructed at CERN to help sort out installation issues on the surface.

PERFORMANCE

The ZDC sections have been studied with both stand-alone simulations and CMS GEANT4 simulations. These studies have been directed both towards optimization of the calorimeter design and towards understanding the detector performance. Following the simulation studies, we have fixed the main version of the converter for the EM section to be 33 tungsten plates with a thickness of 2 mm (~$19X_0$) and quartz fibers with a core diameter of 0.6 mm. We required that the energy resolution should be better than 10% for 50 GeV photons and that the calorimeter should be linear within 2% in the energy range from 10 GeV to 100 GeV. The simulation shows that a calorimeter with this geometry meets our requirements. The resulting energy resolution is displayed in Fig. 3.

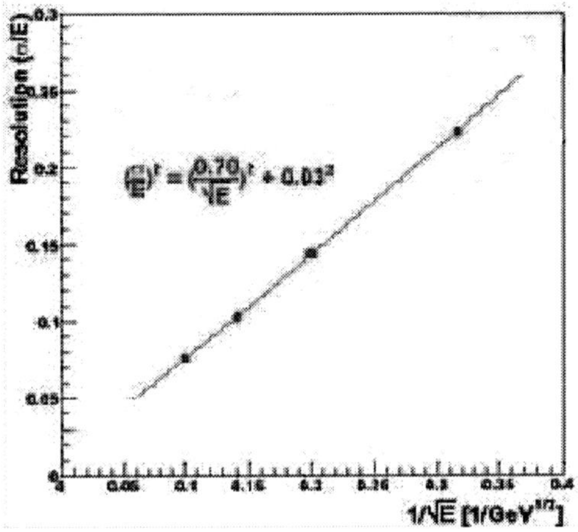

FIGURE 3. Energy resolution of the ZDC electromagnetic section.

The transverse cell size (in the horizontal plane) for the electromagnetic calorimeter is fixed at ~18 mm. This ensures that the main energy deposition from a single particle occurs in one cell. The 50 GeV photon shower profile is given in Fig. 4. The width of the shower is ~5 mm and the separation of two particles with a relative distance of about 2 cm in the horizontal plane at the entrance to the calorimeter is easy to distinguish. The standard technique of shower-profile fitting can be applied to separate closer showers. In practice, this position resolution will allow measurement of the beam crossing angle with a resolution of ~10 mrad.

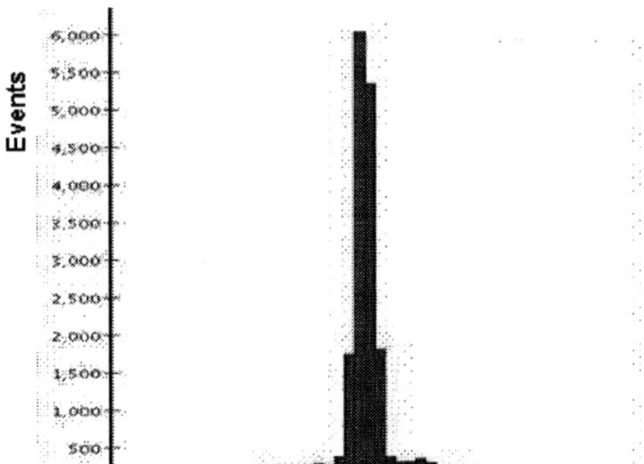

FIGURE 4. The 50 GeV photon shower profile in the EM section of ZDC.

The ZDC hadronic section will be in the very forward region, and thus the calorimeter needs to be quite dense so as to absorb hadrons with an energy of a few TeV. The choice of tungsten as the passive material meets this requirement. We have studied different tungsten plate thickness varying from 5 mm to 20 mm and we did not find significant changes in energy resolution for energies in the region from 500 GeV to 3 TeV. For quartz fibers with core diameter of 0.6 mm the energy resolution will be in the range 10% ± 2%, which is better than our requirement of 15% for 3 TeV. For practical reasons, such as the number of tungsten plates, the number of fiber ribbons, and the limited space, we chose 15.5 mm tungsten plates as the absorber in the HAD section. The energy resolution for a calorimeter with 10 mm-thick tungsten plates is presented in Fig. 5.

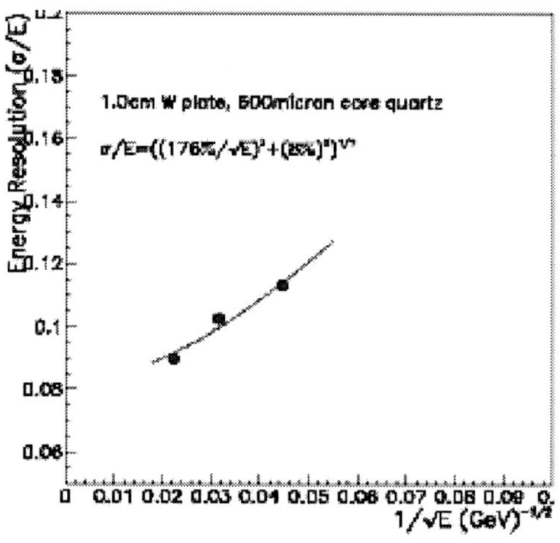

FIGURE 5. The energy resolution of the hadronic section of ZDC.

ELECTRONICS

The ZDC electronics includes the four major functions necessary for the functioning and data reading from the detector: front-end electronics, trigger electronics, data acquisition, and the high voltage system needed for operation of the phototubes.

In the nucleus-nucleus running mode, the ZDC is a central detector for Level-1 triggering. The standard ZDC minimum bias trigger uses a coincidence of the signals from the two calorimeters (right and left) on both sides of the interaction point. In addition, a single (or double) signal from one (or both) ZDCs will be used to trigger on ultraperipheral electromagnetic nucleus-nucleus interactions leading to the dissociation of one (or both) colliding nuclei. These coincidences are processed by the ZDC electronics in the counting room.

The front-end electronics, including signal processing, digitization, buffering, formation of an event trigger, and the first level of readout, is located in HCAL VME crates as close to the ZDC as possible. Because of the time delays introduced these crates located in the adjacent underground counting room (USC55).

There are a total of 18 readout channels for the two ZDCs. The signals from the ZDCs are transmitted through a long (210 m) coaxial cable to the counting room. Signals from the EM channels will go directly to the QIE ASIC. The signal from each HAD channel will be split into two signals; one signal will go to the QIE and the other will be summed with the other three HAD channel signals. The analog sum of the four signals related to each HAD section, proportional to the total energy deposition in each detector and proportional to the total number of spectators coming out from the interaction, will be sent to the splitter. One output signal of the splitter will go to the QIE to provide a Level 1 trigger for several centrality intervals, and other signal will go to the input of the discriminator and then to a logic unit for coincidence. The coincidence of two signals from both sides of the interaction point is sensitive to most of the nuclear and electromagnetic cross section. Information from scalers will be used for tuning the interaction of beams and for defining the real-time luminosity.

INSTALLATION AND PLANS

The ZDC detector configuration necessary to accomplish all of the physics goals includes 2 ZDCs (two hadron sections and two electromagnetic sections). The Day 1 detector implementation will provide physics results. The installation operation is scheduled to start early September of 2006. After all connections are complete, each section will be tested and commissioned. Testing and commissioning of the detectors will mainly involve HV (high voltage) system tests and LED (light emitting diode) and laser calibrations. Each ZDC section will have been calibrated in the test beam. We foresee all functionality of the read-out to be tested and operational shortly after the cables are connected and the electronics in the counting room is functioning.

ACKNOWLEDGMENTS

We thank S. J. Sanders for useful discussions and support. D. Macina, E. Tsesmelis and A.-L. Perrot of CERN TS/LEA provided invaluable help for the integration and installation of the ZDC. D. d'Enterria acknowledges support by the 6th EU Framework Programme (contract MEIF-CT-2005-025073). We are grateful to the entire US Heavy Ion group for support. This project was supported by the HENP Divisions of the U.S.DOE and the U.S.NSF.

REFERENCES

1. C. Adler et al., *Phys. Rev. Lett* **89**, 272302 (2002); A. Denisov et al., *Nucl. Phys.* **A698**, 551 (2002).
2. Y. Onel, 10th *International Conference on Calorimetry in High Energy Physics* (CALOR 2002), Pasadena, CA, 25-30 March 2002; N. Akchurin and R. Wigmans, *Rev.Sci.Instrum* **74**, 2955-2972 (2003).
3. N. V. Mokhov et al., FERMILAB-FN-732.

The LHCf experiment at the LHC accelerator

L. Bonechi [1*], O. Adriani*, M. Bongi*, G. Castellini†, R. D'Alessandro*, A. Faus**, M. Haguenauer‡, Y. Itow§, K. Kasahara¶, T. Mase§, K. Masuda§, Y. Matsubara§, H. Matsumoto§, H. Menjo§, Y. Muraki§, P. Papini*, T. Sako§, T. Tamura‖, K. Tanaka§, S. Torii††, A. Tricomi‡‡, W. C. Turner§§, J. Velasco**, H. Watanabe§ and K. Yoshida‖

*University of Florence and INFN Section of Florence, Italy
†IFAC-CNR, Firenze, Italy
**IFIC, Centro Mixto CSIC-UVEG, Valencia, Spain
‡Ecole-Polytechnique, Palaiseau, France
§Solar-Terrestrial Environment Laboratory, Nagoya University, Japan
¶Shibaura Institute of Technology, Japan
‖Kanagawa University, Japan
††RISE, Waseda University, Japan
‡‡University of Catania and INFN Section of Catania, Italy
§§LBNL, Berkeley, California, USA

Abstract. The claimed discovery of atmospheric shower induced by cosmic-ray with energy beyond the GZK cutoff by the AGASA experiment in 1994-1995, although not confirmed by other important experiments like Fly's Eye and Hi-Res, together with the poor knowledge of the composition of cosmic rays around and beyond the Knee region, have highlighted the necessity of new experiments that should increase our present knowledge of HECR and UHECR. For this reason big efforts have been addressed to the development of new experiments, like Auger, TA and EUSO, for a systematic study of the UHE atmospheric showers with increased capabilities with respect to the previous experiments. Moreover complementary experiments should allow a precise calibration of the methods used for the reconstruction of cosmic-ray showers in atmosphere. Their aim is the measurement of quantities that are used in these procedures and that are not yet precisely known. Under this perspective the LHCf experiment is a compact experiment which has been proposed for the study of neutral pion and gamma production at high energy in proton-proton interaction in the very forward region of the LHC accelerator. It will help calibrating the algorithms that are used to reconstruct the atmospheric shower events for energy beyond the Knee. The LHCf apparatus and the results of the first beam test, held in 2004, are shortly discussed in this work.

Keywords: neutral pion and gamma, nuclear interaction, calorimetry, high energy cosmic rays
PACS: 13.75.Cs, 29.40.Vj, 96.50.sb

1. INTRODUCTION

A lot of interest in particle astrophysics has been raised up in the previous decades, mainly due to the amazing puzzle of the possible existence of Super-GZK particles in cosmic rays hitting the Earth atmosphere. The GZK-cutoff is an upper limit on the energy of cosmic rays coming at Earth from distant sources, that was computed in 1966 by the physicists K. Greisen, G. Zatsepin and V. Kuzmin. This effect predicts that particles

[1] Corresponding author: lorenzo.bonechi@fi.infn.it

CP867, *XII International Conference on Calorimetry in High Energy Physics*,
edited by S. R. Magill and R. Yoshida

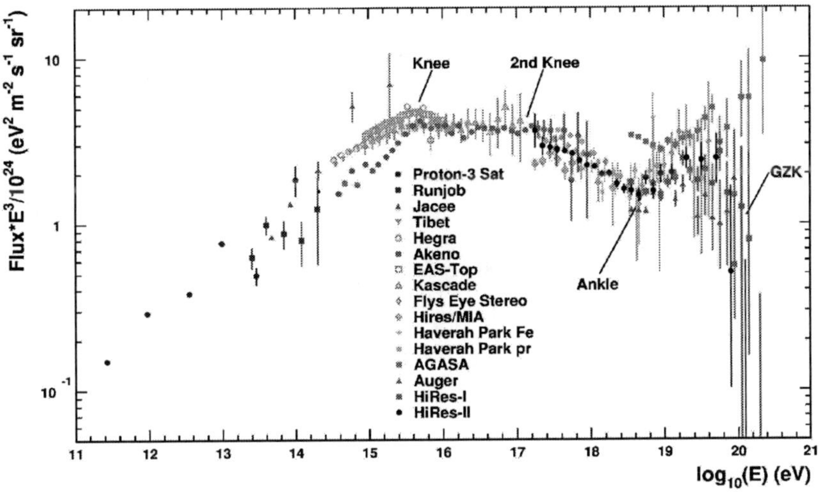

FIGURE 1. Spectrum of cosmic radiation[4].

produced in distant sources with energy over the threshold value of 5×10^{19} eV cannot be observed at Earth due to their interaction with cosmic microwave background radiation to produce pions (pion photo-production $p + \gamma_{3K} \rightarrow \Delta^{+} \rightarrow N + \pi$). The AGASA experiment [1] has claimed the observation of a few events at energy greater that the GZK-cutoff. This could be explained assuming the existence of extremely energetic sources in our galaxy, never observed up to now, or, following some new theories, with the decay of very massive particles. A different explanation can be considered if we look at the AGASA experimental results together with the results obtained by other collaborations, like Hi-Res [2] and Auger [3]. In figure 1 the measurements of the cosmic-ray spectrum up to ultra high energy by these experiments are compared. It can be observed that the results of the AGASA experiment are higher than those of the other experiments over all the investigated energy range, even if the statistical errors around the GZK-cutoff are too high to allow any conclusions. This discrepancy could then in principle be solved if we assume the AGASA energy reconstruction procedure affected by a systematic overestimate around 15-20%. Because the method used by the AGASA and Hi-Res experiments for energy reconstruction depend on the composition of cosmic rays and on the considered hadronic interaction model in atmosphere, that are not precisely known, a calibration of the simulations that are used to give an estimate of the shower energy is necessary.

2. THE LHCF EXPERIMENT

The LHCf experiment, officially approved as a CERN experiment in may 2006, has been proposed for the first time in 2004[5, 6]. Its aim is the measurement of neutral pion and

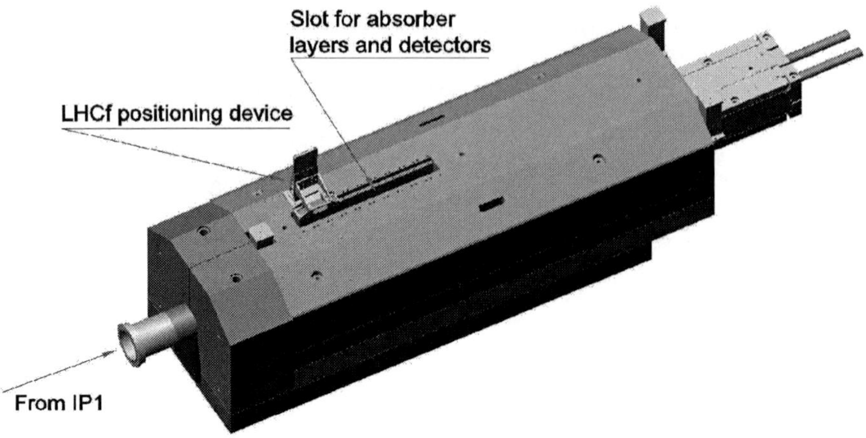

FIGURE 2. Absorber for neutral particle debris, located at $\pm$ 140 m from IP1.

gamma production spectra in proton-proton interactions in the very forward region of LHC. A precise knowledge of these spectra at very high energies is of great interest because it may help choosing the best among the existing hadronic interaction models that are commonly used for simulations at extreme energies (QGSJet, DPMJet, Sibyll etc.), in such a way to better reproduce the development of atmospheric showers induced by UHECR. The neutral pion production spectrum at high energy has been previously measured at the SPS accelerator by the UA7 collaboration only [7], for an equivalent energy of 10^{14} eV in the laboratory system. This energy is far below the energy at the Knee of cosmic-ray spectrum, that is between 10^{15} eV and 10^{16} eV. The LHCf collaboration, descendant of the UA7 collaboration, has proposed a new measurement at the LHC accelerator at CERN. The LHC energy (7 TeV+7 TeV in the CM system) will allow a study of neutral pion and gamma components at an equivalent energy around 10^{17} eV in the laboratory system, just between the energy at the Knee and the GZK-cutoff.

The LHCf apparatus is composed by two small independent detectors which will be installed, at the beginning of 2007, on both sides of the ATLAS interaction region (IR1), 140 m far from the interaction point (IP1). In these regions, known as "recombination chambers", the single straight beam pipe coming from IP1 splits into two separate smaller pipes, 96 mm far from each other, where protons moving in opposite directions travel separately. At IR1 the two recombination chambers are surrounded by two big absorbers (TANs) used to protect the successive cryomagnets from neutral particle debris (figure 2). Thanks to a $9.6\,\mathrm{cm}^{w} \times 60.7\,\mathrm{cm}^{h} \times 100\,\mathrm{cm}^{l}$ wide cavity, the structure of the TAN allows the insertion of material just between the two beam pipes, along the beam line as coming from IP1. Ten slot, 10 cm wide, can be used both for absorber layers (9.4 cm thick copper layers are ready to be installed) and for detectors. While the fourth slot will be used for a luminosity monitor, the first three copper layers of each TAN will be replaced by the LHCf detectors, which should take data during the first phase of the LHC running, when the luminosity is lower than $10^{30}\,\mathrm{cm}^{-2}\,\mathrm{s}^{-1}$ and the number of bunches is limited to 43. A few hours data taking should be enough for the achievement of the physics goals, being the expected rates in LHCf around 500 Hz and 6 Hz respectively for gamma and neutral pion events.

2.1. The LHCf detectors

The two LHCf detectors that will be installed in the TANs are similar, but not exactly identical. They are both sampling electromagnetic calorimeters, made of tungsten absorber layers, plastic scintillator layers for the longitudinal sampling of the shower energy, and tracking layers for the reconstruction of the impact point of gammas. Both the detectors are made of two independent calorimeter towers. They are necessary to reconstruct π^0 events by separately detecting the two gammas coming from decay $\pi^0 \longrightarrow 2\gamma$. Each tower has a total tungsten thickness of 44 radiation lengths and includes 17 scintillator layers (EJ-260 by Eljen Technology Co.). In the first part of the towers (22 radiation lengths), where the electromagnetic showers develop almost completely, we have a 3 mm sampling scintillator layer for each 7 mm thick tungsten layer (2 radiation lengths). In the other part, interested only by hadronic showers, sampling is done every two absorber layers.

Apart from these common parts, some important differences, both in the tracking instrumentation and in the geometry of the sensible regions, characterize the two calorimeters. Let's consider them separately.

2.1.1. Detector I

In figure 3 a projection of detector I on a plane transverse to the beam line is shown. The towers have square sections of $20 \times 20\,\text{mm}^2$ and $40 \times 40\,\text{mm}^2$ and are disposed

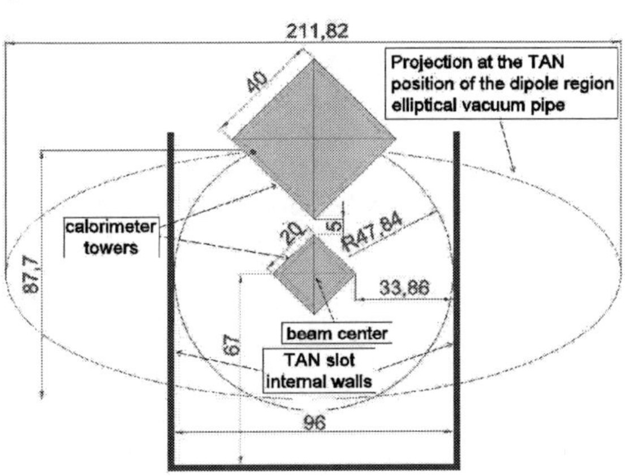

FIGURE 3. Projection of detector I on a plane transverse to the beam line.

with their diagonals aligned along the vertical direction. Four tracking modules, made of 1 mm × 1 mm square scintillating fibers (KURARAY SCSF-38) arranged in two layers for the measurement of two ortogonal coordinates, are placed at 8, 10, 30 and 42 radiation lengths inside each calorimeter towers. The positions of the first two layers

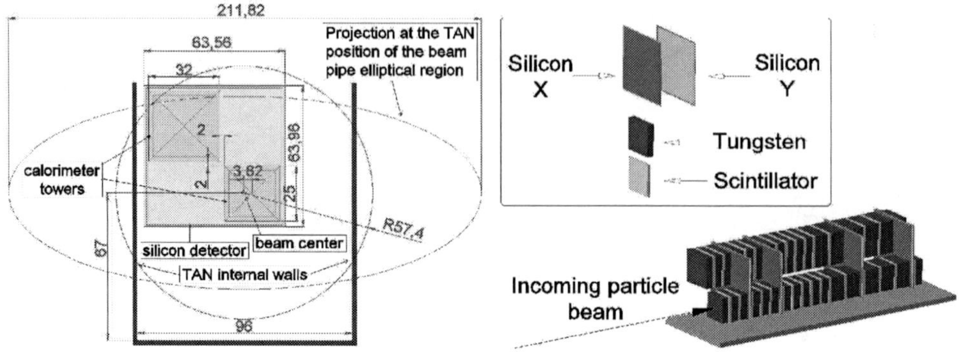

FIGURE 4. Left: projection of detector II on a plane transverse to the beam line. Right: structure of the calorimeter towers.

have been optimized for electromagnetic showers, while the other two layers will be used for neutron studies.

The light produced inside a scintillator layer is collected using a bundle of light guide fibers and readout by a Hamamatsu R7400U photomultiplier. The scintillating fibers of the tracking layers are connected to multi-anode photomultipliers (MAPMTs) Hamamatsu H7546 (64 anodes). High dynamic range VA32HDR14 chips are used for the front-end boards.

The relative position of the towers has been chosen to allow the several scintillating and light guide fibers to be pulled to the top of the detector without tight bendings. The same configuration minimizes energy contamination between the towers when two showers develop simultaneusly in both of them.

The LHCf geometrical acceptance is limited by the aperture of the D1 dipole magnets that are located in the region between IP1 and the TANs. The $20 \times 20\,\mathrm{mm}^2$ tower is usually centered along the beam line; due to the geometrical cut the good region of the big tower is only its lower part. For this reason detector I is designed to be allowed to move up and down using a remotely controlled positioning tool, in such a way that, if the machine in IP1 runs with a not null beam crossing angle, a greater kinematical region may in principle be accessed by simply moving the detector a few centimeters.

2.1.2. Detector II

In figure 4 a projection of detector II on a plane transverse to the beam line is shown. While the longitudinal structure of absorber and scintillator layers is the same as for detector I, some important differences concern the geometry of the two calorimeter towers. Differently from detector I, detector II has been optimized under the hypothesis of data taking in a fixed position, in case we have no possibility of moving it up or down. To cover the greatest kinematical region the towers have been placed with their sides (and not their diagonals) aligned with the horizontal or vertical directions and shifted along the horizontal direction as shown in figure 4. Their size are $25 \times 25\,\mathrm{mm}^2$ and $32 \times 32\,\mathrm{mm}^2$ and they cover an area up to 57 mm far from the beam center (to be

FIGURE 5. Setup of 2004 beam test with a prototype of LHCf detector I. Calorimeter towers are contained in the metal box on the left. Between the beam pipe termination (on the right) and the LHCf box, five silicon microstrip modules from the ADAMO experiment [11] have been used to reconstruct the impact point of particles on the calorimeter.

compared to 48 mm of detector I). This disposition also simplifies pulling the light guide fibers, because they are now naturally aligned along the vertical direction. No bending is thus necessary to connect them to the photomultipliers located in the upper part of the detector. The other big difference with respect to detector I is the use in detector II of silicon microstrip sensors in place of scintillating fibers. Due to the limited available space, the first two tracking modules will be placed at 6 r.l. and 12 r.l. to allow routing the light guides and silicon readout cables in the region between them. The expected spatial resolution in the gamma impact point reconstruction is better than $100\,\mu$m in the energy range relevant for LHCf. Silicon sensors are readout using fast electronics developed at CERN for the CMS experiment [8]. The front-end is based on the large dinamic range PACE3 chip [9, 10].

3. FIRST BEAM TEST RESULTS

In 2004 a prototype of detector I, composed of two calorimeter towers ($20 \times 20\,\text{mm}^2$ and $40 \times 40\,\text{mm}^2$), both provided with 3 pairs of scintillating fibers layer, was built to be tested with a beam of particles at CERN SPS accelerator. The experimental setup is shown in figure 5. The main aim of this test was the study of energy resolution of the calorimeter towers. In particular it was requested by the LHC Committee to analyze the dependence of energy resolution as a function of the impact point of particles on the towers, focusing the attention to the smallest one, whose linear dimensions (20 mm)

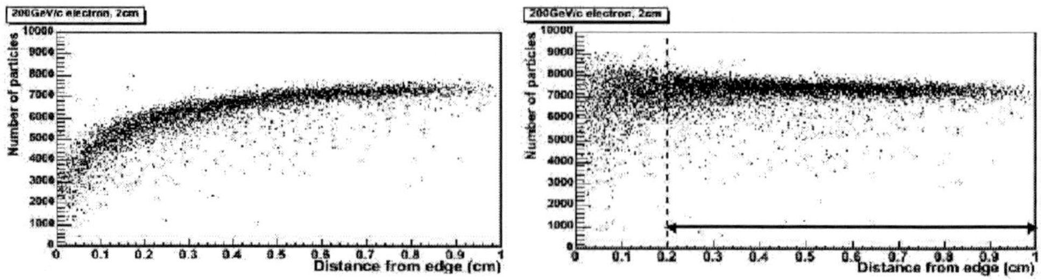

FIGURE 6. Beam test results with 200 GeV/c electron beam. Left: rough energy collection in the small tower as a function of distance of impact point from the edge of the tower. Right: energy collection corrected for the impact point position.

are comparable to the Moliere radius of e.m. showers in tungsten (9 mm). It had been previously shown by simulations that a correction algorithm to energy loss depending on the impact point can be successfully used up to 2 mm distance from any edge of the towers [6]. A microstrip silicon detector tracker from the ADAMO experiment [11] was therefore used to accurately reconstruct the impact point position on the towers. The spatial resolution of this tracker is 5 μm along the horizontal and 13 μm along the vertical direction. Figure 6 shows the results of data analysis for collected energy in the small calorimeter tower using 200 GeV electron beam. On the left the rough total collected energy is reported as a function of the distance of the impact point from the edge of the tower. The right picture shows the same data after correcting them with a function of the energy which takes into account the position of impact points on the tower. It can be seen that this correction gives good results for showers initiated by particles hitting the tower more than 2 mm from its edge. It is interesting to observe that the shape of the correcting function does not depend on the energy of the incoming particle as it can be verified by Monte Carlo simulations. Let's refer to figure 7. Upper picture shows the simulated results for energy collection in both the calorimeter towers, for different energies of the incoming particle, as a function of distance of impact point from the edge. After a normalization of all curves to the maximum energy collection for a 200 GeV/c incoming electron, in the lower picture the ratio between normalized curves and the curve for 200 GeV/c electrons is shown. The results is a flat function, meening that the dependence of energy loss from the impact point position is the same over all the considered energy range of incoming electrons.

4. CONCLUSIONS AND SCHEDULE FOR THE FUTURE

The two LHCf detectors are almost completed and will be installed inside the TANs in the beginning of 2007. A beam test a CERN SPS in 2004 allowed to study the performance of a prototype of detector I. A new beam test has been completed between August and September 2006. During this test the final version of detector I has been used, together with a partial version of detector II, which had only one over four silicon tracking modules. Results will be published soon. Next milestone is the completion of

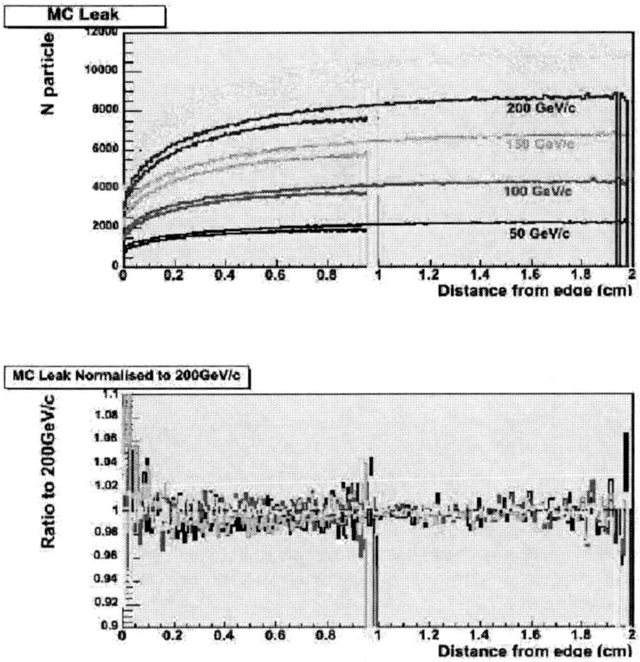

FIGURE 7. LHCf simulation results. Upper: energy collection for detector I towers as a function of the distance of impact point from the edge of the towers. Lower: ratio between energy collection curves normalized to 200 GeV/*c* maximum energy collection, and 200 GeV/*c* curve.

the tracking modules of detector II. The LHCf apparatus will begin data taking at LHC at the end of 2007.

REFERENCES

1. M. Takeda et al., Phys. Rev. Lett., 81 (1998) 1163.
2. P. Sommers et al., Proc. of 29th Int. Cosmic Ray Conf., Pune (2005).
3. S. Yoshida,Rapporteur paper, Proc. of 29th Int. Cosmic Ray Conf., Pune (2005).
4. G. Thomson, presentation at ICHEP conference, 2006.
5. Letter of Intent to CERN LHCC, 4 May 2004.
6. Technical Proposal for the CERN LHCf Experiment, CERN-LHCC-2005-032, CERN-LHCC-P-007, 12 October 2005.
7. E. Pare et al., *Phys. Lett.*, **B242** (1990) 531.
8. LHCf Experiment: Technical Design Report, CERN-LHCC-2006-004, LHCF-TDR-001, 7 February 2006.
9. P. Aspell, *PACE3 - Digital specification, version 4*, http://cmsdoc.cern.ch/~aspell/PACE3/
10. P. Aspell et al., *PACE3: A large dynamic range analogue memory ASIC assembly designed for the readout of silicon sensors in the LHC CMS Preshower*, 10th Workshop on electronics for the LHC experiments (LECC2004).
11. L. Bonechi et al., Proceedings of XXIX ICRC, Pune (India), session HE2.4, vol. 9, pag. 283–286

Hadronic Calibration of the ATLAS Calorimeter

Rolf Seuster

University of Victoria
Department of Physics and Astronomy
3800 Finnerty Road
Victoria, B.C. V8P 5C2
Canada

(on behalf of ATLAS colaboration)

Abstract. A novel approach by the ATLAS collaboration for hadronic calibration of their non-compensating calorimeters is presented, where the detector dependent calibration of hadronic signals is as much as possible decoupled from physics dependent corrections in jet reconstruction. A special feature of this approach is that first the basic hadronic calibration is applied using only the less biased calorimeter signal itself, rather than a specific jet context. This is expected to allow better control systematic uncertainties, with indications of a possible better energy resolution and a smaller jet energy scale error. All results reported here are based on detector signal simulations and should be considered preliminary.

Keywords: detector,hadronic calorimetry,calorimeter signal weighting
PACS: 42

INTRODUCTION

Jets will certainly play a key role in the hadronic final state representation in pp collisions at the soon to be commissioned Large Hadron Collider (LHC) at CERN. The enormous design luminosity ($10^{34}\text{cm}^{-2}\text{s}^{-1}$) will result in data samples of unprecedented size, with a huge number of jets which have to be reconstructed with very high precision. For example, the total uncertainty on the mass of the top quark can be lowered to about 1 GeV with a jet energy scale error of order 0.5%, while the uncertainty for the W boson mass could be decreased to about 20 MeV with a relative precision of about 0.025%.

These stringent requirements on the jet energy scale errors require a very good understanding of especially the hadronic signal in the calorimeters. The ATLAS [1] experiment at LHC features non-compensating calorimeters with typical e/π ratios between 1.2 and 1.6. They provide energy measurements for pseudo-rapidities η up to 5, with electromagnetic liquid argon/lead accordion sampling calorimeters in the $\eta < 3.2$ range, surrounded by hadronic scintillator/iron tile calorimetry within $\eta < 1.7$, and hadronic liquid argon/copper parallel plate sampling calorimeters in the end caps $1.4 < eta < 3.3$. The liquid-argon/copper or tungsten forward calorimeters extend the acceptance by covering $3.2 < \eta < 5.0$. The total depth of the ATLAS calorimeter system across the full η range is typically 10 hadronic interaction lengths (λ) (for more details see [1]).

CP867, *XII International Conference on Calorimetry in High Energy Physics,*
edited by S. R. Magill and R. Yoshida

The non-compensating character of the ATLAS calorimeters introduces a challenge for jet calibration. Jets are mostly made out of hadrons, most of which are pions, which means that a correction for $e/\pi > 1$ must be applied to correctly calibrate their signals. In addition, energy losses in inactive materials in front and between the calorimeters, including cryostat walls and internal structures inside the cryostats, for example, have to be corrected.

The ATLAS calorimeter group has developed a calibration model commonly referred to as *local hadronic calibration* to compensate and correct for these detector inefficiencies, as an alternative to calibration attempts explicitly using the jet context, see [2]. This model is based on initial topological calorimeter cell clustering leading to three-dimensional "energy blobs" in the calorimeter, followed by an attempt to classify these clusters according to their shape as of electromagnetic or hadronic origin. Finally, specific hadronic calibration functions are applied to cells in non-electromagnetic clusters, followed by additional inactive material corrections applied to all clusters. The resulting locally calibrated clusters are then input to jet finding and to further calibration refinement in the jet context, again see [2].

This note starts with a brief description of the basic signal used as a base for hadronic energy calibration. It is followed by a discussion of the various steps involved in local hadronic calibration approach, and some discussion on further steps towards the jet energy scale.

BASIC CALORIMETER SIGNAL

The basic calorimeter cell signal is reconstructed on the electromagnetic energy scale, which is derived from the electron (and photon) response. At ATLAS, this energy scale can be validated and its quality refined to high precision using $Z_0 \rightarrow e^+e^-$ decays, which occur at a rate of about 1 Hz at initial (low) collision luminosity ($\sim 10^{33}\mathrm{cm}^{-2}\mathrm{s}^{-1}$). Within only a few days of running, enough data can be accumulated to have a statistical uncertainty of less than one permille, with the valid assumption that the electron response is symmetric in azimuth.

The initial calibration for the electromagnetic energy scale is found in test-beam experiments with electrons of known energy. For this reason all ATLAS calorimeters, even the hadronic modules, have been subjected to electron beams, from 1 to 300 GeV [3].

Charged pion signals, as jet signals, are typically reconstructed too low, depending on the effective e/π of a given calorimeter region or module. Additional calibrations can be applied to correct for this signal deficiency. These either use cell or calorimeter sampling signals in jets [2] and are mostly motivated by signal weighting techniques similar to the ones developed by the H1 collaboration [4]. This approach folds the physics dependence of the response with the calorimeter signal dependence on the nature of the energy deposit, thus producing hadronic calibration functions depending on jet definitions and collision event environments like signals from underlying multiple interactions.

LOCAL HADRONIC CALIBRATION

The already mentioned local hadronic calibration approach presently studied for the ATLAS calorimeters explicitly decouples the detector dependent calibrations from the final jet energy scale calibration absorbing physics event environment and jet algorithm dependencies.

Utilising the fine granularity of the ATLAS calorimeters, clusters of topologically connected calorimeter cells are formed. The connection is given by signal correlations in neighbouring cells, as introduced by the (hadronic or electromagnetic) shower development. Especially the clusters in the very finely segmented electromagnetic calorimeters can then be classified according to the showers generating the signals. Clusters classified as electromagnetic at this step are then only subjected to signal corrections for inactive material energy losses, if needed.

After classifying a cluster as hadronic, weights are assigned to its cell signals depending on how much of the energy deposit is due to electromagnetic or hadronic interactions. The corresponding weighting functions are discussed further below.

Additionally to the classification and weighing scheme, and because clusters in ATLAS are allowed to span over regions of different cell granularity, and even different calorimeters, the energy loss in inactive material between these detectors, or inside of one of these, must also be accounted for. This is also described in more detail below.

This classification and the weighting scheme obviously strongly depend on how clusters are formed. Therefore, the standard procedure for general purpose clustering in ATLAS is shortly reviewed in the next section.

CLUSTER BUILDING

Different types of clusters are used in the ATLAS collaboration. For triggering purposes, clusters are build withing "region of interest". Reliably identifying clusters caused by electrons or photons, a sliding window algorithm has been developed. For general purpose analysis, however, a topological clustering algorithm has been implemented.

Initial cluster formation

In the topological cluster algorithm, clusters are built out of neighbouring cells which are even allowed to reside in different calorimeters or regions. In a first step, seed cells are found. A seed cell is any cell with an absolute signal $|E_{cell}|$ of more than S (default $S = 4$) σ significance over the total noise of this particular cell. The total noise is the electronic noise in the cell folded with luminosity-dependent physics noise from pile-up events, if applicable.

In the second step, all neighbouring cells with $|E_{cell}| > N\sigma$ (default $N = 2$) are connected to the seed cell. Neighbours are all cells sharing (all or part of) an edge, side, or corner with the seed cell, in all possible directions (full three-dimensional clustering).

Finally, another layer of neighbouring cells is connected to the cluster, with $E_{cell} > L\sigma$ (default $L = 0$), providing a guard ring against possible signal losses around a cluster.

This procedure has proven to perform well in terms of noise suppression for various applications.

The reconstructed clusters at this stage are typically too big, as spatial signal structures in them are ignored in the initial formation. A splitting step, as discussed in the following section, is introduced to shape the topological clusters more along the shower development. On the other hand, the initial clusters provide efficient noise suppression with very little signal loss by cell selections, because small signal cells survive in the presence of a neighbour with a more significant signal.

Cluster splitting

Particles in jets which deposit energy close by will most likely produce overlapping clusters. Therefore a splitting of the clusters is needed to achieve a good representation of the spatial energy distribution in the calorimeters generated by the incoming particles.

Each initial cluster is examined for local signal maxima, which are the core of the new clusters after splitting. One characteristic variable used to define a local signal maximum is the transverse energy density measured by $E_{t,cell}/\sqrt[3]{V_{cell}}$, where $E_{t,cell}$ is the transverse energy in the cell, and V_{cell} is the cell volume. This is only one example for a useful variable, of course.

Cells in between local maxima are allowed to be in two clusters, their energy will be shared between the two clusters so that the sum of these fractional cell energies is the original cell signal. The energy of the new clusters is then recomputed from the cells contributing to this new clusters including the fraction of energy contributing to the cluster.

The objective of this procedure is to have about one cluster per electron or photon with somewhat high efficiency, but possibly several clusters per pions, if the corresponding hadronic shower is scattered accordingly in the calorimeter.

CLUSTER CALIBRATION

The clusters formed according to the prescription provided above are then classified according to the nature of their signal, at least as much as possible within the granularity of the calorimeters. Mainly this means that clusters classified as electromagnetic can really only be found where cell sizes are small with respect to shower sizes, i.e. the electromagnetic ATLAS calorimeters.

Calibration hits

The classification criteria and the signal weighting functions are derived using *calibration hits* in single electron and charged pion, and even jet simulations in Geant4 [5]. These hits are available in special simulation samples, where the additional information of the nature of an inelastic interaction of primary or secondary particles inside a shower is recorded in individual cells in the ATLAS calorimeters. Each interaction inside the calorimeters is classified as one of following energy deposits:

- (EM) electro-magnetic, energy deposits originating from electro-magnetic interactions
- (non-EM) non-electro-magnetic, energy deposits originating from non-electro-magnetic interactions
- invisible, excitations of hadrons which release their energy well outside of the integration window of the pulse shape
- escaped, energy escaping the calorimeter via e.g. neutrinos.

The calorimeter signal is only generated by the electromagnetic and the non-electromagnetic energy deposits in active material, whereas the invisible and escaped energy have no signal equivalent and give rise to $e/\pi > 1$, as they only significantly occur in hadronic showers. In addition to above classification, calibration hits are also recorded not only in active material, but also in the absorber structure of a cell, and the inactive material. They are also a very useful tool to understand the details of the modeling of hadronic showers in Geant4.

Cluster classification

The classification determines which clusters are more likely to be of hadronic nature, therefore in need of weights to determine the correct energy response, or which ones are more electromagnetic, needing less correction. It is done in two steps: First, the EM fraction f_{em} of a cluster is estimated from the calibration hits (see above) in single particle simulations, as function of the average cluster energy density $\rho_{cluster}$ and the depth λ of the cluster inside the calorimeter (other variables are under study). f_{em} are tabulated in look-up tables in $(\rho_{cluster}, \lambda)$ for various calorimeter regions. If $f_{em} > 0.9$, the cluster is classified as being electromagnetic and therefore no additional weighting will be applied. Clusters with $f_{em} < 0.9$ are classified as hadronic if their total signal is greater 0, and the subsequent signal weighing technique will be applied. Clusters with negative energies are classified as *unknown* and can be considered as noise. In QCD di-jet events with an energy range of the hard interaction of $280\mathrm{GeV}/c^2 \leq p_T \leq 560\mathrm{GeV}/c^2$ about 15% of the clusters are identified as being of EM nature. For the barrel, about 90% of pions and electrons are identified correctly, however the performance in the end-caps are somewhat degraded, due to the coarser cell granularity.

Hadronic signal weights

For every cluster identified as being of hadronic nature via the above mentioned classification, a signal weighting scheme is applied. The corresponding weight functions $w(\vec{x})$ have again been determined from calibration hits in charged pion simulations by cell-by-cell calculating the ratio of the true deposited energy over the visible energy (signal on electromagnetic energy scale) as function of a set $\vec{x}$ of experimentally accessible

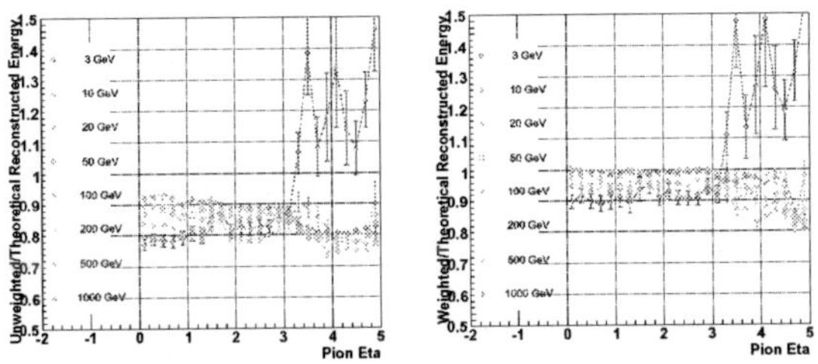

FIGURE 1. The performance of the classification and the weighting scheme on single pions Monte Carlo for the scale. No dead material correction has been applied.

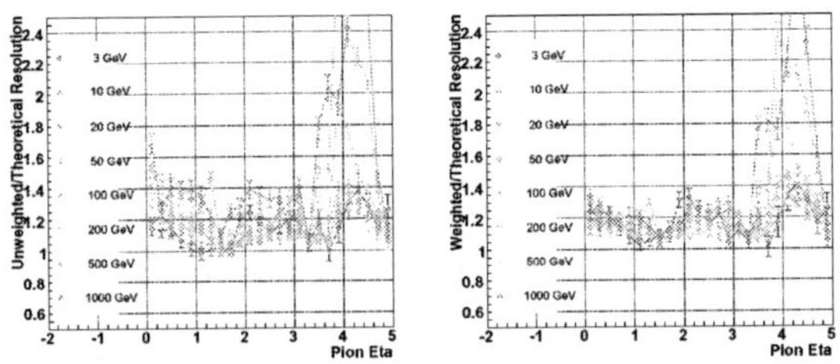

FIGURE 2. The performance of the classification and the weighting scheme on single pions Monte Carlo for the resolution. No dead material correction has been applied.

variables. Presently under study is $\vec{x} = (E_{cluster}, \rho_{cell})$, where $E_{cluster}$ is the cluster energy on electromagnetic energy scale, and $\rho_{cell} = E_{cell}/V_{cell}$ is the cell signal density.

The corrected energy of each cell inside the cluster then is related to the electromagnetic energy scale signal as:

$$E_{cell}^{corr} = w(\vec{x})E_{cell}, \quad \text{with} \tag{1}$$

$$w(\vec{x}) = \frac{E_{em} + E_{non-em} + E_{invisible} + E_{escaped}}{E_{em} + E_{non-em}} \tag{2}$$

The weights $w(\vec{x})$ as defined above are of course not accessible in the experiment. They have been determined from simulations and tabulated in look-up tables as functions of the measurable variables $E_{cluster}$ and ρ_{cell}.

The performance of the current classification and weighing scheme for single pions of various incoming energies is shown in Figures 1 and 2 before (left) and after reweighting has been applied. No correction for energy losses in front and inside of clusters has been applied yet. A clear improvement in scale is visible, see Figure 1. The scale for single pions with energies above 50 GeV reaches close to unity, only for $|\eta| > 3$ the deviation is larger, about 5% until the very forward region. For lower energy single pions, the scale improves significantly, however, residual signal non-linearities remain at a level of 90%. This can be understood by the almost constant energy loss in front of the calorimeter in the inner tracking detectors, showing relatively larger effects for lower energies.

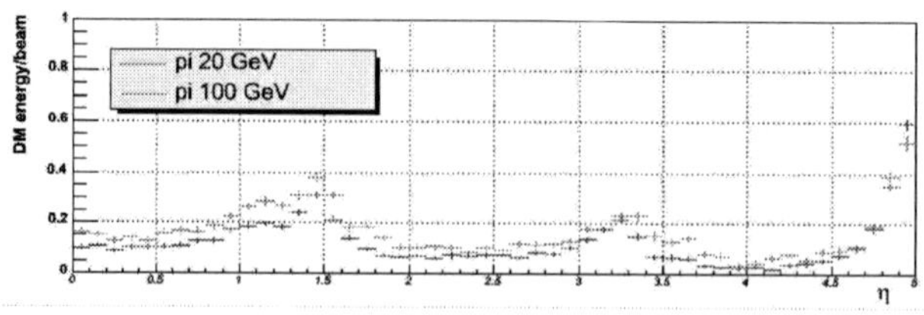

FIGURE 3. The fraction of energy loss in different η regions of the calorimeter for two energies of the incoming single pion. Clearly visible are the transition regions between the different calorimeter subsystems at $|\eta| \approx 1.5$ and $|\eta| \approx 3.2$

Inactive material corrections

The last ingredient to the local hadronic calibration is the correction for energy loss in inactive material inside and in front of the calorimeter, which can be significant. To determine this correction, the same simulated charged pions with calibration hits in the inactive material outside calorimeter cells are used.

Clearly visible in the correction are the transition regions between the different calorimeters. Larger corrections for 20% to 40% energy losses are needed for the region $\eta \approx 1.5$, the transition from the barrel to the end-cap, and for the region $\eta \approx 3.2$, the transition from the end-cap to the forward calorimeters, see Figure 3. The corrections for lower energy particles are relatively larger, because the energy loss itself is basically constant.

The correction factor is parametrised in terms of the measured energy in close-by calorimeter cells, e.g. the energy loss E_{loss} between the central barrel electromagnetic calorimeter and the tile calorimeter behind it is strongly correlated with the energy in the last layer of the barrel $E_{Barrel,3}$ and the first layer in the tile calorimeter $E_{Tile,0}$, as $E_{loss} \sim \sqrt{E_{Barrel,3} * E_{Tile,0}}$.

JET ENERGY CORRECTIONS

In the reconstruction of jets in physics events taken at the ATLAS detector, additional corrections need to be applied. Several approaches are under study, mostly using experimentally accessible calibration signals, as discussed below.

Physics environment and jet algorithm dependent corrections are applied in a separate step. Among the effects from the physics environment are fragmentation effects, especially differences in the fragmentation of light quarks and heavy quarks, signal contributions from underlying event activity, and possible additional signal fluctuations introduced by pile-up events.

Jet energy scale corrections introduced by jet algorithms are missing signals due to limited jet cone sizes, including losses introduced by showering of the jet particles in the calorimeters. These have to be addressed in connection with the chosen jet algorithm choice and its specific configuration.

In addition to this simulation based normalisation, a few calibration channels for jet calibration are available in the pp collisions at LHC. Paramount are the $Z \rightarrow e^+e^-$ or γ co-produced with a jet, where a well calibrated electromagnetic system can be balanced in transverse momentum with the hadronic final state jet. Other examples are events with t-quarks, where the W in the top decay chain can decay fully hadronically itself, thus providing a precise invariant mass reference for the two resulting jets. As b-quarks are very rarely produced in these W decays, this tests the energy scale for light quark jets. For b-quark jets, one can directly use the bs from the top decays. QCD di-jet events can also be used for cross check calibrations between different calorimeter regions. Some of the possible jet energy scale corrections experimentally accessible this way can be very sensitive to the particular physics environment in which they have been determined.

SUMMARY

The approach followed by the ATLAS collaboration for correcting energy deposits in the non-compensating calorimeters are described. All ingredients from building and splitting a cluster are described. The corrections applied to a cluster are listed. These include classification, correction for different energy responses of electrons and hadrons and energy losses before and inside clusters. For single pions, an improvement of the response in resolution and scale is clearly visible. The application to jets is currently being worked on and results will be available for comparison soon.

ACKNOWLEDGMENTS

I like to thank my colleagues in the hadronic calorimeter calibration and the jet reconstruction working group in ATLAS for providing important input to this topic. Many thanks also to the organisers of CALOR06 to create such a nice and productive atmosphere. Last but not least I like to acknowledge the continuous support by all involved funding agencies.

REFERENCES

1. ATLAS Coll., *ATLAS Detector and Physics Technical Design Report*, CERN/LHCC/99-14/15 (1999).
2. S. Jorgensen Roca, these proceedings.
3. W. Lampl, these proceedings.
4. H1 Coll., Nucl.Instrum.Meth.A336, 460-498 (1993).
   H1 Coll., Nucl.Instrum.Meth.A386, 310-347 (1997).
5. GEANT4 Coll., Nucl.Instrum.Meth.A506, 250-303 (2003).

Radiation-Hard Quartz Čerenkov Calorimeters

U. Akgun and Y. Onel (for CMS Collaboration)

Department of Physics and Astronomy, The University of Iowa, 203 Van Allen Hall, Iowa City, IA, 52242, USA

Abstract. New generation hadron colliders are going to reach unprecedented energies and radiation levels. Quartz has been identified as a radiation-hard material that can be used for Čerenkov calorimeters of the future experiments. We report from the radiation hardness tests performed on quartz fibers, as well as the characteristics of the quartz fiber and plate Čerenkov calorimeters that have been built, designed, and proposed for the CMS experiment.

Keywords: Quartz, Čerenkov, calorimeter, radiation.
PACS: 07.20.Fw, 29.40.Vj, 42.70.Ce,

INTRODUCTION

The future colliders are going to reach unprecedented energies and the experiments will experience huge particle fluxes. This requires developing and using radiation hard detector system technologies. For the CMS Forward Calorimeter (HF) we have used quartz fibers as the active material, in light of a series of radiation-hardness tests performed on quartz fibers. We report results from these tests.

In this paper we summarize the test beam results on Pre-Production Prototype (PPP-1) and HF, as well as the Geant4 simulations. The CMS detector has the CASTOR calorimeter behind HF to cover the forward region better. CASTOR has been designed to use Čerenkov light made in quartz plates. We review the design properties of CASTOR calorimeter. As a possible upgrade to the End-Cap Hadron calorimeter for the SuperLHC (SLHC) era, we have proposed to replace existing scintillators with radiation hard quartz plates. In the last section of this paper we give the preliminary results from R&D studies for maximizing the Čerenkov light collection efficiencies.

RADIATION DAMAGE STUDIES ON QUARTZ FIBERS

The simulations show that the CMS HF calorimeters are going to absorb, on average, 760 GeV energy per event. This is huge compared to only 100 GeV for the rest of the main detector. Moreover, this energy is not uniformly distributed, but has a pronounced maximum at the highest rapidities. At $\eta = 5$ and integrated luminosity of 5×10^5 pb^{-1} (~ 10 year of LHC operation), the HF will experience ~1 GRad [1]. The neutron rates ($E_n \leq 14$ MeV) at the front of the detector will be in the order of 10^8 Hz/cm^2. The charged hadron rates will also be extremely high, especially at the

CP867, *XII International Conference on Calorimetry in High Energy Physics*, edited by S. R. Magill and R. Yoshida

shower maximum of the HF where the rate will reach 10^{13} to 10^{16} Hz/cm^2. This hostile environment presents a unique challenge in particle detection techniques.

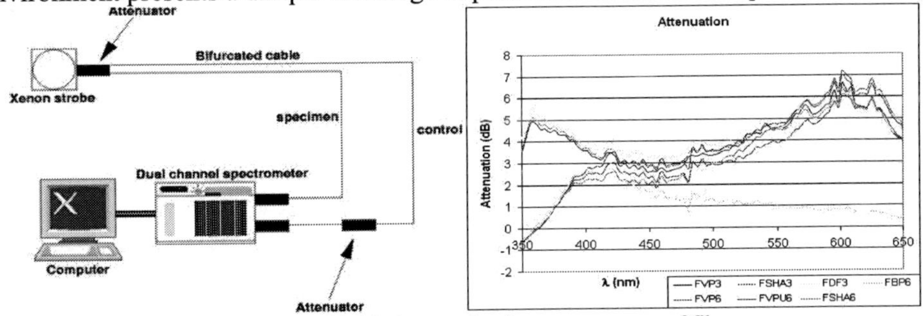

Figure 1: (left) Experimental setup, (right) Attenuation for seven groups of fibers.

Initial radiation damage studies on quartz fibers were performed at two different facilities; a 10 kW experimental reactor, UTR-10 at ISU in Ames, Iowa, and the MGC-20E cyclotron at ATOMKI in Debrecen, Hungary. The details of these tests can be found elsewhere [2]. The initial tests showed some radiation recovery for most silica fibers. For some quartz fiber types < 10 MRad gamma irradiation seems to have generated a similar type of optical damage as neutron irradiation at fluence of 10^{15} n/cm^2.

After these initial tests we performed more radiation damage tests on nine different high OH^- quartz fibers by irradiating them with 500 MeV electrons from the Linac Injector of LEP (LIL) at CERN. The transmission of Xe light was measured *in-situ* in the 350-800 nm range. The induced attenuation at 450 nm is found to be 1.52 ± 0.15 dB/m for a 100 Mrad absorbed dose. We observed some darkening on the fibers, but the radiation did not change the tensile strength of the quartz fibers [3].

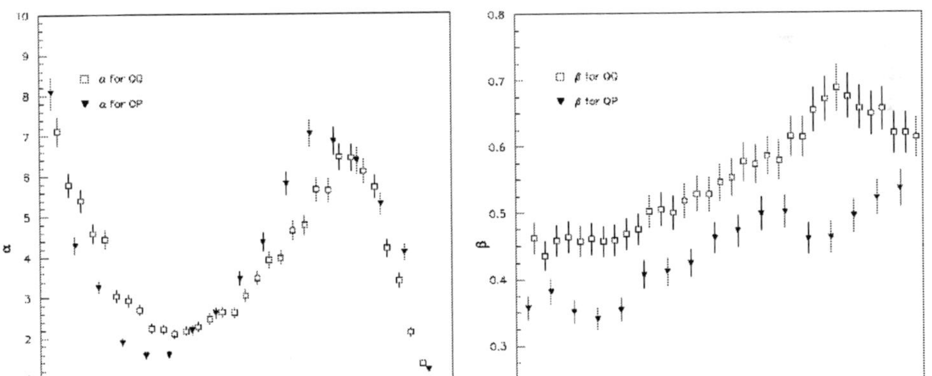

FIGURE 2. (a) α parameters for qq and qp fibers versus wavelength, where α is the attenuation in dB/m at 100 Mrad. (b) β parameters for qq and qp fibers versus wavelength.

In 2004, quartz fibers were exposed to strong neutron and photon radiation and tested for optical degradation. The fibers were irradiated with pulses of high-energy neutrons produced by the Intense Pulsed Neutron Source (IPNS) at Argonne National Laboratory for a two-week period. Damage to the fibers was determined after 7.42 x 10^{17}n/cm^2 of fast neutrons, 17.6 Mrad of neutron and 73.5 Mrad of gamma radiation. The spectra of the irradiated fibers look similar to those from the non-irradiated fibers but were noticeably attenuated. The level of damage was wavelength specific (See Figure 1).

Recently, to complete our knowledge of radiation damage of the quartz fibers we initiated a high level of irradiation with 24 GeV protons at CERN. The quartz fibers with 0.6 mm core diameter were irradiated with 4.6 10^{16} protons/cm^2 corresponding to a 1.2 Grad dose.

The attenuation measurements were performed *in-situ*. We tested quartz fibers with quartz cladding (qq) and with plastic cladding (qp). The light attenuation in the fiber is defined as

$$A(\lambda, D) = \alpha(\lambda)\left[\frac{D}{D_s}\right]^{\beta(\lambda)} \tag{1}$$

Where α and β are parameters for qq and qp fibers that are determined by fitting the ratios as a function of wavelength and dose.

$$\frac{I(\lambda, D)}{I(\lambda, 0)} = e^{\left[-\left(L/4.343)\alpha(\lambda)\left(\frac{D}{D_s}\right)^{\beta(\lambda)}\right)\right]} \tag{2}$$

Choosing a scale factor D_s = 100 Mrad, α is the attenuation at 100 Mrad, in dB/m, L being expressed in m (see Figure 2). After 1.2 Grad of proton irradiation, the results are similar to the electron radiation tests [4]. Up to 0.8 Grad we do not observe a significant difference in attenuation versus dose between qq and qp fibers. Above 0.5 Grad there is no recovery of damage below 380 nm, and in the range of 580-650 nm quartz becomes opaque. We monitored the recovery over 4 months. The decrease of signal in fibers is very fast at the beginning: 30% loss after 20 Mrad at 455 nm.

QUARTZ ČERENKOV CALORIMETERS OF THE CMS EXPERIMENT

This part of the paper will cover three different radiation-hard Čerenkov calorimeters for the CMS experiment (left, Figure 3); i) Pre-Production Prototype (PPP-1) of the Forward (HF) calorimeter, ii) The Forward (HF) calorimeter iii) The CASTOR calorimeter. The PPP-1, and HF use quartz fibers while CASTOR uses quartz plate as the active volume (right, Figure 3).

Pre-Production Prototype (PPP-1)

The PPP-I is an iron absorber matrix with embedded quartz fibers which serve as the active material. The iron matrix is composed of 2.5 mm thick iron layers with grooves every 2.5 mm. The embedded fibers run parallel to the beam and constitute the active component of the detector. The length of the absorber is 165 cm, 8.3 nuclear interaction length (λ_{int}), with a cross sectional area of 18 cm × 18 cm. A single quartz fiber is inserted In each groove. In PPP-I, there are three different lengths of fibers to achieve a longitudinal segmentation. They are called electromagnetic (EM), hadronic (HAD), and tail catcher (TC). Long fibers sample all shower components while the shorter fibers are biased to the hadron component as the hadronic showers tend to penetrate deeper. There are two EM fibers for each HAD and TC fiber. The PPP-I is divided into nine physical regions called towers, each with a 6 cm × 6 cm cross sectional area. At the center of each tower, a radioactive wire source-tube groove exists for calibration [5].

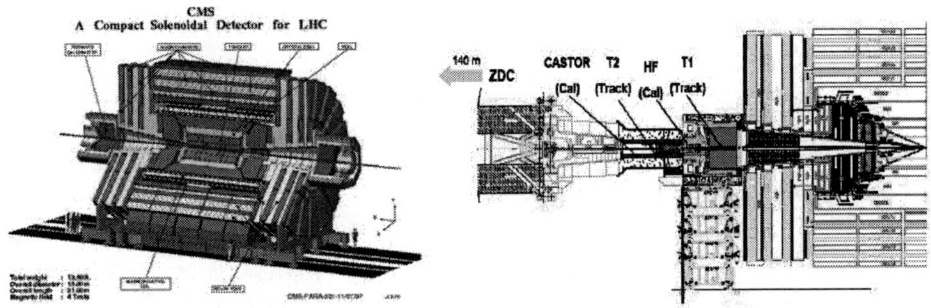

FIGURE 3. The major components of the central part (left) and the forward region (right) of the CMS detector.

At the test beam the detector was placed on a platform that could move in three dimensions with respect to the H4 beam line at CERN. One of the factors that affect the calorimeter's response is the non-uniformities in its structure. In PPP-I, this non-uniformity is the fiber spacing. In order to study the spatial uniformity of the PPP-I, a 120 GeV electron beam was moved with 1.0 cm steps across the face of the detector. The energy resolution of a calorimeter, in general, can be parametrized as

$$\left(\frac{\sigma}{E}\right)^2 = \left(\frac{a}{\sqrt{E}}\right)^2 + \left(\frac{b}{E}\right)^2 + c^2 \quad (3)$$

The first term is the sampling term and characterizes the statistical fluctuations in the signal generating processes. The second term corresponds to noise and includes the energy equivalent of electronic noise as well as pileup. The third term is the constant term and is related to the imperfections of the calorimetry, signal generation and collection non-uniformity, calibration errors and fluctuations in the energy leakage from the calorimeter. In quartz fiber calorimeters the electromagnetic energy resolution is completely dominated by photoelectron fluctuations, therefore we drop the noise term for electromagnetic resolution definition (See Figure 4). The electromagnetic energy resolution was measured to be

$$\left(\frac{\sigma}{E}\right)^2 = \left(\frac{197\%}{\sqrt{E}}\right)^2 + (8\%)^2 \quad (4)$$

Where, E has units of GeV. The hadronic energy resolution was determined to be 18% at 1 TeV.

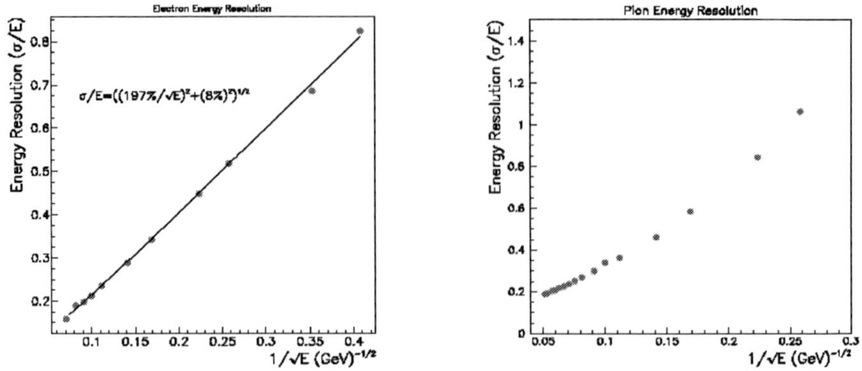

FIGURE 4. The electromagnetic (left) and hadronic (right) energy resolution of PPP-1 as a function of $1/\sqrt{E}$.

The Forward (HF) Calorimeters

The HF calorimeters consist of two modules, located symmetrically at about 11 m from either side of interaction point. They cover the pseudorapidity range 3-5. The length along the beam is 1.65 m or 10 λ_{int}. Each calorimeter consists of a large steel block that serves as the absorber. Embedded quartz fibers in the steel absorber run parallel to the beam and constitute the active component of the detector. In order to optimize energy resolution for E and E_T flows and forward jets, the calorimeter is effectively segmented longitudinally by using two different fiber lengths.

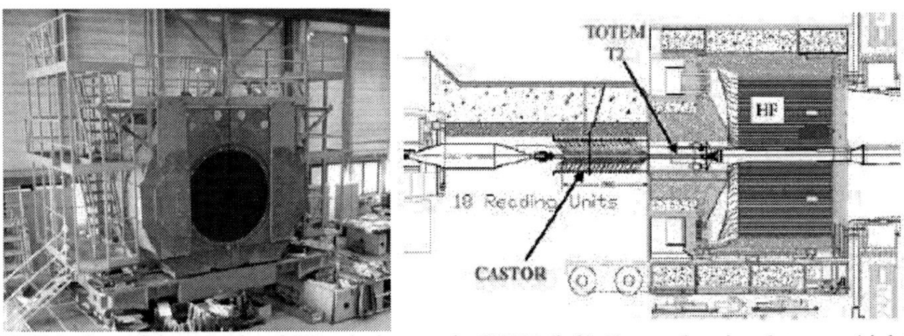

FIGURE 5: One of the completed HF calorimeters in CERN (left). Forward region detectors (right).

In this report we will summarize the results from tests performed on HF modules at the H2 test beam line at CERN, in 2004. And, the Geant4 simulation efforts to verify the test beam.

The HF wedges were individually placed on a remotely controlled table that could move horizontally and vertically with respect to the beam with better than 1 mm precision. The angular positioning of the detector on the vertical plane with respect to the beam line was also remotely controlled and varied from $0^o$to 6° to mimic the particle path in the CMS experiment. The purity of the beam depended on energy; in particular, high energy electron beams were contaminated with pions and muons [6].

The simulations of the test beam, the physics of the electromagnetic and hadronic processes, are implemented by using GEANT4 physics applications. Although electromagnetic physics used in the simulation is standard GEANT4 electromagnetic physics, four different physics lists are used for the hadronic physics, LHEP, QGSP, QGSC, and FTFP [7]. Table 1 and Table 2 summarize the comparison of different Geant4 simulations with the test beam.

The test beam results showed that for the electromagnetic energy resolution (from Eqn. 3), the stochastic term "*a*" is 198% and the constant term "c" is 9%. The hadronic energy resolution is largely determined by the fluctuations in the neutral pion production in showers, with a = 280 % and c = 11%.

TABLE 1. Test beam electromagnetic energy resolution and Geant4 results.

Data	30 GeV	50 GeV	100 GeV	150 GeV
TB04	0.396 ± 0.028	0.315 ± 0.022	0.231 ± 0.016	0.204 ± 0.014
LHEP	0.421 ± 0.001	0.329 ± 0.001	0.239 ± 0.001	0.201 ± 0.001
QGSP	0.420 ± 0.001	0.328 ± 0.001	0.240 ± 0.001	0.201 ± 0.001
QGSC	0.421 ± 0.001	0.328 ± 0.001	0.239 ± 0.001	0.203 ± 0.001
FTFP	0.422 ± 0.001	0.328 ± 0.001	0.238 ± 0.001	0.203 ± 0.001

TABLE 2. Test beam hadronic energy resolution and Geant4 results.

Data	30 GeV	50 GeV	100 GeV	150 GeV
TB04	0.514 ± 0.036	0.407 ± 0.029	0.309 ± 0.022	0.267 ± 0.019
LHEP	0.495 ± 0.001	0.380 ± 0.001	0.280 ± 0.001	0.241 ± 0.001
QGSP	0.467 ± 0.001	0.368 ± 0.001	0.275 ± 0.001	0.234 ± 0.001
QGSC	0.477 ± 0.001	0.373 ± 0.001	0.276 ± 0.001	0.236 ± 0.001
FTFP	0.466 ± 0.001	0.372 ± 0.001	0.274 ± 0.001	0.232 ± 0.001

CASTOR Calorimeter

CASTOR [8] is a calorimeter that will measure the energy of particles emitted at very small angles, covering the polar angles from 0.1° to 0.7° with respect to the beam direction. The detector is an octagonal cylinder, approximately 1.5 m long with an outer diameter of 36 cm and an inner diameter of 3.8 cm (to contain the beam pipe). It is subdivided into 16 sections in azimuth and 18 sections in depth. This fine segmentation allows it to image the 3 dimensional "texture" of energy deposition for different particles. Because the energy of the incoming particles is very high, the CASTOR spectrometer has to be very dense, which is achieved by using tungsten plates. For sensitivity to the angle of the incoming particles it is desirable to measure only the very fast positrons and electrons produced at small angles with respect to the initial direction. This is done by using the Čerenkov effect, which results in emitted light created in an approximate electromagnetic equivalent to the shock front that causes sonic booms. In its proposed location, the detector is expected to experience radiation exposure in the range of 10-100 MGy. The quartz is arranged in thin plates, tilted at 45° in order to efficiently capture the Čerenkov light. In order to increase the amount of light collected, many layers of tungsten and quartz are stacked together.

The light from seven quartz plates is collected by an air filled light guide, covering 22.5° in azimuthal angle (see Figure 6).

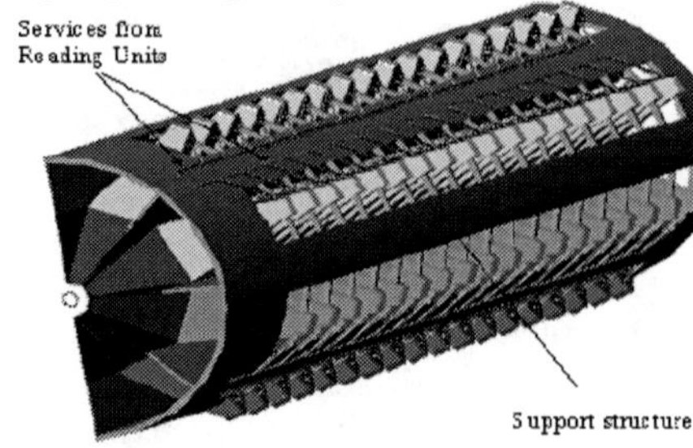

FIGURE 6: The design of CASTOR.

The electromagnetic particles, photons and electrons, penetrate less deeply into the detector than the hadronic particles, such as pions and protons. Therefore, the first 11.5 cm of the detector is segmented slightly differently with tungsten and quartz thicknesses of 3.0 and 1.5 mm, respectively, compared to 5.0 and 2.0 mm, in the remaining 1.25 m hadronic section of the device. With this design of quartz detectors, the diameter of the showers of electrons and positrons produced by hadrons is of the order of 1 cm for tungsten, compared to values orders of magnitude larger in other calorimeters.

CASTOR will provide a very fast, high resolution measure of the azimuthal distribution of energy in a region very close to the beam direction. The longitudinal segmentation will allow this energy to be differentiated into that carried by hadronic and electromagnetic particles and to allow study of exotic objects that have been predicted to occur in this angular range [9]. In events with smaller occupancy, the energy of individual particles, as well as their hadronic or electromagnetic nature, can be determined with high accuracy.

CMS HCAL DETECTOR UPGRADE R&D FOR SLHC

The Large Hadron Collider (LHC) is designed to provide, 14 TeV center of mass energy with proton-proton collisions every 25 ns. After a few years of run, the conditions of LHC will be upgraded to operate at 10 times higher luminosity ($L=10^{35}cm^{-2}s^{-1}$) allowing new physics discoveries. This period is called SLHC. In the current design, Hadronic EndCap (HE) calorimeters of the CMS experiment use Kuraray SCSN81 scintillator tiles, and Kuraray Y-11 double clad wavelength shifting fiber. These materials have been shown to be moderately radiation hard up to 2.5 MRad [1]. The scintillation photons are collected by wavelength shifting fibers which have the geometry shown in the first picture at Figure 7. Under the SLHC conditions the lifetime radiation dose at the HE will increase from 2.5 Mrad to 25 Mrad. The scintillator tiles used in the current design of HE will lose their efficiency due to high radiation. As a solution to the radiation damage problem with SLHC conditions, we propose to replace the scintillators with quartz plates. They will not be affected by the high radiation, but with quartz plates, the light comes from Čerenkov radiation which produces 100 times less light than scintillation. Our aim is to find an efficient way to collect light from quartz plates. At the University of Iowa and Fermilab, we tested and simulated different sizes of quartz plates with different fiber geometries embedded in them to obtain maximum light.

After the test beams and simulations of different fiber geometries for their light collection efficiencies and uniformities we have constructed a quartz plate calorimeter prototype. The Geant4 simulations for the calorimeter prototype are still undergoing. We have test beams scheduled at Fermilab and CERN for the fall of 2006.

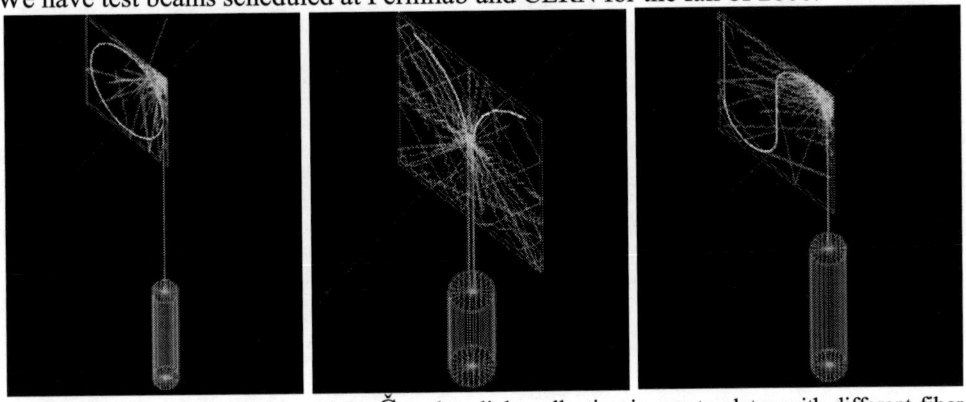

FIGURE 7: Geant4 simulations of the Čerenkov light collection in quartz plates with different fiber geometries.

ACKNOWLEDGMENTS

We would like to thank D. Green, J. Freeman, and A. Skuja of HCAL management, P. Bruecken, K. Cankocak, J.P. Merlo, T. Yetkin, Iowa–Turkish Universities Raddam Group. E. Norbeck, A.D. Panagiotou of the CASTOR group. This research is funded by DOE (DE-FG02-91ER40644), Iowa-Fermilab MOU for CMS construction and M&O projects, and The University of Iowa office of VP of Research and Graduate College.

REFERENCES

1. "The Hadron Calorimeter Project, Technical Design Report", CERN/LHCC 97-31.
2. A.S. Ayan, "The CMS Forward Calorimeter Prototype Design Studies and Ω_c^0 Search at E781 Experiment at Fermilab", *Ph.D. Thesis*, The University of Iowa, 2004.
3. I. Dumanoglu et al., "Radiation-hardness studies of high OH- content quartz fibres irradiated with 500 MeV electrons", Nucl. Instrum. Meth. A 490, 2002, 444-455.
4. J.P. Merlo and k. Cankocak, "Radiation-hardness studies of high OH^- content quartz fibers irradiated with 24 GeV protons", 2006, CMS CR 2006/005
5. A.S. Ayan et al., "Energy resolution and the linearity of the CMS forward quartz fibre calorimeter", J. Phys. G: Nucl. Part. Phys., 30, (2004), N33-44.
6. G. Baiatian et al., "Design, Performance, and Calibration of the CMS Forward Calorimeter Wedges", CSM NOTE 2006/044.
7. T. Yetkin, "Search for SUSY in Missing Transverse Energy Plus Multijet Topologies at $\sqrt{s}$ = 14 TeV and Geant4 Simulation of the CMS Hadronic Forward Calorimeter in the 2004 Test Beam", Ph.D. Thesis, Cukurova University, 2006.
8. CASTOR: Centauro And Strange Object Research, http://cmsdoc.cern.ch/castor
9. E. Gladysz-Dziadus, "CASTOR: Centauro and Strange Object Research Exotic Aspects of Forward Physics at the LHC", Act. Phys. Pol. B, 37, 2006, 153-160.

SESSION VI

LHC CALORIMETRY—COMMISSIONING

Convener: Stanislav Nemecek

The ATLAS Liquid Argon Calorimeter: Construction, Integration, Commissioning

Martin Aleksa
on behalf of the ATLAS LAr Collaboration

PH-Department, CERN, 1211 Geneva, Switzerland

Abstract. The ATLAS liquid argon (LAr) calorimeter system consists of an electromagnetic barrel calorimeter and two end caps with electromagnetic, hadronic and forward calorimeters. The liquid argon sampling technique, with an accordion geometry was chosen for the barrel electromagnetic calorimeter (EMB) and adapted to the end cap (EMEC). The hadronic end cap calorimeter (HEC) uses a copper-liquid argon sampling technique with flat plate geometry and is subdivided in depth in two wheels per end-cap. Finally, the forward calorimeter (FCAL) is composed of three modules employing cylindrical electrodes with thin liquid argon gaps.

The construction of the full calorimeter system is complete since mid-2004. Production modules constructed in the home institutes were integrated into wheels at CERN in 2003-2004, and inserted into the three cryostats. They passed their first complete cold test before the lowering into the ATLAS cavern. Results of quality checks (e.g. electrical, mechanical, ...) performed on all the 190304 read-out channels after cool down will be reported. End 2004 the ATLAS barrel electromagnetic (EM) calorimeter was installed in the ATLAS cavern and since summer 2005 the front-end electronics are being connected and tested. Results of this first commissioning phase will be shown to demonstrate the high standards of quality control for our detectors.

Keywords: LAr, ATLAS, calorimeter, commissioning, electromagnetic calorimeter, hadronic calorimeter
PACS: 29.40.Vj

INTRODUCTION

The Large Hadron Collider (LHC), currently under construction at CERN, will collide protons on protons with a beam energy of 7 TeV, extending the available center-of-mass energy by about an order of magnitude over that of existing colliders. Together with its high collision rate, corresponding to an expected integrated luminosity of $10-100\,\mathrm{fb}^{-1}$/year, these energies allow for production of particles with high masses or high transverse momenta or other processes with low production cross-sections. The LHC will search for effects of new interactions at very short distances and for new particles beyond the Standard Model of particle physics (SM). An excellent knowledge of the electron or photon energy is needed for precision measurements of, for example, couplings within and beyond the SM, or to resolve possible narrow resonances of new particles over a large background. A good energy resolution and a good linearity need to be achieved for energies ranging from a few GeV up to a few TeV. The most prominent example is the possible discovery of the Higgs boson, which in the SM provides an explanation of how the elementary particles acquire mass.

CP867, *XII International Conference on Calorimetry in High Energy Physics,*
edited by S. R. Magill and R. Yoshida

Physics Requirements for the ATLAS LAr Calorimeters

The potential of dicovering the Higgs boson determines most of the performance requirements for the EM calorimetry in ATLAS. The decay channels $H \to \gamma\gamma$ and $H \to 4e^{\pm}$ have been the benchmark processes for the design of the ATLAS LAr EM calorimeter: The necessity of the largest possible acceptance led to the accordion calorimeter, a concept that allows a uniform acceptance over the whole azimuthal range without any gaps for services. The requirement to cover a large dynamic range (20 MeV to 2 TeV) made it necessary to design front end electronics with three different gains (separated by approximately a factor of ten). The stochastic term of the energy resolution for electrons and photons is required to be around $10\%/\sqrt{E}$. A very good overall constant term of about 0.7 % calls for very precise mechanics (lead and LAr gap thicknesses) and a very precise electronics calibration system that can calibrate the response of the calorimeter at the 0.25 % level. In order to be able to perform W-mass precision measurements the linearity of the EM calorimeter response must be around 0.1 %. To achieve this goal a presampler and the longitudinal segmentation of the detector help to correct for dead material in front of the calorimeter (layer weighting). Fine granularity of the detector is also needed for particle identification (e.g. $e^{\pm}$ – jets and $\gamma - \pi_0$). Fine strips and longitudinal segmentation are a prerequisite for precision position and angular measurements down to $50\,\mathrm{mrad}/\sqrt{E}$.

The requirements for the ATLAS LAr hadronic calorimeters are: Almost full 4π acceptance for the measurement of E_t^{miss} for the discovery of SUSY particles and a jet resolution of $\sigma_E/E \approx 50\%/\sqrt{E}$ for pseudo-rapidities η below 3 and $\sigma_E/E \approx 100\%/\sqrt{E}$ for pseudo-rapidities up to $|\eta| < 4.9$. Due to the fact that the ATLAS calorimeters are non-compensating, granularity and longitudinal segmentation are very important to allow the application of software weighting techniques offline.

In order to suppress pile-up, short signal peaking times of around 40 ns are used for the pre-amplification and shaping of the calorimeter signal.

The performance of the LAr calorimeters has been tested in several test beams at CERN. The results of these test beams are the subject of other contributions to this conference. More details on the test-beam performance of the LAr calorimeter can also be found in [1], [2], [3], and [4].

The ATLAS Liquid Argon Calorimeter

Figure 1 shows how the ATLAS LAr calorimeter wheels are located in the three aluminum cryostats. Informations on the design of the cryostats and detectors can be found in [5]. The EM barrel calorimeter covers a pseudo-rapidity range of $|\eta| < 1.475$ and each wheel consists of 16 modules or altogether 1024 accordion shaped lead absorbers (with stainless steel sheets glued on their surfaces), interleaved with read-out electrodes. The drift gap on each side of the electrode is 2.1 mm, which corresponds to a total drift time of about 470 ns for an operating voltage of 2000 V. The sketch on the left of Fig. 2 shows the segmentation of the barrel calorimeter, fine strips (in η), necessary for the required high position and angular resolution, are followed by a long

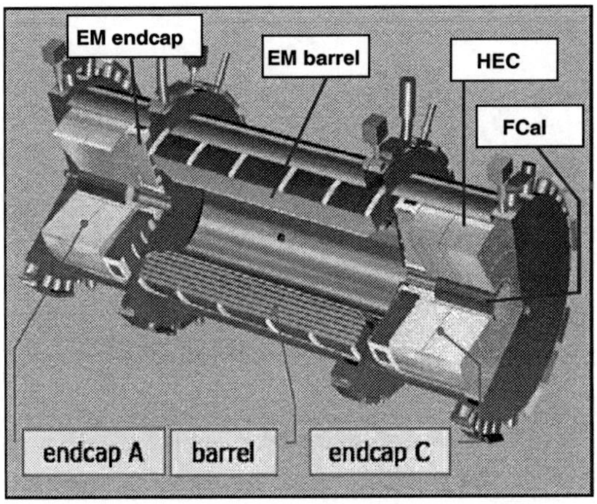

FIGURE 1. Schematic drawing of the ATLAS LAr calorimeter system. The barrel cryostat houses two EM calorimeter wheels and each end cap cryostat contains one electromagnetic, two hadronic and three forward calorimeter wheels.

middle section ($\Delta\eta \times \Delta\varphi = 0.025 \times 0.025$) and a small back compartment, necessary to estimate the back leakage of the calorimeter. The total thickness of the LAr barrel calorimeter is at least $22\,X_0$. In [6] the design and the construction of the EM barrel calorimeter are described in detail. The EM end cap calorimeter wheels use similar accordion shaped lead absorbers and consist of 8 modules each covering $1.4 < |\eta| < 3.2$. The HEC wheels are made of 32 modules per wheel (see the sketch in the middle of Fig. 2), and consist of copper absorbers (25 mm and 50 mm thickness for the first and second wheels respectively) interleaved with electrodes that split the gap between two absorbers into four sub gaps, each of width 1.85 mm (see the drawing on the right of Fig. 2). The two HEC wheels of one end cap have a total thickness of $10\,\lambda$. The wheels of the forward calorimeter (FCal) consist of a tungsten matrix (copper for the first wheel) housing cylindrical electrodes, consisting of copper rods (see Fig. 3). The LAr gaps are $250\,\mu$m, $375\,\mu$m, and $500\,\mu$m for the first, second and third FCal wheel, the thicknesses are anti-correlated with the expected particle rates. The ATLAS FCal covers a pseudo-rapidity range of $3.2 < |\eta| < 4.9$ and has a total thickness of $10\,\lambda$. The ATLAS LAr calorimeter is complemented by a hadronic barrel calorimeter made of iron absorbers and scintillating tiles as active material, this calorimeter was the subject of other contributions to this conference and will not be treated here.

CALORIMETER WHEEL INTEGRATION INTO CRYOSTATS

The detector modules of the LAr calorimeter were produced in institutes of the collaboration between 2001 and 2004. The modules were then delivered to CERN, where they were assembled into calorimeter wheels. All the EM and HEC wheels were assembled – with their axes vertical – on dedicated assembly tables inside clean rooms. After each wheel was finished it was turned to the axis horizontal position, surveyed and pushed

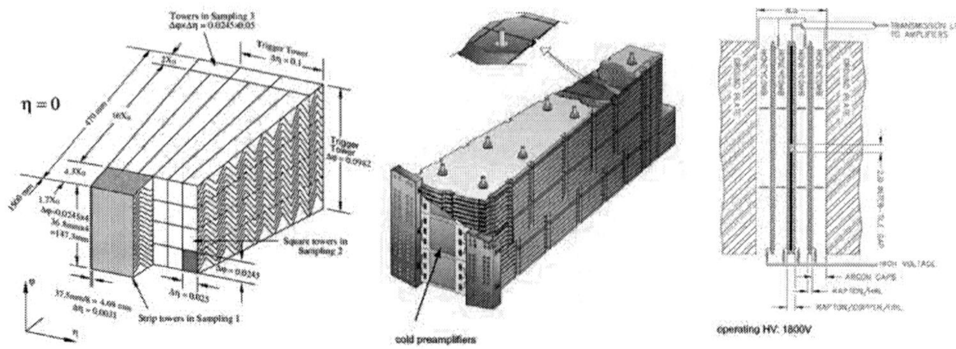

FIGURE 2. Left: Segmentation of the LAr barrel calorimeter. Middle: One hadronic end cap module. Right: The drift gap of the hadronic end cap.

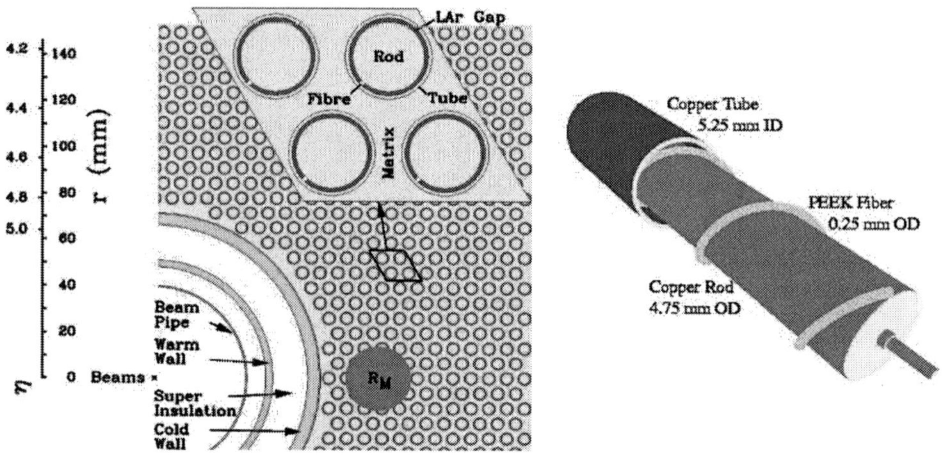

FIGURE 3. The sketch on the left explains the metal matrix housing the rods of the forward calorimeter. On the right the copper rods including the spacers are shown.

into the cryostats. After each step, measurements were performed to check the functionality of all channels. Figures 4 and 5 show the wheel assembly and their integration. The assembly and integration of the calorimeter wheels lasted from 2002 to 2004.

COLD COMMISSIONING ON THE SURFACE

Before being installed in the ATLAS cavern all three cryostats were commissioned at LAr temperatures (87K) and filled with LAr on the surface. The goals of the tests performed during this cold commissioning are listed below:

- In order to check the connectivity and integrity of the connections of all the detector

FIGURE 4. The picture on the left shows the completion of one EM barrel calorimeter wheel. One can see the accordion like absorber structure. The photo on the right depicts one barrel wheel after the insertion into the cryostat.

FIGURE 5. The picture on the left upper corner shows the assembly of one HEC calorimeter wheel. The picture below depicts the completed HEC wheel after rotation before the insertion into the cryostat. The photo on the right shows one EM end cap wheel after the insertion into the cryostat.

read-out channels, signals were injected via the calibration lines and read back over the read-out lines. Reflectrometry measurements were carried out to assure the integrity of the calibration lines.

- For testing the integrity of all HV channels, all channels were ramped up slowly to nominal voltage and kept at nominal voltage during several weeks. In order to verify the connectivity of the HV lines low frequency signals were applied on the

HV lines, which induced a signal on the read-out channels via capacitive coupling in the LAr gaps.

- In order to achieve the required 0.25 % precision of the calibration system all the cell capacitances (in fact product *LC* of the inductance *L* of the signal lines on the electrodes and the capacitance *C* of the detector cells) and the calibration resistors used to inject the precise calibration pulse were measured inside the LAr bath.
- To verify the noise level of the read-out electronics the noise and coherent noise were monitored using ATLAS production electronics boards.

During a total period of 24 weeks the LAr calorimeter system was tested inside the LAr bath, i.e. 190304 read-out channels, 14592 calibration lines and 4248 high voltage channels in all three cryostats. Table 1 shows a summary of the results of the cold commissioning and demonstrates the excellent state of the LAr detectors before their installation in the ATLAS pit. Due to a high redundancy more than 99.9 % of the detector acceptance is operational. A few percent of channels need corrections, only one HV sector in one of the EMEC wheels is broken. Ways to recover this sector have been found and will be applied if the problem persists after the final filling in the ATLAS pit.

TABLE 1. Status of the LAr calorimeter after the cold commissioning. The columns denote the following (from left to right): The detector part, the number (percentage) of bad channels, the number (corresponding percentage of the acceptance) of bad calibration channels, the number (percentage) of dead signal channels, the number (corresponding percentage of the acceptance) of channels that will need a HV correction, and the number (corresponding percentage of acceptance) of dead HV sectors. Over 99.9 % of the detector work.

	Bad channels number (%)	**Bad calib.** number (acceptance %)	**Dead signal ch.** number (%)	**HV corr.** number (acceptance %)	**Dead HV sect.** number (acceptance %)
EMEC C	40 (0.13)	1 (0.04)	6 (0.02)	25 (5.00)	0 (0.00)
HEC C	3 (0.11)	3 (0.37)	3 (0.11)	12 (7.50)	0 (0.00)
FCal C	10 (0.70)	0 (0.00)	0 (0.00)	8 (1.80)	0 (0.00)
EMEC A	20 (0.06)	4 (0.16)	8 (0.03)	35 (8.75)	1*(0.25)
HEC A	0 (0.00)	1/3 (0.8)	3 (0.11)	13 (8.10)	0 (0.00)
FCal A	9 (0.63)	0 (0.00)	0 (0.00)	11†(4.19)	0 (0.00)
EM Barrel	49 (0.04)	1 (0.01)	31 (0.03)	8.5 (1.90)	0 (0.00)

* Possibility to repair in situ, but waiting for final filling of the cryostat.
† Tests on how to repair those HV channels are underway.

INSTALLATION IN THE ATLAS CAVERN AND COMMISSIONING

All three cryostats have already been installed in the ATLAS cavern, the barrel cryostat was lowered in October 2004 (see left picture of Fig. 6), the end cap C followed in December 2005 and end cap A was brought down in April 2006. Since then the hadronic barrel calorimeter has been built around the three cryostats. The barrel cryostat was already moved to its final position around the interaction point in November 2005 (see right picture of Fig. 6). Its alignment with respect to the barrel hadronic calorimeter and the interaction point was measured to be better than 1 mm. Once the cryostats

FIGURE 6. The picture on the left shows the lowering of the barrel cryostat. On the right the ATLAS barrel calorimeter can be seen before its insertion into the ATLAS toroid.

were fitted in the cavern, the installation and connection of the front end electronics started. This operation is finished on the barrel cryostat, and started 2 months ago on the end caps. In parallel to that the back-end read out electronics are being installed in the underground service area. The HV system for the barrel has been completed and is operational (although there are still some modifications foreseen for the HV power supplies).

Since several months, regular calibration runs on readily commissioned front end crates are being taken. At the time this paper is written five front end crates are connected to the data acquisition. The pedestals, noise, coherent noise, autocorrelation matrix, gains and pulse shapes of all channels are being monitored regularly and tested for stability. Starting in April 2006 the barrel cryostat has been cooled down to the operating temperature. Figure 7 shows the temperature of the barrel calorimeter during cool down, the speed of the cool down was limited by stringent conditions on the maximum gradients allowed in the detector wheels. After the operating temperature was reached the cryostat was filled with LAr by condensing gaseous Argon. This procedure was chosen to avoid any turbulences in the LAr bath during filling and hence minimizing the risk of creating additional HV problems. It is planned to record cosmic muons in the calorimeters in the next months using the barrel hadronic calorimeter as a trigger. This cosmic run will be a valuable opportunity to practice the continuous operation of our detectors.

CONCLUSIONS

The ATLAS LAr calorimeter has been constructed in modules, assembled into wheels, integrated into three cryostats and installed in the ATLAS cavern during the last five years. The cold commissioning on the surface demonstrated the excellent condition of

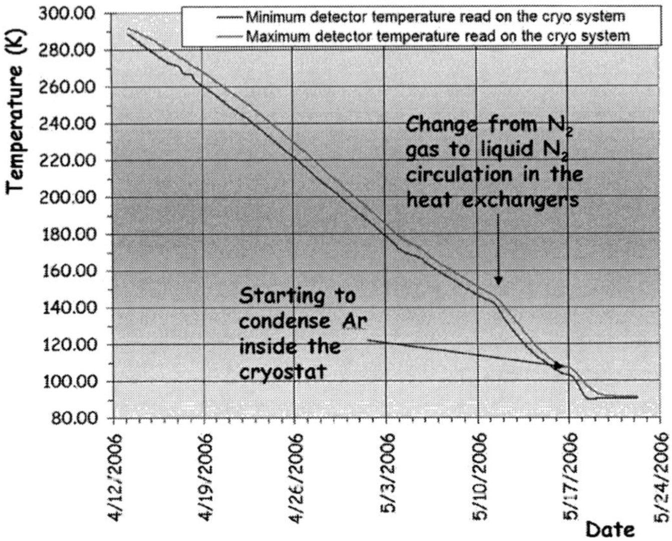

FIGURE 7. The plot shows the temperature of the detector during cool down of the barrel cryostat. The operating temperature of $\approx 87\,\mathrm{K}$ was reached on May 18, 2006.

our calorimeter. By now the calorimeter is being equipped with front end electronics and connected to the data acquisition. During the coming months the LAr calorimeter will be in operation recording cosmic muons. All the commissioning data suggest that the LAr calorimeter is working according to the requirements.

ACKNOWLEDGMENTS

The design and construction of the ATLAS liquid argon calorimeter wouldn't have been possible without the excellent contributions of our technicians. We are indebted to their great implication into the construction work. The installation of the ATLAS LAr calorimeter in the cavern was under the responsibility of the ATLAS technical coordination, we are grateful for this very good collaboration.

REFERENCES

1. M. Aharrouche et al., *Nucl. Instr. and Methods A* **Submitted** (2006).
2. B. Dowler et al., *Nucl. Instr. and Methods A* **490** (2002).
3. B. Aubert et al., *Nucl. Instr. and Methods A* **500** (2003).
4. M.I. Ferguson et al., *Nucl. Instr. and Methods A* **383** (1996).
5. The ATLAS collaboration, Liquid Argon Calorimeter Technical Design Report, Tech. Rep. CERN/LHCC/96-41, CERN (1996).
6. B. Aubert et al., *Nucl. Instr. and Methods A* **558** (2006).

Commissioning of the CMS electromagnetic calorimeter and operational experience

Paolo Rumerio

CERN, CH-1211 Genève 23, Switzerland
On behalf of the CMS Electromagnetic Calorimeter Group

Abstract. The barrel part of the CMS electromagnetic calorimeter consists of 61,200 lead-tungstate crystals optically coupled to twice as many avalanche photodiodes. The on-detector read-out electronics comprises about 20,000 printed circuit boards and 5000 optical links. The integration and commissioning phase of the 36 supermodules, 1700 crystals each, is flowing steadily at CERN in compliance with the experiment installation schedule. Given the high complexity of the apparatus, each integration step is validated by specific quality controls. It follows a week of pedestal, test-pulse and laser light monitoring runs, and 10 days of cosmic ray calibration. An automatic analysis of the data provides immediate feedback for possible intervention and verifies the stability and performance of the detector. In addition, a subset of supermodules is undergoing an electron beam calibration while two supermodules are being operated, within a segment of the CMS detector, in a CMS-like environment during the magnet test.

The strategy and procedure aimed to ensure a zero-fault tolerance in the commissioning of such a highly sophisticated and performing apparatus will be described, along with the operational experience accumulated so far.

Keywords: CMS, ECAL, Electromagnetic Calorimeter
PACS: 07.20.Fw, 29.40.Vj

INTRODUCTION

The Compact Muon Solenoid (CMS) [1] is a general purpose detector being installed at the Large Hadron Collider (LHC) of CERN. The electromagnetic calorimeter (ECAL) [2] of CMS is a hermetic, homogeneous and compact detector designed to operate in a 4 Tesla magnetic field, a 1-2 kGy/year radiation environment and at the LHC bunch crossing rate of 40 MHz. This challenging environment and the energy resolution needed to detect the two photon decay of the postulated Higgs boson have driven the design of ECAL. It consists of a central (or barrel) part and two end-caps made of lead-tungstate ($PbWO_4$) crystals [3]. Avalanche photodiodes (APDs) in the barrel and vacuum phototriodes in the end-caps are used as photodetectors. In order to achieve the design energy resolution, the crystal and APD temperature has to be kept constant to a high precision, and the crystal transparency, which varies with irradiation, has to be monitored and corrected for [4]. The availability of radiation tolerant 0.25 μm CMOS technology has allowed to design an on-detector electronic system that performs several of the trigger and data acquisition functions.

For the barrel part of ECAL, the integration, commissioning and first operation of the detector will be described with emphasis on the on-detector electronics.

CP867, *XII International Conference on Calorimetry in High Energy Physics,*
edited by S. R. Magill and R. Yoshida

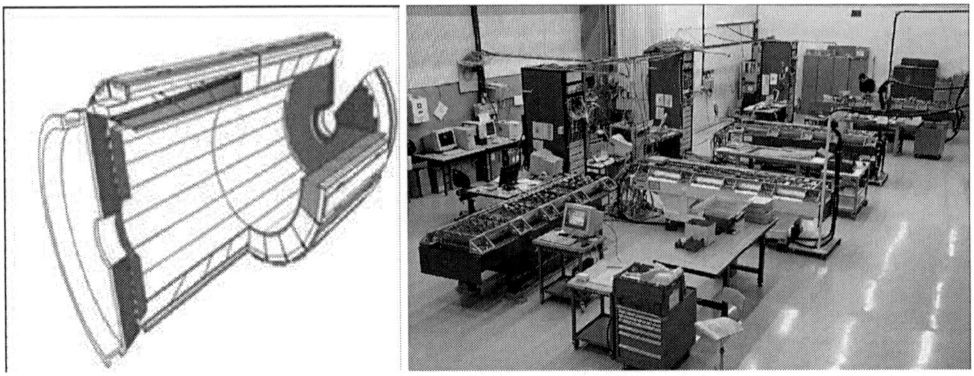

FIGURE 1. Sketch of ECAL barrel and end-caps (left): the barrel part is divided in 36 supermodules. At the electronics integration center (right), up to four supermodules can be hosted for installing the on-detector electronics.

THE BARREL ELECTROMAGNETIC CALORIMETER

A sketch of ECAL is represented in Fig. 1-left. The barrel part of ECAL is made of 61,200 $PbWO_4$ crystals arranged in a cylindrical geometry with axis along the beam line. Its acceptance covers the pseudorapidity region $|\eta| < 1.48$. $PbWO_4$ is a fast scintillator with short radiation length (8.9 mm) and small Moliere radius[1] (22 mm). The crystals have a front face of approximately 2.2×2.2 cm^2 and a length of 23 cm^2. Their axes point toward the nominal interaction point with a slight displacement (3° in polar and azimuthal angle) so to misalign inter-crystal cracks and particle trajectories. Two APDs (Hamamatsu Photonics S8148) are glued to the back face of each crystal and connected in parallel to the read-out electronics. For one every ten crystals, a thermistor is placed in the capsule that contains the two APDs.

A matrix of 5×2 crystals is held in a glass fiber alveola. Groups of 40 or 50 alveolae are assembled into modules and four modules into a supermodule. There are 36 supermodules in ECAL barrel, 18 in the positive and 18 in the negative η region. Each supermodule, completed with its cooling system and on-detector electronics, is a self-contained unit which can be operated in concert with other supermodules or in a stand-alone fashion.

THE ECAL BARREL ELECTRONIC SYSTEM

The electronic system is divided in an on-detector and an off-detector part which are connected through optical links.

The modularity of the on-detector electronics is a grid of 5×5 crystals, or trigger tower (Fig. 2-left) [6]. In a trigger tower, each one of the 5 Very Front End (VFE) cards reads

---

[1] More details on crystal characteristics can be found in [5].

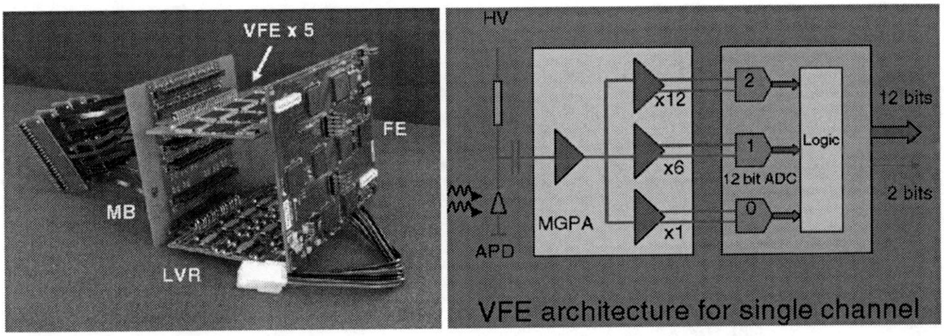

FIGURE 2. The read-out electronics of a trigger tower (left) and the architecture of one of its 25 channels (right).

out a strip of 5 crystals along the azimuthal direction. A Front End (FE) card controls and receives the data from the 5 VFEs. A Low Voltage Regulator (LVR) card stabilizes and distributes the low voltage to FE and VFEs. A Mother Board (MB) distributes the high voltage to the APDs and provides the connections to the VFEs.

On a VFE, each of the 5 crystal signals is amplified by a Multiple Gain Pre-Amplifier (MGPA) [7] with gains 1, 6 and 12 (Fig. 2-right). The 3 analog outputs are digitized in parallel by 3 of the 4 channels of a 40 MHz, 12-bit ADC (AD41240) [8]. The highest, non-saturated signal is selected by the ADC integrated logic.

On the FE card, the data are stored during a 3-μs latency of the trigger Level 1 Accept (L1A) signal, while the trigger primitives, which include quantities such as the energy deposited in the trigger tower, are calculated and transmitted to the off-detector electronics. To this purpose, two Gigabit Optical Hybrids (GOH) links, one for the data and one for the trigger primitives, are used. Each FE hosts a Clock and Control Unit (CCU) integrated circuit which serves as I^2C interface to control the 68 trigger towers[2] of the supermodule through a system of 8 token rings. A redundancy path guarantees the functionality of a token ring in case of non-operational FEs, provided that these FEs are not consecutive.

The off-detector electronics (Fig. 3) has a modularity of one supermodule served by three 9U, VME64x compliant boards with main functionalities described below. A Clock and Control System (CCS) board initializes and controls the on-detector electronics. It also receives the LHC clock, the control signals and the trigger L1A, and distributes them to the on-detector electronics and the other off-detector boards. A Trigger Concentrator Card (TCC) encodes the trigger primitives, transmits them at 40 Mhz to the Regional Calorimeter Trigger (RCT), and stores them awaiting for a L1A signal. Also, upon receipt of a L1A signal, it transmits to a Selective Read-out Process (SRP) a classification of the trigger towers according to their energy deposits. A Data Concentrator Card (DCC) receives the supermodule data at the trigger L1A rate, performs integrity controls, reduces the data size via zero suppression and selective read-out, formats the

[2] Plus two laser monitoring modules, or MEMs.

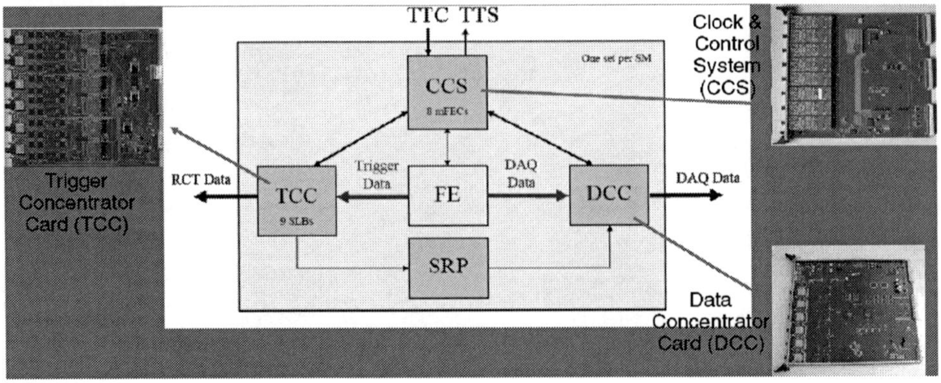

FIGURE 3. Schematic of the off-detector electronics.

data including the trigger primitives retrieved from the TCC, and transmits them to the central data acquisition.

INSTALLATION OF THE ON-DETECTOR ELECTRONICS

The on-detector electronics is installed and tested at an integration center at CERN (Fig. 1). Given the high complexity of the system and the fact that subsequent installation items cover the previous ones, each integration step is validated before proceeding with the next one. First, all MBs are installed and their connectivity is verified. On top of each MB, the 5 VFEs and the LVR are mounted, followed by the FE. As a trigger tower is completed, it is connected to a dedicated 6U VME read-out and its full functionality is tested as follows. The CCU ID, which matches the trigger tower number, is read out. All devices connected to the 16 I^2C interfaces of the CCU are scanned. The registers of FE and VFEs are initialized and read back. A pedestal scan is performed, values are set to $\sim$200 ADC counts and means and RMS are controlled. Mean amplitudes and RMS of injected test-pulses are measured. The LVR settings, the APD leakage currents, the APD temperatures and the temperature of the electronic cards are measured by Detector Control Unit (DCU) chips installed on the LVR and VFEs. Intervention in case of malfunctioning components takes place before proceeding with the integration of the following trigger tower.

Once all trigger towers are completed, the token rings are installed. Default and redundancy path are tested. Then, the GOHs for data and trigger links are connected, as well as distributed fiber patch panels. Every trigger tower is then re-tested with its final links. Fig. 4-left shows a few FE cards with the GOHs and token rings. At this point, the final low voltage distribution system is installed, covering completely the on-detector electronics, as shown in Fig. 4-right.

FIGURE 4. FE cards with token rings and GOHs (left), which are subsequently covered by the LV distribution system (right).

COMMISSIONING AND OPERATION OF THE DETECTOR

Once a supermodule is completed with the final on-detector electronics, one week of automatic operation is performed still at the electronic integration center. The standard off-detector electronic system is employed and is controlled with the latest, stable version of the CMS-ECAL data acquisition software. The data taking consists of sequences of the following run types:

- PEDESTAL SCAN - to determine the proper pedestal settings (~200 ADC counts).
- PEDESTAL RUN - to measure, with the above settings, mean value and RMS of the 3 gains of every channel (Fig. 5).
- TEST-PULSE RUN - to measure the amplitudes of the pulses injected at the input of the amplification chain.
- LASER RUN - to measure the amplitude of laser pulses injected in each crystal.

A monitoring program automatically executes control tasks on the entire supermodule data at the completion of each run. These controls include a check of the data integrity, pedestals and RMSs, and laser and test-pulse amplitudes. Summary plots and outputs of these checks are automatically produced into web pages that provide immediate feedback for possible intervention.

The CMS Detector Control System (DCS) and ECAL Safety System (ESS) are used to monitor at all times low- and high-voltage, temperatures and cooling system, so to allow safe and unattended operation of the supermodule.

Measurements of environmental variables via the DCU chips are also performed. They include the APD temperatures (Fig. 6), the APD leakage currents, and the temperature and low-voltage settings of the on-detector electronics. The access to these variables has provided important clues for understanding system features and solving

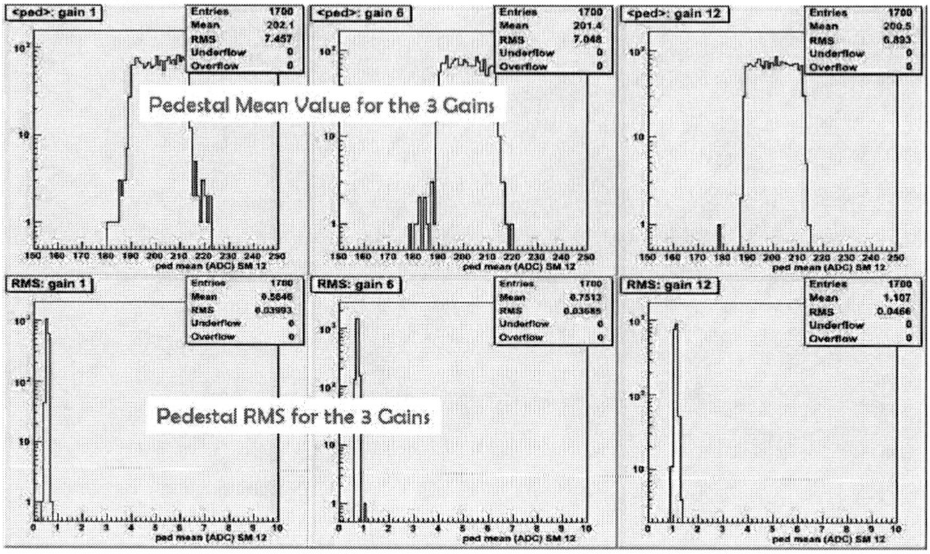

FIGURE 5. Pedestals of all 1700 channels of a supermodule as output of the monitoring program.

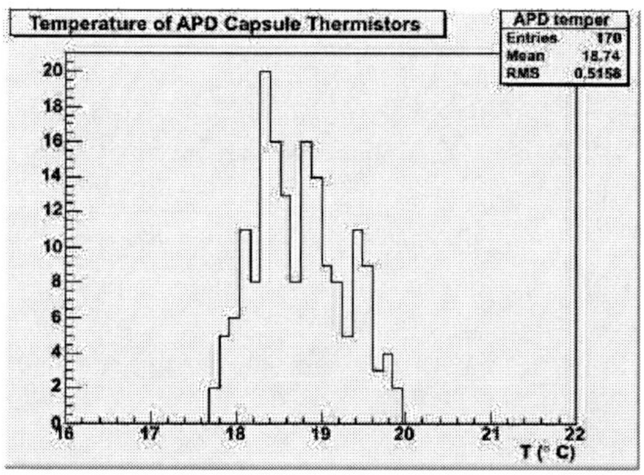

FIGURE 6. Temperature of the APDs (one thermistor every ten crystals). The spread is due to uncalibrated DCUs. Measurement of temperature variation is important: precision and stability are 0.01-0.02°C.

problems during the integration and commissioning phases.

Status and Plans

The integration and commissioning of the supermodules of ECAL barrel is proceeding in compliance with the CMS installation schedule. Fig. 7 shows the commissioned supermodules at the time of the CALOR conference, early June 2006, and the planned

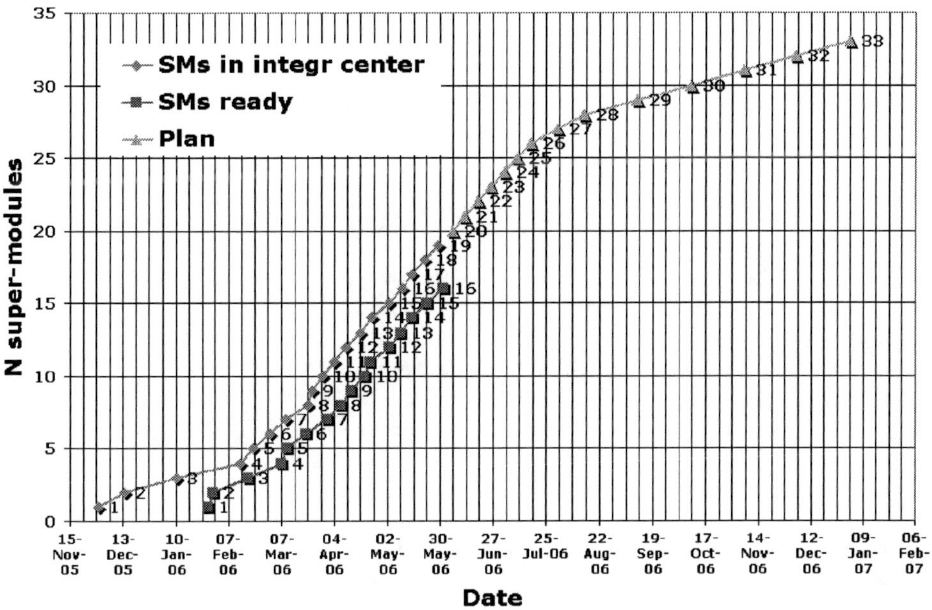

FIGURE 7. The integration progress, showing the number of supermodules arrived to the electronic integration center and the number of completed supermodules. The plan for the following months is also shown.

schedule. Ten of the commissioned supermodules underwent cosmic data taking (described in the next section). About 15 of the total 17,000 channels were not working and other 15 channels were either noisy or had one or two of the three gains not functioning. The channels failure amounts then to a few per-mills.

COSMIC RAY AND OTHER DATA TAKINGS

After one-week commission at the electronic integration center, each supermodule is transported to a cosmic ray calibration stand. Ten days of cosmic ray data are acquired at the nominal CMS temperature and with stable environmental conditions. The supermodule is operated 24 hours a day with no shifters, supervised by the mentioned DCS and ESS. A scheduled sequence of pedestal, test-pulse and laser runs is recorded between cosmic runs. Again, each run is automatically analyzed by the monitoring program upon completion and the data are subsequently analyzed to extract crystal inter-calibration constants. The main goal of the cosmic data taking is to achieve a further calibration (more details can be found in [9]) in addition to laboratory and test-beam calibrations. An important added value is a period of continuous running of all supermodules during which detector stability and performance, in addition to functionality, are probed.

At the time of the writing of this proceedings, further supermodule operations are on-going at CERN. A subset of commissioned supermodules is being calibrated with an electron beam at the H4 line in the North Area, while another one is taking part in a combined hadronic and electromagnetic calorimeter test-beam at the H2 line. At the surface of the CMS location, two other supermodules together with a segment of the CMS detector are participating in the CMS magnet test and cosmic challenge.

CONCLUSIONS

The commissioning of CMS ECAL barrel is flowing steadily in compliance with the CMS installation schedule. The procedures for the integration of the on-detector electronics are well established and have improved with the accumulated experience.

The standard CMS-ECAL off-detector electronics, acquisition software and detector control and safety systems are heavily used during commissioning and cosmic ray calibration, and recently at test-beam calibrations and CMS magnet test. The cosmic ray calibration adds a valuable, continuous operation of each supermodule in a stable environment, where detector performances, other than functionality, are probed and monitored. A level of channel failure down to a few per-mills has been achieved.

The ECAL barrel supermodules are being tested on different fronts, validating functionality and performance both as stand-alone entities and in concert with other CMS sub-detectors.

REFERENCES

1. C. M. S. Collaboration, *CERN-LHCC-94-38* (1994).
2. C. M. S. Collaboration, *CERN-LHCC-97-33* (2002).
3. A. A. Annenkov, M. V. Korzhik, and P. Lecoq, *Nucl. Instrum. Meth.* **A490**, 30–50 (2002).
4. A. Bornheim, *AIP Proceedings for CALOR 2006, Chicago* (2006).
5. R. Paramatti, *AIP Proceedings for CALOR 2006, Chicago* (2006).
6. M. Hansen, *CERN-LHCC-2003-041* pp. 78–82 (2003).
7. M. Raymond, *CERN-LHCC-2003-055* p. 83 (2003).
8. G. Minderico, *CERN-LHCC-2003-055* p. 88 (2003).
9. G. Franzoni, *AIP Proceedings for CALOR 2006, Chicago* (2006).

Installation and Commissioning of the ATLAS LAr Readout Electronics

Jingbo Ye

Southern Methodist University, Dallas TX 75275
For the ATLAS Liquid Argon Calorimeter Collaboration

Abstract. The Liquid Argon Calorimeter (LAr) is a subsystem in the ATLAS experiment which is being assembled at the Large Hadron Collider (LHC) at CERN. The readout electronics system for the LAr is discussed here. The front-end of the readout will work in radiation environment. This special requirement has led to the development of a large number of rad-hard ASICs. The back-end of the readout employs DSPs to process data to reduce transmission bandwidth. The installation and commissioning of this readout is also discussed here.

Keywords: ATLAS, LHC, Liquid Argon, Readout Electronics, ASIC.
PACS: 29.40.Vj

INTRODUCTION

The ATLAS experiment [1] is designed to study the proton-proton collision from the LHC at a center-of-mass of 14 TeV to uncover and explore the physics behind the electroweak symmetry breaking. The Higgs searches through $H \rightarrow \gamma\gamma$ and $H \rightarrow 4$ *leptons* (here *leptons* $= e, \mu$) and other new physics (ex.: $W' \rightarrow e\nu$, $Z' \rightarrow ee$) require efficient identifications, precision energy and spatial measurements of photons and electrons. SUSY searches using missing E_T require precision measurements of visible energies such as photons, leptons and jets. These requirements from physics dictate the design and construction of the ATLAS detector and its readout electronics.

The Liquid Argon Calorimeter (LAr) in ATLAS identifies photons and electrons and measures their energies. As shown in Figure 1, the LAr comprises three parts: the Electro-magnetic (EM) Accordion Calorimeter which forms the barrel part and the front portion of the endcap, the Hadronic End Cap Calorimeter (HCAL) and the Forward Calorimeter (FCAL). Together with the EM endcap part, the HCAL and the FCAL complete the endcap and the LAr has one barrel and two endcaps. This calorimter is almost hermetic: $|\eta| \leq 4.9$. The EM Accordion Calorimeter has a dynamic range from 20 MeV to 2 TeV to maximize the discovery potential, an energy resolution for photons and electrons designed to be $\sigma(E)/E(GeV) = 0.1/\sqrt{E} \oplus 0.007$. In order to achieve high spatial resolutions, the LAr has high granularity: about 195,000 readout cells. To match the detector performance, the readout electronics system has large channel numbers (195,000 channels), large dynamic (16 bits) range and excellenent linearity. The front-end of this readout situates on the detector and will be operated in radiation environment. This special requirement has led to the

CP867, *XII International Conference on Calorimetry in High Energy Physics,*
edited by S. R. Magill and R. Yoshida

development of large number of rad-hard ASICs through various technologies available to different collaborating institutions in the LAr.

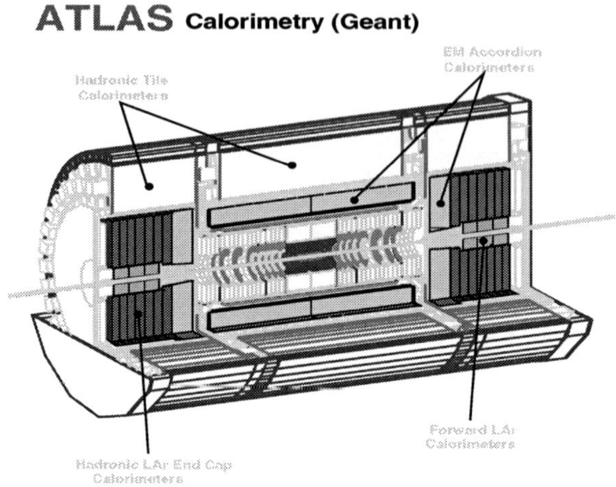

FIGURE 1. The ATLAS Calorimetry. Three parts compose the liquid argon calorimeter: the EM barrel and two endcaps. The EM barrel is placed in the barrel cryostat. The EM endcap, together with the HCAL and the FCAL fill the two endcap cryostats.

The ATLAS experiment is planned to be operational in year 2007. The LAr is being installed and will be commissioned through cosmic rays. In this note we will concentrate on the discussion of the LAr readout electronics system.

THE READOUT ELECTRONICS SYSTEM FOR THE LIQUID ARGON CALORIMTER

The LAr readout system is divided into two parts: the front-end and the back-end. The front-end electronics mounts on the detector and therefore has to withstand the radiation generated from the proton-proton collisions. The back-end receives data from the front-end, process them using a farm of 720 MHz digital signal processors (DSPs) to reduce the transmission bandwidth. Shown in Figure 2 is the block diagram of the LAr readout system. In the front-end crate, there are 28 FEBs (most cases). Each FEB reads out 128 detector channels. These FEB boards are calibrated by the Calibration Board which generates the calibration pulses through a 16 bit digital-to-analog signal converter. The Tower Builder Board (TBB) collects analog signals from each FEB and constructs energy deposition towers to provide the Level 1 trigger with information necessary to trigger the events. The trigger information comes back to the Controller Board in the front-end crate through the TTC fiber, together with the LHC clock that is used to clock the FEB. Triggered events are digitized on the FEB through a 12 bit ADC and are sent to the back-end electronics through a dedicated fast optical data link that operates at 1.6 Gbps serial data rate. In the back-end crate, the FEB data arrive at a board called the Read Out Driver (ROD) that collects data from 8 FEBs. On the ROD there are the 720 MHz DSP based processing units (PU) that process them and send the processed data to later stage electronics via the CERN developed Slink.

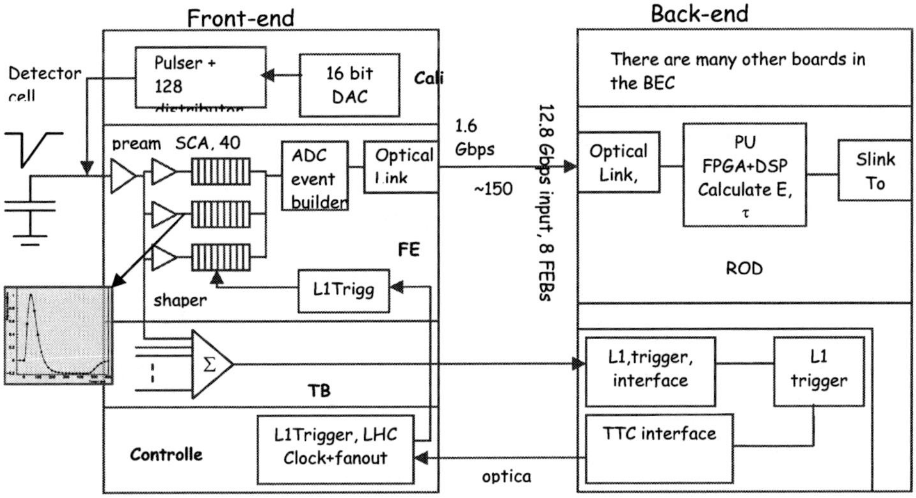

FIGURE 2. The ATLAS Liquid Argon Calorimeter readout electronics system block diagram.

Shown in Figure 3 is the more detailed block diagram of the FEB board.

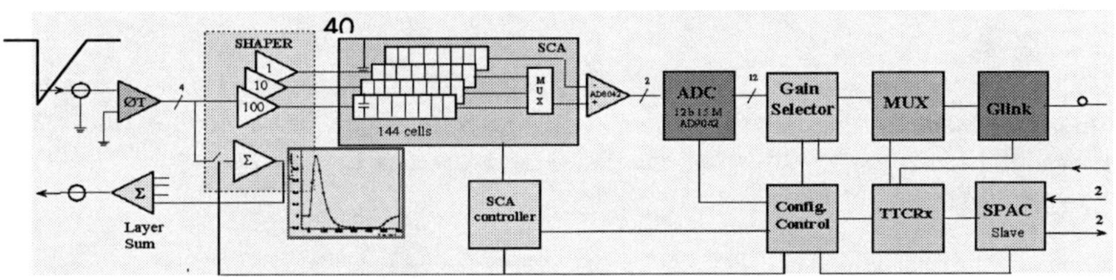

FIGURE 3. The FEB block diagram.

The detector cell generates a negative triangular current pulse from the energy deposition of a passing particle. This signal is amplified and shaped into a bipolar signal to eliminate electronics pileups. This signal is then sampled at 40 mega-sample per second (Msps) and stored in an analog memory (the SCA) waiting for the Level 1 trigger. Once triggered, the event stored in the SCA is sent to the analog to digital converter (ADC) to be converted into digital data format. Typically five samples around the positive peak of the bipolar signal are converted. Optimum Filtering method is used to obtain the peak amplitude which is proportional to the energy deposited in the detector cell. There are three gains from the shaper which amplifies the signal by factors of 1, 10 and 100. There is a gain selection mechanism built in the FEB to select the proper gain. Together with this gain selection and the 12 bit ADC, a

16 bit resolution is achieved. A Glink chip based optical data link for each FEB converts the data from the ADCs into a 1.6 Gbps serial data stream and sends the data through a 50 micron core graded-index (GRIN) fiber to the ROD which sits in the back-end crate. Shown in Figure 4 is a picture of the top side of the FEB with its key components labeled.

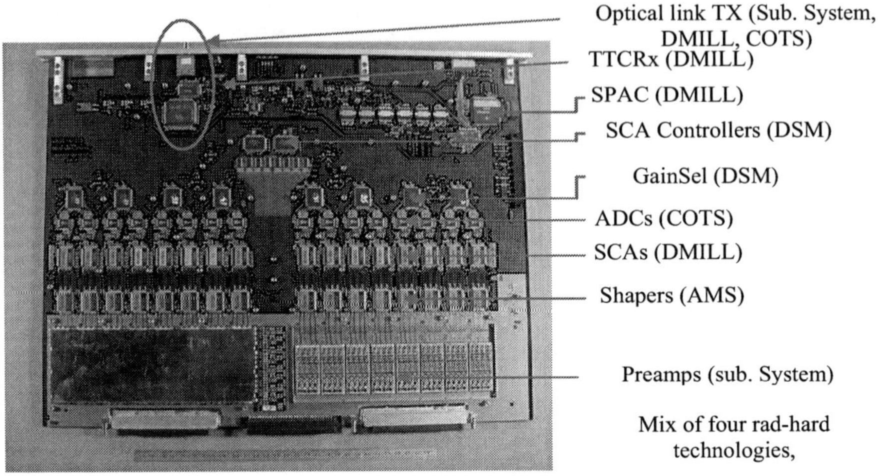

FIGURE 4. A picture of the front-end-board (FEB) and the key components on it. All components on the FEB have to be rad-hard. Four rad-hard ASIC technologies have been used to produce ASICs , The COTS on the FEB have been verified to be rad-hard.

The back-end system consists of several boards. Of them the ROD is the most important and complicated one. The block diagram of the ROD in the back-end is shown in Figure 5. A picture of the ROD and the PU that plugs onto the ROD is shown in Figure 6.

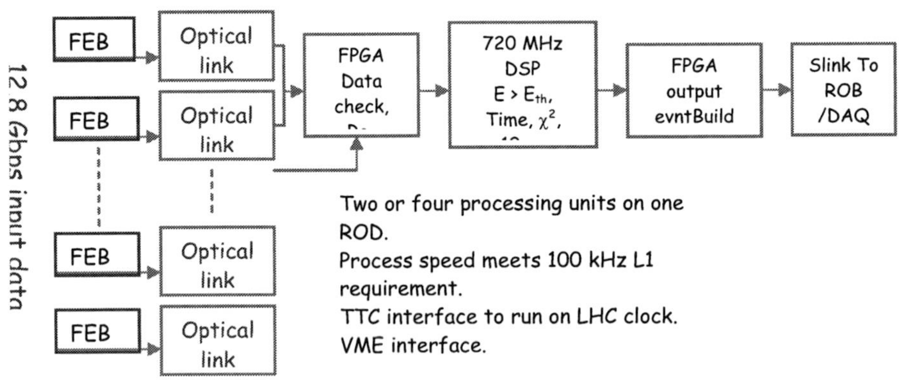

FIGURE 5. Block diagram for ROD in the back-end.

FIGURE 6. A picture of the ROD mother board and the PU that plugs onto it.

The ROD input data bandwidth is 12.8 Gbps, realized with 8 fibers carrying data from 8 FEBs. The data are preprocessed by FPGA chips. After this stage, the data are sent to the PU to be processed by the DSPs. Energy and other quantities, together with histograms are obtained by the DSPs. The energy is calculated using the Optimum Filtering method. Events with energy above a set threshold are sent to the output FPGA and then further shipped off the ROD through the Slink to a later stage in the data acquisition system.

THE DESIGN AND PRODUCTION CHALLENGES FOR THE LAR READOUT ELECTRONICS SYSTEM

The design challenges in the LAr readout system are mostly in the front-end, especially on the FEB, hence it is used as an example here. Once installed and commissioned, there will be very little access to these boards, so reliability becomes a serious issue because maintenance of these boards will only be possible during long period of LHC machine shutdowns. Each FEB reads out 128 detector channels, with 1524 FEBs to cover more than 195,000 detector channels. Although each readout channel consumes only 0.8W on the FEB, one FEB dissipates about 100 W in operation. Highly densely packed in the front-end crate, the FEBs are water cooled. Large 16 bit readout dynamic range dictated by physics is realized by a 12 bit ADC with 3 gains in the amplification/shaping channel. The analog signals are sampled at 40 MHz with 5 samples per pulse. The sampled data are stored in the SCA for Level 1 trigger latency of up to 2.5 μs, or 100 bunch crossings. The digitization of triggered event is carried out on the FEB, meeting the requirement of 100 KHz Level 1 trigger rate. After the digitization, events are sent off the FEB through an optical data link running at 1.6 Gbps serial data rate. The optical data link is used to reduce the coherent noise. The fact that there are small analog signals together with MHz to GHz digital signals on the same PCB makes the design and layout of the FEB very difficult. High speed serial data transmission requires ultra low clock jitter. In the case of the 1.6 Gbps optical link on the FEB, the clock jitter is required to be less than 20 ps. The clock jitter from the 40 MHz LHC clock distribution system is about 1 ns. Special rad-hard clock jitter cleaning circuit developed by CERN is employed on the FEB and an average of 10 ps clock jitter is achieved.

The boards in the front-end crate have to withstand radiation of over 10 years of LHC operation, estimated to be 50 Gy of ionizing radiation and 10^{12} n_{eq}/cm^2 of hadronic radiation. This requirement has led to a large number of rad-hard ASIC development and extensive rad-hardness evaluations of both ASICs and COTS. Listed in Table 1 are the ASICs developed for the FEB.

TABLE 1. ASIC development for the FEB

Technologies	Components	Number used in one FEB
AMS BiCMOS	Shaper	32
DMILL	SCA	32
	SMUX	1
	TTCrx	1
	CFGCTRL	1
	SPAC slave	1
Deep sub micron (0.25 μm)	GainSele	8
	QPLL	1
	CLKFO	1
	SCAC	2
	DCU2	2
RHBip 1	VREG	19

Production challenges lie in component, system integration and final FEB assembly levels. Example of the component level is the OTx, the electrical to optical signal converter that is used in the optical data link. The OTx is a custom designed subassembly that is 100% screened for quality assurance (QA) [2]. Since only 2000 OTx are produced, we do not have the luxury to have a preproduction of a few thousand modules to debug the production procedure. Problems in the production are actually solved during these QA measurements. The optical data link has been developed as a subsystem in lab before it is integrated onto the FEB. In a stand alone mode, the link is clocked by a crystal based oscillator which has a very small jitter. During the integration, jitter from the system clock is found not suitable for high speed serial data transmission. A special click jitter cleaning circuit has to be developed to cope with this problem and that is the QPLL chip from CERN, introduced to the FEB at a rather late stage in its development. The FEB itself is subject to a Highly Accelerated Stress Screening (HASS) during production. The HASS consists of heating and cooling cycles over 6 hours, simulating the first few weeks of LHC operation. This is designed to find problems in an early stage (the "infant mortality" due to problems in production and weak components). Component level and circuit level problems (ex: OTx and QPLL stability) were identified during early stage in the FEB production and were fixed. The FEBs are then tested at collaborating institutions for the full digital and analog functionality to assure that each FEB produced meets its design specifications.

INSTALLATION AND COMMISSIONING STATUS

The LAr readout system has been tested together with detector modules in beam tests. Shown in Figure 7 are preliminary results of electrons at 20 GeV and 180 GeV from beam tests, indicating that the detector and its readout system are working although the absolute calibration is off by a few GeV.

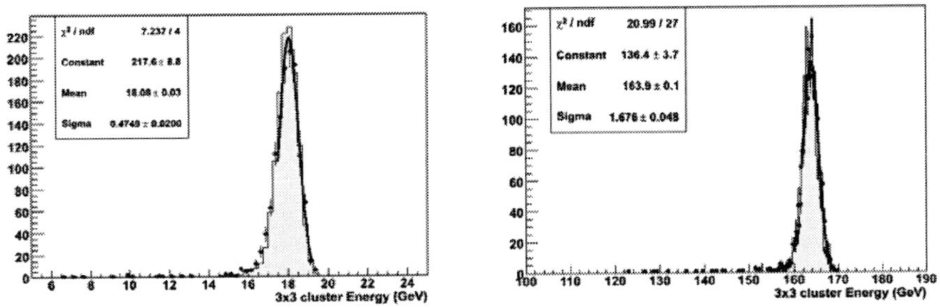

FIGURE 7. Preliminary results on electrons from beam tests indicating that the LAr detector and its readout system work.

The installation of the LAr detector and its readout electronics is taking place now. At time of the conference, the EM barrel front-end crates were fully equipped and tested. Endcap front-end crates were just started to be equipped. There was a lot of work on cabling, fiber splicing, cooling and other services taking placing in the ATLAS experimental hall. Back-end electronics installation matched well with the front-end. The commissioning of the detector with its readout system will be carried out later part of the year with cosmic rays.

CONCLUSIONS

The readout electronics system for the ATLAS LAr calorimeter is discussed in this note. The radiation environment in which the front-end electronics will be running imposes special rad-hard requirement for its components that has led to the development of a large number of rad-hard ASICs and lengthy evaluation of them, together with other COTS that are identified to be rad-hard and are used on the front-end electronics boards. This readout electronics system has been produced, tested in beam tests, and installed onto the ATLAS detector. A great deal of experience has been gain in component production, system integration and final board level assembly. The commissioning of the system together with the detector using cosmic rays is being carried out. This is a crucial step to ensure that the ATLAS experiment will be ready for the LHC colliding beam.

ACKNOWLEDGMENTS

The author of this note wishes to thank the whole ATLAS LAr collaboration. Special thanks go to the following people who provided guidelines, suggestions and materials for this note: Ryszard Stroynowski, John Parsons, Dominik Dannheim, Isabelle Wingerter-Seez, Martin Aleksa and Stefan Simion.

REFERENCES

1. ATLAS collaboration, *ATLAS Detector and Physics Performance*, Technical Design Report, CERN/LHCC 99-14, 1999.
2. Tiankuan Liu *et al*, *Quality Control of Optical Transmitters for the ATLAS Liquid Argon Calorimeter*. ATL-LARG-PUB-2005-004.

Scintillator Pad Detector: Very Front End Electronics. Design and Pre-Series.

S. Luengo[1]*, J. Riera[1], S. Tortella[1], X. Vilasis-Cardona[1], D. Gascón[2], A. Comerma[2], L. Garrido[2]

1 Dept. Electrònica, Enginyeria i Arquitectura La Salle, Universitat Ramon Llull, Barcelona, Spain
2 Dept. Estructura i Constituents de la Matèria, Universitat de Barcelona, Barcelona, Spain

Abstract. The SPD (Scintillator Pad Detector) is a part of LHCb calorimeter which is designed to distinguish electrons and photons for this first level trigger. This detector is a plastic scintillator layer, divided in about 6000 cells of different size to obtain better granularity near the beam. Charged particles will produce, and photons will not, ionisation on the scintillator. This ionisation generates a light pulse that is collected by a Wavelength Shifting (WLS) fibre that is twisted inside the scintillator cell. The light is transmitted through a clear fibre to the readout system. For cost reduction, these 6000 cells are divided in groups using a MAPMT of 64 channels for receiving information in the readout system. The signal outing the SPD PMTs is rather unpredictable as a result of the low photostatistics, 20-30 photoelectrons per MIP, and the response of the WLS fibre, which has low decay time. Then, the signal processing must be performed by first integrating the total charge and later subtracting to avoid pile-up. The SPD Readout system is performed by an ASIC which integrates the signal, makes the pile-up compensation, and compares the level obtained to a programmable threshold (distinguishing electrons and photons), an FPGA which programmes the ASIC thresholds and pile-up subtraction and finally LVDS serializers, in order to send information to the first level trigger system. The design of the VFE unit takes into account not only mechanical constraints, as a result of the little space for the readout electronics but also the radiation quote expected in the environment and the distance between the VFE electronics and the racks were information is sent.

Keywords: Photomultiplier, Readout, Electronics, Scintillator Pad Detector, LHCb
PACS: 29.40.

INTRODUCTION

LHCb is one of the four detectors to be built on LHC, the future proton-proton accelerator in construction at CERN (European Center for Nuclear Research) in Geneva (Switzerland). It is expected to be running by 2007. LHC will be an intense source of hadrons containing quark b: a production of 10^{12} bb is expected, working at a luminosity of $2\cdot10^{32}$ $cm^{-2}\cdot s^{-1}$ at 40 MHz frequency. The LHCb detector is designed to make precise studies of CP asymmetries and of rare decays in the B-mesons systems in the LHC proton-proton collider.

LHCb structure contains a calorimeter whose main purpose is the identification of hadrons, electrons and photons and the measurement of their energies and trajectories.

CP867, *XII International Conference on Calorimetry in High Energy Physics,*
edited by S. R. Magill and R. Yoshida

This is the basis of the trigger system containing information to study B physics and enables the reconstruction of the interactions produced in the collider.

The LHCb calorimetry [1,2] has four elements: a hadronic calorimeter (HCAL), an electromagnetic calorimeter (ECAL), a Preshower detector (PS) and a Scintillator Pad Detector (SPD).

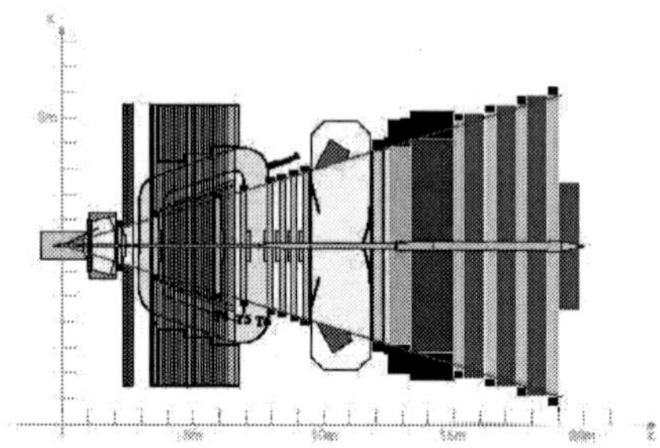

FIGURE 1. Spectrometer of LHCb. .

SPD REQUIREMENTS AND VFE SPECIFICATIONS

The SPD [6] consists of a plane of about 6000 scintillator cells of several sizes sitting in front of the lead plate of the PreShower. Its mission is to increase first level Trigger efficiency by distinguishing charged particles (mainly electrons) from neutral ones (mainly photons).

Such distinction is made by measuring the energy deposited in the scintillator cells. A coil of Wavelength Shifting Fiber captures the emitted photons and takes them out the pad. A further connection to a clear fiber brings these photons into the readout system. To lower cost and space, 64 channel multi-anode photomultipliers are used to convert light into electrical signal. The expected output is one bit for every pad telling whether a charged particle passed by the cell in the current bunch crossing.

Pulsehape Characteristics

Ideally, only charged particles generate a relevant signal. However, high energy photons can create an electron through secondary processes such as Compton Effect or pair production. Applying a threshold at 1.4 MeV (0.7 MIP) is a good compromise to reject photons with an acceptable efficiency for the trigger [7] . The resolution of the discriminator must be better than 0.05 MIP in order to keep the energy resolution given by photostatistics in this threshold area.

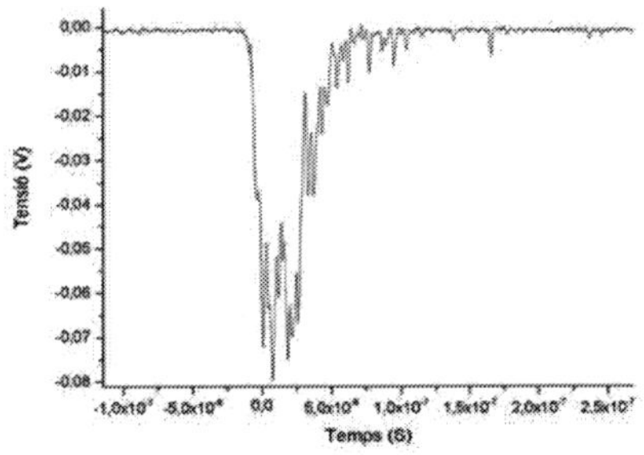

FIGURE 2. Pulseshape characteristics.

The efficiency of the light collection system and the PMT [5] photocathode lays low photostatistics, around 20-30 photoelectrons. This generates irregular pulseshapes lasting longer than the bunch crossing period of 25 ns, see figure 2. Cosmic ray studies and simulations recommend integrating the total signal and performing pile up correction [5]. Besides, the PMT gain is limited by the maximum average current that can be used to avoid PMT premature aging [5]. For the cells with higher occupancy the MIP signal will be cut down to 100fC. Active bases shall be used to keep acceptable gains at low high voltage.

Readout Electronics

SPD Readout system is performed by an ASIC [6] which integrates the signal, makes the pile-up compensation, and compares the level obtained to a programmable threshold (distinguishing electrons and photons), an FPGA which programs the ASIC threshold and pile-up subtraction and finally LVDS serializers, in order to send information to the first level trigger system.

From those elements, functional design for the SPD readout is split into: a Very Front End board hosting a PMT with its corresponding 8 ASICs; a Regulator Card which feeds the Very Front End card; and a control board sitting at the calorimeter front end crates and interfacing the experiment control system and the boards, as depicted in figure 3. Very Front End boards shall be located in boxes sitting on the top and the bottom of the detector, limiting the area of the VFE board to 9 × 9cm2.

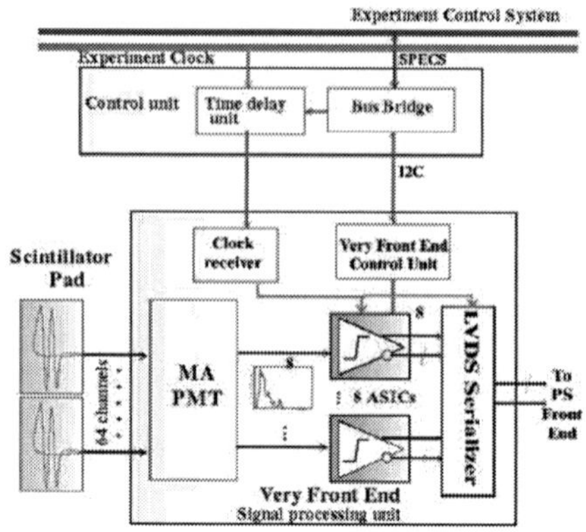

FIGURE 3. SPD Readout System.

In order to save connector space data shall be sent to the PreShower front end card by a multiplexed LVDS link.

VFE DESIGN SOLUTION

Very Front End Cards are the heart of the readout electronics since they are in charge of the pad signal processing and the sending of the resulting bits to the Preshower Front End Card [1].

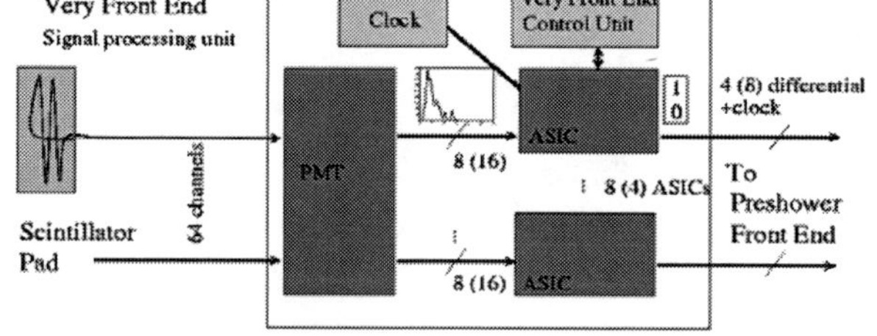

FIGURE 4. VFE Diagram Block.

The most important components are:

- The multianode photomultiplier that is in charge of converting light to analog signal,
- ASICs, that convert the analog signal to a digital one,
- The FPGA (Control Unit), its function is to communicating with Control Card and programming thresholds and pile-ups corrections,
- LVDS Transceivers, converting LVDS signal to CMOS ones,
- LVDS Serializers that prepared the ASICs digital outputs to be transmitted to the high speed LVDS link (25 meters long).

All components have been irradiated in order to pass the radiation qualification [3,4] (Single Event Effects and Cumulative Effects). Some useful components had to be discarded because they do not support the required levels of radiation (E.g. DS90CR494, 64-channel LVDS multiplexer).

One of the challenges in the design of VFE board is the voltage levels of components. For e.g. ASICs voltage supply is (+/-1.65V), but the output of the board should be 3.3V (in order to communicate the PreShower [1] FE board to SPD VFE board). This implies that it has to be a voltage level adaptation between both boards. This adaptation is made by a passive circuit (2 PIN diodes and 2 resistors).

As a first attempt, VFE Unit was designed to be in a single card. After this first design, it was realized that the measure of the card was big enough to not fitting in the space available in the cavern [1].

Actually the design of the VFE board has been splitted into three different boards:

- PMT Base Board
- ASICs Board
- Input Stage/ Control Unit / Output Stage Board

PMT Base Board

This board is the same as PMT Base Board of Pre-Shower [1].

ASICs Board

This board contains the photomultiplier, the eight ASICs and its circuitry and the part of threshold reference and subtractor reference.

It is a ten-layer card, with components on both layers, top and bottom layer, class6 and made with FR-4.

Input Stage/ Control Unit / Output Stage Board

The last board contains all the input stage (connectors, transceivers, and adaptation), the control unit (FPGA) and the output stage (asic's output adaptation, serializers and output high speed connector).

It is an eight-layer card, with components on both boards, top and bottom layer, class6 and made with FR-4.

TESTS

Most important tests made on the VFE Unit are:

- ***Test Beam Tests***: during some runs and prototypes.

RUN3 (Jan 2002)
  - New tunable substractor
  - 1 full channel with digital control
  - On-chip DAC to program thresholds

RUN4 (Sep 2002)
  - 1 Complete processing channel
  - Separate blocs + digital control
  - Working at 3.3 V to reduce power consumption

RUN5 (March 2003)
  - Final Version

- ***Laboratory Tests***: they showed the good performance of the final prototype. Some tests are enumerated below:
  - *Power Consumption*
  - *Programming FPGA*
  - *Noise*
  - *Offset*
  - *I^2C protocol. Read and Write*
  - *ASIC's digital part. Read and Write internal thresholds*
  - *Calibrating the cards: pedestals*
  - *Critical signals'shape (such as 40Mhz clock or LVDS link)*

The noise on the card is around 2mV, which is a current figure with a 470Ω input resistor.

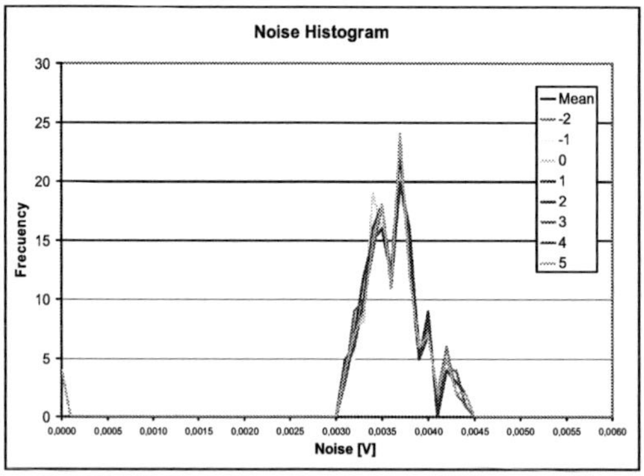

FIGURE 5. Noise Histogram of 64 channels.

- ***Validation Protocol***: For VFE prototype's acceptance a validation protocol was defined. The most important figures:
  - *Noise*
  - *Offset*
  - *Power Consumption*

- ***Cooling Tests***: They had been done in Clermont-Ferrand at LPC (Laboratoire de Physique Corpusculaire) between July and September 2005 and during July 2006 in Barcelona at UB (University of Barcelona).

The cooling is made by a cool water circuit around boards. A conductive material of heating is put on the cards and it is in contact with an aluminium platform that contains the water circulating on.

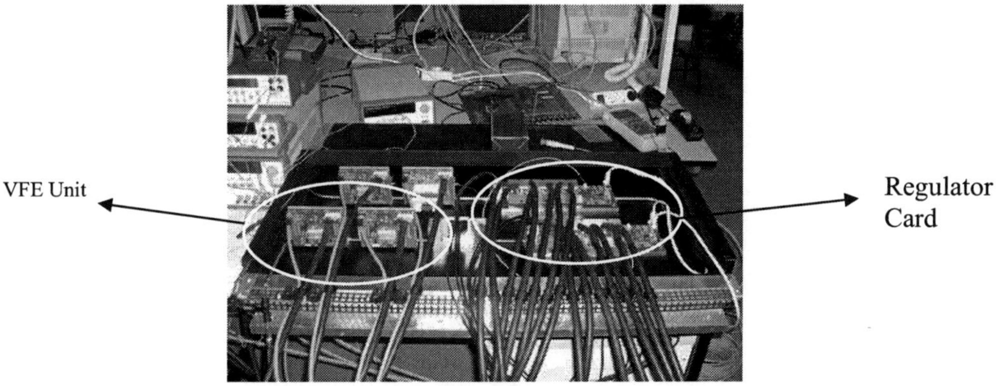

FIGURE 6. Cooling Tests July 2006.

- ***Pre-series tests***: The results of this pre-series will show the good performance of the card, and it will be used to test all the automatic set-up prepared for the series test. Some of the tests done were:
  - *Noise*
  - *Offset*
  - *Critical signals' shape*
  - *LVDS link*
  - *I^2C read and write*

- ***Burn-in Tests:*** These tests are done by Applus+, a Certification Technological Center. The main goal of these tests is avoiding the enfant mortality.

The temperature range and time cicles are:

8 cycles of 8h15min between 0 and 50 Celsius degrees with a 2h20 stop at 0 degrees, 40 degrees, and 50 degrees, with slopes of 2 degrees per minute in 'montée' and 1 degree per minute in 'descent'.

FUTURE WORK

The pre-series is fully tested up to now. During next months, the series will arrive and has to pass some of the test mentioned before: burn-in tests in order to avoid infant mortality, validation protocol…

The commissioning is planned to be done during this winter, and taking the first results during next spring.

ACKNOWLEDGMENTS

Work supported by CICYT under projects FPA2005-06889-c02-01 and FPA2005-06889-c02-02 and by Funitec under project PGR-PR-2004-03.

S. Tortella acknowledges an FI fellowship from Generalitat de Catalunya.

REFERENCES

1. D. Breton "The Front-End Electronics for LHCb calorimeters", X Int. Conf. on Calorimetry in Part. Phys., CALOR 2002, Pasadena.
2. S. Amato et al., "LHCb Technical Design Report", CERN/LHCC/2000-0036, 2000.
3. F. Faccio, " COTS for the LHC radiation environment: the rules of the game". Geneva, CERN 2000.
4. M. Dentan "Overview of the ATLAS policy on radiation tolerant electronics" Geneva, CERN, 2000.
5. O. Dechamps et al, "Study of multianode photomultipliers for the electromagnetic calorimeter preshower read out of the LHCb experiment", 3th Int. Conf. on New Developments in Photodetection, Beaune, 2002
6. D. Gascon et al, "Discriminator ASIC for the VFE SPD of the LHCb Calorimeter", LHCB Technical Note, LHCB 2004-xx.
7. L. Garrido et al.,"Backsplash studies for the scintillator pad detector of LHCb in a tagged-photon test beam", Nucl. Inst. and Methods A, 484/1-3 2002.

SESSION VII

LHC CALORIMETRY—TEST BEAM

Convener: Calin Alexa

Study of the ATLAS central calorimeters response to pions in 2004 combined test beam

V. Giangiobbe

Laboratoire de Physique Corpusculaire de Clermont-Ferrand

Abstract. During summer 2004, a testbeam for the ATLAS experiment was performed at CERN. A full slice of the ATLAS detector was tested with the final version of the acquisition electronics. In the talk presented at Calor06 conference, the ATLAS calorimeters response to high and low energy pions was presented.

Keywords: ATLAS; Testbeam; Calorimetry; Pions
PACS: 01.30.Cc; 29.40.Vj

DESCRIPTION OF THE 2004 COMBINED TEST BEAM OF THE ATLAS DETECTOR

Central calorimetry in ATLAS

The whole calorimetry in ATLAS is contained within a 12 m long cylinder with an outer radius of about 4.2 m. This volume is divided into 3 part along η. The central part called barrel, covers $-1.7 < \eta < 1.7$. The electromagnetic calorimeter (LAr calorimeter) uses lead as absorber and the active material is liquid argon. For the hadronic calorimeter (TileCal) the absorber is iron and the active part consists of scintillating tiles. For mechanical reason the central hadronic calorimetry is separated in 3 parts : 1 long barrel for $-1.0 < \eta < 1.0$ and 2 extended barrels for $0.8 < |\eta| < 1.7$. Figure 1 shows the calorimetry in ATLAS. The central calorimetry which will be the subject of the present document has a light-grey color. The dark-grey circled region represents the part that have been exposed to the beam during the 2004 combined test-beam.

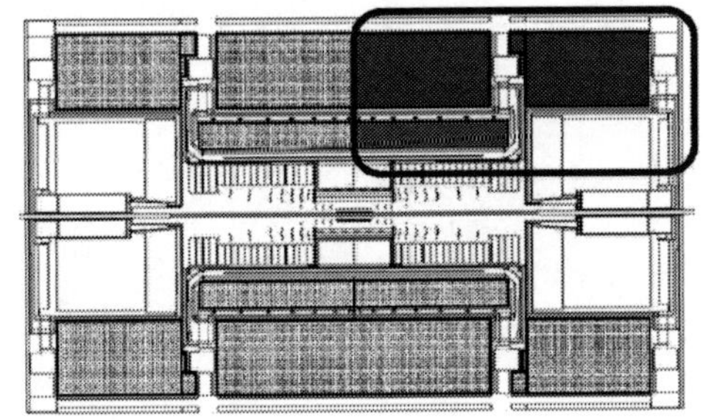

FIGURE 1. The calorimetry in ATLAS experiment.

CP867, *XII International Conference on Calorimetry in High Energy Physics,*
edited by S. R. Magill and R. Yoshida

Central calorimetry in 2004 combined test beam

The 2004 combined test-beam is the first test of a full ATLAS slice (inner detector, calorimetry and muon spectrometer). In the following only the calorimeter response will be discussed. A top view of the calorimetry setup is shown on Figure 2. The LAr calorimeter is on the bottom. Half a barrel of TileCal and 1 TileCal extended barrel are also shown. The modules were laying on a rotative Table giving the possibility to shoot the beam at various η positions (represented in Figure 2). Comparing with Figure 1 it can be observed that the test-beam geometry is similar to the ATLAS one. Figure 3 shows a side view of the calorimetry on the Table. We can see the LAr calorimeter within its large cryostat, and 3 TileCal modules. The ϕ-coverage is 0.4 in LAr and 0.3 in TileCal.

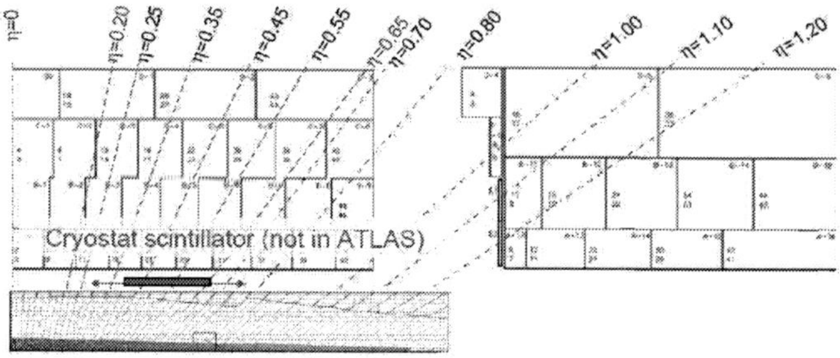

FIGURE 2. Top view of the calorimetry at the 2004 combined test-beam.

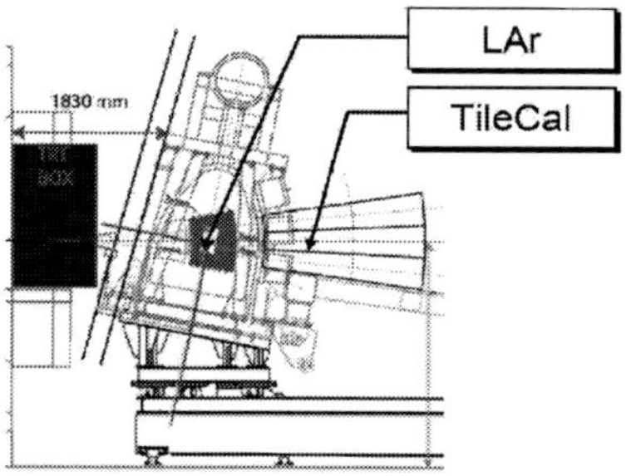

FIGURE 3. Side view of the calorimetry at the 2004 combined test-beam.

Data sample and energy reconstruction

During summer 2004, central ATLAS calorimeter modules have been exposed to muons, pions and electron beams with different energies. In the following only the results obtained using the pion beams are reported. The High Energy (HE) pions had the following energies : 50, 100, 150, 180, 200, 250, 320 and 350 GeV. They had an impact angle with respect to the surface of the detector corresponding to the pseudo-rapidity values : η=0.20, 0.25, 0.35, 0.45, 0.55, 0.65. The Very Low Energy (VLE) pion beams had energies ranging between 1 and 9 GeV and the same η values.

In LAr calorimeter the energy scale is obtained using the features of a discharging capacitor. For TileCal, the energy scale is determined using the response of the detector to electron beams. In the present analysis the energy that is deposited by pions in calorimeters is reconstructed summing the energy of cells in a limited window around the direction of the beam. It has been checked that the pion showers are well contained in $\Delta\eta \times \Delta\phi = 0.4 \times 0.4$ volume in LAr and $\Delta\eta \times \Delta\phi = 0.4 \times 0.3$ volume in LAr.

PIONS THAT DO NOT INTERACT IN THE ELECTROMAGNETIC CALORIMETER

Around 20% of the high energy pions reach the hadronic calorimeter without producing a shower in the electromagnetic section. These pions constitute an interesting probe to check the performances of TileCal standalone and to establish the absolute energy scale in TileCal. In this section the response (ratio between the reconstructed and the beam energy) and the resolution of TileCal will be reported for high energy pions.

Event selection

Figure 4 shows the correlation between the deposited energy in the electromagnetic and hadronic calorimeters for a 180 GeV beam. It can be seen on this plot that the high energy pion beam contains a significant fraction of muons and electrons. Muons give a very small energy in both calorimeters while electrons deposit all their energy in LAr. Muons and electrons can be easily rejected looking at the longitudinal profile of the energy in calorimeters. Pions showering in both calorimeters are visible in the center of Figure 4 (anti-correlation between LAr and TileCal). Pions interacting only in TileCal can be selected requesting a signal compatible with the one of a minimum ionizing particle (mip) in each of the 4 longitudinal compartments of the LAr. This is not enough to select pions reaching TileCal without early interaction. Indeed pions passing through the LAr at their minimum of ionization can interact in the outer cryostat wall. In order to reject such events an additional scintillator (called cryostat scintillator) has been placed between the outer cryostat wall and the front face of TileCal (see Figure 2). A mip-like signal is then requested also in the cryostat scintillator. Figure 5 shows the energy distribution in TileCal for 180 GeV pions that pass the cuts described above (ie mip-like signal in each LAr layer and in the cryostat scintillator). A Gaussian fit has been applied in order to estimate the mean reconstructed energy E_{raw} and the width σ of the distribution. The fit was performed in a region of 2 sigmas around the peak value.

Response and resolution

Figure 6 shows the ratio between E_{raw} and the beam energy for different η positions. Two different beams have been used during the data taking. One has a negative polarity and is expected to contain only negative pions while the other one is a positive beam

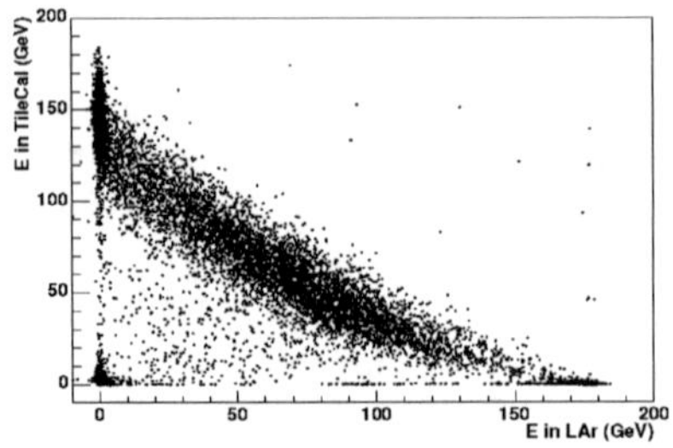

FIGURE 4. Correlation between LAr and TileCal energy

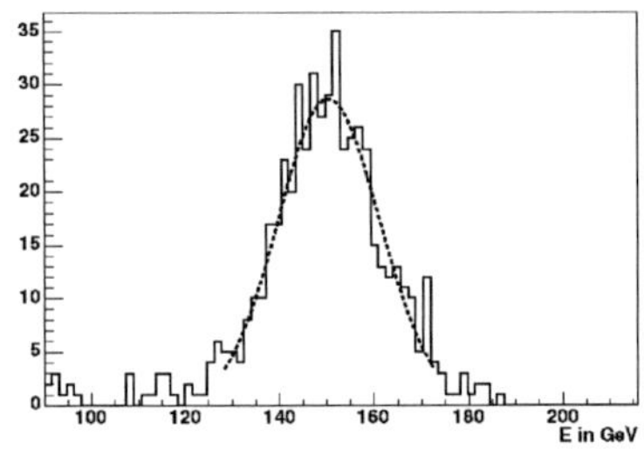

FIGURE 5. Response to 180 GeV pions that do not start showering before reaching TileCal.

for which a proton contamination is expected. It is a well known fact that protons produce a lower response than pions in a non-compensating calorimeter like TileCal [1]. For this reason and because it was not possible to separate pions and protons in 2004 combined test-beam, the response for positive and negative beams have been represented separately in Figure 6. It can be noticed that, as expected, the response for the positive beam, (150 and 180 GeV) is systematically lower than the negative beam one at the same energy. The effect is around 1%. Looking at Figure 6 it can be noticed that the calorimeter response is increasing with the beam energy. This behavior is typical of a non-compensating calorimeter. At high energy (320 and 350 GeV), due to the transversal and longitudinal leakage, the response decreases slightly. So far two sources of systematic effect have been considered. The beam incident energy is known with a 0.5% precision. The influence of the selection cuts has been estimated to be of the order of 0.2%. This gives a global systematic uncertainty of around 0.5%. The statistical error on the reconstructed energy is around 1%. It means that the statistical error dominates the estimation of the TileCal response to pions.

The evolution of the resolution defined as σ/E_{raw} as a function of $1/\sqrt{E_{beam}}$ it shown in Figure 7. This study has been done using the negative beam (no protons) except at 50 and 100 GeV. A fit function has been applied in order to estimate the sampling term a and the constant term b :

$$\frac{\sigma}{E_{raw}} = \frac{a}{\sqrt{E_{beam}}} \oplus b$$

TABLE 1. Sampling term a and constant term b for pions in TileCal.

	Sampling term a ($GeV^{1/2}$)	**Constant term b (%)**
$\eta = 0.25$	0.51±0.03	5.9±0.2
$\eta = 0.35$	0.52±0.02	5.7±0.3
$\eta = 0.45$	0.48±0.02	5.8±0.2
$\eta = 0.55$	0.44±0.03	5.5±0.2
$\eta = 0.65$	0.49±0.03	5.1±0.1

The obtained parameters are given in Table 1. It has to be stressed that these results have been obtained without any correction for leakage or non-compensation. For this reason the resolution is not the best that can be achieved. The results of Table 1 are in good agreement with results obtained in the same conditions during previous TileCal standalone test-beams [2].

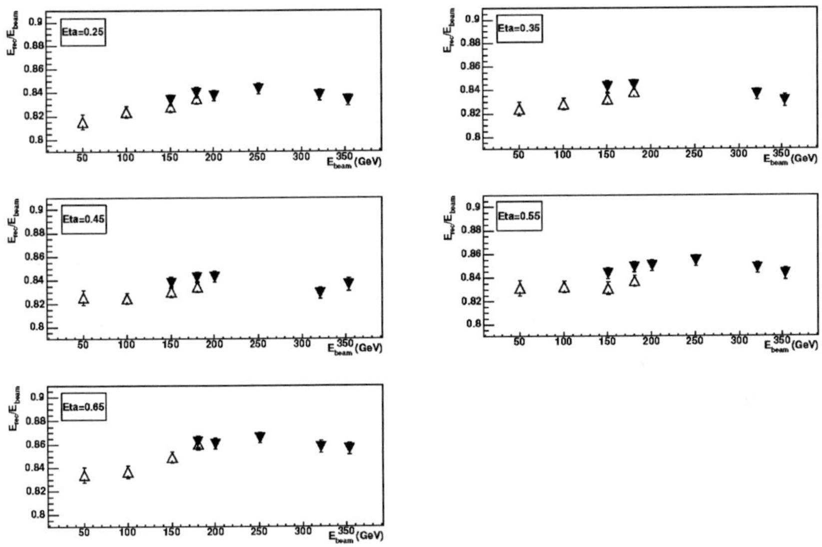

FIGURE 6. Ratio between the reconstructed energy and the beam energy for various η positions. Black triangles represent negative beam results and white triangles represent positive beam ones (see text).

COMBINED CALORIMETRY RESPONSE TO PIONS

Energy reconstruction

The study of combined calorimetry response to pions is done only using beams with a negative polarity in order to get rid of any proton contamination. The energy

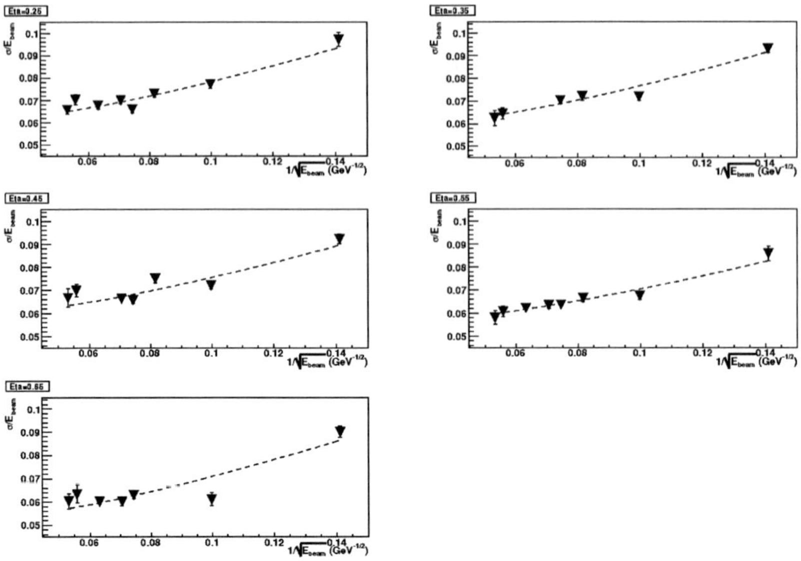

FIGURE 7. Resolution (σ/E_{raw}) as a function of $1/\sqrt{E_{beam}}$.

in calorimeters is computed as the sum of energy in LAr and energy in TileCal. No correction for non-compensation, dead material or leakage has been applied. The aim of this analysis is to provide the combined response that can be used to fix the absolute hadronic scale and that can be reproduced by the simulation. Figure 8 shows the energy distribution in LAr, TileCal, and the sum of the energies in the two calorimeters for 180 GeV pions at $\eta = 0.35$. The reconstructed energy for pions that start showering in TileCal (called penetrating pions) is also shown (bottom left). As before a Gaussian fit is applied on the total energy distribution and the mean reconstructed energy E_{rec} is obtained from the fit.

Combined response

Figure 9 shows the ratio between the mean reconstructed pion energy and the beam energy for different η positions. Again the response is showing the typical behavior of a non-comensating calorimeter : the response increases as the beam energy increases (due to the increase of the fraction of π^0 in the showers). The error on the pion response is around 0.5%. It comes from the uncertainty on the beam energy.

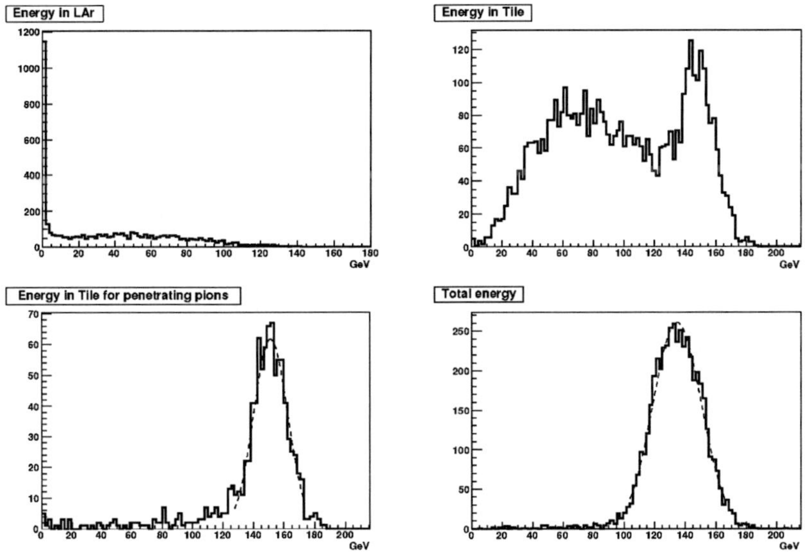

FIGURE 8. Energy distributions for 180 GeV pions at $\eta = 0.35$ Top left : energy in LAr; top right : energy in TileCal; bottom left : energy in TileCal for pions that start showering in TileCal; bottom right : energy in LAr+TileCal for all pions.

CALORIMETERS RESPONSE TO VERY LOW ENERGY PIONS

Motivations

In addition to the high energy program, very low energy data have been taken during the 2004 combined testbeam. From the calorimetry point of view this is motivated by the fact that very low energy particles (namely E<10 GeV) have an important contribution in jets. Figure 10 shows a simulation of the fraction of energy brought by charged particles with E<10 GeV as a function of the jet energy. One observes that about 25% of the energy in a 150 GeV jet is brought by particles with less than 10 GeV. Therefore we need a good knowledge of the calorimeters response to very low energy particles in order to check and tune the simulation of jets.

Selection of the very low energy pions

Very low energy (VLE) data have been taken at η =0.2, 0.25, 0.35, 0.45, 0.55, 0.65 for energies between 1 and 9 GeV. The VLE beam is a mixture of low energy pions, electrons and muons from pion decay. Due to the configuration of the beam line, a large contamination by high energy muons is also expected.

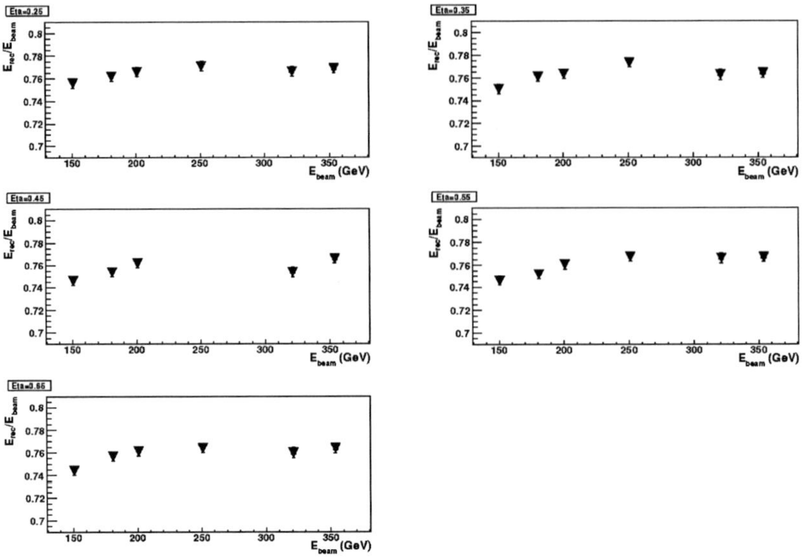

FIGURE 9. Combined response to pions for various η positions.

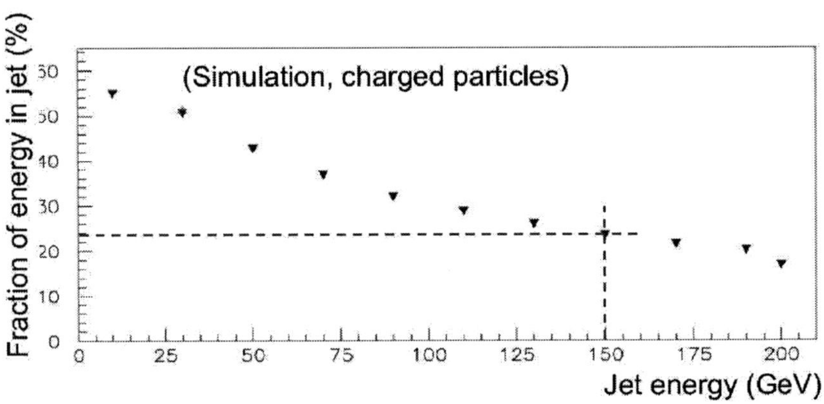

FIGURE 10.

The different particle separation has been performed using a Cerenkov counter[1] and the Transition Radiation Tracker[2] (TRT) that was in the beam line during the 2004 combined test-beam. Both TRT and Cerenkov counter were used to separate pions from

---

[1] Not an ATLAS detector.
[2] TRT is an ATLAS detector.

electrons. The scatter plot in Figure 11 shows on the x-axis the signal (in adc channel) in the Cerenkov counter and on the y-axis the number of hits in TRT. In TRT, pions are expected to produced less transition radiation X-ray than electrons. In the Cerenkov counter a lower signal is expected for pions than for electrons. Electrons and pions (together with muons) can be clearly identified on Figure 11. The following cuts have been applied to select pions : *Cerenkov* < 630 *adc* and *number of hits* ≤ 2 (grey lines on Figure 11).

Once electrons have been rejected we still have to deal with the contamination of high energy muons. High energy muons give a higher signal than very low energy pions in the last TileCal layer. Indeed VLE pions are expected to lose most of their energy before reaching the end of TileCal. High energy muons can be rejected asking less than a given energy in the last TileCal layer. Above 5 GeV this cut can introduce a bias because some pions can reach this layer. Nevertheless such a cut is easy to reproduce in the simulation. Figure 12 shows the distribution of the total energy as a function of the energy in the last TileCal layer. A large cloud with a total energy around 2 GeV is observed. This is the expected signal for high energy muons. The following cut has been chosen to select pions : *E in the last TileCal layer < 150 MeV* (grey line in Figure 12). Once pions have been separated from electrons and muons, the energy E_{raw} is determined fitting the total energy distribution with a Gaussian function. E_{raw} corresponds to the mean of the function. Again no corrections have been applied to take into account the dead material effects.

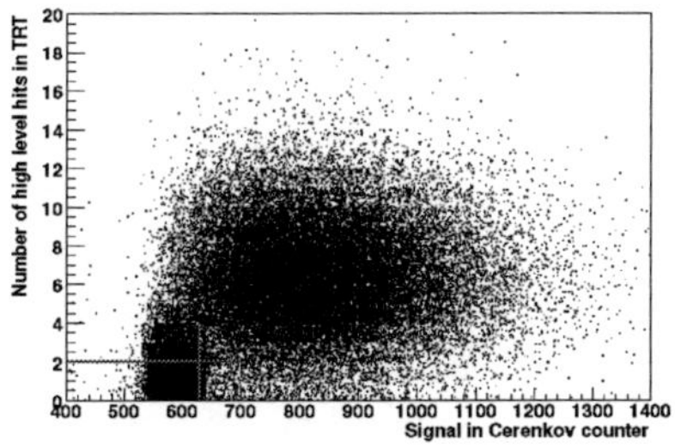

FIGURE 11. Correlation between the Cerenkov signal and the number of hits in TRT.

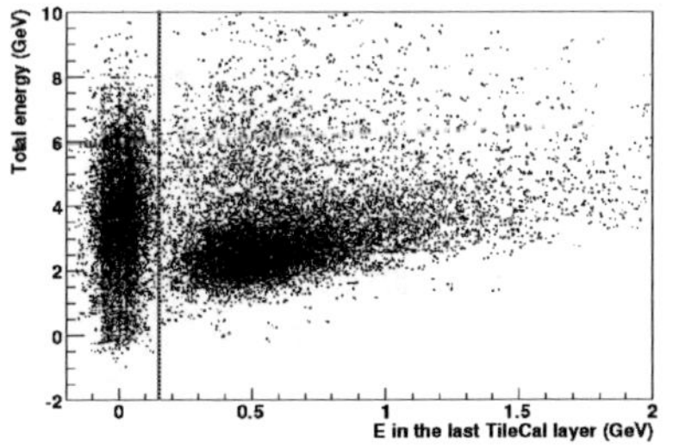

FIGURE 12. Correlation between the total energy and the energy in the last TileCal sample. The beam energy is 7 GeV.

Calorimeters response to VLE pions

Very low energy pions have been selected as described above for various energies and η positions. Figure 13 gives the dependence of the response (E_{raw}/E_{beam}) as a function of η. Due to the low number of pions available, the uncertainty on the response is dominated by the statistical error on the mean reconstructed energy. This statistical error varies from 1% at 9 GeV to 4% at 5 GeV. The error on the knowledge of the beam energy is 0.3%. Systematic effects of the selection cuts have to be studied.

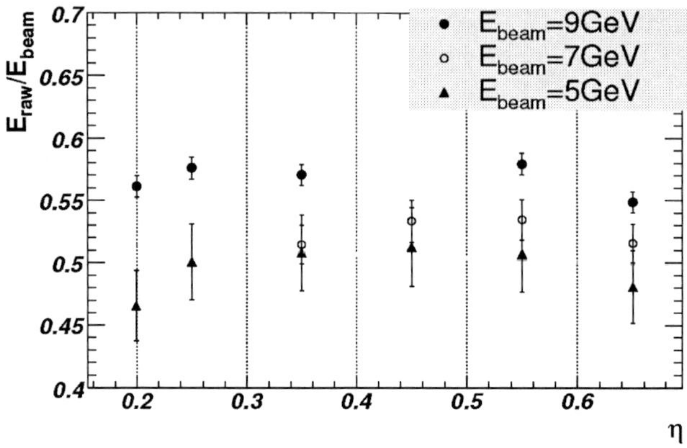

FIGURE 13. Response as a function of η for very low energy pions.

CONCLUSIONS AND OUTLOOKS

In this note results concerning the response of ATLAS central calorimeter to pions have been presented. Data collected during a CERN testbeam were used. The results were obtained studying :

- all high energy pions (energy larger than 50 GeV)
- high energy pions that start showering in TileCal
- very low energy pions (energy lower than 9 GeV)

The study of pions which start to produce a shower in TileCal has been performed in order to determine the hadronic absolute scale for TileCal and to compare TileCal performances in 2004 with the results obtained during previous standalone TileCal test-beams. From this analysis one can conclude that the resolution obtained in 2004 is compatible with the values obtained during previous test-beams. The uncertainty on TileCal response to pions is around 1% (dominated by the statistical error).

For the combined LAr+TileCal response to high energy pions, the uncertainty is less than 1%. It is dominated by the error on the beam energy.

For very low energy pions, the error on the LAr+TileCal response ranges from 1% at 9 GeV to 4% at 5 GeV. Work is still in progress for the determination of the systematic error introduced by the selection cuts (pion/electron separation, pion/muon separation, beam profile...).

At the end of the talk possible jet calibration strategy in ATLAS were discussed. The response of the ATLAS calorimeters has been obtained (in the 2004 testbeam configuration). The next step, which is already in progress, is to compare these results to the simulation in the same conditions (simulation of the test beam geometry). If needed the simulation will be tuned according to the testbeam results (amount of dead material, shower development...). Once a good agreement will be achieved, correction for dead material, leakage, e/π will be derived from simulation. Algorithms could be proposed in order to improve the calorimeters performances for single pions (like weighting techniques to improve linearity and resolution). The final objective is to obtain the jet absolute energy scale with a good linearity and resolution.

REFERENCES

1. S. Constantinescu, S. Dita, A. Henriques, M. Nessi, R Stanek, *Comparative analysis of the ATLAS tile hadronic calorimeter response to pions and protons*, TILECAL-2001-005, CERN, (2001)
2. V. Castillo, S. González, *Analysis of the 1998 test of the TileCal barrel prototype*, TILECAL-99-020, CERN, (1999)

Recent Results From The ATLAS Endcap Combined Test Beam

Peter Schacht

Max-Planck-Institut für Physik, Werner Heisenberg Institut, Föhringer Ring 6, 80805 Munich, Germany

Abstract. The ATLAS detector at the LHC at CERN is entering the last phase of construction, with the ATLAS calorimeter passing already the first steps of commissioning. The endcap region in the range of pseudorapidity $2.4 < |\eta| < 4.0$ is a particular complex one, with the electromagnetic, hadronic and forward calorimeters. In a dedicated beam run this region has been studied using the individual calorimeter modules. The set-up was as close as possible to the real ATLAS calorimeter, including all details like inactive support material structures. The goal is to validate the MC simulation for the different regions and to cross check the intercalibration of the various calorimeters. Electron and pion data have been taken in an energy range $6 < E < 200$ GeV with special emphasis on lateral and vertical scans to study the transition regions in detail.

Keywords: Calorimetry, Liquid Argon Calorimeter, LHC Detector.
PACS: 07.20.Fw

INTRODUCTION

Beam tests of calorimeters provide an important step in the calibration procedure required for the data analysis. The goals are:

- Obtain intercalibration constants for electron and pions in the energy range accessible. In particular the comparison with Monte Carlo (MC) simulation is of high importance in order to be able to extrapolate safely to different event topologies and energies.
- Study in detail the transition regions of complex structure to validate the MC simulation and to understand the impact on the determination of missing energy in the ATLAS detector.
- Test algorithms and constants for the hadronic calibration with real particles.

Calibration runs have been performed previously (2002) in the endcap region in the pseudorapidity range $1.6 < |\eta| < 1.8$ [1] and more recently (2004) in the pseudorapidity range $2.4 < |\eta| < 4.0$. These tests have been carried out at CERN in the H6 beam line with electrons and pions in the energy range $6 < E < 200$ GeV.

CP867, *XII International Conference on Calorimetry in High Energy Physics,*
edited by S. R. Magill and R. Yoshida

THE ATLAS CALORIMETER

The ATLAS calorimeter employs the liquid argon (LAr) technology except for the hadronic part in the central and outer forward region. Fig. 1 shows a schematic overview.

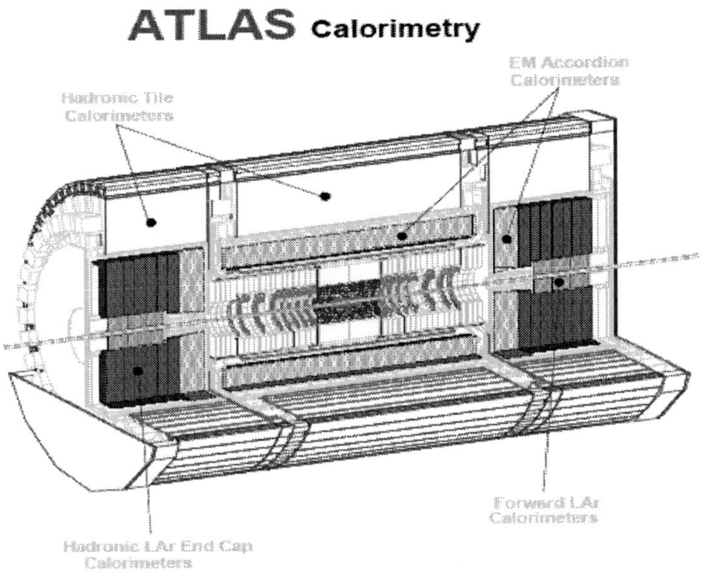

FIGURE 1. The ATLAS calorimeter, which is based on liquid argon technology except for the central barrel and outer forward region.

One barrel and two endcap cryostats comprise the LAr calorimeters [2]. The electromagnetic calorimeter consists of accordion shaped lead absorbers interleaved with electrodes and covers the region $|\eta| < 3.2$. The hadronic endcap calorimeter (HEC) is together with the electromagnetic endcap calorimeter (EMEC) and the forward calorimeter (FCAL) in the endcap cryostat. The absorber material for the HEC and the electromagnetic part of the FCAL is copper. For the hadronic sections of the FCAL tungsten is used as absorber. The pseudorapidity region covered by the HEC is $1.5 < |\eta| < 3.2$ and $3.2 < |\eta| < 4.9$ for the FCAL. The central (TILE barrel) and outer forward (TILE extended barrel) hadronic calorimeter is an iron scintillator sampling calorimeter. The typical depth of the ATLAS calorimeter is in the range 10-13 λ.

More details of the structure in the endcap region are shown schematically in Fig. 2. The separation between the outer and inner EMEC at $|\eta| = 2.4$ and details of the support structure for the FCAL are clearly visible.

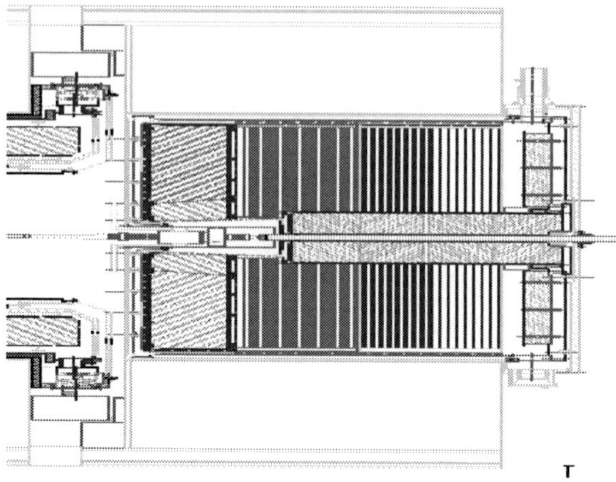

FIGURE 2. The endcap cryostat with the detailed structure and support of the electromagnetic (EMEC), hadronic (HEC) and forward (FCAL) calorimeters.

THE TESTBEAM SET-UP

The beam test carried out in 2004 corresponds as close as possible to the ATLAS region around $|\eta| = 3.2$, the complex region of overlap of the three endcap calorimeters. The EMEC (inner wheel), HEC and FCAL are positioned as in ATLAS, including all details of cryostat walls and supports (inactive material). The Fig.3 shows the schematic of the set-up in the cryostat in the H6 beam line at CERN. One quarter of the HEC1 (first wheel) and HEC2 (second wheel) wheels have been assembled, but with small modules which cover in size only the forward η region. Similarly, one EMEC inner wheel module (1/8 of the EMEC wheel) and ¼ of the FCAL1 and FCAL2 have been assembled. The set-up has been tilted to reproduce the average impact angle in ATLAS corresponding to $|\eta| = 3.2$ (see ATLAS axis). Also shown are typical vertical scan positions accessible with beam and within the cryostat window acceptance.

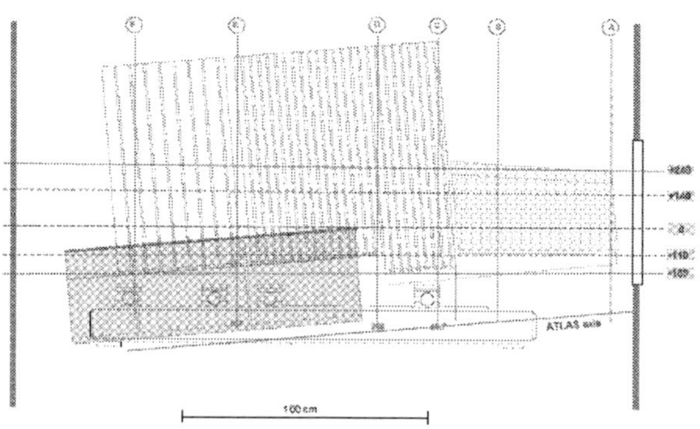

FIGURE 3. Schematic of the set-up in the H6 cryostat: shown are (from right to left) the inner EMEC wheel module, the HEC1 and HEC2 modules and the FCAL1 and FCAL2 modules. The beam enters

from the right. Also shown is the cryostat window and typical vertical scan positions (in mm) as well as the nominal ATLAS axis.

Fig. 4 shows the front view of the set-up with the geometrical structure of the individual read-out channels of the HEC, EMEC and FCAL. The lines indicate the typical horizontal and vertical scan ranges, the squares the regions with high statistics energy scans and local spatial scans.

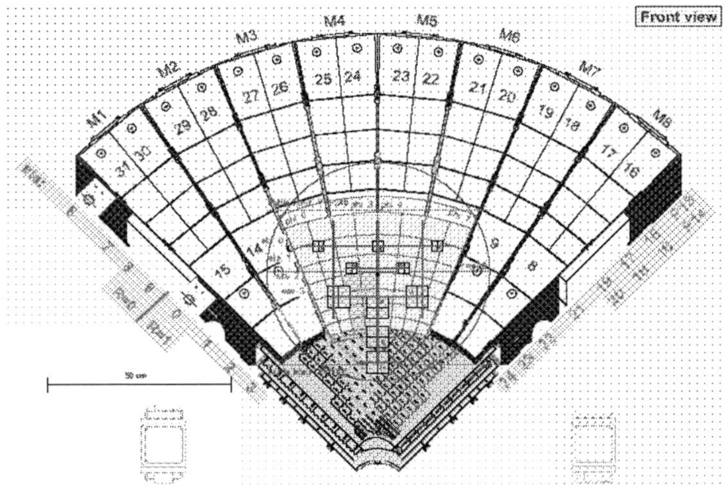

FIGURE 4. Front view of the testbeam set-up showing the individual EMEC, HEC and FCAL read-out structure. The lines and marked squares show the most important scan regions.

ELECTRON RESULTS

Electron data are used not only to assess the detector performance but also to obtain the electromagnetic scale for the related calorimeter. Also details of the inactive material in the crack between the EMEC and FCAL can be verified using MC simulation. Therefore large statistics samples of electron data have been taken in the accessible energy range $6 < E < 200$ GeV.

Electrons in EMEC

An important feature of the EMEC is the variation of the LAr gap size with η due to the specific electrode structure. Partially this is compensated by a variation of the high voltage (HV), but only in defined steps in η. For the inner EMEC wheel there are two regions of constant HV: $2.4 < |\eta| < 2.8$ and $2.8 < |\eta| < 3.2$. Within each zone a cell dependent correction has to be established, based on known models of the electric field variation:

$$E_{corr} = E_{cell} * \frac{\beta}{1 + \alpha * (\eta_{cell} - \eta_0)}$$

From a large number of electron runs covering many readout cells and different energies these parameters have been defined from a fit, very close to model expectations: α = 0.55 and β = 1.0 (high |η|) and β = 1.04 (low |η|). To reconstruct the signal response the energy deposited in a fixed size cluster in Δη×Δφ has been used: either a cone of ΔR = 0.15 or ΔR = 0.25. Fig. 5 shows the energy dependence of the energy resolution for these clusters (upper points for ΔR = 0.15, lower points for ΔR = 0.25). The solid points show the data, the open points the MC prediction. Parameterizing the energy dependence according to $\sigma/E = a / \sqrt{E} + b$ yields typically 10 % for the sampling term a and about 1 % for the constant term b. The larger cluster, which picks up more of the tails of the energy deposition of the electron, gives a slightly better resolution. But additional corrections, e.g. due to the φ variation of the response (absorber structure), dead material in front or cluster leakage corrections, are not yet applied, resulting in the non negligible constant term of 1%. The MC prediction, where the electronic noise is neglected, yields a somewhat better resolution. Fig. 6 shows the corresponding linearity of the response (upper points for ΔR = 0.25, lower points for ΔR = 0.15). Again, the solid points show the data, the open points the MC prediction. The larger cluster, which has been used for the normalization, stays close to one, except for low energies where the dead material in front results in a variation of 2-3%. The MC prediction is rather close to the data.

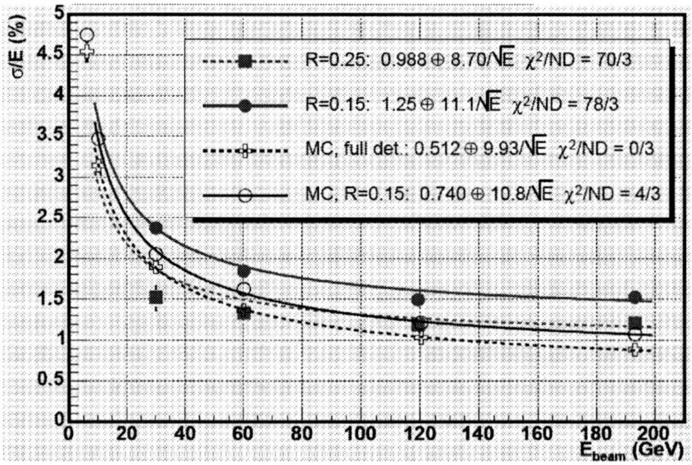

FIGURE 5. Energy dependence of the energy resolution for fixed size electromagnetic clusters in the EMEC. The solid points show the data, the open points the MC prediction. The upper points are for a cone size in Δη×Δφ of ΔR= 0.15, the lower for ΔR= 0.25. The lines show the fit using the parameterization $\sigma / E = a / \sqrt{E} + b$.

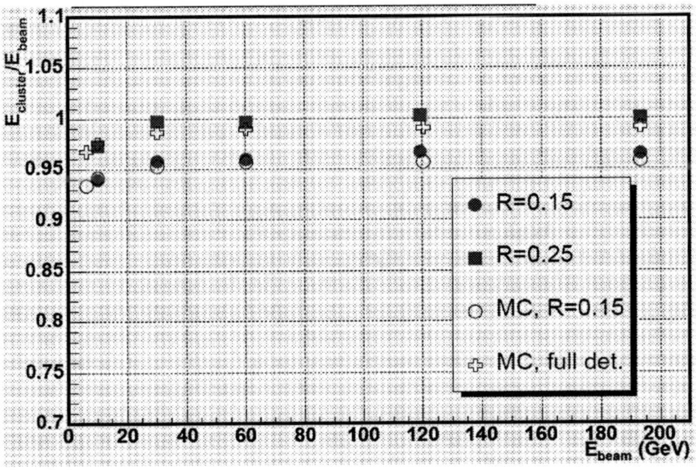

FIGURE 6. Linearity of the response for fixed size electromagnetic clusters in the EMEC. The solid points show the data, the open points the MC prediction. The upper points are for a cone size in $\Delta\eta\times\Delta\varphi$ of $\Delta R= 0.25$, the lower for $\Delta R= 0.15$.

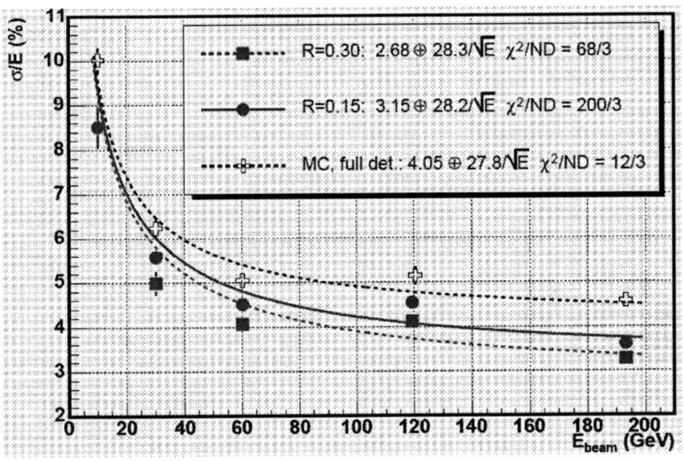

FIGURE 7. Energy dependence of the energy resolution for fixed size electromagnetic clusters in the FCAL. The solid points show the data, the open points the MC prediction. The upper points are for a cone size in $\Delta\eta\times\Delta\varphi$ of $\Delta R= 0.15$, the lower for $\Delta R= 0.30$. The lines show the fit using the parameterization $\sigma / E = a / \sqrt{E} + b$.

Electrons in FCAL

In a similar way the response of electrons has been studied in the FCAL. Again, clusters have been reconstructed in a cone of $\Delta\eta\times\Delta\varphi$, not applying any corrections due to cluster leakage or dead material in front. Fig. 7 shows the energy dependence of the energy resolution for a typical FCAL impact point for the two cone sizes of $\Delta R = 0.15$ (upper points) and $\Delta R = 0.30$ (lower points). Again, the solid points show the data, the open points the MC prediction. Parameterizing the energy dependence of the energy resolution by $\sigma/E = a / \sqrt{E} + b$ yields typically 28 % for the sampling term a and about 2.7-3.2 % for the constant term b. This result is slightly better than the MC expectations (shown for the full detector only). Fig. 8 shows the corresponding linearity of the response (upper points for ΔR = 0.30, lower points for ΔR = 0.15). Again, the solid points show the data, the open points the MC prediction for the full detector. The data are well described by the MC prediction: at low energies the dead material in front of the calorimeter starts to reduce the response.

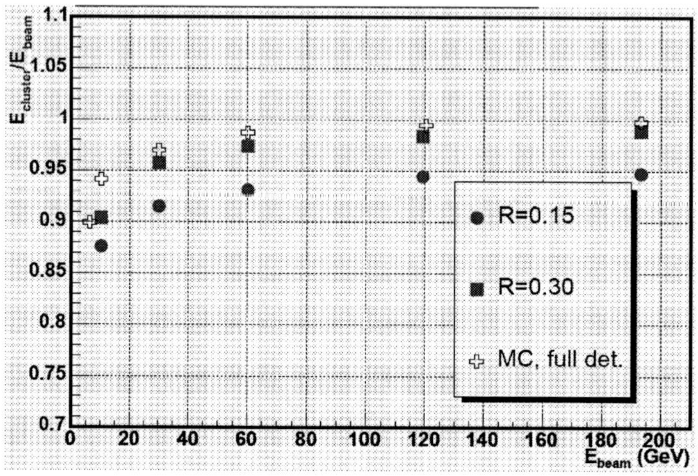

FIGURE 8. Linearity of the response for fixed size electromagnetic clusters in the FCAL. The solid points show the data, the open points the MC prediction. The upper points are for a cone size in $\Delta\eta\times\Delta\varphi$ of ΔR= 0.25, the lower for ΔR= 0.15.

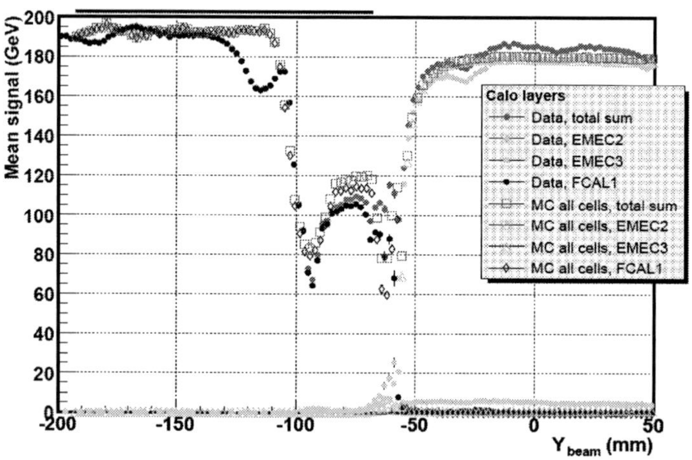

FIGURE 9. Vertical scan with electrons of 193 GeV from the EMEC (y > 50 mm) to the FCAL (y < - 150 mm). The data (solid points) are compared with the MC prediction (open squares). The highest

points correspond to the total sum, the lower to the first and second longitudinal section of the related calorimeter.

Vertical Scan with Electrons in the Crack between EMEC and FCAL

Finally the variation of the response of electrons has been compared to MC simulation in the full range from the EMEC, the crack between the EMEC and FCAL, and the FCAL. The vertical scan has been carried out at the horizontal position x=0 with electrons of energy E=193 GeV. Fig. 9 shows this comparison between the data (solid points) and MC prediction (open points). Shown is the total energy reconstructed (electromagnetic scale) as well as the energy deposited in the individual longitudinal sections of the EMEC and FCAL. For the cluster reconstruction the cone size of $\Delta R = 0.25$ in $\Delta\eta\times\Delta\varphi$ has been used. The MC simulation (full detector) describes the data well in the EMEC (y>50) and FCAL (y<-100) regions. The deviations of the data from the MC prediction in few regions can be traced back to specific problems of a weak response for these channels. In the crack region geometrical details, like exact position of LAr excluders and details of additional support structures need still to be improved.

PION RESULTS

Pion data have been taken in the full energy range available. The most important impact points used are indicated by squares in Fig.4. In addition large statistics data with horizontal and vertical scans (see Fig. 4) have been taken as well, mostly at high energies. Given the limited acceptance of the set-up, lateral as well as longitudinal leakage are an important issue. Typically 10 % leakage is expected at 200 GeV, increasing to more than 20 % at the lowest energies. Therefore the data have been compared to MC simulation wherever possible. In the MC the noise has been ignored.

Pions in EME/HEC

For the energy reconstruction of pions a cluster size of typically $\Delta R = 0.30$ or $\Delta R= 0.40$ in $\Delta\eta\times\Delta\varphi$ has been used. For the MC also the full calorimeter acceptance has been studied. Fig. 10 shows the pion response on the electromagnetic scale as function of energy in the region of EMEC/HEC. Shown are the data in the impact point D (see Fig. 4). The MC prediction (open points) yields a somewhat larger prediction in comparison to the data (solid points). Clearly visible is the reduction of the signal from full calorimeter acceptance to the cone size of $\Delta R = 0.4$. In a next step the energy resolution has been compared to the MC prediction. To start with, the 'baseline approach' has been applied, defining one calibration constant per calorimeter type from the fit minimizing the energy resolution. The obtained energy dependence is shown in Fig. 11. For the given cone size of $\Delta R = 0.30$ the data show a somewhat worse resolution in comparison to the MC simulation. The difference could be due to differences in the energy leakage, which are for this setup substantial. For the MC at least the effect due to the lateral leakage is seen when going from the full acceptance to the limited cone in $\Delta\eta\times\Delta\varphi$. More important is the longitudinal leakage, where we would expect from previous testbeam studies[3] some difference between data and MC. At low energies the contribution due to electronic noise, which has not been simulated in the MC, might also have a strong influence. Further studies rejecting events with strong leakage and finally applying the proposed hadronic weighting schemes[1] are being carried out presently. Therefore we expect the energy resolution to come very close to the prediction obtained for the full ATLAS simulation.

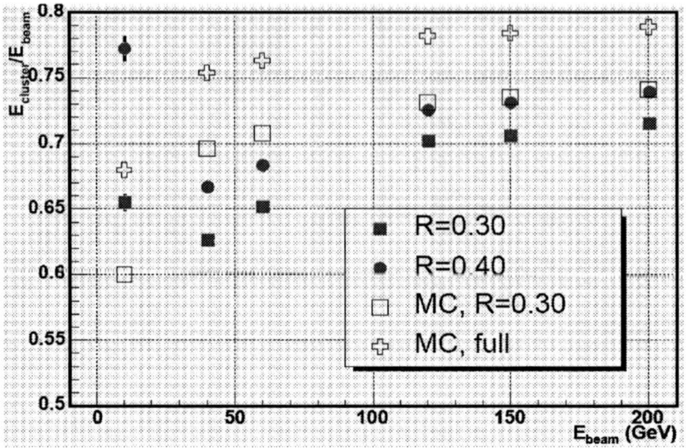

FIGURE 10. Energy dependence of the pion response (electromagnetic scale) in the EMEC/HEC region. Shown is the ratio of the cluster to the beam energy for different cone sizes ΔR in Δη×Δφ. The MC simulation (open points) yields a somewhat larger prediction in comparison to the data (solid points), eventhough the trend of the data is reproduced well.

Pions in FCAL

Again, the response of pions on the electromagnetic scale has been compared to MC simulation. The Fig. 12 shows the energy dependence of the ratio to the beam energy for the impact point H (see Fig. 4). The MC (open points) yields a good description of the data (solid points), except at very low energies where the noise contribution cannot be neglected. The energy dependence of the energy resolution is shown in the Fig. 13. Again, the 'baseline approach' (see above) has been used to compare the data with MC simulation. The MC prediction, where the electronic noise contribution is ignored, describes the data rather well, except for low energies. Also here the substantial difference between the full calorimeter response and the cone in Δη×Δφ is clearly visible in the MC data. Again, applying hadronic weighting schemes, as proposed for ATLAS, we expect a considerable improvement of the energy resolution.

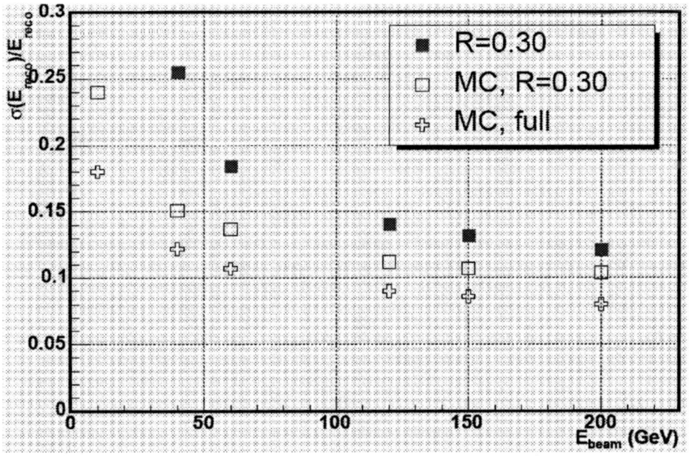

FIGURE 11. Energy dependence of the energy resolution for pion for a cluster size of $\Delta R = 0.30$ in $\Delta\eta\times\Delta\varphi$. The data (solid points) in the EMEC/HEC region are compared with the MC prediction (open points). For the MC prediction the results for the full calorimeter acceptance are given as well.

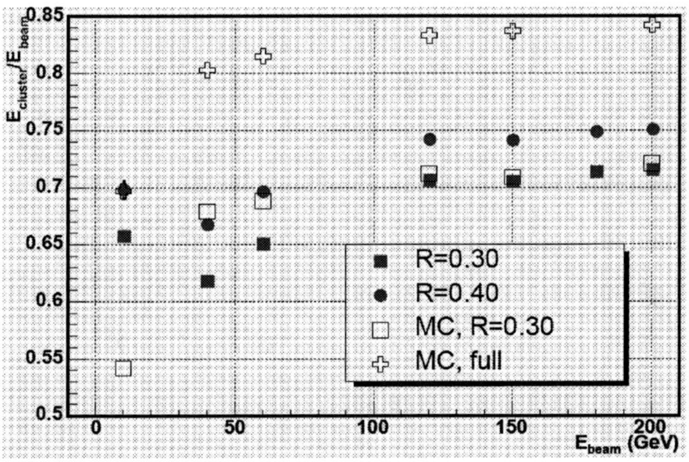

FIGURE 12. Energy dependence of the pion response (electromagnetic scale) in the FCAL. Shown is the ratio of the cluster to the beam energy for different cone sizes ΔR in $\Delta\eta\times\Delta\varphi$. The MC simulation (open points) describes the data (solid points) well, except for low energies where the noise contribution cannot be neglected.

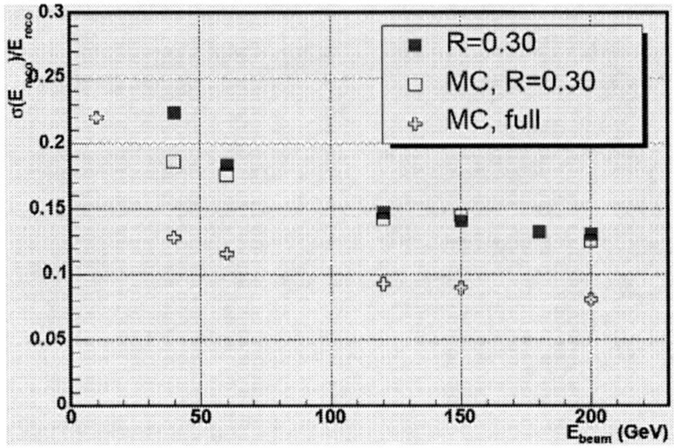

FIGURE 13. Energy dependence of the energy resolution for pions for a cluster size of ΔR = 0.30 in Δη×Δφ. The data (solid points) in the FCAL region are compared with the MC prediction (open points). For the MC prediction the results for the full calorimeter acceptance are given as well.

Vertical Scan with Pions in the Crack between EMEC/HEC and FCAL

Vertical scans across the crack between EMEC/HEC and FCAL have been carried out with pions in a similar way as with electrons. The Fig. 14 shows the total response of pions of 200 GeV on the electromagnetic scale summed over all calorimeters. For the MC simulation the full calorimeter acceptance has been used, for the data only the signal within a cone of ΔR = 0.30 in Δη×Δφ. Therefore some care has to be taken when comparing the data with simulation. The differences seen are partially masked by this cut, in addition to some difference in the pion response on the electromagnetic scale in the two calorimeter regions. But most important, the typical signal losses in the critical crack region are not exceeding the level of 20 %- 25 %.

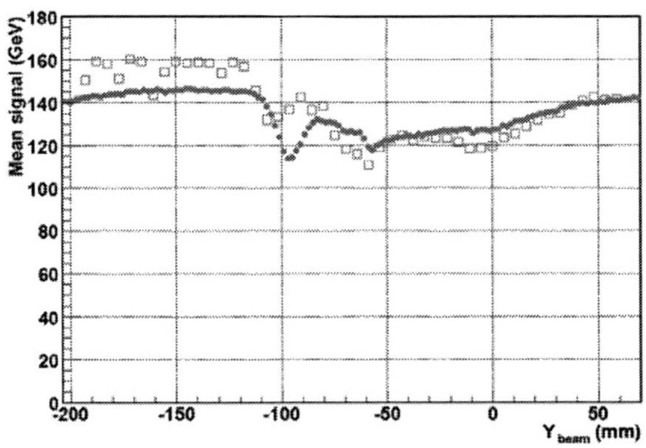

FIGURE 14. Vertical scan with pions of 200 GeV from the FCAL region (left) across the crack up to the EMEC/HEC region (right). Shown is the total response on the electromagnetic scale, where for the

MC simulation the full calorimeter acceptance has been used but for the data only the response within a cone of $\Delta R = 0.30$ in $\Delta\eta\times\Delta\varphi$.

Conclusions

The performace of electrons and pions in the ATLAS endcap region of pseudorapidity $2.4 < |\eta| < 4.0$ has been studied in beamtests in an energy range up to 200 GeV. The set-up with specially built modules was reflecting this particular complex region, with the electromagnetic, hadronic and forward calorimeters in full detail. The data have been compared to the MC prediction. The preliminary results show a rather good agreement in all η-regions, but more detailed studies are ongoing.

ACKNOWLEDGMENTS

This work is based on many studies done by the ATLAS H6 combined testbeam analysis group. We greatfully acknowledge the many contributions, in particular we would like to thank M. Bieri, P. Cavalleri, A. Minaenko, W. Shaw and P. Strizenec for their help and their effort in the analysis of the testbeam data.

REFERENCES

1. C. Cojocaru et al., Hadronic calibration of the ATLAS liquid argon end-cap calorimeter in the pseudorapidity region $1.6 < |\eta| < 1.8$ in beam tests, Nuclear Instruments and Methods A, NIM A 531, 2004, pp. 481-514.
2. ATLAS Liquid Argon Calorimeter Technical Design Report, CERN/LHCC 96-41, 1996.
3. A.E. Kiryunin et al., GEANT4 physics evaluation with testbeam data of the ATLAS hadronic end-cap calorimeter, Nuclear Instruments and Methods A, NIM A 560, 2006, pp. 278-290.

Test beam results on the performance of the CMS electromagnetic calorimeter

Alexandre Zabi

Imperial College, London, UK. Alexandre.Zabi@cern.ch.
On behalf of the CMS ECAL group

Abstract.
The performance of the CMS electromagnetic calorimeter has been studied with the test beam data recorded with a fully equipped barrel supermodule. An efficient method for signal amplitude reconstruction has been implemented and tested on these data. Results on energy resolution and electronics noise show excellent performance of the apparatus. Effects attributed to shower containment in crystal arrays were investigated and a set of corrections has been formulated and used to provide results on energy resolution.

Keywords: CMS, ECAL, Test beam, amplitude, containment, resolution
PACS: 29.40.Vj

1. INTRODUCTION

1.1. CMS electromagnetic calorimeter

The Compact Muon Solenoid (CMS) [1] detector is a general purpose detector to be installed at the 14 TeV proton-proton collider, LHC, under construction at CERN and due to start operation in 2007. The CMS electromagnetic calorimeter (ECAL) [2] is a hermetic homogeneous calorimeter made of 61,200 scintillating lead tungstate ($PbWO_4$) crystals [3] mounted in the central barrel part, closed by 7,324 crystals in each of the two end-caps. The barrel consists of 36 supermodules, each covering half the barrel length and 20° in azimuth, containing 1700 crystals together with the associated front end electronics.

1.2. Test beam experimental set-up

Test beam data taken in the H4 beam line at CERN in October and November 2004 using electron beams with a range of momenta between 20 and 250 GeV/c, have been used to investigate the performance of the ECAL. A supermodule fully equipped with the final readout electronics [4] was installed on a movable table which allowed the beam to be directed at any part of it. Electrons were incident at an angle of $\simeq 3°$ to the direction of the crystal axis in both transverse directions reproducing the average incident angle of particles emerging from the collision region in LHC running. Plastic scintillator counters were used to trigger the readout. The trigger defines an area of 20×20 mm^2. This area is only slightly smaller than the transverse section of the crystals which have a front section

CP867, *XII International Conference on Calorimetry in High Energy Physics,*
edited by S. R. Magill and R. Yoshida

of $\simeq 22 \times 22$ mm^2 and a rear section of $\simeq 25 \times 25$ mm^2. The position of the incident electrons in the transverse directions was determined by four planes of scintillating fibre hodoscopes.

1.3. Readout electronics

The electrical signal from the photodetectors is shaped by a preamplifier and amplified by a multi-gain amplifier (MGPA) before digitization by ADCs at a frequency of 40 MHz [5]. It provides a dynamic range from about 35 MeV to saturation at 1.7 TeV in the barrel. The data read out consists of a series of 10 consecutive digitizations, corresponding to a sequence of samplings of the signal at 40 MHz. The signal amplitude must be reconstructed using these samplings.

2. RECONSTRUCTION OF THE SIGNAL AMPLITUDE

2.1. The weights method

The method used to reconstruct the amplitude from the digitized samples is based on a digital filtering technique. The signal amplitude, $\hat{\mathscr{A}}$, is computed from a linear combination of discrete time samples (see Equation 1).

$$\hat{\mathscr{A}} = \sum_{i=1}^{N} w_i \times S_i \tag{1}$$

where w_i are the weights, S_i the time sample values in ADC counts and N is the number of samples used in the filtering with the index, i, running over the time samples. The weights w_i are obtained by minimizing the variance of $\hat{\mathscr{A}}$ [6]. When considering the simplest case (no noise correlations between time samples) the optimal weights are given by the formula $w_i = f_i / \sum_{j=1}^{N} f_j^2$, where f_i is the value of the function, $f(t)$, describing the time development of the signal pulse in time, t, at the time t_i of sample i. A form of the function, $f(t)$, that provides a good description of the electronics signal is a digital representation (profile histogram) directly built from the test beam data (see Fig. 1). The determination of the optimum function to be used is addressed in Section 2.3.

Various implementations of the weights method have been tested using test beam data and their ability to reduce the noise contribution is used to discriminate between them.

2.2. Measurement of the electronics noise

In the following, we compare the final electronics noise obtained when implementing a "5-weights" and a "3+5-weights" method. The 5-weights method uses the sample on the peak, the one before and the 3 after while the 3+5-weights also uses the 3 pre-samples (before any signal appears). The first implementation requires the pedestal to

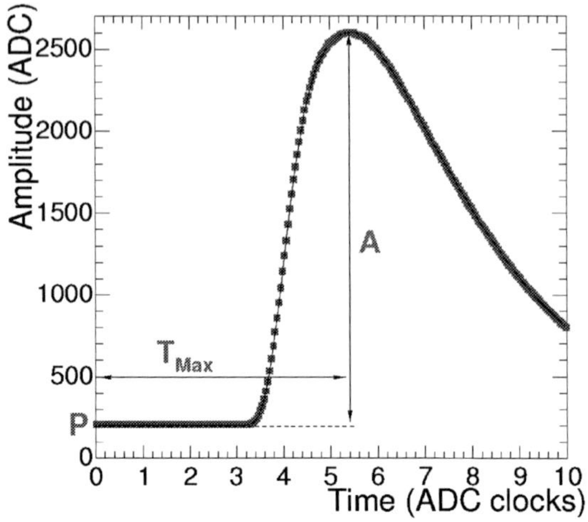

FIGURE 1. Profile of the signal pulse from a crystal of the supermodule using an electron beam of 120 GeV. The peaking time T_{Max}, the pedestal P and the amplitude of the signal A are shown.

be determined independently and subtracted prior to reconstruction, while the second allows a subtraction of the pedestal on an event by event basis (the so-called "pedestal subtracting weights"). For completeness, a 3+1 weights pedestal-subtracting implementation has also been studied. The single weight is then applied to the sample on the peak. This allows the demonstration of the impact of dynamic baseline subtraction in its simplest form. Table 1 summarizes the ECAL noise measured by examining the variation of the reconstructed amplitude for data taken with random triggers, when no signal is present.

TABLE 1. Noise in ADC counts measured in a single channel and arrays of 3×3 and 5×5 channels for three different implementations of the amplitude reconstruction method, and for a single sampling. 1 ADC count ≃ 37 MeV.

Method	Noise (ADC counts)		
	1×1	**3×3**	**5×5**
Single sampling	1.20±0.01	3.7±0.04	6.5±0.1
5-weights	1.11±0.02	3.8±0.1	6.7±0.1
3+1 Pedestal subtracting weights	1.13±0.03	3.4±0.1	5.7±0.1
3+5 Pedestal subtracting weights	1.07±0.02	3.2±0.1	5.4±0.1

The results given in the table show that the lowest noise is achieved by the pedestal-subtracting weights. This implies the presence of low frequency noise which is removed by the dynamic pedestal subtraction. Comparing the noise seen in a single channel with that seen in sums of 3×3 and 5×5 channels shows a small contribution of noise correlations between channels. This is entirely removed by the use of the pedestal subtracting weights. This implementation is used for reconstructing the signal amplitude. Using five samples in the signal, rather than one, results in a slightly lower value of noise as compared to the 3+1 weights.

2.3. Implementation of the reconstruction method

As mentioned in Section 2.1, the determination of the weights requires knowledge of a signal representation $f(t)$. The impact of the choice of this function on the energy resolution has been studied with test beam data. Two parameters characterize $f(t)$: the signal timing T_{Max} and the parameter C which represents the width of the pulse.

In the test beam the scintillation signals have a random phase with respect to the 25 ns ADC clock interval, while during LHC operation the ADC clock will be synchronous with the bunch crossing. The relevance of the choice of $f(t)$ has been investigated in both cases. To simulate synchronous running, test beam data have been analyzed taking only events in a single 1 ns bin of phase. Figure 2 shows the energy resolution measured when the weights used to reconstruct the amplitude are derived from signal representations with peaking times and widths which differ from those of the actual signal.

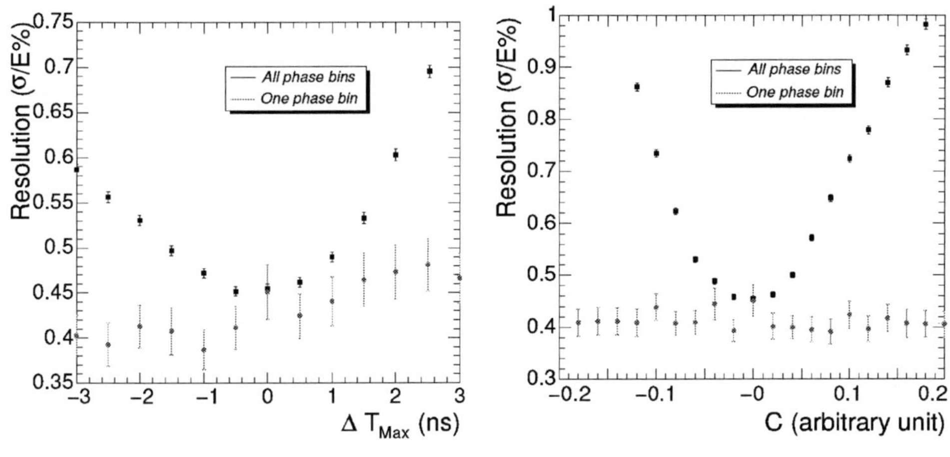

FIGURE 2. Left: Energy resolution in a 3×3 matrix of crystals measured with a 120 GeV electron beam, as a function of the timing difference (ΔT_{Max}). Right: Energy resolution as a function of the contraction factor, C ($t \to t + \mathrm{C} \times (t - T_{\mathrm{Max}})$). The black square points correspond to the summing of all bins in phase as in the asynchronous data taking (Test beam). The red dots correspond to the synchronous mode (CMS) which is reproduced by selecting a single 1 ns bin in phase.

These results suggest that, in the case of synchronous running (such as in CMS operation), the same reference signal representation can be used to derive a single set of weights to be used for the signal amplitude reconstruction of all channels. The uniformity of the readout electronics allows this important simplification. Reconstructing the signal pulse amplitude in test beam is sensitive to the individual channel signal timing and shape due to the asynchronous nature of the data taking. When the weights are derived using the signal description specific to each channel, the resolution is as good as the one obtained in a single 1 ns bin.

3. SHOWER CONTAINMENT VARIATION WITH POSITION

The energy contained in 3×3 and 5×5 matrices of crystals depends on the particle incidence position on the central crystal. The variation of the energy containment with respect to this position was studied with electron test beam data. A correction based on the balance of energy between crystals can be applied to recover a uniform energy response [7].

3.1. Dependence of the measured energy with the position

Figure 3 displays the reconstructed energy summed in 3×3 and 5×5 arrays of crystals centered on three adjacent crystals, as a function of the η coordinate measured by the hodoscope. The maximum response has been normalized to the beam energy (120 GeV) in each case. The peak to peak variation of the measured energy contained in a matrix decreases from 2.5% for a sum of 9 to 1.6% for a sum of 25 crystals. Similar distributions can be obtained when plotting the energy as a function of ϕ.

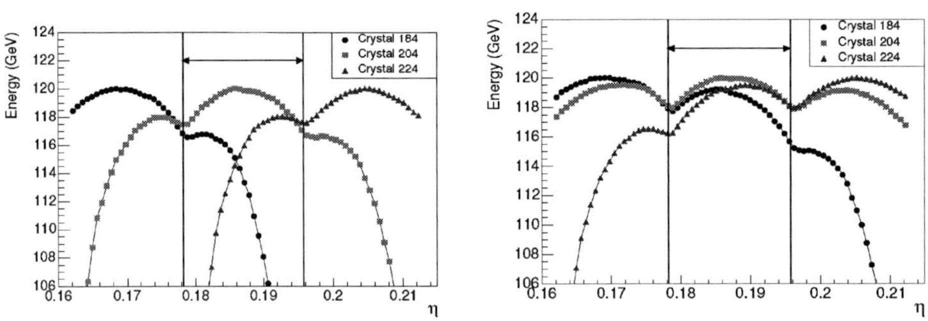

FIGURE 3. Left: Energy in 3×3 matrices centered on crystals 184 (dots), 204 (squares), 224 (triangles) as a function of η; Right: same for 5×5 matrices. The double arrow shows the extent of a single crystal.

3.2. Parametrization of the response with position

The energy balance between crystals can be used to determine the position of incidence. As illustrated by Fig. 4 (left), this method uses the ratio of the energy contained in the side rows (or columns) of crystals to the energy contained in the rest of the matrix (including the central row/column).

Figure 4 (right) shows the energy variation parametrized as a function of position, measured by the variable ln(E2/E1), using test beam data. In order to better determine this parametrization, data have been recorded with the beam directed on the edge of the centre crystal. These functions are used to correct the energy on an event by event basis. An asymmetry can be observed between the distributions in the regions close to the two opposite edges which is due to the 3° off-pointing of the crystals. A single correction function was used for all energies and for all regions in the supermodule.

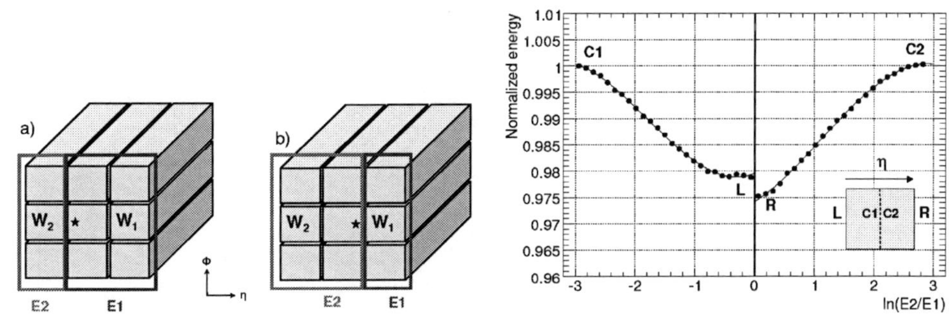

FIGURE 4. Left: Definition of clustered crystal energy into E1 and E2 in the case where W1<W2 (a) and W1>W2 (b) (W1 and W2 are the energies of the crystals on either side of the centre crystal). Right: Normalized sum of energy measured in 9 crystals versus ln(E2/E1) in the η views for crystal 204 ($\eta = 0.187$). The crystal 204 has always the maximum energy. The lines are third order polynomial functions fitted to the measured distributions (independently for positive and negative values of ln(E2/E1) because of the asymmetry mentioned before).

4. ENERGY RESOLUTION

Results on energy resolution are given for both electrons in a small region of incidence and for electrons with uniform incidence over the calorimeter surface. The electron-induced shower energy is reconstructed in matrices of 3×3 crystals centred on the crystal containing the maximum energy. Corrections for variations in the shower containment are applied in the case of uniform incidence.

4.1. Intrinsic energy resolution

As explained in Section 3, the electron shower energy contained in a crystal matrix depends on the shower position. Investigating the energy resolution in the central incidence region, defined as a limited 4×4 mm^2 region around the point of maximum containment, gives a measure of the intrinsic resolution of the calorimeter. In this region, the containment is almost constant.

In the following, results are given for runs with 30 000 events taken in crystal 704, located at pseudorapidity $\eta = 0.62$. The energy spectra for 120 GeV electrons reconstructed in a 3×3 matrix centred on crystal 704 is shown in Fig. 5 (left). The energy scale in these plots is obtained by normalizing the mean signal amplitude to the beam momentum. The crystals were intercalibrated with constants determined directly from the test beam data.

The energy dependence of the resolution is studied for data taken at a series of beam energies: 20, 30, 50, 80, 120, 180 and 250 GeV. The energy dependence is fitted by the following functional form :

$$\left(\frac{\sigma}{E}\right)^2 = \left(\frac{S}{\sqrt{E}}\right)^2 + \left(\frac{N}{E}\right)^2 + C^2 \tag{2}$$

where S is the stochastic term, N the noise and C the constant term (E is in GeV). The fit is performed with the noise term fixed at its measured value. The energy resolution values for a 3×3 matrix centred on crystal 704 is shown in Fig. 5, together with the fitted resolution function.

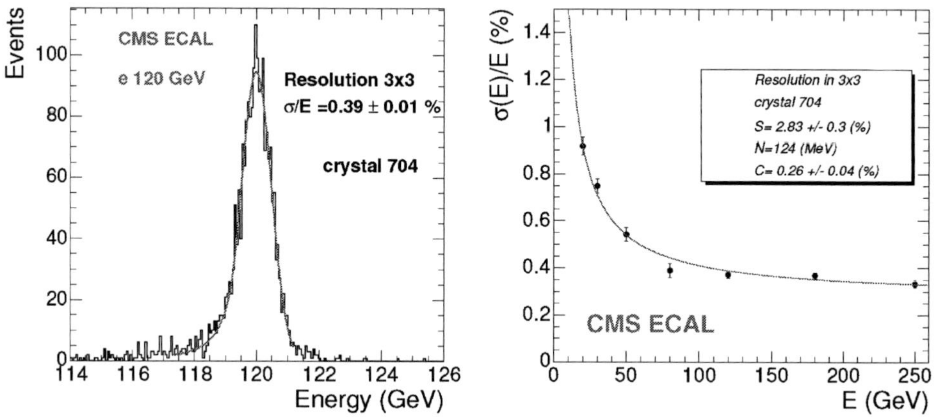

FIGURE 5. Left: Distribution of energy reconstructed in a 3×3 crystal matrix centered on crystal 704 for electrons at 120 GeV. Right: Resolution as a function of the energy reconstructed by summing for the same matrix after hodoscope cuts of 4×4 mm^2.

4.2. Energy resolution for uniform impact

Figure 6 shows the effect of shower containment corrections summing data from many runs where the incident beam is impacting at many locations on an array of 9 crystals. This configuration reproduces quite well a uniform coverage of an array of 9 crystals. The measured relative resolution at 120 GeV is 0.50%.

5. CONCLUSION

Test beam data taken in 2004 allowed exhaustive performance studies to be made on one CMS ECAL supermodule. Reconstruction of the signal amplitude in the CMS ECAL can be performed by a digital filtering technique. An implementation with dynamic pedestal subtraction gives a noise in a single channel of 40 MeV, and coherent noise

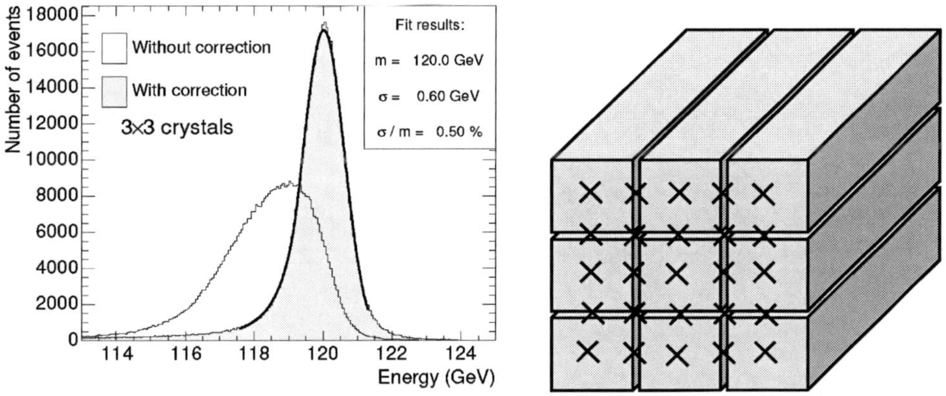

FIGURE 6. Distribution of energy before (unshaded) and after shower containment corrections (shaded) when the incident beam is centered at many locations of an array of 3×3 crystals as shown by the black crosses on the right side of the Figure.

between channels is effectively removed. In synchronous running, as will be the case at LHC, a single set of weights can be used to reconstruct the amplitude in all channels.

Consistent energy resolution is obtained for electrons with central incidence in different areas of the calorimeter. The resolution obtained using the energy reconstructed in a matrix of 3×3 crystals is $0.40\pm0.01\%$ for electrons of 120 GeV. A typical energy dependent resolution for central incidence gives a stochastic term of 2.8% and a constant term of 0.3%. For a uniform incidence, shower containment can be effectively corrected to give a resolution better than 0.5% for electrons of energy greater than 100 GeV. The resolution measured for the supermodule is compatible with design expectations.

REFERENCES

1. CMS Collaboration, *"The Compact Muon Solenoid Technical Proposal"*, CERN/LHCC 94-38, 1994
2. CMS Collaboration, *"The Electromagnetic Calorimeter Technical Design Report"*, CERN/LHCC 1997-033, 1997
3. A.A. Annenkov, P. Lecoq and M.V. Korzhik, *"Lead Tungstate scintillation material"*, Nucl. Instr. and Meth. **A490** (2002) 30
4. CMS Collaboration, *"CMS Physics Technical Design Report Vol. 1"*, CERN/LHCC 2006-001 (February 2006) 547pp
5. Raymond, M. *et al.*, *"The MGPA Electromagnetic Calorimeter Readout Chip for CMS"*, Proceedings of the 9th Workshop on Electronics for the LHC Experiments, CERN-LHCC-2003-055, 2003
6. CMS ECAL Group, *"Reconstruction of the signal amplitude of the CMS electromagnetic calorimeter"*, The European Physical Journal C, Vol 46, Sup 1, 2006, pp:23–35
7. J. Descamps and P. Jarry, *"Performance of the cluster containment correction for the CMS electromagnetic calorimeter barrel"*, CMS Note 2006/045.

Recent Results of the ATLAS Barrel Combined Test Beam

Marco Delmastro[a,b]
(on behalf of the ATLAS Liquid Argon Collaboration)

[a] *CERN, Geneva, Switzerland*
[b] *Università degli Studi and INFN, Milano, Italy*

Abstract.
In summer 2004 a full slice of the ATLAS detector – including all detector subsystems from the inner tracker, the calorimetry to the muon system – was exposed to particle beams (electrons, pions, photons, muons, protons) with different energies (from 1 GeV to 350 GeV). The aim was to study the combined performance of the different detector subsystems in ATLAS-like conditions.

We present the electronics calibration scheme of the electromagnetic calorimeter (EMC) and its complete implementation. The calibrated response of the EMC is compared to the Monte Carlo simulations, showing that a very good description of the data has been achieved. Results of the combined data analysis are presented, focusing on the combined reconstruction of converted photons using the EMC and the ATLAS inner tracker.

Keywords: ATLAS Calorimetry, Electronics Calibration, Signal Reconstruction, Combined Performances.
PACS: 29.40.Vj

1. THE ATLAS BARREL 2004 COMBINED BEAM TEST

Goals of the 2004 ATLAS combined testbeam (CTB) were to demonstrate that all the ATLAS sub-detectors in a barrel slice can acquire and analyze combined data using ATLAS-like read-out electronics and online/offline software, and to confirm the previously measured subdetectors' performance.

The ATLAS inner detector (ID) components, the liquid Argon (LAr) electromagnetic calorimeter (EMC), the Tile hadronic calorimeter and the muon spectrometer system were accommodated on the CERN H8 beam line at the North Area. The set-up of the combined testbeam is shown in Figure 1 and described in details in [1].

The CTB was an unprecedented occasion for the ATLAS LAr collaboration to test the full read-out chain (front-end electronics, back-end electronics, online software, DAQ system, offline reconstruction tools) of the EMC using the final ATLAS components. The complete setup and readout of the LAr EMC module used during the CTB is described in details in [2].

2. ELECTRONICS CALIBRATION OF THE LAR EMC

Charged particles in the electromagnetic or hadronic shower developing in the LAr EMC ionize the liquid argon as they traverse the active gaps. The electron-ion pairs are

CP867, *XII International Conference on Calorimetry in High Energy Physics,*
edited by S. R. Magill and R. Yoshida

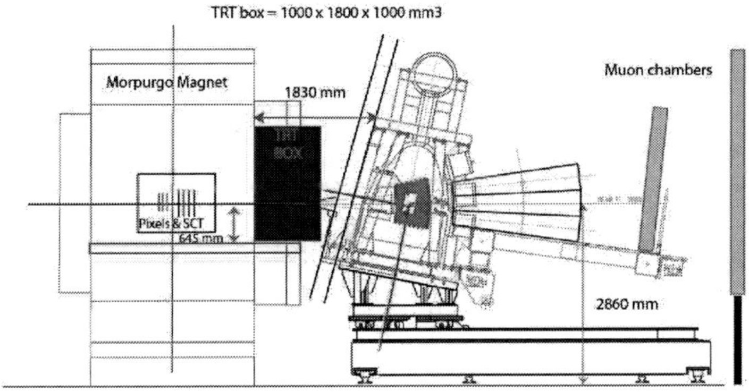

FIGURE 1. Schematic view of the CTB setup, including the inner detector components and the LAr and Tile calorimeters. The muon setup is located 20 m further down on the beam, on the right outside of the scope of the drawing.

separated by a high-voltage; the electrons drifting toward the anode induce a triangular signal at the signal electrode (Figure 2 (left)). The amplitude of the signal is directly proportional to the energy deposit [3].

2.1. From digits to "raw" energy

The signal collected on the readout electrodes is passed through a Pre-amplifier and a bipolar shaper, then sampled at the 40.08 MHz LHC bunch crossing frequency by a 12 bit ADC in 3 gains (Figure 2(b)). In contrast to what is planned in normal ATLAS data-taking, during the CTB energy was reconstructed offline, thus the digitized samples were transmitted by the back-end electronics and recorded.

The EMC is equipped with an electronic calibration system [4], that allows to promptly measure the actual gain and electrical response of each readout cell. This system is based on the ability to inject into the detector a calibration signal of known amplitude that mimics the ionisation pulse generated by the particles hitting the detector (Section 2.4).

The energy deposit in each LAr readout cell is reconstructed from the digitized samples as:

$$E = F \sum_{j=1,2} R_j \left[\sum_{i=1,5} a_i (s_i - p) \right]^j \quad (1)$$

where:

- s_i are the digitized samples, measured in ADC counts;
- p is the pedestal level of the readout ADC for the given cell (see Section 2.2);
- a_i are the *optimal filtering* coefficients used to reconstruct the signal amplitude (see Section 2.3);

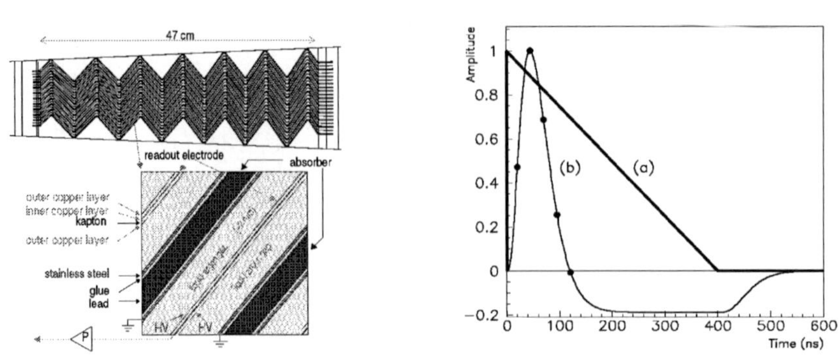

FIGURE 2. Left: detailed view of a EMC LAr gap section. Right: form of the ionization signal (a) and the shaped ionization signal (b); the samples are also indicated [5].

- the R_j coefficients parametrize the electronic gain, converting the signal amplitude from ADC to current (see Section 2.4);
- F is the conversion factor between current and energy (see Section 2.4).

2.2. Pedestals and noise

Dedicated runs without beam or injected charge are used to monitor the noise level in the EMC, as well as to measure the pedestals p and noise time autocorrelation in a given cell needed for the energy reconstruction. Pedestal and noise levels were measured regularly with this approach every 8 hours. Alternatively during the standard data taking, random triggers were used to obtain the same information.

Stability of pedestals was very good (better than 1 ADC count over 4 months of data taking, see Figure 3 (left), corresponding to $\sim$10 MeV in the central compartment of the EMC), small temperature variations were easily corrected for. Noise and coherent noise were as expected.

2.3. Optimal Filtering Coefficients

The *optimal filtering* (OF) method is an elegant way to compute the peak of the ionization signal by minimizing the electronics noise contribution and the fluctuations introduced by jitter in the readout electronics. The signal amplitude is obtained as the linear combination of the digitized samples (see equation (1)), the optimal filtering coefficients (OFC) a_i being computed through a Lagrange multiplier technique [6].

In order to perform this computation, the knowledge of the normalized shape of the ionization signal, its derivative and the noise time autocorrelation is required for each cell. The noise time autocorrelation is computed from the data acquired during the pedestal runs; the normalized ionization pulse is predicted from the corresponding calibration profile, as obtained by pulsing the readout cell with a signal that mimicks the ionization current with an exponential pulse, and introducing a variable delay between

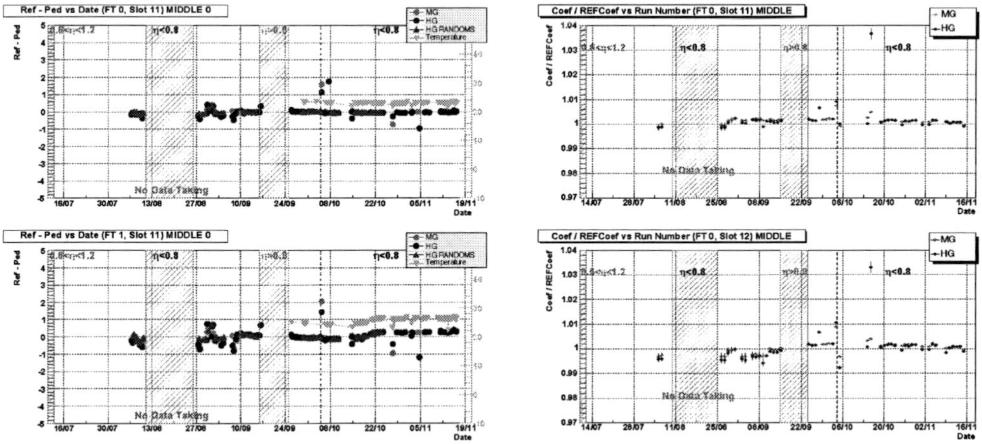

FIGURE 3. Left: example of stability of pedestals (in ADC counts) over time, averaged over the 128 channels of a EMC readout Front End board (FEB), for 2 FEBs. Reference runs are acquired on October 24 and 25, 2004. Right: example of stability of R_1 coeffi cients over time for 2 FEBs. Reference runs are acquired on November 14, 2004. The blue vertical lines indicate a change in the trigger timing. The blue hashed areas are the period where there was no data taking. The red hash areas indicate periods with changes in the EMC cabling.

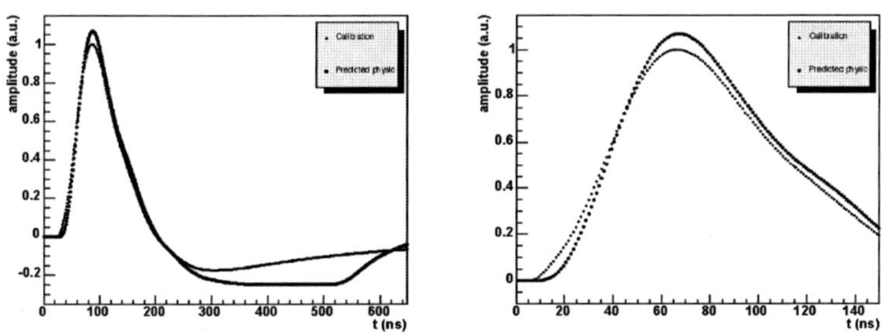

FIGURE 4. Example of calibration signal and corresponding predicted ionization signal for a the EMC central sampling (Middle) cell at (left: full pulses; right: detail of peaks' region.

the calibration pulser and the data acquisition system in order to reconstruct the cell response with a fine time granularity [2].

The prediction of the ionization pulses is based on the electrical model of the readout cell; the predicted pulses include the correction for the distortion introduced by the different electrical properties of each readout channel, and for the shape difference between the calibration and the ionization currents [7] (Figure 4).

Since at the beam test particles are asynchronous with respect to the central clock, several OFC sets were computed, corresponding to different portions of the ionization signal being digitized. The proper OFC set was chosen according to an external timing

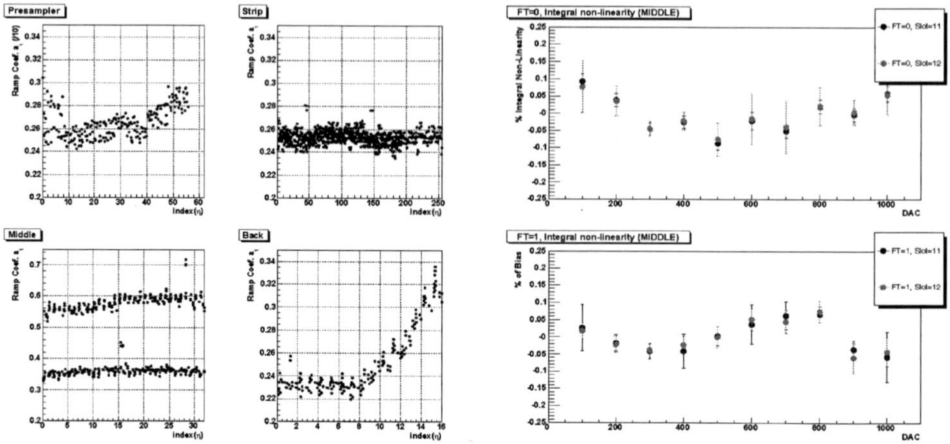

FIGURE 5. Left: coefficient R_1 as a function of the η index for the Presampler and EMC longitudinal compartments in high gain; note the variation related to the detector capacitances (bottom right), and the different populations related to different preamplifier types (bottom left). Right: integral non-linearity (in %) measured on the calibration pulses of different amplitude for the EMC central sampling (Middle) in high gain.

information, provided by a TDC measuring the time different between the main trigger and EMC readout clock.

2.4. ADC $\rightarrow$ MeV conversion

The relation between the OF-reconstructed amplitude of the signals and the currents collected in the EMC is measured on a regular basis by an electronic calibration system. Each readout cell is pulsed with different current settings covering the whole dynamic range of the readout electronics, mimicking the triangular ionization signal with an exponential pulse. The response of the system is reconstructed, the electronics gain is then parametrized with a second order polynomial, in order to take possible non-linearities of the readout into account; the R_j coefficients in equation (1) account for this parametrization (Figure 5 (right)) .

The electronic gains were very stable: variations up to a few permil were observed over 4 months of data taking (see Figure 3 (right)). A clear dependence of the gains on the temperature measured on the readout Front End boards (FEB) was observed, and systematically corrected for in offline reconstruction, thanks the excellent time granularity of the conditions database. Integral non-linearity of the readout remained below 0.1%, as required by the ATLAS performances (Figure 5 (left)).

The conversion of the reconstructed current and the corresponding deposited energy is accounted by the factor F in equation (1):

$$F = \frac{\mathrm{M_{phys}}}{\mathrm{M_{cali}}} \cdot F_{\mathrm{DAC}\rightarrow\mu\mathrm{A}} \cdot F_{\mu\mathrm{A}\rightarrow\mathrm{MeV}} \tag{2}$$

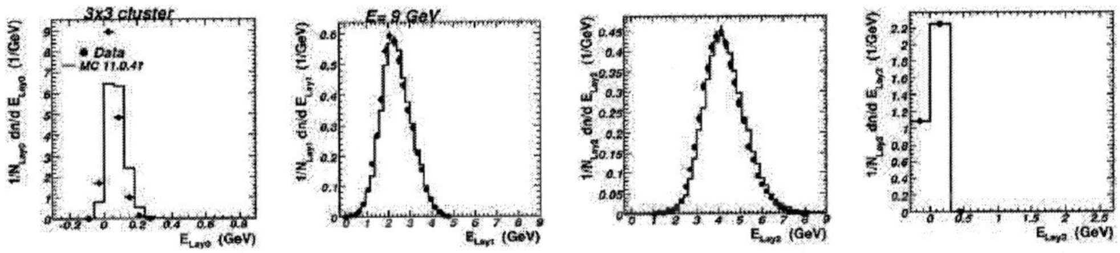

FIGURE 6. Response to 9 GeV electrons of in the ATLAS Presampler and EMC longitudinal compartments, as simulated for the VLE 2004 ATLAS CTB setup (solid) and reconstructed from data (points).

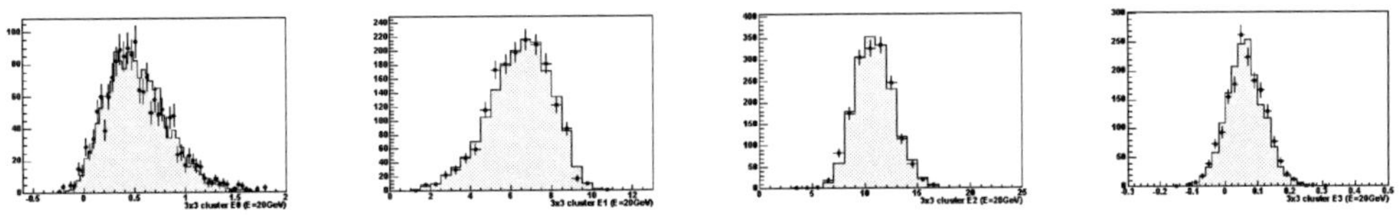

FIGURE 7. Response to 20 GeV electrons of in the ATLAS Presampler and EMC longitudinal compartments, as simulated for the HE 2004 ATLAS CTB setup (solid) and reconstructed from data (points).

where:

- the $\frac{M_{phys}}{M_{cali}}$ factor takes into account the different distortions showed by an ionization and a calibration pulse corresponding to the same amount of currents – distortions related to the different generation points in the detector, and by the pulse shape differences (ionization: triangular; calibration: exponential) [8, 7];
- the $F_{DAC \to \mu A}$ factor is related to the calibration pulser specificities [4];
- the $F_{\mu A \to MeV}$ factor converts current to energy, and takes into account the ionization potential into LAr, the electric field applied on the EMC LAr gap, the recombination effects, the ionization electrons drift velocity, the sampling fraction for electrons which translates the energy deposited in LAr into the total deposited energy in the EMC [2, 9].

3. RESPONSE OF THE ELECTROMAGNETIC CALORIMETER TO ELECTRONS

A good description of the energy deposits in the EMC is crucial to obtain the proper energy scale calibration [9]. The response to electrons was extensively studied with the simulation of the two different beam–line setups: Very Low Energy (VLE, 1-9 GeV), where the momentum selection were done very close to the beam-test setup trigger, and High Energy (HE, 9-180 GeV).

Extensive comparisons between real an simulated data have been performed: a very good description of energy deposits in each EMC layers is obtained for the VLE setup (Figure 6), as well as for the HE setup (Figure 7).

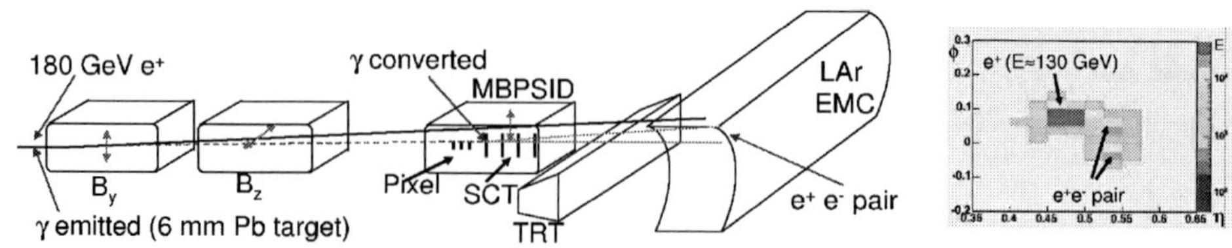

FIGURE 8. Left: CTB photons setup. Right: example of clusters distribution in the EMC for a CTB photon event.

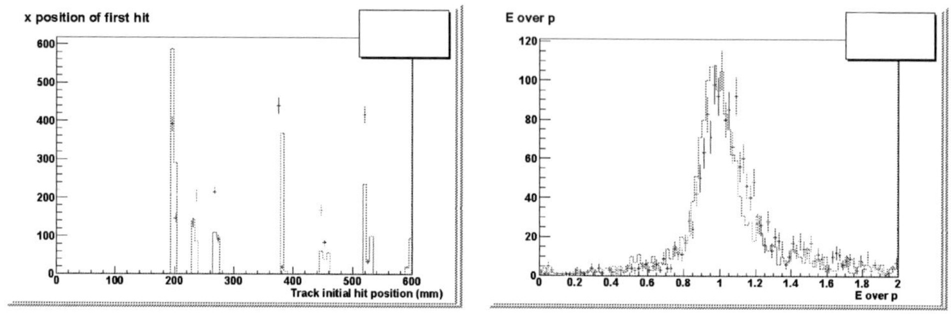

FIGURE 9. Left: positions of the photons' extrapolated conversion points (solid: simulation: points: data). Right: E/p distribution of $e+$ and $e-$ from converted photons (solid: simulation; points: data).

4. COMBINED RECONSTRUCTION OF CONVERTED PHOTONS

4.1. Photons run setup

Precise reconstruction of photons will be an important task of the ATLAS detector at LHC. A crucial issue would be the combined reconstruction of photons converting before reaching the EMC.

At the 2004 CTB a special setup was used to study photons, consisting a 6 mm Pb target on which a 180 GeV e^+ beam was shot. The target was followed by magnets which deflected the incoming e^+ from the beam trajectory, that would instead be followed by photons eventually emitted at the target. The field was chosen so that both the primary e^+ and the emitted γ could reach the EMC module acceptance (Figure 8 (left)).

4.2. Backtracking of converted photons

Topological clustering was used to reconstruct 3 objects in EMC: the cluster associated to the primary e^+, and the two clusters corresponding to the e^+e^- pair from the converted photons (Figure 8 (right)). A back-tacking algorithm was developed to associate the clusters with tracks reconstructed by the ID, then to extrapolate the γ conversion point.

The positions of the extrapolated conversion points nicely point to the internal layers of the ID: a good association between between clusters in EMC and conversion points is found (Figure 9 (left)). A first combined measurement of E/p is obtained, p being measured by the ID and E by the EMC, along with a good agreement between combined data and combined description of the setup (Figure 9 (right)).

5. CONCLUSIONS

The 2004 ATLAS CTB was an unprecedented occasion to exercise the electronics calibration of the LAr EMC. The full electronic calibration chain was implemented, the performances of the ATLAS LAr final electronics met the requirements for the ATLAS physics. All the EMC electronics calibration procedures have been implemented in the ATLAS reconstruction software, the system is ready for the commissioning of the full EMC, that will start in Summer 2006,

The response of the detector to electrons is very well understood: very good agreement between data and Monte-Carlo in the different beam-line setups is obtained, the simulation can be used to finalize the response calibration of the detector at different energies..

Combined studies are ongoing, the first results are very encouraging.

ACKNOWLEDGMENTS

We are particularly grateful to all the people who made the 2004 ATLAS combined test-beam a success. The work presented here has been performed within the ATLAS collaboration, and we thank all the collaboration members for the helpful discussions. We are indebted to the CERN accelerator division and in particular to Ilias Efthymiopoulos for their help with the H8 beam line.

REFERENCES

1. B. Di Girolamo et al., *ATLAS Internal Note* **ATL-TECH-PUB-2005-001** (2005).
2. M. Aleksa et al., *ATLAS Internal Note* **ATL-LARG-PUB-2006-003** (2006).
3. W. Willis and V. Radeka, *NIM A* **120** (1974).
4. J. Colas et al., *ATLAS Internal Note* **ATL-LARG-2000-006** (2000).
5. The ATLAS collaboration, Liquid Argon Calorimeter Technical Design Report, Tech. Rep. CERN/LHCC/96-41, CERN (1996).
6. W.E. Cleland and E.G. Stern, *NIM A* **338** (1984).
7. D. Banfi , M. Delmastro and M. Fanti, *ATLAS Internal Note* **ATL-LARG-2004-007** (2004).
8. L. Neukermans, P.Perrodo and R. Zitoun, *ATLAS Internal Note* **ATL-LARG-2001-008** (2001).
9. W. Lampl, "Studies of the Linearity of the ATLAS EM Barrel Calorimeter," in *Proceedings of CALOR 2006*, Chicagi, IL-USA, 2008.

Studies of the Linearity of the ATLAS EM Barrel Calorimeter

Walter Lampl
on behalf of the ATLAS Liquid Argon Collaboration

University of Arizona, Tucson, AZ, 85721

Abstract. The ATLAS experiment uses an accordion shaped lead/liquid argon sampling calorimeter for electromagnetic calorimetry. During beam test campaigns in summer 2002 and summer 2004 dedicated linearity scans were carried out. During the 2004 run, not only the beam energy but also the material in front of the calorimeter has been varied.

This contribution discusses methods to achieve an excellent linearity and resolution even in the presence of a significant amount of upstream material. These methods were developed and tested on data obtained from the described beam tests and on a Geant4 simulation of the beam test setup.

Keywords: ATLAS Calorimetry Linearity Testbeam
PACS: 29.40.Vj

1. INTRODUCTION

The ATLAS Experiment currently under construction at CERN in Geneva uses a lead-liquid argon sampling calorimeter in accordion geometry for electromagnetic calorimetry. As shown in figure 1 the calorimeter is subdivided into three longitudinal compartments with different granularity in η and ϕ. The total active thickness is at least 22 radiation lengths. The whole structure is enclosed by a double-walled aluminum cryostat. A presampler is installed between the cryostat wall and the accordion calorimeter. In ATLAS, the total material upstream of the accordion amounts to 2 to 3 X_0 from which about 0.5 X_0 are downstream of the presampler, just before the accordion calorimeter. The main contributors to the upstream material are the cryostat walls and the solenoid coil that provides the magnetic field for the inner detector. More details about the design and the construction of the calorimeter can be found in [1, 3] and are subject of other contributions to this conference [5, 6].

The performance goals given by the envisaged physics program of ATLAS include an energy resolution with a sampling term of about $10\%/\sqrt{E[GeV]}$ and a linearity of 10^{-4} [2].

It was found that the relatively large amount of upstream material has a deteriorating effect on the linearity if not properly calibrated. In order to develop and test calibration methods that achieve the required linearity while preserving the resolution, a series of beam tests have been carried out. This contribution will discuss results from electron beam tests done in summer 2002 and 2004.

CP867, *XII International Conference on Calorimetry in High Energy Physics,*
edited by S. R. Magill and R. Yoshida

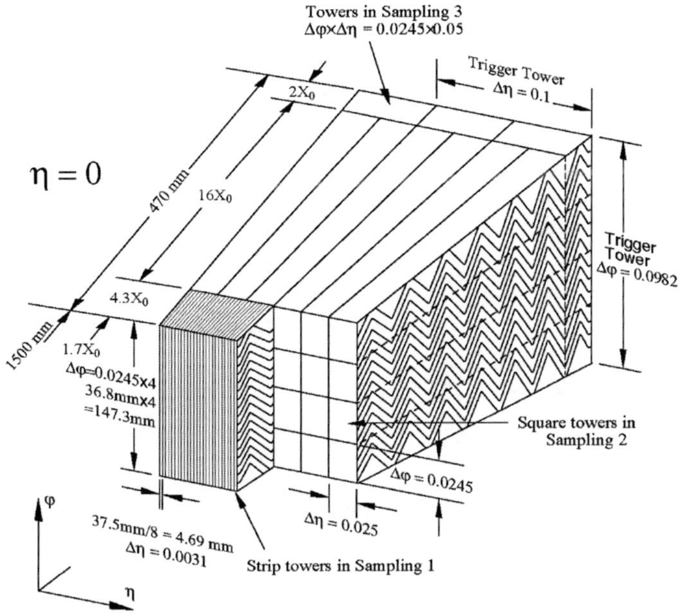

FIGURE 1. Segmentation of the ATLAS EM Barrel Calorimeter

1.1. Calibration Strategy

The Monte-Carlo simulation is a very useful tool to understand in detail the effects caused by the inactive regions around the calorimeter. It has been used to develop the calibration methods presented in the following sections.

The excellent agreement of the Monte-Carlo simulation of the beam test setup with the data (see later in this paper) encourages the usage of the simulation to derive the necessary constants.

2. THE 2002 AND 2004 BEAM TEST CAMPAIGNS

All ATLAS Liquid Argon electromagnetic barrel calorimeter beam tests have been carried out in the H8 beam line of the CERN SPS. This beam line can provide a wide variety of particles and beam energies. In 2002 [8], the beam line was extended by dedicated measurement devices that allowed to control the beam energy up to a precision of σ_E=11 MeV+3.4$\cdot$ 10^{-4} E. The linearity scan that was performed consists of 15 energy points at an impact point that would correspond to a pseudo-rapidity of η=0.687 and azimuthal angle of ϕ=0.282 in ATLAS. In the 2004 run [7], this data set was complemented by an energy and material scan. The material scan was realized by placing aluminum plates in the beam line just before the cryostat. Data was taken with four different material configurations with 0, 25, 50 and 75 mm aluminum. Including other upstream material like the cryostat walls at the given impact point of η =0.4 and ϕ=0, this leads to 2.4, 2.7, 3.0 and 3.3 X_0 upstream of the presampler. For each material configuration, 6 energy points have been taken. The precision equipment used in 2002 to control the beam energy was not available in 2004.

The simulation of both setups is based on Geant4.

3. CALIBRATION OF THE ELECTRON ENERGY MEASUREMENT

3.1. Upstream energy loss

In this section, we concentrate on the energy lost in inactive material upstream of the presampler. The presampler is a thin (11 mm) layer of liquid argon that is instrumented to be read-out. To first order, one could regard the ensemble of inactive material upstream followed by the presampler as very basic sampling calorimeter and calibrate it by a weight or "inverse sampling fraction" that is given by the MIP (minimum ionizing particle) energy loss in the two regions. This assumes that the same number of particles traverses the active and the inactive material. However, this approach is too simple. The beam electron does loose energy not only by ionization but also radiates bremsstrahlung. In some cases, the created photons will convert to an $e^+ e^-$ pair that passes only through a part of the upstream material or the conversion might event take place in or after the presampler. In many cases the bremsstrahlung-photon has too little energy for pair-production and will dissipate its energy through other processes without creating a signal in the presampler.

Figure 2 (a) shows a plot of the upstream energy deposit versus energy deposit in the presampler obtained from the Monte-Carlo simulation. The plot shows a clear correlation between the two quantities and an offset. We conclude from this plot that the proper calibration for the presampler is given by:

$$E_{Upstream} = a + b \cdot E_{PS} \tag{1}$$

Where the offset a and the weight b can be derived fitting the correlation shown in figure 2 (a). The offset a is interpreted as energy loss by the initial electron towards particles that don't reach the presampler. The weight b is smaller than the MIP sampling fraction because of the effects discussed above.

Applying only a weight and omitting the offset leads to either a good linearity or a good resolution - depending on how the weight was optimized - but such an approach can't optimize both.

3.2. Energy loss in the gap between Presampler and Accordion

The region between presampler and the accordion is occupied by support structures as well as cables and electronics boards immersed in liquid argon. The situation in this region is different form the upstream region because the shower has already somewhat developed at this point. The energy deposit is neither well correlated with the presampler measurement nor with the energy deposit in the front compartment.

Again using the Monte-Carlo simulation, a suitable correlation has been found between the energy deposit in the gap region and the square-root of the product of the energy deposits in the adjacent active compartments. Figure 2 (b) shows this correlation for a simulation of 100 GeV electrons. Using this empirically found correlation, the energy deposit in the gap region can be estimated as:

$$E_{Gap} = c\sqrt{E_{PS} \cdot E_1} \tag{2}$$

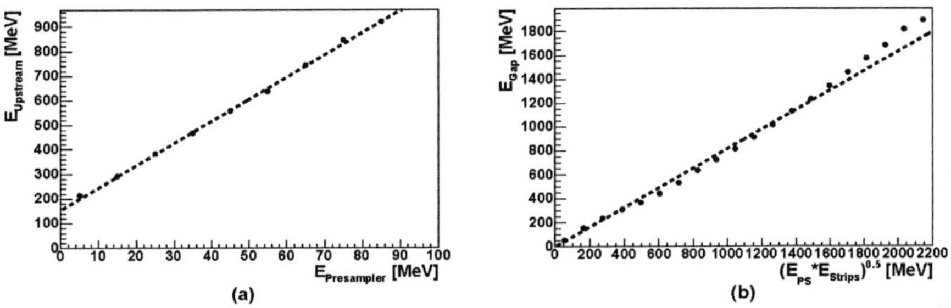

FIGURE 2. (a) Correlation of the energy deposit upstream versus the energy deposit in the presampler. (b) Correlation of the energy deposit in the gap between presampler and accordion versus $\sqrt{E_{PS} \cdot E_1}$. Both figures where obtained from a simulation of the 2004 beam test setup with 100 GeV electrons.

3.3. Calibration of the energy measured in the Accordion

The energy of a single cell of the accordion calorimeter is reconstructed by multiplying the peak of the ionization signal (measured in ADC counts) with a factor that incorporates the electronic gain, the inverse sampling fraction and the current-to-energy conversion factor. Eventually, this factor will be fine-tuned using in-situ calibration in ATLAS.

Neighboring cells are summed up to clusters that contain the electromagnetic shower. The cluster size is a trade-off between noise and lateral leakage. For the beam test analysis, clusters consisting of 3×3 middle cells (built around the hottest middle cell) plus the adjacent front, back and presampler cells were used. The value used as sampling fraction at cell-reconstruction level is an average number that needs to be corrected depending on the cluster size and also on the energy. This correction factor includes also lateral out-of-cluster corrections.

The exact sampling fraction, the fraction of the energy of an electromagnetic shower that gets deposited in the active layers of the calorimeter, depends on the shower composition. The bulk of the energy is deposited by low-energy particles that have a very short range, much shorter than the thickness of layers of the calorimeter. To understand the sampling fraction, one has to take into account also the production mechanism of such particles. The dominant process at low energy is the photoelectric effect. The resulting electron will most likely be absorbed very close to where it was produced. Since the cross-section for the photoelectric effect depends strongly on the Z of the material ($\approx Z^n$ where n is between 4 and 5), low-energy photons have a much higher probability to get absorbed in the lead of the absorber than in the liquid argon.

At the start of the shower, there are also high-energetic charged particles that indeed traverse several layers of active and inactive material, while the tail and outer radii of the shower are made up exclusively of soft particles. This means that the beginning of the shower a larger fraction of the energy gets sampled than towards the end of the shower. Figure 3 (b) illustrates how the sampling fraction decreases with the depth along the shower. Figure 3 (a) shows the longitudinal profiles of electron showers in the ATLAS EM calorimeter.

Electrons with higher energy penetrate deeper and cause a shower with a higher fraction of high-energetic particles in the accordion calorimeter compared to showers induces by electrons of lower energy.

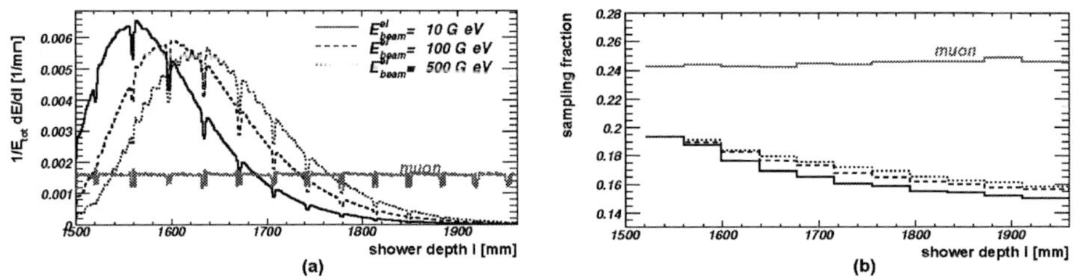

FIGURE 3. (a) Longitudinal profile of an electron-induced shower in the accordion calorimeter for 10, 100 and 500 GeV electrons as well as for 10 GeV muons. The spikes come from the accordion folds. (b) Sampling fraction versus depth in the shower for the same energies. A clear decrease of the sampling fraction for electrons is visible, while the muon sampling fraction stays constant. Both figures have been obtained from a simulation of the 2002 beam test setup.

3.4. Final Calibration Formula

Summarizing the calibration strategies lined out in the previous sections leads to the following final calibration formula:

$$E_{cl} = a + b \cdot E_{PS} + c \cdot \sqrt{E_{PS} E_{Front}} + d \cdot E_{Acc} + e \cdot E_{Back} \quad (3)$$

The last term was used in the 2004 analysis to correct for longitudinal leakage. In 2002, this was not necessary because the data were taken at a different η-position. The total thickness of the calorimeter at this point exceeds 30 X_0 and a correction by an average value was sufficient. The total thickness of the calorimeter changes significantly with η and in general a leakage correction is necessary [4].

The parameters a to d in equation 3 depend on the impact point and slightly on the energy but this dependence can be easily parametrized. Figure 4 shows how these constants develop versus the beam energy.

When more upstream material is added, all corrections except the back leakage are rising. This can be easily understood, since an earlier shower leads to higher energy

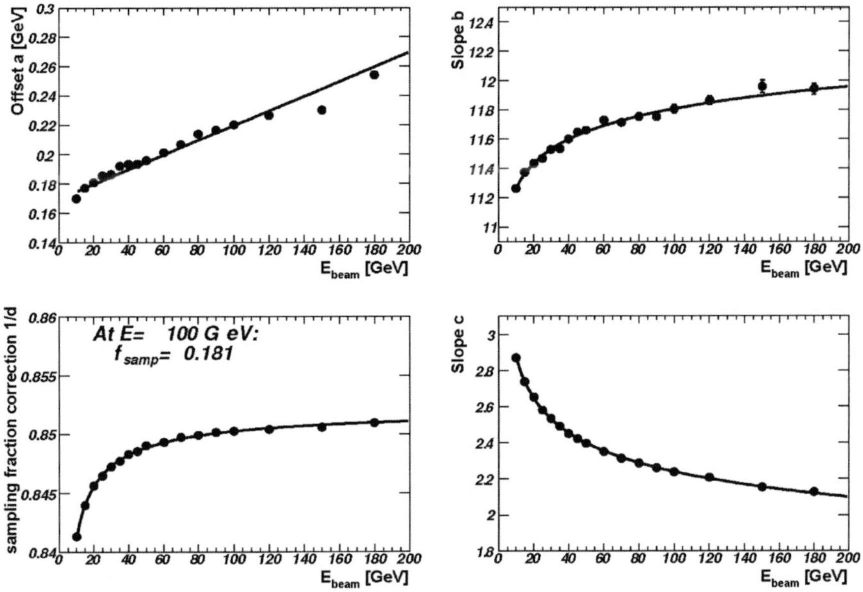

FIGURE 4. Calibration parameters used in equation 3. They have been derived from the 2002 beam test simulation. The calibration parameter for the accordion is plotted as correction to the sampling fraction. This is the inverse of the parameter d in equation 3.

deposit in inactive regions upstream and the accordion samples a later part of the shower that contains more soft photons.

4. COMPARISON OF DATA AND SIMULATION

Since we use the Monte-Carlo simulation to derive cluster correction constants, the accuracy of the simulation is crucial. To ensure the validity of the simulation, extensive comparisons have been carried out. It turned out, that the accurate description of the material distribution in the beam line as well as the cables and electronics boards in the region between presampler and accordion is the most difficult aspect. All plots presented here use 'equivalent' material: thin aluminum plates that have been placed the simulation of the beamline to account for air and beam pipe windows far upstream of the calorimeter.

Figure 5 compares the distribution of the energy fraction deposited in the presampler and the three compartments of the accordion for electrons with and energy of 10 and 100 GeV. The plot has a logarithmic scale to focus on the tails of the distribution which are most sensitive to upstream material.

For the 2004 run, figure 6 shows the ratio of the mean energy deposit in the presampler and the three longitudinal compartments of the accordion. It is clearly visible that the comparison for the presampler is much worse than for the accordion. This can be

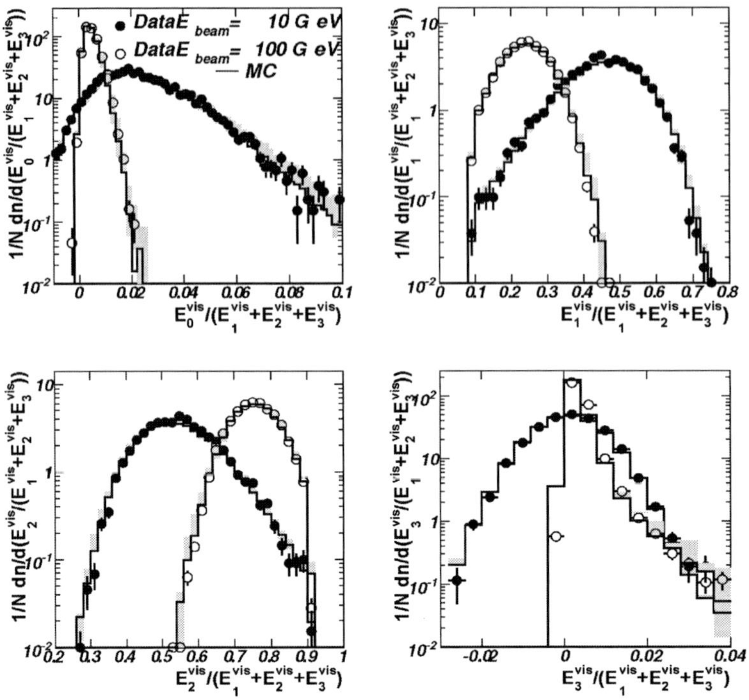

FIGURE 5. Visible energy fraction distribution for 10 and 100 GeV electrons in the presampler and the three longitudinal compartments of the accordion calorimeter. Shown are data (circles) and a Monte-Carlo simulation (line). The band indicates the uncertainty in the Monte-Carlo simulation due to the far upstream material.

explained by the fact that the impact point of ϕ=0 corresponds to a boundary of two presampler modules and the stray-field in the gap is difficult to simulate. Most of the other points match within 2%. The RMS of all points (6 beam energies and 4 material configurations) is 0.75%.

5. RESULTS

Applying the calibration method outlined in section 3 to the linearity scan performed during the 2002 beam test yields a linearity of better than $\pm 0.1\%$ for energies bigger than 10 GeV (see figure 7). The resolution can be fitted with a sampling term of $10.1\%/\sqrt{E[GeV]}$ and a constant term of 0.17%.

Figure 8 (a) shows the linearity achieved on 2004 beam test data for all energies and material configurations. The solid lines show the accuracy of the beam energy. Almost all data points are within these errors, even for large amounts of additional upstream material. The sampling term of the resolution is about 11% with an 0.5% constant term.

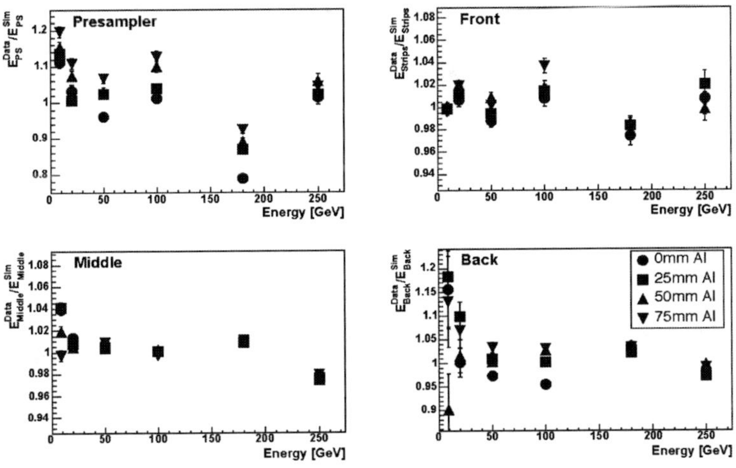

FIGURE 6. Ratios of the mean energy deposit in data and in simulation in the presampler and each compartment of the accordion calorimeter for all energies and all material configurations measured in the 2004 beam test. For low energies, the signal in the back compartment is dominated by noise since the shower does not penetrate deep enough.

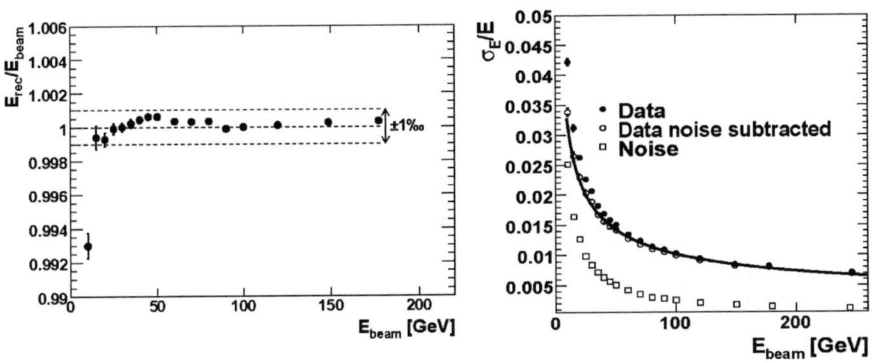

FIGURE 7. Linearity and Resolution achieved for the 2002 beam test using the calibration formula 3. The values are normalized to the 100 GeV point.

Figure 8 (b) shows the robustness of the method against wrongly estimated upstream material. For energies above 50 GeV, an error on the upstream material estimation of 0.3 X_0 would cause 1% error on the energy measurement.

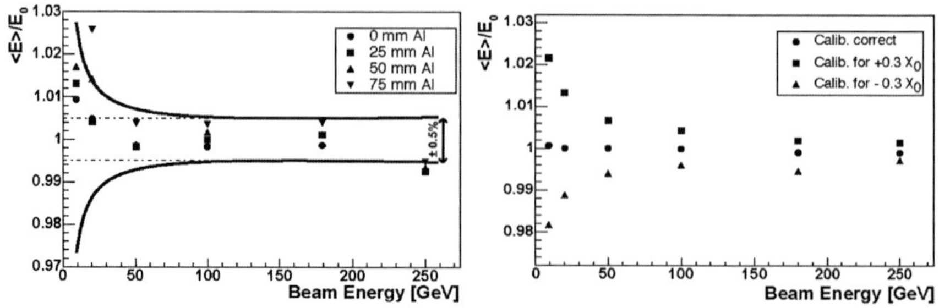

FIGURE 8. (a) Linearity achieved on 2004 beam test data. using the calibration formula 3.The values are normalized to the 100 GeV/25 mm point. The solid lines indicate the error on the beam energy. (b) Estimate of the robustness of the calibration model using the beam test simulation. The circles give the linearity using the correct set of constants while the squares denote the linearity that would be achieved when the upstream material is overestimated by 0.3 X_0 or 25 mm of aluminum. The triangles show the linearity if the upstream material is underestimated by the same amount

ACKNOWLEDGMENTS

The design and construction of the ATLAS Liquid Argon calorimeter as well as setting up and running the beam tests was a collaborative effort of many people and institutes. We wish to thank the technicians and engineers that contributed to this great project.

We are indebted to the CERN accelerator division and in particular to Ilias Efthymiopoulos for their help with the H8 beam line.

REFERENCES

1. The ATLAS Liquid Argon Collaboration, *Liquid Argon Calorimeter Technical Design Report*, CERN/LHCC/96-41, 1996
2. The ATLAS Collaboration, *ATLAS Calorimeter Performance Technical Design Report*, CERN/LHCC/96-40, 1997
3. Aubert, B. et. al, *Construction, assembly and tests of the ATLAS electromagnetic barrel calorimeter*, Nuclear Inst. and Methods in Physics Research A, 558, 2006, pp. 338–428
4. Nikolic, I., *Uniformity in the ATLAS electromagnetic calorimeter measured in test-beams*, Contribution to this conference
5. Aleksa, M., *The ATLAS Liquid Argon Calorimeter: Construction, Integration, Commissioning*, Contribution to this conference
6. Delmastro, M., *Recent Results from the ATLAS Barrel Combined Test Beam*, Contribution to this conference
7. Lampl, W., *Optimizing the Energy Measurement of the Atlas Electromagnetic Calorimeter*, Ph.D. Thesis, Vienna University of Technology, 2006
8. The ATLAS Liquid Argon Collaboration, *Energy Linearity and Resolution of the ATLAS Electromagnetic Barrel Calorimeter in an Electron Test-Beam*, in preparation, 2006

Uniformity in the ATLAS electromagnetic calorimeter measured in test-beams

Irena Nikolic-Audit
on behalf of the ATLAS Liquid Argon Collaboration

LPNHE, IN2P3-CNRS, Universités Paris VI et Paris VII

Abstract. The construction of the ATLAS liquid argon electromagnetic calorimeter has been completed and all the modules are assembled and are in the cryostat in the ATLAS cavern. During the construction, three barrel modules and three endcap modules were exposed to beam test in order to assess the uniformity performance and to verify the production reproducibility of the detector. The energy reconstruction method and the performance with electron beams are presented here. An overall constant term below 0.7 % is obtained.

Keywords: Calorimetry, Uniformity, ATLAS, Testbeam
PACS: 29.40.V,06.20.f

1. INTRODUCTION

The liquid argon electromagnetic calorimeter in ATLAS was primarily designed to meet the goals of the Higgs search at the LHC and to optimally detect the Higgs boson decaying into photons or electrons. In order to meet the physics requirements at the LHC, the global constant term of the energy resolution of the calorimeter must be less than 0.7 %.

The constraints on the calorimeter constant term are presented in section 2, the test beam setup is presented in section 3, the energy reconstruction method is presented in section 4, and the uniformity results are presented in sections 5 and 6.

2. CONSTANT TERM IN THE LARG CALORIMETER

The ATLAS electromagnetic calorimeter description, the signal collection and reconstruction methods have been presented elsewhere in this conference [1, 2] and in [3, 4, 5]. Only the features relative to the uniformity study of the detector are given here. The physics program in ATLAS, in particular the search of $H \rightarrow \gamma\gamma$ events, requires a constant term below 0.7 % over the full calorimeter acceptance relevant for precision physics ($|\eta| < 2.5$). Mechanical, geometrical, calibration or material effects are all potential sources of non uniformities. The numerous contributions to the constant term can be separated in two components: the "local" constant term C_{loc} and the "long range" constant term C_{LR}.

- The local constant term can be measured with testbeams, and those measurements will be presented here. The requirement from the ATLAS detector performance

CP867, *XII International Conference on Calorimetry in High Energy Physics,*
edited by S. R. Magill and R. Yoshida

TDR [6] is to have the C_{loc} below 0.5 % in a calorimeter region of the size of $\Delta\eta \times \Delta\phi = 0.2 \times 0.4$, which corresponds to 128 cells in the middle compartment.[1]

- The long range variations will be measured in situ with physics events. There are 440 regions of $\Delta\eta \times \Delta\phi = 0.2 \times 0.4$ in the electromagnetic calorimeter. With 250 electrons in each of these regions, a C_{LR} below 0.4 % can be obtained with 10^5 $Z \rightarrow e\ e$ events. This will be the case after a few days at the low LHC luminosity.

In order to maintain the local term C_{loc} as small as possible, a large effort was done during the detector construction to have all the calorimeter modules as reproducible as possible. For example, the variations of the lead and absorber thicknesses induce a variation in the signal response. In the calorimeter, the dispersion of lead sheets has been measured to be equal to 9 microns, resulting in a contribution to the constant term of $< 0.2\%$.

3. EXPERIMENTAL SETUP IN TESTBEAMS

During the years 2001 and 2002, three barrel and three endcap modules making the final detector were exposed to electron beams at the CERN H8 and H6 facilities in order to assess the calorimeter performances and the detector response uniformity. Among other measurements such as the energy resolution, linearity, muon response, timing, the uniformity of the detector as a function of the electron impact point has been studied by performing a position scan in the η and ϕ directions over entire modules at fixed beam energies (245 GeV for the barrel and 120 GeV for the endcap). The read out electronics used during those test beams was very close to the final one.

4. ENERGY RECONSTRUCTION

The energy reconstruction method used here is the same one as in the energy resolution and linearity studies of the electromagnetic calorimeter (see [2] in these proceedings). It is a Monte Carlo based method, with corrections factors derived from the simulation. The correction factors have been determined for one ϕ position only, and apply to all ϕ angles. The simulation of the shower developments was done with a 4.7 and 4.8 versions of GEANT, and using a full ATLAS reconstruction package with the testbeam geometry.

The particle impact position is estimated with energy weighted barycenters, taking advantage of the fine granularity of the front sampling along the η direction for the barrel and for the outer endcap. For the inner endcap and in the ϕ direction, middle cells are used.

Most of the energy of electromagnetic interacting particles is deposited in the calorimeter. A small fraction is deposited in non instrumented regions such as the cryostat, the region between the presampler and the beginning of the accordion, the cables and also after the back sampling. The agreement between the test beam data and the Monte Carlo for the energy deposited in all three samplings of the calorimeter and in

---

[1] The cell size in the middle is 0.025×0.025.

the presampler is excellent, as shown in Figure 1. Many small effects, such as the cross talk, the differences between physics and calibration signals ... are present in the data. The dedicated test beams were a unique opportunity to study the calorimeter response in detail and precisely tune the Monte Carlo to implement all these effects.

Calorimeter cells form a 3×3 cluster around the most energetic middle cell. The reconstructed energy E^{rec} is evaluated from the measurements of the visible energies E^{meas} in the PS and the three accordion (acc) compartments of the calorimeter (front, middle and back), and from the η and ϕ angles of the cluster with the following scheme:

$$E^{rec} = [(a_{\eta,E} + b_{\eta,E} E_{PS}^{meas} + c_{\eta,E} \sqrt{E_{PS}^{meas} E_{front}^{meas}} + d_{\eta,E} \sum_{i=1}^{3} E_{i\ acc}^{meas}) + E^{leak}] \cdot f_{CI} \cdot f_{TR(\eta,\phi)}$$

where η and ϕ are the measured energy weighted barycenters of the reconstructed cluster respectively along the longitudinal and azimuthal directions. The total energy deposited in the material upstream of the calorimeter ($\simeq 1.5X_0$) is reconstructed by the $a + b \cdot E_{PS}$ term. The energy lost in the cables, electronics and the support structure between the PS and the front sampling of the accordion is corrected by the empirically determined square root term. The d term contains the finite cell size and the sampling fraction dependence in the energy and in η. The variation of the a, b, c and d coefficients as a function of η expressed in middle cell units is shown in Figure 1.

E^{leak} is the correction for the longitudinal leakage of the electromagnetic shower. The shower leakage at the back of the calorimeter is clearly a function of η as the depth varies by $\simeq 8X_0$ with η. A parametrization of the leakage has been derived as there is a linear dependence between the leakage normalized to its average value and the energy deposited in the back compartment of the calorimeter : $E_{leak}/ < E_{leak} >= \alpha + \beta \cdot E_{back}$. α and β coefficients have been extracted as a function of η (see Figure 2). If this leakage were not corrected for, it would be a dominant contribution to the non uniformity of the detector, with a 0.6% contribution to the constant term.

The f_{CI} corrects for transverse leakage in a 3×3 cluster and for geometrical effects originating from local non uniformities. f_{TR} corrects for the drop of the signal in the transition region between two types of electrodes at $\eta = 0.8$. The lead thickness is different between the two regions in order to maintain a constant sampling term in the energy resolution.

5. BARREL UNIFORMITY

The ATLAS em calorimeter is made of two half barrels. For ease of construction, each one is made of 16 modules of size $\Delta\eta = 1.475 \times 0.4$. Three of the modules (M10, P13 and P15) were exposed to beam particles. Fine η and ϕ scans were performed with 245 GeV electrons. Results will be given for modules P13 and P15 and only for an extension of 0.15 in ϕ because parts of modules were readout with a poor quality equipment. The final energies for the tested modules are shown in the Figure 3.

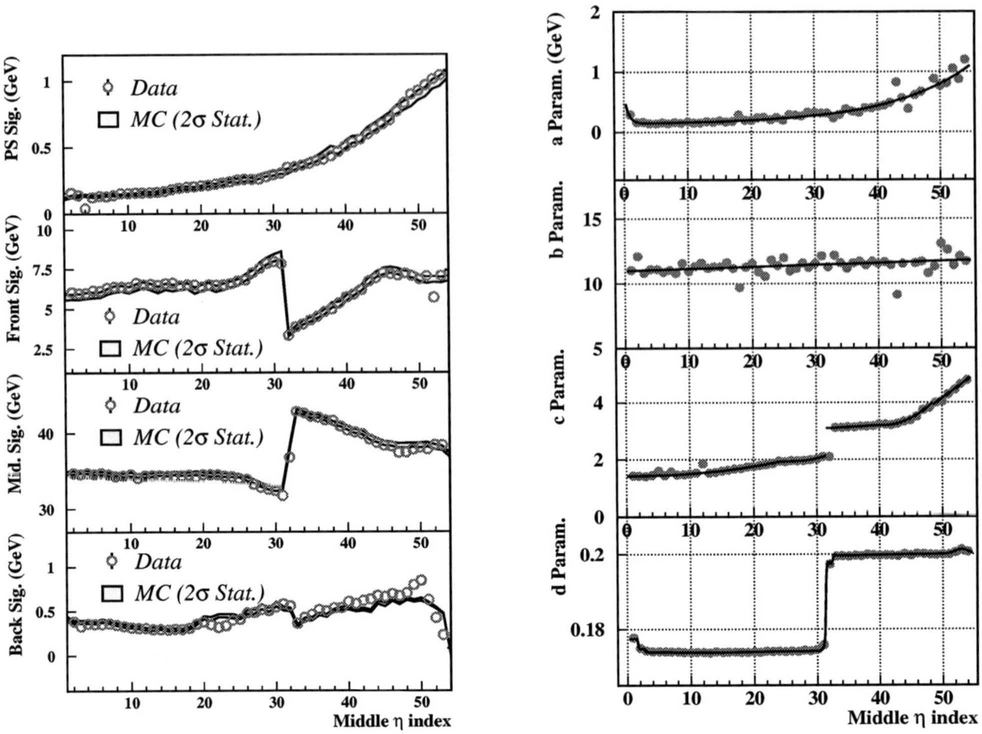

FIGURE 1. Left: Deposited energies as a function of η in the presampler and the 3 calorimeter compartments before applying the correction factors for 245 GeV electrons. In the horizontal axis is the η value in units of the middle sampling cell number. 56 cells in the middle sampling correspond to 1.4 in η. Right: Coefficients a,b,c,d as a function of η. The step at $\eta = 32$ in the d coefficient is is due to the change of the lead thickness.

5.1. Barrel results

The dispersion (RMS/mean energy value) of the two barrel modules, with over 300 cells per module in the analyzed region, is given in table 1. The energy resolution varies between 0.7 % and 0.9 %. The absolute energy scale is known at a level of 5 10^{-4}. The signal dependence with the liquid argon temperature has been corrected.

5.2. Understanding the barrel uniformity

For a better understanding of the barrel uniformity, a detailed analysis of a barrel region $\Delta\eta \times \Delta\phi = 0.8 \times 0.15$ with $\simeq$ 190 cell was done. The requirements from the

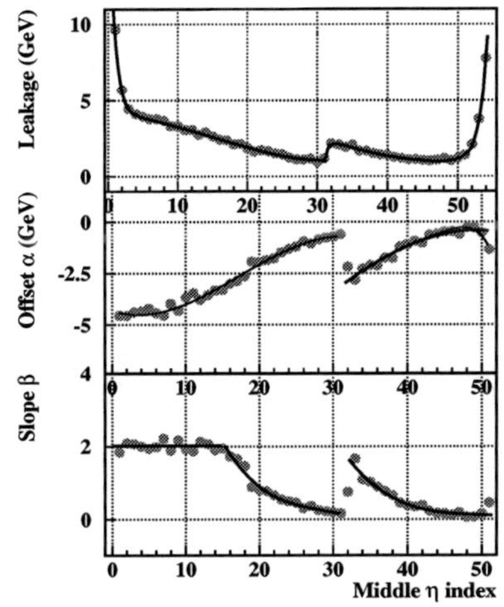

FIGURE 2. Longitudinal energy leakage and the relevant correction parameters as a function of the middle η index.

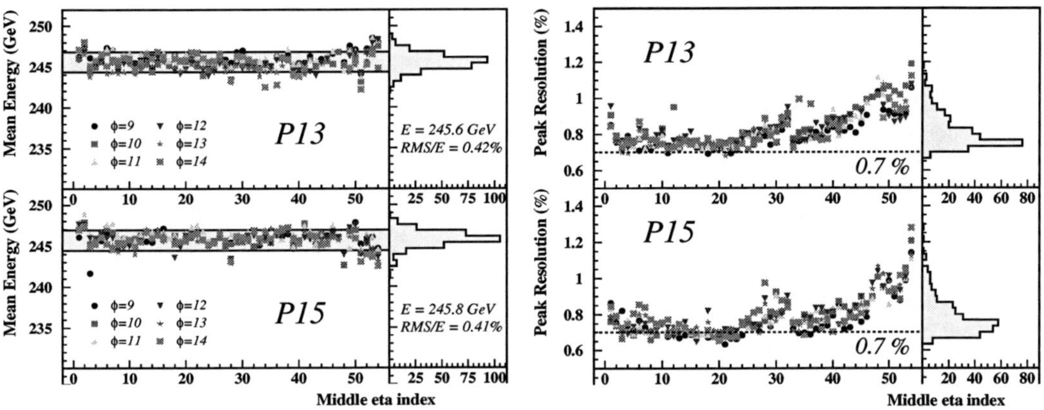

FIGURE 3. The mean cell energy after all corrections and the energy resolution as a function of the middle cell η index for various ϕ lines in barrel modules. The dotted line shows the 0.7% energy resolution.

TABLE 1. Mean energy and non-uniformity of 2 barrel modules over the whole analysis region.

Module	P13	P15
$< E >$	245.6 GeV	245.8 GeV
RMS/< E >	**0.42%**	**0.41%**
global constant term	0.62 %	0.56 %

TABLE 2. Requirements from the ATLAS TDR on the local constant term.

Source	Contribution to uniformity	
Calibration: amplitude + stability	< 0.25 %	correlated in testbeam
Signal reconstruction + inductance	< 0.3 %	correlated in testbeam
Φ modulation + longitudinal leakage	< 0.25 %	correlated in testbeam
Mechanics: Pb + Ar gap	< 0.25 %	uncorrelated in testbeam

ATLAS TDR [3] on various contributions to the constant term are shown in table 2. The non uniformities can be separated into two categories:

- the uncorrelated non uniformities, due to the mechanical construction of the module.
- the correlated non uniformities, common to two modules. Their origins are stated in the first three lines of the table 2, the total contribution being 0.45%. P13 and P15 have been tested consecutively with the same electronics. The same reconstruction scheme has been derived from Monte Carlo.

The dispersions for restricted barrel regions are 0.34 % for both P13 and P15 modules. This term contains the correlated and the uncorrelated non uniformities. The dispersion of the ratio of the P13/P15 cell energies is 0.24 %, with only the uncorrelated non-uniformities contributing. From these numbers, we find:

- 0.29% for the correlated non-uniformity in P13 and P15, to be compared to 0.45%.
- 0.17% for the uncorrelated non-uniformity in P13 and P15, to be compared to 0.25%.

Those numbers are well below the specifications, in a region larger than specified in the ATLAS TDR.

6. ENDCAP RESULTS

The endcap extends to $|\eta| = 3.2$ and consists of 2 wheels. Each endcap consists of the outer wheel ($1.4 < |\eta| < 2.5$) and the inner wheel ($2.5 < |\eta| < 3.2$). The position scan was performed with 120 GeV electrons on 3 (ECC0, ECC1 and ECC5) out of 16 modules of the endcap in the H6 beamline. Due to the accordion geometry of the endcap [7] and the continuous modification of the gap value between the electrodes and the absorbers, the high voltage has to vary accordingly. The endcap is divided into 8 high voltage (HV) sectors. This is illustrated in Figure 4. For technical reasons, the HV variation is done by steps along the η angle. The HV effect is corrected for by weighting the energy in each cell depending on its η position and its sector. The corresponding correction parameters are shown in Figure 5. As in the barrel, the corrections for geometrical effects and for the energy leakage have been applied on energies at the cell and the cluster level. Contrary to the barrel, the corrections need to be parametrized both in η and in ϕ.

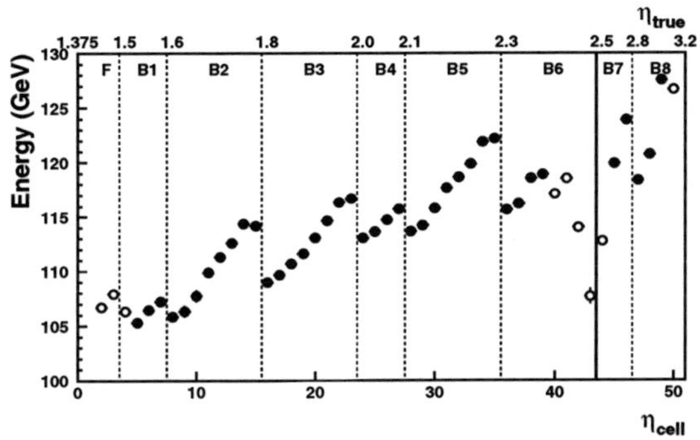

FIGURE 4. Energy averaged over Φ for one module as a function of η before high voltage correction. The vertical dashed lines separate the high voltage sectors and the full line at $\eta = 2.5$ separates the outer and the inner wheel. The full (empty) dots correspond to cells inside (outside) the analysis region.

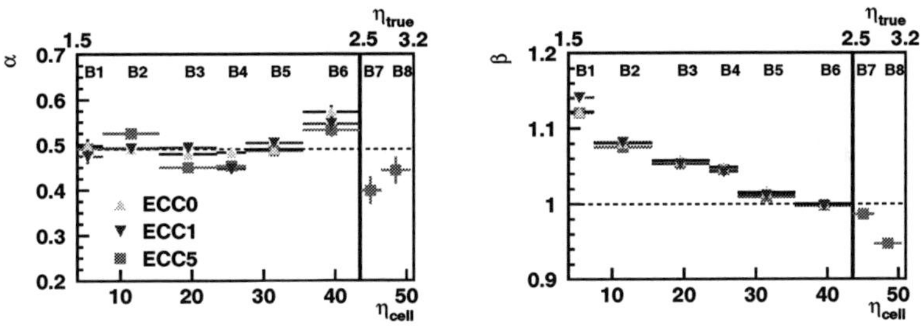

FIGURE 5. High voltage correction parameters α (left) and β (right) obtained in the eight scanned HV sectors for the three tested modules. Results have been averaged over Φ.

The mean reconstructed energy after all corrections is shown in Figure 6 for the three tested modules. The dispersion of cell energies are given in table 3.

7. CONCLUSIONS AND ACKNOWLEDGMENTS

The uniformity of the ATLAS electromagnetic calorimeter was tested in 6 production modules in dedicated test beams. Using a Monte Carlo based energy reconstruction method, the achieved performances are well within expectations: global uniformity over

TABLE 3. Mean energy and non-uniformity of the three tested modules over the whole analysis region.

	Outer wheel			Inner wheel
Module	ECC0	ECC1	ECC5	ECC5
$< E >$	118.8 GeV	119.1 GeV	119.3 GeV	119.1 GeV
RMS/$<$ E $>$	**1.27%**	**1.28%**	**1.22%**	**1.26%**
global constant term	0.70%	0.72%	0.61%	0.78%

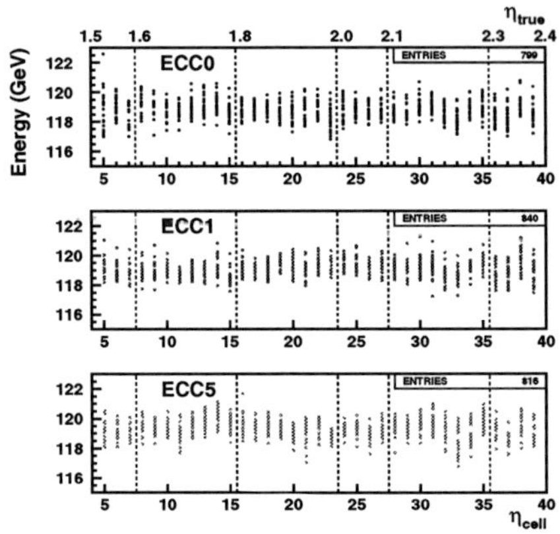

FIGURE 6. Endcap energies as a function of the η impact point.

a large area of a barrel module is 0.44 % and the energy scale between modules is known at a per mil level.

I would like to thank all my colleagues from ATLAS Larg community, and in particular Marumi Kado and Laurent Serin for their help in this presentation.

REFERENCES

1. M. Aleksa, contribution to these proceedings.
2. M. Delmastro and W. Lampl, contributions to these proceedings.
3. ATLAS Collab. "Liquid Argon Calorimeter technical Design Report", CERN/LHCC/96-41, 1996
4. B. Aubert *et al.* "Performance of the ATLAS electromagnetic calorimeter barrel module 0," Nucl. Instrum. Meth. A **500**, 202 (2003)
   B. Aubert *et al.* "Performance of the ATLAS electromagnetic calorimeter end-cap module 0," Nucl. Instrum. Meth. A **500**, 178 (2003).
5. B. Aubert *et al.* "Construction, assembly and tests of the ATLAS electromagnetic barrel calorimeter," Nucl. Instrum. Meth. A **558** (2006) 388.
6. ATLAS Collab. "Technical Design Report, Detector and Physics Performance", CERN/LHCC/99-14, 1999
7. F. Hubaut and C. Serfon, "Response uniformity of the ATLAS electromagnetic endcap calorimeter" ATLAS note **ATL-LARG-2004-015.**

SESSION VIII

LHC CALORIMETRY—CALIBRATION AND MONITORING

Convener: Ambreesh Gupta

Calibration and monitoring of the ATLAS Tile calorimeter

J. Carvalho

LIP-Coimbra, Departamento de Física, Universidade de Coimbra, 3004-516 Coimbra, Portugal
E-mail: jcarlos@lipc.fis.uc.pt
(for the ATLAS Tilecal Collaboration)

Abstract.
The calibration and monitoring systems of the ATLAS hadronic calorimeter, the TileCal, are presented. Special attention is given to the experience gained so far and perspectives for the future, alongside with the results obtained from the analysis of data acquired during the testbeam periods.

Keywords: hadron calorimetry, calibration, monitoring
PACS: PACS numbers: 07.20.Fw, 06.20.Fn

1. INTRODUCTION

The ATLAS hadronic sampling calorimeter is built from iron absorbers and scintillator tiles. It is divided into three parts along its axis, a central and two extended cylindrical barrels, each part divided in 64 wedges. The TileCal covers a pseudorapidity up to $|\eta| = 1.7$. The good hermeticity is ϕ was obtained using a vertical placement of the scintillating tiles in the $r - \phi$ plane. Each tile is readout by two wavelength shifting fibers, coupled to photomultipliers. The fibers are joined in bundles in such a way to form projective towers pointing to the interaction region, to reach the desired segmentation, divided in three longitudinal samples (with different interaction lengths), with a granularity of $\eta*\phi$=0.1*0.1 [1].

Several systems are used to calibrate and monitor the calorimeter performance. The charge injection system is used to calibrate the relative response of the photomultiplier readout electronics, across all the calorimeter, and to track any variation with time. The laser system is designed to calibrate and monitor the response of the photomultipliers, in particular the stability of their gains, their global linearity, and for timing studies. A cesium radioactive source system is designed to measure the quality of the optical response of each calorimeter cell, to equalize the signal response from all the cells and to monitor it with time. Finally the minimum bias current will be used to continuously monitor the calorimeter response during physics runs. About 12% of the modules were calibrated in a testbeam with electrons, muons and pions [2].

2. CALIBRATION AND MONITORING OF THE TILECAL

Each TileCal cell can be divided, for monitoring and calibration purposes, into three parts: the optical part (the scintillating tiles where the light is produced by the charged

CP867, *XII International Conference on Calorimetry in High Energy Physics*,
edited by S. R. Magill and R. Yoshida

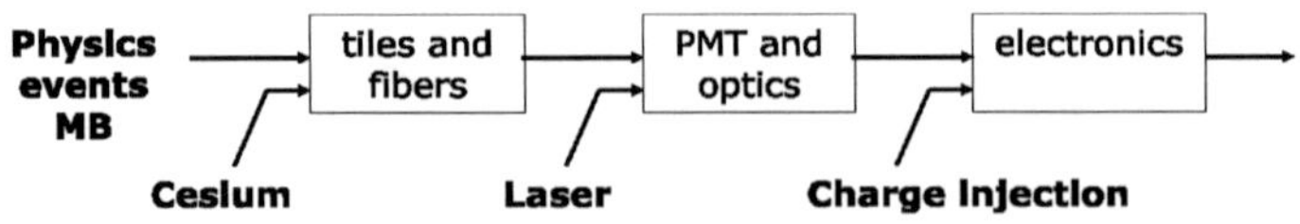

FIGURE 1. Diagram of the optical and electronic readout, and the calibration and monitoring systems.

particles and the wavelength shifting fibers that transport the light into the photomultipliers); the photomultiplier tube (PMT) that converts and amplifies the optical signal; and the readout electronics that shapes, amplify and digitizes the signal from the PMTs [3]. For each one of these parts a dedicated monitoring and calibration system was designed and built, which are described in the next sections. Figure 1 shows a diagram of the sequence of the TileCal optical and electronic readout, and the location of the calibration and monitoring systems inputs.

2.1. Charge injection system

The charge injection system (CIS) is designed to calibrate the relative response of the pulse readout electronics for all calorimeter PMTs and to monitor the variations with time. The objective is an accuracy of 1%. Each PMT channel has two analogue paths (bi-gain), the high and low gain (82 counts/pC and 1.3 counts/pC, respectively), digitized by a 10-bit ADC, covering a range of 800 pC (corresponding to a energy deposition of about 700 GeV). The CIS constants allows for the conversion of ADC counts to energy (in units of pC).

Each channel is equipped with calibration capacitors, charged from a high precision voltage source and discharged into the input electronics. The resultant waveform is similar to the one produced by the PMT for a given charge, except for an amplitude 10% larger and a FWHM 10% smaller. Dedicated CIS runs are taken, scanning the full range of charges for both gains, allowing for the determination of the ADC counts/pC ratio for each channel. The pulse amplitude is reconstructed, from ADC samples taken every 25 ns, with a 3-parameter fit. An example of this fit is shown in fig. 2.

The response as a function of the injected charge is shown in fig. 3 over the full range of both gains. Both the total response and the residuals from a linear fit are shown, and the deviation from linearity is no more than 2.2 counts over the active range of both gains.

The channel-to-channel variation on the measured gains has a RMS of 1.5% before applying any correction, comparable with the 1% variation in the calibration capacitors. During a 4-month period the corrected gains were stable at a RMS level better than 0.2%.

In ATLAS there will be periodic CIS runs over the full dynamic range during beam-off periods (between LHC fills and maintenance periods). Interleaved with data are also mono-CIS events, where a fixed amplitude signal is injected during missing bunches in

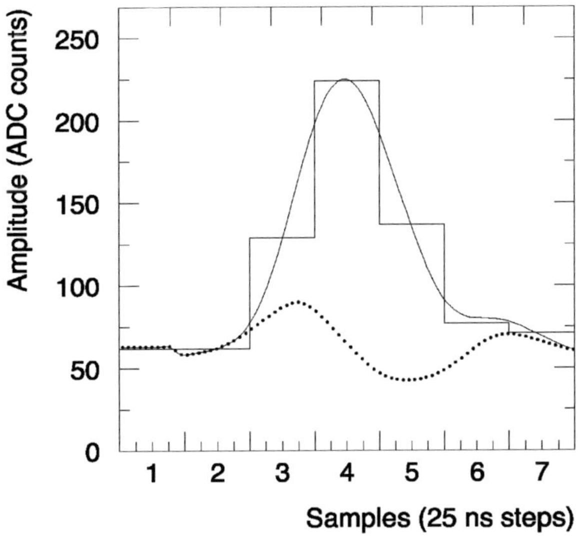

FIGURE 2. CIS pulse of a given channel. The histogram shows the ADC samples taken in 25 ns steps and the solid line corresponds to the 3-parameter fit. The contribution of the switch capacitance is shown as the dotted curve.

the LHC beam structure.

2.2. Laser system

The laser system was designed to calibrate and monitor the PMTs response with an accuracy better that 0.5%. It is used to check the PMTs linearity, for studies on pulse saturation recovery and in calorimeter timing, and also for debugging and diagnostics tests.

A frequency doubled Nd:YVO4 laser is used to produce pulses with a wavelength of 532 nm and a $\sim$ 10 ns width, synchronized with the bunch-crossing clock. Close to the laser the light is split, with part of it being sent to a set of photodiodes for relative intensity monitoring, and the remaining being delivered, through a light distribution system and clear optical fibers, to each calorimeter PMT.

The short and long term monitoring of the PMTs gain stability is the most important role of the system. This is done in special calibration runs from the ratio of the charge measured by each PMT, for a given laser pulse, to the photodiodes response. This ratio provides a reference for the relative gain of each PMT. The results obtained in testbeam conditions show that the PMTs are stable and that the laser system accuracy matches the design goal.

The linearity of the PMTs can be studied varying, in a known range, the intensity of the light delivered. Although the laser intensity itself can be varied only by a factor 10,

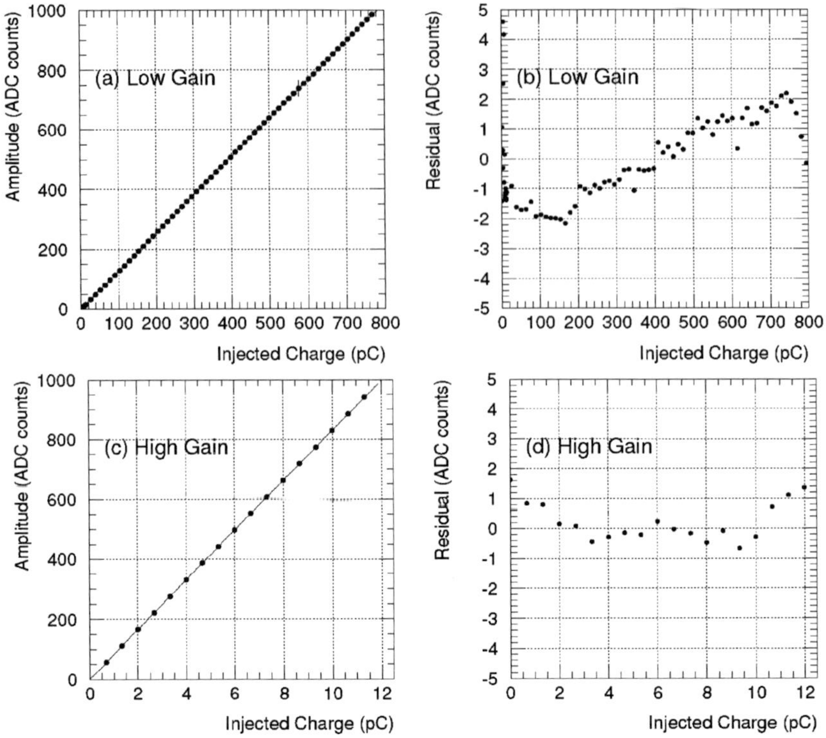

FIGURE 3. Amplitude of the response of the low gain, top, and high gain, bottom, paths of a single channel as a function of the injected charge. The right plots show the residuals from a linear fit.

an automated set of filters allows an extra dynamic range of over 1000. The global PMTs non-linearity was found to be better than 1% in the 0.7-5 pC range, 0.7% in the 5-80 pC range and 0.3% in the 80-700 pC range.

The laser system is also used to measure the absolute gain of each PMT, and the photoelectron yield can be obtained from the combination with the measured calorimeter response in pC/GeV. It takes advantage of the fact that the pulse-to-pulse intensity fluctuations of the laser are accurately tracked by the monitoring photodiodes. The laser system is also useful in timing studies, as taking into account the differences in propagation of signals, the timings done with projective particles and with laser pulses can be easily correlated, as shown in fig. 4. High intensity laser pulses can be sent to the PMTs in order to saturate the readout response and then apply saturation recovery strategies.

In ATLAS there will be special multi-pulse laser runs, linearity runs over the full dynamics ($\sim$ 60000), but also mono-pulse events inside empty bunches. In addition there will be saturation studies (with pulses well above the 800 pC limit ($\sim$ 1.4 TeV/cell)).

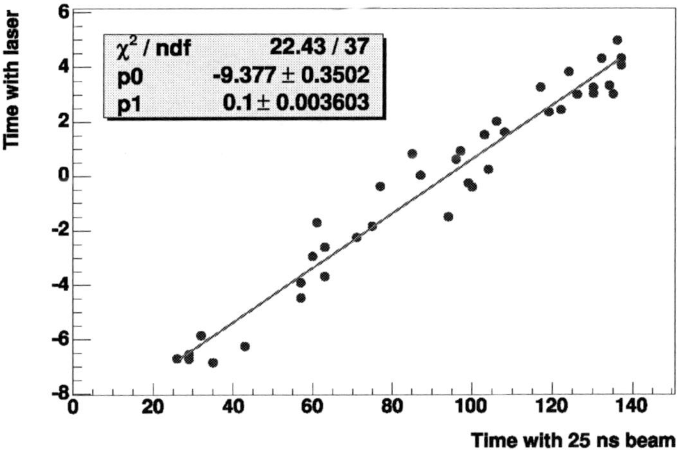

FIGURE 4. The timing with the laser system as a function of the timing with the 25 ns beam is shown.

2.3. Cesium source system

The cesium calibration is designed to determine the quality of the optical response of each calorimeter cell, to adjust the PMTs high voltage in order to equalize the response from all cells and to monitor it with time [4]. The objective is to keep the stability of the energy calibration at the 0.5% level. The system uses a ^{137}Cs γ source which moves, orthogonally to the tile planes, through a hole in the scintillating tiles. This source is transported by a hydraulic system in a series of straight paths along the calorimeter modules, so it passes through holes in every single scintillating tile and absorber plate. The current from each PMT is measured with an integrator which samples the signal at a 90 Hz rate [5]. The response of the individual tiles is clearly seen as the mean path of the γs is comparable with the periodic 18 mm separation between tiles. Figure 5 shows the scan of three adjacent cells where the response of the individual tiles is clearly seen and the effect of a poor optical coupling of a tile to the fiber is evident.

Two methods are used to measure the calorimeter response. In the first of them, the "integral method", the area of the curves for each cell, as in fig. 5, is evaluated and divided by the appropriate cell width. The cells at the calorimeter boundary need a special procedure to take into account the unobserved tail, and the cross-talk between cells is also corrected for. The accuracy of this method is about 0.2%. In the "amplitude method" the signal of each individual tile is fitted, taking into account the tails, and it yields the signal amplitude for every tile in the calorimeter. The accuracy of a single tile response is better than 2% and for the average amplitude of a cell better than 0.3%. The integral method is faster so it is used for a fast equalization of the cells responses (by adjusting the PMTs gains). The amplitude method is used for detailed scans to check the module quality.

In the procedure to equalize the response of every calorimeter cell, the high voltage

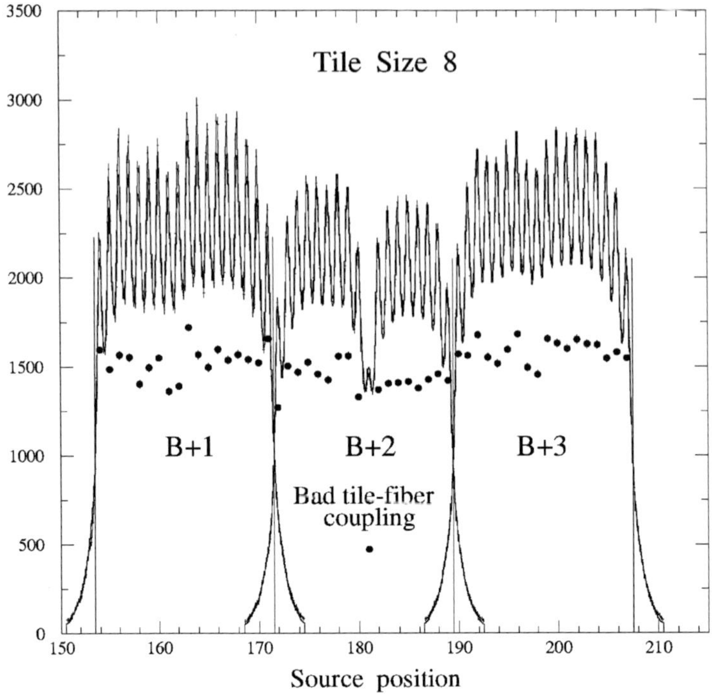

FIGURE 5. The PMT current as a function of the source position for three adjacent cells.

(HV) of each PMT is adjusted in a iterative process to reach a target value of the gain. The new HV values are calculated with the formula:

$$HV_{new} = HV_{old} \left(\frac{\text{target response}}{\text{measured response}} \right)^{1/\beta} \tag{1}$$

The parameter β changes slightly for each PMT, but a single value, $\beta = 7$, gives a convergence to within 0.5 V in the HV after three iterations. The target response, the ratio of the measured PMT charge to the deposited electromagnetic energy, was established to be in the range 1.04-1.1 pC/GeV, depending on the angle of the incident particle.

In testbeam the overall cell-to-cell non-uniformity of the calorimeter after the Cesium equalization, as seen by muons and electrons, is less than 3%. It is worse than the Cesium measurements because the muon and electron beams and the Cesium source deposit energy in different parts of the cells, and the scintillating tiles are not identical (a 5–8% variation was observed during instrumentation). A hadronic shower spans over many tiles and the response non-uniformity for single pions is at the level of 1.3%.

In ATLAS the Cesium system will be used to monitor the long term stability of the calorimeter. It allows not only to monitor the PMTs stability but also to detect bad tile-fiber couplings, scintillator aging and optical problems in general. Between the Cesium runs the PMTs stability is monitored by the laser system.

2.4. Minimum bias events

The minimum bias (MB) events in ATLAS are inelastic *pp* collisions with low momentum transfer, whose rate is proportional to the LHC luminosity [6]. Twenty three MB events per bunch crossing are expected in high luminosity runs. These events produce a non-negligible occupancy of the TileCal cells, with rates uniform in the azimuthal angle ϕ and moderately dependent on the pseudo-rapidity η. Since the MB current, averaged over milliseconds, is almost constant and proportional to the interaction rate, it can be used to monitor the calorimeter response, the relative luminosity and the beam quality during physics runs. Dead and hot channels are also identified in MB data.

For each PMT channel the anode output is DC coupled to an integrator circuit based in an operational amplifier. This has a fixed RC time constant of 10 ms and a gain selectable from six predefined values (as the MB current changes with the position of the cell on the calorimeter, as well as with the luminosity). All integrator outputs, multiplexed to a 12-bit ADC, are readout every 2 s. In ATLAS, in a high luminosity run, are produced about 8 million MB events during the integration time.

The MB monitoring system was tested with pion beams, at different rates, in order to reproduce the conditions expected in ATLAS and check the system's performance and dynamic range. The results are consistent with the expectations, based on the spills time profile, the average energy deposited, the beam sampling and the MB system design.

3. CONCLUSIONS

The ATLAS hadronic calorimeter monitoring and calibration systems were presented. The CIS, laser and Cesium allow to calibrate and to monitor the calorimeter response with a 0.5–1% precision. After HV equalization the overall cell-to-cell non-uniformity, as measured with electron and muon beams, is better than 3%. This non-uniformity for hadronic showers is at the 1–1.5% level. Other TileCal monitoring systems, like the HV, the low voltage or the cooling systems, were not presented in this paper.

After the testbeam and the commissioning phases, the different calibration and monitoring systems are ready, waiting for the first ATLAS data taking.

REFERENCES

1. ATLAS/Tile Calorimeter Collaboration "Tile Calorimeter - Technical Design Report" CERN/LHCC 96-42, 1996.
2. ATLAS/Tile Calorimeter Collaboration, "Testbeam Summary Paper", in preparation, 2006.
3. F. Sarri, "Calibration of the ATLAS Tile Calorimeter", proceedings of the "XI International Conference on Calorimetry in High Energy Physics – CALOR2004", published in "Calorimetry in Particle Physics", World Scientific (2005).
4. E. Starchenko et al., "Cesium monitoring system for ATLAS Tile Hadron calorimeter", Nucl. Instr. an Meth. A494 (2002) 381.

5. N. Shalanda et al., “Radioactive source control and electronics for the ATLAS Tile Calorimeter cesium calibration system", Nucl. Instr. an Meth. A508 (2003) 276.
6. X. Portell et al., “Distributions of Minimum Bias current measurements in TileCal", ATL-TILECAL-03-010, CERN, 2003.

Cosmic ray calibration of the $PbWO_4$ crystal electromagnetic calorimeter of CMS

G. Franzoni

University of Minnesota, 116 Church Street S.E., Minneapolis

Abstract. The Compact Muon Solenoid experiment at the CERN LHC features a high precision $PbWO_4$ crystal electromagnetic calorimeter. Each crystal is first precalibrated with a radioactive source and by means of optical measurements. After the assembly, each supermodule (1700 crystals) is exposed to comics rays.

The comparison between intercalibration obtained from cosmic muons and electrons from test beam was performed at the end of 2004 for an initial set of 130 channels and showed that a precalibration with a statistical precision of 1 to 2 % can be achieved within approximately one week. An important aspect of the comics muons analysis is that it is entirely based on the calorimeter data, without using any external tracking device.

We will present the setup and results from the 2004 test as well as recent data recorded on many supermodule.

Keywords: LHC, CMS, Electromagnetic calorimeter, Calibration
PACS: 07.20.Fw, 29.40.Mc

THE COMPACT MUON SOLENOID

The CMS experiment is one of the two multi-purpose detectors that will take data at the Large Hadron Collider (LHC) at CERN starting from 2007. The physics goals of the experiments operating at the LHC are diverse, ranging from understanding the origin of mass (searches of the Higgs boson) and physics beyond the Standard Model to precision measurements on known particles exploiting the huge production rates. The detectors need to have excellent performance in identifying leptons, reconstructing masses of heavy decaying particles and particle discrimination. To achieve this and to detect neutrinos through missing energy, the calorimeters need to be as hermetic and accurate as possible [1]. CMS is composed of different subdetectors, each optimized for different complementary measurements. Placed closest to the interaction point, is a tracking system (pixel layers and silicon strips) which will perform charge and momentum measurements using the 4T magnetic field. Outside the tracker, but still within the magnetic coil there are the two calorimetric subdetectors. The first, a homogeneous electromagnetic calorimeter (ECAL [2]), based on scintillating $PbWO_4$ crystals which measures the energy of high energy electrons and photons. Secondly, a sampling hadronic calorimeter consisting of layers (17 in the barrel, 19 in the endcaps) of brass absorbing material interleaved with plastic scintillator. The outermost subdetector is dedicated to the muon detection. The barrel muon system relies on drift tubes in the central region, and on cathode strip chambers in the endcaps. Resistive plate chambers are used throughout the central and forward region as additional, fast, muon triggering detectors.

CP867, *XII International Conference on Calorimetry in High Energy Physics,*
edited by S. R. Magill and R. Yoshida

THE CMS ELECTROMAGNETIC CALORIMETER

The CMS electromagnetic calorimeter is composed of 75848 $PbWO_4$ scintillating crystals. Their shape and size slightly changes along the pseudorapidity range and a typical crystal volume is $24.7 \times 24.7 \times 230$ mm^3 . The crystals' length corresponds to 26 radiation lengths. Both halves of the ECAL barrel consist of 18 supermodules (wedges in ϕ), divided in four modules. To avoid cracks aligned with particle trajectories, the axes of the crystals are tilted by 3° with respect to the direction leading to the interaction point both in *eta* and in *phi*. The crystals are grouped in subunits of 5×2 held together by a glass fiber alveolar structure. Two photodetectors are glued on the back of each crystal embedded in a plastic capsule. For the barrel, avalanche photodiodes (APD's) produced by Hamamatsu are employed. Such devices have an 80% quantum efficiency and have been chosen since they are able to operate in the 4T magnetic field with a gain equal to 50 and are able to withstand the severe irradiation of ten years LHC running. Just outside the APD's, there is a thick aluminum plate (the grid) which holds the weight of the crystals and in which the connectors for the read-out electronics are embedded. The two endcaps consist of identically shaped crystals, shorter (220 mm, 25 X_0) but slightly larger than the barrel crystals. The endcaps are divided into mechanical units of 5×5 crystals held together by an alveola. The crystals are off-pointing by angles between 2 and 5. The electronics and the cooling system are placed just outside the grid. The electronic read-out chain follows a modular structure whose basic elements are groups of 5×5 crystals which corresponds, in the barrel, to a trigger tower. Each APD is connected to a preamplifier which is followed by an ADC, and five of such channels sit on the same Very Front End board (VFE). Groups of five VFE's are plugged into the same motherboard which hosts also the Low Voltage Regulator board (LVR) supplying them.

CMS ECAL PRECALIBRATION

Before installing the supermodules in CMS, each channel of ECAL is precalibrated by means of laboratory measurements [3] performed at CERN and in Rome (INFN-Enea Casaccia). The scintillation light yield of any crystal is measured positioning a radioactive source next to it and performing a scan along the crystal length in steps of 1 cm. The light is collected with a photomultiplier.

Besides that, the laboratory measurements comprise also a study of the longitudinal optical transmission which is determined at 11 different wavelengths. The transmission has a plateau for high wavelengths and a steep rise at 360 nm.

It has been found that there is a correlation between the longitudinal transmission at 360 nm, which can be measured in a simpler and more precise way, and the light yield. Such correlation is used to provide the best laboratory intercalibration. By comparing the laboratory intercalibrations to the intercalibration obtained by test beam data, they are found in agreement at 4.2% level.

Presently (summer 2006), a test beam campaign at CERN is occurring that will provide intercalibrations from 120 GeV electrons for some of the 36 supermodules. These intercalibrations will be the most accurate but will not be available for all the crystals.

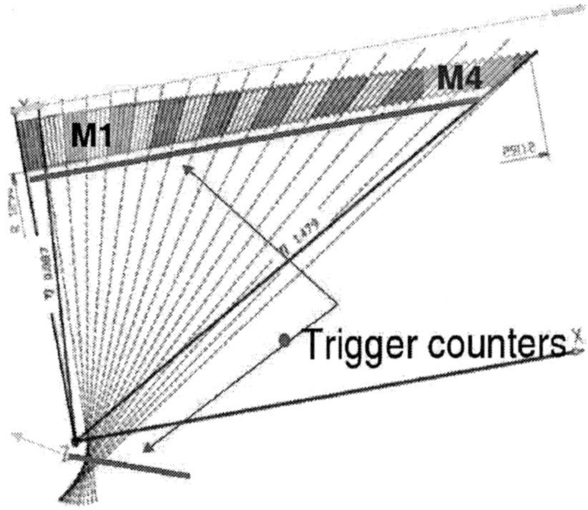

FIGURE 1. Schema of the CMS ECAL cosmic calibration stand. The supermodule tilted of 10 degrees and the two trigger scintillator planes can be seen.

An independent strategy of precalibration of the ECAL channels using cosmic muons has as proposed in 2004 will be described here [4]. The goal is providing an intercalibration method alternative to the laboratory measurements which can give information for all the channels of the CMS ECAL barrel and have a better accuracy.

THE COSMIC MUON CALIBRATION STAND

A dedicated stand for the cosmic muons data taking has been put in place in the CERN North Area. A structure which can hold one supermodule has been developed such that the supermodule can take data inclined of 10° on the horizontal, as shown in figure 1. This allows the axes of the crystals to be, on average, closer to the vertical direction where the cosmic muons flux is the most copious. In order to enhance the signal amplitude, the APD's are operated at gain 200, a factor 4 higher than what it will be *in situ* at CMS. The trigger is provided by the coincidence of two planes of scintillators. One is located just below the crystal and covers the whole $\eta - \phi$ range, while the other is smaller and placed at the point where the LHC interaction point would be *in situ*. The trigger geometry selects muons which traverse the ECAL crystals along an almost pointing direction, resulting in a rate of 7 Hz.

MUONS IN ECAL AND SELECTION OF THE SAMPLE

Cosmic muons aligned to the crystal axis provide a reference signal for the calibration, depositing on average 250 MeV of energy in the $PbWO_4$. This corresponds to a signal of 27 ADC counts, to be compared to the pedestal spread which is on average 1.3-1.5 ADC counts.

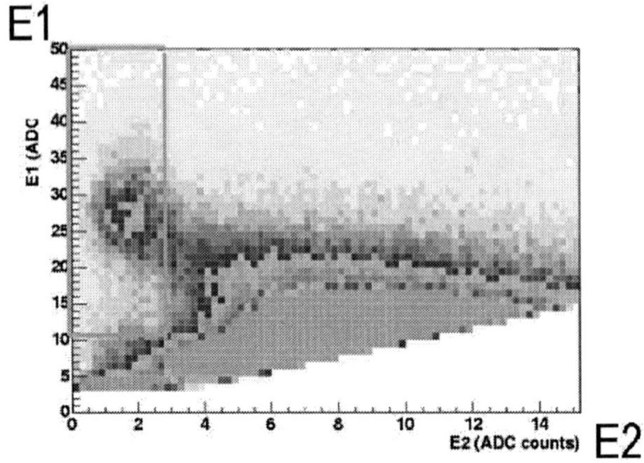

FIGURE 2. E_1 versus E_2 for all the crystals of one supermodule. See text for definition of E_1 and E_2.

The actual ratio of the APD 200, as used at the cosmic stand, to the value that will be used at CMS, 50, is measured very accurately using laser events, the spread of the distribution of such ratios being 2.5%.

Pointing muons are selected based on calorimetric cuts. In each event, the crystal with the maximal energy deposition E_1 is sought and required to have an amplitude above 10 ADC counts. Furthermore, a veto is imposed: the maximum of the energies deposited over its eight neighbors E_2 has to be below 3 ADC counts. Events that fulfill these two cuts (7% of the total) are likely to have had a muon traversing mainly one crystal and are used for the calibration.

The distribution of E_1 versus E_2 for all the crystals of one supermodule is shown in figure 2. The rectangle shows the region of the selected events.

The number of selected events collected in a crystal in a day is shown in figure 3. There are about 4 times as many events in the region with maximal occupancy (between module 1 and 2) as in module 4, due to the different orientation of the axis of the crystals with respect to the vertical direction. The tree peaks among the modules come from a reduced effectiveness of the veto ($E_2 < 3ADC$) due to the 3 mm inter-module cracks.

INTERCALIBRATION PROCEDURE

The intercalibration coefficients are determined from position of the peak of E_1 distribution. The shape of the E_1 distribution mildly depends on the η coordinate of the crystal. The plots in figure 4 show the E_1 distributions integrating over all the crystals of a given module as obtained from the data collected in 14 days.

For the intercalibration, 17 η-dependent reference E_1 distributions are built from a simulated sample merging events of 5 rings in η. Such reference distributions are used to perform an unbinned maximum likelihood fit to data of each individual crystal, with an energy scale factor (the intercalibration) and a normalization as free parameters. This

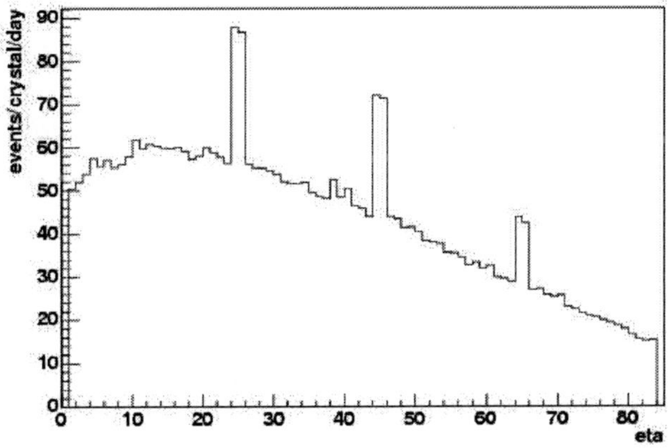

FIGURE 3. Number of selected events per day per crystal as a function of pseudorapidity.

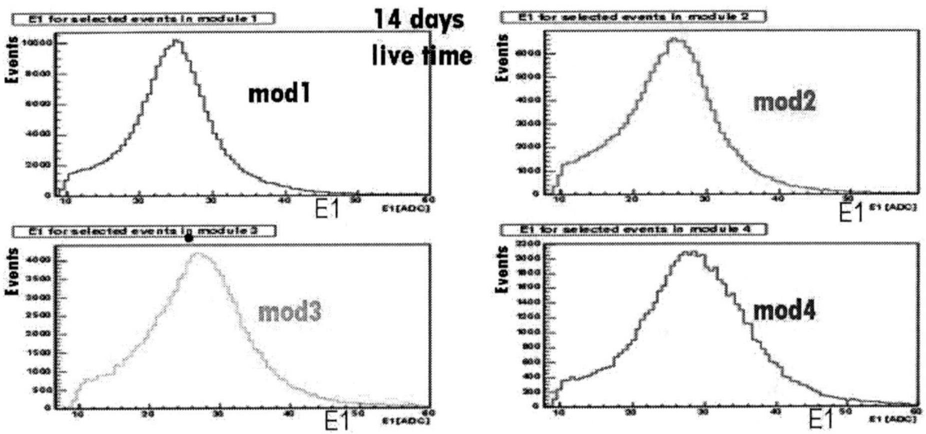

FIGURE 4. E_1 distributions integrating over all the crystals of a given module.

method has proven robust and allows intercalibration of all the crystals of a supermodule except those located at the supermodule boundaries where the lack of neighbors makes the calorimetric cut inefficient. Studies dedicated to the edge crystals are on going and are not presented here.

STATISTICAL PRECISION

Studies have been carried out to estimate the statistical precision of the method. Comparisons of the intercalibrations as obtained from two equally numerous sub samples of

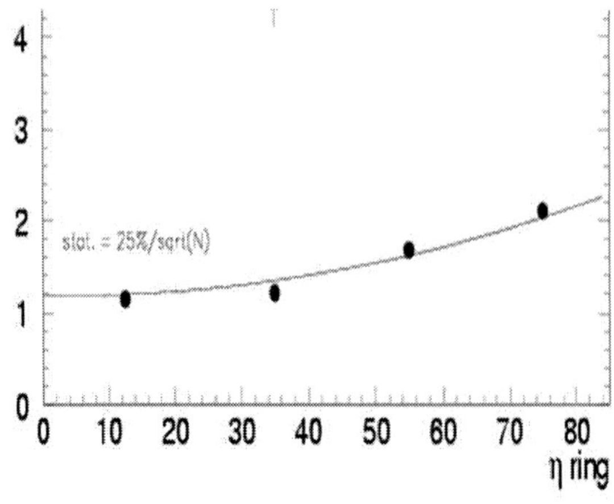

FIGURE 5. Statistical precision of the intercalibration method.

the overall available statistics have been performed for several supermodules resulting in what shown in figure 5. The accuracy is better in the regions at low η, 1.2%, than at high η, $\sim$ 2%. This is due to the larger data sample available for those crystals. Fitted to the points a curve $25\%/\sqrt{N}$, N being the number of selected events and 25% the ratio of the width of the distributions of figure 4 to their mean.

ACCURACY

Direct comparison of the cosmic muons intercalibrations to test beam results for several supermodules will be possible shortly, thanks to the data collected at the 2006 testbeam campaign.

One supermodule was exposed to cosmic rays just after having been intercalibrated with electrons during the last test beam in 2004. Using a prototype trigger telescope, muon data were collected for 41 hours. The comparison of the comic muons intercalibrations to test beam ones was performed for 130 channels in module 1 (boundaries excluded): a scatter plot and a histogram with the differences are shown in figure 6.

The 3.2% spread is the result of three factors. Firstly the statistical uncertainty of the cosmic intercalibration, 1.8%. Secondly the spread in the distribution of the ratios between gain 200 and gain 50 (the test beam intercalibrations are obtained with the same APD gain as *in situ*). This contribution will be accounted for in the final procedure employing the measured ratios. Finally, neglecting the uncertainty on the test beam intercalibrations, the systematic effects between the two methods have to be taken into account. By subtraction in quadrature, the 2004 data indicate that the latter effect is at the level or below 2%. Such value, obtained by a simple direct comparison, could be reduced once extensive studies will be possible on the correlations between the two methods and more experience will be gained on possible biases of one method with respect to the other.

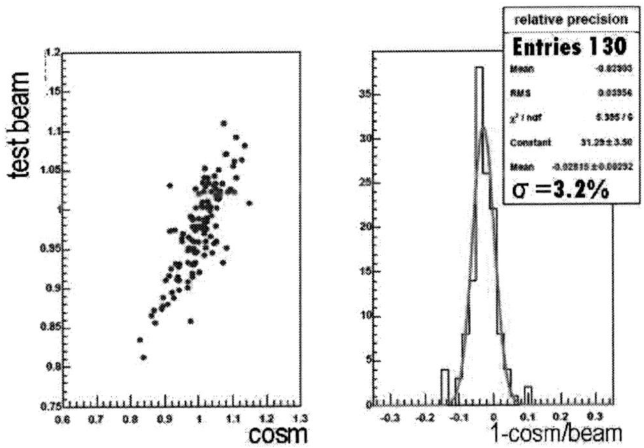

FIGURE 6. Comparison of intercalibration obtained from cosmic muons and test beam on the supermodule which was intercalibrated at the 2004 test beam.

CONCLUSION

We have presented a newly proposed method to intercalibrate all the 36 barrel supermodules of the CMS electromagnetic calorimeter based on cosmic muons data. The data collection is performed at a dedicated stand equipped with a scintillator telescope and allowing the supermodule to be inclined of 10° with respect to the horizontal position. By means of a calorimetric veto, a sample of muons that have traversed crystals quasi aligned to their axis is selected for each crystal not located on the supermodule edges (for the calibration of such crystals studies are on going). The intercalibrations are obtained by fitting reference distributions to the distribution of the amplitudes of each crystal.

The statistical accuracy of the method ranges from 1.2% for crystals between module 1 and 2, and $\sim$ 2% for crystals in module 4.

Comparison of the cosmic muons intercalibration to the test beam results of 2004 indicate that systematic differencies are at the level or below 2%. Such value, obtained by a simple direct comparison, could be reduced once extensive studies will be possible on the correlations between the two methods and more experience will be gained on possible biases of one method with respect to the other.

REFERENCES

1. *The CMS Technical Proposal*, CERN/LHCC 94-38
2. *The Electromagnetic Calorimeter Project. Technical Design Report*, CERN/LHCC 97-33
3. F. Cavallari *et al*, *CMS ECAL intercalibration of ECAL crystals using laboratory measurements*, CMS NOTE-2006/073
4. M. Bonesini *et al.*, *Inter-calibration of the CMS electromagnetic calorimeter with cosmic rays before installation*, CMS NOTE-2005/023

CMS ECAL Calibration Strategy

Georgios Daskalakis

Imperial College, London, UK

Abstract. The CMS Electromagnetic crystal Calorimeter (ECAL) must be precisely calibrated if its full potential performance is to be realized. Inter-calibration from laboratory measurements and cosmic ray muons will be available for all crystals and has been demonstrated to give good pre-calibration values at the start-up; some crystals will be also inter-calibrated using an electron beam. *In-situ* calibration with physics events will be the main tool to reduce the constant term of the energy resolution to the design goal of 0.5%. In the following the calibration strategy will be described in detail.

Keywords: CMS, ECAL, calibration, calorimeter, crystals, in-situ, electrons, lead tungstent
PACS: 06.20.Fn, 07.20.Fw

The ECAL Detector

The ECAL [1] is made out of 75848 lead tungstate ($PbWO_4$) crystals (Fig. 1a). They are arranged into a Barrel (61200 crystals), covering the central rapidity region ($|\eta| < 1.5$) and two Endcaps (7324 crystals each) which extend the coverage (up to $|\eta| <$ 3.0). Due to the high density (8.28 g/cm^3) and the small radiation length (X_0 = 0.89 cm) of $PbWO_4$, the calorimeter is very compact and can be placed inside the magnetic coil [2] needed for precise momentum measurements with the Tracker [3] and the Muon [4] systems. The small value of the Molière radius (2.2 cm) matches well the very fine granularity needed by the high particle density of the events at LHC. Crystals are organized in a quasi-projective geometry [1], which improves the hermeticity of the detector.

The electrons lose energy via bremsstrahlung in the Tracker material (Fig. 1b) while the photons are converted into electron pairs. In addition, the strong magnetic field bends the electrons causing the radiated energy to spread in φ. Both effects impact electron/photon energy resolution making the *in-situ* calibration of ECAL a challenge. Special reconstruction algorithms [5] have been developed to recover the radiated energy from electrons and to reconstruct the converted photons.

ECAL Calibration

The physics reach of the ECAL, in particular the discovery potential for a low mass Standard Model Higgs boson in the two photon decay channel [6], depends on its excellent energy resolution.

CP867, *XII International Conference on Calorimetry in High Energy Physics,*
edited by S. R. Magill and R. Yoshida

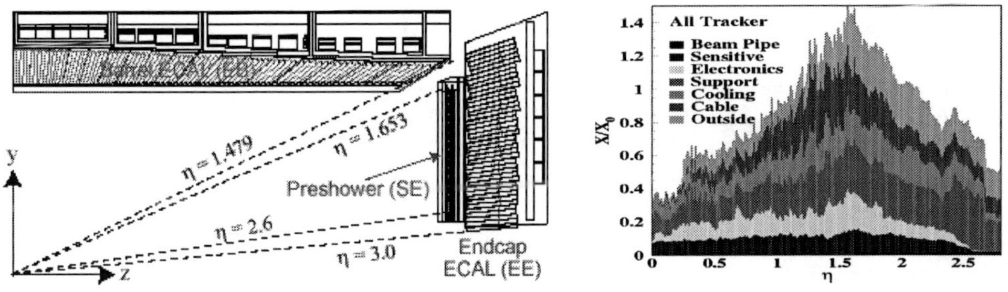

(a) (b)

FIGURE 1. (a) A slice through a quadrant of the CMS Electromagnetic Calorimeter (b) the Tracker material distribution in front of the ECAL.

The intrinsic ECAL energy resolution measured in Testbeam (by summing the deposited energy in a 3×3 array of crystals around the crystal in beam) is expressed by the following parameterization (E is in GeV):

$$\left(\frac{\sigma}{E}\right)^2 = \left(\frac{2.9\%}{\sqrt{E}}\right)^2 + \left(\frac{125(MeV)}{E}\right)^2 + (0.30\%)^2 \quad (1)$$

which matches the design resolution for a perfectly calibrated detector. Mis-calibration will directly affect the constant term, degrading the overall ECAL performance.

The goal of the calibration strategy is to achieve the most accurate energy measurement for electron and photons. Schematically, the reconstructed energy can be decomposed into three factors [5]:

$$E_{e,\gamma} = G \times F \times \sum_i^{Cluster} c_i \times A_i \quad (2)$$

where G is a global absolute scale and F accounts for energy losses due to bremsstrahlung and containment variations. The c_i factors are the inter-calibration coefficients while the A_i are the signal amplitudes, in ADC counts, which are summed over the clustered crystals. The main source of channel-to-channel response variation in the Barrel is the crystal-to-crystal variation of scintillation light yield which has an r.m.s. of ~15%. In the Endcaps, the crystal signal yield and the product of the gain, quantum efficiency and the photocathode area of the VPTs, have an r.m.s. variation of almost 25%.

The target inter-calibration precision can only be achieved using physics events. Over the period of time in which the physics events used to provide an inter-calibration are taken, the calorimeter response ideally should remain stable and constant to high precision. One source of significant variation is changes in crystal transparency caused by irradiation and subsequent annealing. The changes are tracked and corrected using a laser monitoring system [7]. In addition, the sensitivity of both crystal and photo-detector response to temperature fluctuations requires a precise control of the temperature stability. The water-cooling system guarantees a long-term temperature stability of the crystal volume and the APDs below the 0.1 °C level [7], in order to be able to meet the target values for the energy resolution.

Calibration Roadmap

There are two distinct periods during which calibration can be performed. The first period is before the installation of the detector. During that period ECAL crystals can be calibrated in Testbeam (around 10k crystals only, due to the restricted beam time), using light yield measurements in the laboratories that check the crystals quality and by using cosmic muons. When the detector will be fully operational, minimum bias events and/or Level-1 jet triggers could be used in order to achieve a fast crystal inter-calibration. Precise *in-situ* inter-calibration can be achieved with isolated electrons from sources like $W \to e\nu$ or $Z \to e^+e^-$ decays. The absolute calibration scale as well as other calibration tasks can be achieved by using the mass constraint for electrons from $Z \to e^+e^-$ decays. There is also the possibility to achieve a fast and accurate calibration using $\pi^0, \eta \to \gamma\gamma$ or $Z \to \mu\mu\gamma$ decays.

Calibration in Testbeam

In the Testbeam, supermodules [1] are mounted on a rotating table that allows rotation in both the η and φ coordinates and are fully scanned with high-energy electron beams. The incident electron positions are measured with a set of hodoscopes. The response of a single crystal to electrons depends on the electron incident position. The dependence on the two lateral coordinates can be factorized and corrected. The inter-calibration coefficients c_i for crystal i are defined as the ratio of the mean value of the corrected response to a reference value.

The statistical uncertainty remains negligible (less than 0.1%) provided that at least 1000 events are taken per crystal. The inter-calibration precision, when these constants are used *in-situ*, is expected to be limited by variations occurring in the time between their determination in the Testbeam and their utilization in the installed detector.

Calibration from Laboratory Measurements

The crystal calorimeter is being assembled in 2 regional centers: at CERN and at INFN-ENEA Casaccia near Rome. During the assembly phase, all the detector components are characterized and the data are saved in the construction database. From these data it is possible to estimate the inter-calibration coefficients c_i of each channel i as [7]:

$$\frac{1}{c_i} \sim LY \cdot \varepsilon_Q \cdot c_{ele} \cdot M,$$

where LY is the Light Yield of the crystals, M and ε_Q are respectively the gain and quantum efficiency of the photo-detectors and c_{ele} is the calibration of the electronics chain. The crystal LY is measured in the laboratory with a photo-multiplier tube, exciting the crystal with a ^{60}Co source that emits photons with energy of 1.2 MeV. This gives an average LY_{PMT} for the $PbWO_4$ crystals of 10 pe/MeV at 18 °C. The

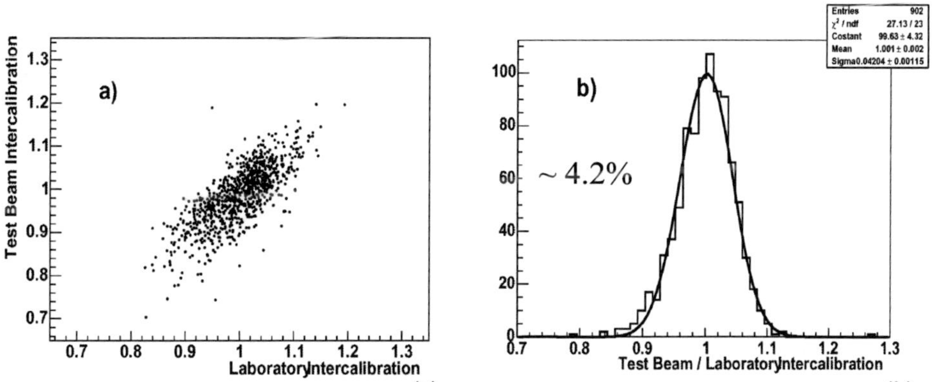

(a) (b)

FIGURE 2. (a) Inter-calibration coefficients obtained in Testbeam versus those obtained from laboratory measurements (b) ratio between Testbeam and laboratory inter-calibration coefficients.

measurements span about 7 years of crystal production.

In order to establish the inter-calibration precision achieved with the laboratory measurements, the inter-calibration coefficients are compared with those from Testbeam measurements (Fig. 2a). As can be seen from their ratio (Fig. 2b) an inter-calibration precision of about 4.2% can be obtained from laboratory measurements.

Calibration with Cosmic Ray Muons

Inter-calibration coefficients for Barrel crystals are also obtained using cosmic muons that are well aligned with the crystal axes [8]. For this measurement the APD gain is increased by a factor 4 with respect to the gain to be used during normal data taking by increasing the bias voltage. This improves the signal to noise ratio and allows the selection of muons passing through the full length of crystals by vetoing on signals in surrounding crystals. An overall precision of 3.0-3.5% should be achievable in one week of data taking. The statistical contribution to the overall uncertainty was estimated to be around 2%.

Calibration with the φ – uniformity Method

The proposed technique makes use of the φ-uniformity of deposited energy to inter-calibrate crystals within rings at constant η. Due to the symmetry of the ECAL about $\eta=0$, crystals with same $|\eta|$ are folded. In the Barrel, there are 85 pairs of rings with 360 crystals per ring. In the Endcaps, there are 39 pairs of rings and the number of crystals per ring varies with η.

Inter-calibration is performed by comparing the total energy deposited in each crystal with the mean of the distribution of total energies for all crystals in a ring. For the moment, two choices of event trigger have been investigated: random bunch crossings [9], and Level-1 jet triggers [10]. The inter-calibration precision for a

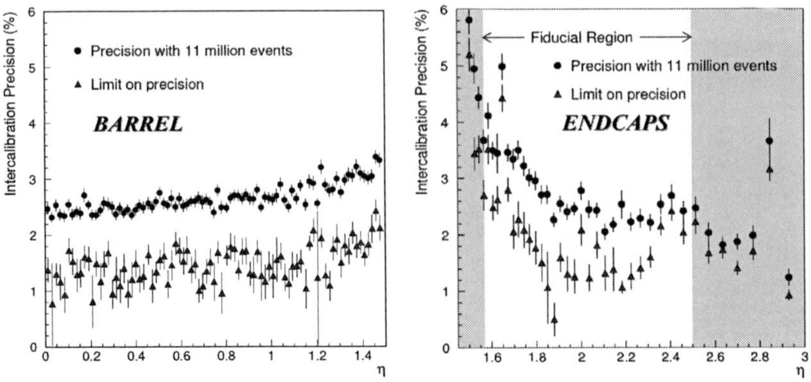

(a) (b)

FIGURE 3. Inter-calibration precision achieved with the φ-uniformity method (a) in the ECAL Barrel and (b) in the ECAL Endcaps, obtained with 11 millions Level-1 jet trigger simulated events (circles). The expected limit on the inter-calibration precision is also shown (triangles).

given $|\eta|$ is obtained from the Gaussian width of the distribution of ΣE_T for the pair of rings of crystals at that value of $|\eta|$.

A limit on the precision arises due to non-uniformities in φ, primarily from the in-homogeneity of Tracker material, but also from geometrical asymmetries such as the varying off-pointing angle of Endcap crystals, and the boundaries between Barrel supermodules.

Results based on a Level-1 jet-triggers simulated sample are shown in Fig. 3a for the Barrel and in Fig. 3b for the Endcaps. It can be seen that without using any knowledge about the material distribution in the Tracker, the limit on the precision is close to 1.5% throughout the Barrel and between 3.0% and 1.0% for the fiducial region of the Endcaps. It is expected that the limit on the precision will be closely approached with a few tens of millions of events. This is equivalent to about 10 hours of data taking, under the assumption that 1 kHz of Level-1 bandwidth is allocated to single jet triggers, and that the calibration software has access to this rate, either running on the Filter Farm, or more probably, running offline on a highly compacted data stream (a few tens of channels stored per event).

Calibration with $Z \rightarrow e^+e^-$

The Z mass constraint in $Z \rightarrow e^+e^-$ decays is a powerful tool for calibration. A number of different uses are envisaged, from tuning the corrections of the electron reconstruction algorithms as shown in [11], to the inter-calibration of regions of the ECAL, for example as a complement to the φ-uniformity method at the start-up.

For a preliminary estimate of the inter-calibration factors between rings, electrons that radiated little were chosen since their reconstructed energy shows the least dependence on the Tracker material, and hence η. The method has been tested taking the calorimeter regions as rings of crystals (at fixed η) in the ECAL Barrel. The results obtained when starting from a 5% mis-calibration between rings and a 2% mis-calibration between crystals within a ring, using events corresponding to an integrated luminosity of 2.0 fb^{-1}, corresponds to 0.6% ring inter-calibration precision.

Calibration with Isolated Electrons

Once the Tracker is fully operational and well aligned, inter-calibration of crystals can be performed using the momentum measurement of isolated electrons [12]. The main difficulty in using electrons for inter-calibration is that electrons radiate in the Tracker material in front of the ECAL, and both the energy and the momentum measurement (P) are affected. Moreover the average amount of bremsstrahlung varies with Tracker material thickness. The ECAL energy will be measured by summing the energy deposited in the 5×5 array of crystals (S25) around the crystal with the maximum signal. The energy in the 5×5 array does not require the complexity of a single crystal containment correction and helps in cleanly separate the inter-calibration from the corrections required by the super-clustering algorithms. In the Endcap, the energy measured in the Preshower and associated with the electron cluster is added to the energy summed in the crystals.

In order to extract the inter-calibration constants the individual crystal contributions must be unfolded, while minimizing the difference between the energy and momentum measurements. Two algorithms to achieve this minimization have been tested: an iterative technique that was used for the *in-situ* calibration of the BGO crystals in the L3/LEP experiment and a matrix inversion algorithm. The results, both in terms of precision and in terms of speed of algorithm, are similar, and show no dependence on the technique used. The event selection was based on variables that are sensitive to the amount of bremsstrahlung emission, chosen to select events with little bremsstrahlung.

Due to the variation of the average value of S25/P with pseudorapidity, caused by the variation of the amount of material in front of the ECAL, the calibration task will be divided into two steps. In the first step crystals in small regions in η, over which the average value of the S25/P is rather constant, will be inter-calibrated. In the second step the small regions will be inter-calibrated with each other.

The calibration precision versus η achievable for a fixed integrated luminosity follows the Tracker material budget distribution (Fig. 4a). The simulated data used to obtain these results correspond to about 5(7) fb^{-1} in the Barrel(Endcaps). This estimation uses the PYTHIA cross section for the W-production with no k-factor. The calibration precision was also extensively studied in different φ-regions keeping the same η interval. There is no evidence of any φ-dependence.

The calibration precision achievable is strongly dependent on the number of electrons collected per crystal (HLT output). In Fig. 4b the inter-calibration precision versus the number of electrons per crystal is shown for three different areas of the ECAL Barrel. The curves, from bottom to top, represent the accuracy for low, middle and high η regions in the ECAL Barrel. As can be seen, an inter-calibration precision of 0.6% averaged over the Barrel can be achieved with 10 fb^{-1} of integrated luminosity.

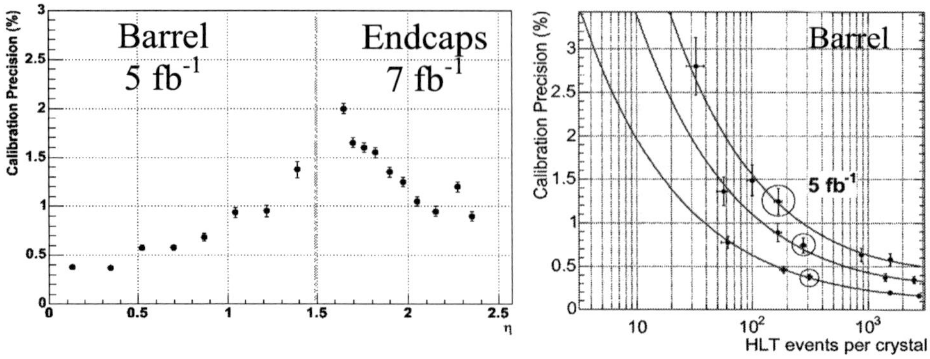

(a) (b)

FIGURE 4. (a) Calibration precision versus η using isolated electrons (b) Calibration precision versus HLT events per crystal for different η Barrel regions. Upper curve: the last 10 crystals in the Barrel ($1.305<\eta<1.479$); middle curve: 10 crystals in the middle of the Barrel ($0.783<\eta<0.957$); lower curve: the first 15 crystals in the Barrel ($0.0<\eta<0.261$). The third point along each line gives the precision of 5 fb^{-1} of integrated luminosity.

After crystals within regions of η are inter-calibrated, the regions have to be calibrated among themselves. This task is accomplished by selecting electrons with minimum energy loss due to bremsstrahlung. After this selection, the resulting values of the peaks of the S25/P distributions that are found are consistent with the pseudorapidity dependence of shower containment.

The rate of the jet background in the single electron trigger stream (HLT output) is estimated to be 27Hz at low luminosity out of which 16Hz are expected in the Barrel. The residual background has been investigated for the Barrel case. After the calibration selection is applied, the surviving background corresponds to a rate of 2.3 Hz. One third of this rate comes from b/c→e semileptonic decays. Such decays might be useful in the calibration process, increasing the overall calibration statistics. If required, the background can easily be further reduced by a factor 10, using isolation cuts, with only a small effect on the signal efficiency.

Calibration with $\pi^0,\eta\rightarrow\gamma\gamma$ and $Z\rightarrow\mu\mu\gamma$

The possibility of inter-calibrating the ECAL using the reconstructed mass of $\pi^0,\eta\rightarrow\gamma\gamma$ is being investigated. These low mass particles could provide an important additional calibration tool which is useful for relatively rapid inter-calibration of all crystals, study of the effects of crystal transparency corrections from the laser monitor and rapid check-out and monitoring of detector performance.

The inter-calibration obtained from low-energy $\pi^0\rightarrow\gamma\gamma$ is not sensitive to Tracker material if unconverted photons are selected. The only effect of the Tracker material is a rate loss at larger η values due to photon conversions.

It has been shown that π^0s, useful for calibration, can be located within events using the ECAL Level-1 trigger information, requiring very little processing time to extract the small amount of information relevant for calibration. In the ECAL Barrel, the π^0 mass peak, with relatively little background, has a mass resolution of about 8%. Around 1.4% of the Level-1 trigger events have a usable π^0 in the Barrel and almost all of them are tagged by the isolated electron Level-1 trigger. With an assumed Level-1 global trigger rate of 25 kHz, about 100 π^0s per crystal can be obtained in a running period of less than 5 hours.

Events from $\eta\rightarrow\gamma\gamma$ are also being studied. The signal has a much lower rate once the background is reduced sufficiently, but the mass resolution is about 3%. The $\eta\rightarrow\gamma\gamma$ decay should be a useful calibration tool at higher energy and may prove very useful in the Endcaps, although it will take longer.

A significant rate of high-P_T photons with very little background and an energy that can be known independently of the ECAL, is available in radiative decays of $Z\rightarrow\mu\mu$. These photons are being investigated as a valuable tool for various calibration related tasks, as well as a probe for measuring photon reconstruction efficiency. They can be used, for example, to inter-calibrate different regions of the ECAL (coefficient c_i of Equation 2), and to tune the various cluster correction algorithms (coefficient F) and the overall energy scale (coefficient G). They can also be used to relate the energy scale of unconverted photons to that of electrons (from converted photons). For an integrated luminosity of only 1 fb^{-1}, an average of nearly 1 such photon per crystal will be collected.

Summary

The calibration of the CMS crystal calorimeter will be performed before and after the assembly of the detector. Before the assembly, crystals will be calibrated in the Testbeam, using laboratory measurements and with cosmic muons. After the assembly, crystals will be calibrated using physics events. At startup, the φ-uniformity inter-calibration technique will provide a precision of around 2% in a couple of hours. The design precision of 0.5% will be achieved using the E/P ratio of isolated electrons mainly from W and Z decays. The mass reconstruction of π^0,$\eta\rightarrow\gamma\gamma$ and $Z\rightarrow\mu\mu\gamma$ will provide important additional calibration tools.

REFERENCES

1. CMS Collaboration,"The Electromagnetic Calorimeter Technical Design Report", *CERN/LHCC* **97-033**.
2. CMS Collaboration,"The Magnet Project Technical Design Report", *CERN/LHCC* **97-010**.
3. CMS Collaboration,"The Tracker Project Technical Design Report", *CERN/LHCC* **98-006**.
4. CMS Collaboration,"The Muon Project Technical Design Report", *CERN/LHCC* **97-032**.
5. CMS Collaboration,"The CMS Physics Technical Design Report, Volume I", *CERN/LHCC* **2006-001.**
6. CMS Collaboration,"The CMS Physics Technical Design Report, Volume II", *CERN/LHCC* **2006-021.**
7. P. Adzic et al., "Results of the first performance tests of the CMS e/m calorimeter", *Eur. Phys. J.* **C44**, s02, 1-10 (2006).
8. M. Bonesini et al. "Inter-calibration of the CMS e/m calorimeter with cosmic rays before installation", *CMS Note* **2005/023**.
9. D. Futyan et al. "Intercalibration of ECAL crystals in φ Using Symmetry of Energy Deposition", *J. Phys.* **G29** (2003) 1229-1236.
10. D. Futyan, "Intercalibration of the CMS e/m calorimeter using jet trigger events", *CMS Note* **2002/031**.
11. R. Paramatti et al. "Use of $Z\rightarrow e^+e^-$ events for ECAL calibration", *CMS Note* **2006/039**.
12. L. Agostino et al. " Intercalibration of the CMS e/m calorimeter with isolated electrons", *CMS Note* **2006/021**.

The CMS ECAL Laser Monitoring System

Adolf Bornheim, for the CMS ECAL Collaboration

California Institute of Technology, 1200 East California Blvd. , Pasadena, CA 91125, USA

Abstract. The CMS detector at LHC will be equipped with a high precision lead tungstate electromagnetic calorimeter. To ensure the stability of the calorimetric response at a level of a few per mille, every channel of the detector is monitored with a laser system. This system allows to correct for fluctuations in the detector response with high precision, in particular the expected radiation induced changes of the crystal transparency, which are on the level of several per cent at nominal luminosity. Here we report results from long term tests of the system on fully equipped supermodules of the CMS ECAL which are performed after the final step of the construction but before insertion of the modules in CMS. The in situ monitoring strategy for the CMS ECAL is discussed in detail and the performance of the full monitoring procedure as achieved in the test beam is illustrated.

Keywords: Lead Tungstate Crystal Calorimeter, CMS ECAL, Monitoring System
PACS: 29.40.Vj, 29.40.Mc,

INTRODUCTION

One of the main physics goals of the CMS detector at LHC is the search for the Higgs boson. In the mass range around 125 GeV, the decay $H\rightarrow\gamma\gamma$ is of particular interest because of its clean final state. This mode however requires a very good energy resolution for the electromagnetic calorimeter to take advantage of the expected narrow decay width of the Higgs boson in this mass range. The design resolution for this calorimeter foresees a constant term in the resolution formula of 0.55 %. To fulfill this requirement, the CMS experiment features a high resolution crystal calorimeter utilizing lead tungstate (PWO) crystals [1]. These crystals do experience a very slight, dose rate dependent decrease of their transparency under irradiation, which recovers to a large extend in irradiation free periods. In CMS, the variation of the crystal transparency under irradiation and the corresponding decrease in light out put of the crystals will be monitored with a high precision light monitoring system. In this paper we describe the experience with the final design monitoring system in the test beam at CERN and discuss the strategy for correcting the variation in the crystal response during LHC operation.

CP867, *XII International Conference on Calorimetry in High Energy Physics,*
edited by S. R. Magill and R. Yoshida

TRANSPARENCY CHANGE OF LEAD TUNGSTATE CRYSTALS UNDER IRRADIATION

The mass produced lead tungstate crystals used for the CMS ECAL are radiation hard to high integrated dosage, but they experience a dose rate dependent transparency loss, as shown in Figure 1 (left). Although there is no damage to the scintillation mechanism, color centers are created during irradiation. These radiation induced color centers may annihilate at room temperature. During irradiations both annihilation and creation processes coexist and the color center density reaches an equilibrium at a level depending on the dose rate applied.

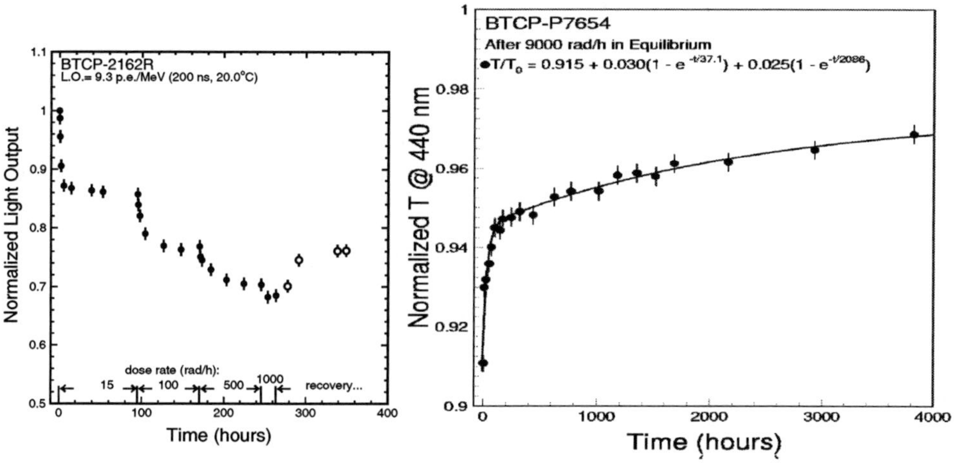

FIGURE 1. The left plot shows the change in the crystal transparency at different dose rates. Note that the transparency change reaches an equilibrium at each rate under continued irradiation. The dose rate during the first 100 hours of 15 rad per hours corresponds to the expected dose rate at LHC during low luminosity running. Note also that the relative change in the transparency is largest at small dose rates and becomes increasingly smaller at larger doserates. The right plot shows the recovery of the transparency loss after an irradiation. The recovery can be described by two time constants.

When the irradiation stops, the color center density decreases until it reaches an equilibrium again and thus the transparency change recovers. The time constants for the damage and recovery process are a crystal property and can vary to some extend. The recovery can be described by two time constants. An initial recovery phase takes place on a time scale of approximately 10 hours, followed by a slower recovery on the time scale of 1000 hours, as illustrated in Figure 1 (right). This has to be compared of the typical time constants of the LHC accelerator cycle. The luminosity operation will last for about 12 hours, followed by a refill of the machine which takes of the order one hour. The interplay between the recover time constants and the LHC operational cycle will lead to a distinct pattern in the transparency change, as illustrated in Figure 2. Since the crystals will not fully recover during the refill period of the accelerator the maximal transparency change will be limited. This will reduce the systematic effects related to the transparency correction since it will reduce the size of the correction on average.

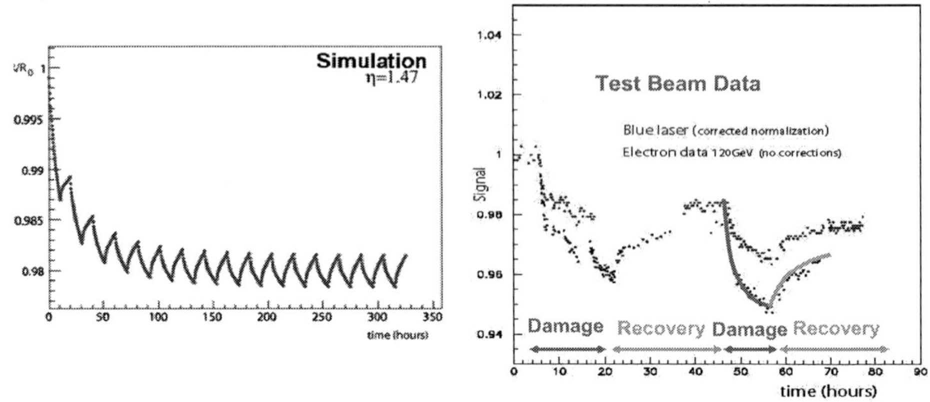

FIGURE 2. The diagram on the left from an simulation illustrates the interplay between the two time constants of the crystal recovery and the operational cycle of the LHC accelerator. Since the recovery time is much longer than one LHC cycle, the transparency change will not fully recover during refilling of the LHC. This will limit the overall fluctuations of the transparency change. The diagram on the right shows two such irradiation cycles mimicked at the test beam by repeatedly irradiating a final design CMS ECAL super module. The expected behavior can clearly be seen.

THE CMS ECAL MONITORING SYSTEM

The laser monitoring system must track the change in the transparency precisely enough to maintain the constant term in the ECAL resolution of 0.55%. This requires a measurement of the transparency with an accuracy of better than 0.2%. The transparency in each crystal will have to be measured approximately every 30 minutes during LHC operation to account for the fast initial change at the beginning of an irradiation cycle.

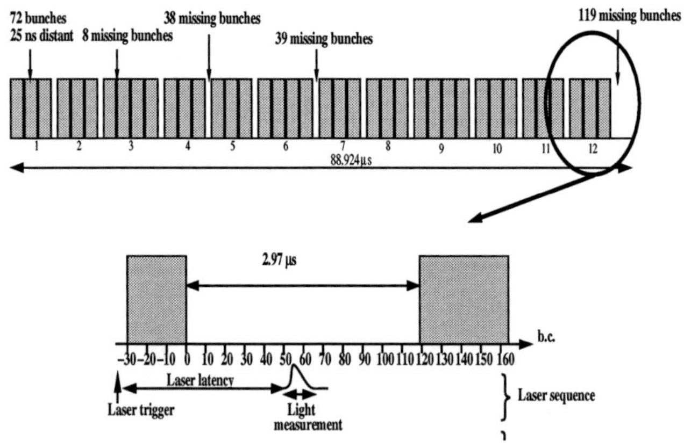

FIGURE 3. The diagram illustrates the continuous laser monitoring during LHC operation. Taking advantage of a 3 μs gap in the LHC bunch train laser pulses are injected at a rate of 100 Hz.

To achieve this, the monitoring has to be performed while the data taking of the detector is ongoing. This can be realized by taking advantage of gaps in the LHC bunch train which occur once every 90 μs. In Figure 3 the time structure of the LHC bunch train is illustrated. The laser system will be synchronized with the LHC clock cycle via the CMS trigger system and send laser pulses during the extended gaps. To allow this operational mode the laser source has to fulfill a number of design parameters. In particular the laser pulse width must be shorter than 40 ns and stable to better than 10% to match the shaping of the ECAL electronics. Furthermore must the pulse timing be stable to a few ns to allow a synchronization which the LHC clock and the detector readout. Finally, the pulse energy must be on the order of 1mJ to provide sufficient light to large groups of crystals to ensure that the entire ECAL can be monitored in 30 minutes. The pulse energy also has to be stable to better than 10% to prevent possible non-linear effects in the reference system against which the signals of the individual crystals are compared. To achieve all these performance requirements a dual-stage laser system is used [3,4]. Two such laser systems provide two different wavelengths, on at the peak of the scintillation light wavelength at 440 nm, one far away in the red wavelength range to allow for systematic cross checks. These two systems plus an spare backup system for the blue laser light source distribute the light pulses via a system of optical switches to the ECAL super modules, as shown in Figure 4.

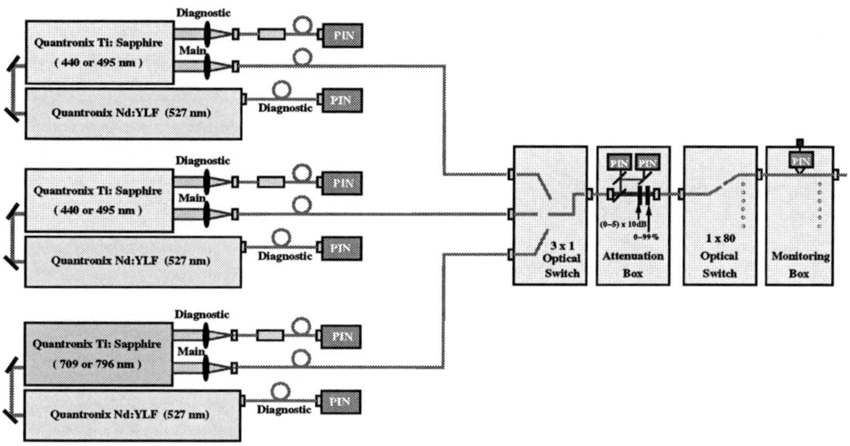

FIGURE 4. The diagram shows the layout of the CMS ECAL laser monitoring source. Two laser systems provide light at 440 nm and 800 nm as primary and secondary monitoring wavelengths. The 440 nm system has an identical spare system as a backup. A system of optical switches allows to distribute the light to one half of a super module at a time. All half super modules of the entire ECAL will be scanned once within 30 minutes.

Inside the super module the laser light is distributed via optical fibers and a fanout system to each individual crystal. The optical fibers enter the crystals from the front in the barrel part of ECAL and from the back in the endcaps. In parallel light is sent to reference photo diodes inside the supermodule. The relative difference seen in the photodetectors of the ECAL with repect to the reference diodes measures the relative transparency change in the crystals. More details of the on-detector part of the monitoring system can be found in reference [5].

TRANSPARENCY CORRECTION WITH THE CMS ECAL LASER MONITORING SYSTEM

The feasibility of the radiation damage monitoring with the desired precision has been demonstrated in the test beam [6]. One important measure of the performance is the stability of the monitoring system response over time in the absence of any radiation. This is a measure of the resolution of the system which sets the scale for the minimal transparency change which can be detected. The design goal for the CMS ECAL monitoring system is to be able to track transparency changes on the level of better than 0.2 %. The second benchmark is the ability of the system to correct back transparency changes induced by irradiation. This is verified in the test beam by irradiating crystals in a high energy electron beam with dose rates similar to the ones expected in CMS. During the irradiation the crystals are calibrated with the electron beam in regular time intervals, followed by laser monitoring runs during which the transparency is measured. The response to the electrons is then corrected using the transparency measurements of the laser system. Aside from the stability of the monitoring system there are several other issues which impact the performance of this correction procedure. The most important one is the relation between the relative change of the laser monitoring signal and the relative change of the response to electrons. This relation can be described by a linear function for small transparency changes, as shown in the left plot of Figure 4.

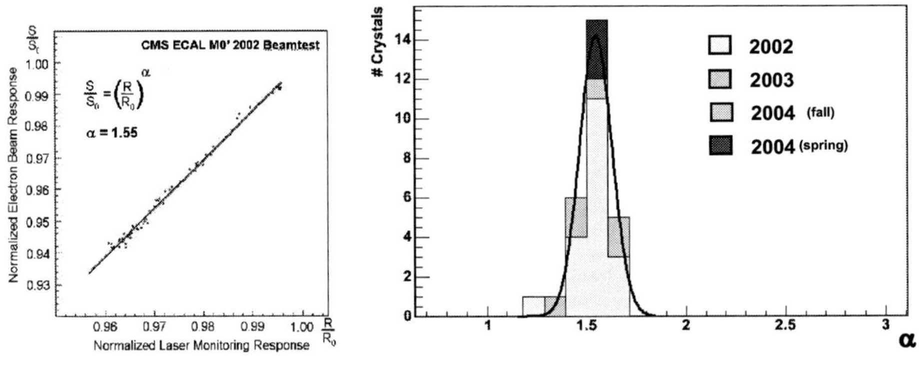

FIGURE 5. The left plot shows the linear correlation between the relative loss of the laser monitoring response and the relative loss of the response to electrons. The slope α of this linear correlation is quite uniform for crystals produced under the same conditions. The plot on the right shows the slope from 28 crystals produced by BTCP, the main supplier of crystals for the CMS ECAL. The distribution has an RMS of about 5 %.

Thus, only a scaling factor is needed to determinc the correction factor from the relative change as measured by the laser system. This scaling factor α is found to be the same within about 5 % for crystals produced under the same conditions, as shown in the right plot of Figure 4. For a total of 28 crystals the parameter α has been measured in various test beam campaigns and found to be the same in all crystals within 5 %. There are indications that the orientation of the crystal optical axis with respect to the crystal geometrical shape do impact in the scaling factor α. At CMS the transparency change in the LHC low luminosity running will be on the level of a few percent. At this level it is sufficient to know the average value of the parameter α as determined in the test beam. Procedures to extract the parameter α in situ using physics events are under consideration.

PERFORMANCE OF THE MONITORING SYSTEM

In Figure 5 we show an example of an irradiation cycle in the test beam. A final design super module was irradiated with 120 GeV electrons at does rates equivalent to four times the expected does rate at LHC in low luminosity operation. During the irradiation the crystals exposed to the beam were calibrated repeatedly with the electron beam. In parallel the transparency was monitored with the laser monitoring system. Over a period of six hours we observe a decrease of the raw response to electrons of about 2 %.

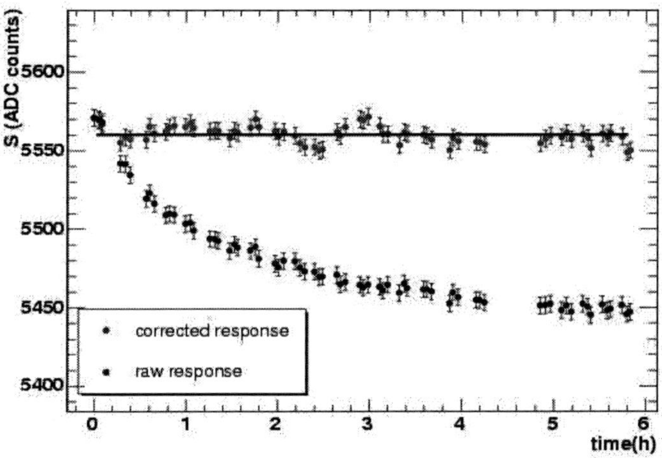

FIGURE 6. The raw ECAL response to 120 GeVelectrons during an irradiation period on a super module in the test beam. The response is the sum of nine crystals centered around the crystal hit by the beam. The beam rate was set such that the radiation level corresponds to four times the expected radiation level at the LHC. The crystal response drops by about 2 percent and then approaches a plateau. The data has then be corrected using the laser monitoring data. The individual calibration points are within 0.15 % of the initial calibration.

After applying a correction based on the measurements from the laser monitoring system the individual calibration points are within 0.15 % of the initial calibration before irradiation [7]. This value is achieved for a set of four crystals all irradiated in the same was as shown in the example in Figure 5. For this correction the average value for the parameter α as determined from Figure 4 was used. This is to be compared to the design value for the constant term of the energy resolution of the CMS ECAL which is 0.55 %. The effect of the transparency change correction is well below other effects contributing to the overall constant term, in particular the intercalibration to be performed in situ [8].

SUMMARY

A final design laser monitoring system of the CMS ECAL has been tested in the test beam. With this monitoring system the radiation induced transparency change in the crystals of the CMS ECAL can be measured and corrected for. The residual effect after the correction of the transparency change is found to be less than 0.15 %. This is well within design specifications and negligible compared to other contributions to the constant term of the resolution of 0.55 %.

ACKNOWLEDGMENTS

I would like to thank the entire ECAL collaboration for their outstanding effort, R.-Y. Zhu and the CALTECH team for their superb work on the laser system and the organizers of the XII International Conference on Calorimetry in High Energy Physics at Chicago.

REFERENCES

1. "The Electromagnetic Calorimeter Technical Design Report", CMS Collaboration , CERN/LHCC 97-33, 1997
2. "Radiation Induced Color Centers and Light Monitoring for Lead Tungstate Crystals", X. Qu, L. Zhang, R.-Y, Zhu, IEEE Trans. Nucl. Sci. NS-47 (2000).
3. "Performance of the Monitoring Light Source for the CMS Lead Tungstate Crystal Calorimeter ", IEEE Trans. Nucl. Sci.Vol. 52 No. 4 (2005) 1123--1130.
4. "ECAL Monitoring Light Source at H4", D. Bailleux, A. Bornheim, L. Y. Zhang, K. J. Zhu, R.-Y. Zhu, and D. Liu, CMS IN 2003/045.
5. "Laser monitoring system for the CMS lead tungstate crystal calorimeter", M. Anfreville et. al., to appear in *Nucl. Instr. And Meth.*
6. "The CMS ECal laser monitoring system.", A. Bornheim, in proceedings of the 9th ICATPP Conference on Astroparticle, Particle, Space Physics, Detectors and Medical Physics Applications, Como, Italy.
7. "Analysis of the response evolution of the CMS electromegnetic calorimeter under electron and pion irradiation*", A. Ghezzi et. al.,* CMS NOTE-2006/038
8. ***"CMS ECAL Calibration Strategy", G. Daskalakis, these Proceedings***

Test beam results on the Proton Zero Degree Calorimeter for the ALICE experiment

R. Arnaldi*, E. Chiavassa*, C. Cicalò†, P. Cortese**, A. De Falco†, G. Dellacasa**, N. De Marco*, A. Ferretti*, M. Floris†, M. Gagliardi*, M. Gallio*, R. Gemme*, A. Masoni†, P. Mereu*, A. Musso*, C. Oppedisano*, A. Piccotti*, F. Poggio*, G. Puddu†, E. Scomparin*, S. Serci†, E. Siddi†, D. Stocco*, G. Usai†, E. Vercellin* and F. Yermia*

**Università di Torino and INFN, Torino, Italy*
†Università di Cagliari and INFN, Cagliari, Italy
***Università del Piemonte Orientale and INFN, Alessandria, Italy*

Abstract. The proton Zero Degree Calorimeter (ZP) for the ALICE experiment will measure the energy of the spectator protons in heavy ion collisions at the CERN LHC. Since all the spectator protons have the same energy, the calorimeter's response is proportional to their number, providing a direct information on the centrality of the collision. The ZP is a spaghetti calorimeter, which collects and measures the Cherenkov light produced by the shower particles in silica optical fibers embedded in a brass absorber. The details of its construction will be shown. The calorimeter was tested at the CERN SPS using pion and electron beams with momenta ranging from 50 to 200 GeV/c. The response of the calorimeter and its energy resolution have been studied as a function of the beam energy. Also, the signal uniformity and a comparison between the transverse profile of the hadronic and electromagnetic shower are presented. Moreover, the differences between the calorimeter's responses to protons and pions of the same energy have been investigated, exploiting the proton contamination in the positive pion beams.

Keywords: Sampling calorimeter; Quartz optical fiber; Radiation hardness
PACS: 29.40.Vj; 42.81.Wg; 42.70.Ce; 42.88.+h

INTRODUCTION

The ALICE experiment will study ultrarelativistic heavy ion collisions at the CERN LHC[1]. Its primary aim is to investigate the physics of strongly interacting matter at extreme energy density, where the formation of the Quark Gluon Plasma (QGP), a state of matter where quark and gluons are deconfined, is expected. Many QGP signatures manifest themselves as a threshold behaviour of certain observables as a function of the energy density reached in the interaction, which is estimated through the centrality of the collision. In the ALICE experiment the Zero Degree Calorimeters (ZDCs) will measure the collision's impact parameter by measuring the zero degree energy, namely the energy carried away by non-interacting nucleons (spectators) [2]. Two identical sets of calorimeters will be located at opposite sides with respect to the beam interaction point IP2, 116 m away from it, where the two LHC beams circulate in two different pipes (Fig. 1). Each set consists of two hadronic calorimeters placed at zero degree with respect to LHC axis. One of them, the neutron calorimeter (ZN), is positioned between the two beam pipes to intercept the spectator neutrons [3]. The other one, the

CP867, *XII International Conference on Calorimetry in High Energy Physics,*
edited by S. R. Magill and R. Yoshida

proton calorimeter (ZP), is placed externally to the outgoing beam to collect the spectator protons, which will be separated from the ion beams by means of the dipole magnet D1.

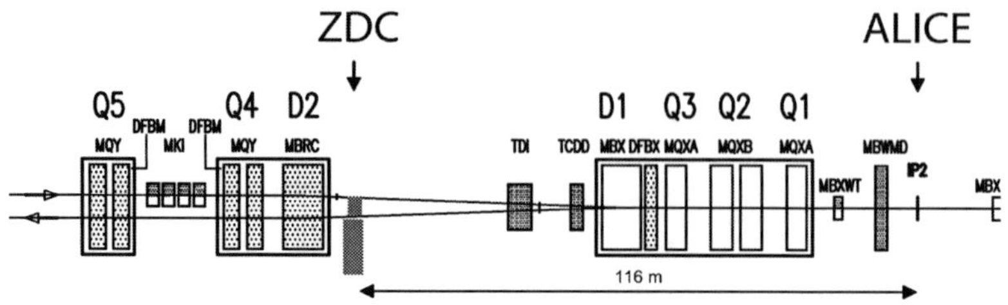

FIGURE 1. Schematic representation of the beam line and ZDC location.

ZDCs are sampling calorimeters with silica optical fibers as active material, embedded in a dense absorber [4]. The principle of operation is based on the detection of Cherenkov light produced in the fibers by the shower's charged particles. This technique has several advantages like a fast response, important for triggering purposes, and a reduced lateral dimension of the detectable shower, which allows to make compact detectors, as required by LHC integration technical constraints. Moreover the high radiation hardness, characterizing the quartz fiber calorimetry, will enable the ZDCs to operate in a highly radioactive environment [5], where the estimated dose is of the order of $\sim 10^4$ Gray/day, at a luminosity L=10^{27} cm^{-2}s^{-1}.

The energy resolution of the ALICE ZDC should be comparable with the fluctuations on the number of spectator nucleons at a fixed impact parameter, which ranges from $\approx 20\%$ for central events to $\approx 5\%$ for peripheral ones, according to simulations that use HIJING as event generator.

FIGURE 2. Photo of the calorimeter at the end of the assembly.

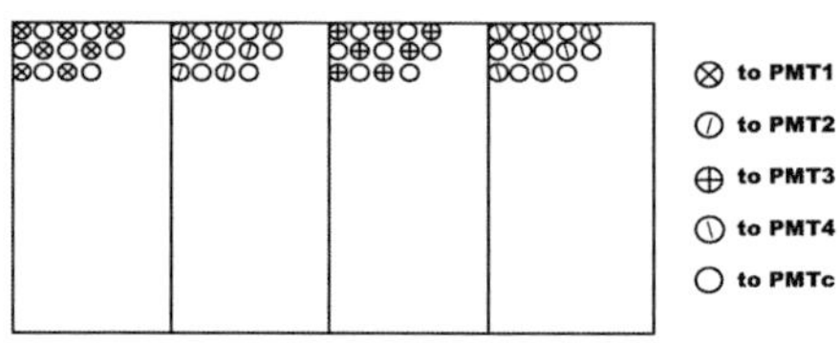

FIGURE 3. Schematic connection of the fibers to the PMTs.

PROTON CALORIMETER DESCRIPTION

The Proton ZDC for the ALICE experiment is a detector conceptually similar to the ZDC used by NA50 experiment [6]. The ZP calorimeter (Fig. 2) has been built by stacking 30 grooved brass plates, each of them 4 mm thick to form a parallelepiped of 22.8x12x150 cm^3. The grooves, inside the absorber matrix, run parallel to the beam axis and host 1680 quartz fibers with a pitch of 4 mm corresponding to a filling ratio, i.e. the ratio of active volume to the absorber volume, of 1/65. The fibers are of HCG-M-550-U type (manufactured by SpecTran Specialty Optics Company, USA) and have a pure silica core, silica fluorinated cladding and a hard polymer coat with a diameter of 550, 600 and 630 μm, respectively. The numerical aperture is 0.22. The fibers come out from the rear face of the calorimeter and are read by 5 photomultipliers Hamamatsu R329-02, with a quantum efficiency around 25%. One out of every two fibers is sent to a single photomultiplier (PMTc), while the remaining fibers are connected to four different photomultipliers (PMT1 to PMT4) [7], collecting the light from four vertical towers, as can be seen in Fig. 3. This segmentation allows to check the relative behaviour of the different towers and to give a rough localisation of the spectator protons' spot.

THE TEST BEAM EXPERIMENTAL SETUP

The Proton Calorimeter has been tested at the H6 beam line of the CERN SPS. The detector has been irradiated with secondary and tertiary hadron and electron beams of momentum ranging from 50 to 200 GeV/c.

A schematic layout of the test set-up is shown in Fig. 4. A plastic scintillator telescope S1-S4 provided the trigger. A MWPC has been installed in front of the calorimeter to localize the incident beam particles on the ZP front face. Two plastic scintillators MU1 and MU2, placed beyond an iron wall, allowed to detect muons. To be able to scan the whole calorimeter front face, the calorimeter has been mounted together with the MWPC on a movable platform.

A Lecroy 2249W ADC module has been employed to digitize the analog signals from the PMTs of the calorimeter.

The PMTs high voltages were equalized at the beginning of the test by sending the beam in the centre of each tower and of the calorimeter front face. A fine tuning of the ADC signals for each channel was made offline to compensate possible small differences.

FIGURE 4. Schematic view of the experimental set-up.

EXPERIMENTAL RESULTS

Response to hadrons and electrons

Fig. 5 shows an example of ADC spectrum obtained as the sum of the signals from the 5 PMTs when a 120 GeV/c pion beam hits the centre of the calorimeter's front face. The data have been fitted with this function

$$f(x) = Ne^{\frac{-(x-\mu)^2}{2\sigma^2(x)}} \tag{1}$$

where the term

$$\sigma(x) = \sigma_0 + \sigma_1 \frac{x-\mu}{\mu} \tag{2}$$

accounts for the asymmetry of the calorimeter's response to hadrons, which is due to the energy fluctuations of the electromagnetic component of the hadronic shower.

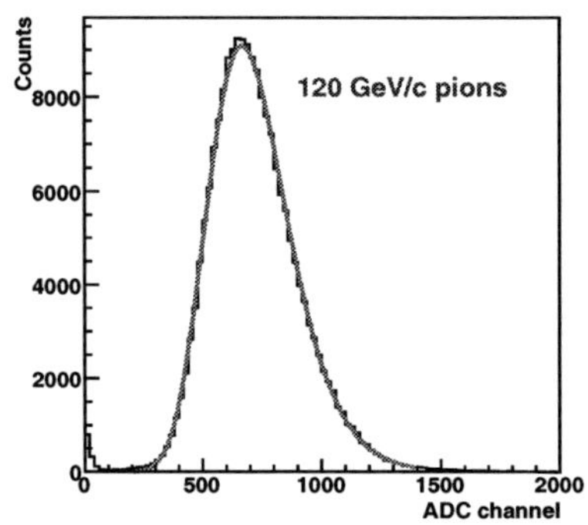

FIGURE 5. ZP response to 120 GeV/c pion beam.

Fig. 6 and 7, respectively for hadrons and electrons, show, as a function of the beam energy, the ratio between the ADC response and the corresponding beam energy, normalized to 120 Gev/c. ZP response to electrons is linear within 3%, while the response to hadrons is not linear. In fact, since the calorimeter is not compensating, it is sensitive to the electromagnetic component of the hadronic shower, which increases with the beam energy. However, in the ALICE experiment, the ZP will measure the number of spectator protons, which all have the same fixed energy as the beam nucleons. The charged particles produced at IP will be bent by D1 separator magnet outside ZP acceptance and the energy due to neutral particles hitting the calorimeter is found to be negligible with respect to the energy carried by spectator protons. Therefore the fact that the calorimeter is not compensating will not create any problem in ALICE.

The measured e/π ratio depends on the beam energy and ranges from $\sim$2.0 at 50 GeV/c to $\sim$1.6 at 180 GeV/c. The electromagnetic shower is completely contained in the calorimeter, while for the hadronic shower we estimated, from the simulation, a lateral leakage of about 15%. Taking into account this correction factor for the experi-

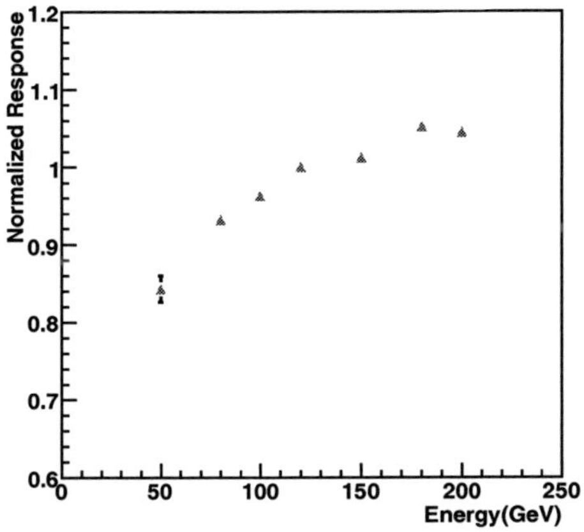

FIGURE 6. ZP normalised response to hadrons as a function of the beam energy.

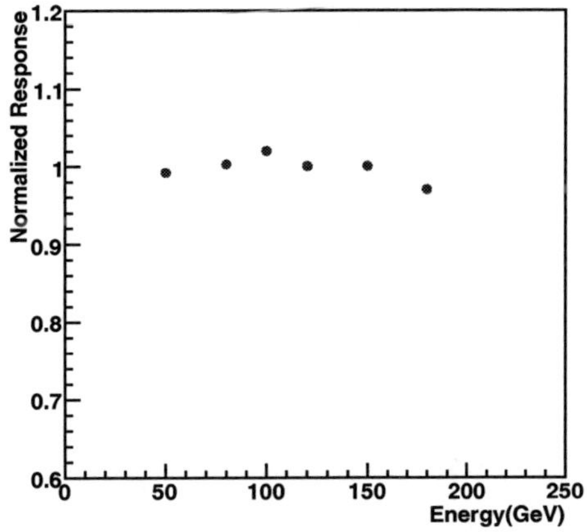

FIGURE 7. ZP normalised response to electrons as a function of the beam energy.

mental data, we could evaluate *e/h*, the ratio of the calorimeter's responses to the e.m. and non-e.m. components of the hadronic shower [8] [9]. We found an *e/h* value of ~ 3.

Energy resolution

The test allowed to measure the calorimeter's energy resolution as a function of the beam energy, for both hadron and electron beams; the results are shown in Fig. 8.

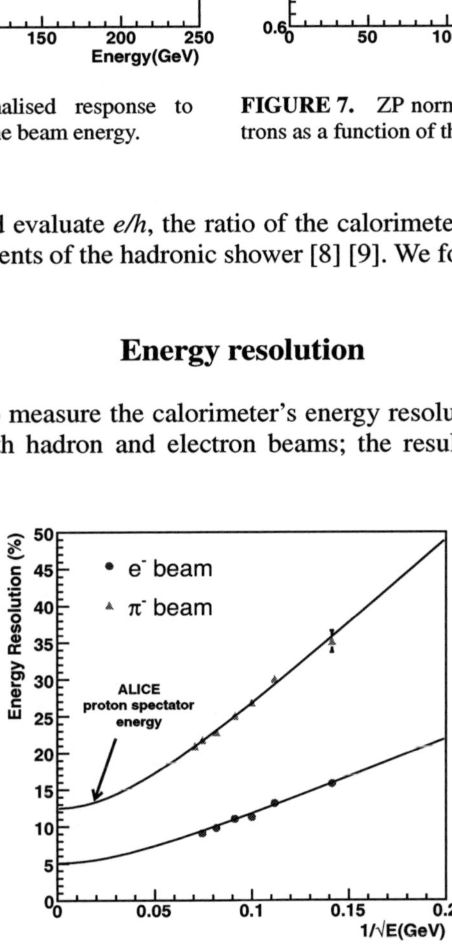

FIGURE 8. ZP energy resolution as a function of $1/\sqrt{E(GeV)}$.

The energy resolution has been calculated as the ratio σ_0/μ (see Eqs. (1) and (2)); the relationship between the FWHM of the ADC spectra and σ_0 is:

$$FWHM \simeq 2.4 \cdot \sigma_0 \tag{3}$$

very similar to that of a Gaussian distribution.

The data have been fitted with the function:

$$\frac{\sigma}{E} = \frac{a}{\sqrt{E(GeV)}} \oplus b. \tag{4}$$

We obtained $a = (237. \pm 2.)$ % $GeV^{1/2}$, $b = (12.5 \pm 0.2)$ % for pions and $a = (106.8 \pm 0.9)$ % $GeV^{1/2}$, $b = (5. \pm 0.3)$ % for electrons. By making an extrapolation to the ALICE single spectator proton energy (2.76 TeV) we find a resolution value of about 13%, in agreement with the ALICE experiment requirements [2].

Response's uniformity and detectable shower's transverse profile

With a 120 GeV/c pion beam we have performed a complete horizontal scanning leaving the vertical beam position fixed at the centre of the calorimeter front face, in order to study the uniformity of the response. In Fig. 9 the responses of the four adjacent towers as a function of the beam impact horizontal coordinate are represented. It is interesting to note that only the towers 3 and 4 can see the quartz fiber periodicity. This is due to the fact that the sampling fraction is slightly higher when the beam particles hit the calorimeter in a fiber plane than when they hit the absorber [10]. The vertical beam position was not the same during the whole scan, as can be seen in Fig. 10, where we plot the vertical coordinate of the beam particles detected by the MWPC. This beam displacement is due to the fact that the complete scanning was done at two different periods and we couldn't reproduce exactly the same beam conditions.

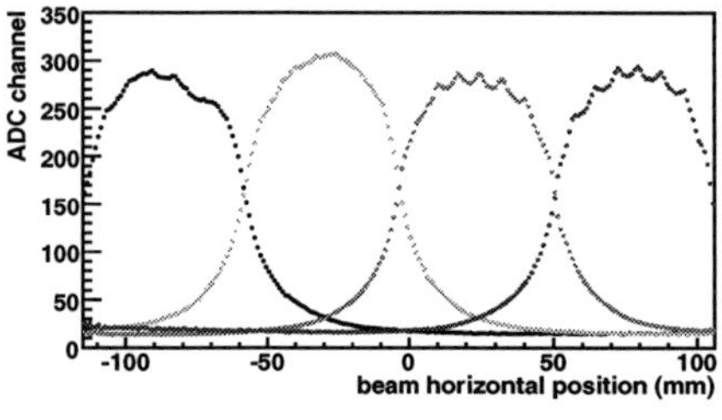

FIGURE 9. Signals of the four ZP towers T1, T2, T3, T4 (from left to right) as a function of the beam impact horizontal coordinate.

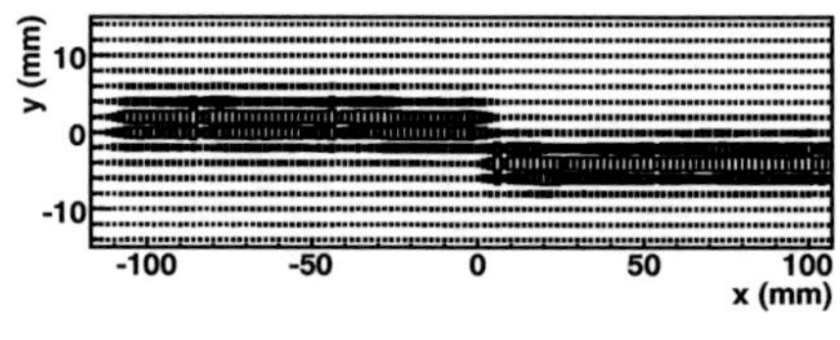

FIGURE 10. Beam impact point on the calorimeter front face defined using the MWPC.

Taking the derivative of a single ZP tower response as a function of the beam impact point in a boundary region between two towers, we can estimate the transverse size of the detectable shower. The FWHM of the derivative is $\sim$ 13 mm for the hadronic shower, while we obtained a value of $\sim$ 6 mm for the electromagnetic shower, in an analogous study performed with an 100 GeV/c positron beam.

Difference in the response to p and π

During the test, for three beam momenta (100, 120, 150 GeV/c) the calorimeter behaviour has been investigated with both negative and positive hadron secondary beams. While the negative beams are composed almost enterely by pions, the positive beams are a mixture of pions and protons. As can be seen in Table 1, the fraction of pions depends on the beam energy and on the selected production angle on the target [11]. Since the fraction of pions is known, it is possible to separate the π^+ and proton contributions in the positive beam spectra, assuming that the shape of π^+ spectra is equal to that of π^-, scaled by a proper factor. Subtracting from the positive hadron spectrum the π^+ contribution, we could obtain the calorimeter's response to protons.

TABLE 1. Pions' fraction in positive beams, for the investigated beam momenta (CERN SPS data).

Beam momentum (*GeV/c*)	**Production angle** (*mrad*)	**Fraction of** π
100	-5.46	61%
120	-0.2	61%
150	5.	31%

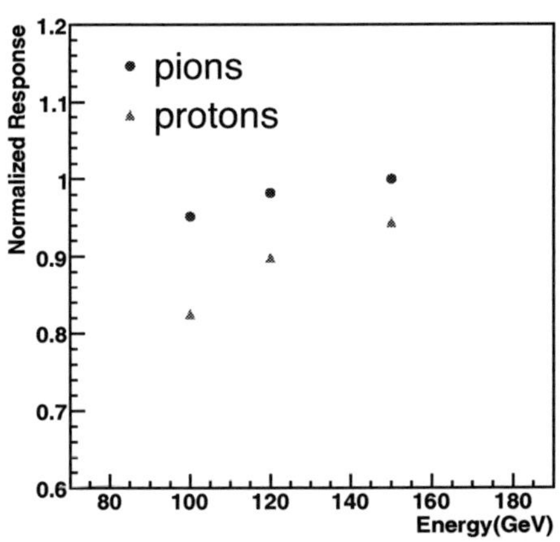

FIGURE 11. Normalized ZP response to pions and protons.

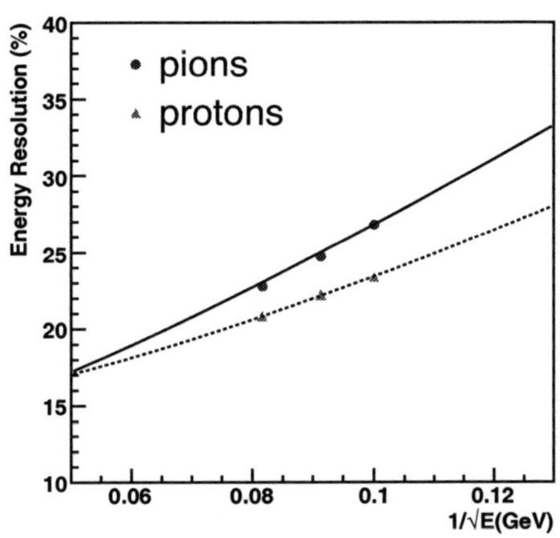

FIGURE 12. ZP resolution for pions and protons as a function of $1/\sqrt{E(GeV)}$.

In Fig. 11, for both pions and protons, we plot the normalized response, i.e. the ratio between the ADC response and the corresponding beam energy, normalized to the 150 Gev/c pion beam. From this plot the ZP response to protons is found to be, on average, $\sim$ 9% lower than the response to pions of the same energy. In fact, due to the baryon number conservation law, the proton interactions have a smaller fraction of energy going into π^0's, and therefore into the e.m. component of the shower, than pion interactions [12]. The ZP calorimeter, being non-compensating, is sensitive to this different e.m. component in the shower development.

Finally, in Fig. 12 we show the calorimeter resolution for pions and protons as a function of the beam energy; it can be noticed that the energy resolution for protons is better than for pions, despite the fact that the ZP response to protons is lower than the response to pions. The data relative to proton resolution have been fitted with the function described in Eqn. (4). For the fit parameters we get $a = (185. \pm 4.)$ % $GeV^{1/2}$ and $b = (14.3 \pm 0.4)$ % for protons, while the curve describing the points relative to pion resolution is the one shown in Fig. 8.

CONCLUSIONS

After an intensive R&D phase the final proton Zero Degree Calorimeter of the ALICE experiment was built and tested. The ZP has a linear response to electrons in the investigated energy range. The calorimeter is non-compensating, with an *e/h* value of ~ 3, but this fact will not create any problem in the ALICE experiment. The energy resolution is well described by the function $a/\sqrt{E(GeV)} \oplus b$ and fulfills the ALICE physics requirements. Moreover we could verify that the chosen sampling frequency is adequate for hadronic calorimetry. Finally the test has shown that the detector is sensitive to the different electromagnetic component of the hadronic shower developed by pions and protons.

ACKNOWLEDGMENTS

We wish to thank M. Arba and M. Tuveri from INFN Cagliari and G. Alfarone and R. Farano from INFN Torino, who provided an essential technical support. We greatly appreciated the help of I. Efthymiopoulos in the SPS beam operations.

REFERENCES

1. *"ALICE Technical Proposal"*, CERN/LHCC, 95-71.
2. *Zero Degree Calorimeter Technical Design Report*, CERN/LHCC, (1999).
3. R. Arnaldi et al., *NIM*, **A 564**, 235 (2006).
4. P.Gorodetzky et al.,*NIM*, **A 361**, 161 (1995).
5. R. Arnaldi et al., *NIM*, **A 456**, 248 (2001).
6. R. Arnaldi et al., *NIM*, **A 411**, 1 (1998).
7. R. Arnaldi et al., *Proc. Nuclear Science Symposium, Rome, Italy, 16-22 October 2004*, IEEE, (2004).
8. R. Wigmans, *NIM*, **A 265**, 273 (1988).
9. D. E. Groom, *Proc. Seventh Inter. Conf. on Calorimetry in High Energy Physics, Tucson, Arizona, 9-14 November 1997, ed. by E. Cheu, T. Embry, J. Rutherfoord, R. Wigmans.*, World Scientific, (1998).
10. A.S. Ayan et al., *J.Phys.G:Nucl.Part.Phys.*, **30**, N33-N44 (2004).
11. http://cern.ch/gatignon/partprod.html
12. N. Akchurin et al., *NIM* , **A 408** , 380 (1998).

SESSION IX

LHC CALORIMETRY—SIMULATION

Convener: Ren-Yuan Zhu

Jet Performance and Jet Energy Scale Determination at CMS

Anwar A Bhatti
for CMS Collaboration
Experimental Physics Laboratory, The Rockefeller University, 1230 York Ave, New York NY 10021, USA

Abstract. We describe the jet response of the CMS calorimeter which will be used to study pp collisions at Large Hadron Collider at CERN, Geneva, Switzerland at $\sqrt{s} = 14$ TeV. The electromagnetic section of calorimeter consists of lead tungstate crystals which gives an excellent resolution for electrons. The hadron section is brass-scintillator sampling calorimeter read by wavelength shifting fibers in the central region ($|\eta| < 3.0$) and steel/quartz-fibers in the forward ($3.0 < |\eta| < 5.0$) region. Extensive test beam calibration data has been collected. A GEANT-based calorimeter simulation has been tuned to reproduce the test beam measurements. The calorimeter response to jets has been determined using this tuned simulation. We describe the calorimeter response to jets, the jet energy resolution, and the procedure we plan to use to establish the jet energy scale from a combination of test beam and pp data when we start taking data in September 2007.

Keywords: CMS,calorimeter,jet energy calibration
PACS: 07.60.Dq,07.20.Fw,06.20.Fn,13.85.-t

1. INTRODUCTION

Identification of jets and their accurate energy determination is important for most physics studies at hadron-hadron collider. Precise energy determination of jet energy is essential to reconstruct any particle, such as Higgs boson, into hadrons. Missing transverse energy, calculated by vectorially summing the transverse energy deposits in the calorimeter, is a signature of particles which do not interact with the detector. Such particles are predicted in Supersymetry, an extension of Standard Model. An accurate determination of missing transverse energy is also essential to observe Higgs boson in $H \to ZZ \to \nu\nu l^+l^-$ and $H \to W^+W^- \to l^+\nu l^-\nu$ channels where neutrinos leave the detectors without any energy deposition. Jet are also important to discover the contact interactions as an excess in jet production rate at very high E_T, provided the contact interaction parameter Λ is moderately large. In addition to measuring the jets and missing transverse energy, hadron calorimeter is also used to identify leptons. The leptons form a very clean signal for certain physics channels e.g. $H \to ZZ \to \mu\mu\mu\mu$. Thus understanding of calorimeter is essential for most if not all the physics at a hadron collider.

At LHC, to study pp interactions at $\sqrt{s} = 14$ TeV, the P_T range of the interesting jets range from 20 GeV, essential to identify top quark events, which form background to many important signals, to a few TeV to measure the jets coming from decay of a new heavy particle. At LHC design luminosity, there are, on average, 20 minimum bias interactions in the same bunch crossing which produce ~ 1000 particles. Time interval

CP867, *XII International Conference on Calorimetry in High Energy Physics,*
edited by S. R. Magill and R. Yoshida

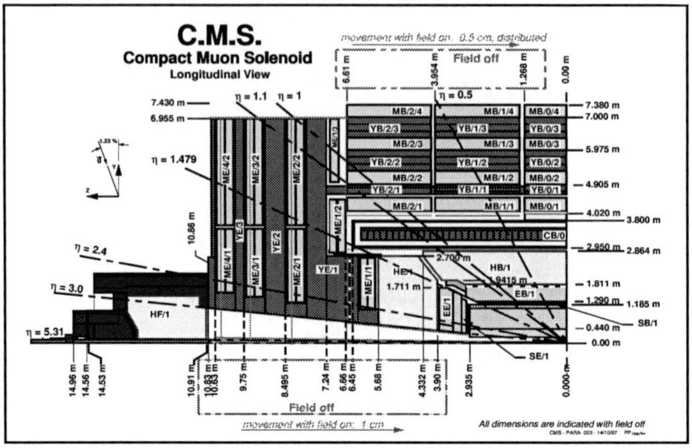

Fig. 1.1.2(color): Longitudinal view of one quadrant of the CMS detector.

FIGURE 1. One quadrant of the CMS detector.

between two bunch crossing is 25 *ns*. The required large dynamic range and short digitization time and size of the CMS magnet puts strong constrains on the calorimeter design choices.

In this note, after a brief description of the CMS detector with emphasis on the hadron calorimeter in Section 2, we describe the calorimeter response to single particles and jets in Section 3; followed by a description of the procedure to calibrate the jet response and verify it using collider data in Section 4. Finally, in Section 5, a method to improve the jet energy resolution using only the calorimeter information is described.

2. CALORIMETER DESIGN AND CONSTRUCTION

The Compact Muon Solenoid (CMS) detector consists of a very large magnet with solenoid field of 4 Tesla and radius of ~ 3 meters with an excellent silicon tracker inside and second-to-none muon system outside. Central calorimeter is located inside the magnet and is divided into barrel ($|\eta| < 1.5$) and endcap calorimeters ($1.5 < |\eta| < 3.0$) consisting of electromagnetic (EB, EE) and hadronic (HB, HE) sections . The electromagnetic calorimeter is constructed with 0.0175×0.0175 ($\Delta\eta \times \Delta\phi$) granularity, lead tungstate ($PbWO_4$) crystals which are 26 (25) radiation length long in $|\eta| < 1.479(> 1.479)$. The central hadron calorimeter is a sampling calorimeter consisting of brass and scintillator plates. There are 17 active scintillators layers which are divided in projective towers. The section of barrel hadron calorimeter which resides in the magnet is 5.9 absorption length (λ) thick at $\eta = 0$. The endcap is 10λ thick. To detect the part of signal from very high energy particles which leaks out of HB, a 3.9λ deep calorimeter is installed outside the magnet coil. This HO calorimeter is two active layers in $|\eta| < 0.435$ region and only one active layer for $0.435 < |\eta| < 1.45$.

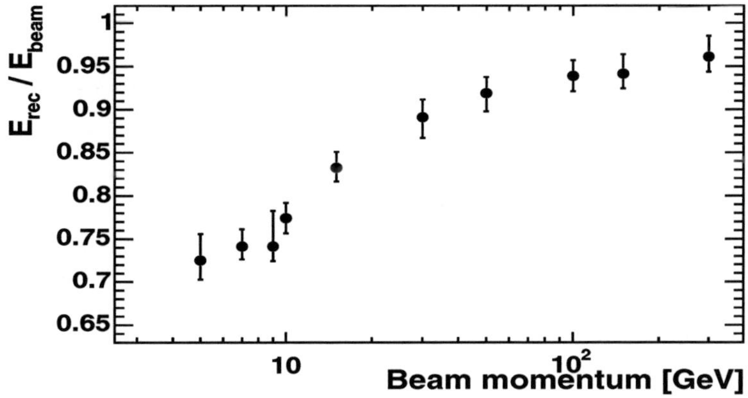

FIGURE 2. Calorimeter response to charged pions: the simulated data is compared to test beam measurements

There is 0.4 to 1.4 radiation lengths thick material in front of EM calorimeter. As a results 35% (10%) of electrons radiate more than 70%(95%) of their energy before the calorimeter. For EM section, the optical signal is converted to electrical signal by Avalanche Photon Diodes (APD) which are mounted on back of the crystals. For hadron sections, the light produced by shower particles is collected by wave length shifting fibers and converted to electronic signal by APDs and digitized by custom built electronics. The readout chain has noise which equivalent to a few hundred MeV of energy for hadron sections. In forward region, an iron-scintillating fiber calorimeter covering $3.5 < |\eta| < 5.5$ consists of short and long fibers which are read out separately. These signals are combined appropriately to determine the electromagnetic and hadronic content of the energy cluster.

3. CALORIMETER RESPONSE

The CMS calorimeter has been calibrated using test beam data and simulated using GEANT4 program. This GEANT4-based simulation describes the test beam data adequately [2]. The calorimeter response to charged pions as a function of particle momentum is shown in Fig. 2. The response increases from $\sim 72\%$ at 5 GeV to 95% at 300 GeV. The signal produced by a 50 GeV charged pion which has its first inelastic interaction in the hadronic section is defined to be equivalent to 50 GeV energy. The calorimeter resolution defined as σ_E/E, where E is the sum of energy deposited in electromagnetic and hadronic section and σ_E in the RMS of E, is shown in Fig. 3. The electronic noise in the calorimeter is suppressed by applying tower level energy thresholds. These thresholds decrease the noise contribution to a jet of cone of radius 0.5 to less than 1 GeV with a small loss in the jet energy. This loss in energy has small effect on the jet energy resolution. Using these thresholds, less than one in 10 events in QCD sample have a fake jet with $P_T > 10$ GeV; a fake jet is a jet without a corresponding generated (physical) jet.

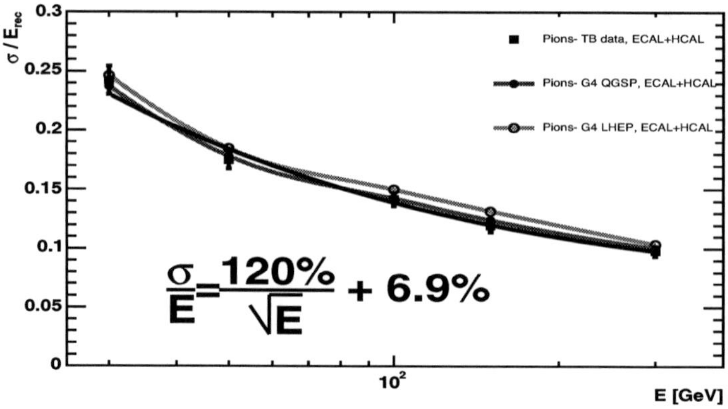

FIGURE 3. Charged pions Resolution: the simulated data is compared to test beam measurements

Jet reconstruction efficiency of 100% can be obtained at ~ 40 GeV particle jet P_T with some appropriate cut on reconstructed jet P_T.

CMS is currently using an iterative cone algorithm to reconstruct the jets with clustering cone radius R=0.5 in $\eta - \phi$ space. Starting with the highest P_T tower, the algorithm combines the towers above a threshold within a cone into a cluster. The centroid of the cluster is calculated and process is iterated until the geometric center of the cone is same as the centroid of the energy cluster. The towers associated with a cluster are removed from the list and clustering is repeated using the remaining towers. In this algorithm, although geometric cones of two reconstructed jets can overlap, jets do not share towers by construction. Thus, there is no splitting or merging phase. CMS collaboration has also implemented Midpoint Algorithm[3] but all the studies reported in this paper have used iterative cone algorithm.

The calorimeter response to a jet, defined as the ratio of P_T of the calorimeter jet to the P_T of the corresponding particle jet, is shown in Fig. 4 as a function of psuedorapidity for various P_T ranges. The calorimeter response is reasonably uniform in $|\eta| < 1.3$ region with a $\sim 5\%$ drop at the boundry between barrel and endcap sections. The endcap region show a stronger P_T dependence of the response. The higher response in $|\eta| > 3.0$ region is due to different calibration scheme used in the HF calorimeter. The jet energy resolution is shown in Fig. 4. Different curves correspond to different noise suppression schemes.

4. JET ENERGY CALIBRATION

The CMS collaboration plans to determine the jet energy scale through a combination of data-driven and Monte Carlo based techniques. A section of calorimeter comprising full η range but only four towers (20°) in azimuthal direction has been calibrated using charged pions and electrons. The calorimeter response to a radio-active source has been measured and cross-calibrated using test beam measurements. These calibration will be

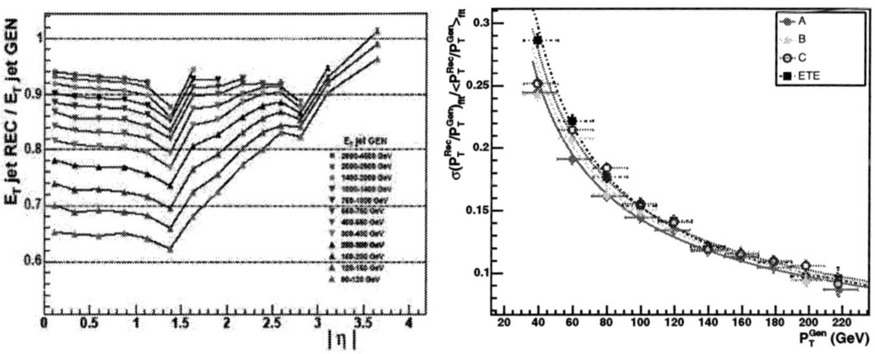

FIGURE 4. The ratio of calorimeter jet P_T to the particle jet P_T and the jet P_T resolution

used at the start of the run to set the initial calorimeter energy scale. The tower to tower variation in response to the same radio active source is less than 5%. To the extent that magnetic field effect on the calorimeter response can be accurately accounted for, the initial energy scale is expected to be correct to 5%.

After accumulation of *pp* data, the azimuthal symmetry in the ECAL will be established using photons. The uniformity of the hadron calorimeter response will be established by measuring the tower occupancy in minimum bias data and forcing azimuthal symmetry at each η annulus modulo any differences expected due to variation in material before the calorimeter. Alternatively, the muon signal can be used to establish the azimuthal symmetry in hadron calorimeters. The most probable value of the energy deposited by muons in HB section is $\sim$ 1.8 GeV and HE section is $\sim$ 3 GeV. However, muons and hadrons sample the depth of the tower differently. During construction, the tile to tile scintillator response variation of less than 10% was maintained and thus differences between muon and pions response should be small. In addition muons are less sensitive to the material in the tracker.

The jet response as a function of η can be made uniform by scaling the P_T of the jets at η such that it is, on average, equal to jets within $|\eta| < 1.0$ in dijet events. In these events, in $2 \to 2$ approximation, the transverse momentum of two outgoing partons should be the same. Even at particle level where the jet is reconstructed from the stable particles, the exact P_T balance may not hold if the out-of-cone radiation has η dependence. The underlying event energy and pile-up energy contributions may also depend on η. In addition, the electronic noise is an additive contribution and is different for different sections of calorimeter. However, these effects are small and are being studied.

The initial scale determined using test beam measurement will be confirmed by comparing the jet transverse energy to the photon in γ-jet events which is a $2 \to 2$ process in LO QCD. The photon transverse energy is measured in electromagnetic calorimeter which can be calibrated using electrons from $Z \to ee$ or photons from π° decays. In addition to out-of-cone energy effect, the photon-jet balancing suffers from the large dijet background where one of the jet fakes a photon. In addition the photon-jet sample is a mixture of quark and gluon jets. The quark and gluon jets have different particle multiplicity and thus due to non-linearity of calorimeter response to hadrons, their calorimeter response differs by $\sim 15\%$. These effects can be evaluated by judicious use of calibrated detector simulation and a good physics model of parton fragmentation.

CMS plans to measure the calorimeter response to single particle, observed in pp collisions and test beam data and, tune the calorimeter simulation. Similar program has been successfully used by CDF Collaboration[5]. CMS is currently collecting test beam data using pions, electrons and muons with special emphasis on the low energy pions. Combining the in-situ single particle and the test beam measurements, with a significant overlap region, one should be able to achieve jet energy scale to $< 3\%$ for particle jets. At this stage, the jet energy scales of real data and simulated data can be independently determined from photon-jet data to achieve even better precision.

After taking care of the calorimeter non-linearity, the $W \to jj$ mass peak in top quark events will be used to confirm the absolute calorimeter calibration obtained by photon-jet balancing and Monte Carlo based calibration. At LHC, the expected statistical uncertainty on the top mass determination is $\sim 1\%$. The systematic uncertainty is dominated by the pile-up effects. In my opinion, it will be difficult to disentangle the effects of calorimeter energy (mis)calibration and the out of cone radiation in multijet environment and one will have to use on some physics model for parton radiation and hadronization.

5. JET ENERGY RESOLUTION

The calorimeter jet energy is a convolution of single particle response with the spectrum of particles in a jet. Normally jets are corrected as a single object and thus the uncertainty in the jet energy scale arises from the fluctuations in the jet fragmentation and fluctuations in the calorimeter showering. The fragmentation fluctuations can be factored out if each individual particle can be identified and corrected based on its type and momentum. In addition, the effect of shower fluctuations can be reduced if the longitudinal shower profile can be used to separate early showering particles from late showering particles and corrected accordingly.

The single particle response measured in the test beam has been used to derive the corrections based on energies deposited in EM and HAD sections. The energy deposited by a hadron in a calorimeter [6] is

$$E = (e/\pi)_{em} E_{em} + (e/\pi)_{had} E_{had} \tag{1}$$

where E_{em} is energy observed in EM section and E_{had} is the energy observed in HAD sections of the calorimeter. The two correction factors are:

$$(e/\pi)_{em} = \frac{(e/h)_{em}}{[1 + ((e/h)_{em} - 1)f_0]} \tag{2}$$

$$(e/\pi)_{had} = \frac{(e/h)_{had}}{[1 + ((e/h)_{had} - 1)f_0]} \tag{3}$$

The intrinsic electron to hadron response in EM ($(e/h)_{em}$) and and HAD ($(e/h)_{had}$) sections can be parameterized as

$$(e/h)_{had} = A_1[1 + A_2/(E_{em} + E_{had})] \tag{4}$$

$$(e/h)_{em} = A_3[1 + A_4/(E_{em} + E_{had})] \tag{5}$$

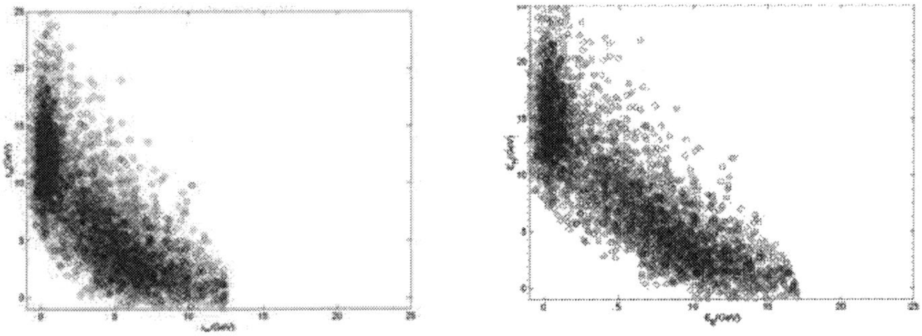

FIGURE 5. Correlation between energy observed in EM and HAD sections of calorimeters (left) before corrections, (right) after corrections

where A_1 is fixed at 1.39 and A_2, A_3 and A_4 are free parameters.

The π^0 fraction of the hadronic shower, f_0, depends on the energy observed in the calorimeter, and is parameterized as

$$f_0 = A_5 \log(E_{em} + E_{had})^{A_6} \tag{6}$$

where A_5 and A_6 are free parameters. These five parameters were determined from charged pions response measured in test beam at momenta $p = 10, 30, 100$ and 300 GeV. The correlation between E_{had} and E_{em} before and after applying these corrections for 15 GeV pions is shown in Fig. 5. A clear improvement in linearity is observed.

To identify the individual particles within a jet, we cluster the EM and HAD sections of calorimeter separately with clustering size $R = 0.03$ and $R = 0.15$ respectively. The EM and HAD clusters are matched in space. The clusters are identified as "electromagnetic", "hadronic" and "mip-like hadronic" based on the E_{had}/E_{em} where E_{had} and E_{em} are the energies in HAD and EM sections respectively. Preliminary studies in the barrel region ($\eta < 1.5$) show an improvement of $\sim 10\%$ in resolution. Further studies are in progress.

6. CONCLUSIONS

CMS Collaboration has built a calorimeter with 0.5% resolution for electrons and $120\%/\sqrt{E}$+6.9% resolution for charged pions. Current calibration scheme will ensure $< 5\%$ tower to tower variation in the calorimeter response at start of the data taking. A jet calibration scheme based on collider measurements and accurate simulation of the calorimeter response is being finalized. The work to improve the jet resolution based on local clustering is promising.

REFERENCES

1. CMS Collaboration, *"The CMS Physics Technical Design Report, Volume 1. CERN/LHCC (2006)"*.

2. Jordan Damgov, *CMS HCAL Testbeam Results and Comparison with GEANT 4 Simulation* this proceedings.
3. G. Blazey *et al.* "Run II Jet Physics: Proceedings of the Run II QCD and Weak Boson Physics Workshop", hep-ex > hep-ex/0005012.
4. G. Grindhammer, M. Rudowicz and S. Peters, Nucl. Instrum. Methods. and Phys. Res. A **290**, 469 (1990).
5. CDF Collaboration, A. Bhatti *et al.* "Determination of Jet Energy Scale at the Collider Detector at Fermilab", *Accepted by Nucl. Instument. Meth.*, FERMILAB-PUB-050470(2005), hep-ex/0510047.
6. Dan Green, private communication.

Jet Reconstruction and Calibration in the ATLAS Calorimeters

Sigrid Jorgensen Roca

IFAE, Universitat Autonoma de Barcelona, UAB E-08193 Bellaterra, Spain
E-mail: sigrid@ifae.es

(on behalf of ATLAS collaboration)

Abstract. Many physics studies in ATLAS require precise reconstruction and calibration of particle jet kinematics. Among these are the reconstruction of the top quark mass, the search for the Higgs boson, and possible supersymmetric particles. The ATLAS calorimeter system has been designed to meet these requirements across a wide acceptance in pseudorapidity ($|\eta|$<5). Different calorimeter technologies are applied in different rapidity regions to optimize the performance with respect to coverage, containment, highest possible spatial granularity, and the best possible energy resolution, in the difficult and changing experimental conditions characteristic for each of these regions.

In this talk we briefly illustrate the ATLAS calorimeter features most relevant for the jet measurement. The general approach to calorimeter jet calibration is two-fold. First, the jet signal shape is used to correct for detector effects such as non-compensation and energy losses in inactive materials. This followed by corrections for biases introduced by the jet clustering algorithms and effects from the collision physics environment. We intend to discuss this calibration procedure and the different strategies available to implement it, in the context of the evaluation of the jet reconstruction performance for various available jet clustering algorithms, including a fixed cone and the Kt algorithm. A focus in this discussion is on the expected initial run condition at ATLAS start-up.

Keywords: ATLAS, calorimetry, calibration, jets
PACS: 29.40.Vj

ATLAS JET RECONSTRUCTION AND CALIBRATION STRATEGY

Various effects play a role in the chain that goes from the partons produced in the pp collisions to the signals detected in the calorimeter: physics effects (as parton shower and fragmentation, underlying events (UE), etc.), detector effects (as non compensation, dead material, electronic noise, etc.) and clustering effects (as "out-of-cone" energy losses, etc.). The ATLAS jet calibration procedure is designed to factorize the corresponding corrections as much as possible. It consists of two steps. First, the calibration correcting detector effects is applied. This provides the particle level energy of the clustered jet. Second, the process dependent corrections for physics effects are applied, which are intended to recover the kinematics of the scattered parton.

Characteristics of ATLAS calorimetry

The ATLAS hadronic (Tile and Liquid Argon) and electromagnetic (Liquid Argon) calorimeters have a pseudorapidity coverage $|\eta| < 5$, and a fine lateral and longitudinal

CP867, *XII International Conference on Calorimetry in High Energy Physics,*
edited by S. R. Magill and R. Yoshida

granularity for performant resonance resolution, particle identification and energy calibration [1]. The level of non-compensation (e/h) is about 1.3 to 1.5, depending on the calorimeter. This introduces a non-linearity of the energy response to hadrons which can be corrected with cell signal weighting techniques, with significant improvement of the resolution, as demonstrated with test beam data [2, 3].

Jet response uniformity

The calorimeter response to jets depends on the level of non-compensation, dead material and gaps. (Fig. 1, left). In the precision region ($|\eta| < 2.5$) the response is quite uniform. The barrel-end cap transition is visible around $|\eta| = 1.5$. The forward region shows a more pronounced effect of the transition between the end-cap and forward calorimeter at $|\eta| = 3.2$.

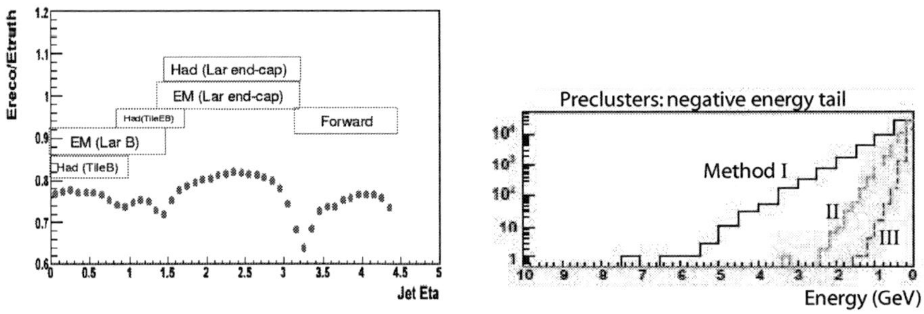

FIGURE 1. Jet energy calibrated at EM scale normalized to the particle level jet energy as a function of jet $|\eta|$ (left). Negative energy left in preclusters (right) showing that CaloTopoCluster is the best method (see text).

Reconstruction and calibration scheme

Different strategies to do the jet reconstruction and calibration are under study. The current baseline uses calorimeter cell signals at the electromagnetic (EM) energy scale, with two options for cell clustering prior to jet finding: the cells can either be collected into projective towers (CaloTowers) in a $\Delta\eta \times \Delta\phi = 0.1 \times 0.1$ grid, or into three-dimensional topological cell clusters (CaloTopoClusters) [2]. The main difference between the two approaches is the calorimeter noise suppression, as discussed later. In any case, the jet finder then merges theses towers or clusters into jets, which still need to be calibrated.

The alternative strategy under development will allow to correct for detector effects before jet clustering: it builds CaloTopoClusters, estimates the nature of the corresponding energy deposit (electromagnetic or hadronic) using their shape and position, and applies the appropriate calibration to them, thus correcting detector effects, see [4]. Jet clustering algorithms are then applied to the calibrated CaloTopoClusters, and final jet energy scale corrections are applied for precise particle level calibration. Finally, parton level calibrations are applied.

Preclustering and noise treatment

Three methods for pre-clustering and noise suppression are presently implemented for the ATLAS calorimeters. The currently best tested option is based on CaloTowers (method I). Before actually running any jet finding algorithm, CaloTowers with negative signals are combined with neighbours in a window of 7×7 towers until the total signal becomes positive. If there is no positive signal sum within this window, the CaloTower energy is set to 0, thus introducing a bias in the jet energy scale. The second less biased method uses CaloTopoClusters. The algorithm starts from a seed, a cell with energy $|E_{cell}| > 4\sigma$ of the noise, aggregates neighbours with $|E_{cell}| > 2\sigma$, and ends summing all directly connected cells. Unclustered cells are not used further. A third method, mostly used in missing E_t reconstruction, simply removes all cells with $|E| < 2\sigma$. Fig. 1 (right) shows the performance of all three methods.

Jet clustering algorithms

In this study we use three representative algorithms. The seeded cone with splitting and merging [5] is used with two optional cone sizes $R = \sqrt{\Delta\eta^2 + \Delta\phi^2} = 0.7$ to avoid fragmentation losses for low E_t jets, and $R = 0.4$ to optimize jet separation in resonance decays and at high luminosity. The K_t [6] algorithm is also available. It merges signals based on radial distance and transverse momentum with a jet size parameter $D = 1$.

Jet energy calibration to particle level

The particle level jet calibration is constrained by signal linearity and resolution optimization, while correcting for dead material and non-compensation. There are various strategies, most prominently: apply weights to individual cell signals, or to calorimeter sampling layer energy sums. The latter is simpler and fast, but less performant. Both take into account η and jet energy dependencies in different ways.

In the current default calibration method in ATLAS, a weight is applied to each cell [3, 7]. It depends on the cell energy density and the sampling S to which the cell belongs, with the reconstructed jet energy $E_{rec} = \sum_i^{cells} W((E_i/V_i), S)E_i$. An hadronic shower typically contains lower energy density cell signals, so that W is usually larger for these cells ($W \sim 1.3 - 1.6$). Higher signal density cells are more likely from electromagnetic deposit, thus $W \sim 1$.

Additional corrections $R(E_t, \eta)$ are applied to correct for residual non-linearities, typically amounting to very few percents.

Jet energy resolution and linearity

Fig. 2 shows the linearity and resolution as a function of energy at EM and calibrated scale. The calibration coefficients have been obtained for $R = 0.7$ cone jets from simulated QCD di-jet samples within $|\eta| < 0.7$. Note that the non-linearity reduces from 15% at EM scale to 4% when the calibration is applied. The resolution at EM scale is $\frac{\sigma(E)}{E} = \frac{83\%}{\sqrt{E(GeV)}} \oplus 5\% \oplus \frac{2.3}{E(GeV)}$ and $\frac{\sigma(E)}{E} = \frac{67\%}{\sqrt{E(GeV)}} \oplus 2\% \oplus \frac{4.3}{E(GeV)}$ after calibration, with a significant improvement especially in the constant term. The increase of the noise term is due to the enhancement of small signals by the weighting procedure. It is still a small contribution at typical LHC jet energies.

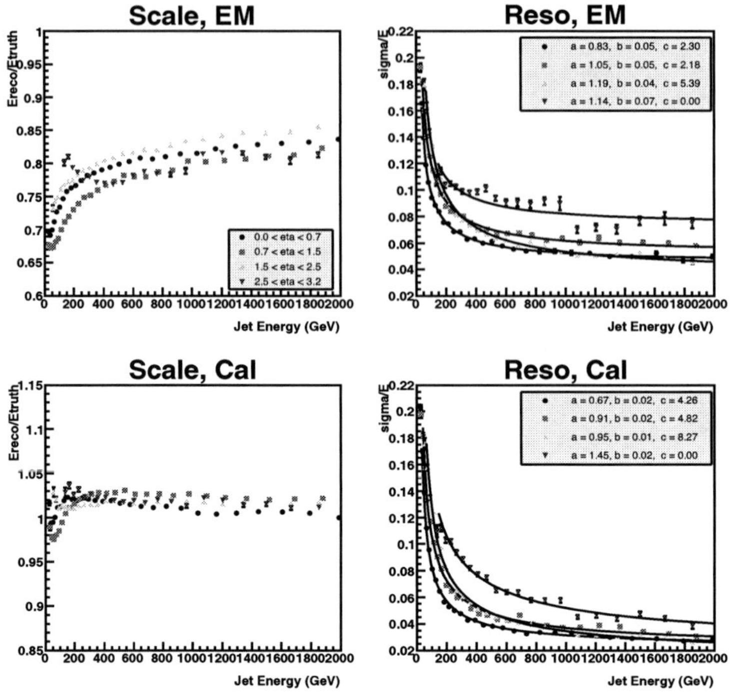

FIGURE 2. Response linearity (upper left) and resolution (upper right) at EM scale.The linearity and resolution at calibrated scale are shown in the lower plots.

In situ physics processes

In situ processes give the possibility to calibrate jets to parton level and are also very useful to validate simulations, especially concerning the physics effects. In ATLAS, QCD di-jets, $\gamma/Z(\rightarrow ll)+jet$ and $W \rightarrow jet\,jet$, have been studied. Di-jet events are mainly used to cross-calibrate different parts of the detector, while $\gamma/Z(\rightarrow ll)+jet$ can be used to perform parton level calibration, jet clustering, UE studies, etc. The main advantage is that the γ and Z are well understood EM references recoiling against the hadronic system. Statistics available at luminosity $L = 10^{33} cm^{-2} s^{-1}$ will be large in ATLAS. For transverse momentum p_t in the range $20\text{GeV}/c < p_t < 60\text{GeV}/c$, for example, the $Z + jet$ rate is about 2 Hz, while the $\gamma + jet$ rate is about 0.1 Hz due to the prescaled sinle γ trigger. For higher $p_t > 60\text{GeV}/c$, the $Z + jet$ rate is reduced to 0.1 Hz, while the $\gamma + jet$ rate is going up to 2 Hz. $W \rightarrow jet\,jet$ can be used for parton level calibration using the (precise) W mass constraint, at an expected rate of a few hundred events per day after selections.

Two complementary variables are used to study the recoil of the jets against the EM systems (γ or Z). One is the p_t balance, which is calculated as the relative difference Δp_t between the most energetic recoiling jet and the EM system, with $\Delta p_t = \left(p_t^{jet} - p_t^{\gamma}\right)/p_t^{\gamma}$. Δp_t is sensitive to out of cone showering, gluon radiation, UE and detector effects.

The missing E_t projection method calculates the recoil of the complete hadronic system against the photon as a response function $R = 1 + \frac{\vec{E_t^{miss}} \times \vec{E_t^\gamma}}{\vec{E_t^\gamma} \times \vec{E_t^\gamma}}$, $\vec{E_t^{miss}} = -\sum_{calo} \vec{E_t^{calo}}$. It is essentially only sensitive to the particle response.

Jet algorithm studies with γ + jet events

Fig. 3 shows the most probable values of the p_t balance as a function of the average of true γ and parton energy. Here the most energetic jet opposite to the photon has been selected, thus allowing unbiased comparative jet algorithm studies.

Energy leakage out of cone and UE affect the p_t balance. For $R = 0.7$ cone jets at particle level, both effects cancel. At reconstruction level, i.e. calibrated reconstructed jets from detector signals, the bias is $0 - 2\%$ showing that the calibration procedure works very well. The large negative bias at particle level using the $R = 0.4$ cone jet algorithm is due to the larger energy leakage out of cone. For reconstructed jets the bias is even worse due to the calibration, since the reference $R = 0.7$ cone jets have a different composition than the $R = 0.4$ jets, where basically only the core of the jet with more energetic particles is reconstructed. The particle level p_t balance shown in fig. 3 indicates that the K_t algorithm collects too much energy coming from UE.

The balance obtained with real data have to be compared with simulations to validate both the description of experimental and physic effects.

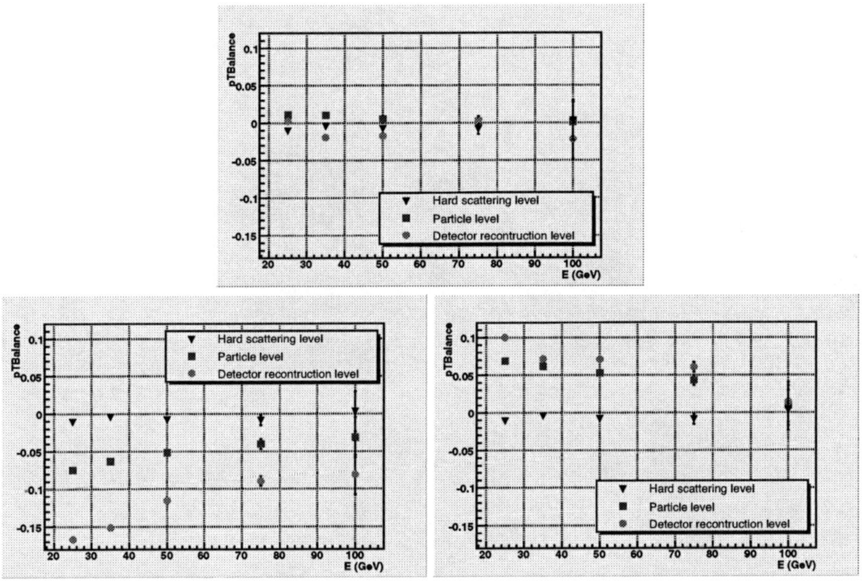

FIGURE 3. Most probable values of the p_t balance of jets as a function of the average of true γ and parton energy from simulations, for $R = 0.7$ cone jets (top), $R = 0.4$ (bottom left), and the k_t algorithm (bottom right).

In situ W $\rightarrow$ jet jet

$W \rightarrow jet\,jet$ in $t\bar{t}$ events is a well known resonance reconstructed with little physics background. These events can be used to calibrate jets to parton level as well, especially for the smaller cone jets ($R = 0.4$). Residual sensitivity to the boost of the W system limit the kinematic range for which these decays can be used. Also, the fact that these jets are from a colourless source may limit the validity of jet energy scale corrections derived this way, especially with respect to QCD jets.

CONCLUSIONS

Different strategies for signal reconstruction and jet energy calibration are presently studied with the ATLAS calorimeters, indicating that the requirements from physics analysis on the jet energy measurement error can be met, if not exceeded for certain physics channels. A strategy for the use of first pp collision data is under design, including the preparation of measurements of important event contributions to the jet energy scale.

ACKNOWLEDGMENTS

I would like to thank Martine Bosman, Chiara Roda, Peter Loch and all the members of the ATLAS Jet/ETmiss group who have helped in the preparation of this talk.

REFERENCES

1. ATLAS Collaboration, *Detector and Physics Performance Technical Design Report*, CERN/LHCC/99-14, Vol I, 25 May 1999.
2. C. Cojocaru et al., *Hadronic calibration of the ATLAS liquid argon end-cap calorimeter in the pseudorapidity region* $1.6 < |\eta| < 1.8$ *in beam tests*, Nucl.Instrum.Meth. A531 (2004) 481.
3. S. Akhmadaliev et al., *Results From a new Combined Test of an Electromagnetic Liquid Argon Calorimeter with an Hadronic Scintillating-Tile Calorimeter*, Nucl.Instrum.Meth., A449 (2000) 461.
4. R. Seuster in these proceedings
5. G. C. Blazey et al., *Run II Jet Physics: Proceedings of the Run II QCD and Weak Boson Physics Workshop*, http://arxiv.org/abs/hep-ex/0005012
6. S. D. Ellis, and D. E. Soper, *Successive combination jet algorithm for hadron collisions*, Phys. Rev. D 48 3160 (1993).
7. I. Abt et al., H1 Collaboration, Nucl. Instrum. Meth. A386(1997) 348.

The Simulation of the ATLAS Liquid Argon Calorimetry

V. Niess

(on behalf of the ATLAS collaboration Liquid Argon Calorimetry Group)

CPPM –IN2P3/CNRS – U. Mediterranée
163, av. de Luminy – case 902 –13288 Marseille cedex 09 - France

Abstract. The liquid argon calorimeter of the ATLAS detector is aimed at measuring electromagnetic energy deposit from pp collision produced at the Large Hadron Collider. It was designed to allow rare electromagnetic final states detection and jet spectroscopy in a high-background environment. The simulation of the calorimeter has a long history and is now a component of an integrated GEANT 4 based simulation of ATLAS. Its sophistication has increased during its development. At present it models many effects which significantly affect shower development and response. Step by step, the accuracy of the simulation is tested and tuned to beam test data accumulated in dedicated runs from 2001 to present. The framework in which it runs allows users to switch simulation configuration between pp collisions, cosmic rays and single particle data taking, with the full ATLAS detector or with the various beam test setups. For the purpose of hadronic and electromagnetic calibration, a detailed accounting scheme of energy flow, allowing energy deposits tracking in upstream dead material for example, is implemented. Fast parameterization of electromagnetic shower development has also recently become available. The software is developed by many experts around the world and works at a very high level of reliability because of the constant scrutiny it receives from a large community of developers and users, as well as a dedicated validation team.

Keywords: Simulation; Calorimetry; Particle Physics.
PACS: 07.05.Tp ; 29.40.Vj.

INTRODUCTION AND GENERAL CONTEXT

The ATLAS collaboration[1] is building the largest ever man made particles physics detector for the Large Hadron Collider (LHC), to start operations at CERN late 2007. The detector has been designed to fully exploit the great physics potential provided by 14 TeV pp collisions at the LHC and to cope with a very generalist physics panorama, oriented toward the discovery of new energetic and rare processes. Key features are the search for the Higgs boson and physics beyond the Standard Model of particles physics.

In ATLAS, calorimetric measurements are provided by several sub-detector systems one of which being the Liquid argon (LAr) calorimeter[2]. This sub-detector is a sampling calorimeter relying on liquid argon as active medium and metallic sheets as

CP867, *XII International Conference on Calorimetry in High Energy Physics,*
edited by S. R. Magill and R. Yoshida

passive media. The ionization in liquid argon, resulting mostly from cascading particles, is collected with a 10 kV/cm high voltage electric field. It provides full coverage electromagnetic calorimetry but also hadronic calorimetry in the end-cap regions.

The LAr calorimeter is composed of four sub-systems immersed in liquid argon. Design specificities, from one sub-system to another, allow a $|\eta| \leq 5$ pseudo-rapidity coverage. The core of the LAr Calorimeter is a cylindrical barrel, 6.3 m long and 4 m outer diameter, aligned along the beam line and surrounding the primary vertex. The barrel is purely designed for electromagnetic calorimetry. It covers a $|\eta| \leq 1.4$ range with a 25 to 35 radiation length depth. The absorber plates are composed of lead sheets, a few millimeters thick, maintained by stainless steel and folded in a peculiar accordion shape to improve the calorimeters hermeticity. Both extremities of the barrel are closed by end-caps made of an electromagnetic end-cap (EMEC, $|\eta| \in [1.4;3.2]$) with 'accordion' like absorbers, followed by the hadronic end-cap (HEC, $|\eta| \in [1.5;3.2]$) with stacked copper absorber plates. Thc very forward region is complemented by a forward calorimeter systems (FCAL, $|\eta| \in [3.1;4.9]$) incorporated in the end-caps, as can be seen on figure 1.

The simulation activity of LAr components has started with design studies early 1994, using the GEANT 3 code. The LAr simulation now stands on an object oriented architecture, using a GEANT 4 core, and it has been integrated into the ATLAS's common analysis framework ATHENA[3] since May 2004. Motivations for a simulation of the LAr calorimetric systems, and more-over a fully detailed simulation, are mostly driven by the stringent calorimetric performances required for physic analysis. The ATLAS detector discovery potential, in searching for rare final states in a high-background environment, requires an accurate understanding of the calorimeters response. For example, the significance of the *light* Higgs boson search into its electromagnetic final state modes is limited by the instrumental resolution on its invariant mass reconstruction. Accuracies better than 1 % are required[1]. At these 100 GeV high energies, the resolution becomes limited by fine systematic effects. Therefore, a proper understanding and estimate of these is required. This goes through a detailed description and simulation of the detectors response.

In the following sections we give a general overview of the strategies used in the LAr simulation covering two main aspects which are: the description of matter distributions in the detector elements and the simulation of the detectors response in both sensitive and non-sensitive regions. We then discuss the performances of the simulation: Its accuracy and resource usage. In this paper we focus on the particular case of the EMEC test beam, the author has been most dealing with. Other test beam results are covered by other papers[4-8] presented in these conference proceedings. Finally we conclude with achievements and future scopes the simulation has to deal with.

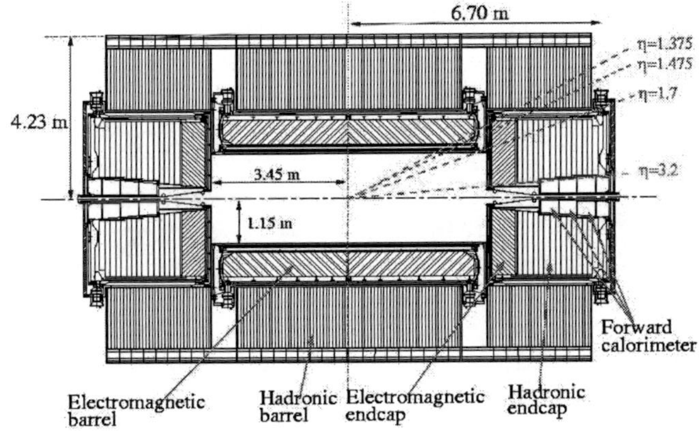

FIGURE 1. A general view ATLAS Calorimeters components. The Liquid Argon components are embedded in the hadronic Tile calorimeter system.

OVERWIEW OF THE LIQUID ARGON SIMULATION

The ultimate goal of the LAr simulation is to run a GEANT 4 Monte-Carlo simulation of particles showering in the detector systems. However, for the very end user this simulation has to be integrated in a 'transparent' way in the offline analysis software of ATLAS, ATHENA, which relies on python scripting languages. Therefore a dedicated interface framework was developed for controlling and running the GEANT 4 simulation from ATHENA: The G4AtlasApps package[9]. From this very high user level one can manage the GEANT 4 core features: range cuts, particle guns, physics lists, as well as pre-defined simulation options provided to the end user via a set of flags. The G4AtlasApps framework allows users to switch simulation configuration between pp collisions, cosmic rays and single particle data taking, with the full ATLAS detector or with the various beam test setups with the same 'mechanics'. This, as well as relational data-base usage, participates in a general effort to centralize the key information used in the simulation.

The lower level of the simulation stands on layers of C++ coded components organized in separated packages. These components describe the two main features involved in the simulation which are: the building of the detectors geometry and the simulation of its response. This later is done via the management of particle 'hits' which are the quanta of information issued by the simulation.

The Geometry Description

A key point in simulating the calorimeters response is an accurate description of the energy deposit. This involves two main aspects which are the accuracy of GEANT 4 algorithms in modeling energy loss processes and a precise description of the matter distribution in the detector. Considering the first point, we stand at a user level. We can only check the various GEANT 4 releases to test beam data and feed back to the author. However, for what concerns the geometry description, we are granted full responsibility and therefore many cares are taken.

The description of the detectors geometry goes through successive layers of C++ codes. In order to disentangle information on the geometry from the GEANT 4 code, a dedicated C++ layer, called GeoModel, was built. This layer allows the definition of geometric components: boxes, tubes, as well as custom shapes, atomic elements and composed materials, and this is done outside of the GEANT 4 simulation code. The GeoModel information is then translated to GEANT 4 definitions for the need of the simulation. But it can also be shared with other tools, like reconstruction or dedicated visualization algorithms (HEPvis), without the need to go down to the GEANT 4 level. The GeoModel layer is also granted additional features, like control over the global alignment of sub-detector elements and handling of time dependent schemes.

The description of the various detectors components follows a hierarchized structure with successive inclusions. A complete setup is built by assembling single components elements defined in GeoModel so that, for example, the EMEC wheel description is used in the full ATLAS detector simulation as well as in the simulation validation on test beam data.

Energy Flow Schemes

The goal of the GEANT 4 simulation is to provide an estimate of the energy deposited in the various detector elements. Because of the large number of individual tracks involved in a full GEANT 4 simulation, the information from the tracking level is binned and collected over a grid of cells, resulting in so called *hits*. This collection of hits serves two purposes. It is used to simulate the calorimeters output current response, by branching dedicated 'sensitive detectors' during the GEANT 4 tracking processes, and it is also used to estimate energy deposition flows for calibration schemes.

Visible Energy In Liquid Argon And Charge Collection

In the first case, hits are associated to the readout cells of the detector elements. The cell response is computed from the various tracks ionizing the sensitive area. More precisely, the peak output current, I_0, is estimated by summing up individual track contributions, ΔI_0, according to the following equation :

$$\frac{\Delta I_0}{\Delta E} = K \cdot (1 - r(\varepsilon)) \cdot v_d(\varepsilon) \cdot \frac{\varepsilon}{V} \cdot S(\vec{r}) \quad (1)$$

where, ΔE, is the energy deposited by the track in the cell. The previous equation lies on strong physics basis. Based on energy conservation law, it is obtained by equalizing the mechanical power of a charge, *q,* drifting in an electric field, ε, at a drift velocity, v_d, with the resulting electric power $\Delta P = \Delta I_0 \cdot V = q \cdot \varepsilon \cdot v_d$, where, V, is the applied voltage. We also take into account ions recombination effects by a $r(\varepsilon)$ screening factor on the electric charge. In addition, the global apparatus function, including the readout electronics response and corrections from cell calibration is simulated by an average suppression factors, S, depending on the track mid-location, $\bar{r}$. Tracks located too close to the collection electrode have a small contribution to the calorimeters signal, due to the non instantaneous electronics impulse response.

The simulation only provides a relative estimation of the calorimeters response. There is a global energy scale factor, K, which is tuned on test beam data. This global energy scale accounts for the effective charge resulting from ionization loss as well as for conversion factors used during the signals reconstruction.

Energy in Dead Materials and Calibration

A detailed scheme was developed to track the 'real' energy deposited in the calorimeter parts, in particularly including the non sensitive 'dead materials' area. The GEANT 4 deposited energy is recorded in so called 'Calibration hits' with a classification in four sub-categories. The deposited energy can be: of 'EM' origin, of 'Non EM' origin, hence hadronic, 'not visible' because lost in nuclear transitions and 'escaped' from the detector, as neutrinos for example. For dead materials a grid divided in angular sectors around the beam line is used, with cells mapping pseudo-rapidity values, η, and angular sectors, ϕ.

The simulation is used in calibration schemes. The knowledge of both the visible energy and the non visible energy allows to develop dedicated schemes to recover in real data this non visible loss, from the visible energy[4,10]. These calibration schemes are investigated and tuned according to energy deposition estimations given by the simulation.

PERFORMANCES OF THE LIQUID ARGON SIMULATION

A very first achievement of the ATLAS integrated GEANT 4 simulation is that the code has proved to be robust via massive usage, in large scale exercises like the DC2 data challenge with more than 8.5 Mega events fully simulated, as well as from a world-wide distributed community of physicists: the ATLAS collaboration. In addition, the accuracy of the simulation is cross-checked on test beam experiments[4-8] carried out from 2001 to the present. This provides very stringent guidelines for tuning the simulation.

Comparison to Test Beams and Accuracy

There are numerous test beam setups that allow comparing the simulation to real data, components by components or through combined test beams. As a general results, the LAr simulation provides a good agreement with the test beam data, despite some fine effects left to understand. In this paper we present results for the EMEC stand alone test beam. Some simulation results for other components are reported in [4,8] for the stand alone barrel, [4,7] for the combined one and [6] for the combined end-cap test beam.

The electromagnetic end-cap has a complex geometry compared to the barrel. It is contained in a 4 m outer diameter disk with a 51 cm width and divided in two concentric wheels: the Inner and Outer wheels, with a $|\eta| = 2.5$ separation. The accordion like fan absorbers are arranged radially. Consequently the gap between electrodes increases with the radius. To maintain a roughly homogeneous electric field between electrodes, the high voltage is also increased by steps over so called 'HV sectors'. This results in a succession of ramps in the EMECs response, as can be seen on figure 2.

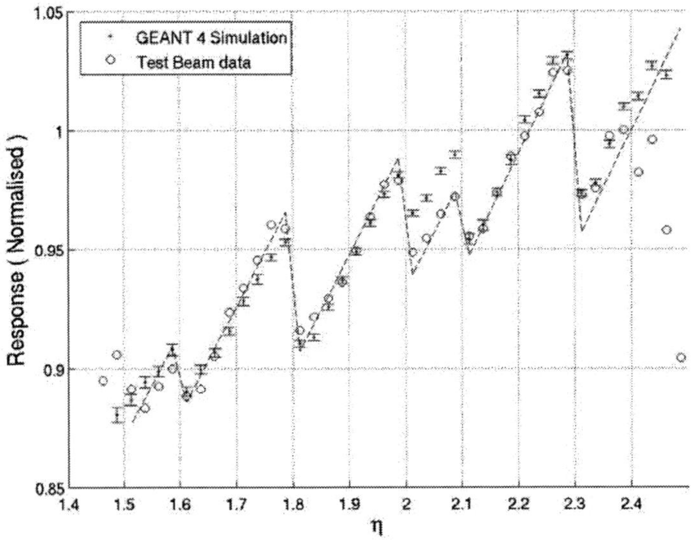

Figure 2. Normalized response of the outer wheel of the Electromagnetic end-cap calorimeter as a function of the pseudo-rapidity coverage, η, for 120 GeV electron showers.

EMEC Global Accuracy

The simulated response of the EMEC stand alone test beam was confronted to real data analysis[11,12] for 120 GeV electron showers. A detailed charge collection scheme was used in the simulation with a finely interpolated electric field map. The scale of the simulation was fitted according to a chi-square criteria. The overall agreement, over various η ranges, can be seen on figure 2. A 2 % relative accuracy is achieved in the central region of the EMEC Outer wheel. Leakages are observed in border regions that are still under investigation. In addition, a 2 % systematic can be observed in the

HV sector $\eta \in [2.0;2.1]$. No particularity was found in the simulation, in this η range, that could explain the observed bias. This systematic would be consistent with a resistive loss in real test beam of ~50 V over 1500 V nominal value. This later hypothesis needs further investigation.

EMEC Non-Uniformities and ϕ-Modulations

At the level of readout cells, the EMEC response can vary significantly by up to 3% due to geometric non-uniformities. Non uniformities results from the design specificities, as for example the so called ϕ-modulations[10-12] which are generated by the accordion like shape of the absorbers. But non uniformities can also be hazardous. For example the precise width of the absorber plates needs to be controlled to the level of 10 μm during manufacturing processes[2,8]. Control is given in the simulation to tune various parameters in the simulation that affect the non uniformity.

The observation of ϕ-modulations allows very fine tuning of the simulated geometry of the EMEC. The amplitude and shape of the modulations vary significantly with η. This dependency is also sharply affected by variations of the slant angle in the absorbers plates, to the level of 1° as well as by the deformation caused by the sagging of the absorber plates during assembly. The tuning of these parameters in the simulation, according to test beam data, is currently under investigation.

Dedicated muons runs allowed a fine scanning of the EMEC structure. Particularly, one could see the specificities of the electrode printout scheme in the global response to muons[13]. According to this, a detailed description of the readout cells geometry was developed to reproduce these features at the level of the simulation.

Resource Usage

The performances of the LAr simulation are tested according to benchmark jobs scripts[9]. The memory consumption is of 600 Mo with one half linked to GEANT 4 processes. The CPU time, however, is mostly driven by GEANT 4 performances. With the currently in use version 4.7 of GEANT, it is of the order of 10 min KSI2K for 100 GeV electrons in a full setup. For a stand alone test beam line like the EMEC one, with less instrumentation it can go down to 2 min KSI2K. Simulation time can be marginally reduced by using looser tracking range cuts. Going from the standard value in use of 30 μm to 1 mm, the simulation duration is reduced by a factor of ~1.5. However, with the latest version 4.8 of GEANT, where the accuracy of scattering processes involved in shower lateral development has been improved, the simulation time is increased by a factor of ~2.

The most promising feature for reducing simulations duration is the use of fast shower parameterizations[14] within GEANT 4. Such schemes have recently been implemented within the LAr simulation for electron showers in the 0.5 to 100 GeV range. They allow reducing the duration of the simulation by up to a factor 100 for 100 GeV electron showers. Therefore, there is a balance to find between the simulation accuracy and its speed. The implementation of these fast parameterization tools may allow to find it and to fit the every users need. The simulation can cope with both 'private' analysis on individual laptop, for example, and more accurate massive data challenge productions.

CONCLUSION AND OUTLOOK

The Liquid argon simulation is in good shape for full ATLAS data taking. It has proved to be stable and reliable in many exercises. It provides a good over-all description of the detector elements response. Its flexibility allows the handling of many geometries within a single mechanics. Hence, the code to be used for full ATLAS analysis is validated by confrontation to numerous test beam data within the same frame.

Further steps focus on tracking down fine systematic effects and on providing an 'as built' detector. With the commissioning of the ATLAS detector there is also an ongoing work in estimating the sensitivity to cosmic rays[15] via simulation studies.

ACKNOWLEDGMENTS

We thank the ATLAS liquid Argon calorimetry group for the opportunity to report this work and particularly the CPPM team. We also thank the Organizing Committee for financial support.

REFERENCES

1. ATLAS Collaboration, "Technical Proposal for a General purpose pp Experiment at the Large Hadron Collider at CERN", CERN/LHCC/94-93.
2. ATLAS Collaboration, "Liquid Argon Calorimeter Technical Design Report", CERN/LHCC/96-41.
3. The ATHENA framework: http://atlas.web.cern.ch/Atlas/GROUPS/SOFTWARE/OO/architecture/General/Documentation/
4. W. Lampl, "Studies of the Linearity of the ATLAS Electromagnetic Calorimeter Response", CALOR 2006 proceedings.
5. V. Giangiobbe, "Studies of the Response of the ATLAS Barrel Calorimeters to Pions using 2004 combined Test Beam Data", CALOR 2006 proceedings.
6. P. Schacht, "Recent Results of the ATLAS Endcap Combined Test Beam", CALOR 2006 proceedings.
7. M. Delmastro, "Recent results of the ATLAS Barrel Combined Test Beam", CALOR 2006 proceedings.
8. I. Nikolic, "Recent Results on the Uniformity of the Liquid Argon Calorimeter Measured in Test Beams", CALOR 2006 proceedings.
9. G4AtlasApps: http://atlas-computing.web.cern.ch/atlas-computing/packages/simulation/geant4/G4AtlasApps/doc/
10. R. Seuster, "Hadronic Calibration of the ATLAS Calorimeter", CALOR 2006 proceedings.
11. B. Aubert et al, "Performances of the ATLAS electromagnetic calorimeter end-cap module 0", Nucl. Inst. Meth. A 500 (2003) 178-201.
12. F. Hubaut, C. Serfon, "Response uniformity of the ATLAS electromagnetic end-cap calorimeter", ATL-LARG-2004-015.
13. see the talk corresponding to this abstract : http://www.hep.anl.gov/CALOR06.
14. G. Grindhammer, S. Peters, "The Parameterized Simulation of Electromagnetic Showers in Homogeneous and Sampling Calorimeters",hep-ex/0001020 (1993).
15. M. Aleksa, "The ATLAS Liquid Argon Calorimeter : Construction, Integration, Commissioning", CALOR 2006 proceedings.

Electron/Photon Reconstruction Performance with the ATLAS Detector

K. Benslama, on behalf of the ATLAS Collaboration

Physics Department, University of Regina, Saskatchewan, Canada

Abstract. The calibration of electromagnetic clusters is one of the key issues of ATLAS (and the LHC) in 2007. The cluster algorithm starting from electronically calibrated calorimeter cells will be described. Local position and local energy variations are corrected for. As a last step, longitudinal weights are applied to correct for energy loss upstream of the calorimeter. Strongly influenced by testbeam studies, the new longitudinal weighting optimizes simultaneously energy resolution and linearity. Methods to derive the final calibration parameters from the physics events have been developed. The electron and photon identification methods and their performance will then be discussed.

Keywords: Identification of electron and photon
PACS: 29.40.Vj

INTRODUCTION

The Large Hadron Collider (LHC) at CERN is a proton-proton collider with 14 TeV energy in the center of mass and a design luminosity of $10^{34} cm^{-2}s^{-1}$. The ATLAS experiment is one of the two major multi-purpose detectors currently under construction at the LHC. Its inner detector consists of tracking detectors enclosed in a solenoidal magnet with 2T field. From the inner radius to the outside radius, it consists of pixel detectors, silicon strip detectors (SCT) and transition radiation drift tubes (TRT), covering the pseudo-rapidity region $|\eta| < 2.5$.

The inner detector is surrounded by a sampling electromagnetic calorimeter based on lead and liquid argon (LAr) technology and a hadronic calorimeter based on LAr in the end-caps and iron/scintillator tiles in the barrel. The global detector dimensions are defined by a large air-core muon spectrometer, providing precision measurements of high-p_T muons over $|\eta| < 2.5$.

The LHC physics program [1] ranges from the search for the Higgs boson, searches for physics beyond the standard model, such as supersymmetric particles, new additional W and Z bosons, and also precision studies, such as measurements of the t quark and W boson masses and unexpected signals from unpredicted physics scenarios.

ELECTRONS AND PHOTONS RECONSTRUCTION AND IDENTIFICATION

Events with electrons and photons in the final state are important signatures for many physics analyses at the LHC, since electrons and photons are relatively easy to measure

CP867, *XII International Conference on Calorimetry in High Energy Physics,*
edited by S. R. Magill and R. Yoshida

precisely and to trigger on. Isolated high p_T electrons and photons are not easy to identify at the LHC because of the large QCD background from high-p_T jets, which results in an electron/jet ratio of about 10^{-5} at the LHC (to be compared to about 10^{-3} at the Tevatron) for isolated electrons from W/Z decays, and to a photon/jet ratio of 10^{-4} (to be compared to about 10^{-3} at the Tevatron). Final states containing electrons or photons such as $H \longrightarrow 4e$ or $H \longrightarrow \gamma\gamma$ decays provide convincing discovery channels [1].

Electrons and photon reconstruction mainly uses data coming from the electromagnetic calorimeter (EM) and the inner detector (ID) systems. Electromagnetic objects can be identified by looking at the transverse and longitudinal shower shapes and at isolation variables. For electrons, a track is then required to match in position and energy that measured in the electromagnetic calorimeter. For photons, no track is required (except in the frequent case of converted photons) but γ/π^0 separation criteria is required using some of the unique features of the electromagnetic calorimeter.

Energy Measurement

Four longitudinal samplings are available over most of the EM calorimeter coverage for the energy measurement. Where available, the presampler is used to correct for the energy lost in the upstream material. The strip section provides information on the early part of the shower, besides contributing to the energy measurement. Most of the shower energy, however, is deposited in the middle section sampling, while the back sampling is used for high energies.

The energy response of the calorimeter is potentially affected by the following effects:

- upstream energy losses;
- ϕ-modulations and azimutal gaps between presampler sectors;
- lateral leakage outside the cluster;
- longitudinal leakage behind the EM calorimeter.

Efficient corrections are applied for most of these effects, which allows to preserve a good energy resolution.

In general, the energy reconstructed in the EM calorimeter can be written as:

$$E_{tot} = W_{glob}(b + w_{ps}E_{ps} + E_{str} + E_{mid} + E_{back}) \tag{1}$$

Where w_{glob} is a global calibration factor, b is an offset with units of energy, motivated by recent test beam analyses. This offset was found to optimize simultaneousely electron energy linearity and resolution. $E_{ps}, E_{str},$ *and* E_{back} are the energies measured in a given cluster of cells in the presampler, strip section, middle and back samplings of the calorimeter respectively. The mean fractional energy deviation from truth (data points) and the energy resolution (error-bars) for 100 GeV electrons after application of longitudinal weights is shown in Fig.1.

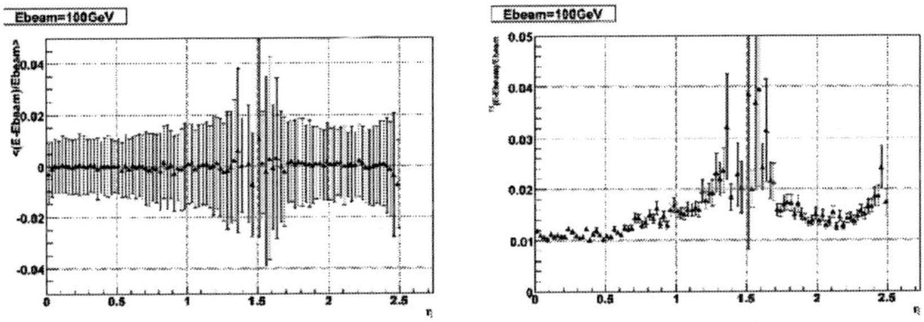

FIGURE 1. The mean fractional energy deviation from truth (data points) and the energy resolution (error-bars) for 100 GeV electrons after application of longitudinal weights.

TABLE 1. Electron identification efficiency ε_e for single electrons with $p_T > 25$ GeV and jet rejection of the offline analysis at low luminosity [2].

	$\varepsilon_e(\%)$	**rejection** $(\times 10^3)$
Calo	91.2 ± 0.4	1.3 ± 0.1
ID	81.3 ± 0.5	19.5 ± 2.5
ID-Calo	76.4 ± 0.6	57.0 ± 7

Electron/jet Separation in ATLAS

The identification of isolated electrons with $p_T > 20$ GeV/c will be essential for many physics searches at the LHC. A challenging task will be to identify electrons in the presence of high QCD backgrounds, which is $\sim 10^5$ times higher, as in the case of W and top decays.

To separate electrons from jets, cuts were developed to maintain a reasonable electron identification efficiency even with pile-up at high luminosity, while removing a high fraction of jets events. The cuts include Level-1 and High Level Trigger cuts, shower shape and isolations cuts in the calorimeter, cuts on track in ID, cuts on ID-Calo matching in position and energy and TRT cuts. The effect of applying all these cuts one after the other to both single electrons and jets samples is shown in Tab. 1. After calorimeter selection, the dominant background consists of photons from π^0 and η decays. This is significantly reduced by requiring the presence of a high p_T track. After the ID-Calo match, charged hadrons remain the main background. The signal-to-background ratio is 2:1 for a QCD-jet rejection of 0.6×10^5. The signal is equally from semileptonic decays of heavy quarks and isolated electrons from W and Z decays. The QCD jet rejection can be improved by using the transition radiation rejection of the TRT. An electron efficiency of about 70% is obtained while a QCD jet rejection above 10^5 can be achieved. Finally, removal of photon conversions by direct reconstruction, would allow the identification of a pure electron inclusive sample with a jet rejection of around 10^6. Further improvements can be achieved by using multivariate techniques.

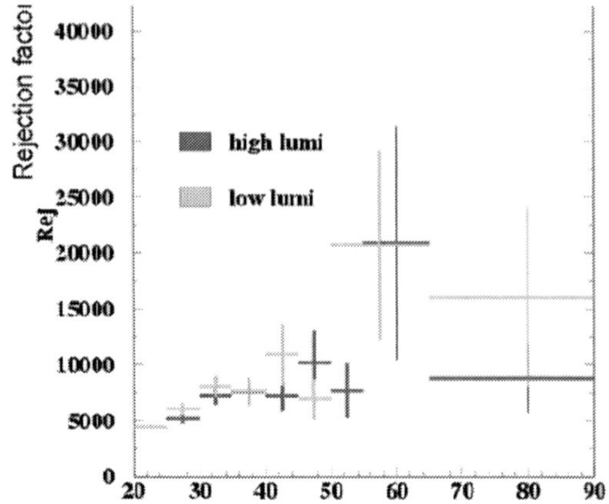

FIGURE 2. Jet rejection as a function of the transverse energy of the jet for a photon identification efficiency of 80% at low and high luminosity [3].

Photon/jet Separation in ATLAS

Given the amount of material in front of the calorimeter, about 40% of the photons from e.g $H \rightarrow \gamma\gamma$ decays convert into e^+e^- pairs before depositing their energy in the calorimeter. Since the $H \rightarrow \gamma\gamma$ is small, it is important to recover these conversions to maintain its efficiency as high as possible. Conversions are first searched for in the Inner Detector, and the energy deposited in the calorimeter in a 3×7 $\eta \times \phi$ window is computed, since this window size allows to recover a significant fraction of the initial photon energy. Using an estimated dependence between the conversion radius, E_T and the photon energy, it is possible to reconstruct the initial photon energy.

If the photon has not been reconstructed as a conversion, the photon/jet separation relies on the search for electromagnetic objects, with cuts including the Level-1, High Level Trigger cuts, shower shapes and isolation cuts in the calorimeter, with the requirement that no track is found in the ID within a $\Delta\eta \times \Delta\phi$ region of size $\pm 0.1 \times \pm 0.1$ around the calorimeter cluster. Fig. 2 shows the jet rejection after photon selection cuts as a function of the jet transverse energy E_T. A rejection factor better than 7000 can be obtained for $E_T > 40 GeV$, both for low and high luminosity.

Electron/pion Separation in ATLAS

The efficient tagging of low energy electrons is an important tool for B-physics, as well as a complementary method to b-tagging. Separating low energy electrons from pions by analysing the energy deposits in the calorimeter alone is not an easy task because these electrons are within or near to jets. A better option is to use the ID to seed the calorimeter clustering. This strategy consists of several steps. First, tracks with $p_T > 2$ GeV are found in the ID and then one looks for the EM calorimeter regions hit

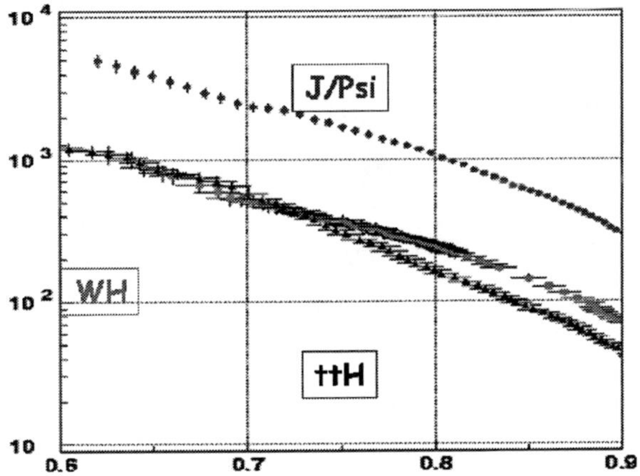

FIGURE 3. Pion rejection as a function of electron identification efficiency for low energy electron [4] [5].

by the tracks. By combining various shower shapes estimators and the E/P value and the information from the TRT, it is possible to get the pion rejection versus the electron identification efficiency curves of Fig. 3.

For the J/ψ sample, a pion rejection factor of 1000 is achieved for an electron identification efficiency of 80%. Electrons from $WH \rightarrow b\bar{b}$ events are located inside jets. Thus, their identification is more difficult. For 80% electron identification efficiency, the pion rejection factor is ~ 250. This soft electron identification could then be used for b-tagging as a complementary method to the standard vertex-based tagging.

CONCLUSION

The ATLAS collaboration has developed powerful electron and photon identification algorithms and tuned them over the past years on detailed Monte Carlo simulation. While maintaining high electron and photon identification efficiency, these algorithms achieve a very high QCD jet rejection. This strong identification ability coupled with a good performance of the ATLAS detector are crucial for many discovery channels.

ACKNOWLEDGMENTS

The author is greatful to the ATLAS Collaboration.

REFERENCES

1. ATLAS detector and physics performance technical design report, ATLAS TDR 14 & 15, CERN/LHCC/99-14 & 15.

2. F. Derue, C. Serfon, "Electron/jet separation with DC1 data", ATLAS Note ATL-COM-PHYS-2005-016.
3. M. Escalier et al., "Photon/jet separation with DC1 data", ATLAS Note ATL-COM-PHYS-2005-048.
4. T. Bold et al., "Pile-up studies for soft electron identification and b-tagging with DC1 data", ATLAS Note ATL-COM-PHYS-2005-027.
5. F. Derue et al., "Reconstruction of DC1 $J/\psi \to e^+e^-$ decays and use for low energy calibration of the ATLAS electromagnetic calorimeter", ATLAS Note ATL-COM-PHYS-2005-022.

Missing Transverse Energy Reconstruction in the ATLAS Calorimeter

Ambreesh Gupta

Enrico Fermi Institute, University of Chicago,
5640 S. Ellis Ave, Chicago IL, U.S.A

Abstract. A very good missing transverse energy (ETMiss) measurement is a crucial requirement for the study of many physics channels at the LHC, including but not limited to Standard Model W or top production, production of Higgs bosons decaying to taus, and SUSY events, for example. The most important contribution to this measurement in the ATLAS detector comes from the calorimeters, which provide near hermetic energy reconstruction up to pseudo-rapidities of |eta|<5. The transverse energy components reconstructed in individual calorimeter cells and clusters of cells, in addition to corrections for losses in inactive materials and cracks, are the basis of the EtMiss calculation. Additional refinement can be achieved by including specific calibrations depending on calorimeter signal shapes used in hadronic calibration approaches, and/or identified particle and jet signals, for example, all applied at the level of the cell and cluster signals. In this talk we present comparisons between several different strategies on how to include these calibrations, all judged by the EtMiss reconstruction performance. The discussed relevant variables are signal linearity, missing transverse energy resolution, and tails in the EtMiss distribution. The corresponding studies have been performed with high statistics full simulations of the most important physics channels generating missing transverse energy at LHC.

INTRODUCTION

Weakly interacting particles do not leave a visible signature in the detector. Such particles are invisible to the detector and measured by measuring the missing energy in the event when the event is required to satisfy energy-momentum conservation. In a proton-proton collider detector like ATLAS, kinematics of particles in transverse plane is used to reduce dependence on boost and activity in forward region. Henceforth transverse missing energy is referred with ETMiss. In the Standard Model of particle physics neutrinos interact only weakly and occur in processes involving W- and Z- Boson. They are expected to form an important part of Higgs boson signature at the Large Hadron Collider (LHC) energies. Figure 1 (left) shows the production of Higgs in vector boson fusion channel and then decaying to two tau leptons. In order to reconstruct the tau pair mass a good measurement of neutrino from tau is needed. The missing transverse energy of neutrinos involved in this case is ~50 GeV [1].

CP867, *XII International Conference on Calorimetry in High Energy Physics,*
edited by S. R. Magill and R. Yoshida

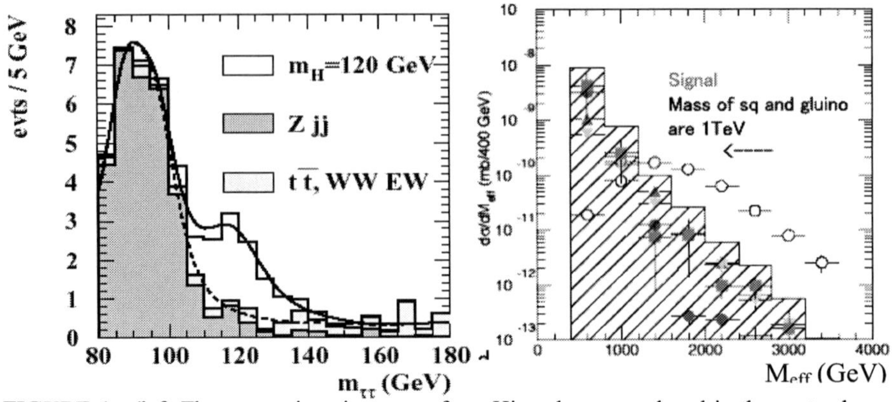

FIGURE 1. (lef) The tau-tau invariant mass from Higgs boson produced in the vector boson fusion channel. A good ETMiss measurement is needed to sharpen the difference between signal and background. (right) Distribution of Meff variable, which is the sum of four jet Et and ETMiss in the event.

Weakly interacting particles also play important role in physics beyond the Standard Model. In supersymmetric (SUSY) theories of the lightest supersymmetric particle (LSP) is weakly interacting and escapes detection. Figure 1 (right) shows the distribution of Meff variable, which is defined as the sum of four highest jet ET and ETMiss in the event. These events tend to have large ETMiss from LSP that gives a good discrimination between signal and background. Such events are required to have a ETMiss of 100 GeV or larger [2]. Also in theories related with large extra dimension if gravitons are involved in the signature, they will interact with the detector very weakly and produce large ETMiss.

Measuring ETMiss is challenging because apart from the true ETMiss from neutrino, LSP, gravitons, etc., any miss-measurement in the event will produce ETMiss. The second kind of ETMiss is normally referred as fake ETMiss. Source of fake ETMiss are particles lost in gaps and dead material, dead and noisy calorimeter cells, noise/pile-up suppression, detector resolution and errors in energy scale. A good ETMiss measurement will minimize the fake ETMiss component relative to the true ETMiss component. In general one would veto events with badly reconstructed pieces and understand the response of the calorimeter to electrons and hadrons precisely to minimize the affect of miss-measurement.

ATLAS has a hermetic calorimeter with coverage up to $|eta| < 5$. Figure 2 (left) gives a schematic view of the different sections of the calorimeter. The electromagnetic calorimeter is a Pb/LAr sampling calorimeter. Between barrel and extended barrel, it covers an $|eta| < 3.2$. The central hadron calorimeter is an Fe/Scintilator sampling calorimeter that covers $|eta| < 1.7$. Beyond that there are the hadron calorimeters (HCAL), which are Cu/LAr sampling calorimeters and cover $1.7 < |eta| < 3.2$. In the very forward region $3.2 < |eta| < 5.0$ are the forward calorimeter (FCAL), which are Cu/LAr and Wu/LAr sampling calorimeters. In the intermediate regions of these calorimeter sections there are regions with gaps and dead materials for service outlet and support structure. There also are dead materials associated with the cryostat that houses LAr calorimeter. The complexity of the calorimeter system can be seen in the material distribution of the calorimeter in Figure 2 (right). A good ETMiss measurement requires an understanding of the nature of energy deposit and their correction in the different sections of the calorimeter.

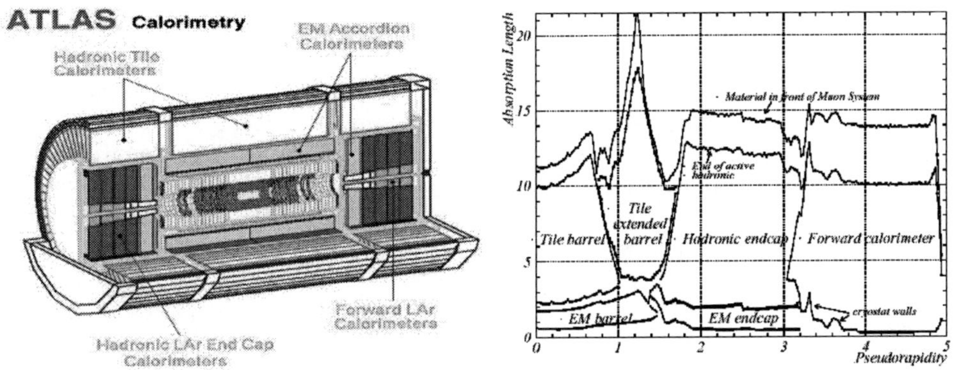

FIGURE 2. (left) Schematic view of the ATLAS calorimeter. (right) Material distribution in the calorimeter as a function of pseudorapidity.

Extra energy in the event other than the hard scattering process will also contribute to the fake ETMiss component described earlier. At LHC proton bunches will cross every 25 ns with about 23 interactions per crossing at peak luminosity. These extra interactions per crossing and electronics pile-up, which is due to the electronics shaping time being larger than 25 ns crossing time, add extra energy to the event. The ATLAS Calorimeter has about 200K readout channels, majority of which are from the LAr calorimeter readout. The design noise level is fairly low, 35-50 MeV per channel in the central calorimeters, but the large number of channels produces a large amount of event-by-event energy fluctuation. There are also stringent requirement on the coherent noise in the calorimeter. These extra sources of energy have to be suppressed via effective cuts and algorithms, and properly accounted for a good ETMiss measurement.

ETMISS RECONSTRUCTION IN ATLAS

ETMiss is reconstructed using the energy deposit in the calorimeter cells. Cells from the full calorimeter coverage of |eta| < 5 are used for best resolution and scale determination. Calorimeter cells are calibrated according to the calorimeter section, eta region and the energy density in the cell. The energy density is used to distinguish between electromagnetic energy, which are dense and of higher energy density, from hadronic energy deposits, which tend to be more spread out and have smaller energy density.

Before the calorimeter cells can be used for ETMiss reconstruction they have to be selected so as to suppress noise as much as possible while keeping maximum amount of signal. Two techniques have been studied. First is a simple two sigma symmetric cut on the energy in a cell based on the expected noise level in the calorimeter cell. Present noise estimates are based on test-beam data. A more sophisticated technique is to use the calorimeter cells that are part of three-dimensional clusters. Energy deposit from signal will be close together and form a cluster whereas random energy deposit from noise for example will tend not to form a cluster. As will be seen in the next section the ETMiss performance improves with the use of cluster noise suppression technique.

Once the calorimeter cells are selected, corrected and their four-vector summed, the next step is to account for losses in the dead material. For example the losses in the cryostat are estimated using the energy deposit in the layer before and after the cryostat. Finally muons in the event are taken into account to get the final missing-ET estimate of the event. Further optimization of the above ETMiss calculation can be done by using reconstructed/calibrated electron, tau, photon and jet, in the event to refine the corrections applied to the calorimeter cells.

ETMISS PERFORMANCE

The reconstructed ETMiss is tested for different kind of physics events, with different amount of total summed transverse energy and amount of ETMiss. Figure 3 shows the reconstructed ETMiss compared to the true ETMiss for samples with Z- and W- boson, Higgs, Top and SUSY. The reconstructed samples are Monte Carlo generated physics events that pass through GEANT4 detector simulation. In general a good agreement with true ETMiss in the event is seen and the reconstructed ETMiss is within 5% of true. At low true ETMiss affect from fakes start becoming larger. One of the SUSY points, SU2, has very large ETMiss and is plotted in the figure with scale written for convenience. The blue points, which use cluster noise suppression technique are in general better agreement with truth ETMiss than by selecting calorimeter cells based on two sigma symmetric noise cut.

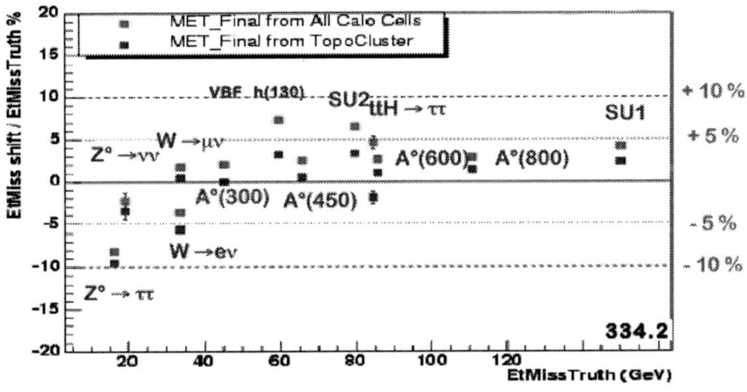

FIGURE 3. Reconstructed missing-ET in different physics event compared to the true ETMiss in the event. Events are detector simulated with GEANT4.

The ETMiss resolution is well modeled with the following formula –

$$\sigma(\text{ Ex(y)Miss }) = \text{p0}.\sqrt{\text{ SumE}_T}$$

where p0 is an arbitrary parameter. Figure 4 shows ETMiss resolution in different physics event as a function of the SumET in the event. It can be seen the simulated events follow the above equation model fairly closely. The fitted value for parameter for three physics samples, Z, Dijet and SUSY Higgs are also shown.

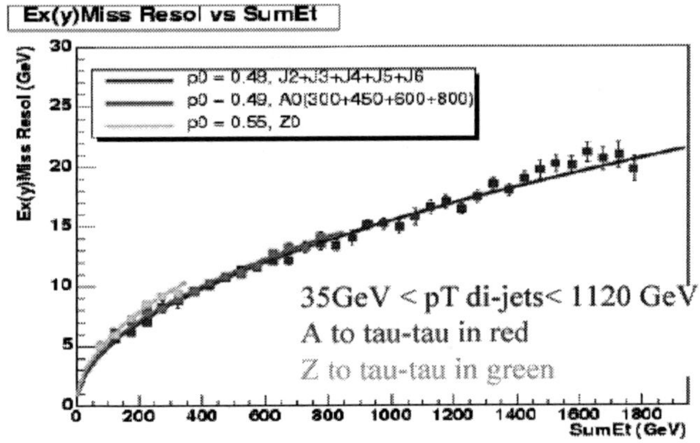

FIGURE 4. ETMiss resolution as a function of the SumET in the event.

IN–SITU ETMISS CALIBRATION

Events with Z boson that decay to two tau leptons can be used for in-situ calibration of ETMiss, where one of the tau decays to lepton. Under the following three assumptions – the mass of tau is small, neutrinos in tau decay are coincident with other decay products and that the decay products are not back-to-back - the tau-tau invariant mass can be written as,

$$m_{\tau\tau} = \sqrt{2(E_1 + E_{\nu 1})(E_2 + E_{\nu 2})(1 - \cos\theta)}$$

where E1 and E2 are the energy of the decay products and θ is the angle between the decay products. Eν1 and Eν2 are the neutrino energies. The resolution of the measured tau-tau invariant mass is related to the ETMiss resolution by the formula –

$$\sigma\,(m\tau\tau) = \sigma\,(\mathrm{ETmiss}) \,/\, |\sin(\Delta\phi)\ \mathrm{prod}_1\ \mathrm{prod}_2\,|$$

where Δϕ is the azimuthal angle between the decay products. By measuring Z mass from tau decays ETMiss scale can be determined. Figure 4 shows the dependence of Z mass measurement on missing-ET scale. Z mass measured to a precision of 3% will give a ETMiss scale to a precision of 10%. This is a feasibility study with idealistic assumptions on the detector and does not include any background.

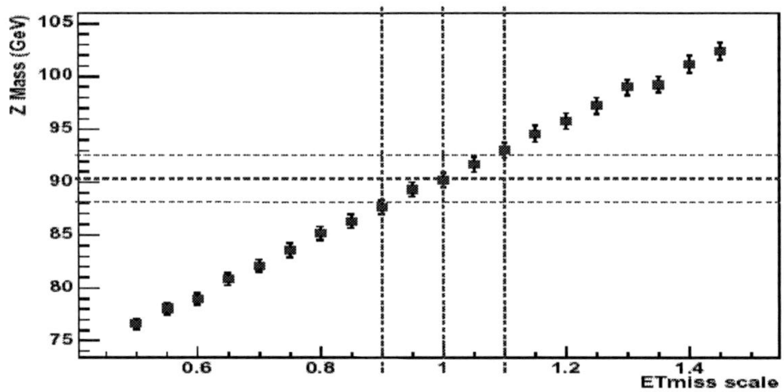

FIGURE 5. Dependence of ETMiss scale on the precision of measured Z boson mass; when Z boson decays to two tau leptons.

The main background in this analysis are expected from W+jet, t-tbar events with semi-leptonic decay of the top to electron or muon. b-bbar with leptonic decays of b. About 900 Z decaying to tau lepton pairs is expected in ATLAS for 1fb-1 of data. The signal and background level depends on the efficiency of the event selection. For typical selection - jets and lepton with pt larger than 20 GeV, within 2.5 eta range, at least 50 GeV of ETMiss - the expected background level is about 20-30%.

CONCLUSION

ATLAS will be taking data in very near future and a good ETMiss measurement will be crucial for both standard model and beyond the standard model physics. Good linearity and resolution is achieved by ETMiss reconstruction algorithms on GEANT4 detector simulated Monte Carlo samples. Studies are under way to use Z events decaying to two tau leptons as an in-situ calibration sample for ETMiss.

REFERENCES

1. S. Asai. et. Al.,"Prospects for the Search for a Standard Model Higgs Boson in ATLAS using Vector Boson Fusion" , Euro. Phys. J
2. Atlas Physics and Detector Performance Technical Design Report, CERN-LHCC-99-15.

SESSION X

CALORIMETER SIMULATION

Conveners: Adam Para and Norman Graf

Geant4 hadronic physics validation with ATLAS Tile Calorimeter test-beam data

C. Alexa, S. Constantinescu and S. Diţă

IFIN-HH, Bucharest, România

Abstract. We present comparison studies between Geant4 shower packages and ATLAS Tile Calorimeter test-beam data collected at CERN in H8 beam line at the SPS. Emphasis is put on hadronic physics lists and data concerning differences between Tilecal response to pions and protons of same energy. The ratio between the pure hadronic fraction of pion and the pure hadronic fraction of proton F_h^π/F_h^p was estimated with Tilecal test-beam data and compared with Geant4 simulations.

Keywords: Calorimeters particle physics
PACS: 29.40.Mc; 06.20.F

INTRODUCTION

Geant4 hadronic physics lists validation presents a special interest for all LHC experiments. In this paper, we report new results on Geant4 [1] hadronic physics lists validation based on comparisons of the simulations with test-beam data collected with ATLAS Tile Calorimeter (Tilecal) [2]. Preliminary results were presented in [3] and [4].

The experimental data used in our analysis allow us to observe differences between Tilecal response to pions and protons of the same energy. Tilecal data gives us the opportunity to perform a quite detailed validation of the Geant4 hadronic physics lists since, as it has been mentioned in several papers [5, 6], neutral pion production processes are different in pion showers and in proton showers. Mainly, the production of leading neutral pions, in showers initiated by pions, are plays a crucial role in the observed differences. Because Tilecal is a non-compensating calorimeter, it responds differently to electromagnetic and pure hadronic components of the showers and the comparison of this data with Geant4 predictions is a leading topic for the hadronic physics lists validation.

Tilecal beam tests were performed in H8 beam line of CERN-SPS where in pion beams we had a significant amount of protons while in positron beams we had pions and protons.

The simulated data were produced with two Geant4 releases, Geant4.5.2 with QGSP2.7 and Geant4.7.1 with QGSP_GN.

Geant4 predictions and test-beam data were mainly compared for the following quantities: the ratio between the pion and proton response - π/p, the ratio between positron and pion - e/π as well as the ratio between positron and proton response - e/p. Using Tilecal test-beam data, the values obtained for the ratio between the pure hadronic fraction of pion and the pure hadronic fraction of proton are shown against Geant4 predictions.

This paper organized as follows. In the next two sections Tilecal test-beam set-up

CP867, *XII International Conference on Calorimetry in High Energy Physics,*
edited by S. R. Magill and R. Yoshida

is described and the main features of the performed Geant4 simulation are presented. We continue with the presentation of the results of this study and we finalize with conclusions.

TEST-BEAM SET-UP DESCRIPTION

The ATLAS Tile Calorimeter[2] is a sampling device made of steel and scintillating tiles, as absorber and active material respectively. The detector consists of a cylindrical structure with an inner radius of 2280 mm and an outer radius of 4230 mm. It is subdivided into a 5640 mm long central barrel and two 2910 mm long extended barrels. Each cylinder is made of 64 ϕ-modules. The scintillating tiles lie in the $r-\phi$ plane and span the width of the module in the ϕ direction. Wavelength shifting (WLS) fibers running radially collect the light from the tiles along their two open edges. Readout cells are then defined by grouping together a set of fibers into a photomultiplier (PMT), to obtain a three dimensional segmentation. Radially, the calorimeter is segmented into three layers, approximately 1.4, 3.9 and 1.8 interaction lengths thick at $\eta = 0$. The $\Delta\eta \times \Delta\phi$ segmentation is 0.1×0.1 in the first two layers and 0.2×0.1 in the last radial layer. The two PMT of each calorimeter cell were equalized with a cesium source. The equalization uniformity over a whole module is within 1%. In each event the PMT signals are sampled in nine time slices at 40 MHz. The first time slice defines, for each event and channel, the pedestal level, which is directly subtracted from the energy calculation. The total energy released by a particle is obtained by summing PMT signals. For each geometry and beam energy, the energy scale was set using electrons.

The data presented in this paper were taken with two different setups. In 2002 the test-beam setup was built with two extended barrel modules, one barrel module and one pre-series barrel module, while in 2003 with two barrel modules and the same pre-series barrel module.

Tilecal beam tests were carried out in the H8 beam line of SPS. From the test-beam data obtained with prototypes, a significant amount of protons in pion beams has already been seen using the H8 beam line Cerenkov Counter [8] for pion - proton separation, as well as a strong energy dependence of the amount of protons in pion beams was observed. For example, at 50 GeV the amount of protons was 30% while at 180 GeV we had 60% protons.

To separate protons from pions, we have used, as in the beam tests with prototypes, the H8 beam line Cerenkov Counter. The pressure of this Counter was set to a value: a) lower than the proton threshold for the current incident energy and b) assuring a high efficiency of Cerenkov Counter to pions. For an efficient separation between protons and pions both conditions have to be fulfilled and as a consequence the separation was possible only for the beam energy interval $50 \leq E_{beam} \leq 180$ GeV.

First, we separated two groups of particles: I) the group of particles with Cerenkov signal near zero which were identified as protons (and some negligible background with low energy deposition) and II) the group of particles with a higher Cerenkov signal - pions, muons and positrons.

Secondly, to eliminate muons we have used the deposited energy in calorimeter. Particles with high Cerenkov signal and an energy deposition less than 10 GeV were identified as muons. For the separation between pions and positrons we have performed a cut in the fraction of the total energy deposited in the first Tilecal layer of the central module. The cut values were established by Geant simulations for each rapidity and incident energy. From simulations $\sim 97\%$ of positrons ($\sim 98\%$ of pions) were identified by imposing that more (less) than about 0.76 of the total energy was deposited in the first layer.

GEANT4 SIMULATION

Geant4 has four physics lists [9] for high energy physics calorimetry: LHEP, QGSP, QGSC and FTFP. They differ mainly in the physics of the initial reaction in the primary hadron inducing the shower. Previous studies [3] have shown that test-beam data seem to indicate a slightly better overall agreement with QGSP [10] which is a theory-driven modeling for the reactions of energetic pions, kaons, and nucleons.

For this study QGSP physics list was used. The simulated data were produced with Geant4.5.2 with QGSP 2.7 and Geant4.7.1 with QGSP_GN.

A Geant4-based simulation package which describes in full detail Tilecal modules and the test-beam layout has been developed [7]. The effects of the beam smearing is also included in the simulations. The range cut was set to 1 mm. The contributions of the electronic noise and photo-statistics were considered at the analysis level. Electronic signal response reconstruction was recently implemented in Tilecal software environment, only results obtained with Geant4.7.1 contain this new feature. For example, for an 80 GeV pion, before the implementation we obtained $E_{mean} = 66.21 GeV$ and $\sigma = 6.205 GeV$ while after implementation $E_{mean} = 64.87 GeV$ and $\sigma = 6.475 GeV$. As a consequence, electronic signal response reconstruction produced a decrease of the energy mean value and a slightly increase of the standard deviation σ.

In simulated data, the starting point for energy reconstruction is the total energy released by a shower in the active material, which is calibrated to the incident energy using electrons as for the real data.

RESULTS

π/p ratio

Gaussian fits over a $\pm 2\sigma$ range to raw pion and proton deposited energy spectra [11] have been performed and we have obtained the mean values of the deposited energies for pions and protons. Using these mean values of the deposited energies in calorimeter, we have calculated the π/p ratio - the ratio between the pion and proton response.

To avoid effects introduced by leakage, we selected only events satisfying the following relation:

$$\frac{E_{first\ layer} + E_{second\ layer}}{E_{tot}} > 0.7 \qquad (1)$$

were selected in our the analysis.

In Figure 1, the energy dependence of the π/p ratio is represented. Higher signal for pions than for protons makes the ratio π/p greater than unity for all studied incident energies. Since Tilecal is a non-compensating calorimeter with $e/h > 1$, π/p ratio greater than unity indicates a smaller pure hadronic fraction for pions than for protons.

For comparison, π/p ratio obtained with the CMS quartz fiber calorimeter [12], an extreme non-compensating calorimeter, are also represented. The higher values obtained with the quartz fiber calorimeter are explained by its much higher non-compensation $e/h \sim 5$ in comparison with Tilecal $e/h = 1.36 \pm 0.11$. All the experimental values obtained for π/p ratio with Tilecal are inside the interval $1.02 < \pi/p < 1.05$.

It can be seen that the π/p ratio obtained by Geant4 simulations has higher values than the test-beam data for all the incident energies. Geant4.7.1 has the highest π/p ratio for all beam energies.

A slow decrease of the π/p ratio with increasing energy can be observed for both test-beam data and simulation values represented in Figure 1.

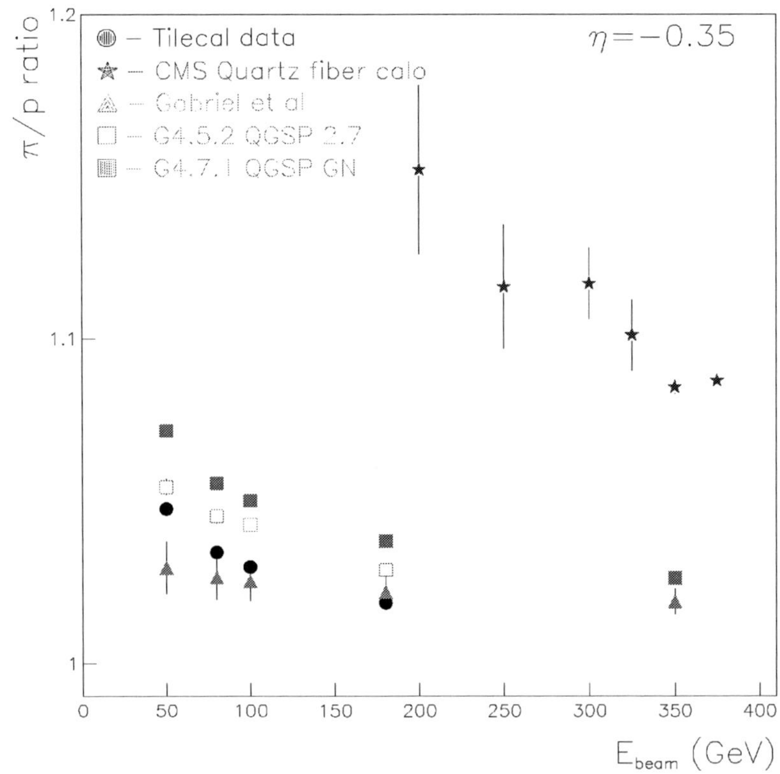

FIGURE 1. The beam energy dependence of the π/p ratio.

Preliminary results showing that pion and proton shower profiles are different have been obtained. They say that protons have wider lateral shower profiles than pions, in agreement with results published in [6].

e/π and e/p ratios

Performing the separation of pions, protons and positrons from the positrons beams, we were able to estimate e/π and e/p ratios for each incident energy.

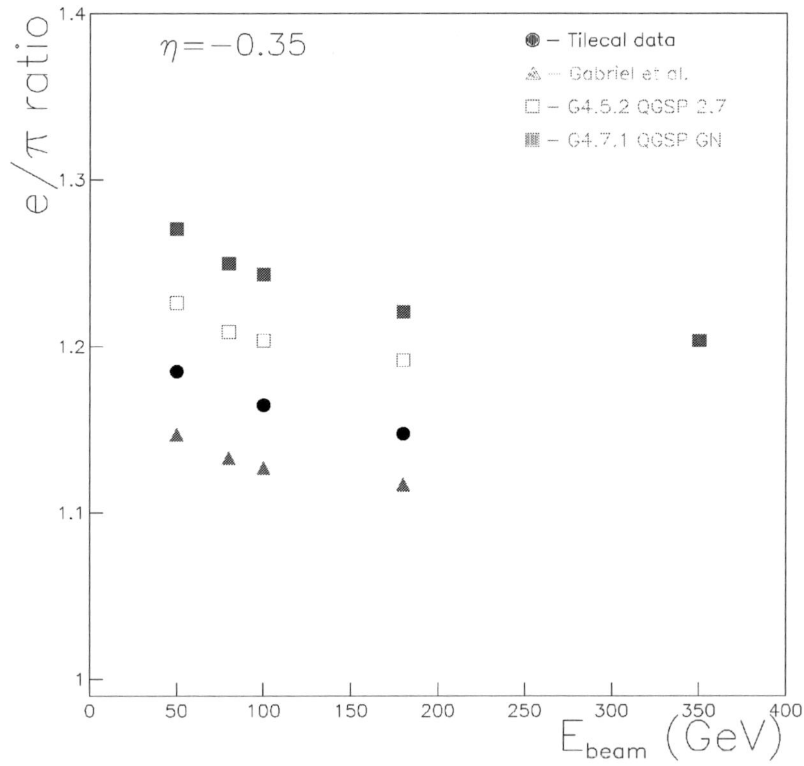

FIGURE 2. The energy dependence of the e/π ratio.

In Figure 2 and Figure 3 we present the energy dependence of the e/π and e/p ratios obtained from data and simulations. Also plotted are the values of these ratios calculated using Groom parametrization of the pure hadronic fraction for pions and protons with parameter values corresponding to an iron absorber [5].

The highest values were given by Geant4.7.1 with *QGSP_GN*, while smallest values were obtained using Groom parametrization of the pure hadronic fractions for pions and protons. The slow decrease with increasing energy of the e/π and e/p ratios is observed in both data and simulations.

Due to higher Tilecal response for pions than for protons of the same energy, the e/p ratio is higher than the e/π ratio.

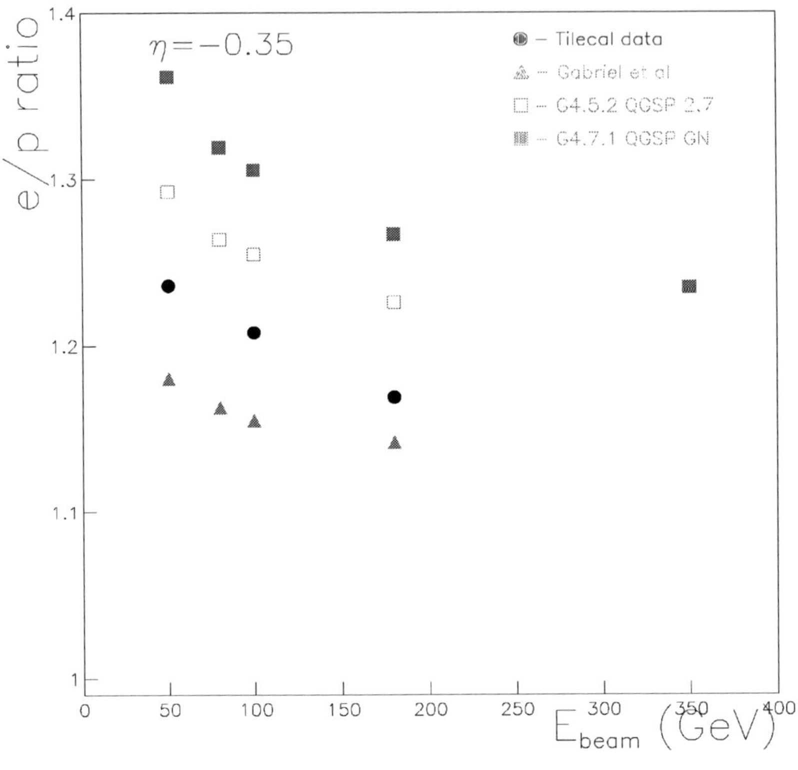

FIGURE 3. The beam energy dependence of the e/p ratio.

F_h^π / F_h^p ratio

Data collected with positron beams offer us the posibility to get the energy dependence of the ratio between the pure hadronic fraction of pions and the pure hadronic fraction of protons.

By simple calculations, we have obtained the following relation:

$$\frac{F_h^\pi}{F_h^p} = \frac{<E_e> - <E_\pi>}{<E_e> - <E_p>} \tag{2}$$

and we used it to compute the ratio F_h^{π}/F_h^{p} from the mean values of deposited energy, obtained experimentally for positrons, pions and protons.

In Figure 4, the energy dependence of the ratio F_h^{π}/F_h^{p} is represented. The line shown in the plot represents F_h^{π}/F_h^{p} calculated with Groom parametrization. Also plotted are the values given by both Geant4 versions.

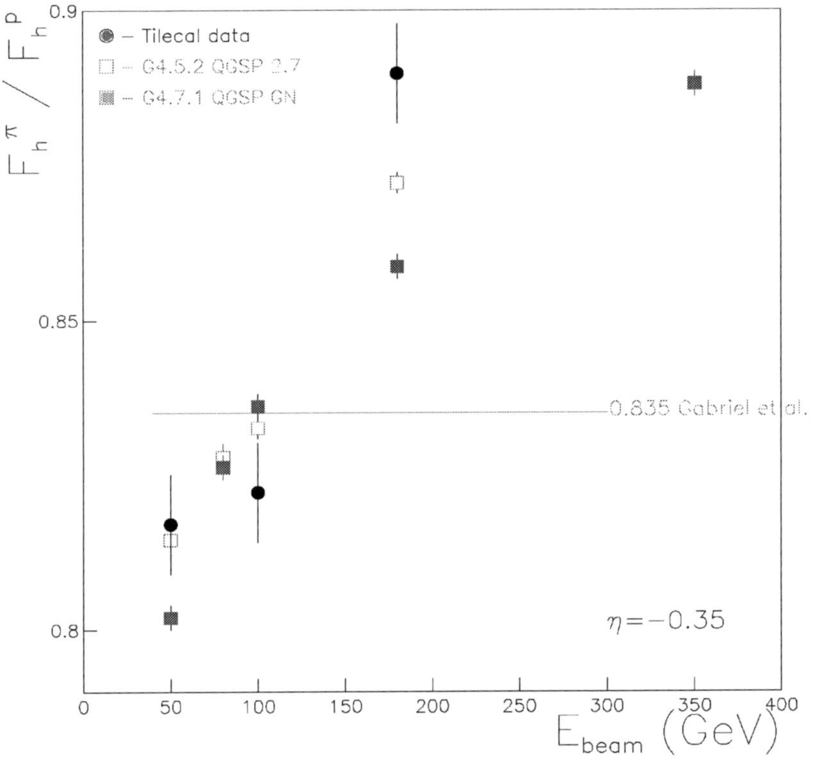

FIGURE 4. The beam energy dependence of the F_h^{π}/F_h^{p} ratio.

The test-beam data values and both Geant4 versions predictions show an increase of the ratio F_h^{π}/F_h^{p} with energy, in contrast with the constant behaviour given by Groom parametrization.

Both test-beam data and simulations suggest that the differences between Tilecal response to pions and protons decrease with increase of energy.

CONCLUSIONS

The conclusions of the study presented in this paper can be summarized as follows:

- π/p ratio values are greater than unity for all incident energies used in this study for both test-beam data and simulations;
- the decrease with the energy of the π/p ratio is observed in both data and simulations;
- the highest values for π/p ratio were obtained by Geant4.7.1 *QGSP_GN*
- the decrease of e/π and e/p ratios with the beam energy is well described by both Geant4 predictions;
- values less than unity were obtained for the F_h^{π}/F_h^{p} ratio for both test-beam data and simulations, in agreement with Gabriel et al [5];
- the increase of the F_h^{π}/F_h^{p} ratio with beam energy, observed in both test-beam data and Geant4 simulations, is in contrast with the constant behaviour predicted by Gabriel et al. [5].

ACKNOWLEDGMENTS

This work is supported by the Romanian R&D Agency through CORINT no.3 project.

REFERENCES

1. S. Agostinelli *et al.*, Geant4 Collaboration, Nucl. Instrum. Meth. A **506**, 250 (2003).
2. ATLAS Collaboration, *Tile Calorimeter Technical Design Report, CERN/LHCC 96-42 (1996).*
3. C. Alexa *et al.*, *Geant4 hadronic physics validation with LHC test-beam data: first conclusions*, CERN-LCGAPP-2004-10.
4. C. Alexa, *Geant4 hadronic physics validation with LHC test-beam data*, Calor2004, Perugia, Italy
5. T. A. Gabriel *et al.*, Nucl. Inst. and Meth. A 338 (1994) 336
6. R. Wigmans, Ann. Rev. Nucl. Part. Sci. 41 (1991) 133.
7. A. Solodkov and V. Tsulaia, *ATL-TILECAL-2003-002*
8. S. Constantinescu, S. Dita, A. Henriques, M. Nessi and R. Stanek, *Comparative analysis of the ATLAS Tile Haronic Calorimeter Response to Pions and Protons, ATL-TILECAL-2001-005.*
9. Geant4 physics lists for calorimetry, *http://geant4.web.cern.ch/geant4/physics_lists/*
10. A. B. Kaidalov, K. A.Ter-Martirosyan, Phys. Lett. **B117** 247 (1982); Sov. J. Nucl. Phys. **39** (1984) 979.
11. F. Ariztizabal *et al.*, Nucl. Inst. and Meth. A 349 (1994) 384
12. N. Akchurin *et al.*, Nucl. Instrum. Meth. A **408** (1998) 380-396.

CMS HCAL Testbeam Results and Comparison with GEANT 4 Simulation

Jordan Damgov

Fermi National Accelerator Laboratory
Batavia, IL 60510, USA [1]

for The CMS HCAL Collaboration

Abstract. A study of the CMS hadron calorimeter has been performed using pions, protons, electrons, and muons in the H2 test beam in the North Area at CERN in 2004. We present some of the measured properties of the HCAL barrel calorimeter. These include the detector energy resolution, linearity, and longitudinal shower profile as a function of the incident particle momentum. The detector configuration that was used included a mock-up of electromagnetic calorimeter (a 7X7 array of $PbWO_4$ crystals) and two hadronic barrel modules . The momentum range of the beam particles was from 5 to 300 GeV/c. A validation of the GEANT4 was done by comparing the predictions of the simulation with the test beam data.

Keywords: Calorimeters,Scintillation detectors,Monte Carlo
PACS: 07.20.Fw;29.40.Vj;29.40.Mc;02.70.Uu

INTRODUCTION

The Compact Muon Solenoid (CMS) [1] is one of the new general purpose detectors now under construction for the new Large Hadron Collider (LHC) at CERN. The CMS hadron calorimeter (HCAL) is a sampling calorimeter with alternating layers of brass absorber and plastic scintillator tiles. It plays a fundamental role in most anticipated LHC physics through its role in generation of level-1 trigger primitives [2], the high level trigger [3], and the offline reconstruction of jets, taus, and missing transverse energy .

Like all high energy experiments, CMS depends critically on the accuracy of the description of the behavior of the detector by the simulation programs. An accurate description of hadronic interactions and hadronic showers over a broad energy range is one of the most challenging parts of any software package that is used to simulate the environment of an experiment at the LHC.

For the first time, the data from the 2004 HCAL expand the low-energy measurements down to 5 GeV/c^2, and also measures the longitudinal shower profile in a production HCAL module. This provides a unique opportunity to validate the hadron physics in the simulation program down to the low momenta of the incident beam particles.

---

[1] On leave at Institute for Nuclear Research and Nuclear Energy, Bulgarian Academy of Science, Sofia, Bulgaria

CP867, *XII International Conference on Calorimetry in High Energy Physics,*
edited by S. R. Magill and R. Yoshida

beam

FIGURE 1. HCAL test beam 2004 setup.

2004 HCAL TEST BEAM SETUP

The 2004 HCAL Test Beam studies were conducted in the H2 test beam in the North Area at CERN. Two HCAL barrel wedges (HB) and one HCAL end-cap module (HE) (see Fig. 1) were placed on a movable table. The movable table was designed in such a way that it appears as if the beam is coming from the interaction region as it would at the CMS interaction point in the LHC. The table allows movement in both directions - η and ϕ. The HB is a sampling calorimeter covering the pseudo-rapidity range $-1.3 < \eta < 1.3$. The active material is 17 layers of plastic scintillator, divided into 16 η sectors, resulting in a segmentation $(\Delta\eta, \Delta\phi)$=(0.087,0.087). The absorber material is brass (70% Cu, 30% Zn), with a steel plate in front, and on the back of the calorimeter. The total absorber thickness at $\eta = 0$ is 5.82 absorption lengths ($\lambda \equiv 1.0$ absorption lengths). The effective thickness increases with $|\eta|$, such that at $|\eta| = 1.3$, the effective thickness is 10.3 λ. Most of the data for this study was taken at $\eta = 0.566$, where the thickness is 6.8λ.

An ECAL module was placed in front of the HCAL on the movable table. This ECAL module simulated the ECAL electro-magnetic calorimeter modules that will exist in the final CMS detector, It consisted of a 7x7 matrix of lead tungstate ($PbWO_4$) crystals. Each crystal is a quadrilateral prism with a depth of 23 cm, a square front surface 20.5 x 20.5 cm^2, and a square rear surface, 23.8 x 23.8 cm^2. Each crystal is 25.8 radiation lengths thick, for electro-magnetic particles, and 1.1λ absorption lengths for hadrons. An aluminum block was placed between the ECAL and HCAL modules to mimic the material in the ECAL electronics at that location in the CMS detector.

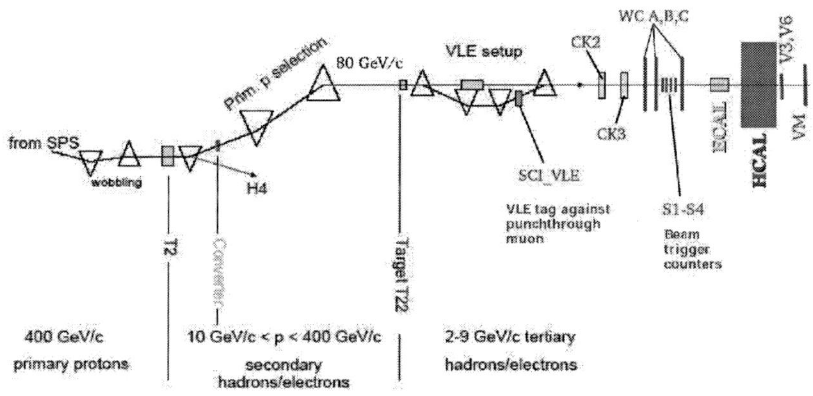

FIGURE 2. H2 beam line setup for high and low energies.

H2 Beam Line in the North Area and Particle Identification.

Two configurations of the H2 beam line were used. For particles with momenta in the range 10-300 GeV/c, the normal magnet configuration of H2 was used. For particles with momenta in the range 2-9 GeV/c, a special configuration (VLE) was used with additional elements and a secondary target. A schematic view of the normal and the VLE beam line configuration is shown on Fig. 2. The primary proton beam with momentum 400 GeV/c coming from SPS accelerator strikes the primary target T2. Downstream from the primary target, a secondary hadron or electron beam is made in the momentum range from 10 to 400 GeV/c. This was the beam configuration used for the high energy data. For the low energy data, an additional target T22 and a beam stop were inserted into the beam line. Additional beam focusing and bending elements were introduced along the VLE path in order to make a narrow beam with momentum from 2 to 9 GeV/c from particles produced at the target T22.

Particle identification was performed with Cerenkov counters CK2 and CK3, and muon scintillators VM, V3 and V6. The muon scintillators were located behind the calorimeters to identify muons since they rarely interact and are not absorbed in the material of the calorimeters. The gas pressure in Cerenkov counter CK2 was adjusted to identify electrons. Pions with momentum below 20 Gev/c emit no Cerenkov light in CK2. Cerenkov counter CK3 was dedicated to identifying the protons in the positively charged pion beams. The gas pressure in CK3 was adjusted in such a way that positively charged pion beams with momenta from 5 to 9 GeV/c should produce signals in the CK3 Cerenkov counter only for pions and muons, and not for kaons and protons.

Particles that interacted along the beam line before the calorimeters were identified by multiple hits in a set of three wire chambers, each with active area 10x10 cm. In the VLE configuration of the beam line, an additional scintillator *SCI_VLE* was included to assure that the beam particle followed the VLE path during the very low energy data taking. This was necessary to eliminate any muons created in the beam dump just after the T22 secondary target, that were capture by the downstream beam optics. Any remaining uncertainty in the beam contamination and interaction in the beam line is added to the systematics of the measurement.

SIMULATION OF THE TEST BEAM CONFIGURATION

For the Monte Carlo simulation of the test beam is GEANT4 [4] version 6.2.p01.

Geometry The implementation of the geometry of the two HCAL barrel wedges was fully detailed (Fig. 3. An aluminum block was implemented to represent the magnet coil behind the wedges, followed by HO scintillator layers. The ECAL calorimeter was implemented as a single $PbWO_4$ crystal with a geometrically-calculated readout structure representing the 7x7 crystal matrix. In addition, a 8 cm Al block was located between the ECAL and HCAL calorimeters representing material in the ECAL electronics. The trigger scintillators S1-S4 and wire chambers were also implemented in order to study the interactions along the beam line.

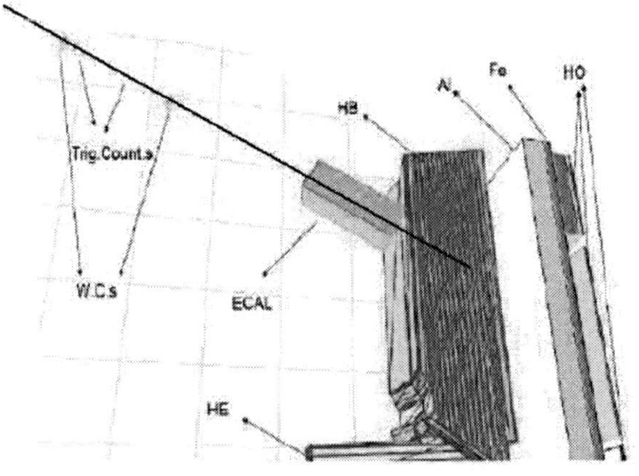

FIGURE 3. Visualization of the G4 test beam geometry.

Physics Lists In the development of a GEANT4-based application, it is the user's responsibility to decide which physics processes are required, and then to include them in the physics list. In realistic applications, the physics list can become long and involved. Also, the large number of physics processes available can make the choice difficult. For this reason the GEANT4 group at SLAC has developed and supports a number of physics lists which can be easily integrated into applications.

- LHEP: LEP/HEP parametrized models for inelastic scattering.
- QGSP: Quark Gluon String model for the " Punch-through " interactions.
- QGSC: QGSP + Chiral invariant phase-space decay.
- FTFP: diffractive string excitation, similar to FRITJOF and Lund.

The HCAL test beam data was compared with the LHEP 3.7 and QGSP 2.8 physics lists as a representatives of the groups which are based on either parametrized or theoretical models.

ENERGY RESPONSE OF THE CALORIMETER

The energy response to pions with momentum from 5 to 300 GeV/c and electrons with 100 GeV/c was measured for the ECAL + HCAL combined system, and for HCAL-only.ECAL energy scale is calibrated with 100 GeV electron and HCAL alone is calibrated with 50 GeV pions.

ECAL plus HB Combined System

The energy response of the combined ECAL and HCAL system is presented here for the barrel HCAL wedges and the ECAL crystals. The configuration included a 8 cm Al block, located between the ECAL and HCAL calorimeters representing material in the ECAL electronics.

Calorimeter Response to Pions

Fig. 4 compares the energy spectra in the combined ECAL + HCAL system for the beam data and GEANT4 simulation.

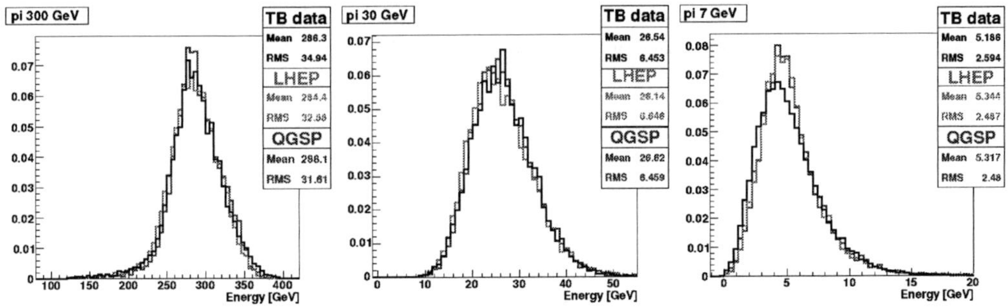

FIGURE 4. dN/dE spectra. Test Beam vs. G4 LHEP and QGSP for pion beam with momentum 7,30 and 300 GeV/c.

One of the main characteristic of many calorimeters is that they respond to electrons and hadrons differently. Calorimeters with a differing response (often quantified by the ratio e/h different then 1.0) are called non-compensating. However, this characteristic is usually measured experimentally by the different calorimeter response to electrons and pions, as a function of incident particle energy, using the ratio π/e. For this study, the absolute calibration constant for ECAL was adjusted using a beam of electrons with momentum 100 GeV/c such that $E_{meas}/E_{beam} = 1.0$. Using this particular choice for calibration of ECAL over the whole energy range, a measurement of π/e then is equivalent to measuring the linearity of the calorimeter response to pions. Fig. 5 shows the ratio of measured energy to beam momentum for pions as a function of the beam momentum, for the the test beam data, and for the GEANT4 simulation.The GEANT4 simulation shows good agreement with the data within the measurement precision.

However, there is a noticeable discontinuity in the 7-10 GeV energy range. This may result from the transition between underlying high and low energy physics models in the simulation program.

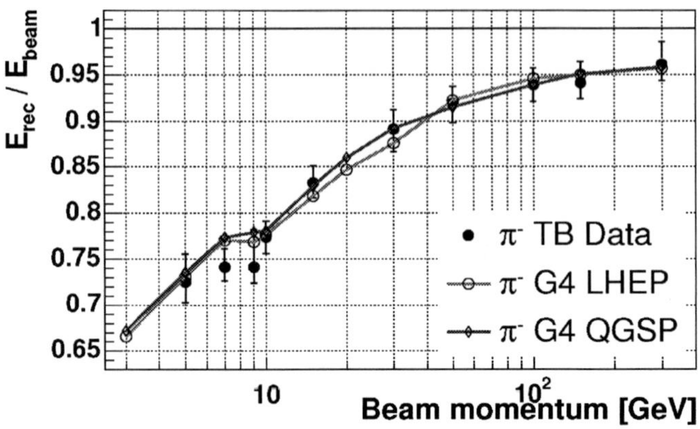

FIGURE 5. e/π for ECAL +HCAL. Test Beam vs. G4 LHEP and QGSP physics lists.

Fig. 6 shows the energy distribution between the electromagnetic and hadronic part of the calorimeter system. The deviation from a strait line of the energy distribution is an illustration of the non-compensating of this calorimeter.

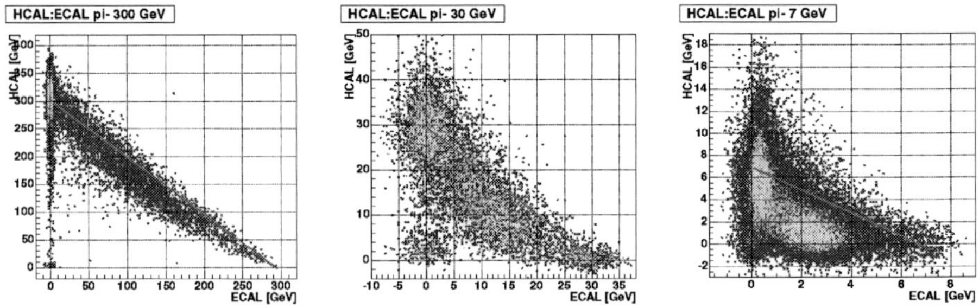

FIGURE 6. ECAL vs HCAL energy distrinution.

Calorimeter Response to Protons

Fig. 7 shows the ratio of proton to pion response (E_{proton}/E_{pion}) for the combined HCAL+ECAL system . The proton response is 15-20% lower than the pions in 5 to 9 GeV/c momentum range and this difference is well reproduced by the simulation.

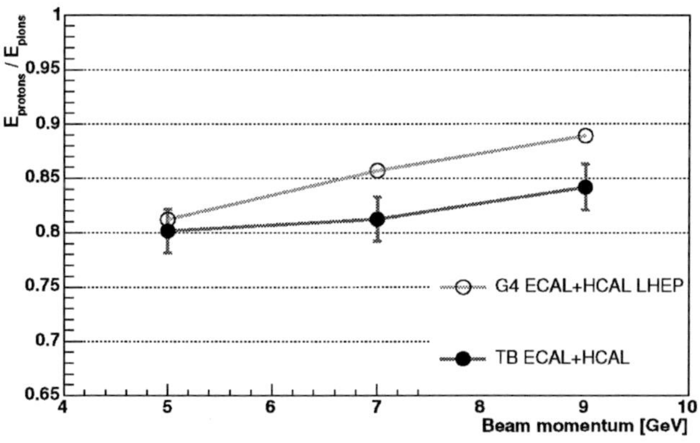

FIGURE 7. Proton over pion response ratio.

HCAL-only Response to Pions

The linearity of the HCAL-only response to pions is shown on Fig. 8 for beam momentum from 10 to 300 GeV/c. At high energies, the LHEP physics list reproduces the test beam data better. The QGSP physics list has less leakage out the back of the HCAL, and thus gives a high value for the energy response.

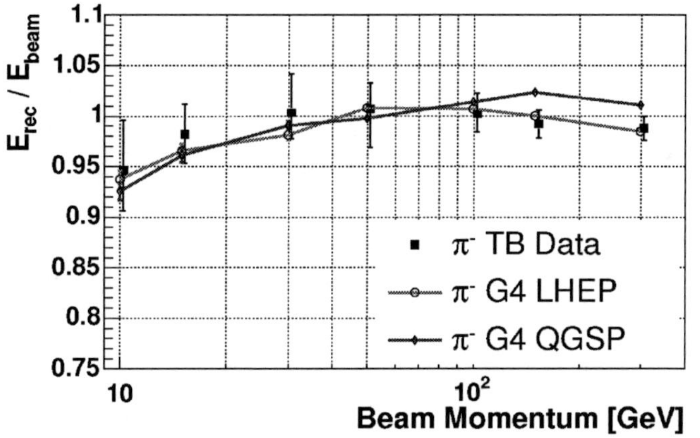

FIGURE 8. Linearity of HCAL-only. Test Beam vs. G4 LHEP and QGSP.

LONGITUDINAL SHOWER PROFILE

The special configuration of the HCAL wedge 2 (HB2) was used to measure the longitudinal shower development. Data with a 100 GeV/c electron beam and 5 to 300 GeV/c pion beams was used in this measurement. Fig. 9 shows the longitudinal shower profile for 100 GeV electrons in HB2 alone. This plot shows both the uncorrected data, and the data corrected for HPD cross-talk. The corrected data agree very well with simulation.

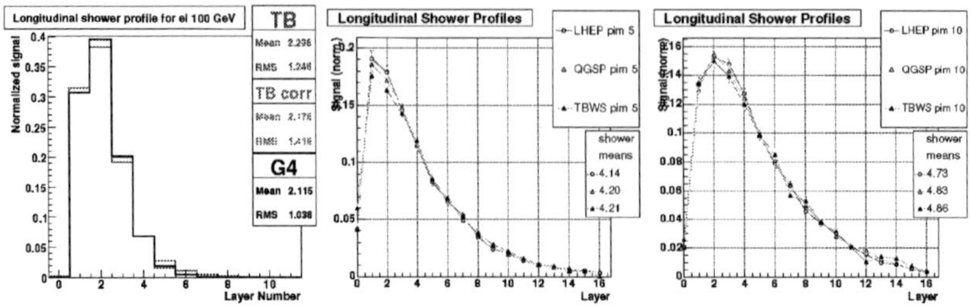

FIGURE 9. Longitudinal shower profile from data and G4 simulation for 100 GeV electrons, 5 and 10 GeV pions.

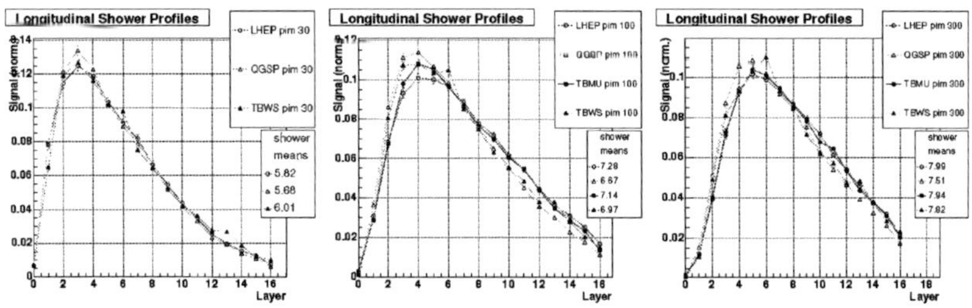

FIGURE 10. Longitudinal shower profile from data and G4 simulation for 30,100 300 GeV pions.

The electromagnetic shower from electrons and photons does not penetrate very deeply in HCAL. The first three layers contain 91% of the measured energy. Four layers are enough to collect 98% of the energy. The longitudinal shower profiles for pions was measured for a wide range of beam momentum, and compared with the GEANT4 for LHEP and QGSP physics lists. Fig. 9 and Fig. 10 show the profiles for beam momenta 5 to 300 GeV/c. Overall, the GIANT4 simulation reproduces well the test beam data. The QGSP physics list shows a shorter shower profile at 100-300 GeV/c beam momentum, which explains why the reconstructed energy is larger from smaller leakage out the back of the HCAL. At momenta below 50 GeV/c, both LHEP and QGSP show good agreement with the test beam data and with each other.

REFERENCES

1. *The Compact Muon Solenoid Technical Proposal*, CERN/LHCC 94-38, 1994.
2. *The Hadron Calorimeter Technical Design Report , CMS TDR 2*, CERN/LHCC 97-31, 1997.
3. *CMS, The TriDAS Project, Technical Design Report, Volume 1: The Trigger Systems*, CERN/LHCC 2000 - 38, 2000.
4. S. Agostinelli, and et al., *GEANT4: Asimulation Toolkit*, Nucl. Instrum. Meth. A 506, 2003, pp. 250–303.

Recent Developments and Validations in Geant4 Hadronic Physics

Dennis H. Wright, Tatsumi Koi*, Gunter Folger, Vladimir Ivantchenko, Mikhail Kossov, Nikolai Starkov†, Aatos Heikkinen** and Hans-Peter Wellisch‡

Stanford Linear Accelerator Center, Menlo Park, California, USA
†CERN, Geneva, Switzerland
***Helsinki Institute of Physics, Helsinki, Finland*
‡Geneva, Switzerland

Abstract.
The Geant4 hadronic models cover the entire range of energies required by calorimeters in new and planned experiments. The extension and improvement of the elastic, cascade, parameterized and quark-gluon string models will be discussed. Such improvements include the extension to more particle types, a review and correction of cross sections, and a better treatment of energy and momentum conservation. Concurrent with this development has been a validation program which includes comparisons with double differential cross sections. An ongoing hadronic shower validation will also be discussed which includes the examination of longitudinal shower shapes and the performance of the above models as well as their interaction with electromagnetic processes such as multiple scattering.

Keywords: simulation, Geant4, validation
PACS: 24.10.Lx, 02.70.Uu, 01.30.Cc

INTRODUCTION

For a given calorimeter simulation using Geant4 [1, 2], a number of physical processes is assigned to each particle type. To each of these processes several models and cross section data sets are assigned. For the hadronic processes, the user must choose the models which are most appropriate to the energy range and level of detail required in his simulation. For calorimetry in LHC and ILC detectors, hadronic physics requirments are similar, so that a common set of processes and models may be used. These have been included in a few "physics lists" which users may invoke in their applications. Validation and subsequent improvement of these physics lists and their constituent processes and models is currently underway.

Elastic scattering, cascade, high energy and parameterized models will be discussed in particular, followed by selected highlights of the test-beam validation of showers.

IMPROVEMENTS IN ELASTIC SCATTERING

Elastic hadron scattering is important in the simulation of shower shapes and calorimeter response. The current elastic scattering model in Geant4 is derived from the GHEISHA

CP867, *XII International Conference on Calorimetry in High Energy Physics,*
edited by S. R. Magill and R. Yoshida

code of Geant3 [3]. It does a reasonable job of reproducing the scattered angular distributions for angles smaller than about 15° . Beyond that it tends to over-estimate the large angle scattering. A better model has been developed which is relativistically correct, includes coherence effects such as diffraction minima due to the nuclear radius, and removes charge exchange as a separate reaction. In also includes improved p and n scattering from p, d , and He targets, as well as high precision neutron elastic scattering from isotopes.

IMPROVEMENTS IN THE GEANT4 PARAMETERIZED MODEL

Like the elastic scattering model, the Geant4 parameterized models are derived from the GHEISHA models of Geant3. They are based on fits to experimental data over a broad energy range and employ theoretical guidance when little data is available. They are computationally fast, and may be used for all long-lived hadrons, at all incident energies.

This versatility is balanced by the approximate nature of these models. Originally formulated as a shower code, they do not conserve energy, momentum, charge or other quantum numbers on an event-by-event basis, but instead seek to conserve these quantities on average.

The approach of Geant4 hadronics is to use these models only where speed is important, or when no better models exist. Current physics lists depend on the parameterized models over the incident particle energy range from $\sim$5 GeV to $\sim$20 GeV, and at all energies for anti-baryons. Because of this dependence, and because many calorimeter studies employ energy flow methods, these models were recently improved with respect to energy conservation. An example of this improvement is shown in Fig. 1. The plot shows the total energy recovered from a 50 MeV anti-neutron coming to rest in a perfect calorimeter. The sum of all the annihilation energies and masses should be 1.93 GeV. Using the original version of the model, the recovered energy has its mean at around 1.9 GeV, but the broad tails above and below the mean are problematic for any serious energy accounting. The new model enforces the 1.93 GeV upper limit and narrows the energy distribution considerably.

Charge and baryon number conservation was also improved in the new version of these models, especially for hadron collisions with light nuclei.

THE CASCADE MODELS

In Geant4 the energy range from 0 to 5-10 GeV is handled by the so-called binary and Bertini cascade models. Both of these models are more detailed, and depend more heavily on theoretical arguments, than the parameterized models discussed previously. They also implicitly conserve energy-momentum and quantum numbers, and thus are good candidates to replace the parameterized models in this energy range.

The binary cascade model is valid for incident p, n, π^+, π^- from 0 to 3 GeV. It is based on hadron collisions with nucleons which form resonances. The resonances then decay according to their quantum numbers. This model was recently extended to include

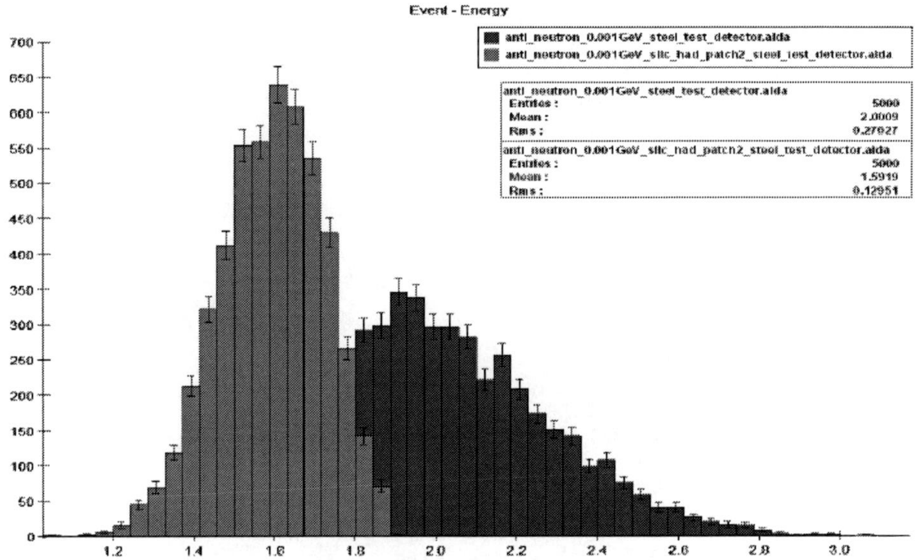

FIGURE 1. Recovered energy due to an annihilating anti-neutron in a perfect calorimeter. The dark gray histogram is the result of the original parameterized model and the light gray histogram is the result of the improved parameterized model. The horizonal axis is in GeV. Plot courtesy of Ron Cassell (SLAC).

ion-ion collisions, with the restriction that either the projectile or the target must have $A \leq 12$.

The Bertini cascade is based on the Bertini INC [4] code and is valid for p, n, π^+, π^- from 0 to 10 GeV. Hadron collisions with nucleons are assumed to proceed according to the free-space partial cross sections and final state distributions measured for the incident particle types. This model was recently upgraded to handle elastic scattering, and the inelastic scattering of incident kaons and hyperons.

Both the Bertini and binary cascade model the target nucleus in detail, taking into account density variations and tracking in the nuclear field.

Both models have undergone extensive validation over the energy range 100 MeV to 1 GeV. One such validation for incident 800 MeV protons is shown in Fig 2. The binary cascade does a better job at reproducing the 7.5° data, while the Bertini cascade does better at larger angles.

THE HIGH ENERGY MODELS

For incident particles of energy 15 GeV to $\sim$ 10 TeV, three models are available in Geant4. The high energy parameterized (HEP) model is derived from the high energy part of GHEISHA, which was discussed previously. The quark gluon string (QGS) model is almost entirely a theory-based model which contains diffractive string excitation and string decay into hadrons. The Fritiof fragmentation (FTF) model is an alternate

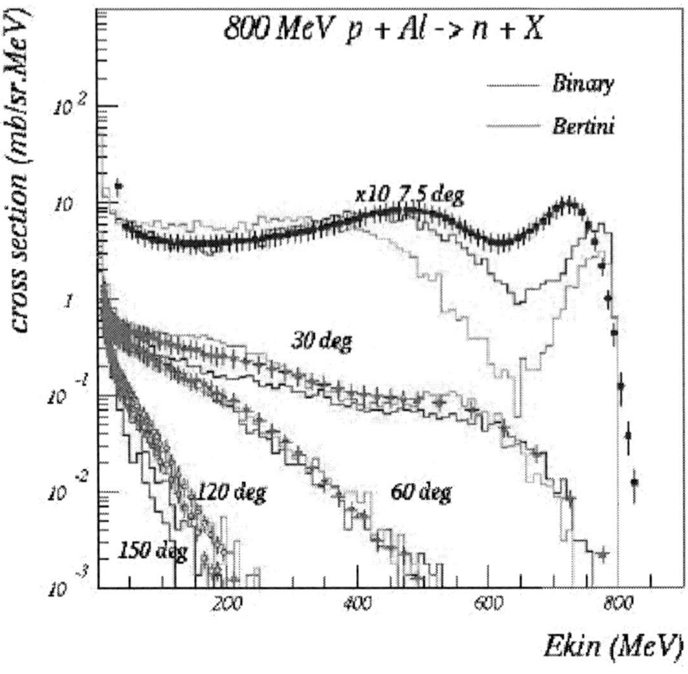

FIGURE 2. Neutron production from 800 MeV protons on Al. The double differential cross section (mb/sr/MeV) vs. neutron kinetic energy (MeV) is shown at 7.5, 30, 60, 120, and 150 °. The dark gray histogram represents the Binary cascade and the light gray histogram represents the Bertini cascade. The data are from Ref. [5].

theoretical model with different string fragmentation functions. Of the two theoretical models (QGS and FTF), QGS works better in most cases. Most high energy calorimetry physics lists use either QGS or HEP models.

Validation of these models is underway. An example of the comparison of the QGS and HEP models is shown in Fig. 3. Here the π^+ rapidity is examined for 320 GeV/c incident π^-. There are clear differences between the models: QGS produces a smooth curve with a central plateau, while HEP produces a three-peak structure. Both models agree reasonably well with the data, but one cannot currently be preferred over the other.

SHOWER SHAPE STUDIES

As mentioned in the introduction, a realistic Geant4 simulation is built by combining physics processes and models into a physics list. Two such physics lists, named LHEP and QGSP, are the most used and most tested in high energy calorimetry. Shower shapes provide especially good integration tests of these physics lists and data from several

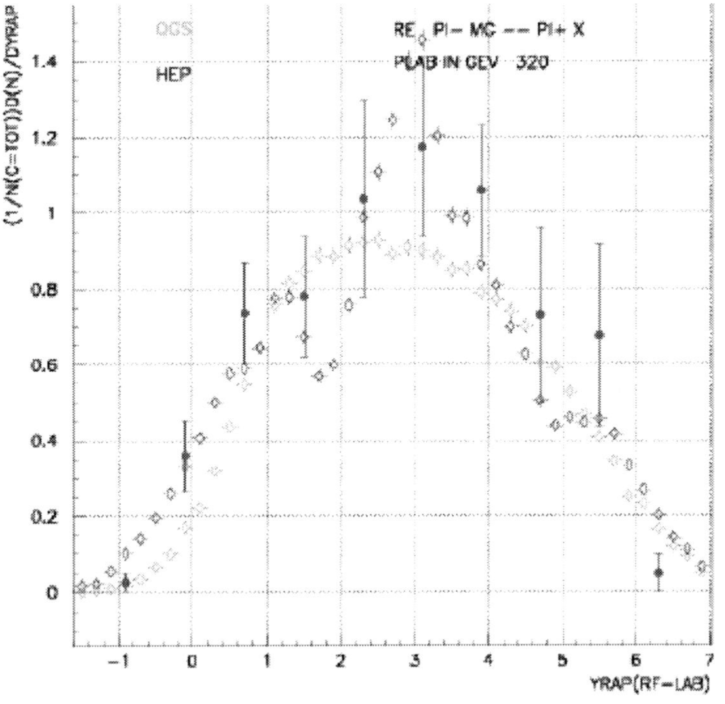

FIGURE 3. π^+ yield as a function of rapidity for 320 GeV/c π^- incident upon Mg. Open diamonds represent the HEP prediction and open circles represent the QGS prediction. The data are from Ref. [6].

test-beam experiments have been compared to their predictions.

The LHEP physics list consists of the low energy parameterized (LEP) models and the high energy parameterized (HEP) models, plus the Geant4 standard electromagnetic package. The QGSP physics list consists of the quark gluon string (QGS) model, the precompound nucleus model, some of the LEP models, and the standard electromagnetic package.

An example of the shower shape studies using these physics lists is shown in Fig. 4. The data are from the CMS test-beam setup of 2004, which consisted of a 7×7 array of $PbWO_4$ crystals and two wedges of alternating brass absorber and scintillator. Pion beams of between 2 and 300 GeV were incident upon the detector. Longitudinal shower shapes, pion energy spectra and recovered energy were examined. From the longitudinal shower shape it is seen that the LHEP physics list does a good job of reproducing the profile. The QGSP physics list, however, produces a significantly shallower profile. The reasons for this are currently under study.

For lower beam energies, such as 5, 10 and 30 GeV, both QGSP and LHEP physics lists show good agreement with the shower shape data.

Pion energy spectra measured in the calorimeter are in good agreement with both models at all energies.

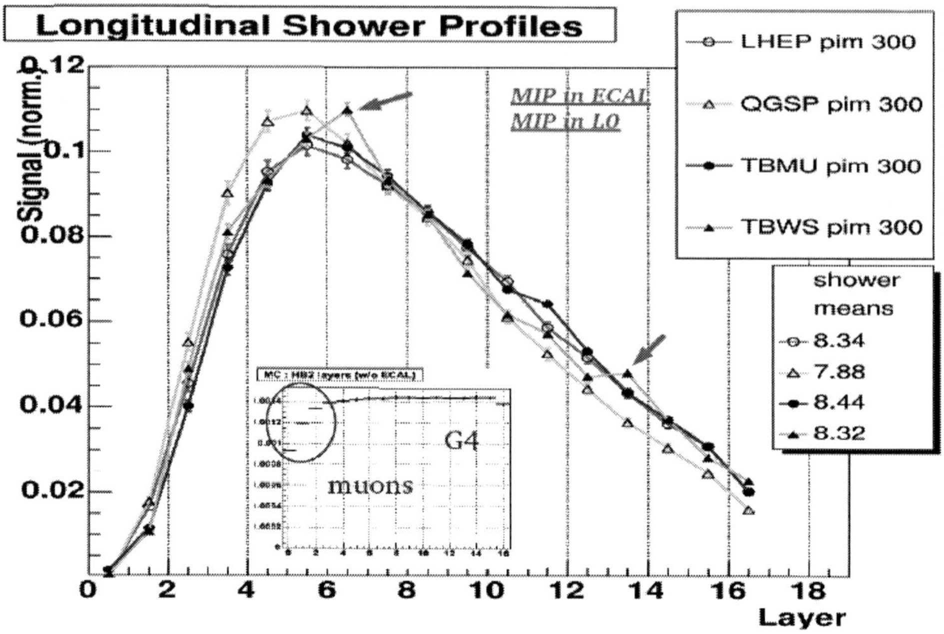

FIGURE 4. Longitudinal shower profile in the CMS test-beam calorimeter. Data are shown as filled triangles and filled circles connected by lines to guide the eye. Predictions of the LHEP physics list are shown by open circles and predictions of the QGSP physics list are shown by open triangles. The vertical axis is the normalized energy deposit and the horizontal axis is the hadronic calorimeter layer in which the energy is depositied. Plot courtesy of J. Damgov and the CMS collaboration.

In other measures, such as recovered energy, again the agreement with both physics lists is good, as shown in Fig. 5.

OTHER DEVELOPMENTS AND VALIDATIONS

Models dealing with the energy range 1-15 GeV have until recently been difficult to validate due to a scarcity of data. A number of test-beam experiments have been performed which now provide more data in this range. For example, the HARP experiment has recently published data for 12.9 GeV/c incident hadrons [7]. This will be used to validate and tune the LEP models in Geant4.

Shower shapes and calorimeter response are certainly dependent on the interaction cross sections. A review of all hadronic cross sections in Geant4 is now underway. Already the cross sections internal to the QGS model have been improved. This may lead to improved shower shapes at high energies.

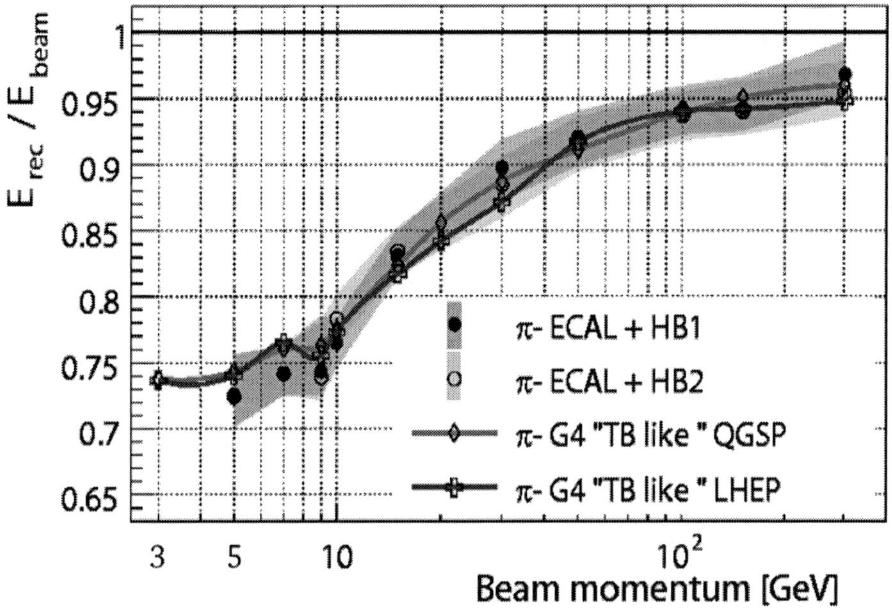

FIGURE 5. Recovered energy in CMS test-beam calorimeter as a function of incident particle momentum (GeV/c). Data are shown as filled and open circles (with error bands). Predictions of the LHEP physics list are shown as open crosses and predictions of the QGSP physics list are shown as open diamonds. Plot courtesy of J. Damgov and the CMS collaboration.

CONCLUSION

Many improvements to the Geant4 hadronic models are being made in order to improve calorimeter response and shower shape agreement. Elastic scattering is important in shower development and new models are now available which are more physical. The low energy and high energy parameterized models of Geant4 are still widely used and improvements have been made to them with respect to energy conservation. However, new, more physical models are being substituted where appropriate. The binary and Bertini cascade models are being used and improved as is the quark gluon string model. These models have been combined into the LHEP and QGSP physics lists, which are in the process of being validated against test-beam data from various high energy calorimeters. Agreement with CMS test-beam data is good for the most part, but some deviations of the QGSP list are being examined.

ACKNOWLEDGMENTS

This work was supported by the Stanford Linear Accelerator Center (SLAC) through a grant from the U.S. Department of Energy. It was also supported by the European

Organization for Nuclear Research (CERN) and the Helsinki Institute of Physics.

REFERENCES

1. S. Agostinelli et al., Nucl. Instr. Meth. Phys. Res. A 506 (2003) 250-303.
2. J. Allison et al., IEEE Trans. Nucl. Sci. 53 No. 1 (2006) 270-278.
3. GEANT - Detector description and simulation tool, CERN Program Library Long Write-up W5013, CERN Geneva.
4. H.W. Bertini and P. Guthrie, Nucl. Phys. A169 (1971).
5. W.B. Amian et al., Nucl. Sci. and Eng. 112 (1992) 78.
6. J.J. Whitmore et al., Z.Phys. C62 (1994) 199.
7. The HARP collaboration, Nucl. Phys. B732 (2006) 1.

Performance of the CDF Calorimeter Simulation in Tevatron Run II

Pedro A. Movilla Fernández
(*for the CDF Collaboration*)
Lawrence Berkeley National Laboratory, One Cyclotron Road, Berkeley, CA 94720, U.S.A.

Abstract. The CDF experiment is successfully collecting data from $p\bar{p}$ collisions at the Tevatron in Run II. As the data samples are getting larger, systematic uncertainties due to the measurement of the jet energy scale assessed using the calorimeter simulation have become increasingly important. In many years of operation, the collaboration has gained experience with GFLASH, a fast parametrization of electromagnetic and hadronic showers used for the calorimeter simulation. We present the performance of the calorimeter simulation and report on recent improvements based on a refined *in situ* tuning technique. The central calorimeter response is reproduced with a precision of 1-2%.

Keywords: CDF, GFLASH, calorimeter simulation
PACS: 07.05.Tp, 07.20.Fw, 13.85.-t

1. INTRODUCTION

Since the start of Run II in 2001, the Collider Detector at Fermilab (CDF) [1] has collected over 1 fb^{-1} of data of $p\bar{p}$ collisions at 1.96 TeV center-of-mass energy. In a variety of resulting publications, the simulation of the calorimeter has proved to be a crucial element for the precision measurement of physical observables, like the mass of the top quark, since it appears as one of the keys to control the jet energy scale systematics. The CDF calorimeter simulation is based on GFLASH [2], a FORTRAN package used for the fast simulation of electromagnetic and hadronic showers. It is embedded in a GEANT3 framework [3] as part of the whole detector simulation and has various advantages w.r.t. the detailed GEANT shower simulation: In CDF it is up to 100 times faster, and it can be flexibly tuned. The calorimeter simulation was initially tuned to test beam data and has been improved due to a steadily refined *in situ* tuning using samples of isolated charged particles, which recently became available over a remarkably extended momentum range up to 40 GeV. Here we report on the GFLASH performance effective for current CDF physics publications and on ongoing improvements contributing to the ambitious Run II physics program.

2. CDF CALORIMETRY

CDF is a general-purpose charged and neutral particle detector with a calorimeter and a tracking system. The calorimeter has a central and a forward section subdivided into a total of five compartments (Fig. 1): the central electromagnetic, CEM [4], the central hadronic, CHA [5], the plug electromagnetic, PEM [6], the plug hadronic, PHA [6], and the wall hadronic, WHA [5]. The calorimeter is of sampling type, with lead (iron)

CP867, *XII International Conference on Calorimetry in High Energy Physics*,
edited by S. R. Magill and R. Yoshida

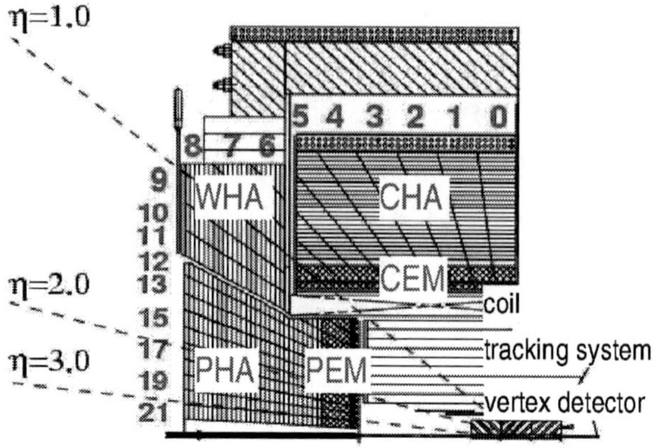

		Central	Plug
EM	thickness	$19\,X_0$, 1λ	$21\,X_0$, 1λ
	sample(Pb)	$0.6\,X_0$	$0.8\,X_0$
	sample(scint.)	5 mm	4.5 mm
	wavelength sh.	sheet	fiber
	resolution	$\frac{13.5\%}{\sqrt{E_T}} \oplus 2\%$	$\frac{14.5\%}{\sqrt{E}} \oplus 1\%$
HAD	thickness	$4.5\,\lambda$	$7\,\lambda$
	sample(Fe)	25-50 mm	50 mm
	sample(scint.)	10 mm	6 mm
	wavelength sh.	finger	fiber
	resolution	$\frac{50\%}{\sqrt{E_T}} \oplus 3\%$	$\frac{70\%}{\sqrt{E}} \oplus 4\%$

FIGURE 1. *Left:* Quadrant view of the CDF calorimeter showing the electromagnetic (CEM, PEM) and hadronic compartments (CHA, WHA, PHA). *Right:* Sampling structure and energy resolutions for the central (CEM, CHA) and the plug part (PEM, PHA).

absorbers for the electromagnetic (hadronic) compartments, scintillating tiles and wavelength shifters. It is subdivided into 44 projective tower groups, each group made of 24 wedges (partially 48 in the plug part) circularly arranged around the Tevatron beam and pointing to the nominal event vertex. The central and plug part together cover the pseudorapidity range $|\eta| < 3.6$. The calorimeter encloses a tracking system consisting of a vertex detector and a cylindrical drift chamber, both situated within a 1.4 T solenoid magnet. It provides a precise measurement of single charged particle momenta serving as an energy reference for the calorimeter simulation tuning.

3. GFLASH IN A NUTSHELL

The CDF simulation uses `GEANT` to propagate particles from the main interaction point through the detector volume. A shower in `GFLASH` is initiated when a particle undergoes the first inelastic interaction in the calorimeter. `GFLASH` treats the calorimeter as one effective medium using `GEANT` geometry and material information. `GFLASH` is ideal for calorimeter modules which may have a complicated but repetitive sampling structure.

Given an incident particle energy, $E_{\rm inc}$, the visible energy in the active medium,

$$dE_{\rm vis}(\mathbf{r}) = E_{\rm inc}\hat{m}\left[\frac{\hat{e}}{\hat{m}}c_{\rm em}f_{\rm em}(\mathbf{r}) + \frac{\hat{h}}{\hat{m}}c_{\rm had}f_{\rm had}(\mathbf{r})\right]d\mathbf{r}\,, \qquad (1)$$

is calculated according to the sampling fractions for electrons ($\hat{e}$) and hadrons ($\hat{h}$) relative to the sampling fraction for minimum ionizing particles ($\hat{m}$), taking their relative fractions $c_{\rm em}$ and $c_{\rm had}$ of energy deposited in the active medium into account. $f_{\rm em}(\mathbf{r})$ and $f_{\rm had}(\mathbf{r})$ are electromagnetic and hadronic spatial energy distributions of the form $f(\mathbf{r}) = \frac{1}{2\pi}L(z)\cdot T(r,z)$, factorizing into a longitudinal profile $L(z)$, which is a function of the shower depth z, and a lateral profile $T(r,z)$, which depends on z and on the radial distance r from the shower center. The showers are treated as azimuthally symmetric.

The longitudinal electromagnetic profiles are assumed to follow a Gamma distribution,

$$L_{\text{em}}(x) = \frac{x^{\alpha-1}e^{-x}}{\Gamma(\alpha)}\,, \tag{2}$$

where $x = \beta z$ and z measured in units of radiation lengths X_0. α and β are correlated parameters generated using two Gaussians, whose means and widths are free parameters subject to tuning, and a correlation matrix hardwired in GFLASH. Longitudinal hadronic shower profiles are a superposition of three shower classes:

$$L_{\text{had}}(x) = f_{\text{dep}}\left[c_{\text{h}}\mathscr{L}_{\text{h}}(x_{\text{h}}) + c_{\text{f}}\mathscr{L}_{\text{f}}(x_{\text{f}}) + c_{\text{l}}\mathscr{L}_{\text{l}}(x_{\text{l}})\right]\,, \tag{3}$$

$$\mathscr{L}_i(x_i) = \frac{x_i^{\alpha_i-1}e^{-x_i}}{\Gamma(\alpha_i)}\,,\ x_i = \beta_i z_i \qquad (i = \text{h},\text{f},\text{l})\,. \tag{4}$$

$\mathscr{L}_{\text{h}}$ is a purely hadronic component (z_{h} given in units of absorption lengths λ_0). $\mathscr{L}_{\text{f}}$ accounts for the component induced by neutral pions from a first inelastic interaction ($[z_{\text{f}}] = X_0$), and $\mathscr{L}_{\text{l}}$ originates from neutral pions occurring in later stages of the shower development ($[z_{\text{l}}] = \lambda_0$). Each subprofile in Eq. (4) is characterized by an individual correlated pair (α, β) analogously to Eq. (2). The coefficients c_i are the relative probabilities of the three classes expressed in terms of the fraction of showers containing a neutral pion (f_{π^0}) and the fraction of showers with a neutral pion in later interactions ($f^{\text{l}}_{\pi^0}$):

$$c_{\text{h}} = 1 - f_{\pi^0}\,,\ \ c_{\text{f}} = f_{\pi^0}(1 - f^{\text{l}}_{\pi^0})\,,\ \ c_{\text{l}} = f_{\pi^0}f^{\text{l}}_{\pi^0}\,. \tag{5}$$

The global factor f_{dep} is the fraction of deposited energy w.r.t. the energy of the incident particle. When a shower is generated, correlations between all parameters are properly taken into account. In total, the longitudinal shower profile is described by 18 independent parameters for the hadronic part (the means and widths of the α's, β's, and of the fractions f_{dep}, f_{π^0} and $f^{\text{l}}_{\pi^0}$), and four parameters for the purely electromagnetic part.

The lateral energy profile at a given shower depth z has the functional form

$$T(r) = \frac{2rR_0^2}{(r^2 + R_0^2)^2}\,. \tag{6}$$

The free quantity R_0 is given in units of Molière radius (for electromagnetic) or absorption lengths (for hadronic showers), respectively. R_0 is an approximate log-normal distribution with a mean and a variance parametrized as a function of the incident particle energy E_{inc} and shower depth z. The mean is given by

$$\langle R_0(E_{\text{inc}}, z)\rangle = [R_1 + (R_2 - R_3 \ln E_{\text{inc}})z]^n\,, \tag{7}$$

where n=1(2) for the hadronic (electromagnetic) case. The spread of hadronic showers increases linearly with shower depth z and decreases logarithmically with E_{inc}. Both shower types have their own set of adjustable parameter values (the R_i plus three independent parameters for the variance), thus giving a total of 12 parameters.

After generating the profiles, GFLASH distributes the incident particle energy in discrete interval steps following the longitudinal profile, and then deposits *energy spots* in the simulated calorimeter volume according to the lateral profile. The number of energy spots is smeared to account for sampling fluctuations. The visible energy is obtained by integrating over the energy spots and applying the relative sampling fractions Eq. (1).

4. GFLASH TUNING METHOD

The tuning of GFLASH mostly relies on test beams of electrons and charged pions with energies between 8 and 230 GeV [7], and has been refined during Run II using the calorimeter response of single isolated tracks in the momentum range 0.5-40 GeV/c measured *in situ*. The tuning of hadronic showers follows a four-step procedure.

Adjusting the MIP peak. Reproducing the response of minimum ionizing particles (MIP) is the first step since it serves as a reference for all other responses, see Eq. (1). The position and width of the simulated MIP peak is tuned to 57 GeV test beam pions in the electromagnetic compartments and involves the adjustment of charge collection efficiencies, which is handled by GEANT at this stage of the simulation.

Setting the hadronic energy scale and shape. Next, the shape of the individual energy distributions in the electromagnetic and hadronic compartment, the sum of both, and the hadronic response of MIP-like particles in the hadronic compartment are adjusted. The tuning is based on 57 GeV test beam pions and involves the most sensitive longitudinal parameters, the means and widths of α_{l} and β_{l} (steering the shower component induced by π^0's from later interactions), of β_{h} (which is related to the purely hadronic component), see Eq. (4), and of the fractions f_{dep} and f_{π^0} in Eqn. (3) and (5).

Fixing the energy dependence. After the longitudinal profile has been adjusted at one energy point, the other test beam samples are employed to parametrize the energy dependence of the parameters, which is typically logarithmic. At this point, also the inclusion of *in situ* tracks is important in order to provide a robust extrapolation into the important low energy region where test beam data are not available. A functional form $f(E_{\mathrm{inc}}) = a + b\tanh(c\ln E_{\mathrm{inc}} + d)$ is used to describe the dependence of the fractions in Eqn. (3) and (5) on the incident particle energy E_{inc}. Distinct parametrizations in the central and plug calorimeter were introduced to account for their different sampling structure. Initial Run II tunes involved samples of single isolated tracks up to 5 GeV/c in the central and plug part. Recently, the energy evolution of f_{dep} and the relative sampling fractions, $\hat{e}/\hat{m}$ and $\hat{h}/\hat{m}$, have been tuned using the total and MIP-like responses of *in situ* tracks up to 40 GeV/c (Sec. 6).

Tuning the lateral profile. The lateral profile is treated almost independently from longitudinal profile details and is tuned solely with Run II data. The parameters R_i of Eq. (7), which have been initially adjusted to describe the profiles in the energy range 0.5-5.0 GeV/c (using default values provided by the H1 collaboration at higher energies), are now tuned up to particle momenta of 40 GeV/c (Sec. 6).

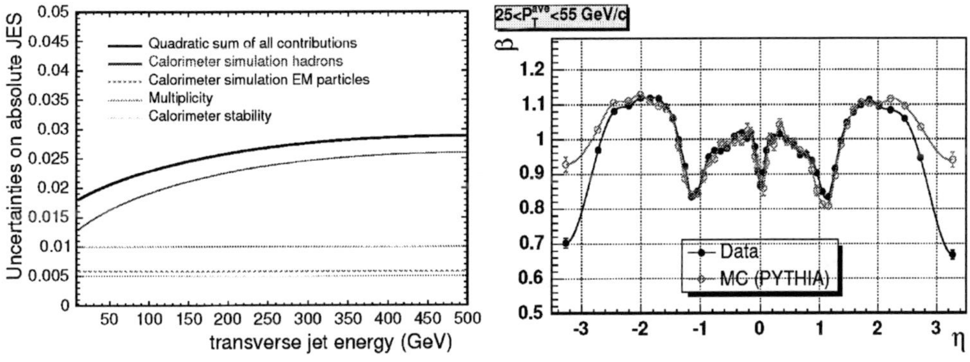

FIGURE 2. *Left:* Jet energy scale uncertainties due to calorimeter simulation and calibration published in [8]. *Right:* Di-jet balance β vs. η in data and simulation for transverse jet momenta 25-55 GeV/c.

5. IMPACT ON CDF JET ENERGY SCALE

We first report on the performance of the early Run II tuning embodied by CDF physics publications to date. Despite the good agreement with test beam data [7], the quality of the simulation used to be hard to control in the intermediate energy region, which is of particular importance since it constitutes a relevant fraction of the particle spectrum of a typical jet in CDF. For the initial tuning of the energy evolution of `GFLASH` parameters, minimum bias data providing tracks only up to 5 GeV were involved. Later checks based on special single track samples at higher momenta revealed a underestimation of the data. The level of agreement in the central was 2% for <12 GeV, 3% for 12 to 20 GeV, and 4% for >20 GeV [8]. Through convolution with a jet's typical particle spectrum, these numbers directly translate into the systematic uncertainties of the jet energy scale used by CDF to date (Fig. 2, *left*). The dominant contribution to the total uncertainties originate from discrepancies between the simulated and measured calorimeter response and from the low statistical precision of early Run II control samples.

The inhomogeneity of the calorimeter response is accounted for by an η dependent tuning. Jet responses in the plug and wall part are re-calibrated w.r.t. the better understood central part using a correction derived from di-jet events, $\beta = p_{\mathrm{T}}^{\mathrm{probe}} / p_{\mathrm{T}}^{\mathrm{trigger}}$, which relates the transverse momentum p_{T} of the non-central "probe" jet to the p_{T} of the central "trigger" jet. A comparison of the simulated and measured di-jet balance (Fig. 2, *right*) shows that the tuning is reproducing many calorimeter particularities along η.

6. IN SITU TUNING PROGRESS

The situation in the central calorimeter has now substantially improved due to the development of dedicated single track triggers with high momentum thresholds up to 15 GeV/c. The samples collected so far (a total of over 20 M events) allow a precise monitoring of the simulation performance in the energy range 0.5-40 GeV. In addition, steadily increased single track samples selected in minimum bias events now allow a consistent tuning of the plug simulation up to incident particle energies of 20 GeV.

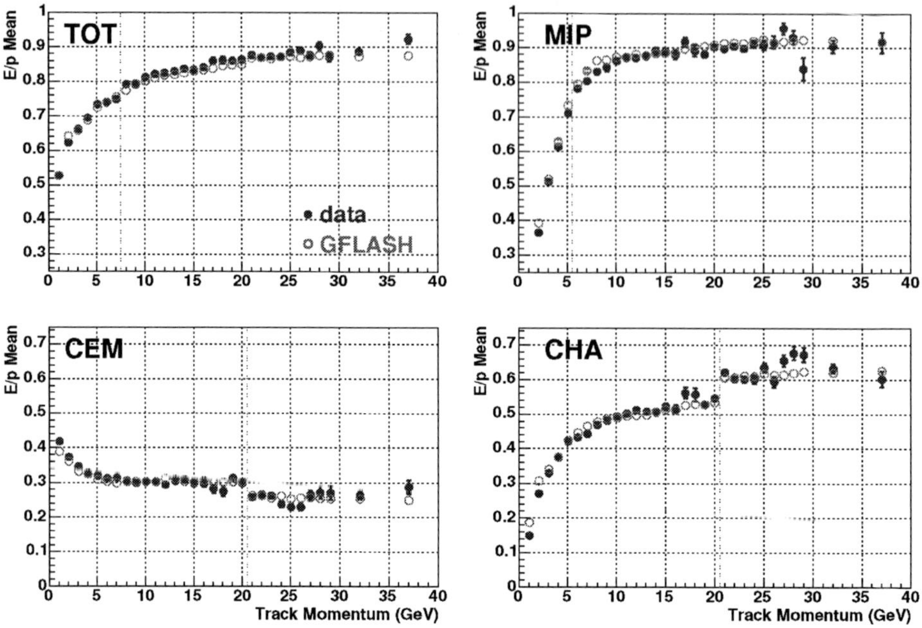

FIGURE 3. Simulated and measured $\langle E/p \rangle$ responses vs. track momentum in the central calorimeters CEM and CHA, the sum of both (TOT), and the CHA response of MIP-like particles (MIP). The vertical dashed lines indicate biases due to momentum dependent track analysis cuts and calculation methods.

Measurement technique. The *in situ* approach employs high quality tracks which are well contained within the target calorimeter tower (given by the track's extrapolation into the electromagnetic compartment) and which are isolated within a 7×7 tower block around the target tower. The signal is defined as the energy deposition seen in a 2×2 block (for PEM and CEM) or a 3×3 block (for CHA and PHA), respectively. For the background estimate, tower strips with the same η range but along the edges of a 5×5 tower group around the target tower are used. The simulation is usually based on a particle gun using a controlled momentum spectrum of a mixture of pions, kaons and protons, plus a minimum bias generator on top for realistic background modeling.

Absolute $\langle E/p \rangle$ response. Fig. 3 shows the comparison of the average E/p responses between data and simulation based on the most recent tuning of the energy dependence of f_{dep}, $\hat{e}/\hat{m}$ and $\hat{h}/\hat{m}$. The basic idea is to adjust simultaneously the simulated total response (TOT) and the CHA response of MIP-like particles with CEM$<$ 670 MeV using *in situ* tracks up to 40 GeV, keeping the test beam tuning at higher energies. Indeed, the simulated TOT and MIP *(top)* as well as the CEM and CHA responses *(bottom)* agree quite well with the data. The total $\langle E/p \rangle$ in the central part, which serves as an important benchmark for the systematic jet energy scale uncertainties, is reproduced with 1-2% precision within 0.5-40 GeV. The new tuning reflects more properly the current condition of the CDF calorimeter, which includes aging effects of the photomultipliers, and partially replaces the former conservative test beam uncertainties of 4 %.

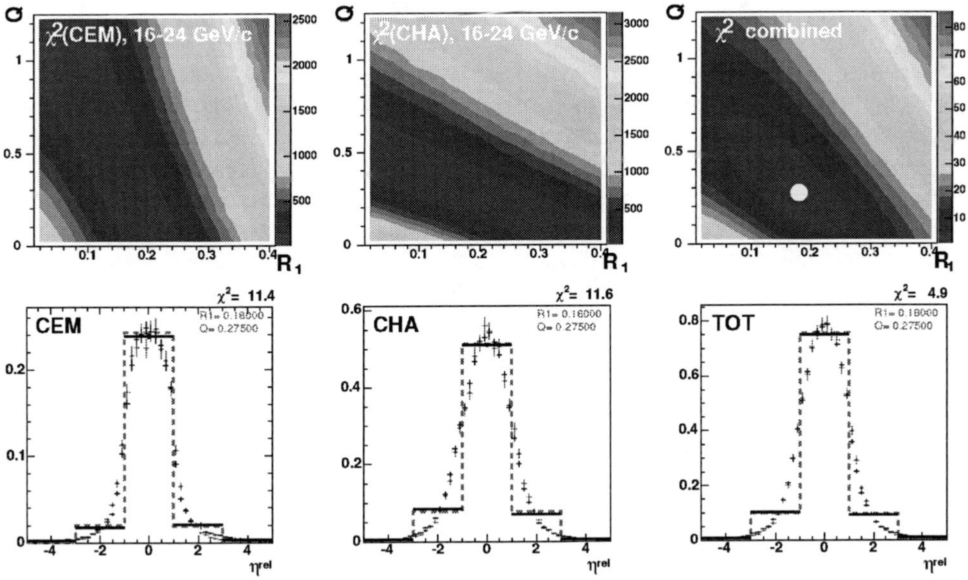

FIGURE 4. *Top*: χ^2 contours versus GFLASH parameters R_1 and Q (see text) obtained from comparisons of simulated and measured $\langle E/p\rangle$ profiles in CEM (left) and CHA (center) for 16-24 GeV/c particles, and a weighted combination of both (right). *Bottom*: Simulated $\langle E/p\rangle$ profiles versus data in CEM, CHA, and the sum of both (TOT) corresponding to the combined minimum χ^2 position (highlighted point).

$\langle E/p\rangle$ profile. An accurate adjustment of the hadronic lateral profile is important for two reasons. First, it controls the energy leakage out of the limited $\langle E/p\rangle$ signal regions. Thus, any profile mismatch between simulation and data causes a bias when tuning the absolute response. Second, leakage effects directly contribute to systematic uncertainties due to the correction of jet energies for the energy flow out of a jet cone [8].

For the tuning, an experimental profile is defined using the individual $\langle E/p\rangle$ responses of five 1×3 tower strips consecutive in η, versus a relative η coordinate normalized to the η of the target tower boundaries ($\eta = 0$ denoting the center of the target tower). The availability of large single isolated track samples allows a straightforward systematic approach: In Eq. (7), the constant R_1 denotes the shower core, while $Q \equiv R_2 - R_3 \ln E_{\rm inc}$ fixes how the profile evolves with shower depth and incident particle energy. Since the electromagnetic and hadronic calorimeter compartments probe different stages of the average shower development, R_1 and Q can be constrained using the profiles measured in the *individual* calorimeter compartments. The top of Fig. 4 shows a comparison of 16-24 GeV profiles between data and simulation in terms of a standard χ^2 estimator. CHA and CEM provide different contours of preferred parameter values, which helps to resolve the ambiguity due to the strong anti-correlation of R_1 and Q by using a combination of both *(top right)*. Generally a good agreement between data and simulation is obtained *(bottom)*. Thus, R_1 has been fixed from 0.5 to 40 (20) GeV in the central (plug) calorimeter, and R_2 and R_3 can be extracted from the linear energy dependence of Q.

7. ELECTROMAGNETIC RESPONSE

The tuning of electromagnetic showers in GFLASH presents less difficulty and is, although important, not detailed in this report. The simulated electromagnetic scale, which has been set using electron test beam data, was validated in Run II at low momenta using electrons from J/ψ decays and at high momenta using electrons from W/Z decays. The simulation reproduces the measured $\langle E/p \rangle$ response with a precision of 1.7%. The uncertainty is dominated by a contribution of 1.6% due to electrons pointing at the cracks between the towers, whereas electrons well contained in the target tower account for less than 1% [8]. The crack response is complicated due to the presence of instrumentation (*e.g.* wavelength shifter) but can be monitored using electron pairs from Z decays. One electron leg is required to be well contained within a target tower and serves as an energy reference, the other leg is used as probe to scan the $\langle E/p \rangle$ profile along the tower up to the crack. The description of the crack response in GFLASH has thus been improved using a correction function applied to the lateral profile mapping of energy spots.

8. CONCLUSION AND OUTLOOK

GFLASH has been tuned to reproduce the average hadronic response in the CDF central calorimeter with a precision of 1-2% within the energy range 0.5-40 GeV/c. The electron response is reproduced with similar precision. An *in situ* tuning technique developed and re-fined in Run II has proved to be crucial to overcome past and current performance limits. GFLASH might also be a promising simulation tool for LHC experiments. It is more flexible than GEANT, it is tunable, and it demonstrates excellent CPU performance.

ACKNOWLEDGMENTS

The main author would like to thank the German Max Planck Society and the Alexander von Humboldt Foundation for their support, and is grateful to the members of the CDF simulation group, in particular S.Y. Jun, S.A. Kwang, Y.S. Chung, G. Yu and D.A. Ambrose.

REFERENCES

1. CDF Coll., "The CDF II Detector Technical Design Report", FERMILAB-PUB-96-390-E (1996).
2. G. Grindhammer, M. Rudowicz and S. Peters, Nucl. Instrum. Meth. **A 290** (1990) 469.
3. R. Brun and F. Carminati, "GEANT Detector Description and Simulation Tool", CERN Programming Library Long Writeup W5013 (1993).
4. CDF Coll., L. Balka *et al.*, Nucl. Instrum. Meth. **A 267** (1988) 272.
5. CDF Coll., S. Bertolucci *et al.*, Nucl. Instrum. Meth. **A 267** (1988) 301.
6. CDF Coll., G. Apollinari, Proceedings of the 4th International Conference on Calorimetry in High Energy Physics, World Scientific, Singapore (1994) p. 200.
7. CDF Coll., C.A. Currat, Proceedings of the 10th International Conference on Calorimetry in High Energy Physics, World Scientific, Singapore (2002) p. 345.
8. CDF Coll., A. Bhatti *et al.*, "Determination of the Jet Energy Scale at the Collider Detector at Fermilab", *accepted by Nucl. Instrum. Meth.*, FERMILAB-PUB-05-470 (2005), hep-ex/0510047.

Simulation of the Hermes Lead Glass Calorimeter using a Look-Up Table

A. Vandenbroucke* and C.A. Miller†

*Department of Subatomic and Radiation Physics, Gent University, Belgium
†TRIUMF, Vancouver, Canada
ON BEHALF OF THE HERMES COLLABORATION

Abstract. This contribution describes the Monte Carlo simulation of the Hermes Electromagnetic Lead-Glass Calorimeter. The simulation is based on the GEANT3 simulation package in combination with a Look-Up Table. Details of the simulation as well as a comparison with experimental data are reported.

Keywords: GEANT3, GEANT4, photo-electron, electromagnetic shower, lead-glass
PACS: 07.05.Tp,07.20.Fw

INTRODUCTION

The Hermes experiment has been running continuously since 1995. Meanwhile, several improvements were made to hardware as well as software to cope with more challenging physics goals. This report details the improved Monte Carlo description of the Lead-Glass calorimeter of the Hermes experiment. In the first section some general ideas about the physics of Hermes will be given, as well as a description of the Calorimeter. Next, the focus will be on the layout of the simulation of the Calorimeter, and finally the performance of the simulation will be shown by comparing with experimental data.

The Hermes Experiment

The principal physics goal of the Hermes experiment is to study the spin structure of the nucleon. It is a fixed target experiment located at the Hera accelerator facility at DESY, Hamburg. Polarized leptons with an energy of 27.5 GeV are scattered off internal (un-)polarized gas targets.

Hermes has already completed some studies of the spin structure function of the nucleon $g_1(x,Q^2)$ via polarized Deep Inelastic Scattering (DIS) [1, 2]. For these measurements only the scattered lepton needs to be detected. Furthermore, via the analysis of polarized Semi-Inclusive Deep Inelastic Scattering (SIDIS), the Hermes collaboration has provided a unique measurement of the polarized quark distribution functions $\Delta q(x,Q^2)$ [3]. In SIDIS a produced pion or kaon is detected in combination with the scattered lepton. These measurements indicated that the spin of a nucleon isn't simply the sum of the individual valence and sea quark spins. Quark orbital angular momenta are expected to play a non-negligible role in the nucleon spin structure. Total orbital angular momenta of

CP867, *XII International Conference on Calorimetry in High Energy Physics*, edited by S. R. Magill and R. Yoshida

TABLE 1. Chemical composition and calorimetric properties of F101 Lead Glass. Cerium is making the Lead Glass radiation hard, while also reducing its transparency.

Chemical Composition F101	weight %
PB_3O_4	51.23
SiO_2	41.53
K_2O	7.0
Ce	0.2
Radiation Length	2.78 cm
Critical Energy	17.97 MeV
Refraction index	1.65
Molière Radius	3.28 cm

quarks can in principle be accessed in exclusive reactions such as Deeply Virtual Compton Scattering (DVCS) in which a photon is created by scattering a virtual photon off a quark.

To comply with the original main physics goals, a spectrometer was built [4] consisting of tracking chambers in combination with a magnet to determine particle momenta and charge. High efficiency electron/hadron separation is done by a Transition Radiation Detector (TRD), a preshower detector and a lead-glass calorimeter. A Ring-Imaging Cherenkov Detector (RICH) was built to distinguish between pions, kaons and protons.

As the interest in DVCS has grown in the last years, and since the produced real photons are measured by the calorimeter only, it became clear that a more detailed Monte-Carlo description of the calorimeter was needed. In (SI)DIS reactions the calorimeter is mainly used for triggering and particle identification.

The Hermes Calorimeter

The Hermes Calorimeter has several functions: First of all it serves as a first level triggering device. Secondly, it separates leptons from hadrons with high efficiency, and finally π^0's can be reconstructed. The design consists of 10 rows and 42 columns of radiation hard F101 lead glass blocks [5]. One calorimeter block surface measures 9×9 cm^2, while the block length is 50 cm, which corresponds to about 18 radiation lengths. The chemical composition and calorimetric properties of F101 are given in Table 1. All blocks are wrapped with an aluminized mylar foil to reflect the Cherenkov photons, and a tedlar foil to make the blocks light tight. To convert the optical photons to an electrical pulse, Photo Multiplier Tubes (PMT) are attached to each block via a two component silicon glue.

In order to improve the electron-pion separation capabilities of the calorimeter, a 1 cm thick lead preshower was put in front of the calorimeter, which is succeeded by a scintillator array for triggering purposes. The preshower detector is depicted in Figure 1 together with the calorimeter. The performance of the Hermes calorimeter within the Hermes experiment in terms of energy resolution and particle identification can be found

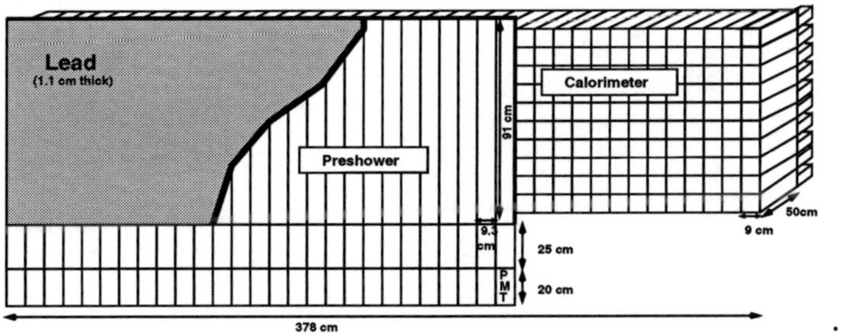

FIGURE 1. Schematic of the Hermes Preshower - Calorimeter combination. The segmentation of the preshower scintillator array and the calorimeter are indicated in the figure.

in reference [6].

SIMULATION OF THE LEAD GLASS CALORIMETER

The Hermes Monte Carlo (HMC) output is intended to mimic decoded Hermes data. It simulates the elementary particle interaction within all relevant detector volumes, as well as the digitization. This approach has the advantage that both Monte Carlo output and Hermes data can be processed by the same tracking and clustering reconstruction code. HMC itself is based on GEANT3, which in the case of the calorimeter is responsible for the simulation of the electromagnetic shower and of the resulting Cherenkov photon production.

Shower description in GEANT3

In order to test the simulation of electromagnetic showers in HMC, the simulated longitudinal shower profile within a lead-glass block was compared to empirical expectations. For various incident energies the profile was fitted by a function of the form:

$$\frac{\mathrm{d}E}{\mathrm{d}t} = E_0 b \frac{(bt)^{a-1} e^{-bt}}{\Gamma(a)}, \tag{1}$$

with E_0 the critical energy, t the depth in radiation lengths, and a and b fit parameters [7]. Electrons of various energies were generated at perpendicular incidence on the calorimeter surface. The function stated above describes the profile very well, with reasonable values of the parameters a and b over a large energy range. This indicated that the GEANT3 based simulation of the shower was functioning properly. The option `IABAN=2` was selected for GEANT3, so that particles are tracked down to a user controlled cut off energy, rather than the default cut off of 10 MeV. The latter value serves to increase computational efficiency. For electromagnetic showers however, this value is too large. The cut off energy was chosen to be 200 keV.

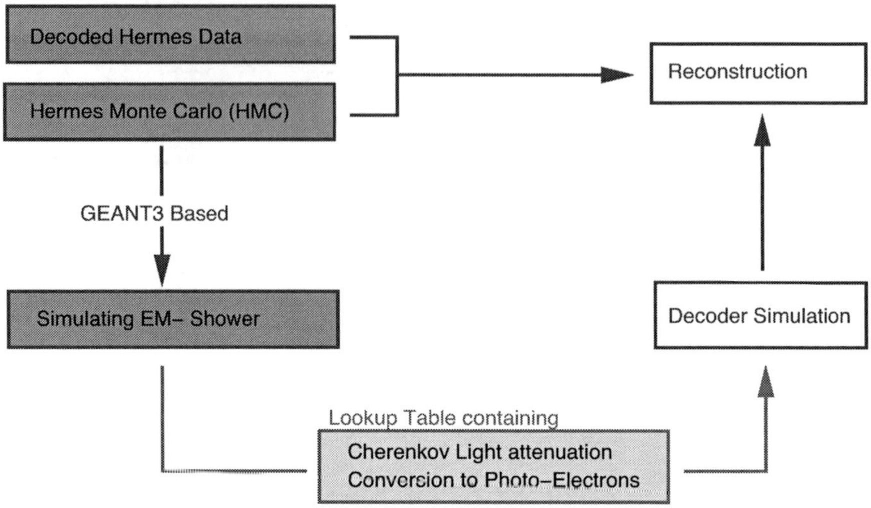

FIGURE 2. Schematic layout of the Calorimeter simulation

During these studies it became clear that the resolution of lead glass calorimeters isn't limited by photo-statistics but rather by stochastic fluctuations in the longitudinal shower profile, in combination with strong attenuation of blue light in the lead-glass. A typical high PMT quantum efficiency for this wavelength, makes this effect even worse. (In reference [8] an improved resolution is reported by making use of a blue light filter in front of the PMT's.)

Optical photon propagation

The produced Cherenkov photons in a particle shower all undergo several reflection and attenuation effects before reaching the PMTs where they are converted to photo-electrons. If this GEANT package would be responsible to track every single Cherenkov photon through the lead glass, the Monte Carlo code would be too slow, even on modern CPUs. Therefore clearly a different approach was needed.

Figure 2 schematically shows the layout of the calorimeter simulation. The method chosen was to use a Look-Up Table (LUT) to represent the photo electron yield per millimeter electron or positron track length. The LUT takes into account the measured reflectivity of the foils, the measured transparency of the lead glass, as well as the refraction at all surfaces, such as the calorimeter block wrapping, and the glue between the PMTs and the lead glass blocks. Moreover, the quantum efficiency of the PMTs was included.

Since Cherenkov photons are generated in the forward as well as in the backward direction, and since their production angle is velocity dependent, the LUT needed to be momentum dependent at least. It also needs to be position dependent, as the geometric collection efficiency depends on position. Moreover, the amount of attenuation clearly is dependent on the longitudinal position. For all of the above reasons the LUT depends on $x, y, z, \theta_p (= \arccos \frac{p_z}{|\vec{p}|}), \phi_p (= \arctan \frac{p_x}{p_y})$ and $\beta (= \frac{v}{c})$.

The most rapid variations were observed for the θ_p dependence of the photo-electron yield. Consequently, the segmentation of the table is finest in θ_p . Rapid variations as a function of z and β were also observed, leading to a $10 \times 10 \times 16 \times 40 \times 10 \times 16$ grid. Due to symmetry reasons the table was generated for one block octant only. A stand-alone GEANT4-based simulation of a 9 block cluster was implemented to simulate electromagnetic showers in the lead-glass. It was impractical to convert the entire calorimeter simulation to GEANT4 because of the large effort involved in redefing the geometry for the entire spectrometer. The propagation of optical photons was simulated in detail including all the effects mentioned above. For each bin in the 6-dimensional table, 1000 electrons were propagated along a one millimeter long track segment at the position and angle corresponding to that bin. The resulting photo-electrons were collected and recorded as the response for that bin. about 50 CPU days at WestGrid in Vancouver for the table to be calculated. A simple coordinate translation is required to convert Hermes coordinates into LUT coordinates.

Table Interpolation

Due to the 'curse of dimensionality', only linear interpolation algorithms have been considered to interpolate the Look-Up Table. Using the most straightforward multidimensional linear interpolation, $2^6 - 1$ interpolations are needed for a six dimensional LUT. The implementation of this algorithm was done using the FINT function from the CERNLIB package[1], which had to be extended from five to six dimensions.

Another interpolation method is the simplex method [9]. Rather than taking an n-dimensional hypercube around the point to be interpolated, this method considers an n-dimensional simplex, which is an n-dimensional analogue of a triangle. As this hypersurface has only $n+1$ vertices, the number of interpolations required reduces to $n+1$. The method is about twice as fast as the FINT method. However, a drawback of the simplex algorithm is that it can only do interpolation, whereas the FINT algorithm is able to extrapolate the table where necessary.

RESULTS OF THE SIMULATION

In order to test the simulation's performance several checks were made. First of all the integration of the Look-Up table together with the interpolation algorithm itself was tested. Later a dedicated comparison between GEANT3 and GEANT4 was done, followed by a comparison with experimental data. This section will report on the aforementioned tests.

---

[1] `http://wwwasdoc.web.cern.ch/wwwasdoc/shortwrupsdir/e104/top.html`

Testing the Integration of the Look-Up Table

In order to test the integration of the LUT interpolation in the context of HMC, high energy muons were generated right in front of the calorimeter. Muons of 100 GeV offer the simplification of eliminating the shower simulations; they keep on going straight without much deflection. The x,y, θ_p, ϕ_p and β thus stayed constant, so that the z-dependence could be tested. Both interpolation methods were able to reproduce the rapid variations in the photo-electron yield given by the Look-Up Table, indicating a good functionality of the algorithms.

Observed differences with GEANT4

The results of the GEANT3 (G3) and GEANT4 (G4) simulations were compared in detail. While muons yielded a good overall agreement, both in terms of Cherenkov photons as well as photo-electrons, some differences were observed for electron initiated showers. In this case, the HMC/GEANT3 based simulation produced about 6 % more Cherenkov photons than the GEANT4 based one, while the photo-electron yield of the GEANT3 based simulation was about 13 % higher. The numbers stated here are the sum of the responses of the central lead-glass block and its eight surrounding neighbours. This is the usual clustering method in the Hermes reconstruction code. In an attempt to understand these differences, the total track length and escaping energy out of the 9-block cluster were calculated. The latter is shown in Figure 3, where HMC/G3 shows no obvious energy dependence, while G4 indicates an increase with energy. G4 also has more fluctuations as indicated by the error bars, which represent the RMS of the distribution. The behavior of G4 can be understood from the fact that the transverse shower development is larger for increasing energies, yielding more escaping energy from a 9-block cluster. These studies indicated that the electromagnetic shower in GEANT4 has a larger transverse spread. Also elsewhere a difference in the electromagnetic shower profile between GEANT3 and GEANT4 has been reported [10]. The observed differences between the GEANT4 and the GEANT3 based simulations are therefore expected to be a consequence of the intrinsic difference between both packages.

Comparison with data

The main goal of any detector response simulation is that it well describes the response of that detector to various kinds of particles and energies. Both the ratio of energy response over incident momentum and the π^0 invariant mass are interesting observables to check the quality of the simulation. Corresponding results will be reported in the next paragraphs.

As electrons lose all their energy in the calorimeter, the ratio of energy as measured by the calorimeter over momentum, as determined with the tracking system, should be close to one. This feature actually is used to calibrate the calorimeter.

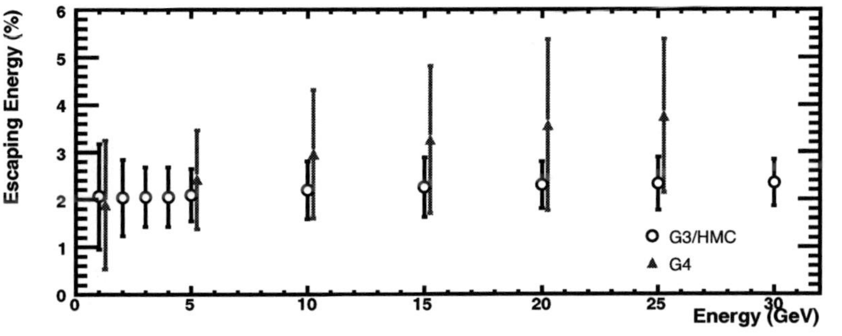

FIGURE 3. Escaping energy from a 9-block cluster for both G4 (triangles) and HMC/G3 (circles), as a function of incident energy. The error bars indicate the RMS of the distributions. For visibility, the G4 points have an offset of $+0.25$.

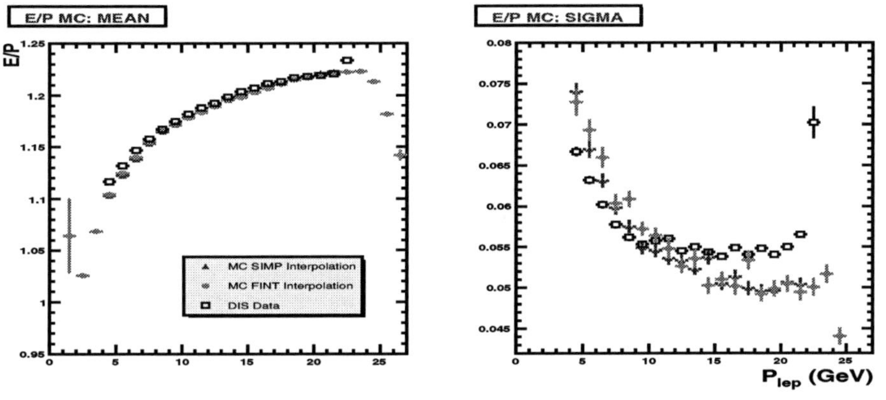

FIGURE 4. Mean (left) and standard deviations (right) of the E/P distributions for electrons in 1 GeV energy bins. Open squares are experimental data, while filled circles (triangles) are Monte Carlo simulations based on the FINT (Simplex) interpolation method. Some of the data points in the left panel are not easily visible because they overlap with the FINT results.

The magnetic momentum analysis for charged particles has been shown to be accurately simulated by HMC. Therefore, a comparison of E/P gives direct information about the quality of the calorimeter simulation.

Both the mean and the width of the E/P distribution are depicted in Figure 4 for data as well as for Monte Carlo. Both the FINT and the simplex method were used for comparison. Sufficient lepton data is available for momenta between 5 and 22 GeV. A good agreement between data and Monte Carlo can be seen. The Simplex method is a bit less precise, due to its inability to extrapolate where needed. The figure illustrates that the calorimeter response is quite non-linear. This is because of both non-linearities in the shower profile and the presence of the preshower detector in front of the lead-glass blocks.

Next to E/P, the two-photon invariant mass spectrum with the visible π^0 peak forms

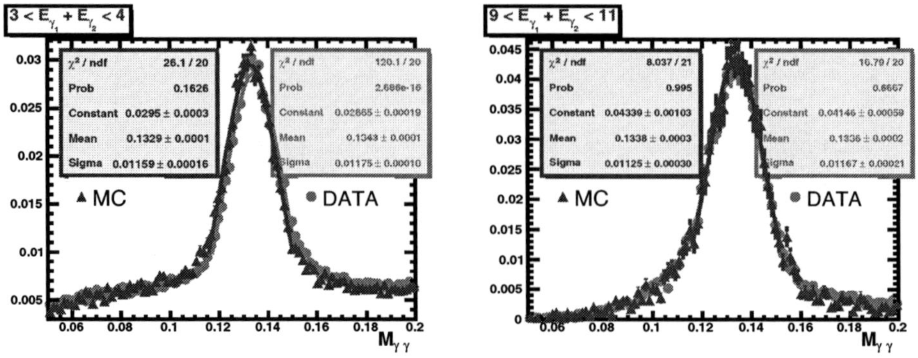

FIGURE 5. Two-photon invariant mass spectrum for two π^0 energies for data (circles) and MC (triangles). The MC is using the FINT interpolation. A Gaussian has been fit to the central peak. Fit parameters are indicated on the right (left) of the peak for data (MC).

an excellent tool for the data to Monte Carlo comparison for photon-induced electromagnetic showers. In order to simulate the combinatorial background in the spectrum, a physics generator was required. Figure 5 shows the Monte Carlo simulation of the invariant mass spectrum as generated with PYTHIA and tracked with HMC, together with the experimental data for two different energy bins. A good agreement is seen. The additional minimal difference in the width could probably be improved by increasing the simulated electronic noise in the digitization routine of the Monte Carlo.

CONCLUSIONS

The simulation of the Hermes Calorimeter was implemented as a combination of the GEANT3 package to simulate the electromagnetic shower, and a Look-Up Table containing the amount of photo-electrons per millimeter electron or positron track. Experimental data is well described by the Monte Carlo code. Previously observed differences between the GEANT3 and GEANT4 based simulation of electromagnetic showers are confirmed.

REFERENCES

1. K. Ackerstaff, et al., *Phys. Lett.* **B404**, 383–389 (1997), hep-ex/9703005.
2. A. Airapetian, et al., *Phys. Lett.* **B442**, 484–492 (1998), hep-ex/9807015.
3. A. Airapetian, et al., *Phys. Rev.* **D71**, 012003 (2005), hep-ex/0407032.
4. K. Ackerstaff, et al., *Nucl. Instrum. Meth.* **A417**, 230–265 (1998), hep-ex/9806008.
5. H. Avakian, et al., *Nucl. Instrum. Meth.* **A378**, 155–161 (1996).
6. H. Avakian, et al., *Nucl. Instrum. Meth.* **A417**, 69–78 (1998), hep-ex/9810004.
7. E. Longo, and I. Sestili, *Nucl. Instrum. Meth.* **128**, 283 (1975).
8. V. Hanin, and E. Stern, *Nucl. Instrum. Meth.* **157**, 455–460 (1978).
9. R. Ravotti, et al., *IEEE Trans. on Comp.* **47**, 894–899 (1998).
10. J. Alvarez-Muniz, et al., *Phys. Rev.* **D68**, 043001 (2003), astro-ph/0206043.

Simulator For The Linear Collider (SLIC): A Tool For ILC Detector Simulations

Norman Graf, Jeremy McCormick

Stanford Linear Accelerator Laboratory, Menlo Park, CA, 94025

Abstract. The Simulator for the Linear Collider (SLIC) is a detector simulation program based on the GEANT4 toolkit. It is intended to enable end users to easily model detector concepts by providing the ability to fully describe detectors using plain text files read in by a common executable at runtime. The detector geometry, typically the most complex part of a detector simulation, is described at runtime using the Linear Collider Detector Description (LCDD). This system allows end users to create complex detector geometries in a standard XML format rather than procedural code such as C++. The LCDD system is based on the Geometry Description Markup Language (GDML) from the LHC Applications Group (LCG). The geometry system facilitates the study of different full detector design and their variations. SLIC uses the StdHep format to read input created by event generators and outputs events in the Linear Collider IO (LCIO) format. The SLIC package provides a binding to GEANT4 and many additional commands and features for the end user.

Keywords: Geant4, Detector Simulation
PACS: 07.05.Tp, 07.05.Wr, 07.20.Fw, 29.40.Vj, 29.85.+c

INTRODUCTION

Detectors for the International Linear Collider (ILC) will require precision tracking, vertexing, and calorimetry in order to meet the ambitious physics goals of the experimental program. Distinguishing between W and Z bosons decaying into quark jets requires a calorimetric energy resolution of roughly $30\%/\sqrt{E}$. The design of the full detector and its sub-detectors will need to be aggressively optimized to meet this performance goal. Monte Carlo simulations are used in this optimization process to study the performance of different proposed detector designs and their variations. In order to conduct these studies and fully explore the phase space of possible detectors, we have developed an executable based on the Geant4 toolkit which is capable of fully defining a detector using runtime input. This frees the physicist from having to worry about details of C++ coding and concentrate on the optimization of the detector performance.

The full detector designs must be optimized for maximum performance while being constrained by a reasonable cost. At the level of the full detector, different combinations of subdetector parameters need to be studied. It may therefore be true that although any particular subdetector may not be optimal, the detector as a whole may have optimal performance. Although the interface targets basic Geant4 shapes, and therefore any detector could be modeled, the focus of this particular presentation is on detectors optimized for the ILC. Although modern collider detectors have

CP867, *XII International Conference on Calorimetry in High Energy Physics,*
edited by S. R. Magill and R. Yoshida

converged on a common cylindrical topology, the exact composition of the trackers and calorimeters varies between concepts.

Since the number of possible full detector designs that need to be modeled by the simulation software is quite large, detector simulation software for the ILC needs to provide capabilities that facilitate an iterative development cycle allowing rapid prototyping of these detector geometries. Changes to the detector geometry must not only be easy to make but also easily to propagate to all packages that depend on this information, from the simulator through reconstruction and analysis algorithms. The detector geometry should be treated as conditions; client programs need random access to full detector designs based on the current physics event.

As an example, the following are some of the basic detector parameters which can be varied at runtime in the geometry of a sampling calorimeter:

- Inner radius of the barrel calorimeter.
- Extent along the beam axis.
- Number of layers.
- Absorber layer material.
- Absorber layer thickness.
- Sensor layer material.
- Sensor thickness.
- Sensor technology.
- Sensor segmentation ("cell size").
- Configuration of dead material / supports.

The above parameters represent only the first level of variations in a subdetector design. The nested geometry hierarchies used in detector simulation can be represented to varying levels of detail. The most complex geometries model the "as designed" or "as built" components as closely as possible. Simplified detector geometries can be used for parametric geometry studies using several full detector designs or variants thereof. For fast studies, tube or cylindrical pieces may be used to model the calorimeters.

SLIC OVERVIEW

The Simulator for the Linear Collider (SLIC[1]) was designed around the requirements of an iterative development cycle. SLIC provides a flexible and powerful geometry system and includes all the necessary packages for a full simulations engine. The "SLIC" name includes a number of different software packages. The GEANT4[2] toolkit provides the geometry and physics simulation engine describing the interaction of particles with fields and materials. LCIO[3] is the IO subsystem for interconnecting different ILC software packages. GDML[4] is an XML-based geometry system. SLIC uses a GEANT4-based library called Linear Collider Detector Description (LCDD) for the detailed and complete description of the detector geometry at runtime. LCDD is an extension of GDML and will be discussed in more detail in the following section. Finally, SLIC is the hub package depending on all seven of the other libraries. It provides concrete definitions of the user classes for geometry, physics, and event generation required by GEANT4. SLIC includes a complete command-line interface and many new Geant4 macro commands to facilitate both batch and interactive work.

LCDD GEOMETRY SYSTEM

The Linear Collider Detector Description (LCDD) is an XML data format for describing detector geometries along with the associated processing infrastructure. The LCDD format is defined by an XML Schema document (XSD) at a standard URL.

```
http://www.lcsim.org/schemas/lcdd/
```

The LCDD schema uses built-in XSD features to import and extend the GDML schema. An LCDD document contains an embedded `<gdml>` block, or sub-document, that must conform to GDML's schema. Applications that need to either produce geometry data or read it in thus have a clear and complete definition of the detector description with these standardized formats.

The GDML format provides bindings to the core geometry classes of the GEANT4 toolkit. The GDML format defines expressions, materials, shapes, volumes, and volume structure. LCDD extends these basic definitions by adding information essential to the full simulation. LCDD adds a header with detector meta-information, an identifier dictionary, sensitive detectors and readouts, regions, physics limits, visualization parameters, and magnetic fields.

Unlike in HTML, where elements can be referenced regardless of order, GDML and LCDD support in-order references, only. An element must have already been defined to be referenced. For this reason, the ordering of the top-level container elements in GDML and LCDD files is important and must conform to the order specified in their respective schemas. For instance, a box solid must be defined before it can be used as the shape for a volume. The primary benefit of this approach is reduction of memory consumption during the processing phase.

The following snippet of pseudo-XML outlines the top-level structure of an LCDD file, including the embedded GDML element.

```
<lcdd>
     <header> … </header>
     <iddict> … </iddict>
     <sensitive_detectors> … </sensitive_detectors>
     <regions> … </regions>
     <limits> … </regions>
     <display> … </display>
     <gdml>
          <define> … </define>
          <materials> … </materials>
          <solids> … </solids>
```

```
            <volumes> … </volumes>
            <structure> … </structure>
            <setup> … </setup>
        </gdml>
        <fields>
    </lcdd>
```

The embedded GDML block may contain additional LCDD tags in the `<volume>` element. This is the only extension to the GDML format made by LCDD. The LCDD processor can also read plain GDML files, so a file with `<gdml>` as the top element is considered valid within the LCDD processing framework.

GDML's `<define>` block contains expressions and definitions read into the CLHEP Expression Evaluator. These expressions may contain double precision numbers, simple arithmetic operators (* / + -), trigonometric functions, and units. The processor predefines a number of standard units for distance, weight, etc. Important primary constants such as the speed of light are also predefined. Rotations and positions that will be referenced later to create physical volumes are also included in this area.

The `<materials>` section has `<material>` and `<element>` elements that bind to the `G4Material` and `G4Element` classes for materials and atomic elements. Materials are defined by atomic or mass composition and density. Material parameter sheets may be attached to provide pre-computed values for dEdx calculations and similar algorithms. The materials defined here are referenced by `<volume>` elements in the `<structure>` area.

The `<solids>` block contains shape definitions that are also referenced by the GDML volumes. Constructive Geometry Solids (CSG) are the most common type of shapes used in GEANT4 geometries. (Boundard Represented Solids (BREPS) are also available but are not supposed by the GDML system.) GDML has bindings to a large and nearly complete subset of the CSG solids defined by the GEANT4 geometry subsystem, including tubes, boxes, trapezoids, circles, ellipses, twisted tubes and boxes, polyhedra, and facetted shapes. Boolean subtraction and addition can be used with these primitives to define arbitrarily complex geometries, such as masks and detectors containing cutouts for ingoing and outgoing beampipes.

The `<structure>` block contains a nested hierarchy of geometric volumes. A volume is composed of a shape plus its material and may contain any number of sub-volumes, defined with the `<physvol>` tag. The volumes in the `<physvol>` elements are called "child" volumes of their "parent" volume. The child volumes must contain a reference to a logical volume, plus an in-lined or referenced position and rotation. The top-level volume or "world volume", typically a large box containing the detector envelope volume, is defined in the `<setup>` block using the `<world>` element.

The `<iddict>` is an identifier dictionary that defines 32 and 64 bit identifiers to uniquely identify geometric elements in the detector. These identifiers are used to write cell identifiers into the LCIO hits. Each identifier has a corresponding `<idspec>` with fields that are defined using `<idfield>` elements. After the dictionary is processed, a GEANT4 runtime object is created for each identifier specification. During hits processing, data is retrieved automatically from the simulation based on the identifier specification. A sample <iddict> with an <idspec> for an ECAL barrel is defined below.

```
<iddict>
     <idspec name="EcalBarrelId" length="64">
          <idfield label="layer" start="0" length="7" />
          <idfield label="sys" start="7" length="3" />
          <idfield label="theta" start="10" length="8" />
          <idfield label="phi" start="18" />
     </idspec>
</iddict>
```

The <idspec> of 64 bits describes an ECAL calorimeter barrel. The <idfield> label sets the source of the value. Possible sources are <physvolid> elements or segmentation values from the subdetector's readout. The "layer" field is the layer number, and it is read from a <physvolid> of the GEANT4 volume that corresponds to the layer volume. The "sys" field is also taken from a <physvolid> and gives a unique identifier for the sub-detector. The "theta" and "phi" fields are not read from the volume hierarchy. These values are provided by the readout of the ECAL sensitive detector. This flexible system allows the geometric identifier information to be transferred easily from the simulation runtime to the LCIO hits. Applications requiring the identifier information at a later time can read it directly from the LCDD file.

The <sensitive_detectors> section lists sensitive detectors that are referenced from the GEANT4 volumes. When a particle passes through a sensitive volume, e.g. one that has an assigned sensitive detector, the hits processing method is called on the sensitive detector. Depending on the characteristics of the energy deposition, a hit may be created that will be transferred into the LCIO output file. The <calorimeter> and <tracker> elements are the two types of sensitive detectors supported by LCDD.

```
<sensitive_detectors>
     <calorimeter name="EcalBarrSD"
                  hits_collection="EcalBarrHits"
                  ecut="0.0"
                  eunit="MeV"
                  verbose="4">
          <idspecref ref="EcalBarrelId" />
          <projective_cylinder ntheta="840"
                               nphi="1680" />
     </calorimeter>
</sensitive_detectors>
```

The calorimeter and tracker classes have some common features. The name attribute is used for referencing the <sensitive_detector> from other LCDD elements, such as the <volume>. The sensitive detector name must be unique. The hits_collection attribute sets the name of the hits collection that is produced and written to the LCIO file. The ecut (and the associated eunit) sets a cut on energy, below which hits will not be recorded. This allows low energy hits to be easily discarded. The verbose attribute turns on debugging output which controls how many messages from the sensitive detector are printed during the event. The <idspecref> gives the name of an <idspec> from the identifier dictionary.

The primary difference between the calorimeter and tracker types is that hits in the calorimeters are segmented into bins. Typical ILC calorimeters contain millions of cells per subdetector, and modeling these readout components as separate geometric objects would be prohibitive in terms of memory usage and simulation performance. So the layers in calorimeters are divided with virtual segmentation. An energy deposition in the calorimeter is added to the total for a single virtual cell. The energy contributions of each particle are tallied and stored into the hit should this information be needed later in reconstruction algorithms. A number of segmentation classes are provided for the end user.

GEANT4 regions are defined using <region> elements in the <regions> block. Regions are used to define the range cut, which determines the number of secondary particles produced. Trajectory storage can be tuned using the regions and physics limits may also be assigned to regions.

Some of the parameters controlling GEANT4's handling of tracks and steps can be tuned using physics limits, which are assignable to regions and volumes. These parameters are grouped into sets using the <limitset> element. Physics limits are referenced from logical volumes or regions using the <limitsetref> element.

```
<limits>
     <limitset name="TestLimitSet">
          <limit name="step_length_max"
                 particles="*"
                      value="1.0"
                 unit="mm"/>
          <limit name="track_length_max"
                 particles="p, pi+, pi-, p0"
                 value="100.0" unit="cm"/>
          <limit name="time_max"
                 particles="*"
                 value="100"
                 unit="ns"/>
          <limit name="ekin_min"
                 particles="*"
                 value="1.0"
                 unit="MeV"/>
          <limit name="range_min"
                 particles="e+,e-,gamma"
                 value="1.0"
                 unit="cm"/>
```

```
        </limitset>
    </limits>
```

The `step_length_max` sets the maximum distance of a single `G4Step`. Since SLIC uses virtual cells, the `step_length_max` can be used to minimize the occurrence of single steps crossing multiple cells. The `track_length_max` specifies the limit at which GEANT4 will kill a track. The `time_max` gives the maximum time a particle can exist after which it is killed. Tracks that are under the kinetic energy given by `ekin_min` will not be created. The `range_min` is duplicated by the region's range cut setting. Particles that will not go farther than the value in `range_min` will not be produced but the parent track's energy will be decremented by the energies of the un-produced secondary tracks (secondaries). The region's setting overrides the setting from the physics limits. Most of these settings can have adverse effects on the physics simulation. For instance, energy is not conserved when the tracks are killed. These settings are generally useful for "expert" users only but are still quite useful for tuning the simulation in certain volumes and regions.

Magnetic fields are defined in the `<fields>` block. The currently available types are simple solenoid, parameterized dipole, RZ field map and full field map.

The GDML `<volume>` is extended by LCDD to support additional reference tags. The GDML volume supports referencing of the solid and material. LCDD adds optional references to sensitive detector, region, limit set, and visualization elements (all previously described). The ref attribute gives the name of a previously defined element in the XML file, with the correct type.

```
    <volume name="EcalBarrelEnvelope">
        <materialref ref="Air"/>
        <solidref ref="EcalBarrelTube"/>
        <sdref ref="EcalSD"/>
        <regionref ref="EcalBarrelRegion"/>
        <limitsetref ref="EcalBarrelLimits"/>
        <visref ref="EcalBarrelLimits"/>
        <physvol>
            <volumeref ref="EcalLayer1"/>
            <physvolid name="layer" id="1"/>
        </physol>
    </volume>
```

This approach allows multiple volumes to reference the same elements, which is quite typical. For instance, each layer in a calorimeter needs to reference the sensitive detector. The other extension made to <volume> by LCDD is the addition of the `<physvolid>` that provides volume "numbering" in the context of the parent volume. Typically these values will correspond to important integer ids in the volume stack, such as layer number, sublayer number, subdetector number, and so forth. The full stack of physical volume identifiers is used to uniquely identify geometric elements using cell IDs, which are written into the hits in the LCIO file.

GEOMETRY COMMANDS

The LCDD geometry file is a mandatory input that must be specified by the user at runtime. Other important settings, such as the physics list and event generator, are given reasonable defaults. The geometry can be read from a URL or local file path. The LCDD files for standard detectors are stored at a URL of this form.

```
http://www.lcsim.org/detectors/[detector_name]/[detector_name].lcdd
```

For instance, the LCDD file for the sid00 detector may be read as follows.

```
slic -g http://www.lcsim.org/detectors/sid00/sid00.lcdd
```

SLIC also accepts a local file url (file://), relative file path, or absolute file path.

Since the LCDD system is an extension of GDML's processing infrastructure, SLIC has the capability of loading plain GDML files.

```
slic -g mygeom.gdml
```

Conversely, the plain GDML processor is able to load a valid GDML file from LCDD by simply ignoring the LCDD elements.

COMPACT DETECTOR DESCRIPTION

From the user's perspective, the LCDD format can be too verbose and complex for hand editing. Typically, researchers want to change a few parameters such as inner radius or layering scheme and have these changes automatically propagated to create the detailed full geometry. The Compact Detector Description, henceforth referred to as the "compact description", is a high-level format designed to facilitate this type of usage. The GeomConverter Java package converts compact descriptions to LCDD. It also supports the HepRep and GODL geometry formats and can create Java runtime objects representing the detector geometry for the org.lcsim reconstruction and analysis framework.

The compact format covers many of the same areas as LCDD, including meta-information, parameters defined by expressions and definitions, detector geometry, regions, and fields. The compact description also has an XML schema, much of which is similar in format to the LCDD schema. Each LCIO file contains a detector tag, such as "sidaug05", and the org.lcsim conditions system can locate and cache zip files containing the compact description and other geometry data sources, such as simple parameter files. The WIRED event display is given a HepRep geometry description produced from the compact description. LCDD files are produced from the compact description; then the LCDD file is used by SLIC. This is an example of the LCDD XML format being used as a "target" by an external application.

The following is a simple example of a compact description of a sampling Si/W calorimeter:

```
<detector id="2"
        name="EMBarrel"
        type="CylindricalBarrelCalorimeter"
        readout="EcalBarrHits">
    <dimensions inner_r="150.1*cm"
                outer_z="208.0*cm" />
        <layer repeat="20">
            <slice material="Tungsten"
                   thickness="0.25*cm" />
            <slice material="Silicon"
                   thickness="0.032*cm"
                   sensitive="yes" />
        </layer>
</detector>
```

EVENT GENERATION

SLIC has several methods for generating physics events. Built-in GEANT4 methods are available, such as the General Particle Source (GPS) and the particle gun. File-based sources are also supported. These are the StdHep binary format, supported by most physics event generators. Events can also be read from the MCParticle collection of an LCIO file.

LCIO OUTPUT BINDING

SLIC creates LCIO files from the simulated GEANT4 events. SimCalorimeterHits, SimTrackerHits, and MCParticles are the three classes comprising SLIC's LCIO binding. A collection of MCParticle objects represents the particles created by the event generator and propagated through the simulation. Typically, an LCIO file will have one collection of MCParticles per event. Each subdetector has a single collection of hits. Hits in the tracking devices are represented by collections of SimTrackerHits. Calorimeter hits are written out to collections of SimCalorimeterHits. LCDD includes a set of "generic" hit classes, so that the sensitive detectors may create hits collections. These intermediate objects are converted to the LCIO format by utility classes in the SLIC package, primarily the `LcioHitsCollectionBuilder`. During this processing stage, the MCParticle contributions to each SimCalorimeterHit may be combined according to user settings.

SOFTWARE DISTRIBUTION

The SLIC binary is built from 8 different software packages. Though these packages are primarily based on the GEANT4 build infrastructure, Xerces, LCIO, and GDML do not use directly the GEANT4 build system.

In order to simplify the distribution of the SLIC binary and manage the inter-package dependencies and different build systems, a build system based on Make and Autoconf is used to automate the tedious, manual configuration of each package. Instead, users need only checkout one CVS package to build SLIC.

```
> cvs -d :pserver:anonymous@cvs.freehep.org:/cvs/lcd co SimDist
> cd SimDist
> ./configure
> make
```

This system has been tested successfully on Linux, Windows, and OSX, and prebuilt binary executables are available for these platforms.

SUMMARY

A fully-featured detector simulation package using the GEANT4 toolkit and runtime detector geometry definition has been developed. This package is being extensively used for detector design optimization for the ILC[5]. The program can be used to simulate arbitrary detectors and is freely available for other experiments.

ACKNOWLEDGMENTS

Work supported by Department of Energy contract DE-AC03-76SF00515.

REFERENCES

1. http://www.lcsim.org/software/slic/
2. http://geant4.web.cern.ch/geant4/
3. http://lcio.desy.de/
4. http://gdml.web.cern.ch/GDML/
5. Graf, et al. "Intelligent Detector design", these proceedings

Extruded scintillator for the Calorimetry applications

A. Dyshkant, V. Rykalin

Northern Illinois Center for Accelerator and Detector Development (NICADD), Northern Illinois University, DeKalb, IL 60115, USA

A. Pla-Dalmau

Fermi Nationa Acceleratorl Laboratory, Batavia, IL 60510, USA

D. Beznosko

State University of New York at Stony Brook Stony Brook, NY 11790, USA

Abstract. An extrusion line has been installed and successfully operated at FNAL (Fermi National Accelerator Laboratory) in collaboration with NICADD (Northern Illinois Center for Accelerator and Detector Development). This new Facility will serve to further develop and improve extruded plastic scintillator. Recently progress has been made in producing co-extruded plastic scintillator, thus increasing the potential HEP applications of this Facility. The current R&D work with extruded and co-extruded plastic scintillator for a potential ALICE upgrade, the ILC calorimetry program and the MINERvA experiment show the attractiveness of the chosen strategy for future experiments and calorimetry. We extensively discuss extruded and co-extruded plastic scintillator in calorimetry in synergy with new Solid State Photomultipliers. The characteristics of extruded and co-extruded plastic scintillator will be presented here as well as results with non-traditional photo read-out.

Keywords: Extruded scintillator, co-extruded plastic scintillator, calorimeter, Solid State Photomultiplier
PACS: 01.10.Fv

CP867, *XII International Conference on Calorimetry in High Energy Physics,* edited by S. R. Magill and R. Yoshida

INTRODUCTION

The reason to choose extruded plastic scintillator among the different types of plastic scintillators for Calorimetry applications is based on the attractiveness of using a simple, inexpensive and robust extrusion procedure [1]. Furthermore, in some cases there is no alternative to the extrusion because of geometry requirements. For instance, if you need to build a detector with long strips of different shapes and plastic scintillator is your choice, you have to utilize an extrusion method. The emerging MINERVA experiment [2] intends to use a triangle-shaped extruded scintillator as an active target and calorimeter active media. The triangular shape of the strips and their length (approximately 4 m long) make the extrusion process an ideal option. In general, the scintillator choice for a given project is based on the compromise between light output, readout system and cost. Cast plastic scintillator may cost between \$40-60 per kg. In contrast, extruded scintillator that has significantly lower price may be used if a wavelength shifting (WLS) fiber readout system is used. With this approach, the MINOS (Main Injector Neutrino Oscillation Search) experiment was able to build two detectors that required almost 300 tons of finished extruded plastic scintillator [3]. The overall cost of the extruded plastic scintillator for MINOS was about \$10 per kg. The ALICE ECAL upgrade at the LHC needs 15 tons of scintillator. Preliminary R&D studies [4] have shown the advantages of using extruded plastic scintillator. In the frame of the ILC R&D program, NICADD proposed using extruded scintillating strips as the active medium for the tail-catcher and muon tracker (TCMT) [4]. NICADD also developed a readout system based on recent development with Solid State Photomultipliers (SSPM) [5]. Coupling with this semiconductor sensor comes in a natural way due to the co-extruded hole and small size of the photo-transducer [6].

In this paper, the basic concept of the extrusion will be illustrated and recent results from measurements testing the response to cosmic muons for different configurations of extruded and co-extruded scintillators will be provided. We hope that the experience accumulated will be useful for different applications in calorimetry and beyond.

EXTRUSION SETUP

The FNAL-NICADD extrusion line has been operating since April 2003. In 2005 it was upgraded with co-extrusion equipment that allows the placement of a TiO_2 white reflective coating on the scintillator strip. The central and vital part of the extrusion line is shown on Fig. 1. From right to left, one sees the 40-mm twin-screw extruder with a die at the end, the co-extruder at 45 degrees above the extruder, and the vacuum cooling tank.

The coating that is a blend of polystyrene and TiO_2 is being injected into the die by the co-extruder. From that die, the melt is directed to the sizing tooling located in the vacuum cooling tank. The polystyrene pellets that are the base for the extruded scintillator are a commercial grade, general-purpose polystyrene. The dopants currently in use are 1% PPO and 0.03 % POPOP, available in a pre-mixed form. This is a blue-emitting scintillator with an absorption cut-off at approximately 400 nm and an emission maximum at 420 nm [7].

FIGURE 1. Main part of the extrusion setup. From right to left: the 40-mm twin screw extruder with a die at the end, the co-extruder at 45 degrees above the extruder, and the vacuum cooling tank.

Extrusion Shapes We Produce

The existing dies produce scintillator strips of different shapes, some with holes for a wavelength shifting (WLS) fiber and some without holes for routing the top surface as needed [7]. The co-extruded hole or groove enables the insertion of the WLS fiber and saves additional expenses. Fig. 2 displays the scintillator profiles with co-extruded holes and white TiO_2 coating and other profiles without the white capstocking. The triangular and rectangular-shaped extruded scintillators were chosen by MINERvA as the basic configuration to build the active target and calorimetric part of the detector, respectively.

The scintillator shape depends on the needs of the experiment and its physics demands. For the TCMT prototype, the extruded scintillator strips were produced in the small amount needed for the experiment [8]. Fig. 3 shows the measurements performed on the scintillator production (nominal size: 100 mm wide by 5 mm thick). The dimensions had an average width of 101.33 ± 0.17 mm evaluated over 300 m with 50 cm interval, and an average thickness of 4.98 ± 0.03 mm evaluated over 300 m with 20 cm interval.

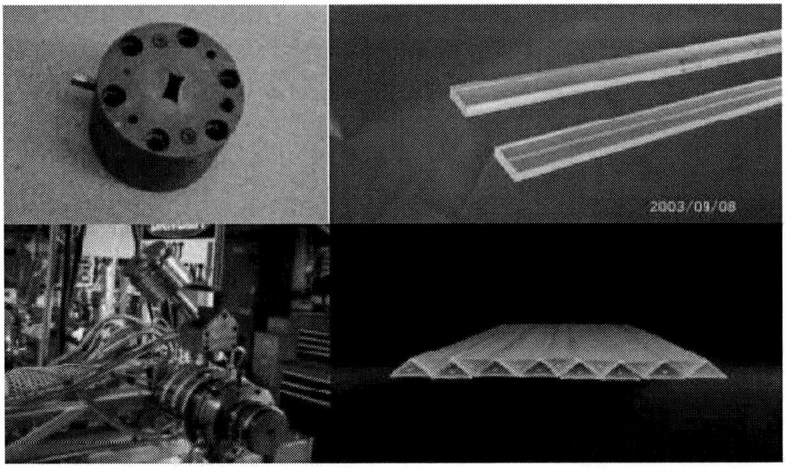

FIGURE 2. Different profiles of scintillator extrusion and dies used to make this particular shape.

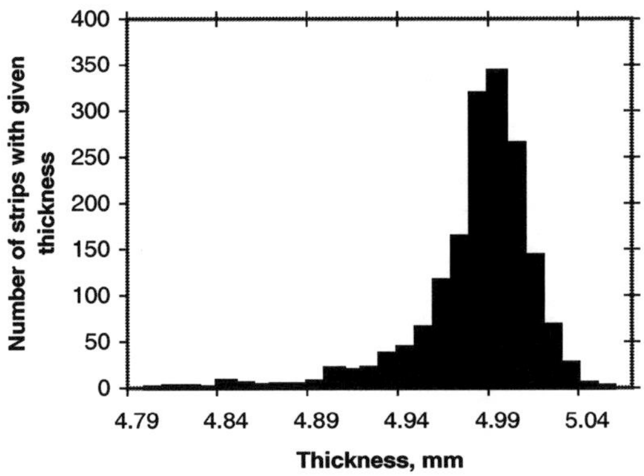

FIGURE 3. Distribution of the strip size with a given thickness.

Experimental results

Attenuation Length Measurements

Measurements of the Light Attenuation Length (LAL) [6] show that the value of the long and short components is 42.15±1.14 cm and 30.52±0.74 cm, respectively. The extruded scintillator samples were wrapped in Tyvek since they did not have the co-extruded reflective coating. In order to evaluate the impact of the TiO_2 coating, a comparative study was performed using 1 m long triangular strips. The results are shown on Fig. 4.

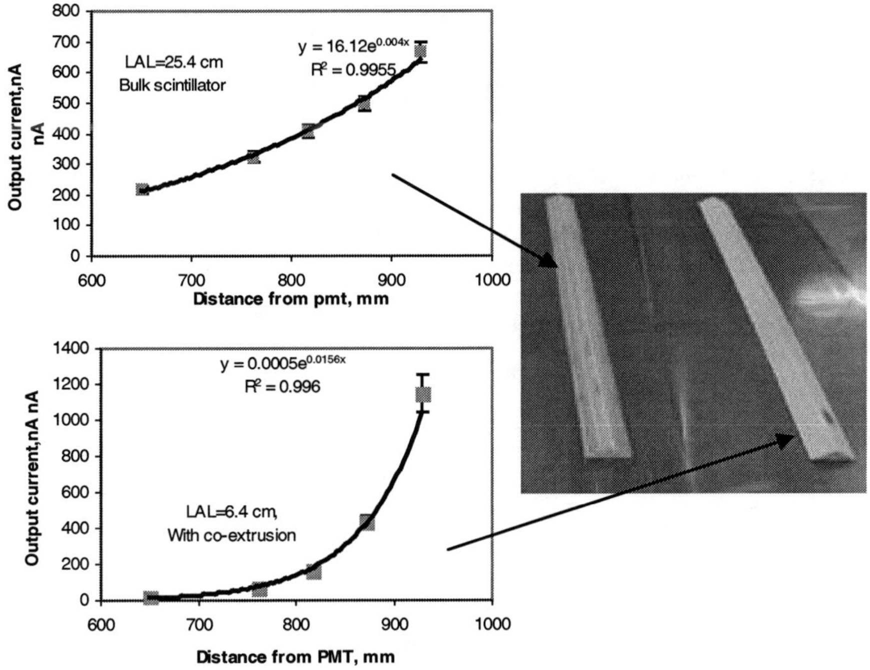

FIGURE 4. Impact of TiO_2 co-extrusion on Light Attenuation Length.

It was observed in this set of measurements that the value of the LAL for the extruded scintillator strips (without co-extruded TiO2 coating) may exceed as much as four times the same value for the TiO2 co-extruded scintillator strips. The origin of this phenomenon is based on the total internal reflection occurring in the extruded scintillator wrapped in Tyvek that is missing in the TiO_2 co-extruded scintillator due to the diffusive character imparted by the coating. It limits a variety of application of this type of scintillator to the use of WLS (Wavelength Shifting) fiber where the LAL is greater than 2 m.

Scintillation Light Yield

As it was shown in [7], the light yield of our extruded scintillator is of the same value as of Kuraray [9] SCSN-81. The light yield was measured from the extruded tiles using cosmic rays. A peripheral sigma-groove was machined in each tile, where Kuraray Y11 (200ppm) MS WLS fiber with diameter of 0.83 mm was inserted. The absolute light was estimated using previous measurements of single photoelectron spectra for the Hamamatsu [10] R6094 Photomultiplier Tube (PMT). The tiles produced from two different extruded scintillators averaged to 8 photoelectrons per MIP with 5% uncertainty on the measurements.

In order to measure if the size of extruded hole would affect the light yield of the scintillator strips, we have performed measurements on a sample with a variety of hole sizes. Regardless the hole size, we inserted the same 1.2 mm Kuraray Y11 WLS fiber and measured the resultant light yield while using a ^{137}Cs gamma source. Fig. 5 (top) shows the approximate ratios of the hole size to fiber diameter, with 0.9 being a closest fit in sample used, and 0.3 being very loose. A total of 25 strips were measured. The same figure (bottom) indicates that a better match (0.7-0.9 ratios) does not usually render a higher light yield. The "far Light Yield" and the "near Light Yield" refer to the measurements done in the position farthest from and nearest to the PMT respectively. This measurement illustrates that the hole size does not have a significant impact on the light yield.

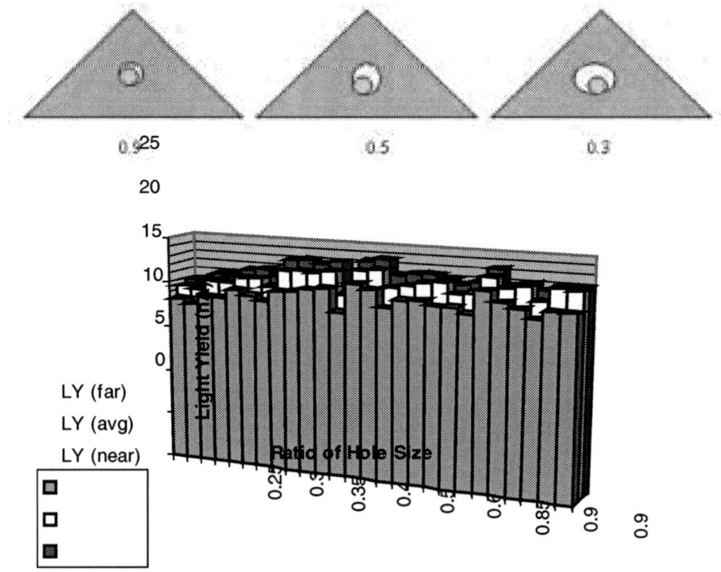

FIGURE 5. Effect on Light Yield due to the Hole Sizes

Several measurements using a Solid State Photomultiplier (SSPM) were also performed. This is a multi-pixel solid-state device with every pixel operating in the Limited Geiger multiplication mode. Avalanche quenching is achieved by means of a resistive layer on the sensor surface. The device has about 1500 pixels per 1x1 mm^2 sensor [5]. The quantum efficiency (QE) of the device is approximately 20% at 500 nm: a value that is comparable to the QE of a conventional green-extended PMT.

The dynamic range of the sensor fits many applications where the linearity issue is of importance. Deviation from linearity at the level of 5% was observed [5] at approximately 2200 photons of input signal and deviation from linearity at the level of 10% was seen with light intensity of about 3200 photons, as shown in Fig. 6.

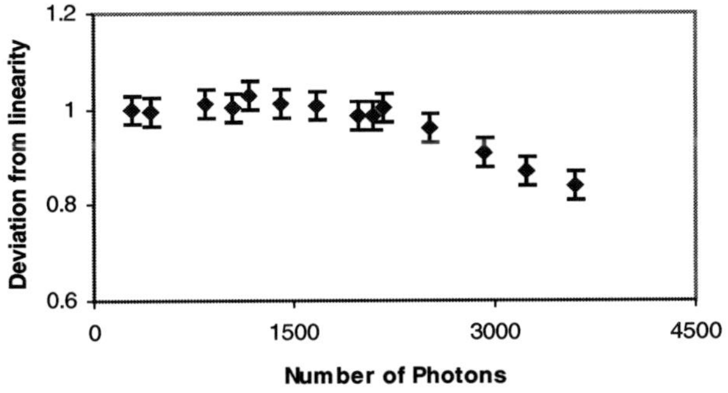

FIGURE 6. Dynamic range of SSPM

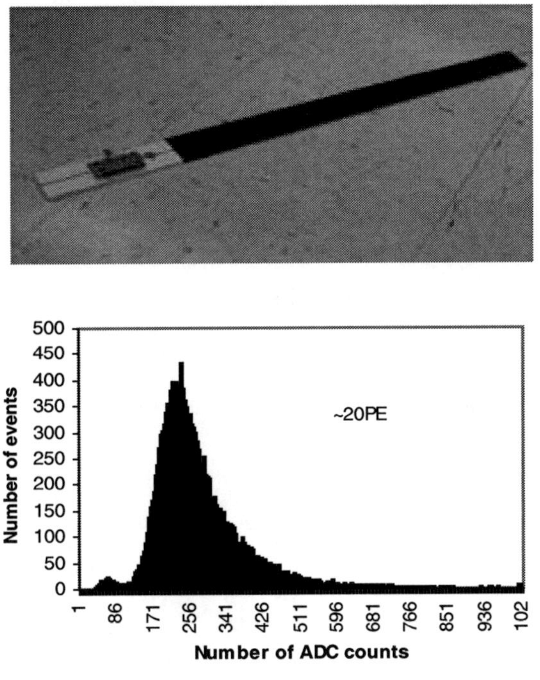

FIGURE 7. Cosmic counter assembly on the top and amplitude spectra from cosmic muons

The coupling of the SSPM and the WLS fiber comes in natural way [6]. Several scintillator counters for a cosmic-ray test stand were built with SSPM photo-readout. For the prototype a MINOS strip was utilized with a WLS fiber in the co-extruded groove. The SSPM sensor and biasing circuit was placed on the same strip for portability (Fig. 7, top). The cosmic-ray signal spectrum is shown on Fig. 7 (bottom). The pedestal is slightly smeared due to the geometry of the trigger. The detector has a perfect single photoelectron (PE) distribution that was measured before the amplitude distribution was accumulated for calibration purpose. The estimated number of photelectrons for this spectrum is about 20 PE. Fig. 7 also indicates that the signal is well separated from the pedestal. Since SSPM is insensitive to strong magnetic fields [11], it is an attractive option for detectors where a strong field is an issue.

CONCLUSION

The FNAL-NICADD Extrusion Line is producing a variety of scintillator shapes and sizes, including co-extruded hole and TiO_2 coating. The scintillator bars with co-extruded capstocking can be used in conjunction with WLS fibers in calorimeters that require long and inexpensive strips, like the MINOS and MINERvA experiments. The use of SSPM with the extruded scintillator yields attractive results for their combined use in future experiments.

ACKNOWLEDGMENTS

The authors would like to thank Chuck Serritella and Elena Baldina for general support and assistance during the R&D extrusion work. We also would like to thank Daniel Ruggiero for providing the Y11 fibers.

REFERENCES

1. D. Beznosko, A. Bross, A. Dyshkant, A. Pla-Dalmau, V. Rykalin, "Production Using FNAL-NICADD Extruder", FERMILAB-PUB-05-344, Jul 29, 2005
2. D. Drakoulakos et al. [the MINERvA collaboration], "Proposal to Perform a High-Statistics Neutrino Scattering Experiment Using a Fine-grained Detector in the NuMI Beam". http://Minerva.fnal.gov
3. MINOS Press Releases, http://www.fnal.gov/pub/presspass/press_releases/minos_3-30-06.html
4. A. Dyshkant, D. Beznosko, G. Blazey, D. Chakraborty, K. Francis, D. Kubik et al., "Towards a Scintillator-Based Digital Hadron Calorimeter for the Linear Collider Detector", IEEE vol. 51, no. 4, pp.1590-1595, Aug. 2004.
5. D. Beznosko, G. Blazey, D. Chakraborty, A. Dyshkant, K. Francis, D. Kubik et al., "Investigation of a Solid State Photodetector", NIM A 545 (2005) 727-737
6. D. Beznosko, P. Polozov, V. Rykalin, G. Sellberg, "Fiber Positioning and MRS Response ", FERMILAB-TM-2247, May 7, 04
7. D. Beznosko, A. Bross, A. Dyshkant, A. Pla-Dalmau V. Rykalin, "Production Using FNAL-NICADD Extruder", FERMILAB-PUB-05-344, Jul 29, 2005
8. A. Dyshkant, E. Baldina, D. Beznosko, G. Blazey, A. Bross, D. Chakraborty et al., "Studies of NICADD Extruded Scintillator Strips", FERMILAB-CONF-05-243, June 20, 2005
9. Kuraray America Inc., 200 Park Ave, NY 10166,USA; 3-1-6, NIHONBASHI, CHUO-KU, TOKYO 103-8254, JAPAN
10. Hamamatsu Corporation, 360 Foothill Road, PO Box 6910, Bridgewater, NJ 08807-0919,USA
11. D. Beznosko, G. Blazey, A. Dyshkant, V. Rykalin, J. Schellpffer, V. Zutshi, "Modular Design for Narrow Scintillating Cells with MRS Photodiodes in Strong Magnetic Field for ILC Detector", accepted to NIM A, Apr 16 2006, NIMA-D-05-00459

SESSION XI

CALORIMETRY AT THE ILC—PFAs

Convener: Jerry Blazey

Intelligent Detector Design

Norman Graf[1], Steve Magill[2], Steve Kuhlmann[2], Ron Cassell[1], Tony Johnson[1], Jeremy McCormick[1]

[1]*Stanford Linear Accelerator Laboratory, Menlo Park, CA 94025*
[2]*Argonne National Laboratory, Argonne, IL 60439*

Abstract. At a future e+e- linear collider, precision measurements of jets will be required in order to understand physics at and beyond the electroweak scale. Calorimetry will be used with other detectors in an optimal way to reconstruct particle 4-vectors with unprecedented precision. This Particle Flow Algorithm (PFA) approach is seen as the best way to achieve particle mass resolutions from dijet measurements in the range of ~30%/√E, resulting in innovative methods for choosing the calorimeter technology and optimizing the detector design.

Keywords: Calorimeter, Particle Flow, Linear Collider
PACS: 07.20.Fw, 29.40.Vj, 07.05.Kf

PARTICLE RECONSTRUCTION GOALS

At the future e^+e^- international linear collider (ILC), precision measurements of jets of particles will be required in order to fully explore and understand the (possibly subtle) effects of the new physics expected at and beyond the electroweak mass scale. A precision measurement of the 4-vector associated with a quark or gluon initiating a jet is best done by fully reconstructing the individual particles produced in the e^+e^- annihilation process. This requires the development of so-called Particle Flow Algorithms (PFAs) which, in combination with an optimized detector, can reconstruct these particles. It is the calorimeter, which must be optimized for the PFA approach, that is the key element which makes possible a precise measurement of both energy and momentum of the particles. For calorimetry, this means that high segmentation and granularity leading to individual particle shower reconstruction is preferred over energy compensation; the loss of good calorimetric energy resolution is more than offset by the gain achieved using the charged particle momenta from tracking.

In this report, the effect of PFA development on calorimeter design for the ILC will be discussed along with a description and results of a particular implementation of the PFA approach for different calorimeter models.

CALORIMETRY REQUIREMENTS FOR PARTICLE FLOW

In the PFA approach to individual particle reconstruction, the goal is to use the best measure of a particular particle's properties – this means that all charged particle momenta are taken from tracking measurements, photon energy is detected and

CP867, *XII International Conference on Calorimetry in High Energy Physics,*
edited by S. R. Magill and R. Yoshida

measured in the electromagnetic calorimeter (ECAL), and neutral hadrons are measured in the ECAL and the hadron calorimeter (HCAL). For this to work, the most important part of the PFA is separation of charged hadron showers from neutral hadron showers in the calorimeter. Therefore, it is 3-D shower reconstruction performance in the calorimeter that drives the calorimeter technology choice and design. A dense absorber is required, especially at the beginning of the calorimeter, in order to best separate the shower starting points of electromagnetic particles from hadrons. As the ratio of hadronic interaction length to radiation length (λ_I/X_0) grows, the starting points of hadron showers are pushed farther into the calorimeter on average, thus providing good longitudinal separation between e.g., photons and neutrons. In a dense calorimeter, the Molière radius (r_M) is small, and to obtain good transverse shower separation, cells should be of the order of r_M in lateral size. A typical choice for the ECAL at the ILC is a 30 layer sandwich of tungsten absorber (~2/3 X_0 per layer) with silicon readout cells of size ~3.5 mm x 3.5 mm [1].

In the HCAL, complete separation of charged and neutral hadron showers requires that high granularity and segmentation be continued. The size of an HCAL cell should be optimized for single particle occupancy of any individual cell, thus allowing high purity shower reconstruction. For most particles, digital readout of the HCAL cells is more than adequate for both shower reconstruction and precise energy determination. Figure 1 shows a comparison of traditional analog calorimetry with a digital method for a 5 GeV π^+. The analog measurement displays Landau broadening of the distribution due to the characteristic minimum ionizing particle (MIP) response and path length effects – both of these are absent in the digital measurement, resulting in improved resolution.

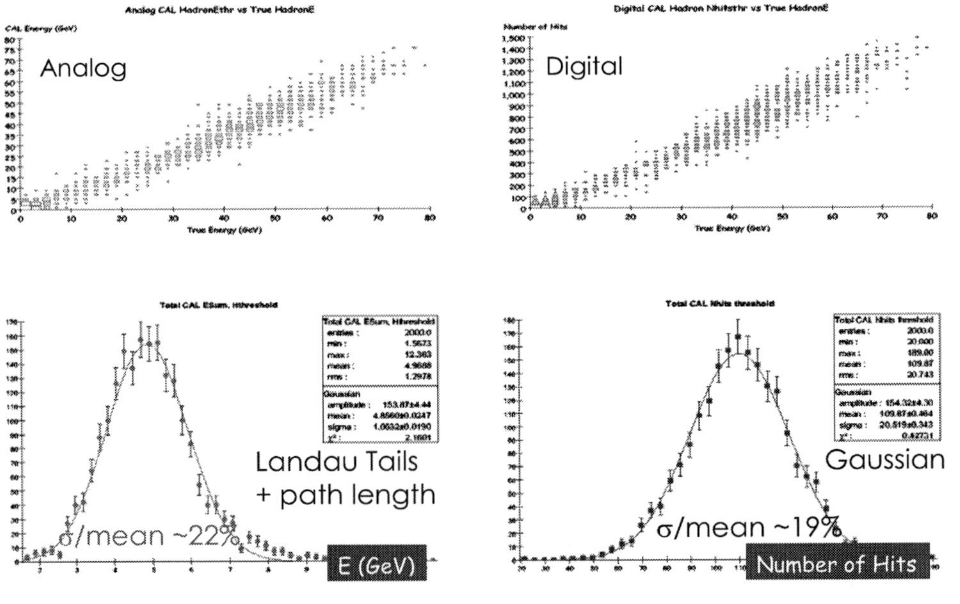

FIGURE 1. (left) Analog calorimeter linearity of response (top) and energy resolution (bottom) compared to (right) digital calorimeter linearity of response (top) and energy resolution (bottom).

Active media choices for the sampling calorimeters under investigation are traditional (but small) scintillator tiles with SiPM readout [2] and technologies employing gas (RPC [3] and GEM [4]). Scintillator is a good neutron detector due to the possibility of detecting elastically scattered protons from incident neutrons as opposed to the gaseous detectors which are somewhat insensitive to neutrals. However, the smaller, better-separated neutral showers in a gaseous detector may enhance the performance of the PFA, more than offsetting the poor neutral energy resolution response of gas. Since the whole calorimeter should be inside a strong solenoidal magnetic field, use of a dense absorber here can keep the magnet bore (and therefore cost) as small as possible.

The PFA approach is used to optimize both jet reconstruction and to optimize the calorimeter design for particle reconstruction. This requires a flexible, fast, full detector simulation with an investigation of many different detector models and options, including variations of detector parameters such as inner detector radius. A complete simulation package has been developed with this in mind [5], and has been used to "build" more than a dozen detector models, each with many single parameter variations. Physics processes at 500 GeV center-of-mass energies and single particle simulations are all included with each detector design variation. Also, for initial development of PFAs, $e^+e^- \rightarrow qq$ events at the Z mass are generated; these are very useful since PFA performance can be evaluated with simple energy sums and no jet algorithm interference. These analyses also provide a point of reference for the LEP and SLC detectors.

Also, since separation of showers in the calorimeter is crucial to PFA performance, the simulation of hadronic showers is of critical importance. Timely test-beam results are needed to validate the hadron shower simulations and verify the feasibility of the PFA approach to detector optimization.

INDIVIDUAL PARTICLE RECONSTRUCTION STRATEGY

The approximate energy content of all jets in e^+e^- annihilation is: 60% charged hadrons, 25% photons, 10% neutral hadrons, and 5% missing energy (neutrinos) as shown in Figure 2 (left). Of course, there are large fluctuations in these quantities per event. In Figure 2 (right), the energy distributions of the particles are shown.

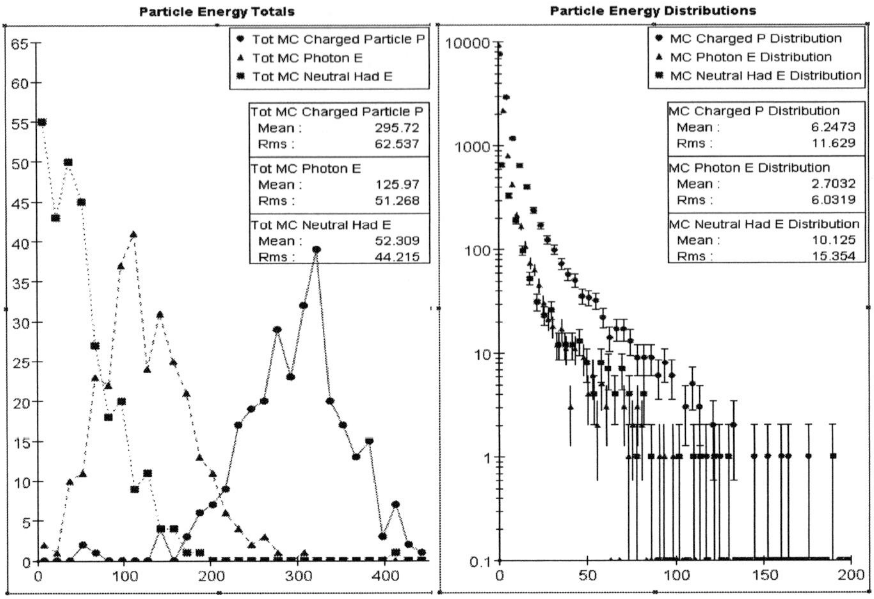

FIGURE 2. (left) Total particle energies and (right) particle energy distributions for typical events in e^+e^- collisions at 500 GeV CM.

Starting with the largest energy fraction, tracks from charged particles are extrapolated into the calorimeter, associating MIP hits until the interaction layer is found – simply characterized as a layer with either no hits or many hits in a small window around the extrapolated track. The calorimeter hits in the resulting MIP "clusters" associated with tracks are removed from further evaluation. Typically, this step results in single-cell roads removed primarily from the ECAL. Next, cells in the ECAL are clustered with a suitable cluster algorithm (cone clusterer, directed tree [6], nearest-neighbor, etc.). All clusters are then tested with an H-Matrix algorithm which forms a chi-squared probability for each cluster based on longitudinal energy correlations which has been trained on real photon clusters. Clusters which pass this test are designated as photons and saved for input to the jet algorithm. The calorimeter hits from photon clusters are then removed from further evaluation. In principle, what is left in the calorimeter now are hit cells from neutral hadrons and the hits of charged hadrons without their MIP tails in both the ECAL and HCAL. These hits are now clustered with a nearest-neighbor cluster algorithm. Continuing the extrapolation of tracks from the end of their respective MIP clusters, the nearest-neighbor cluster fragments are associated with the track if within a small iterated cone. The cone size is gradually increased, adding the energy of cluster fragments until a tuned value of E/p, the ratio of summed calorimeter cluster energy divided by the track's momentum, is reached, typically the value is ~1-2σ_{had}. All hits from clusters associated with tracks are removed from further evaluation. What is now left is due to neutral hadrons plus, of course, any mistakes made in all of the previous algorithms. Additional clustering can be done here, as well as other evaluations of the hits to define the neutral hadron clusters, which are then saved to be evaluated in a jet algorithm along with the photons and tracks. This completes the PFA. A picture illustrating the performance of this algorithm is shown in Figure 3.

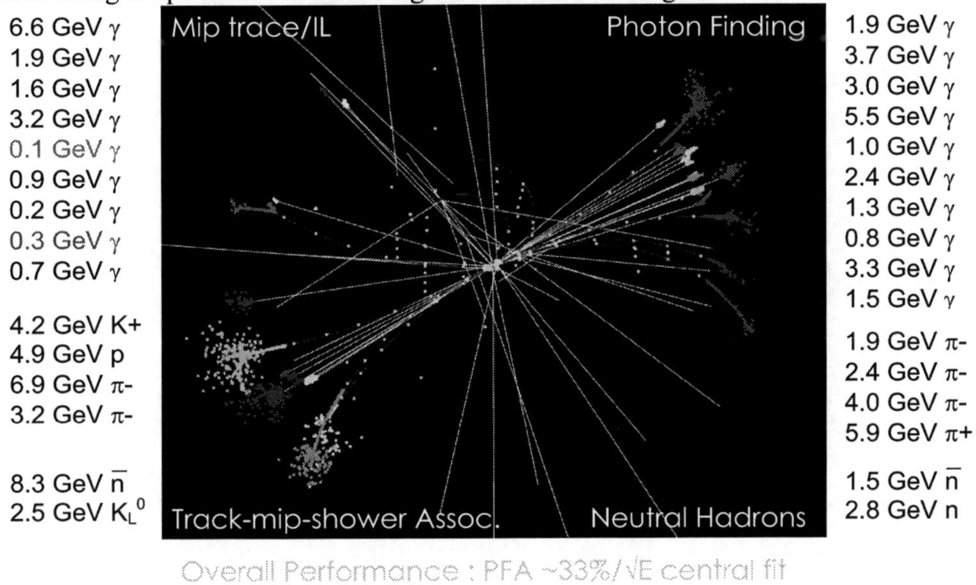

FIGURE 3. Illustration of PFA performance for an e+e- → Z → qq event at 91 GeV.

In this example, all particle showers have been correctly associated with the exception of a 0.1 GeV photon – these photons rarely pass the minimum number of hits requirement for testing with the H-Matrix. A 2-gaussian fit of the response of this PFA at the Z-Pole yields a central peak fit result corresponding to 33%/√E. Also, for Z-Pole events, the "confusion" term – the contribution to the resolution due to algorithm mistakes – is smaller than the contribution from the neutral hadron measurement in the calorimeter.

RESULTS FOR DIFFERENT DETECTOR MODELS

Once the PFA performs at a level in which the calorimeter neutral hadron resolution term is the largest contribution to the total resolution, the algorithm can be used to test different detector configurations. It is assumed that the PFA approach will be sensitive to, among other things, the value of the solenoidal magnetic field and the radius of the calorimeter for a given calorimeter technology.

Magnetic Field Variation

The B-field was varied from its default value of 5 Tesla to 4 Tesla for the same detector design. In this detector, the ECAL was of W/Si sandwich construction, consisting of 30 layers of 2/3 X_0 each with an inner radius of the barrel ECAL of 125 cm. The HCAL was a Stainless Steel/RPC sandwich consisting of 34 layers of 0.12 λ_I each. Figure 4 show the results of this comparison.

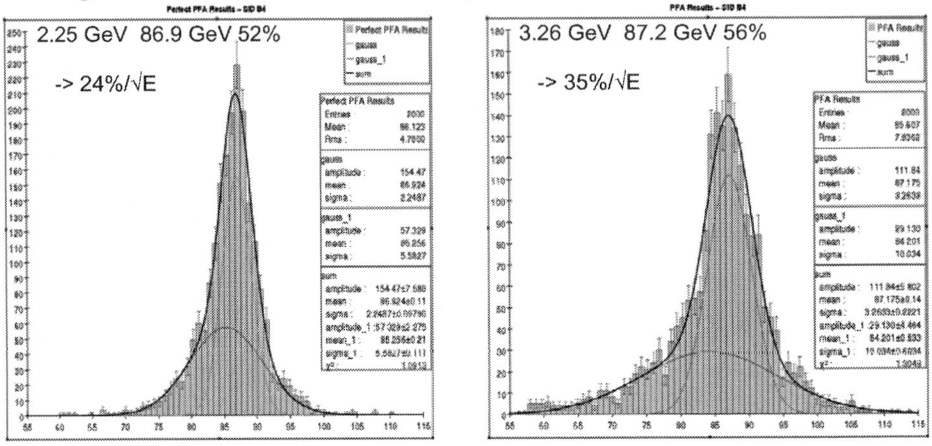

FIGURE 4. (left) PFA result for Z-Pole events in 5 Tesla field detector and (right) in a 4 Tesla detector.

The PFA distribution for the higher B-field model has better resolution than that for the lower field as expected. Not only is the resolution better, but as a consequence, the mistakes made in the PFA are smaller for the high B-field model than for the lower. As expected, increasing the B-field makes the PFA easier to perform and results in a better overall energy resolution.

Calorimeter Radius Variation

Another variation that should affect the PFA result is the distance from the interaction point (IP) to the calorimeter. A detector model was constructed with the inner radius of the barrel ECAL increased from its default value of 125 cm to 150 cm. A further optimization was made in the HCAL, changing from a Stainless Steel/RPC HCAL to one made of W/Scintillator layers. Since tungsten is denser than stainless steel, the similar depth (~4 λ_I) HCAL was ~22 cm thinner than the default one, resulting in the radius of the solenoid coils to only be increased by 1 inch. So, this detector has moved the calorimeter showers farther from the IP without increasing the magnet bore and the cost of the solenoid. Figure 5 shows the results of the PFA comparison of these models.

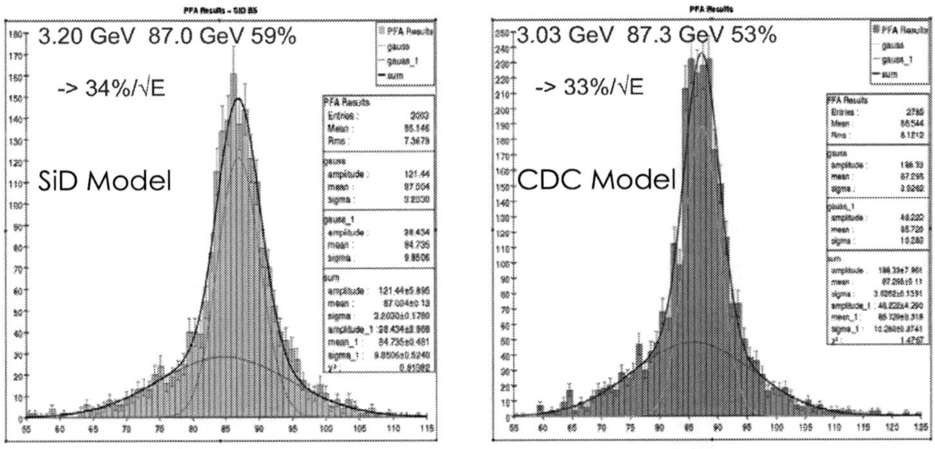

FIGURE 5. (left) PFA result for 125 cm radius ECAL and SS/RPC HCAL (SiD Model) compared to (right) the PFA result for 150 cm ECAL with W/Scintillator HCAL (CDC Model).

The results are similar for these detectors, with a slightly better result for the detector with larger ECAL radius, as expected. This illustrates how PFAs can be used to test different detector models, resulting in an optimized detector configuration which can then be prototyped and placed in a test beam to verify shower characteristics.

PFA IMPLEMENTATION

The PFA approach of reconstructing individual particles in an optimized detector lends itself to modular organization of the reconstruction algorithms. A PFA template has been developed to allow substitution of various clustering and analysis algorithms at each stage of particle reconstruction, thus resulting in faster optimization of a particular set of cluster and analysis codes. It also provides a way of testing individual reconstruction techniques as they are applied. This approach is seen as the best way to combine the efforts of the worldwide community working on a PFA approach to detector design and optimization.

SUMMARY

The PFA approach, combining full simulation with individual particle reconstruction, is a unique approach to detector – and in particular, calorimeter design. This is seen as the best way to optimize the detector configuration at the ILC for precise jet reconstruction. A particular implementation of cluster and analysis algorithms in a full PFA package has been developed and tested on fully-simulated e^+e^- data. The results are approaching the hoped-for goal of ~30%/√E with a non-dominant "confusion" term. First tests of this algorithm on several calorimeter configurations have produced results illustrating how PFAs can be used to optimize the ILC detector.

ACKNOWLEDGMENTS

This work is supported in part by the University of Chicago as Operator of Argonne National Laboratory under Contract No. W-31-109-ENG-38 and in part by Stanford University as Operator of the Stanford Linear Accelerator Center under Contract No. DE-AC03-76SF00515, both with the U.S. Department of Energy.

REFERENCES

1. N. Graf, Proceedings of the XII International Conference on Calorimetry in High Energy Physics (2006).
2. H-G. Moser, Proceedings of the XII International Conference on Calorimetry in High Energy Physics (2006); E. Garutti, Proceedings of the XII International Conference on Calorimetry in High Energy Physics (2006).
3. J. Repond, Proceedings of the XII International Conference on Calorimetry in High Energy Physics (2006).
4. A. White, talk at XII International Conference on Calorimetry in High Energy Physics (2006).
5. N. Graf, Proceedings of the XII International Conference on Calorimetry in High Energy Physics (2006).
6. G. Lima, Proceedings of the XII International Conference on Calorimetry in High Energy Physics (2006).

Development of a Particle Flow Algorithm at Argonne

Lei Xia

Argonne National Laboratory

Abstract. The physics program for the future International Linear Collider (ILC) requires excellent jet energy resolution, which is believed to be obtainable by using particle flow algorithms (PFA) together with a highly granular calorimeter system. In this context we report on our efforts to develop a particle flow algorithm that can demonstrate the required jet energy resolution and be used to optimize the design of an ILC detector. The effort involves development of clustering, photon finding, track cluster association and fraction identification algorithms. We will discuss the current status of the algorithms and our plans for the future.

Keywords: ILC, International Linear Collider, PFA, Particle Flow Algorithm, SiD
PACS: 07.05.Kf

INTRODUCTION

The future International Linear Collider (ILC) will address exciting new physics and perform precision measurements. Many of the interesting physics processes require jet energy and/or jet mass measurements with unprecedented resolution ($\sim 30\%/\sqrt{E}$) to unambiguously determine the hadronic final states. Such a revolution is believed to be achievable by using Particle Flow Algorithms (PFA).

Particle Flow Algorithms (PFAs) achieve the best possible jet energy resolution by measuring the energy of each individual particle in a hadronic jet. In PFAs, the momenta of charged particles are measured with the tracking system, the energy of photons is measured with the EM calorimeter and the energy of neutral hadrons is measured with the combination of EM and hadron calorimeters. The momenta of charged particles (energy of photons) can be measured with high (reasonable) precision. On the other hand, the measurement of neutral hadrons is plagued by poor energy resolution. However, since neutral hadrons carry only 10% to 20% of the energy of a jet, an excellent jet energy resolution can still be achieved, provided the energy deposits of charged and neutral particles in the calorimeter can be identified as such.

Under the assumption that the energy deposits of individual particles can be clearly identified, a so-called *perfect PFA*, which uses the MC information to separate individual particles perfectly, achieves an energy resolution of ~2.3 GeV for di-jet events ($e^+e^- \rightarrow$ qqbar, q = u,d,s) generated at the Z-pole (E_{cm} = 91.0 GeV) and using the SiDaug05_np detector model[1]. The result is shown in Figure 1. However, within PFAs the major contribution to the energy resolution originates from the wrong association of calorimeter deposits to charged and neutral particles. Therefore, when

CP867, *XII International Conference on Calorimetry in High Energy Physics,*
edited by S. R. Magill and R. Yoshida

developing PFAs, the major task is to achieve the best identification of calorimeter energies originating from charged or neutral particles.

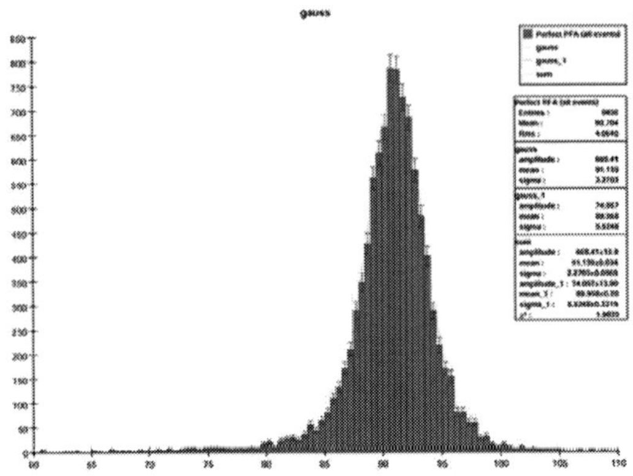

FIGURE 1. Event energy calculated from perfect PFA, for SiDaug05_np detector and Z-pole → di-jet events

PFA COMPONENTS

The PFA described in this paper starts from calorimeter hits and uses a density driven clustering algorithm to group hits into clusters. Clusters originating from photon showers are identified first, by a photon findcr. Then MC charged tracks are extrapolated into the calorimeter and matched with clusters. Clusters matched to tracks are identified as originating from charged particles. In order to identify clusters originating from neutral hadrons, the un-matched clusters are further evaluated according to some geometrical variables. After these steps, the momenta of tracks, energies of photons and neutral hadrons are summed up to provide an estimate of the event energy. So far this algorithm has only been applied to di-jet events at the Z-pole. Therefore, no jet algorithm has been implemented at the moment. The main ingredients of this PFA are described here.

Sampling Fraction Calculation

This is a standalone step, independent of the PFA. The determination of the sampling fractions uses single particle MC samples of the SiDaug05_np detector model. The MC samples include the simulation of single electrons, photons, pions, kaons, neutrons and anti-neutrons originating at the IP and impinging on the calorimeter at $\theta = 90^{\circ}$. The particle energies range from 2 GeV up to 50 GeV. The sampling fraction for a given particle and at a specific energy is calculated by fitting the slope of the 2D distribution of ECal raw energy vs. the HCal number of hits. Since the calorimeters show a linear response for EM showers, the sampling fractions obtained at different particle energies can be averaged and the energy of an EM shower is calculated as,

$$E_{EM}(GeV) = 82.5 \times E_{ECal,raw} + 0.11 \times N_{HCal,hit} \tag{1}$$

The calorimeter response for hadrons is not perfectly linear, leading to an energy dependent sampling fraction. Taking this into account, the energy of a hadron shower is calculated as,

$$E_{hadron}(GeV) = \frac{114. \times E_{ECal,raw} + 0.122 \times N_{HCal,hit}}{1 + 1.34 \times E_{ECal,raw} + 0.00122 \times N_{HCal,hit}} \tag{2}$$

The calorimeter response for different hadron species differs slightly. The sampling fraction shown here is calculated from the kaon sample which assumes a medium value among hadrons. It will be applied to all hadrons in later studies.

Dr. Ron Cassell pointed out an angular dependence of the calorimeter response[2]. In this PFA the angular dependence has been corrected according to the impact position of the particle onto the calorimeter.

MC Particle List, Track List and Perfect PFA Calculation

A list of MC particles is selected to represent all detectable particles in an event while minimizing double counting. In order to calculate the energy resolution with perfect particle separation (perfect PFA), all calorimeter hits are assigned to these MC particles according to their contributing particle. Using appropriate sampling fractions, the particle energy is then calculated from calorimeter hits assigned to each particle. From the MC particle list, those that passed more than 3 layers of the silicon tracker are further selected to form a list to represent measured tracks. In the perfect PFA, the event energy is calculated by summing up the momenta of all tracks and the energies of photons and neutral hadrons as measured by the calorimeter. The result is shown in Figure 1. This represents the theoretical limit for the measurement of the event energy using PFAs and a given detector model. The distribution is fitted with the sum of two Gaussian distributions. The narrow Gaussian contains 80% of the events and has a width of 2.3 GeV, which is significantly smaller than the goal for the jet energy resolution (~3 GeV at Z^0 pole)

Clustering Algorithm

All calorimeter hits are grouped into clusters to represent individual particle showers or fractions of a shower in the calorimeters. The clustering algorithm is based on a hit density calculation. The density of a particular hit 'i' is defined as the sum of contributions from each hit in a group of hits {j}:

$$D_i = \sum_{\{j \neq i\}} \left(e^{-(|\vec{V}_1 \bullet \vec{R}_{ij}|/|\vec{V}_1|^2)} \times e^{-(|\vec{V}_2 \bullet \vec{R}_{ij}|/|\vec{V}_2|^2)} \times e^{-(|\vec{V}_3 \bullet \vec{R}_{ij}|/|\vec{V}_3|^2)} \right) \quad (3)$$

in which, V_1, V_2 and V_3 are linearly independent vectors that start at cell i and stop at its nearest neighboring cells in the same layer (V_1, V_2) and adjacent layers (V_3). R_{ij} is a vector that starts at cell i and stops at cell j. With this density definition the distance between cell i and j is normalized to the local cell density of cell i. To construct a cluster, the clustering algorithm first searches for a cluster seed which is defined as a hit that is not in any existing cluster and has the highest density when {j} runs over all hits in the calorimeters. Neighboring hits of the found seed are then attached to the seed to form a small cluster. A hit, which is not a direct neighbor to the seed cell, is also attached to this cluster if the cell density of that hit, when {j} runs over all hits already in the cluster, is over a given threshold. The clustering for this seed is complete when no more hits can be attached to the cluster. The clustering algorithm then searches for the next seed, until all hits are uniquely assigned to a cluster.

After clustering all calorimeter hits in a Z-pole → di-jet event, 90% of the clusters have hits from exactly one MC particle. Half of the remaining 10% of the clusters have most of their hits from one MC particle and only one or two hits from other particles. The other half of the clusters are merged showers from different MC particles. On average, there will be 1.2 clusters per Z-pole event that contain merged showers. Only when a cluster contains both charged and neutral particles, will it cause difficulties resulting in double counting energy (when the cluster is later identified as a neutral particle) or losing energy (when the cluster is identified as a charged particle).

Photon Identification

After finding all clusters, a photon identification algorithm identifies clusters originating from EM showers by comparing their longitudinal shower profile to the expected shower profile of a photon. The comparison is done with the longitudinal H-matrix as implemented in the org.lcsim framework[3]. The expected shower profile is calculated from a MC sample of single photons with energies from 1 to 10 GeV. A cut is applied on the common logarithm of the probability of a χ^2 calculated from the diagonal elements of the H-matrix for a given cluster:

$$\log(probability(\chi^2_{diagonal})) > -10 \qquad (4)$$

Clusters satisfying this cut are labeled as photons, all others are labeled as hadrons. The efficiency of a real photon passing this cut is shown in table 1.

TABLE 1. Photon identification efficiency

Energy(GeV)	**0.25**	**0.50**	**0.75**	**1.**	**2.**	**5.**	**10.**	**20.**	**50.**	**100.**
Efficiency(%)	43.9	89.7	95.7	96.8	98.4	98.3	96.6	93.9	83.6	52.5

This algorithm can identify >95% of the photon clusters, however, it also mis-identifies a good fraction of the neutral (20-30%) and charged (10%) hadrons as photons. Identifying a neutral hadron cluster as a photon cluster only affects the sampling fraction to be used for that cluster, which results in a small difference in cluster energy. However, if a charged hadron cluster is identified as a photon, the energy of the cluster will be double counted. A better photon identification algorithm is being developed to reduce the misidentified hadron component. As a temporary solution, all identified photon clusters are still subject to track cluster matching algorithm in order to remove charged hadron clusters.

Track Cluster Matching

The trajectory of MC particles identified as charged tracks is extrapolated into the calorimeter assuming a helix. The energy lost along the path of the particle is not yet considered. The extrapolation is stopped when the particle either has traversed the entire calorimeters or starts to curl back. For a given track, the distance between the extrapolated trajectory and each cluster is calculated. The distance is defined as the closest distance between any point on the trajectory and any hit in the cluster. If the distance is smaller than a given threshold, the track and the cluster are matched. All clusters that are matched to a track are label as originating from charged particles.

As pointed out earlier, all identified photon clusters are subject to track cluster matching. The matched photon clusters are relabeled as charged hadrons.

Most of the clusters that match a track are from charged particles. At this step, only 3% of the energy of matched clusters is from neutral hadrons or photons. However, the remaining un-matched hadron clusters have 55% of their energy coming from fragments of charged particles, and only 45% from neutral particles. This shows that a fragment identification algorithm is necessary.

Fragment Identification

Those hadron clusters that don't match any tracks mainly come from neutral hadrons and detached fragments of charged hadrons. Some geometrical variables are used to evaluate these clusters in order to identify the two components. The variables are a) the distance between a cluster and its nearest neighboring track, b) the distance between a cluster and its nearest neighboring 'neutral hadron cluster' and c) the ratio of the two distances. There is no fool proof way to know which cluster originates from a neutral hadron. However, it has been shown that large clusters which don't match any tracks are very likely to be associated to neutral hadrons. So the 'neutral hadron cluster' mentioned above refers to any large enough cluster that doesn't match any track.

For Z-pole → di-jet events, the fragment identification algorithm removes 70% of the energy of charged fragments from clusters not matched to any track, while only 12% of the energy from neural hadrons is misidentified as charged fragments. After this step, clusters labeled as neutral clusters have 70% of their energy coming from neutral hadrons.

PFA Energy Sum

After performing all the above algorithms, a list of MC particles has been selected to represent the measured charged tracks. All the calorimeter hits are grouped into clusters, and each cluster has been labeled as one of the following: photon, charged hadron, charged hadron fragments and neutral hadrons. The final event energy is calculated by summing up track momenta, photon cluster energy and neutral hadron cluster energy.

PFA PERFORMANCE FOR Z^0 → 2 JETS

The particle flow algorithms described above has been used with a MC data sample of $e^+e^- \to Z^0 \to qq$ (q = u, d, s) events. This event sample was generated at a center of mass energy of 91.0 GeV and simulated with SLIC[4] using the SiDaug05_np detector model. After summing up the track momenta, photon energy and neutral hadron energy, the event energy for the Z-pole sample is shown in Figure 2. There is no cut or event selection applied to this MC data sample. The distribution is fitted with the sum of two Gaussian distributions. The narrow Gaussian contains 59% of the events and has a width of 3.4 GeV, and the wider Gaussian contains 41% of the events and has a width of 10.2 GeV. Figure 3 shows the event energy distribution for barrel events which are about 60% of all events. The barrel events are selected with a single cut on the underlying quark direction so that the two jets in these events mainly go to the barrel region of the detector. Using a two-Gaussian fit, the distribution is well fitted by the sum of a narrow Gaussian (3.2 GeV, 59%) and a wide Gaussian (10.0 GeV, 41%). This PFA performs significantly better for barrel events since the barrel cut excluded events which have significant energy leakage into the holes around the beam pipe.

Compared with the perfect PFA (Figure 1), which separates the individual particles in an event perfectly, this PFA is still dominated by misidentifications between charged and neutral particles. The prominent tail on the low energy side is mainly due to photons and neutral hadrons that are misidentified as charged hadrons. Further improvements are needed to reduce this tail.

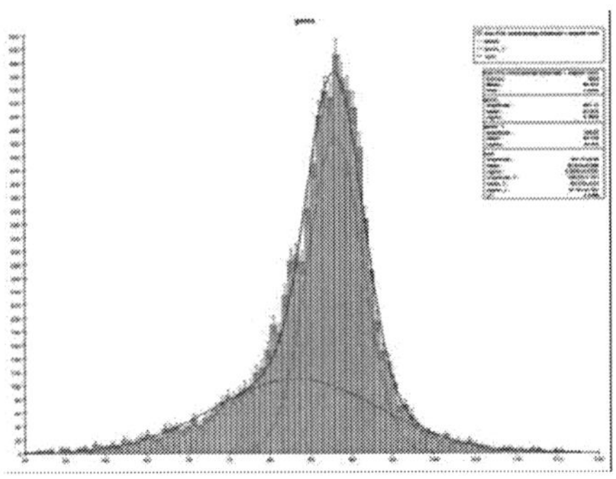

FIGURE 2. Event energy calculated from PFA algorithms, for SiDaug05_np detector and Z-pole → di-jet events. No event selection is applied.

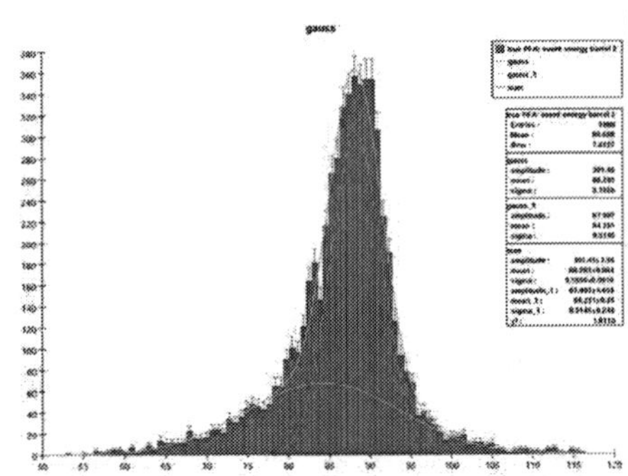

FIGURE 3. Event energy calculated from PFA algorithms, for SiDaug05_np detector and Z-pole → di-jet events. A cut on the quark direction is applied to select barrel events.

FUTURE PLAN

Further developments of this PFA are foreseen in the following areas:

Common PFA Template

The code will be re-written according to a common PFA template[5], which was proposed at the ALCPG Boulder simulation workshop[6]. All the PFA ingredients will be converted into modules with well defined interfaces, to facilitate the exchange and crosscheck of individual routines.

Photon Identification

The photon identification algorithm in this PFA is not optimized yet. It currently only uses a single cut on one variable based on the H-matrix calculation to distinguish photons and hadrons. To maintain a good efficiency for photons, the current algorithm still misidentifies too many hadrons as photons. More variables will be used to reduce the misidentified hadron component. After the photon finding algorithm is optimized, all identified photons will no longer be subject to the track cluster matching algorithm.

Jet Energy Resolution

The current PFA gives a promising jet energy resolution for Z-pole → di-jet events. However, the ILC goal of $\sim 30\% / \sqrt{E}$ has not been achieved yet. There is a prominent low energy tail on the event energy distribution which will affect the jet energy resolution. More studies will be done to optimize the current PFA and improve its performance.

The jet energy resolution will be studied in greater detail with events containing larger numbers of jets and at higher center-of-mass energies. In particular, the energy scale dependence of the PFA performance still needs to be established. New ideas, as they become available, will also be implemented to continuously improve the performance.

Our goal is to develop a PFA that can achieve the required jet energy resolution and use the PFA to improve the detector design.

ACKNOWLEDGMENTS

Work supported by the U.S. Department of Energy, Office of Science, under contract W-31-109-ENG-38.

REFERENCES

1. For details of SiDaug05_np detector model, see https://confluence.slac.stanford.edu/display/ilc/sidaug05_np
2. http://nicadd.niu.edu/cdsagenda//askArchive.php?base=agenda&categ=a0589&id=a0589s1t0/transparencies
3. For details, see https://confluence.slac.stanford.edu/display/ilc/org.lcsim
4. SLIC is a GEANT4 simulator, for details see https://confluence.slac.stanford.edu/display/ilc/SLIC
5. For details, see https://confluence.slac.stanford.edu/display/ilc/lcsim+PFA+guide
6. ALCPG Detector Simulation Workshop, http://www-hep.colorado.edu/cuilc/sim-workshop.html

Particle Flow Performance And Detector Optimization

Predrag Krstonosic

DESY , Notkestrasse 85 , 22607 Hamburg , Germany

Abstract. Different contributions to the resolution of the jet energy as reconstructed following the particle flow approach are being discussed. The achievable performance is studied for a variety of physics processes and compared with that of purely calorimetric methods, where the calibration procedure is found to play a crucial role. *The status of reconstruction software is discussed, and first steps towards a detector optimization using fully simulated and reconstructed events are presented.*

Keywords: particlc flow, calibration, simulation
PACS: 07.20.Fw

Introduction

Physic based requirements for jet energy resolution of $30\%/\sqrt{E}$ for the ILC detector is well established [1]. Initial detector design was made on the base of general requirements - to have the best possible subdetectors (if affordable). All the detector consepts designs were made with intention to be suitable for particle flow algorithm (PFA) type of reconstruction. Performance estimates were made with different level of idealization - perfect particle flow (PPF). In order to estimate real performance of the detector or to optimize it we can make a two side approach from the final reconstruction performance and from calorimeter performance estimate.

Particle Flow Performance

Initial estimate of the reconstruction performance in terms of jet energy resolution was made with the highest level of idealization (taking into account only subdetector resolutions). Initial formulae takes into account resolutions for measurement of single particles (photons, charged particles, neutral hadrons) that are combined with ther mean event abundance, plus the confusion term that absorbs all the unperfectness of the detector and reconstruction :

$$\sigma_{jet}^2 = w_{ph}\sigma_{ph}^2 + w_{had}\sigma_{had}^2 + w_{ch}\sigma_{ch}^2 + \sigma_{conf}^2 \quad (1)$$

where w_i are weights for different kinds of particles ($w_{ph} = 0.26$ for photons, $w_{had} = 0.10$ for neutral hadrons and $w_{ch} = 0.62$ for charged particles), σ_i are measurement resolutions for given type of particle, and σ_{conf} is the confusion term. Resolution of the tracking system is much better then the one of calorimeters and thus could be neglected $\sigma_{ch} = 0$. For the initial consideration the confusion term can be also set to zero, leaving

CP867, *XII International Conference on Calorimetry in High Energy Physics,*
edited by S. R. Magill and R. Yoshida

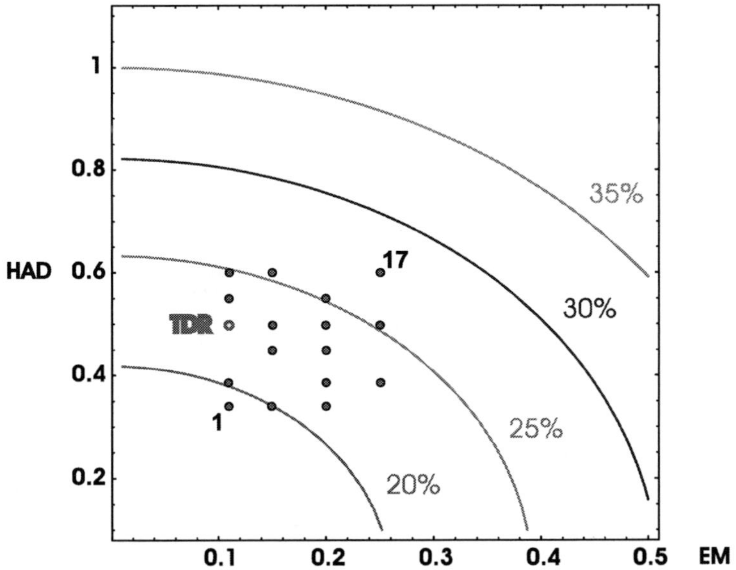

FIGURE 1. Jet energy resolution according to equation (1) in dependence of calorimeter resolutions. Points in te plot represent enetries in Table 1.

TABLE 1. Jet energy resolution according to equation (1) and after energy smearing for each particle for W^+W^- events at 500GeV (in brackets) in dependence of calorimeter resolutions.

$HAD[\%]\backslash EM[\%]$	11	15	20	25
34	12.12 (12.94)	13.19 (13.96)	14.82 (15.50)	
38	13.26 (14.29)		15.76 (16.58)	17.52 (18.21)
45		16.15 (17.08)	17.51 (18.39)	
50	16.77 (17.97)	17.56 (18.76)	18.81 (19.77)	20.31 (21.21)
55	18.27 (19.70)		20.16 (21.55)	
60	19.79 (21.42)	20.46 (21.83)		22.86 (24.06)

only contributions from electromagnetic and hadronic calorimeters. In the figure(1) the results for jet energy resolution were shown in dependence of the calorimeter resolutions (with constant term neglected). It shows that for wide range of detector performances it seam that design goal is within reach leaving large room for choice of the detector components. In order to test the predictions of this simplified approach W^+W^- events at 500GeV were hadronized [2] and single particle energies were smeared according to the detector resolutions. At the end total energy of event was collected and compared with the initial one.

Due to the nongaussian nature of obtained distributions rms was taken as a measure of the spread. In Table 1. it is possible to see that there is rather good agreement with prediction and a spread within 10% around the number obtained with formulae (1).

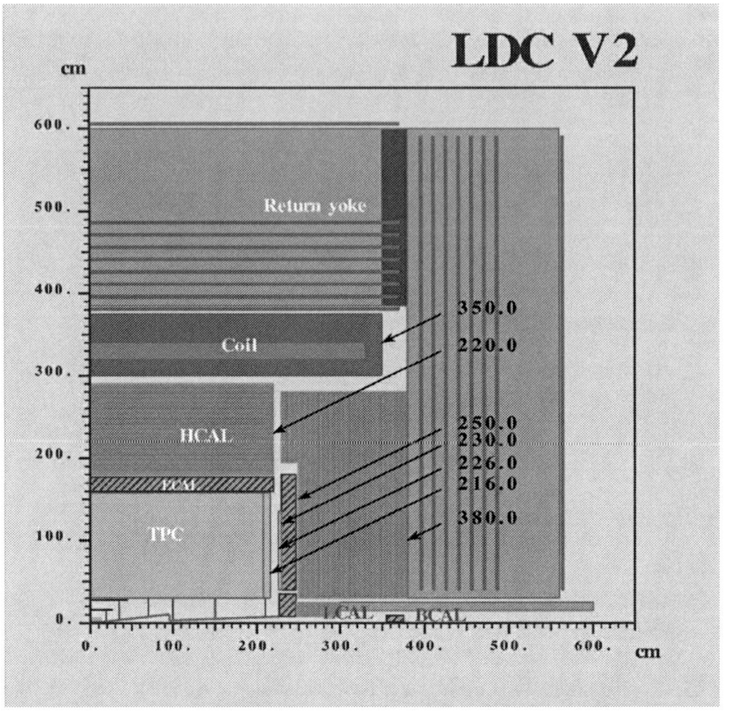

FIGURE 2. Picture to fixed height

After initial excersise we will look into the subject on a bit deaper level. Final jet energy resolution is a convolution of several effects:

$$\sigma_{E\,jet} = \sigma_{PPU} \oplus \sigma_{JFU} \oplus \sigma_{DG} \oplus \sigma_{PFA} \tag{2}$$

where PPU stands for pure physical uncertaintites that is composed of

$$\sigma_{PPU} = \sigma_{\Gamma} \oplus \sigma_{Elumi} \oplus \sigma_{ISR} \oplus \sigma_{E\nu} \tag{3}$$

particle widths, luminosity curve, initial state radiation and energy lost to the neutrinos. DG is detector geometry term which consists of

$$\sigma_{DG} = \sigma_{toBeamTube} \oplus \sigma_{DeadZones} \tag{4}$$

and JFU stands for jet finder uncertainties. Result of the equation (2) will depend on the particular physical process , quality of accelerator (in particular beam spot size and crossing angle), detector geometry and jet finder algorithm applied. But the quality of the PFA has nothing to do with mentioned contributions since it should be only dependent on the sub-detector resolutions. For the PFA quality estimate it is necessary to split the resolution into mentioned independent terms or to remove them froma analysis. Goodness of the Particle Flow or it's comparison is possible only after such splitting.

TABLE 2. Contributions to the width of reconstructed mass distribution from different sources for W^+W^- events at 500GeV. Column joined means that PFA and non PFA contributions were summed sequentialy. Total means that all the contributions were summed. In last colum partial contribution of each effect to the total width is expressed in percent.

Effect	$\sigma[GeV]$ *separate*	$\sigma[GeV]$ *joined*	$\sigma[GeV]$ *total* $(\%/\sqrt{E})$	σ $\%$ *of total*
E_ν	0.70	0.70	0.70 (3.1)	1.50
$\theta^\circ < 5$	2.73	2.82	2.82 (12.6)	22.8
$P_t < 0.36GeV$	1.36	3.13	3.13 (14.0)	5.65
σ_{HCAL}	4.10	4.10	5.16 (23.1)	51.4
σ_{ECAL}	2.17	4.64	5.60 (25.0)	14.4
$M_{neutral}$	1.02	4.75	5.69 (25.4)	3.18
$M_{charged}$	0.60	4.79	5.72 (25.6)	1.10

TABLE 3. Summary of differences between two majos design version for LDC detector consept.

Effect	Z_{pole}	Z	W^+W^-	$t\bar{t}$
$\sigma_{detector}$	1.76	2.76	3.12	3.01
σ_{PPF}	1.63	3.94	4.79	4.38
σ_{total}	2.40	4.81	5.72	5.31
$\sigma\%/\sqrt{E}$	25.1	21.5	25.6	23.7

In order to performe such an analysis full mass of the event was chosen as an estimator since it eliminates use of jet algorithm. Events were generated without ISR and with fixed center of mass energy, thus simplifying equation (2) to :

$$\sigma_{E\,jet} = \sigma_{E\nu} \oplus \sigma_{toBeamTube} \oplus \sigma_{DeadZones} \oplus \sigma_{PFA} \tag{5}$$

Following assumptions were made for detector subcomponents and it's geometry:

- ECAL resolution $12\%/\sqrt{E}$
- HCAL resolution $50\%/\sqrt{E} \oplus 4\%$
- beam tube region up to $\theta = 5^\circ$
- minimal transverse mometum to reach TPC $P_t^{min} = 0.36GeV$
- exact mass assignment for hadrons $M_{charged} = M_{pi^+}$ and $M_{neutral} = M_{KL}$

Fully hadronic WW events at 500GeV were generated and hadronized by PYTHIA [2], and the final particles were first rejected if $\theta < \theta_{BeamTube}$, transverse momentum is $P_t < P_t^{min}$ or they are neutrinos.

Energy of remaining particles were then smeared according to the type, photons with ECAL resolution, neutral hadrons with HCAL resolution and the momentum of charged ones with TPC resolution. Finally mass was assigned according to the above rule. Mass for the event was calculated and the differrence with the initial one was put

TABLE 4. Summary of differences between two majos design version for LDC detector consept.

Detector version		LDC
Tag	00	01
Ecal R_{inner} [cm]	160	170
TPC half Z [cm]	200	250
ECAL		
$N \times W[mm]$	30×1.4	20×2.1
$N \times W[mm]$	10×4.2	10×4.2

into the histogram that was fitted with gaussian distribution. Procedure was repeated in steps to isolate contributions of each effect. Final result is in the table 2. In Table 3. PFA contributions (calorimeter resolutions and mass assignments) versus the non PFA contributions (neutrinos and detector acceptance) are sumarized for different processes. Process specific variations of PFA component contribution to the total width are visible but overall performance is still well within design goal.

Detector and Calibration

LDC [1] detector design hase two major versions with tags 00 and 01 in full Geant4 simulation (Mokka [2]). Calorimeters are tungsten silicon design of the Ecal and stainless stell sicintilator for the HCAL. Most important differencees are sumarized in Table 4. (different structure of ECAL and overall sizes).

In the LDC detector there are three structures with different sampling fractions. It is possible to determine sampling coefficients from the muon simulation for each sampling structure.

$$c_i = E^i{}_{visible}/E^i{}_{whole} \tag{6}$$

where E_{whole} is total lost energy over structure , and $E_{visible}$ is energy deposited in the active part of sampling calorimeter. Simple sum over the calorimeter with proper calibration should give us

$$c_1 E_{vis1} + c_2 E_{vis2} + c_3 E_{vis3} = E_{CM} \tag{7}$$

as you can see from the picture 3. taking into account only sampling fractions is not suficcient. Correlations were conserved but additional tranformation (rotation) is needed to fulfill energy conservation. These rotated coefficients contian in them selves properties of calorimeter as a whole as well as flavour containment of the jets plus angular and magnetic field dependencies. The black line equation is

---

[1] `http://www.ilcldc.org/`
[2] `http://polywww.in2p3.fr:8081/MOKKA`

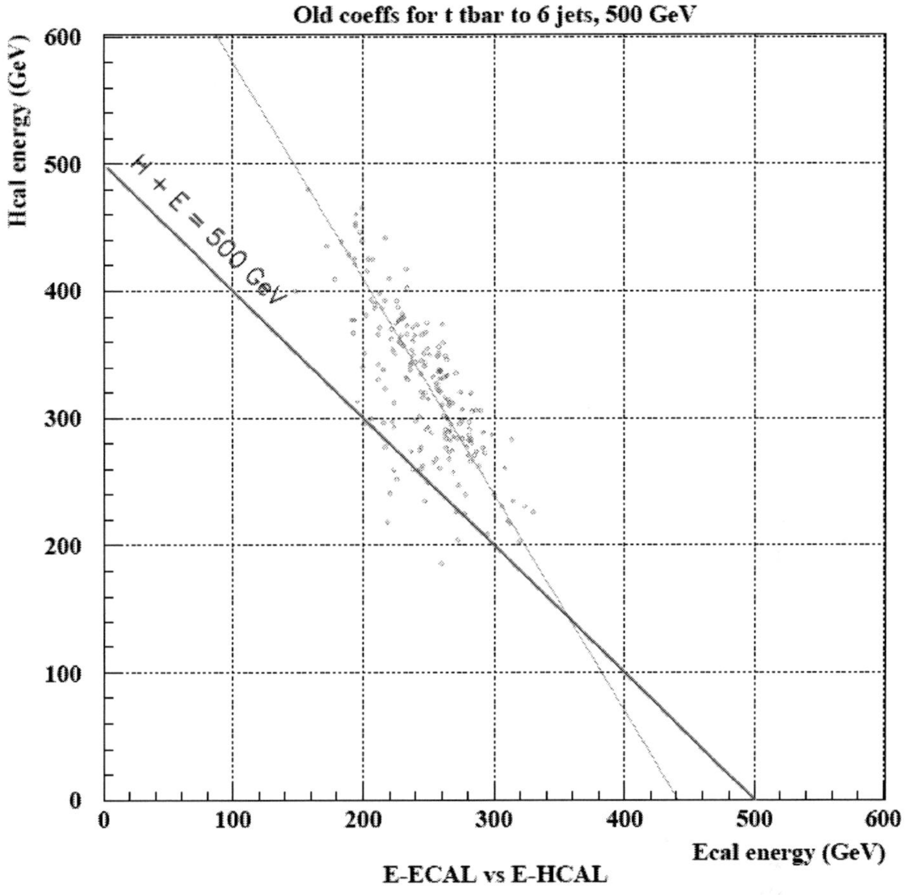

FIGURE 3. Picture to fixed height

$$a_0 E_{Ecal} + E_{Hcal} = a_0(c_1 E_{vis1} + c_2 E_{vis2}) + c_3 E_{vis3} = E_0 \tag{8}$$

where c_1, c_2 *and* c_3 are initial energy conversion coefficients, a_0 is the slope wich gives us the minimala energy width. E_0 is some constant - the line should come through the most probable value of initial energy sum. Let us require $E_0 = E_{CM}$ *and* $a_0 = 1$, then we get new coefficients $c_1{}^{calib} = f a_0 c_1$ where $f = E_{CM}/E_0$ and

$$c_1{}^{calib} E_{vis1} + c_2{}^{calib} E_{vis2} + c_3{}^{calib} H_{vis3} = E_{CM} \tag{9}$$

along the most probable line. Further on new calibration coefficients will be applied on to each hit in the particular sampling region of the calorimeter. Events were generated without luminosity curve and without ISR so the whole sumo of energy in HEP record is

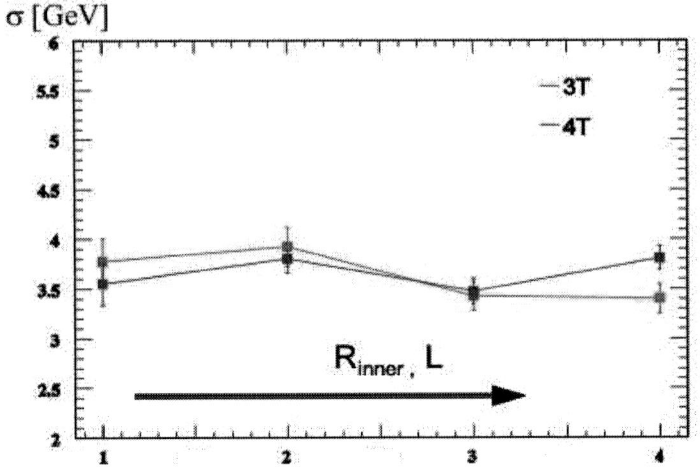

FIGURE 4. Reconstructed width of hadronic Zpole events in dependence of detector configurations and magnetic field. Detector setups are enumerated in the following way : 1. LDC01 -20-20 ,2. LDC01 ,3. LDC00 and 4. LDC00 +20+20.

equal to the center of mass energy exactly. To calculate available energy for calorimeters one should subtract the neutrino energies as well as the energy of particles that are lost due to the acceptance and also the part of energy in muons when they pass through the calorimeter leaving about 1.6GeV per muon. Estimated energy to be measured by the calorimeters for event is then

$$E_{visible} = E_{CM} - E_{neutrinos} - E_{acceptance} - E_{muon} + N_{muons} \times 1.6GeV \qquad (10)$$

Using the reference energy it is possible to make the chek plots, distributions of total calorimeter energy after calibration minus visible energy, and confirm that procedure works in full energy range and without offset (for details see [3]). After calibration check next step in analysis was done full reconstruction of the events using Marlin based reconstruction software.

Optimization

The calibration was proper and the it was time to see are there any dependencies of the full calorimeter responce to the detector dimensions and magnetic field changes. Since two major LDC detector variations are of already significantly different sizes only two additional steps in size were made +20+20 that means +20cm in radius of TPC and +20cm in it's length to the largest detector , and -20-20 on the smallest one. For these four sizes simulation was done in two magnetic fields 3 and 4 Tesla.

Minor differences that could be seen in the table are explainable by second order effects and thus lead to conclusion that optimization on the bases of the calorimeter

TABLE 5. Estimation of the event mass width using different reconstruction approaches.

	PPF	Implementation of PFA	Calorimeter Sum
$e^+e^- \rightarrow$		Sigma [GeV]	
Z pole 91.2GeV	2.4	3.4	4.67
$t\bar{t}$ $500GeV$	5.3	?	12.6
W^+W^- 500GeV	5.7	?	16.9
$q\bar{q}$ $500GeV$	4.8	?	14.8

performance is not possible. Because of this attempt was made to use at the moment available reconstruction software. The process of choice was Z to hadrons at 91.2GeV . As you can see in picture 4. there is practicaly no dependence on the sizes and magnetic fied. Such behaviour is rather understandable due to the low occupation of the detector for this process. On higer center of mass energies performances of reconstruction software was unsatisfying (worser then the calorimeter resolution) so the results were not considered.

Conclusion

At this stage in developement of detector desing and reconstruction algorithms software performance turned to be the limiting factor in search for optimal detector design. But knowlege collected so far puts the strict requirements on the performance. In Table 5. is the example of such requirements. Reconstruction software should be able to fill the gap between the Prefict Particle Flow in the left column and the calorimete responce in the right column. Only such kind of reconstruction with apropriate choice of performance estimators would lead to the correct detector optimization.

REFERENCES

1. Tesla TDR , part IV, 2001, pp. 160–162.
2. T. Sjöstrand, P. Eden, C. Friberg, L. Lonnblad, G. Miu, S. Mrenna and E. Norrbin,PYTHIA V6.221,Computer Physics Commun. 135 (2001), pp. 238
3. V. Morgunov, Calorimeter energy calibration using the energy conservation law, *LCWS06*.

Particle flow reconstruction based on the directed tree clustering algorithm

D. Chakraborty, J.G.R. Lima*, R. McIntosh and V. Zutshi

Northern Illinois Center for Accelerator and Detector Development (NICADD)
Physics Dept., Northern Illinois University, Dekalb, IL 60115

Abstract. We present the status of particle flow algorithm development at Northern Illinois University. A key element in our approach is the calorimeter-based *directed tree* clustering algorithm. We have attempted to identify and tackle the essential challenges and analyze the effect of several different approaches to the reconstruction of jet energies and the Z-boson mass. A number of possibilities have been studied, such as analog vs. digital energy measurement, hit density-based clustering and the use of single or multiple energy thresholds. We plan to use this PFA-based reconstruction to compare some of the proposed detector technologies and geometries.

Keywords: Linear collider, ILC, calorimetry, particle flow, PFA, jet energy resolution
PACS: 07.05.Kf

INTRODUCTION

Particle Flow Algorithms (PFA) have been considered a promising path to realize the requirements posed for the jet energy resolution at the International Linear Collider (ILC). The idea behind PFA is to use the (magnetized) inner tracker to measure the charged particle momenta and the calorimeter for the neutral particle energies in a jet. To do this, one needs to identify the charged and neutral particle showers in the calorimeter, so the charged particles are not counted twice. The challenging PFA goal is a jet energy resolution of about $30\%/\sqrt{E}$ at $\sqrt{s} = M_z$, for good separation of W and Z bosons. This goal requires the tuning of geometrical detector parameters for optimal interplay with the PFA algorithms.

While the momentum measurement of charged tracks is performed with excellent resolution using the tracker system, it is still important to correctly identify and remove their calorimeter showers in order to maximize the energy measurement of photons and neutral hadrons. A high-granularity calorimeter is essential for the proper separation of clusters, but the cost will scale with the number of channels. Sophisticated clustering and jet finding algorithms should play a role in finding the optimal balance between physics performance and costs.

In this paper we describe the status of PFA developments at NICADD/NIU. For this study, we used a sample of $Z \to q\bar{q}$ events at $\sqrt{s}$ = 91 GeV, discarding events with both parton jets in the endcaps ($\theta < 45^{\circ}$ or $\theta > 135^{\circ}$). The data was processed through SLIC[1] (v2r0p0, based on Geant4 v8.0.1[2]), and saved in the LCIO format. The PFA software development is done within the (Java-based) org.lcsim[3] analysis framework.

The detector model used in the current study is *sidaug05_tcmt*, which is a variant of the familiar *sidaug05* model with the projective RPC-based calorimeter-like components

CP867, *XII International Conference on Calorimetry in High Energy Physics,*
edited by S. R. Magill and R. Yoshida

TABLE 1. Details about the electromagnetic (Ecal) and hadronic (Hcal) calorimeters for the *sidaug05_tcmt* detector model.

Subdetector	Layer configuration	# layers	Int./Rad. length	Cell size
Ecal	Si-W, each 3.75mm thick	30	$0.72\,X_0$/layer	4x4 mm^2
Hcal	Sci-Steel, each 28mm thick	34	$0.13\,\lambda_I$/layer	10x10 mm^2
TCMT	Sci-Steel, each 28mm thick	48	λ_I/layer	30x30 mm^2

(hadronic calorimeter and muon system) replaced with similar components based on high-granularity, non-projective scintillator tiles. A 5T solenoid is located between the Hcal and the muon system. Table 1 provides details of the calorimeter components in this model, and full details can be obtained from Ref. [4].

PERFECT PFA

Ignoring constant terms, the jet energy resolution can be decomposed into the energy resolution of its components, as[5]

$$\sigma(E_{\text{jet}}) = \sigma(E_{\text{charged}}) + \sigma(E_{\text{photons}}) + \sigma(E_{\text{neutrals}}) + \sigma(E_{\text{confusion}}) \tag{1}$$

where in average, the charged, photonic and neutral components contribute with respectively 65%, 25% and 10% of the total energy in events with hadronic decays of the Z-boson. At the typical energies involved, $\sigma(E_{\text{charged}})$ is negligible, while $\sigma(E_{\text{photons}})$ and $\sigma(E_{\text{neutrals}})$ are estimated to be about $15\%/\sqrt{E}$ and $60\%/\sqrt{E}$ respectively, so the PFA method becomes feasible when the confusion term becomes smaller than the neutrals term. The so-called perfect PFA is an estimation of the jet energy resolution for a negligible confusion term, so it depends only on the intrinsic detector resolution, or more specifically, on the calorimeter granularity and the detector technology used.

As a first step, we looked at the perfect PFA reconstruction of the total energy of $e^+e^- \to Z$ events (at $\sqrt{s} = M_z$). One minimizes the confusion term by consulting the Monte Carlo information for perfect clustering (perfect association of particles to all their calorimeter hits), as well as perfect cluster ID (e.g. charged, photon or neutral hadron clusters). Cluster energies are then estimated using an analog scheme for photons and a digital scheme for neutral hadrons. In the digital scheme, the cluster energy is based on the number of hits in each cluster. This assumes a linear relationship between number of hits and the energy of the incident particle, which is acceptable for 1x1cm^2 cells but requires some corrections at larger cells for hadron energies above 30 GeV.

For charged particles, a mass assignment is combined with the generated momentum to determine the particle energy. The pion mass was assigned to all charged hadrons, representing a lacking ability to distinguish between pions, kaons and protons.

Figure 1 shows the distributions of $\sum E_{\text{reco}} / \sum E_{\text{gen}}$, from each component. The widths of these distributions provide rough estimates for the contributions from each component in Eq. 1. The calibration was performed in a global way, forcing the distributions for photons and neutral hadrons to be centered around 1. The distribution for charged clusters have a larger mean than for neutrals because, for similar energies, their showers

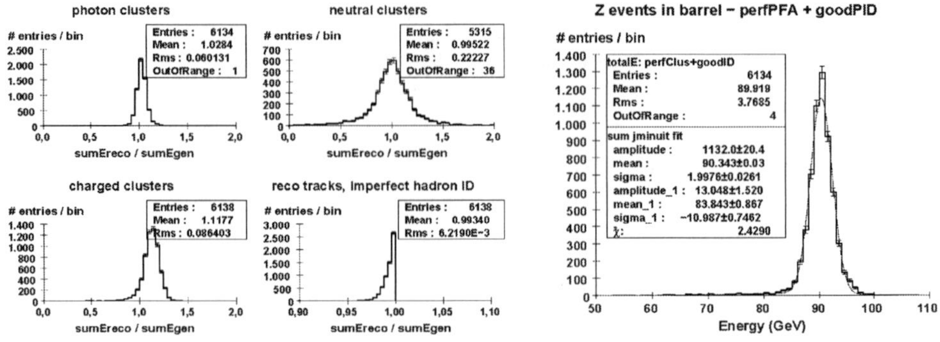

FIGURE 1. Perfect PFA results for $Z \rightarrow q\bar{q}$ events in the barrel of the *sidaug05_tcmt* detector model. The energy distributions are shown on the left plots for perfect clusters from (a) charged particles, (b) photons and (c) neutral hadrons. The effect of using imperfect particle ID for the total charged distribution is shown in (d). The total event energy distribution (Z-mass) is shown on the right plot.

have about the same number of hits, whereas charged clusters usually contain several extra MIP calorimeter hits before showering starts. By using a digital scheme for both charged and neutrals, we find slightly larger energies for charged than for neutrals of equivalent energies. We do not need to correct for this effect, as the energies for charged clusters are not used for the PFA reconstruction (track momenta are used instead).

The plot on the right of Fig. 1 shows the final distribution for the total event energy. We fit this distribution to a double gaussian, and determine the perfect energy resolution to be 1.97 GeV, which corresponds to an energy resolution of $21\%/\sqrt{E}$. As one can see, there is some room to accomodate the contribution from the confusion term to the jet energy resolution, and still keep the latter below $30\%/\sqrt{E}$.

REAL PFA

We developed the PFA algorithm around the directed tree clustering, so that directed tree clusters and reconstructed tracks are the main inputs to the algorithm. The directed tree clustering algorithm is described elsewhere[6]. From now on, directed tree clusters are refered only as *clusters*.

A typical hadronic shower is fragmented into multiple clusters. One usually refers as primary cluster to the one containing most of the shower energy, whereas all others are called fragments. The PFA consists in identifying and correctly classifying the primary clusters for each particle reaching the calorimeters (cluster ID), and then extending these cluster cores by fragment association. These steps are described in more detail in the next subsections.

We are also evaluating the use of some tools to tag showers misidentified at the cluster ID steps. The ordering of all the pieces may affect the final results significantly, and it is part of the optimization procedure.

Cluster ID

The first step in the PFA consists in classifying the input clusters as either charged, photon, neutral or fragment. The main tools for this purpose are track matching for charged clusters, and a longitudinal H-matrix for photon clusters.

Charged clusters: track matching

Charged clusters are tagged by extrapolating the reconstructed tracks into each calorimeter layer, and collecting all nearby hits within a window around the intersection points. The track extrapolation is performed using a simple helix parametrization, hence it does not take into account energy losses or multiple scattering. Typical windows are 5x5 cells in Ecal and 1x1 cells in Hcal.

Once one or more hits from a cluster get matched as above, the whole cluster containing those hits gets associated to the track. Multiple clusters can be associated to any given track. An average of 70% of the energy from charged clusters get correctly associated in this step, with a purity better than 95%.

Photon clusters: longitudinal H-matrix

Clusters containing only Ecal hits (directed tree clusters don't cross calorimeter borders), with at least 8 hits, are fed into a longitudinal H-matrix[7] trained with photons at several energies. The H-matrix variables used are the fraction of cluster energy deposited at each Ecal layer, and also $\log(E_\gamma)$.

Those clusters with a diagonal χ^2 satisfying $\log(\chi^2_{\text{diag}}) > 2.5$, and containing at least one hit in the first 6 layers, are tagged as photons. These requirements have an efficiency of 97.6% with a purity of 74%.

Neutral clusters and fragments

Those clusters with at least 10 hits and not tagged as either charged or photon clusters are taken as neutral showers. Several charged fragments get tagged as neutral at this point, but most of them may get absorbed by charged seeds at the next step before being definitely established as neutral clusters. See next section for more details.

Fragment association

Once the clusters have been identified, they are used as seeds to absorb neighboring fragments. This association is performed according to cluster shapes and directions first, and then to distances between fragments and primary clusters. These steps are refered to as fragment association, or cluster merging steps.

Direction-Based Cluster Merging

In the direction-based merging method, a cluster seed absorbs other clusters (fragments) based on the directions of cluster extremities as seen from the origin. The hits within the first and last 3 layers of each cluster are used to determine direction cosines of cluster extremities with respect to the origin. Opposite extremities (inner vs. outer) of each fragment are compared to those of all seeds, and the fragment is associated to the closer seed if the differences between their direction cosines is smaller than 0.05.

This algorithm is applied first using neutral shower candidates as fragment candidates to charged seeds. Then the remaining untagged clusters are used as fragment candidates to charged and neutral seeds. This algorithm is not used for photons.

Distance-Based Cluster Merging

In the distance-based merging method, all hits in each seed/fragment pairs are searched for the closest pair of hits, based purely on the 3D distance between a fragment hit and a seed hit. Each fragment will be associated to its closest seed, if the closest distance between them is smaller than 200mm. This fixed distance is currently under optimization, and optimal values probably depend on sizes (energies) of fragment and/or seed clusters.

This algorithm is applied after the direction-based merging, only to untagged clusters as fragments to charged, neutral and photon seeds.

At the last last step of fragment association, all remaining untagged are sorted according to sizes (and then energies). The largest one becomes a new neutral seed, and the other fragments are tested against this new seed for a possible distance-based association. This process is repeated until all untagged clusters with 8 or more hits have been processed.

All fragments surviving this last step are currently discarded, in an attempt to minimize the association of hits to incorrect clusters. We are studying the effect of it to the final energy resolution.

Other useful algorithms?

We are also studying some alternative algorithms which may be useful to identify clusters which have been incorrectly tagged as charged, photon or neutral. One such algorithm currently uses statistical correlations to the x,y,z,r coordinates to catch MIP-like stubs in Ecal which miss the track matching, being tagged as either photon or neutral cluster. We also use E/p checks to avoid associating too many hits during the track matching, by comparing cluster energy to track momentum when building the charged clusters.

Another potentially useful tool, not yet implemented at this point, could identify the presence of more than one incoming particle in a given cluster, maybe by identifying the presence of two clear shower starts at the inner part of a reconstructed shower.

Optimization of the algorithms

The performances of the algorithms previously discussed are being analyzed by comparing efficiency and purity for individual showers before and after each step of cluster ID and fragment association. Efficiencies are estimated using $E_{\rm reco}/E_{\rm gen}$ distributions for individual showers. For perfect clustering, these distributions should be gaussian with mean 1 and widths related to the energy resolution for single particles. For cluster purity, we look at the fraction of the hits in each cluster which are compatible with the cluster ID. More details on this optimization procedure will be provided in a longer note, which is currently under preparation.

PFA results and possible improvements

Once all of the cluster seeds and fragments have been merged, the final energy accounting takes place. The energy contributions from each component are determined from the reconstructed clusters. Cluster energies are estimated using a digital scheme for Hcal, while the Ecal is treated as digital for neutral hadrons and analog for photons. The distribution from charged clusters is used for control only, as the tracks rather than the clusters are used for the final PFA result.

Figure 2 shows the current, preliminary results for the charged, photon and neutral components, as reconstructed by our PFA algorithm. The bottom plots can be directly compared to the distributions shown at the left of Fig. 1. We can conclude that the shower reconstruction is working much better for charged particles and photons, but some work is still necessary to improve the performance for the neutral hadrons.

Figure 3 shows the current (preliminary) results for the reconstruction of the Z-mass using our PFA algorithm. The total event energy is obtained by simply adding, for each event, the total energies from the neutral hadrons and photons as reconstructed in the calorimeters, and using the momenta from the reconstructed tracks rather than their cluster energies.

The energy resolution for the Z-mass is measured by a fit to a double gaussian model, which represents well the observed distribution. The inner gaussian contains about 59% of the barrel events, and the width of 3.88 GeV for the Z-mass peak corresponds to an energy resolution of $41\%/\sqrt{E}$.

This energy resolution is still far from the goal of $30\%/\sqrt{E}$, but there is a lot of room for improvements. For instance, it is clear from Fig. 2 that the identification of the neutral clusters is the main problem, and could be improved by a multivariate-based discriminating tool dedicated to neutral core vs. fragment identification. The longitudinal H-matrix also needs to be optimized for different photon energies, as it allows for a significant amount of low energy hadron clusters satisfying the photon criteria.

The fragment association algorithms are performing well, although the specific criteria for deciding when to merge fragments may need some fine tuning. Finally, by investigating the tails of the mass distribution, one frequently finds new insights on improvements to the existing algorithms, or to the development of new ones.

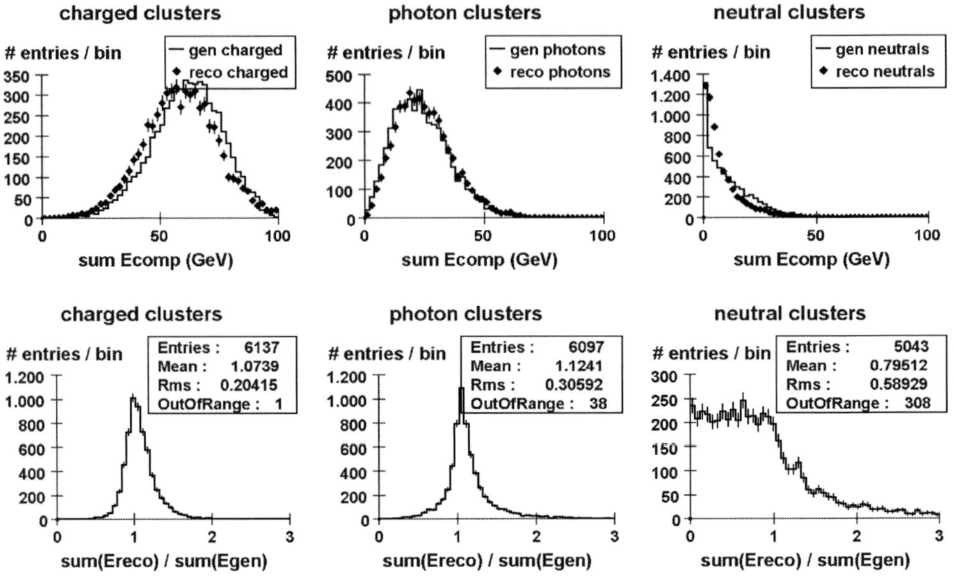

FIGURE 2. TOP: Comparing total generated and reconstructed contributions per event, for charged (left), photons (middle) and neutral hadrons (right). BOTTOM: distribution of event-by-event ratios between the total reconstructed and generated energy contributions, for charged (left), photons (middle) and neutral hadrons (right).

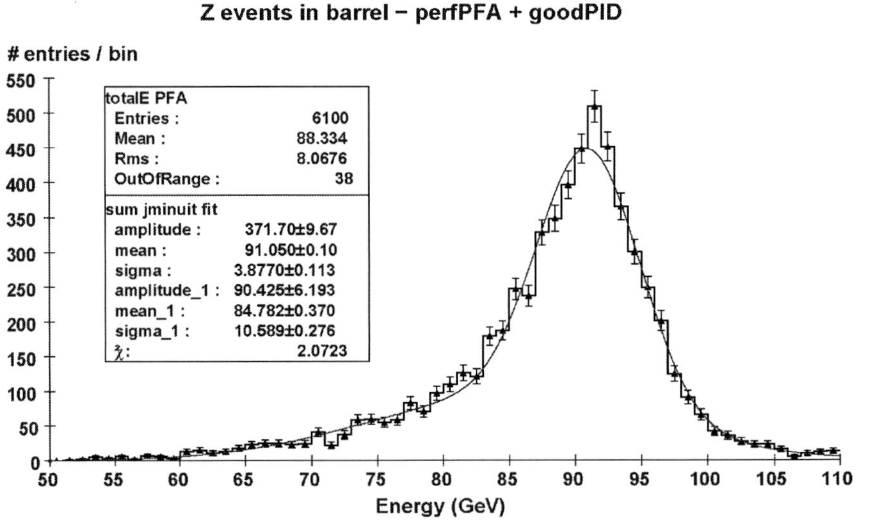

FIGURE 3. Preliminary results for the total event energy, in simulated events containing $Z \to q\bar{q}$ decays at $\sqrt{s} = 91$ GeV. Events where both partons are produced into the endcaps are removed.

The calibration used is another potential source of improvements to the final jet energy resolution (affecting both perfect and real PFA). The results presented here are based on a global calibration, as discussed in Fig. 1. A more sophisticated calibration should account for differences in detector response depending on incident particle type, incidence angles with respect to the calorimeter layers, and also to the calorimeter depth where shower starts.

Finally, the TCMT (tail catcher) data may also be useful to improve the final energy resolution by correcting for punch-throughs surviving the calorimeters and the solenoid.

SUMMARY

In this paper we present the status of the development of a complete particle flow algorithm (PFA) for the International Linear Collider. We have just put all the basic PFA pieces together, and we are currently studying the interplay of all these pieces in order to maximize the performance of the PFA algorithm as a whole. With the current configuration, we have 59% of the events reconstructed with an energy resolution of $40\%/\sqrt{E}$. This result is preliminary, and we have already identified the most important contributions to this poor resolution, and we are working to fix the problems.

Stable PFA algorithms will be a very effective tool for optimizing the calorimeter systems, maximizing the physics capabilities of detectors for a future high energy linear collider. Once the algorithms performance are certified, and their strengths and weaknesses are better understood, the optimization procedure can be streamlined and applied to other detector geometries and/or technologies, as well as variants of the same detector, say different radii, aspect ratios, magnetic fields and calorimeter materials and granularities.

ACKNOWLEDGMENTS

The authors wish to thank the developers and maintainers of the org.lcsim analysis framework. This work is supported in part by grants from the US Departments of Education and Energy, and the National Science Foundation.

REFERENCES

1. See the SLIC web site, at *http://www.lcsim.org/software/slic*
2. S. Agostinelli *et. al.* (Geant4 collaboration), NIM A 506 (2003), 250-303
3. See the *org.lcsim* web site, at *http://www.lcsim.org*
4. *http://confluence.slac.stanford.edu/display/ilc/sidaug05_tcmt*
5. A.Raspereza, "Modular implementation of a Particle Flow algorithm with minimized dependence on detector geometry", in *Proc. of the 2005 ILC Physics and Detector Workshop*, Snowmass, arXiv physics/0601069 and eConf C0508141:ALCPG1101 (2005)
6. ILC-Note, in preparation
7. *http://www.lcsim.org/software/lcsim/apidocs/org/lcsim/recon/emid/hmatrix/HMatrix.html*

SESSION XII

CALORIMETRY AT THE ILC

Conveners: Jerry Blazey and Harry Weerts

GLD Calorimeter

Tohru Takeshita on behalf of GLD Calorimeter group

Shinshu University (Nagano, Japan)

Abstract. The GLD calorimeter, which is under study to aim as the calorimeter of ILC detector with an idea of Particle Flow, is described and discussed. The calorimeter consists of ECAL and HCAL. Both calorimeters are composed of plastic scintillator as the active medium. Fine granularity is required to achieve Particle Flow, so a small scintillator tile technique is developed with wavelength shifting fiber of MPPC read out. The requirements and our solutions for them are discussed and presented.

Keywords: GLD, calorimeter, scintillator and MPPC.
PACS: 07.20.Fw

GLD DETECTOR CONCEPT

At the ILC (International Linear Collider), where electrons and positrons collide at around 1 TeV, the physics events will be dominated by narrow jets which come from fundamental particles. In such environment, we employ an idea called PF (Particle Flow) [1] to resolve jets into the source elementary particles, such as quarks, and intermediate bosons. Those elementary particles carry the interaction information which we search for. In order to identify those fundamental particles in jets, we have to prepare the detector in advance carefully.

GLD is one of the ILC detector concepts, which stress the measurement to be as large as possible to resolve the jets. To reduce overlaps between particles in jets, the calorimeter is set as far as possible from the interaction point. That means that a bigger inner tracker is required to measure the charged track as precisely as possible. The benefit of this calorimeter location and geometry is that the jet components can be more easily separated. In this sense, we expect that the GLD has a smaller confusion term.

Particle Flow (PF)

In the context of Particle Flow (PF), we measure the charged particle momentum by the tracker detector, the photon components by the ECAL (Electromagnetic CALorimeter) and the neutral hadrons by the HCAL (Hadron CALorimeter). On average, the charged particles carry 65% of a jet energy, the photons carry 25%, while the remaining 10% is carried by neutral hadron component.

CP867, *XII International Conference on Calorimetry in High Energy Physics,*
edited by S. R. Magill and R. Yoshida

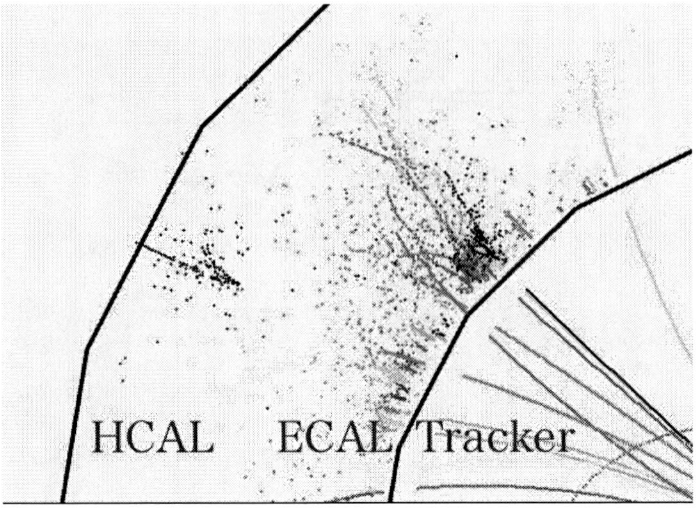

FIGURE 1. An event of closer look at the calorimeter detector at ILC. It shows one of a jet detected by the tracker, ECAL and HCAL.

The jet energy resolution of 30%/√(Ejet) is one of the target values in the linear collider detector development. In order to achieve such performance of the detector, PF algorithms are discussed. The jet energy resolution is dominated by the poorest energy resolution detector, namely by the HCAL normally, even though the energy of the neutral hadrons is only 10% on average. This is due to the error propagation principle. There is another pollution factor in this calculation, the so called confusion factor, which means the miss-assignment of particles in a jet causes wrong measurement of particle energies. This effect is strongly dependent on the segmentation of the calorimeter detector. We have to study the PF at this moment to clarify those effects in mind.

GLD calorimeters

Since the GLD detector concept is "large", the calorimeter volume will be huge. So as to save cost of the calorimeter, however having enough segmentation, we decide to use scintillator strips as a basic component of the calorimeter. To reduce the number of channels, which is the driving term of the cost, length of a strip should be optimized. At the moment, we are studying on the basic design of a strip of 1cm by 4cm with 2mm thick scintillator for ECAL and that of 1cm by 20cm with 5mm thick scintillator for HCAL. In each successive layer, the directions of strips are orthogonal, so as to measure the particle tracking with effective resolution of 1 cm. They are shown in Figure 2 in the cross section picture. In the HCAL we added the tile layer so as to resolve confusions in the track finding.

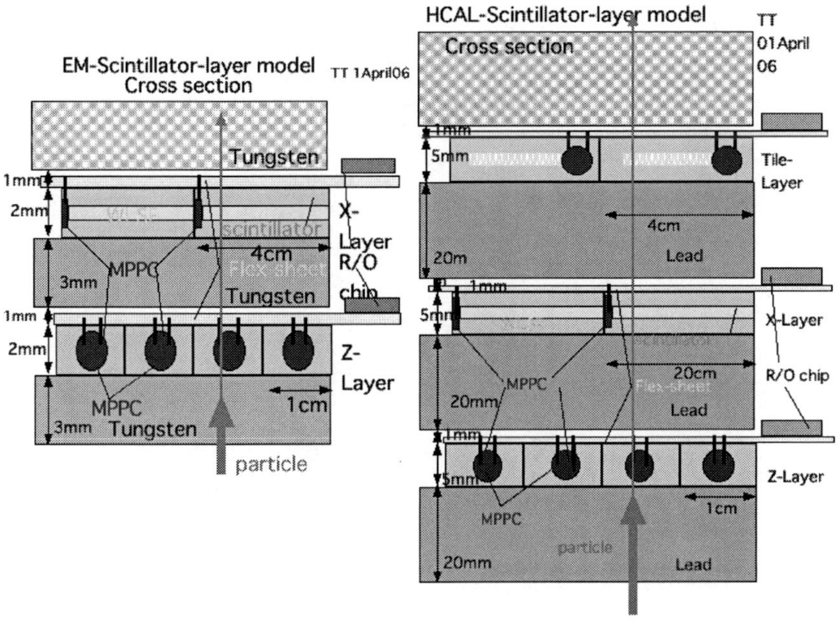

FIGURE 2. Schematical view of GLD- ECAL and HCAL which consist of absorber and plastic scintillator of strip structure with orthogonal direction in each layer.

The absorber materials are Tungsten of 3mm thick for ECAL and Lead of 20mm thick for HCAL, respectively. Even with 4cm long strips, the number of readout channels of GLD-ECAL will be 10 million channels and that of HCAL, which consists of 20 long strips, will be 4 million. Since we try to read out every layer in the longitudinal direction of jets, the number of layers is 33 for ECAL and 47 for HCAL.

Beam test results

We have already constructed and tested the strip calorimeter in the test beam [2]. It consisted of 1cm wide, 20cm long scintillator strips with a WLS fiber read out at the center of a strip. The outer size of 20cm by 20cm was covered by 20 x-strips and 20 y-strips in a layer. There were 24 layers in longitudinal direction. Four successive layers were connected into a multi-anode photo-multiplier (MAPMT) channel by optically connected at the same location of strips. There were in total 480 channels of readout from this detector which is shown in Figure 3.

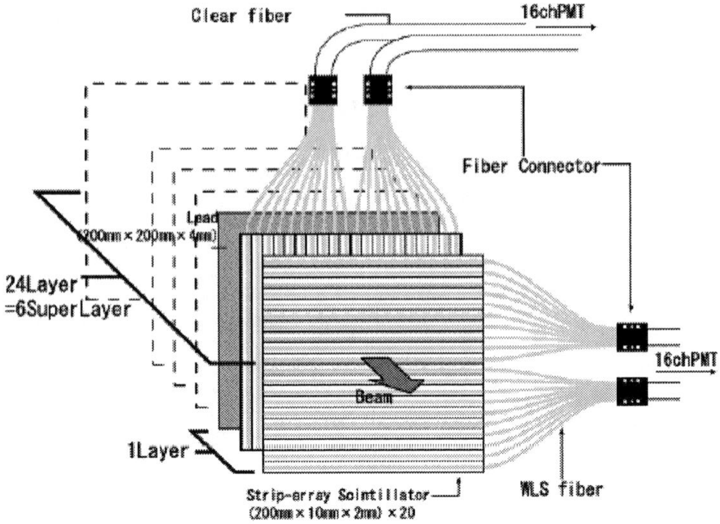

FIGURE 3. A schematic view of the strip ECAL test module, which consists of strips with WLS fiber read out with MAPMT.

One of the nice features is shown in Figure 4 which indicates the pictorial calorimeter for tracking capability. In Figure 5, the lateral electro-magnetic shower profiles are plotted as a function of transverse distance x(cm).

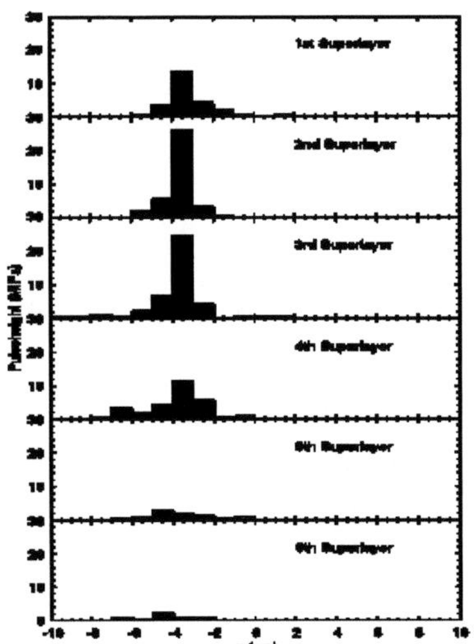

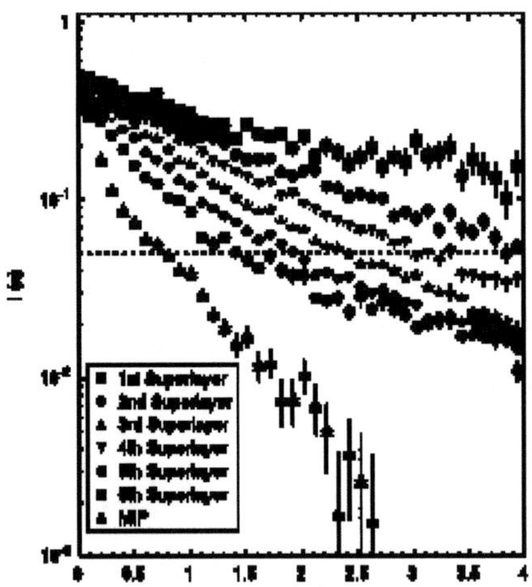

FIGURE 4(left). An electron event display in the longitudinal direction of the detector, which is sub-divided into 6 super layers. A 4 GeV electron is injected into the strip calorimeter.
FIGURE 5(right). The lateral shower profile measured by the ECAL of 1cm wide strips in 6 super layers.

In this beam test of strip ECAL, we maintain that this type of strip structure is able to measure the tracks in the calorimeter and the energy deposit as the calorimeter. There we have tested with 1cm wide strips in orthogonal directions. The strip was 20cm long, because of no other particles getting into the calorimeter in the beam test. The scintillation light was measured through the wave length shifting fiber (WLSF) and fed into the multi-anode photo-multiplier.

We have constructed and tested a fine granular tile calorimeter [3], which is relevant in the GLD- HCAL detector for the strip configuration. Twenty five plastic scintillator tiles of 4cm by 4cm of 1mm thick were arranged by 5 x 5, which composed a layer. In a tile, there was a sigma groove of 0.8mm diameter and 0.7mm WLSF was inserted into it to read out the signal. The schematic view of the tile calorimeter is shown in Figure 6.

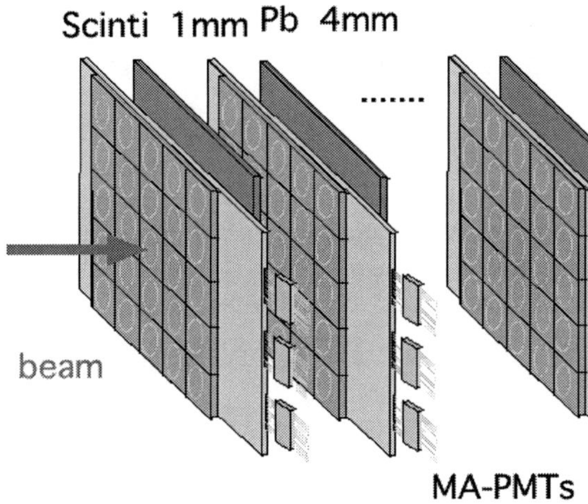

FIGURE 6. A schematic view of the tile calorimeter tested at the beam.

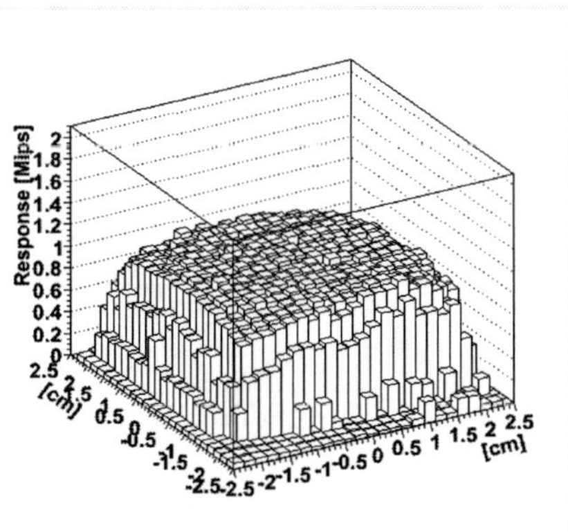

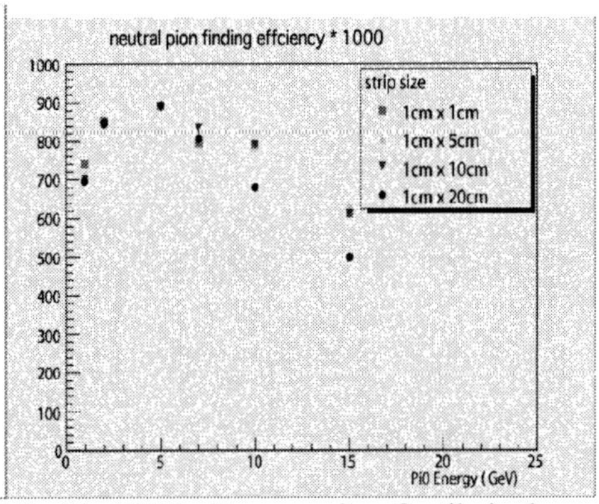

FIGURE 7(left). One of the test results of the tile ECAL. Homogeneity of a tile is shown in a lego plot.

FIGURE 8(right). Measured efficiency of neutral pions as a function of incident pi-zero momentum for the strip ECAL.

With such thin plastic tile of 1mm, we constructed and tested the ECAL satisfactorily. In Figure 7, the homogeneity of a 1mm thick tile is shown. There is enough wide area of flat response in a tile.

ECAL consideration

The most important role of the ECAL is to find photon showers in a jet and if possible, to select neutral pions effectively. In a clean sample, that of pi-zero alone, an algorithm is developed to find neutral pions in the strip ECAL simulation. The invariant mass of pi-zero is calculated from two photon clusters in ECAL. The found efficiency as a function of incident pi-zero momentum is shown in Figure 8. For up to 15 GeV pions, the detector is able to identify pi-zero's with reasonable efficiencies.

HCAL consideration

One of the good points of the scintillator Hadron Calorimeter is the fact that the scintillator is neutron sensitive, since it contains Hydrogen atoms in its contents. This is shown in Figure 9 as an example of a pion injection into a model HCAL of 1cm by 1cm segmentation in 100 layers. We can find many hits far away from the injection position of a pion in this typical event. They are due to the hits made from neutrons traversing the detector and finally scattered with a proton which caused the hits.

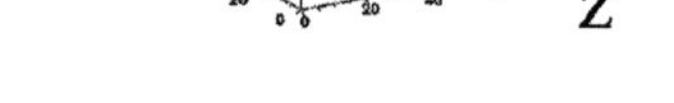

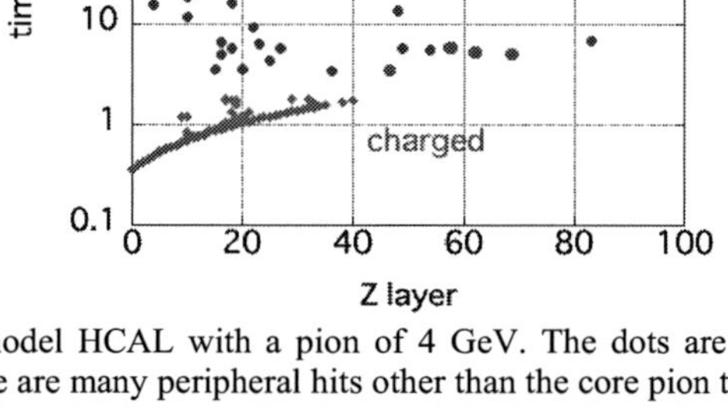

FIGURE 9(left). A typical event in a model HCAL with a pion of 4 GeV. The dots are hits with energy deposit greater than 0.1MIPs. There are many peripheral hits other than the core pion tracks.
FIGURE 10(right). Hit timings as a function of Z layer number in a model HCAL by a pion of 4 GeV. The lighter points are due to neutron hits. There are many peripheral hits other than the core charged hits which are shown in darker dots.

However, the hits due to neutrons could be recognized by some method. For this purpose we have introduced hit time. Figure 9 shows the hit timings as a function of transverse layer number.

Figure 10 indicates that the timing information will help us to select neutron hits from charged hits, if we set a threshold around 5ns. So as to achieve such performance, we need to have timing measurement capability in the readout electronics of the calorimeter with ns resolution.

Photon sensor (MPPC)

For the fine segmented scintillator calorimeter, we need to have semi-conductor based photon sensors, because the strong magnetic field prevents us to operate the normal photon sensor such as photo-multiplier. The photon sensor MPPC named after Multi Pixel Photon Counter is composed of a number of small pixels which work at the Geiger mode or Limited Geiger Mode. The number of photons on the sensor is digitized to the number of pixels fired which is shown in Figure 11 schematically. A photon will be absorbed and produce a pair of a hole and an electron in a certain pixel with high efficiency. The electron will be accelerated in the strong electric field and produce electrons in avalanche. This multiplication will be quenched by their own charges in Geiger Mode. The recovery time of a pixel is supposed to be about 100ns. The number of electrons in a fired pixel will be fairly constant, therefore the signal will be clearly separated and is able to be counted. The MPPC is able to count the number of pixel fired as number of photons into the MPPC. The ADC distribution of a small number of photons shows clear separation from 1 to up to 10 or more.

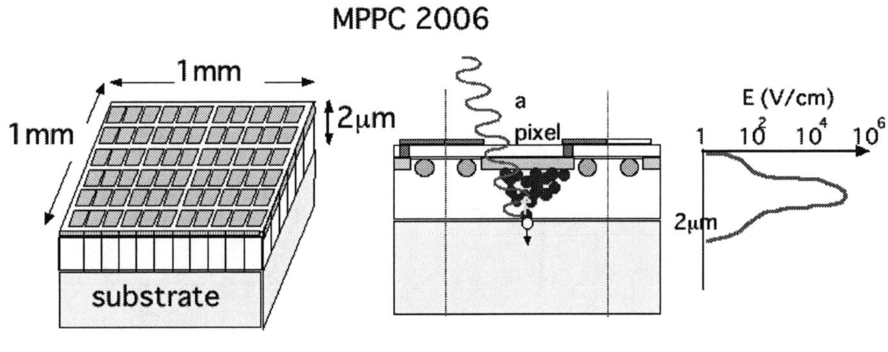

FIGURE 11. A schematic view of MPPC. On the substrate, a very thin layer is divided into many pixels. In a pixel, strong electric field is generated due to its small thickness.

The development of the MPPC is still underway. Here, one of the results of pixel homogeneity is plotted in every pixel of 400 pixels MPPC in Figure 12.

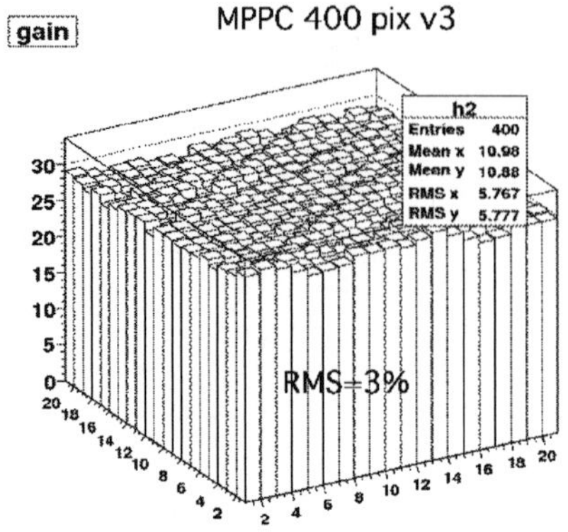

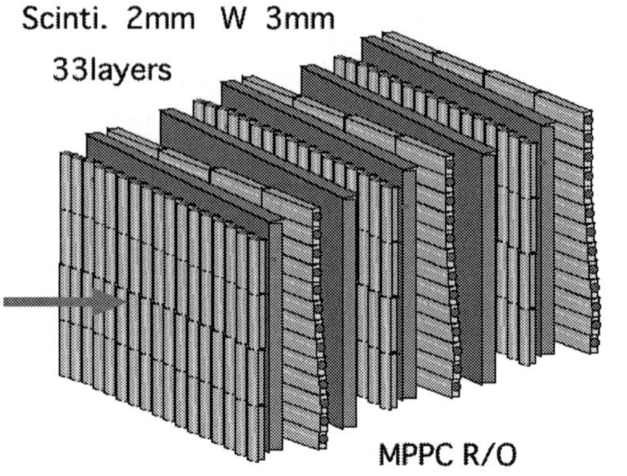

FIGURE 12(left). One of the test results of MPPC of 400 pixels. Gain homogeneity is plotted on each pixel.

FIGURE 13(right). The ECAL which will be tested at the beam. The detector consists of 2000 strips with sandwich of Tungsten plates of 33 layers.

PLANS AND TO DO

We are preparing beam tests of the above detectors with a scintillator ECAL and HCAL. For both cases, MPPC readout is the essential part of the experiment. A schematic view of the ECAL detector which will be constructed and tested in 2006-2007 is shown in Figure 13.

We have to construct a similar structure HCAL and test it at in a high energy hadron beam, in order to clarify models which give many different results to each other.

ACKNOWLEDGMENTS

The author wishes thanks the members of the GLD-CAL group for their collaboration, valuable suggestions and discussions.

This work was supported in part by a Grand-in-Aid for Scientific Research from Ministry of Education, Science and Technology under Contract No. 17340071.

REFERENCES

1. For example, The ALEPH Collab. D. Buskulic et. al., Necl. Insrumt. Meth. A 360 (1995) 481.
2. A.Nagano et. al., Necl. Insrumt. Meth. A 557 (2006) 460.
3. A.L.C. Sanchez et. al., Necl. Insrumt. Meth. A 546 (2005) 535.

The Calice Scintillator HCAL Testbeam Prototype

Gerald Eigen

University of Bergen, Allegaten 55, 5007 Bergen, Norway, representing the Calice collaboration

Abstract. A highly-granular, 38-layer, analog scintillator-tile hadron calorimeter prototype is under construction by members of the Calice collaboration in order to study hadronic shower shapes and to test the calorimetric aspects of particle-flow.

Keywords: Linear collider detector; Analog calorimeter; Semiconductor detectors; Scintillator; High granularity
PACS: PACS: 07.20.Fw; 29.40.Wk; 42.70.Nq; 42.81.Pa; 42.81.Qb

INTRODUCTION

The physics goals at the international linear collider (ILC) impose high demands on calorimetry. The basic idea consists of reconstructing jets with an energy resolution of $30\%/\sqrt{E}$ using the concepts of particle flow [1?]. Thus, it is necessary to separate electromagnetic showers and neutral hadron showers from other particles requiring fine segmentation in the longitudinal and transverse directions both for the electromagnetic (ECAL) and hadron calorimeters (HCAL). At DESY, members of the Calice collaboration are constructing a 1 m^3 analog hadron calorimeter prototype ("physics" prototype) to study hadron shower shapes for different energies and incident angles and to test the calorimetric aspects of particle flow. The design is based upon the experience gained with a technical prototype in a positron testbeam at DESY that operated 108 scintillator tiles read out with wavelength-shifting fibers coupled to SiPMs [3]. The "physics" prototype will be tested in hadron beams together with an ECAL and a tail catcher.

THE PHYSICS PROTOTYPE STACK

The analog hadron calorimeter prototype is a 38-layer sampling calorimeter made of a plastic-scintillator steel sandwich structure with a lateral dimension of about $1 \times 1\text{m}^2$. Each layer consists of 2 cm thick steel absorber plates (including the walls of the scintillator cassettes) and a plane of 0.5 cm thick plastic scintillator tiles housed in a steel cassette. For the first 30 layers, the tile sizes vary from 3×3 cm^2 for 10×10 tiles in the center of the module, to 6×6 cm^2 in the intermediate region (4×24 tiles) and 12×12 cm^2 (4×5 tiles) in the outer ring (see Figure 1). In the last eight layers, the granularity is decreased to 6×6 cm^2 in the central region to save costs. Each tile has a wavelength-shifting (WLS) fiber that is inserted into a groove and that is coupled to a SiPM via an air gap on one end. The other end is covered with a mirror (3M reflector

CP867, *XII International Conference on Calorimetry in High Energy Physics,* edited by S. R. Magill and R. Yoshida

FIGURE 1. Tile layout of a scintillator module for layers 1-30.

foil) to increase the light collection efficiency. In small tiles, grooves are quarter-circle shaped, in all other tiles they have a full-circle shape. All tile side faces are edge-treated (matted) providing a diffuse reflector, while the front and back faces are covered with the 3M super reflector foil. The tiles are positioned and fixed in a frame. The high granularity in the core is suited to test semi-digital readout options.

The prototype consists of 7608 cells and has a thickness of 4.6 interaction lengths. It is placed on a movable platform providing flexibility to select specific tile positions and to choose incident beam angles. Several beam tests are planned together with a highly granular Si-W calorimeter and a tail catcher to address the following objectives:

- study hadron shower shapes at different energies and incident angles.
- validate simulated showers with measured showers.
- study the separability of two adjacent showers.
- gain large-scale, long-term experience with a calorimeter based on SiPM readout.
- develop a calibration strategy.
- identify critical operational issues.

SIPMS AND THEIR PROPERTIES

A SiPM is a multipixel avalanche photodiode operated in the Geiger mode [4]. The detectors are manufactured in Russia by PULSAR enterprise together with the CALICE MEPhI group. Enlarged photographs of an entire detector and the pixel structure are depicted in Figure 2. The physical size of the entire detector is 1.5×1.5 mm^2. The photosensitive area is 1×1mm^2 and holds 1156 pixels, of which each is about 20 μm

FIGURE 2. Photograph of a SiPM (left) and an enlargement of the pixel structure (right).

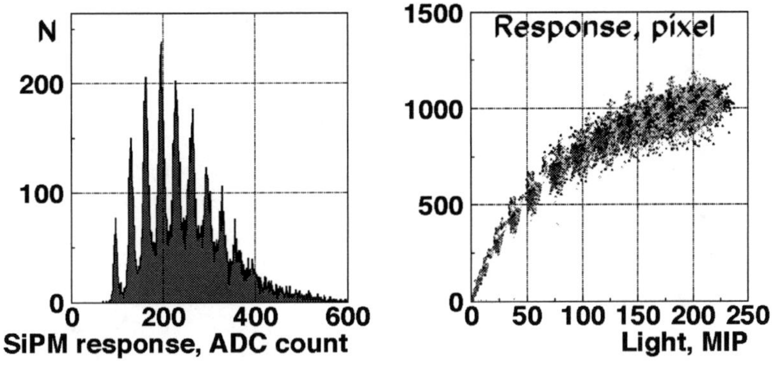

FIGURE 3. A typical SiPM spectrum for low-intensity light, showing up to nine individual pixels (left). The response function of fired pixels versus intrinsic LED light intensity in units of MIPs (right).

wide. SiPMs are reversely biased with a voltage of $U_b \sim 50\ V$ producing gains of the order of 10^6. Once a pixel is fired it triggers a Geiger discharge. The analog information is obtained by summing up the number of fired pixels. Thus, the dynamic range for photon signals is about three orders of magnitude. Each pixel has a quenching resistor of the order of a few MΩ built in, which is necessary to break off the Geiger discharge. The recovery time is less than 100 ns per pixel. Our SiPMs are insensitive to magnetic fields as magnetic-field studies have shown [4]. Due to the Geiger mode operation, SiPMs have a negligible nuclear counter effect. The SiPM detectors are auto-calibrating, however, their response is non-linear (see Figure 3).

Test results of SiPMs

Currently, more than 5000 SiPMs have been produced. Testing is conducted at ITEP Moscow in an automatic setup, where 15 SiPMs are simultaneously illuminated with calibrated LED UV light. During the first 48 hours, the SiPMs are operated at an increased reverse-bias voltage lying 2 V above the nominal operation voltage. Next, gain, noise and relative efficiency with respect to a reference photomultiplier are recorded as a function of U_b. The U_b working point is chosen as that yielding 15 pixels for a MIP-like

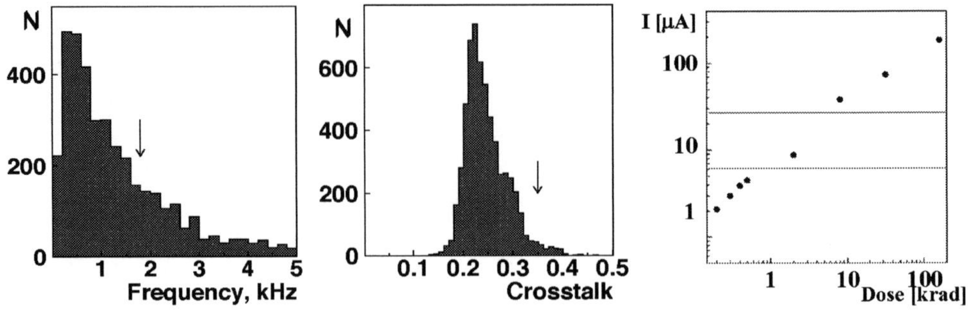

FIGURE 4. The noise distribution of SiPMs at half a MIP threshold (left), the cross-talk distribution (middle) and the SiPM current versus radiation dose for 200 MeV protons.

signal provided by the calibrated LED.

At the working point, several SiPM characteristics are measured. With low-intensity LED light, single-pixel spectra used for gain calibrations are recorded. A typical single-pixel spectrum is shown in Figure 3 (left). The response function of 4000 individual SiPMs, depicting the number of fired pixels versus intrinsic LED light over the entire range (zero pixel to saturation) is plotted in Figure 3 (right). The individual curves all lie within $\pm 15\%$. Noise rate at half a MIP threshold, cross talk and SiPM current are measured in addition. The cross talk here denotes the effect that a Geiger discharge in one pixel triggers a Geiger discharge in one or more neighbor pixels. Figures 4 (left, middle) show the noise rate and cross talk distributions, respectively. For the prototype, we select detectors with a noise rate of less than 1.8 kHz (3.0 kHz for part of the detector), a cross talk probability of less than 0.35 and a SiPM current of less than 2 μA.

SiPM Radiation Hardness and Aging Studies

First radiation hardness tests of SiPMs have been performed with a 200 MeV proton beam at the ITEP synchrotron in Moscow. Figure 4 (right) shows the SiPM current as a function of accumulated dose. The current increase is compatible with that observed in other Si detectors [5], however, SiPMs are more sensitive to radiation damage because of the high amplification ($\sim 10^6$) and a very small initial noise (~ 0.1 photoelectrons). Both properties are essential for separating signals with different numbers of detected photons. This for calibrations important property is lost for doses above 1 krad, since individual pixel peaks are not resolved any longer. Nonetheless, SiPMs are still operable even for radiation doses much higher than 1 krad, but their noise rate is increased. For example, at 5 krad the equivalent noise is about two pixels in a 50 ns wide gate. For MIP signals being 15 pixels on average a two-pixel noise does not increase the width of the Landau distribution and thus does not influence the detection of MIPs.

The radiation hardness of SiPMs is sufficient for operations in a hadron calorimeter at the ILC. Only in the endcaps close to the beam pipe, we expect a neutron flux above $10^{10}/\text{cm}^2/500\ \text{fb}^{-1}$, which would lead to excess currents above 5μA and thus cause a smearing of individual pixel peaks. We have assumed here a standard energy-dependent relative radiation damage efficiency of neutrons and protons [5]. Future tests will study the effect of long-term low-dose irradiation on aging.

We have operated about 650 SiPMs for more than 180 days with the high voltage turned on 50% of the time. During the first three weeks of operation the noise rate increased in six SiPMs, however, these detectors are still operable. After the initial turn-on period, the system remained stable.

Measured Light Yield of the Scintillator Tiles

In order to cover the full dynamic range but still reveal a clear signal above noise, the MIP of a cell should lie in the 10-20-pixel range. For each tile, the MIP response is measured using triggered electrons from a ^{90}Sr source. The mean value of the MIP response varies from tile to tile because of different light collection efficiency. For 407 tiles read out with SiPMs, ITEP measured a mean value of 16.6 pixels with a spread of about 15%.

HCAL READOUT STRUCTURE

For stress relief, the SiPMs pins are soldered to strips that are connected to flexible pads, which in turn are connected via cables with the front-end base boards. Two base boards are needed to read out all 216 tiles of a module. Each base board houses six ILC SiPM chips. The ILC chip produced by the Orsay group [6] and shown in Figure 5 is based on ECAL ASICs. It houses 18 individual channels, each containing a variable-gain charge-sensitive preamplifier, a variable shaper as well as track and hold and multiplexing. Via an 8-bit DAC, the voltage can be adjusted between 1 V and 5 V.

The gains are switchable in 16 steps in the range of 1 mV/pC to 100 mV/pC. High-gain operation is needed for the gain-calibration mode to resolve individual pixels, while low-gain operation ascertains that energies up to 100 MIPS can be measured. In the high-gain mode, a short shaping time of 40 ns is used to keep the noise contributions small. In the low-gain operation, the shaping time needs to be increased to 180 ns due to the beam trigger electronics. Figure 6 shows typical signal shapes for the gain-calibration and MIP-calibration modes.

THE DATA ACQUISITION SYSTEM

For data collection and storage a VME-based data acquisition system (DAQ) is used that was produced by the UK Calice group [8]. The readout board (RO) is a modification of the CMS Si-tracker readout board that was first adapted for the ECAL. The RO board performs different functions, providing the very front-end control, digitization and data buffering. The ADC is a 16-bit device allowing to record energies up to 100 MIPs. Since one RO card contains 8×216 channels, five cards are needed to read out the entire HCAL prototype. The DAQ accepts an instantaneous rate of more than 1 kHz and an

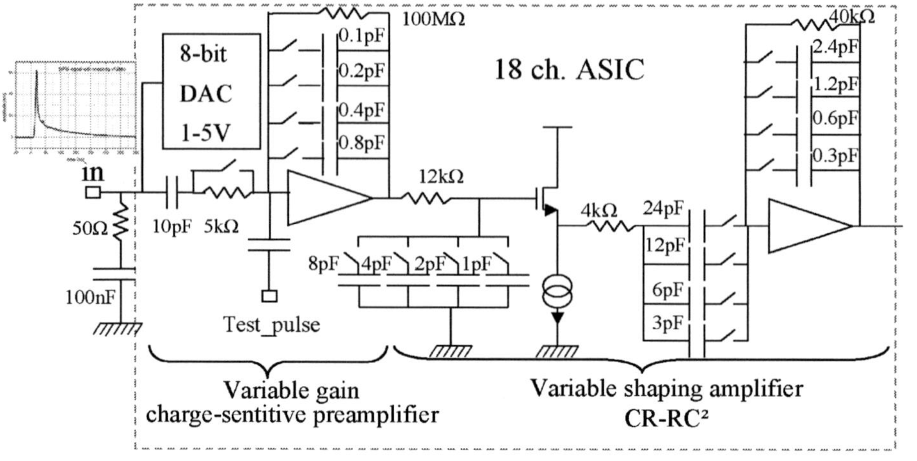

FIGURE 5. Layout of the preamplifier and shaping amplifier in the ILC chip.

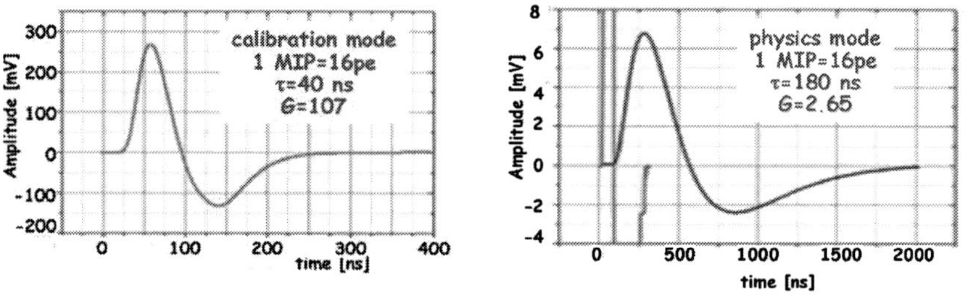

FIGURE 6. Signals in the gain calibration (left) and MIP calibration modes (right).

average rate of 100 Hz. The buffer can store up to 2000 events. Two CPUs are operating, one controlling the functions of the RO board and the other providing monitoring. The data are stored on a 3 TB disk.

Figure 7 depicts the ADC spectra for the gain calibration and the MIP calibration. To record the energy in units of MeV and for comparison with Monte Carlo simulations, intercalibration constants need to be determined. This procedure is discussed in detail in these proceedings by Erika Garutti [7].

THE CALIBRATION-MONITORING SYSTEM

For the test-beam operation, it is important to monitor between MIP calibrations the stability of the entire readout system (scintillator, WLS fiber, SiPM, preamplifier and readout electronics). In particular, we need to monitor the SiPM gain, SiPM response function at fixed LED intensity, and intercalibration constants. This task is achieved

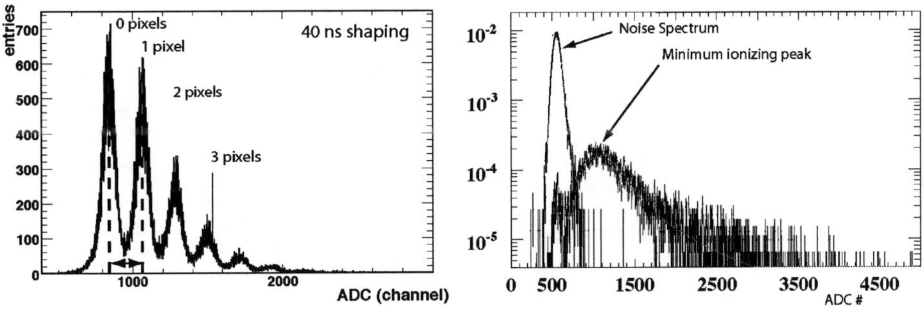

FIGURE 7. Signals in the gain-calibration (left) and the MIP-calibration modes (right).

with an LED/fiber system that distributes UV light from an LED via clear fibers to each tile. 19 fibers are coupled to one LED, of which 18 fibers distribute light to 18 tiles and one fiber is read out by a PIN diode to monitor the LED light output. Each HCAL plane is, therefore, equipped with 12 LEDs and 12 PIN diodes to illuminate all 216 tiles.

The light output of all fibers in one HCAL plane has been measured. For fixed LED intensity, the light illuminating the tiles lies within a factor of two between maximum and minimum intensities. This is sufficient for all our calibration and monitoring tasks. The low-intensity spectra of all channels clearly display the typical multiple pixel peaks that are well separated from the pedestals. In addition, the SiPM response function is measurable over the entire range.

In addition, an independent slow-control system operates that records the temperature of five sensors per module and the reverse-bias voltage of each SiPM. This is necessary, since the SiPM gain has a temperature dependence of $dG/dT \sim -1.7\%/K$ and a voltage dependence of $dG/dV \sim 2.5\%/0.1V$. These effects are further increased for the light collection of the entire cell at our operating conditions, yielding variations in the light output of $dQ/dT \sim -4.5\%/K$ and $dQ/dV \sim 7\%/0.1V$, respectively. From tests we know that the stability of the LED system after PIN diode correction is $< 1\%$.

SOFTWARE FOR TEST BEAM

Figure 9 shows a schematic layout of the analysis and simulation software. The basic idea is to use structured software, which is essential for a distributed effort. The development was conducted in close interaction with world-wide study (WWS) software group. The data format is LCIO and we use the same tools for the test-beam analysis and for the detector optimization.

CONCLUSION

Presently, 15 HCAL modules have been completed and installed at CERN for hadron test beams together with 27 slabs of the ECAL prototype. Data taking starts end of July

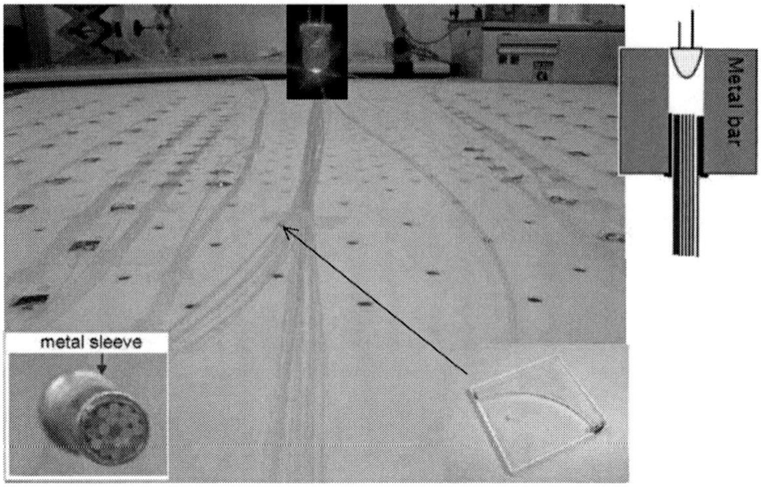

FIGURE 8. The LED-based monitoring system. The top insert shows a UV LED; the top-right insert shows the LED-fiber coupling; the bottom-right insert shows a tile to which the fiber is routed to; the bottom-left insert shows the arrangement of 19 fibers in a metal sleeve.

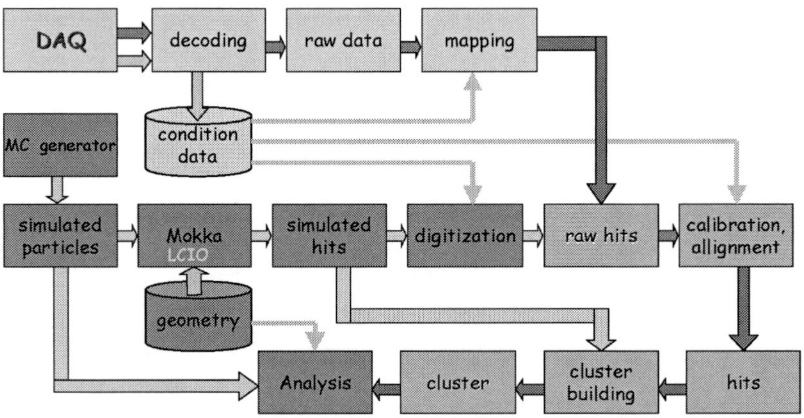

FIGURE 9. Schematics of the analysis and simulation software for test beam data.

for ten days resuming end of August and middle of October for two additional 10-day periods. Another 10-14 HCAL modules are expected to be completed for the October run period. The entire HCAL prototype will be completed by the end of 2006. It will operate 7606 cells of scintillating tiles with WLS-fiber plus SiPM readout, arranged in 38 layers. All tiles have been produced and tested, while about two-thirds of the SiPMs have been produced and tested. First radiation hardness tests look fine for ILC applications. The electronics is designed such that a common DAQ serves for both ECAL and HCAL. A calibration and monitoring system based on monitored UV light that is distributed to each tile via clear fibers allows to correct for temperature and voltage fluctuations and thus provides a stable system.

ACKNOWLEDGMENTS

I would like to thank the Calice collaboration for support, in particular E. Garutti, F. Sefkow and M. Danilov for helpful discussions. This work has been supported by the Norwegian Research Council.

REFERENCES

1. H. Videau, J.-C. Brient, Proc. of the 10th Int. Conf. on Calorimetry in Particle Physics, Pasadena, World Scientific, Singapore, p. 747; V. Morgunov, *ibid*, p 70 (2002).
2. F. Sefkow, Proc. of the 12th Int. Conf. on Calorimetry in Particle Physics, Chicago (2006).
3. V. Andreev et *al*., NIM A540, 368 (2005).
4. G. Bondarenko *et al.*, NIM A 442, 187 (2000).
5. G.Lindstrom NIM A512, 30 (2003).
6. S. Manen et *al*., physics/0501063 (2005).
7. E. Garutti, Proc. of the 12th Int. Conf. on Calorimetry in Particle Physics, Chicago (2006).
8. P.D. Dauncey, hep-ex/0211052 (2002).

CALICE scintillator HCAL commissioning experience and test beam program

Erika Garutti

DESY, Notkestr. 85, 22607 Hamburg, Germany.

Abstract. The active modules of the CALICE scintillator HCAL (or AHCAL for analog hadronic calorimeter) prototype are being commissioned using cosmic muons and the DESY electron test beam. These first data allow to check the functionality of the photo-detectors, the readout chain and the calibration electronics; and to test the performance of the entire system. This proceeding will discuss first operational experience, calibration and correction procedures, data analysis and results. The status of preparations for the first hadron beam measurements to start in summer 2006, and the data taking plans are also presented.

Keywords: silicon photomultiplier, scintillator, calorimeter
PACS: 07.20.Fw, 79.60.Ht

INTRODUCTION

The calorimeter for experiments at the future ILC must be realized as a dense and hermetic sampling calorimeter with very high granularity to allow the separation of the various particles in a jet and to use the best suited detector component to measure their four-momentum. The goal is to reach a jet energy resolution of $\sigma/E \sim 0.3/\sqrt{E}$. The success of this approach will originate more from the higher segmentation (both lateral and longitudinal) than from the stochastic and constant term in the energy resolution, both of which can be moderate. One possible realization of the calorimeter is proposed for the LDC detector and described in [1]. This contribution concentrates on the progresses in the prototype development of the scintillator-steel hadron calorimeter (AHCAL)[1] readout by pixilated avalanche photo-diode operated in Geiger mode, the Silicon-Photomultiplier (SiPM, [2, 3, 4]).
The 38 prototype layers covering about 4.5 interaction length consist each of a matrix of 216 scintillator tiles with on-tile SiPM readout via coaxial cable routing the analog signal to the Very Front-End (VFE) electronics. One such a layer will be referred to as module. A dedicated ASIC chip, developed by LAL (Orsay), is used for multiplexed readout of 18 SiPM. This chip offers various combinations of preamplifier gain and shaping time to best match the SiPM readout and calibration. A more detailed description of the AHCAL prototype structure, and commissioning status is given in [5].
This contribution focuses on a series of tests made during the commissioning phase of the first half of the modules. The aim of the studies was to establish the readout of a

[1] The institutes participating to this project are: DESY, Hamburg U, ICL (London), ITEP (Moscow), LAL (Orsay), LPI (Moscow), MEPHI (Moscow), Northern Illinois U., RAL, UCL (London).

CP867, *XII International Conference on Calorimetry in High Energy Physics,*
edited by S. R. Magill and R. Yoshida

large sample of channels; to develop a calibration procedure applicable to the ~8000 SiPM used to read out the AHCAL; and to test the monitoring of the SiPM responses in time.
The calibration of each tile is based on minimum ionizing particle signals such as cosmic muons or high energetic electrons from the DESY test beam facility. The full calibration of a calorimeter cell, though, requires to account for the non-linearity introduced by the finite number of pixels (1156/mm^2) in the SiPM. For this purpose a versatile UV-LED light distribution system was adopted, capable of delivering light to all tiles with intensity from few photo-electrons to the saturation of SiPM.
Furthermore, the LED system allows to monitor the variations of SiPM gain and signal amplitude; the latest being the product of gain and photo-detection efficiency, both sensitive to temperature and voltage fluctuations. The monitoring system ensures the applicability of calibration factors extracted in dedicated runs at the beginning and end of a data taking period of several days.

CALIBRATION OF CALORIMETER CELLS

Tests with fully equipped modules of AHCAL were performed with cosmics and electron test beam at DESY and with UV light from the LED monitoring system. Up to 4 modules (~900 channels) at a time were read out via the VME DAQ. The procedure to calibrate all calorimeter channels was tested, which requires to account for the signal non-linearity introduced by the limited dynamic range of SiPM.
The charge collected by a SiPM is a multiple of the Geiger mode gain corresponding to the number of pixels fired. To extract the information of light intensity (proportional to the energy deposited on tile) the number of fired pixels is converted into number of photo-electrons using a measurable response function

$$f_{resp} : N_{ph.e} = f_{resp}(N_{pixel}), \tag{1}$$

Fig.1 shows an example of the SiPM response curve from which f_{resp} is obtained. The horizontal axis of the function can be expressed in number of MIP given the light yield (LY) relation for each tile, i.e. the amplitude of a MIP signal expressed in number of pixels fired. The measured LY in the first AHCAL module is in average 15 pixels/MIP with a spread (RMS) of ~20%. This variation is too large to assume the LY value equal for all tiles therefore, the slope of the saturation curve has to be calibrated for each tile individually according to the equation

$$E[MIP] = f_{resp}\left(A[MIP] \cdot LY\left[\frac{pixel}{MIP}\right]\right) \cdot \frac{1}{LY'} \tag{2}$$

where $LY'[ph.e./MIP] = f_{resp}(LY[pixel/MIP])$.
The LY is determined using the SiPM gain (G_{pixel}) extracted from single photo-electron peak spectra. Such spectra are simultaneously obtained for all SiPM by fleshing low intensity LED light to each tile. For this measurement the VFE electronics is operated with the highest possible preamplifier gain (~90 mV/pC). The shortest shaping time

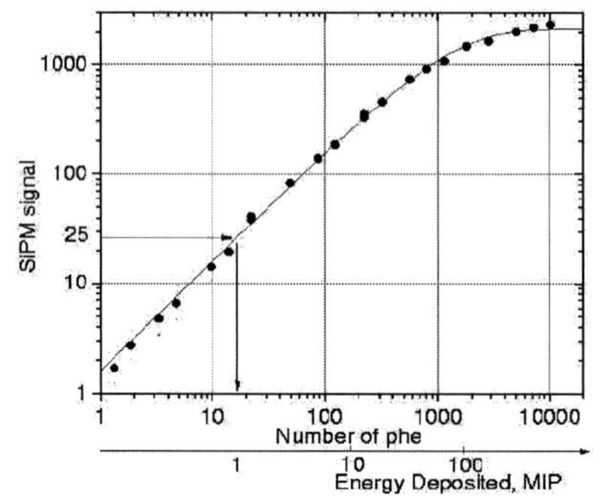

FIGURE 1. Response to increasing intensity LED light of a 1156-pixels/mm^2 SiPM.

of the ASIC chip is selected ($\sim$40 ns) to minimize integration of SiPM dark-rate and optimize single photo-electron peak separation.
Though more favorable from the noise point of view, the short shaping time is not sufficient to accommodate for trigger latency in the test beam setup (typically 150 ns). A shaping time of $\sim$200 ns is used when collecting physics data, in combination with a medium gain ($\sim$8 mV/pC) to match the range of energy deposited on one tile to that of the 16-bit ADC.
When extracting the LY ratio, between MIP and SiPM gain, it is necessary to account for these two modes of operation of the VFE electronics with an inter-calibration factor (I^{calib}_{phys}). Including all corrections Eq. 2 assumes the more complex form

$$E[MIP] = f_{resp}\left(\frac{A[ADC]}{A_{MIP}} \cdot \frac{A_{MIP}}{G_{pixel}} \cdot I^{calib}_{phys}\right) \cdot \frac{1}{f_{resp}\left(\frac{A_{MIP}}{G_{pixel}} \cdot I^{calib}_{phys}\right)}. \tag{3}$$

MIP CALIBRATION

All produced AHCAL modules have been commissioned at the DESY electron test beam and, a sub-sample of 4 modules, also in a dedicated cosmic muons test bench. The scan performed with the e^- beam through the module surface proves to be relatively faster than accumulating the required statistics with cosmic muons. The comparison of the extracted MIP calibration indicates good correlation between the two methods therefore, the faster one has been adopted for the mass test of all modules.
The MIP calibration for each tile is extracted from a fit to the ADC spectrum. Various fitting functions have been tested, for this results a Gaussian fit in the range of $-1.5/+1\sigma$ has been chosen. The fit result for a typical channel is shown in the left panel of Fig.2. The average signal to noise separation for MIP signals is 4.

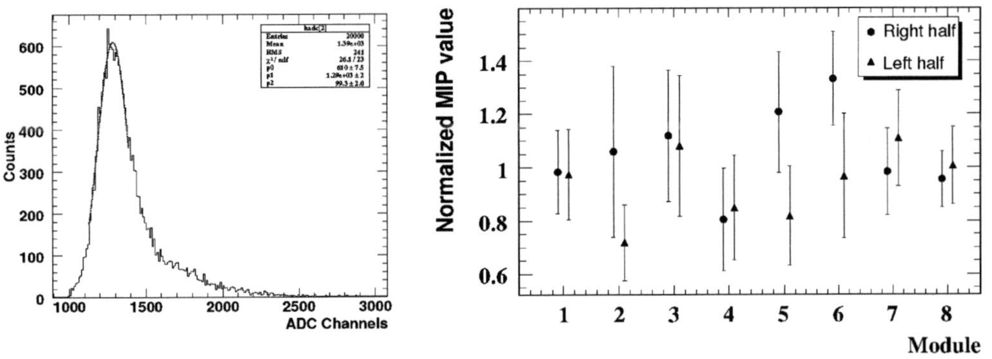

FIGURE 2. Left) A typical MIP spectrum from one AHCAL tile with SiPM readout. Right) Comparison of MIP calibration values for 15 modules. The error bars represent the RMS of 108 tiles per half module.

The right panel of Fig.2 is a compilation of all MIP calibrations for the first 8 commissioned AHCAL modules. The results are presented as average over each half module since in general the two halves can be made out of SiPM from different batches, powered via an independent bias voltage line. The averages are normalized to the mean value of all modules. The error bars represent the RMS of 108 tiles. The error on each MIP calibration is below 2% including the systematic of the fit. The spread of about 25% in MIP calibration for various tiles reflects both the SiPM gain and the LY spread.

UNIVERSALITY OF SIPM RESPONSE FUNCTION

Detailed studies on SiPM saturation correction have been carried out on the first produced AHCAL module in a dedicated setup, which allows to measure energies up to $\sim$40 MIPs per tile and compare the SiPM response with that of a linear photo-multiplier tube (PMT). As shown in Fig.3, a lead absorber plate of variable thickness is positioned on the beam line to initiate an electromagnetic shower. Fig.4 shows the correlation between the energy collected by the scintillator behind the lead plate (PMT2) and the SiPM tile behind it. Data (closed symbol) from a 5 GeV e^- beam showering on 5 X_0 lead are compared to the GEANT3 simulation of the setup (open symbol). Above 40 MIP amplitude the low statistics does not allow to extract sensible information from the data and the simulation was not performed. A clear deviation of the data from simulation is observed already above 15 MIP amplitude (>200 pixels). The largest fraction of the shower maximum, generated after 5 X_0 is in the range between 15 and 35 MIP. For this amplitude the deviation from simulation is $\sim$10-25%. After correcting on an event by event basis the SiPM amplitude according to Eq.3 the data is in very good agreement with the simulation, as it can be seen in the right upper panel of Fig.4.
To study the applicability of a unique SiPM response function to many channels the same test was repeated for a sample of 25 tiles in the core of the first AHCAL module.

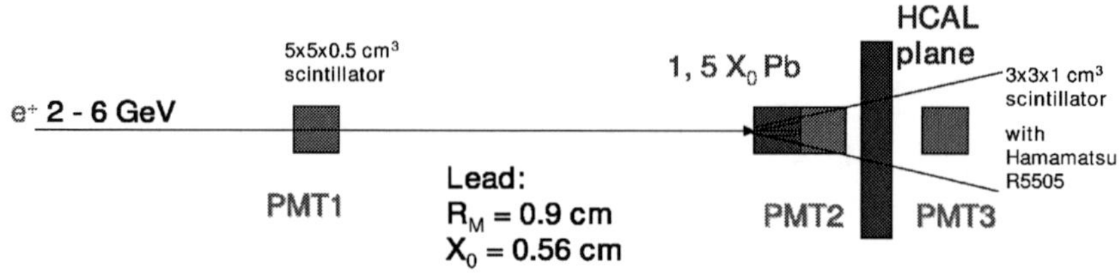

FIGURE 3. Sketch of the setup used for studies of SiPM saturation correction.

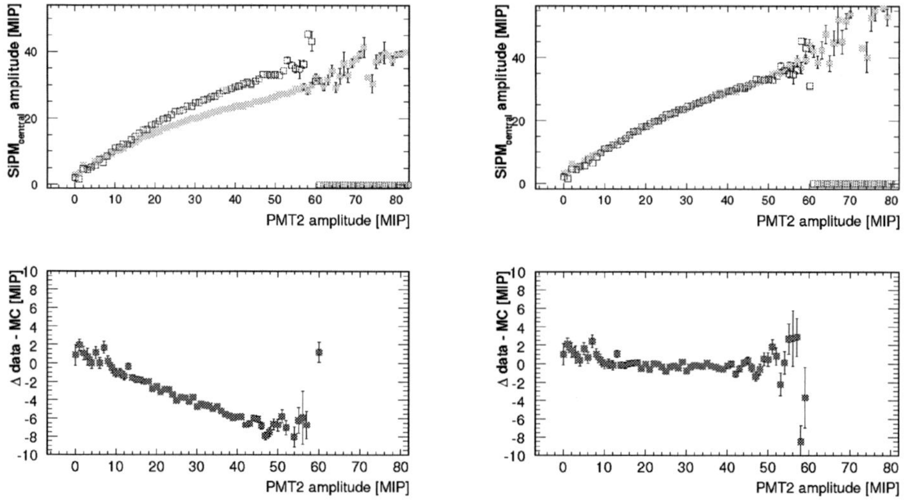

FIGURE 4. Comparison of SiPM and PMT amplitude for a 5 GeV e^- beam showering on 5 X_0 lead. Left) raw SiPM amplitude. Right) SiPM corrected for non-linear response. In the top two panels the data are closed symbol, while the open symbol are a GEANT3 simulation. The bottom two panels show the difference of data and simulation.

The average shower maximum in each tile normalized to the simulation prediction is presented in Fig.5, without (left) and with (right) non-linearity correction. The mean of the distribution for the tested tiles differ from simulation of $\sim$20% when no correction is applied. After correcting each tile response with the same SiPM response function calibrated with the appropriate LY, data and simulation agree very well.
The spread of 5% between tiles reflects the present uncertainty on all the calibration factors, and the fact that no temperature correction has yet been applied to account for the variations during the data taking period.
To correct the non-linearity of SiPM two methods have been used. The parameterization of the response function in Fig.1, measured in a laboratory setup; and a model function inspired by the linearization often applied to PMT ("PMT par"). The model is the first

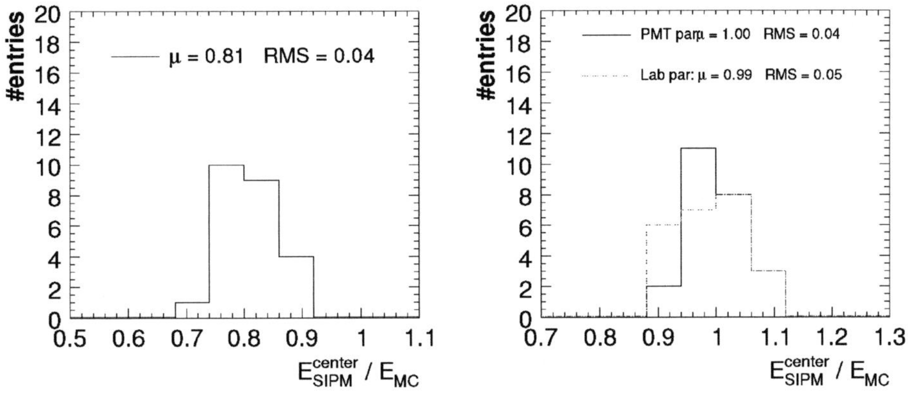

FIGURE 5. Average energy deposited in one AHCAL tile by a 5 GeV e^- beam showering on 5 X_0 lead, normalized to MC expectation, for a sample of 25 tiles. Left) raw SiPM amplitude. Right) SiPM corrected for non-linear response with two functions described in the text.

order expansion of the saturation function

$$f_{PMT\,par} : A_{ph.e} = -N_0 \cdot \left(1 - ln\left(1 - \frac{A_{pixel}}{N_0}\right)\right) \sim \frac{A_{pixel}}{1 - \frac{A_{pixel}}{N_0}}, \tag{4}$$

with effective number of fired pixels N_0 ~1500. In both cases the corrected amplitude agrees with the simulation expectations. The simplified model description works as well as the parameterization of the response function at the moderate amplitudes reached in this test. This observation needs to be corroborated by larger statistics and by measurements at higher amplitude.

UNIFORMITY STUDIES

A study of the uniformity of the calorimeter response has been performed using the same setup as presented in Fig.3 with 1 X_0 lead to initiate an electromagnetic shower. In this configuration the lateral extension of the shower is 99% contained in a matrix of 9 3×3 cm^2 tiles. It is intended to study the homogeneity of one calorimeter layer by comparing the energy deposited in many such matrices, in a region of amplitude where the SiPM saturation is relatively small.
The energy collected in the 9 tiles is in average ~7 MIP (~100 pixels), for which the non-linearity correction is ~10%. The spread between the 45 tile matrices is of the order of 5%, which is in the current uncertainty of these studies. At present this measurements are only indicative that the calorimeter uniformity must be better that 5%. Higher precision is expected when correcting for temperature variations, not yet possible when the data were collected.

MONITORING OF TEMPERATURE VARIATIONS

During the calorimeter operation possible variations in the system are detected by a threefold monitoring system. A slow control system reads the SiPM bias voltage and the temperature of each module with a 5 points interpolation. The SiPM gain is measured at the beginning of each run with low LED light intensity. The total temperature (T) and voltage (V) dependence of the SiPM gain at room temperature as measured by the manufacturer (MEPHI) is $\frac{dG}{dT} = -1.7\frac{\%}{°C}$ and $\frac{dG}{dV} = 2.5\frac{\%}{0.1V}$. Finally, the peak position of a medium amplitude LED signal is monitored. LED light fluctuations are corrected using the light amplitude read out by a photo-diode. Variations in the response to UV-LED light reflect the SiPM signal amplitude dependence on T and V. This dependence is expected to be larger than that of the gain given that the SiPM signal amplitude is the product of SiPM gain, quantum efficiency and Geiger efficiency; and that all three factors depend on T and V. According to the manufacturer the signal amplitude varies according to $\frac{dQ}{dT} = -4.5\frac{\%}{°C}$ and $\frac{dQ}{dV} = 7\frac{\%}{0.1V}$. The combined information from the monitoring system is used to correct the calorimeter response to an expected stability of 1-2%.

The measurements performed on one test AHCAL module over two weeks of continuous monitoring indicate a similar dependence on temperature variations. For this measurement the average of all monitored calorimeter tiles is taken assuming that they all undergo the same temperature fluctuation. Measurements of SiPM gain are repeated through out the day in dedicated low light-intensity runs. Otherwise the module is continuously fleshed with low-rate medium-intensity LED light to observe change in SiPM amplitude. The day-to-night temperature variation was within one degree, but the very high-precision temperature measurement and the low systematic error in the monitored values allow to extract the following results:

$$\frac{dG}{dT} = -1.7 \pm 0.1\frac{\%}{°C}, \frac{dQ}{dT} = -3.7 \pm 0.2\frac{\%}{°C},$$

which are in quite good agreement with the expectation from the manufacturers.

CERN TEST BEAM PLANS

At the moment 15 commissioned AHCAL modules have been installed on the H6 test beam line of the SPS at CERN. Having equipped every other layer of the steel sandwich structure ~80% of the total depth is equipped with half the foreseen granularity. The complete granularity coverage is expected latest for October.

Three data taking periods are planned in which the AHCAL prototype will be tested both as stand alone and with a prototype of the Si-W ECAL in front. The first two runs will have reduced longitudinal segmentation, while for the last run all 38 calorimeter active layers will be installed. For the last data taking period also a tail catcher and muon tagger should be fully equipped in the rear of the AHCAL, to ensure full hadronic shower containment.

The test program is quite rich, both electromagnetic and hadronic showers will be

investigated at energies ranging between 6 and 200 GeV, and under various incident angles to the calorimeters. The beam can be varied between: $\mu, e^{\pm}, \pi^{\pm}, p$, or $\bar{p}$.
While electromagnetic showers are correctly modeled in simulation tools like GEANT3 or GEANT4, the unprecedented longitudinal and transverse granularity of the data collected with the AHCAL prototype will allow a deeper understanding of hadronic showers, and will serve to discriminate among the large variety of existing hadronic models.
Furthermore, the operation of 8000 channels with SiPM readout will be a technological step toward establishing this new photo-detector for calorimetry applications

ACKNOWLEDGMENTS

I wish to thank the people who actually did all the analysis work presented in this proceeding: Nicola D'Ascenzo, Marius Groll, Benjamin Lutz, Hendrik Meyer, Sebastian Schätzel, Yuri Soloviev, Nanda Wattimena, and the many others who contributed to the realization of this project.
This work is supported by DESY and by the Helmholtz-Nachwuchsgruppen fond VH-NG-206.

REFERENCES

1. LDC outline document, http://www.ilcldc.org.
2. G. Bondarenko, P. Buzhan, B. Dolgoshein, V. Golovin, E. Guschin, A. Ilyin, V. Kaplin, A. Karakash, R. Klanner, V. Pokachalov, E. Popova, K. Smir nov, *Nucl. Instrum. Methods* A**442** (2000) 187.
3. P. Buzhan, B. Dolgoshein, L. Filatov, A. Ilyin, V. Kantzerov, V. Kaplin, A. Karakash, , F. Kayumov, S. Klemin, A. Pleshko, E. Popova, S. Smirnov , Yu. Volkov, Proceedings of the 7th Internaional Conference on Advance Technolo gy & Particle Physics, 717 (2002).
4. P. Buzhan, B. Dolgoshein, L. Filatov, A. Ilyin, V. Kantzerov, V. Kaplin, A. Karakash, F. Kayumov, S. Klemin, E. Popova, S. Smirnov, Nucl. Instr um. Meth. A **504** (2003) 48 (2003).
5. G. Eigen, proceeding of this conference.

A Hadron Calorimeter with Resistive Plate Chambers

José Repond

Argonne National Laboratory
Argonne, IL 60439, U.S.A.

Abstract. The application of Particle Flow Algorithms to the measurement of hadronic jets requires calorimeters with extremely fine segmentation. Within the PFA paradigm the role of the hadron calorimeter is to measure the energy of neutral hadrons, i. e. neutrons and K_L^0's. We propose to use Resistive Plate Chambers as active medium for the hadron calorimeter of a detector for the International Linear Collider. We present our plans for the construction of a 1 m^3 prototype section of a hadron calorimeter to be tested in the Fermilab test beam and the current status of the project is reviewed.

Keywords: Calorimetry, Linear Collider, Particle Flow Algorithms, Resistive Plate Chambers.
PACS: 29.40.Vj, 29.40.Cs, 29.40.Mc, 29.40.Wk

PARTICLE FLOW ALGORITHMS AND HADRON CALORIMETRY

Particle Flow Algorithms (PFAs) have been applied to existing detectors, such as ZEUS and CDF, to improve the energy resolution of hadronic jets. The algorithms attempt to measure all particles in a jet (originating from the interaction point) individually, using the detector component providing the best momentum/energy resolution. Charged particles are measured with the tracking system (except for high momenta, where the calorimeter provides a better measurement), photons are measured with the electromagnetic calorimeter (ECAL) and neutral hadrons, i.e. neutrons and K_L^0's, are measure with both the ECAL and the hadronic calorimeter (HCAL). The energy of a jet is reconstructed by adding up the energy of the individual particles comprising the jet. Additional details on PFAs can be found in reference [1].

At previous colliders, such as HERA and the Tevatron, PFAs have been applied to already constructed detectors as an afterthought to improve the jet energy resolution. In preparation for the International Linear Collider (ILC), on the other hand, detectors are being developed with emphasis on the optimization for the application of PFAs [2].

PFAs impose the following challenges and requirements on the design of the HCAL:

CP867, *XII International Conference on Calorimetry in High Energy Physics,*
edited by S. R. Magill and R. Yoshida

- In order to effectively identify energy deposits in the calorimeter belonging to charged or neutral particles, the readout needs to be extremely segmented, of the order of 1 x 1 cm^2 laterally and layer-by-layer longitudinally, thus eliminating the traditional 'calorimeter towers' of past calorimeters.
- The high segmentation of the readout leads to a large number of channels, of the order of 50 x 10^6 for the HCAL alone. In order to reduce the complexity and cost of the readout system, the front-end system needs to be located on the detector and be highly multiplexing.
- To provide excellent momentum resolution for tracks up to high momenta and to reduce the background from low energy particles swamping the inner radius of the detector, a large magnetic field of the order of 3 – 5 Tesla with its direction parallel to the beam axis is required. The magnetic field is to be provided by a superconducting coil with a considerable thickness, corresponding to 1 – 2 λ_I. To preserve the single particle resolution of the calorimeter both the ECAL and the HCAL need to be located inside the solenoid. Technologies which operate in high magnetic fields are required.
- To minimize the radius (and the cost) of the coil, the active medium needs to be as thin as possible, of the order of 10 mm or less.
- Given the difficulty to access the calorimeter in a colliding beam experiment, the active element and its electronic readout system need to perform reliably over the lifetime of the experiment, i.e. 10 – 20 years.
- To equip the entire active area of the HCAL, estimated at approximately 5,000 m^2, a cost-effective medium needs to be used.

RESISTIVE PLATE CHAMBERS

Resistive Plate Chambers (RPCs) are simple in design, robust, quiet, well understood, reliable and cheap to build [3]. The readout can be segmented to match the need of a variety of applications, from large area muon detectors to finely granulated time-of flight counters. In the past both Bakelite and glass has been used as resistive plates. Whereas there are some ageing issues with the Bakelite chambers, no ageing effects have been observed so far with RPCs built with glass plates and operated in avalanche mode [3]. Glass RPCs are therefore an ideal candidate for the active element of a finely granulated hadron calorimeter.

Designs for use in a Hadron Calorimeter

We [4] have investigated a number of design and operational options for the use of RPCs in hadron calorimetry. Chambers with three glass plates (two gas gaps), two glass plates (one gas gap) and only one glass plate (the PCB with the readout pads serves as anode) have been built and thoroughly tested. The design using a single glass plate is particularly attractive for calorimetry due to its reduced depth and greater simplicity. Measurements with two-gap chambers showed no advantage over chambers based on simpler designs. We, therefore, concentrated our efforts on testing chambers with either two or one glass plates, i.e. one gas gap. See Figure 1 and 2 for a schematic of the two default chambers designs.

The plates were regular, commercially available window glass with a thickness of 1.1 mm and a bulk resistivity $\rho \sim 4.7 \cdot 10^{12}$ Ωcm. The glass plates were held apart by fishing lines (from K-mart) to create a gas gap of 1.2 mm. A resistive paint providing a surface resistivity of $\sigma \sim 50$ M$\Omega/\square$ was applied to the outer surface of the glass plates. The thickness of the two-glass (one-glass) chambers, including the boards containing the readout pads and electronics, was approximately 5 (4) mm.

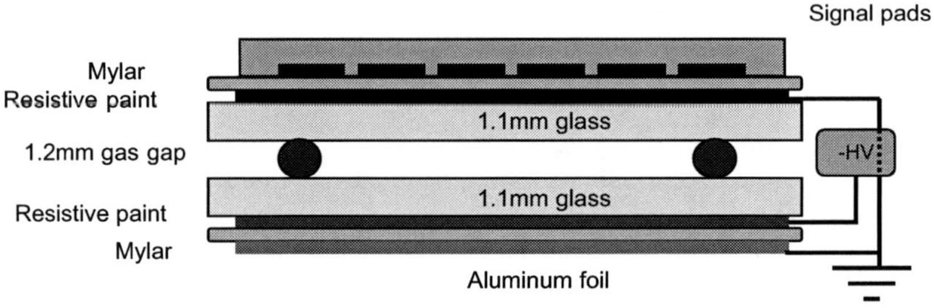

FIGURE 1. Schematic of the two-glass chambers. Not to scale.

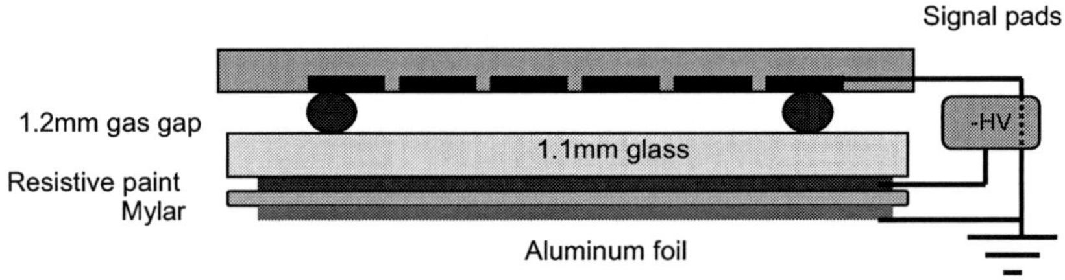

FIGURE 2. Schematic of the one-glass chambers. Not to scale.

The chambers were tested both in streamer and avalanche mode. Since the signals obtained in avalanche mode are quite large, in the range of 100 fC to approximately 5 pC, we saw no advantage in operating the chambers in streamer mode, which would have supplied signals about a factor 10 – 50 larger. We explored a variety of different gas mixtures, but then settled on using a mixture of R134A (94.5), Isobutane (5.0%) and Sulfur Hexa-fluoride (0.5%) as default gas. The readout pads were squares measuring 1 x 1 cm^2. Typically an array of 8 x 8 pads was read out simultaneously.

Measurements with prototype chambers

We performed a large number of measurements with cosmic rays, radioactive sources and with protons and pions from the Fermilab test beam at MT6. The chambers were readout either using a charge integrating data acquisition system with a 14-bit resolution or a so called digital system, with a single-bit resolution per channel. In the following we present selected examples of the measurements which were performed with the two- and one-glass chambers.

Figure 3 shows the single particle detection efficiency and streamer fraction as a function of the applied high voltage as measured with cosmic rays. The efficiency rises to values close to 100% for voltages above 7.6 kV. At values above

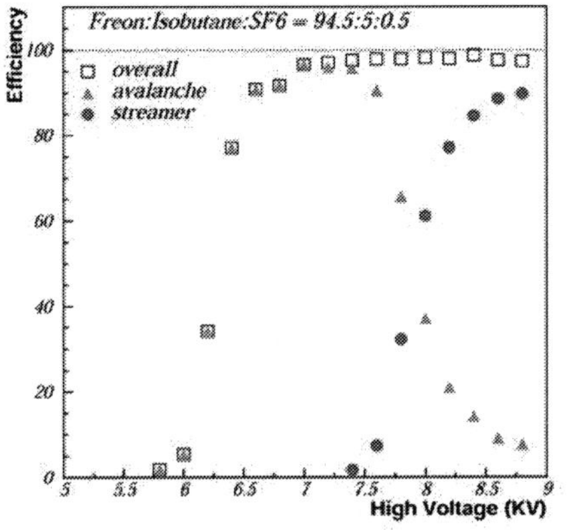

FIGURE 3. Single particle efficiency (open squares) and streamer fraction (red circles) as function of applied high voltage. The measurements were performed with a two-glass chamber.

about 9 kV streamers start to appear. These measurements were performed with a data acquisition system with a 14-bit resolution/channel.

Figure 4 shows the pad multiplicity as a function of the single particle efficiency as measured with the digital readout system. These measurements were performed with a fixed high voltage setting and for each measurement the threshold was adjusted to yield a given value of the efficiency.

It was found that the results are independent of the high voltage setting within the range of operation in avalanche mode. The two-glass chamber showed a strong increase in pad multiplicity with increasing efficiency. For an efficiency of 95% it was found that in average 1.6 pads showed signals above threshold. On the other hand, the hit multiplicity of the one-glass chamber was significantly lower, around 1.1, and appeared to be independent of the single particle efficiency. The higher pad multiplicity of the two-glass chamber is unfortunate, but in simulation studies did not show a negative impact on the overall jet energy resolution. Assuming that the higher pad multiplicity is a pure geometrical effect of the induced charge on the pad, the effect can be reproduced by assuming charge radius of 1.1 mm, which is consistent with measurements performed elsewhere [3].

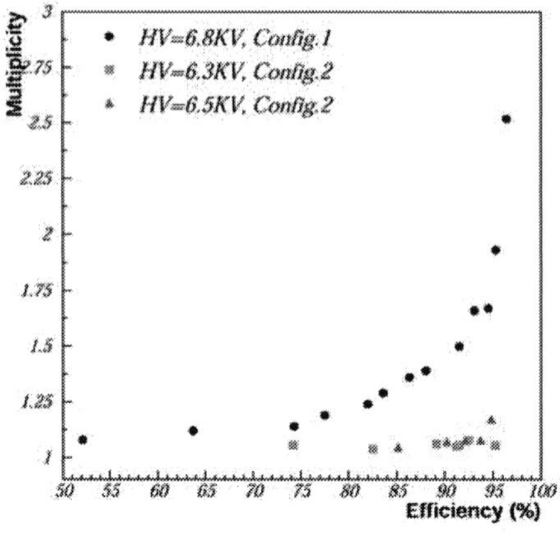

FIGURE 4. Pad multiplicity versus single particle efficiency for the two-glass chambers (black dots) and the one-glass chamber (green squares and red triangles).

In addition to the above measurements the noise rate (< 0.1 Hz/cm^2), the rate capability (> 50 Hz/cm^2), and the signal charges (100 fC to 5 pC in avalanche mode and above 5 pC in streamer mode) were measured as function of applied high voltage and with different gas mixtures. Other groups tested the chambers in high magnetic fields and found no significant effect as function of field strength and/or direction [5].

From these measurements we conclude that RPCs with either two or one glass plates are an excellent choice for a finely granulated hadron calorimeter for an ILC detector. The one-glass option offers the advantage of lower pad multiplicity and reduced thickness. However, to date the design has not yet been fully tested. In particular, long term effect of the gas on the readout pads need to be investigated before this design can be proposed for a larger system with a required life expectancy of decades.

Further R&D on chambers

Further R&D on the chambers is needed to perfect their application as active elements of a hadronic calorimeter for the ILC. In particular, as mentioned above, the one-glass design, with its advantage of low pad multiplicity and reduced thickness, needs to be further explored.

Due to their low density, gaseous active elements have a reduced sensitivity to slow neutrons in the 100 keV to several MeV range. Tests of calorimeter prototypes might indicate the importance of measuring this component of hadronic showers, even within the paradigm of PFAs. Ways of increasing the sensitivity to low energy neutrons need to be developed and tested.

ELECTRONIC READOUT SYSTEM

The large number of channels in a calorimeter optimized for the application of PFAs permits to greatly simplify the readout chain for each channel. Monte Carlo simulation of the response of such a finely granulated calorimeter to single hadrons showed that the number of pads hit (or with energy deposits above a certain threshold) is proportional to the energy of the incoming particle. The single particle energy resolution obtained with a digital readout (single-bit accuracy) was found to be comparable to the one obtained with analog readout (multi-bit accuracy). We, therefore, concentrated our efforts in developing a digital readout system, capable of handling large numbers of channels at a cost of about \$1 - \$2 per channel.

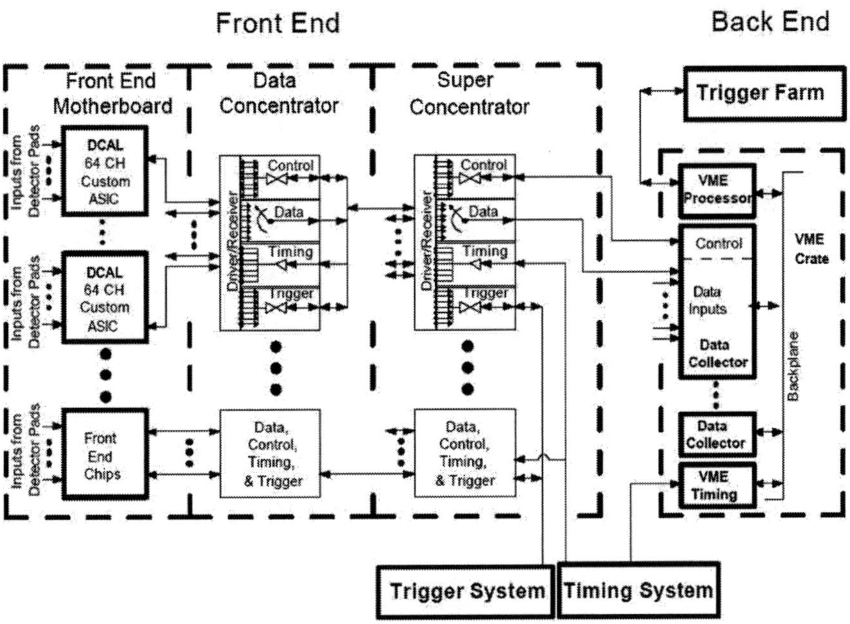

FIGURE 5. Schematic of the electronic readout system.

Concept of the electronic readout system

The electronic readout system consists of several stages: a) the front-end ASIC; b) the front-end boards, c) the data concentrators; d) the super concentrators, e) a VME-based data collector; and e) a trigger and timing system. See Figure 5 for a schematic of the system. In the following we briefly describe the individual stages:

a) The front-end ASIC receives signals from 64 individual pads. The signals are shaped, amplified, and discriminated. The resulting hit patterns are time stamped and stored at a speed of 10 MHz. In triggered mode, an external trigger selects the events to be passed on to the data concentrator. In triggerless mode any hit pattern with at least one hit will be transferred to the back-end of the readout system. The ASICs are located directly on, but on the opposite side of the PC boards containing the readout pads of the chambers.
b) The front-end boards are mounted directly on the outside of the chambers. The boards contain the readout pads and transfer lines to the ASIC (analog signals) as well as the path for the digital output signals and slow control to/from the ASIC. The boards contain 10 – 12 layers, including a number of ground planes for proper shielding of the readout pads from digital noise.
c) The data concentrators each receive data from 12 individual ASICs. They also provide the fast and slow control signals to the front-end ASICs.
d) The super concentrator introduces further multiplexing by reading out six data concentrators. Their design is similar to the data concentrators.
e) The data collector is VME–based and receives the output of the super concentrators. Each card connects to 12 individual data concentrators. The system specifications are very similar to the recently developed system for the MINOS test beam effort.
f) The trigger system distributes the trigger information to the data concentrators. The timing system provides the clocks and clock resets of the readout system.

The Front-end ASIC: the DCAL chip

The front-end ASIC has been designed by a collaboration between Argonne and Fermilab. A first prototype run was submitted in March 2005 and produced of the order of 40 chips. Two of these were mounted on test boards as seen in Figure 6.

FIGURE 6. Prototype DCAL chip mounted on a test board.

The chips underwent extensive testing and performed in all aspects as expected. Measurements with injected charges showed a uniform and linear behavior, the noise rate was found to be very low, the pipeline and time stamp counter worked as specified, etc. Figure 7 shows the average threshold corresponding to 50% efficiency as function of injected charge. The error bars in the figure indicate the RMS of the distribution for the 64 individual channels of the chip.

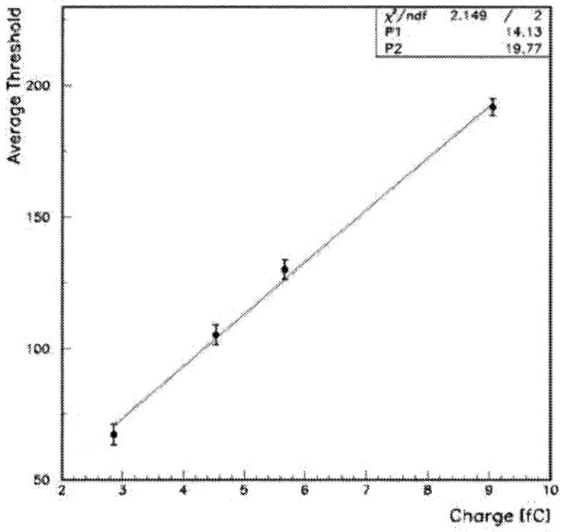

FIGURE 7. Average threshold for 50% efficiency as function of injected charge.

On July 22nd a second prototype run was submitted with two relatively minor modifications. The dynamic range was extended by two orders of magnitude, to be compatible with the large signals from RPCs, and the data transmission clock was decoupled from the readout chip clock.

The remainder of the system

The remaining subsystems of the electronic readout are currently being designed and prototyped. The front-end boards with their mixture of analog (pads) and digital signals (output of ASIC) pose a particular challenge. Prototype boards were used to measure the digital to analog cross – talk and to determine the optimal layout of the digital lines on the board. The cross – talk amounted to approximately 20 fC, to be compared to the signal range of 100 fC to 10 pC for RPCs operated in avalanche mode.

Further R&D on the readout system

Further R&D on the electronic readout system is necessary to optimize the design for the ILC hadron calorimeter. Depending on the results from the test beam (see below), the granularity of the readout might be increased beyond the 1 x 1 cm^2 used so far. The current timing resolution of 100 ns (clock of the front-end chip) might not be sufficient to reduce the background from 2γ events or beam-related backgrounds. The thickness of the front-end board and ASIC is currently approximately 3 mm. This thickness can be reduced to about 1 mm using newer technologies, such as imbedded ASICs. Both the front-end and the back-end of the system needs to undergo further multiplexing to handle the expected $50 \cdot 10^6$ readout channels of the hadron calorimeter of an ILC detector. In particular, the front-end ASICs might be read out via token rings, thus reducing the number of concentrator boards or eliminating their use all together. Finally, taking advantage of the time-structure of the ILC collisions, the front-end ASIC could be power-pulsed to avoid having to actively cool the chips.

PLANS FOR THE NEXT FEW YEARS

The next few years will keep us busy: we are currently preparing a vertical slice test, to be followed by the construction and testing of a 1 m^3 prototype hadron calorimeter. In addition to these activities, further developments of both the chambers and the readout system will take place.

Vertical slice test

The vertical slice test will test the entire readout chain including the RPCs. The test will involve up to eight chambers, each equipped with four ASICs, corresponding to 256 channels. The chambers and all components of the readout system will be as close as possible to the ones designed for the 1 m^3 prototype section. The chambers will be mounted in a mechanical structure containing steel absorber plates. This mini-calorimeter will be tested in Fermilab's MT6 beam line. The tests are planned for January 2007.

1 m^3 prototype section

Following the successful testing of the vertical slice, we will initiate the construction of a 1 m^3 prototype hadron calorimeter. This prototype section will contain 40 layers of steel plates, interleaved with RPCs as active element. Each layer with an area 1 m^2 will be equipped with three RPCs of 32 x 96 cm^2. The total channel count will be close to 400,000. Construction of the section is expected to last for most of calendar year 2007.

Measurements in test beam

The prototype section will be extensively tested in Fermilab's MT6 beam line. The tests will be performed either in standalone mode or with a prototype electromagnetic calorimeter located in front of it. The electromagnetic calorimeter will be provided by the CALICE collaboration [6] and consists of 30 layers of silicon – pad detectors interleaved with Tungsten plates as absorber.

The purpose of the tests will be to validate the approach of digital calorimetry to the measurement of single hadrons, to validate the use of RPCs as active element in a hadron calorimeter and to provide detailed measurements of hadronic showers. The latter are essential to fine tune existing models of hadronic showers, where differences of up to 60 % e.g. in shower radius, have been reported [7].

The tests will include protons with energies up to 120 GeV and pions in a wide momentum range from 1 GeV/c to 66 GeV/c. For a subset of the measurements the section will be tilted compared to the beam axis to study effects of non-vertical incidence on the calorimeter. The tests will last at least 1 – 2 years.

CONCLUSIONS

The application of Particle Flow Algorithms to the measurement of hadronic jets requires finely granulated calorimeters. Resistive Plate Chambers offer an excellent choice as active element of such a calorimeter. R&D on the chambers has been virtually completed with a large number of measurements using radioactive sources, cosmic rays and particle beams. An electronic readout system capable of handling large number of channels at a cost of approximately $1 - $2 has been developed. Preparations for a vertical slice test are ongoing. Assuming its success and sufficient funding, the construction of a 1 m^3 prototype hadron calorimeter will initiate in early 2007 and lead to an extended testing period in the Fermilab test beam.

REFERENCES

1. See session on Particle Flow Algorithms, these proceedings.
2. Three of the currently four ILC detector concepts emphasize the application of PFAs:
   SiD http://www-sid.slac.stanford.edu/
   LDC http://www.ilcldc.org/
   GLD http://ilcphys.kek.jp/gld/
3. See contributions to the proceedings of the VII Workshop on Resistive Plate Chambers, Clermont-Ferrand, France (October 2003), Nucl. Instr. and Meth. **A533**, (2004).
4. The collaboration includes groups from Argonne National Laboratory, Boston University, University of Chicago, University of Iowa, Fermilab and University of Texas in Arlington. The majority of collaborators are members of the CALICE collaboration, see [6].
5. V.Ammosov, private communication.
6. Home page of the CALICE collaboration: http://polywww.in2p3.fr/activites/physique/flc/calice.html
7. G.Mavromanolakis, private communication.

Tail Catcher Muon Tracker For The CALICE Test Beam

Alexander Dyshkant

Northern Illinois University, DeKalb, IL 60115, USA
(For the CALICE collaboration)

Abstract. Results on the construction and commissioning of the CALICE Tail-catcher/Muon Tracker (TCMT) are presented. The ~1 m^3 prototype uses extruded scintillating strips mated to silicon-photomultipliers and is located behind the CALICE hadron calorimeter. The TCMT will provide a snapshot of the tail end of hadron showers, which is crucial to the validation of hadronic Monte Carlo. It will also serve as a prototype muon system for any Linear Collider Detector (ILC) and will facilitate studies of muon tracking and identification within the particle flow reconstruction framework. Additionally, the TCMT will provide valuable practical experience with hadronic leakage and punch-through from thin calorimeters, as are envisaged for the ILC detector, and the impact of the coil in correcting for this leakage.

Keywords: Linear collider detector, scintillation detectors, muon tracker, hadronic leakage.
PACS: 29.40.Vj, 29.40.Mc

INTRODUCTION

To take full advantage of the physics potential of the future electron-positron International Linear Collider (ILC), the ILC detector [1] should have unprecedented performance. For example, hermeticity and resolution constraints require that the calorimeter be placed inside a superconducting coil. Consequently, the hadron calorimeter must be compact and limited in depth to about 1m. The relatively thin hadron calorimeter requires additional calorimetric sampling behind the superconducting coil to estimate and correct for hadronic leakage.

A US-European group is pursuing the construction of a scintillator-steel prototype Tail-Catcher and Muon Tracker (TCMT) to validate a simulation of hadronic showers. This desire is driven by application of particle flow algorithms for the measurement of hadronic jets in the design of a detector for the ILC. In the frame of the CALICE collaboration, the construction of the TCMT involves DESY, Northern Illinois University (NIU), and Fermilab. The measurements obtained in particle beams will be essential for development of a realistic simulation of hadronic showers. CALICE is going to do that using electromagnetic (ECAL) and analog hadronic calorimeters (AHCAL), and TCMT prototypes.

The Northern Illinois Center for Accelerator and Detector Development (NICADD) at NIU proposed and is leading construction of the TCMT prototype. The TCMT (Fig.

CP867, *XII International Conference on Calorimetry in High Energy Physics,*
edited by S. R. Magill and R. Yoshida

1) consists of 8 layers of 2 cm thick and 8 layers of 10 cm thick steel absorber and 16 active medium cassettes, all with dimensions about 1m x 1m.

FIGURE 1. CALICE TCMT has a 142 cm length, 109 cm height, and weight of about 10 tons.

Each cassette has ten extruded scintillating strips as the active medium. Each strip is 5mm thick, 10cm wide, and about 1m long [2]. TCMT cassettes will be installed between the absorber plates with scintillating strips in an alternating x-y orientation. A wavelength shifting (WLS) fiber is inserted into the scintillating strip through a co-extruded hole and is not glued. The WLS fiber is connected to a silicon photo detector, used as readout. NICADD will explore the TCMT prototype performance with test beams.

CONSTRUCTION ELEMENTS

Quality Control For Strips

To estimate the available amount of light, a 1 m long strip with a small co-extruded hole was tested with cosmic rays. The inner diameter of the co-extruded hole was about the outer diameter of the WLS fiber (1.2mm). The WLS fiber was not glued inside the hole. That extruded strip provided about 15 photoelectrons.

Initially TCMT extruded scintillating strips were 1.05 m long, 10 cm wide, and 5 mm thick and had two co-extruded holes along the strip and about 25 mm from the edges. A co-extruded hole's inner diameter was not less than 1.2 mm. At 50 mm from the strip edge, a 1.01 m long and 0.9 mm wide separation groove was milled throughout the strip. The separation groove was injected with white epoxy. The white epoxy was used as reflective material, improved the structure rigidity, and reduced crosstalk between the created strips. The ends of each strip were cut off to the 1.01 m separation groove length.

The strip was wrapped by insertion into a 1 m long sleeve made from one layer of Tyvek. All strips were measured using the same Tyvek sleeve. The strip ends were not painted or wrapped. A 5 mCi Sr-90 radioactive source was positioned on the top at 15mm from the edge of the strip. A Kuraray Y-11, 1.2 mm outer diameter, multiclad, WLS fiber connected to the Hamamatsu PMT R580 was used for readout. The PMT output current was measured with a Keithley 6485 picoammeter, connected to a computer-based data acquisition system.

A few strip responses were significantly lower than the average response. Detailed measurements of the extruded strips and analysis of the strip responses easily identify poor quality strips. The total number of "bad" strips is about 3% for two production runs. For the strips in one cassette the responses are all within 10%.

Quality Control For WLS Fibers

The main goal of the quality control (QC) studies for the TCMT WLS fibers was to identify and remove those fibers with significantly lower response than average. WLS fibers may have lower response because of faulty polishing, faulty mirroring, or fiber problems (such as cracks). After the QC of the extruded scintillating strips [3], the WLS fiber QC was the second step in the TCMT assembly. A double reference method was used in a low background system to test the response of a few hundred WLS fibers.

The TCMT requires 320 WLS fibers. KURARAY Y-11, multiclad, 1.2 mm outer diameter WLS fibers were used. The WLS fiber length is 1035mm, about one inch longer than the strip. The additional fiber length is used for a collar that captures the fiber, provides exact positioning, and holds the fiber end in front of the photo detector active area (1mm^2). The fiber was inserted into the strip via a co-extruded hole. The co-extruded hole position may deviate up to 2mm from the collar center. After the collar, the fiber end displacement from the center of the photo detector active area was within 0.1mm.

Fibers were cut from a reel. The fiber ends were polished using a fly diamond technique. One end of the fiber was mirrored using aluminum sputtering, and the mirror was UV epoxy protected. The UV epoxy protection was in a form of a clear thin film around the mirrored fiber end. The fibers were processed in two groups; the results for the larger group are described here. The other group had similar results.

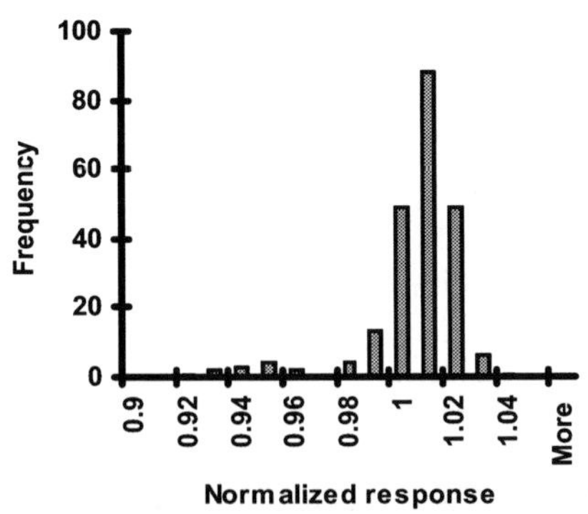

FIGURE 2. Response of 222 WLS fibers normalized to the average value. 210 fibers are in the peak (94%).

The fibers were tested by exciting them with light from a scintillator irradiated with a radioactive source Sr-90 (5mCi). WLS fibers connected to the Hamamatsu PMT

R580 were used as readout. The PMT output current was measured with a Keithley 6485 picoammeter connected to a computer-based data acquisition system [4]. Fig.2 shows the normalized response of a few hundred WLS fibers. The distribution of WLS fibers looks peaked compared to a normal distribution. All WLS fibers used for the TCMT cassettes have responses within 1%.

TCMT Calibration Monitoring System

The silicon photo detector response is bias voltage and temperature sensitive. Each TCMT strip was instrumented with a BIVAR UV LED. The LED driver provides an opportunity to monitor SiPM gain by recording a single photoelectron spectrum at a low light level and a linear range by recording saturation, which is sensitive to the amount of dead micro-cells or pixels at a high-light-level, for each SiPM. The LED driver board is under commissioning and is based on a current driver to reduce temperature sensitivity.

Readout Tests

With one TCMT cassette instrumented with 18 silicon photo detectors, all elements of the readout chain were tested, including common readout with AHCAL and positron beam tests at DESY in November 2005. At DESY the TCMT cassette with electronics was included in the electronics and data acquisition system already in place for the AHCAL test stand.

BEAM TESTS

Positron Beam Tests At DESY

For the November 2005 beam test at DESY 3 GeV/c positrons were used.

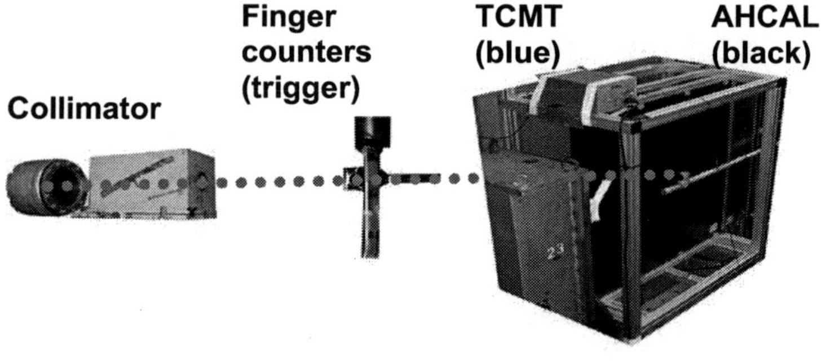

FIGURE 3. Beam test layout at DESY (not to scale).

The TCMT cassette was installed in front of the AHCAL (Fig. 3) on a movable platform. A coincidence between two finger counters was used for triggering. The general dimension of the beam spot was about 5 cm. During the TCMT runs the data for AHCAL was recoded as well (Fig. 4). The platform remote control movement was available in vertical and horizontal directions for full TCMT cassette dimensions. For the transverse scan in Fig. 5 the cassette was moved in the vertical direction in small steps. The response of each strip was normalized to its maximum. Gaps between the neighboring strips are very similar. One of those gaps of a chosen strip is a separation groove. The cross talk through the reflections from the wrapping material is less than 10%.

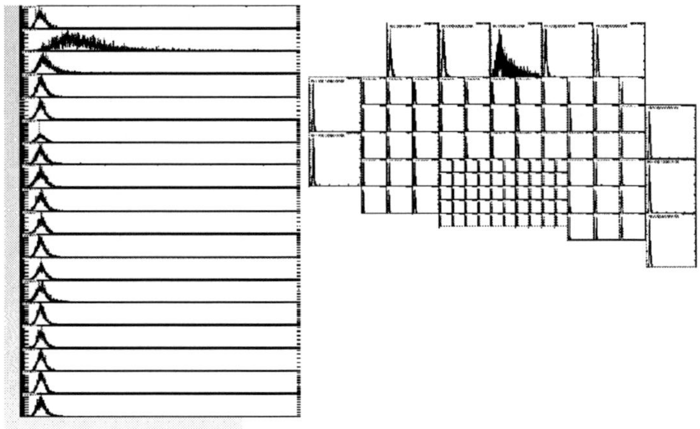

FIGURE 4. 3 GeV/c positron events recoded in TCMT and AHCAL at the same time.

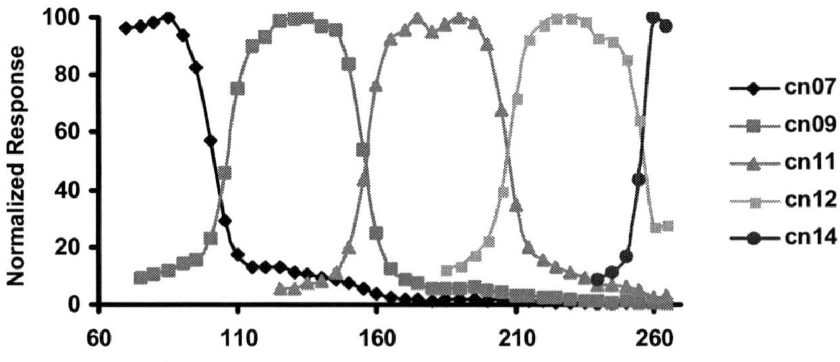

FIGURE 5. Cassette transverse scans with positron beam.

Hadron Beam Tests At Fermilab

On February 2006 a few beam test were performed at Fermilab using a set of hardware and software not used at DESY (Fig. 6). The cassette was exposed to 120 GeV/c proton, 16 GeV/c pion, and muon beams. A coincidence of two finger counters was used for triggering. The electronics box and connecting cables had an additional shielding connected to ground. The length of cables from the cassette to electronic board was 1.5 m. The micro miniature coaxial RADIALL connectors were used.

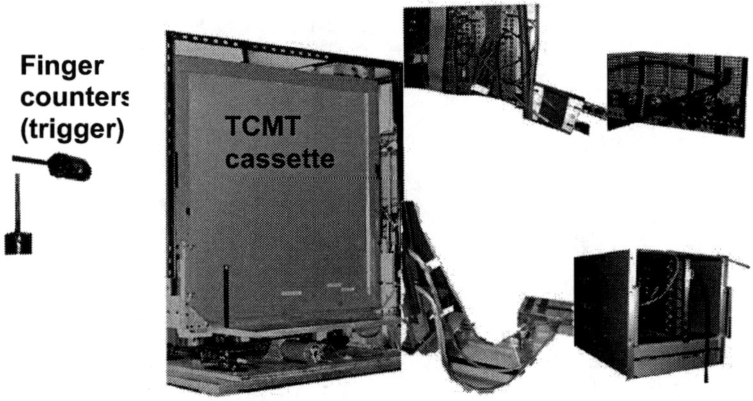

FIGURE 6. Apparatus for beam tests at Fermilab in February 2006. The top right images illustrate the triggering connections. The middle image shows the cassette connected to the electronics board with coaxial cables. The electronic board is connected to the driver panel of the VME crate via the optical cable.

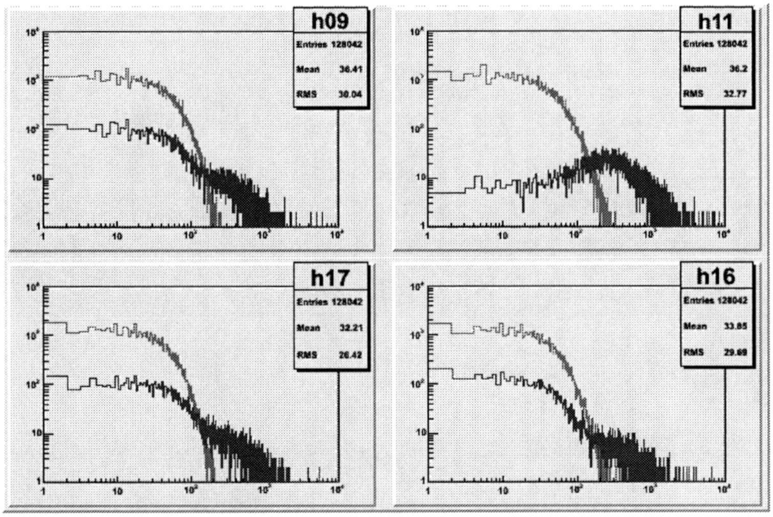

FIGURE 7. Bottom lines represent response of consecutive strips (numbered 09, 11, 17, 16) to 120 GeV/c protons. Beam is in channel 11. Top lines are the pedestals only.

The center contact carries a signal from SiPM. The outer contact carries the operating bias voltage to the SiPM. The example of the cassette response to the hadron beam is in Fig. 7.

SIMULATIONS OF SID WITH TAIL CATCHER FOR ILC

Nowadays the SiD ECAL and HCAL may have 4.6 nuclear interaction lengths (λ) of steel. The solenoid coil and cryostat have 1.27 λ and will sit between 2.37 m radius (outside of HCAL) and 3.5 m radius, where the outer solenoid flux return layers of stainless steel plates with gaps will start. It is at this radius 3.5 m where consideration of a tail-catcher would be natural Fig. 8.

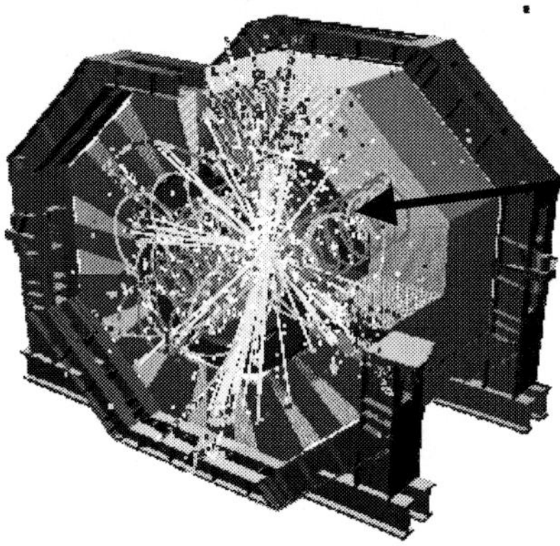

FIGURE 8. In SiD detector a natural tail-catcher would be implemented with modified first layers of the outer solenoid flux return layers of stainless steel plates.

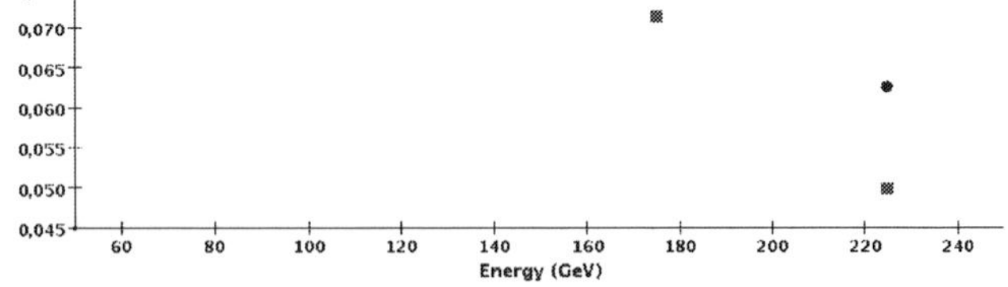

FIGURE 9. b-jets energy resolution simulated by sidaug_tcmt detector model.

A detector model (sidaug05-tcmt)[5], which incorporates a tail-catcher, was used for the simulation of b-jet energy resolution. In this model the ECAL had 5x5 mm^2

active cells; the HCAL had 10x10 mm^2 cells in the active layers and stainless steel plates 20 mm thick in the passive layers; the tail-catcher had 48 layers of 20 mm thick stainless steel plates with 3 mm thick G10, and it had 30x30 mm^2 cells dimensions in the active layers. In Fig. 9, the b-jet energy resolution in the detector with a tail-catcher (squares) is compared with b-jet energy resolution in the "same detector", but not including information from the tail-catcher (dots). The pilot simulation of b-jets in the full SiD model with a tail-catcher demonstrates improvement in b-jet energy resolution.

CONCLUSION

The CALICE TCMT is ready for beam tests. The TCMT cassette readout electronics have been integrated into the AHCAL DAQ chain. The TCMT prototype construction has made good progress with one cassette assembled and all other cassettes ready for SiPM installation. The TCMT will be ready for beam tests at CERN in 2006 with a number of cassettes.

Pilot simulations of b-jets in a full SiD model with a tail catcher demonstrate improvement in b-jet energy resolution.

ACKNOWLEDGMENTS

The authors would like to thank Phillip Stone for excellent mechanical and machining support. This work was supported in part by the U.S. Government of Education grant #P116Z010035, the Department of Energy, the National Science Foundation, and the State of Illinois Higher Education Cooperation Act. The work at Fermilab was supported by the U.S. Department of Energy under contract #DE-AC02-76CHO3000.

REFERENCES

1. ILC Detector Design and R&D, H. Weerts, ILC Meeting at Fermilab, October 27, 2004.
2. A. Pla-Dalmau, A. Bross, V. Rykalin, "Extruded Plastic Scintillator at Fermilab" FERMILAB-Conf-03-318-E, 2003
3. A. Dyshkant, D. Beznosko, G. Blazey et al., "About NICADD Extruded Scintillating Strips" FERMILAB-PUB-05-010-E, 2005.
4. A. Dyshkant, D. Beznosko, G. Blazey et al., "Quality Control Studies of Wavelength Shifting Fibers for a Scintillator-based Tail Catcher Muon Tracker for Linear Collider Prototype Detector" FERMILAB-PUB-06-129-E, 2006.
5. http://confluence.slac.stanford.edu/display/ilc/sidaug05 tsmt

Calorimeters in the Very Forward Region of ILC

Jinlong Zhang
for FCAL Collaboration

University of Colorado
Boulder, CO 80309

Abstract. Two calorimeters, BeamCal and LumiCal, are considered in the very forward region of the International Linear Collider project. BeamCal is important for possible SUSY event detection and fast beam diagnostics. LumiCal is the luminosity detector. Both of them also improve the hermeticity of the detector and shield the central detectors from backscattered particles. Because these two detectors cover the very small polar angles close to the beam pipe, the instrumentation depends on the beam crossing angle and is very challenging. A diamond-tungsten sandwich structure for BeamCal, silicon-tungsten sandwich structure are proposed. The Monte Carlo simulation results to optimize the calorimeter geometry are presented and the performance is estimated. The design difficulties and solutions with respect to different beam crossing angles are discussed. The early R&D progress is reported.

PACS: 29.17.+w, 13.66.Jn, 29.40.Vj, 24.10.Lx

INTRODUCTION

Several concept designs of the detector for the International Linear Collider project (ILC) are being developed. The layout of the very forward region of two of them, SiD [1] and LDC [2], is shown in Figure 1. The detailed configurations are very different but two calorimeters are proposed in common, the beam calorimeter (BeamCal) and the luminosity calorimeter (LumiCal). The following sections are mainly based on the LDC forward region design.

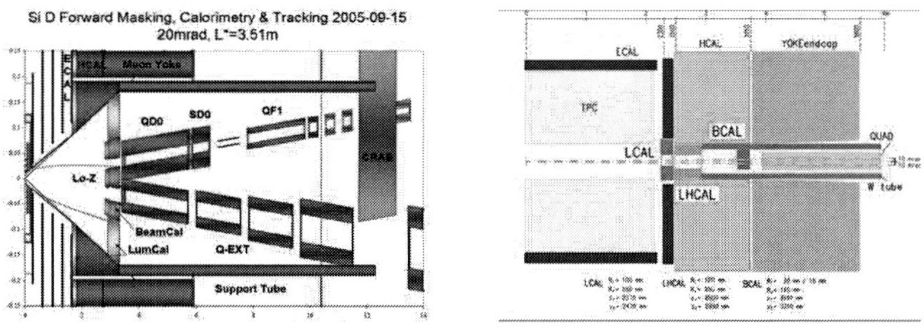

FIGURE 1. The very forward region of SiD (left) and LDC (right).

The precision requirement on the luminosity measurement for ILC can be estimated with the statistics of the interesting physics processes. At $\sqrt{s} = 500$ GeV, for example, the process $e^+e^- \to W^+W^-$ has a cross section of about 10 pb and the process $e^+e^- \to f^+f^-$ about 5 pb. So data samples of $O(10^6)$ events are expected in a few years which require the LumiCal measure the luminosity with a precision of better than 10^{-3}.

CP867, *XII International Conference on Calorimetry in High Energy Physics,*
edited by S. R. Magill and R. Yoshida

In the supersymmetry model (SUSY) the sleptons are heavy particles. Sleptons are pair-produced, then each decays to a regular lepton and a neutralino which escapes with energy. Because the relative orientation of the slepton decay is arbitrary, the two leptons from the slepton pair are usually very acoplanar. In most cases lepton pairs produced via two-photon process have significantly different topology and kinematics, and can be rejected by simple cuts. However, because the cross section of the two-photon process is typically five orders of magnitudes larger than that of the SUSY process, events in the tail of P_t distribution of the spectator electron become important. If the spectator electron of those events is in the acceptance range of the BeamCal, the produced lepton pair could be acoplanar enough to be like a SUSY event. Such events must be vetoed by the BeamCal.

The design of the LumiCal and the BeamCal is not heavily dependent of other subdetectors, but the accelerator parameters like the beam crossing angle. Based on the studies in the world wide community, a few possible beam crossing angles are under consideration, 2 mrad, 20 mrad and 14 mrad. Different versions of LumiCal and BeamCal are being proposed according to these crossing angles.

For a 2 mrad crossing angle the LumiCal performance is similar to the head-on beam configuration. The pair background in the BeamCal is also very similar to that of head-on configuration. The only change to be considered is for the BeamCal to increase the inner radius slightly.

For a 20 mrad crossing angle, the deviation of the beam in the vertical plane at the interaction point (IP) need be compensated in order to preserve the knowledge of the beam polarization. This local compensation of IP angle can be done with the detector integrated dipole (DID) field configuration, which embeds dipoles into the detector coil. The incoming beam will not suffer more synchrotron radiation and spin precession because of DID but the outgoing beam past IP will see increasing field and more pair background dispersion. Subsequently radiation background will increase in the tracking detectors due to backscattering. On the contrary, if the polarity of the detector integrated dipole is reversed (anti-DID), the incoming beam will be followed by more pair background but the outgoing beam not. Depending on which configuration is used, the geometry and the positioning of both LumiCal and BeamCal need be optimized to meet the physics requirement and to survive the background. The instrumentation will be more difficult than that for the head-on beams.

Additionally a 14 mrad crossing angle is proposed to reduce the interaction of the incoming beams with the magnetic field, then to decrease the synchrotron radiation.

LUMICAL

The LumiCal consists of two identical calorimeters positioned at $z = \pm 2.37$ m with respect to IP. Each one covers the polar angle range of $44<\theta<155$ mrad. Longitudinally, each has 30 layers of silicon sensors, interleaving with the tungsten absorber. The active sensor thickness is ~ 500 μm and the absorber thickness is 3.5 mm. One layer corresponds to a depth of about one radiation length (X_0).

Design Optimization and Performance

The relative uncertainty of luminosity is proportional to the relative uncertainty of the Bhabha scattering cross section, thus approximately proportional to the bias of θ measurement from the detector. So a LumiCal design must be able to control the angular bias to $< 10^{-3}$. Two possible designs are under investigating. For a strip design, each sensor layer has either 64 concentric strips or 120 radial strips. The angular bias is at 10^{-7} level. For the pad design, a basic geometry was proposed with each sensor layer divided into 15 rings radially and 24 sectors azimuthally. But the angular bias is two orders larger. In order to improve it the maximum shower peak design can be taken as a good alternate, in which the layers in the vicinity of the maximum shower position are finely granulated. From IP side onward, the sensor layers ate radially segmented as the following:

- 10 rings in the first four layers,
- 60 rings in the middle 15 layers,
- 10 rings in the last 11 layers.

The full Monte Carlo (MC) simulation and reconstruction are performed to study the performance. Bhabha scattering events are generated with the BHLUMI [3] and BHWIDE [4] generators. The CIRCE [5] program is used to generate beamstrahlung loss. The LumiCal detector is fully simulated with the GEANT4 [6] package. The shower position is reconstructed from the energy depositions on the sensors using a weight function. After energy and position have been reconstructed, the following back-to-back cuts are applied:

- energy balance cut $|E_L - E_R| < 0.1E_{min}$, where $E_{min} = \min(E_L, E_R)$ and E_L and E_R are the energy deposited in the two LumiCal calorimeters;
- acoplanarity cut $|\Delta\phi| < 5^o$;
- acollinearity cut $|\Delta\theta| < 1$ mrad.

For the nominal accelerate parameters, the expected performance of this maximum shower peak PAD design and the strip design are summarized in Table 1.

TABLE 1. The expected performance of the two LumiCal designs.

Parameter	**Pad design**	**Strip design**
Energy resolution	25%	
θ resolution	3.5×10^{-5} rad	2.1×10^{-5} rad
ϕ resolution	$10^{-2} rad$	10^{-3}
$\Delta\theta$	1.4×10^{-6}	2.1×10^{-7}
$\Delta L/L$	$< 10^{-4}$	$< 10^{-4}$
Number of electronic channels	25,200	8,000

Both designs meet the required precision. The strip design is more economical in terms of readout electronic channels, but further studies of physics background and sensor properties are needed.

Systematic Effects at Non Zero Beam Crossing Angles

The LumiCal axis aligns with the beam line for the head-on beams. When the beams have a non zero crossing angle there will be a half size angle between the detector axis and the incoming and outgoing beams. Studies have been done to see the systematic effect due to possible geometry shifts of the LumiCal for two possible alignments, with the LumiCal centering on the detector axis or the outgoing beam axis. Certainly if the LumiCal is centered on the detector axis the ϕ symmetry of the Bhabha scattering events will be lost. Then the Bhabha event distribution in the laboratory frame depends on not only the polar angle θ but also the azimuthal angle ϕ due to the boost, furthermore the systematic uncertainties of luminosity become ϕ dependent.

For a possible geometry shift of the LumiCal from its nominal geometry in a particular alignment configuration, the change of the number of reconstructed events after the cuts in the detector acceptance region with respect to the numbers for the nominal geometry provides the systematic change of the luminosity for that alignment configuration. No significant difference in the luminosity changes between the two alignments is observed when the inner radius or the Z position changes. However, a dramatic difference is observed in case of larger crossing angles for radial beam axis shift and LumiCal axis tilt. As shown in Figure 2, for a 0.1 mrad tilt of LumiCal axis, if the LumiCal is centered on the outgoing beam, the relative change of the luminosity is reasonably small and flat; but if the LumiCal is centered on the detector axis, the relative change of the luminosity $\Delta L/L$ becomes strongly dependent of the azimuthal angle ϕ and would change the measured luminosity at $\phi \approx \pi/2$ by about one per cent in the case of a 20 mrad beam crossing angle. An even larger systematic change of $\Delta L/L$ is seen as a function of ϕ for a 1 mm radial displacement of the beam axis in the case of a 20 mrad beam crossing angle if the LumiCal is centered on the detector axis. So the systematic changes of the luminosity for a small tilt or a small radial beam position shift, for certain values of ϕ, are more than an order of magnitude larger than the anticipated precision. Thus in order to control the systematic uncertainty of the luminosity measurement, the LumiCal must be centered around the outgoing beam axis for large crossing angles, with a shift of 23.7 mm along x axis and a radius change from 60-350 mm to 100-350 mm.

Background Suppression

Two-photon events are the major background for Bhabha scattering selection when the additional fermion pairs have low energy.

A sample of 10^5 Bhabha scattering events was generated with electrons and positrons in the LumiCal acceptance region. A sample of 10^5 four-lepton events, $e^+e^- \rightarrow e^+e^-l^+l^-(l = e, \mu)$, was generated as the background with the WHIZARD [7] generator, including all neutral current tree level processes. The background was generated in full polar angle range with the invariant mass of the outgoing leptons greater than 1 GeV and the momentum transferred in photon exchange greater then 1 GeV. Though the cross sections of the signal and background process are comparable, most of the background events can be rejected by typical cuts for Bhabha events described

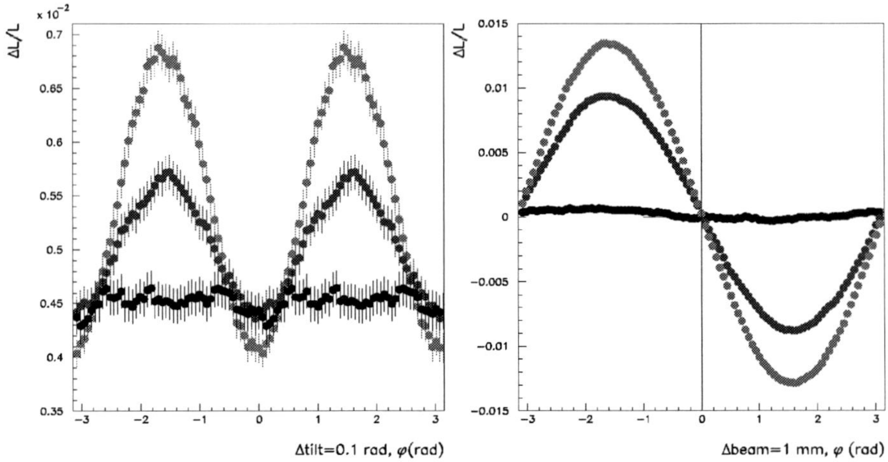

FIGURE 2. The relative change of the luminosity as a function of the azimuthal angle for the tilt(left) and beam position shift(right): the LumiCal aligned on the outgoing beam (black), the LumiCal aligned on the detector axis with 14mrad crossing angle (blue), and the LumiCal aligned on the detector axis with 20mrad crossing angle(red).

before.

The distributions of energy deposition on the front face of the LumiCal are shown in figure 3 for both signal and background, before and after applying the cuts.

The signal efficiency is $\sim 20\%$ and the signal to background ratio S/B is $\sim 10^5$. With looser cuts, the study shows an efficeincy of 98% and a S/B of 10^4 can be achieved. Two-photon processes with hadronic final states have not yet studied, but no dramatic change of S/B is expected.

These preliminary results indicate that the two-photon process background can be rejected at a very high level so that the precision of the luminosity measurement can be obtained. Again a large beam crossing angle complicates the luminosity measurement and the total background will increase due to beam particles and backscattered particles with larger transverse momentum in the magnetic field.

BEALCAL

The diamond-tungsten sandwich calorimetry has excellent electron identification efficiency from the beamstrahlung pair background. Two identical calorimeters locate at $z = \pm 3.65$ m respectively along the beam line, and cover the polar angle range of $5 < \theta < 45$ mrad. Each one consists of 30 layers of diamond sensors longitudinally, interleaving with the tungsten absorber. The sensor has thickness of ~ 0.3 mm to minimize the overall Molière radius (R_M) of the calorimeter. The absorber has thickness of 3.5 mm which corresponds to one X_0.

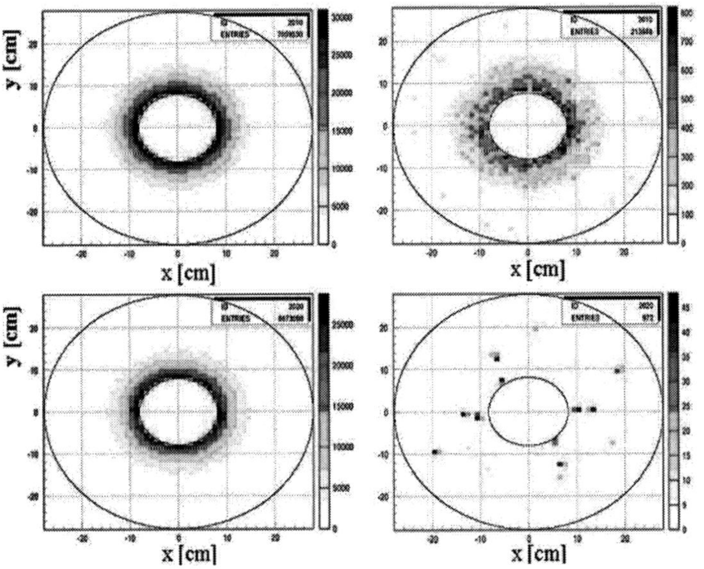

FIGURE 3. Energy deposition at the front face of the LumiCal: signal (left) and background (right), before (up) and after (down) cuts.

Design Optimization and Performance

The electrons and positrons came from the beamstrahlung photon conversion, deposit $\sim$20 TeV of energy into the BeamCal per bunch crossing for the nominal beam parameters. The energy density peaks at a small calorimeter radius and drops with increasing radii, then shows a sharp edge of the affected area and leaves a very small fraction at very large radii. The energy distribution is strongly dependent of the azimuthal angle ϕ because of the flat beam and the helical trajectory of electrons and positrons in the magnetic field.

For SUSY signal search, fine granularity is necessary to identify the localized depositions of high energy electrons from the energy depositions of the huge beamstrahlung background. Using the electron identification efficiency as a benchmark, detailed MC studies have been performed to optimize the BeamCal transversal segmentation.

With the head-on beam configuration and the nominal beam parameters, the beamstrahlung pairs are generated using the GUINEA-PIG generator [8]. Then the single electrons and the pair events are run through the full detector simulation with GEANT4 package. The energy depositions from the pairs and the single electron in the sensor pads are superimposed. A shower reconstruction algorithm is then applied. During the reconstruction, fake electrons can be found because of either a high energetic particle from beamstrahlung or background fluctuations which mimic the electron signal. The current reconstruction algorithm gives a fake rate below 10%.

The inefficiency for the identification of 200 GeV electrons, shown as a function of the sensor size in Figure 4, has a minimum around 5×5 mm^2, corresponding to half

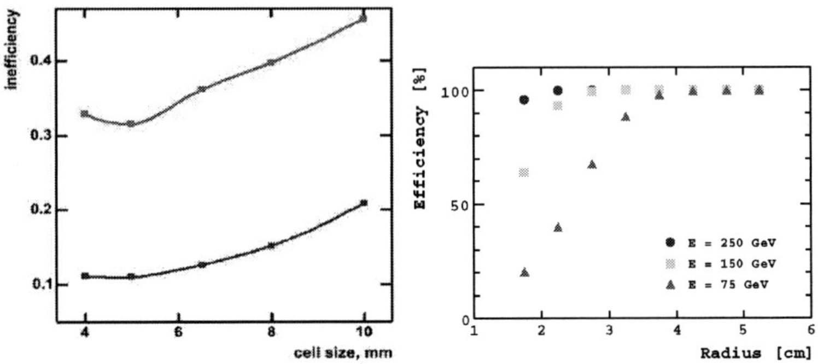

FIGURE 4. The inefficiency vs. the sensor size (left) and the efficiency vs. the radius for different electron energy (right).

R_M. In order to decrease the number of readout channels, an overall ϕ segmentation with sensor size of 8×8 mm^2 may be considered without loosing too much efficiency.

The identification efficiency of high energy electrons is shown in Figure 4 as function of the electron energy. An electron of 75 GeV can only be identified with high efficiency at larger BeamCal radii. An electron of 250 GeV can be identified with almost 100% efficiency even in high background region. The efficiency drops near the innermost radius, partly due to shower leakage.

Instrumentation for Different Beam Crossing Angles

The impact of a non zero beam crossing angle on the BeamCal design and performance is critical, especially due to the different magnetic field configurations such as DID and anti-DID for the large crossing angles. MC simulation has been done for the deflection of the beamstrahlung pairs in both DID and anti-DID cases.

In Figure 5 the radial energy distribution is shown for a 2 mrad design and a 20 mrad design with DID field. The 2 mrad case is very similar to the head-on case and the only change that may needed is a slightly large inner radius (R_{in}). The 20 mrad design with anti-DID field is not shown but also similar to the head-on or 2 mrad case. For the 20 mrad design with DID field case, as seen in Figure 5, the pair events spread significantly and it leads to a need to increasing the angular coverage by increasing the outer radius (R_{out}), otherwise the backscattering background to the LumiCal and the tracking detectors will be tremendous.

In the case of a large crossing angle the incoming and outgoing beams have to pass through individual openings. It is not clear if the inner region between the two beam pipes can be instrumented and the worst case is assumed with a blind area in which no sensor will be implemented. To keep the coverage the BeamCal should be centered on the outgoing beam as well. If the anti-DID configuration need be implemented for the purpose not to disturbing the beam collision too much, 14 mrad will be a better choice technology-wise. But the the blind region will be larger because the beam pipes are

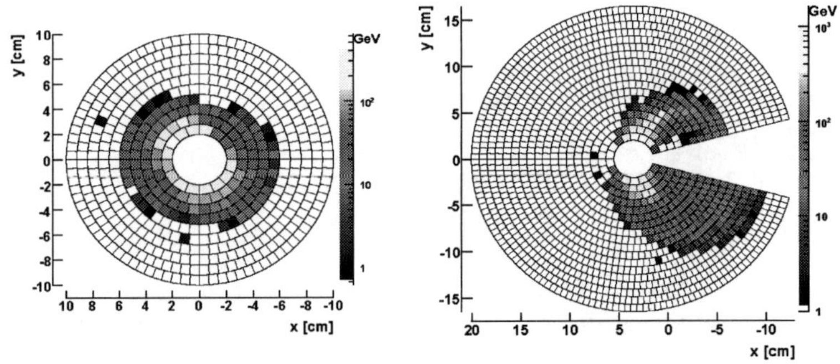

FIGURE 5. Energy deposition at the BeamCal frontface for the 2 mrad design (left) and the 20 mrad design with DID field (right).

closer. Table 2 provides the geometry details for these different configurations.

TABLE 2. BeamCal geometry details for different crossing angles.

Configuration	R_{in} (mm)	R_{out} (mm)	Blind area
head-on	15	100	0
2 mrad	20	100	0
14 mrad	15/20	165	40°
20 mrad	15/20	165	30°

CONCLUSION

Detailed simulation of a silicon-tungsten LumiCal is performed and the required precision of luminosity measurement is achievable. A diamond-tungsten sandwich BeamCal has been shown capable for SUSY event search and fast beam diagnostics, which is not discussed in this document. The instrumentation difficulties from the non zero beam crossing angles and the solutions have been investigated.

Promising results are obtained from laboratory tests of CVD diamond sensors. For both calorimeters the dedicated readout electronics prototypes are being designed. Mechanical structure design is also undergoing and a laser system with a CCD matrix sensor to measure the displacement of the LumiCal detector with respect to the beam pipe flange, is being developed for precise alignment.

ACKNOWLEDGMENTS

The author would like to thank all members of the FCAL collaboration for their excellent work which makes this document possible. The author also thanks the organizers of CALOR06 for the support.

REFERENCES

1. SiD working Group, *SiD Detector Outline Document*, version of 19 May 2006.
2. LDC working Group, *The LDC Outline Document*, version of v1.3.
3. S. Jadach, W. Placzek, E. Richter-Was, B. F. L. Ward and Z. Was, "BHLUMI, A Monte Carlo Event Generator for BhaBha Scattering at Small Angles", Version 4.04.
4. W. Placzek et al., *Phys. Lett.* **B390**, 298 (1997).
5. T. Ohl, *hep-ph/9607454-rev*.
6. GEANT4 Collaboration, S. Agostinelli et al., "GEANT4: A SIMULATION TOOLKIT", *Nucl. Instrum. Meth.* **A506**, 250 (2003).
7. W. Killian, "WHIZARD 1.40: A Generic Monte Carlo integration and event generation package for multi-particle processes", *LC-tool 2002-039*.
8. D. Schulte, Ph. D. Thesis, University of Hamburg 1996.

List of Registered Participants

	Last Name	First Name	Institute
1	Abdo	Aous	MSU - LANL
2	Aleksa	Martin	CERN
3	Alexa	Calin	IFIN, Particle Physics Department
4	Aplin	Steve	DESY
5	Barbi	Mauricio	University of Regina
6	Benslama	Kamal	Columbia University
7	Bhatti	Anwar	The Rockefeller University
8	Blazey	Jerry	NICADD/Northern Illinois University
9	Bonechi	Lorenzo	INFN - Section of Florence
10	Bornheim	Adi	California Institute of Technology
11	Buchanan	Norm	Florida State University
12	Byrum	Karen	Argonne National Laboratory
13	Carvalho	Joao	LIP-Coimbra
14	Chakraborty	Dhiman	Northern Illinois University
15	Chekhtman	Alexandre	Naval Research Laboratory/ George Mason University
16	Chen	Chin-Tu	University of Chicago
17	Connolly	Amy	University of California, Los Angeles
18	Dyshkant	Alexandre	Northern Illinois University
19	Damgov	Jordan	Fermilab
20	Daskalakis	Georgios	Imperial College, London, UK
21	Delmastro	Marco	CERN
22	Derrick	Malcolm	Argonne National Laboratory
23	Dragowsky	Michael	Case Western Reserve University
24	Eigen	Gerald	University of Bergen
25	Ereditato	Antonio	University of Bern
26	Francis	Kurt	NICADD
27	Franzoni	Giovanni	University of Minnesota
28	Freeman	Jim	Fermilab
29	Garutti	Erika	DESY
30	Gemme	Roberto	University of Turin
31	Giangiobbe	Vincent	Laboratoire de Physique Corpusculaire
32	Goncharov	Max	Texas A&M UNiversity
33	Grachov	Oleg	University of Kansas
34	Graf	Norman	Stanford Linear Accelerator Center
35	Gupta	Ambreesh	University of Chicago
36	Hatakeyama	Kenichi	The Rockefeller University
37	Hauptman	John	Iowa State University
38	Hays	Elizabeth	University of Chicago/Argonne National Laboratory
39	Jorgensen Roca	Sigrid	IFAE
40	Jui	Charles	University of Utah
41	Juste	Aurelio	Fermilab
42	Keilhauer	Bianca	University Karlsruhe, Institute for Experimental Nuclear Physics
43	Kistenev	Edouard	Brookhaven National Laboratory
44	Koi	Tatsumi	Stanford Linear Accelerator Center
45	Korbel	Volker	DESY
46	Korolkov	Ilya	IFAE
47	Krasberg	Mark	University of Wisconsin-Madison

48	Krstonosic	Predrag	DESY
49	Kubantsev	Mikhail	ITEP(Moscow)/Northwestern University
50	Kuhlmann	Steve	Argonne National Laboratory
51	Kvita	Jiri	Charles University
52	Lampl	Walter	University of Arizona
53	Le Du	Patrick	DAPNIA CEA SACLAY
54	Lee	Moo Hyun	IPST/University of Maryland
55	Lima	Guilherme	Northern Illinois University
56	Lubrano	Pasquale	INFN Sezione di Perugia
57	Luengo	Sonia	Enginyeria i Arquitectura La Salle, Universitat RamonLlull
58	Marmonier	Carole	PHOTONIS
59	Magill	Stephen	Argonne National Laboratory
60	Mao	Rihua	California Institute of Technology
61	Mattson	Mark	Wayne State University
62	Merkin	Mikhail	Skobeltsin Institute of Nuclear Physics, Moscow State University
63	Mocchiutti	Emiliano	INFN Trieste
64	Moser	Hans-Guenther	Max-Planck-Institut fuer Physik
65	Movilla Fernandez	Pedro	Lawrence Berkeley National Laboratory
66	Musgrave	Brian	Argonne National Laboratory
67	Nikolic	Irena	LPNHE University Paris 7
68	Nemecek	Stanislav	FZU AV CR PRAHA
69	Niess	Valentin	Centre de Physique des Particules de Marseille
70	Nigmanov	Turgun	University of Michigan
71	Norbeck	Edwin	University of Iowa
72	Novotny	Rainer	2nd Physics Institute, University Giessen
73	Oberlack	Horst	MPI fuer Physik, Munich
74	Onel	Yasar	University of Iowa
75	Oreglia	Mark	University of Chicago
76	Para	Adam	Fermilab
77	Paramatti	Riccardo	INFN Rome
78	Peters	Krisztian	University Manchester
79	Proudfoot	James	Argonne National Laboratory
80	Proulx	Joseph	University of Colorado
81	Ragazzi	Stefano	Milano-Bicocca University and INFN
82	Repond	Jose	Argonne National Laboratory
83	Rosner	Robert	Argonne National Laboratory
84	Rumerio	Paolo	CERN
85	Rykalin	Victor	NICADD/Northern Illinois University
86	Schacht	Peter	Max-Planck-Institut fur Physik, Werner Heisenberg Institut
87	Sefkow	Felix	DESY
88	Sekaric	Jadranka	Florida State University
89	Seuster	Rolf	University of Victoria
90	Shutt	Tom	Case Western Reserve University
91	Takeshita	Tohru	Shinshu University
92	Underwood	David	Argonne National Laboratory
93	Vandenbroucke	Arne	Universiteit Gent
94	Velasco	Mayda	Northwestern University

95	Volobouev	Igor	Texas Tech University
96	Wagner	Robert	Argonne National Laboratory
97	Wakely	Scott	University of Chicago
98	Weerts	Harry	Argonne National Laboratory
99	White	Andy	University of Texas at Arlington
100	Whitmore	Juliana	Fermilab
101	Wigmans	Richard	Texas Tech University
102	Wright	Dennis	Stanford Linear Accelerator Center
103	Xia	Lei	Argonne National Laboratory
104	Ye	Jingbo	Southern Methodist University
105	Yoshida	Rik	Argonne National Laboratory
106	Zabi	Alexandre	Imperial College London
107	Zhang	Jinlong	University of Colorado
108	Zhu	Ren-yuan	California Institute of Technology
109	Zutshi	Vishnu	Northern Illinois University

Author Index

A

B

C

D

E

F

G

X

Y

Z